From Five Fingers to Infinity

From Five Fingers to Infinity

A Journey through the History of Mathematics

EDITED BY

FRANK J. SWETZ

OPEN COURT

Chicago and La Salle, Illinois

OPEN COURT and the above logo are registered in the U.S. Patent and Trademark Office.

© 1994 by Open Court Publishing Company

First printing 1994

Library of Congress Cataloging-in-Publication Data

From five fingers to infinity: a journey through the history of
    mathematics / edited by Frank J. Swetz.
            p.   cm.
        Includes bibliographical references and index.
        ISBN 0-8126-9193-8. — ISBN 0-8126-9194-6 (paper)
        1. Mathematics—History.   I. Swetz, Frank.   II. Title: From 5
    fingers to infinity.
    QA21.F76   1994
    510'.9—dc20   94-20802
                                                          CIP

Table of Contents

Gdruckt zů Oppéheim.

NOTE TO
THE READER

The essays in this book are
presented without
alteration to content and
with only occasional minor
stylistic changes. Any
offensive content or
language is included not
for its own sake but for the
purpose of historical
accuracy.

Preface

$\mathcal{T}$HE HISTORY of mathematics is a broad subject spanning some 10,000 years of human activity. Its literature is correspondingly vast. There are many excellent sources of information on the history and development of mathematical ideas, including general texts such as those written by Eves (1983), Burton (1985), and Struik (1967), as well as broad surveys such as those compiled by Kramer (1951), Kline (1962), and Moffet (1977). The difficulty with these books, however, is that they require a prolonged commitment on the part of the reader—a commitment to develop an informed perspective in which the historical information can be understood and readily synthesized. For example, in order to appreciate certain nineteenth-century developments in geometry it is first necessary to know of Euclid's fourth-century B.C. attempt at formalizing the subject and the ensuing historical and mathematical controversies surrounding his efforts, particularly the questions that arose due to his postulate concerning parallel lines. In the usual book on the history of mathematics, such information is typically so scattered that all but the most resolute reader will become discouraged from piecing together a clear, comprehensive picture of these issues in geometry.

The present book represents my attempt to remedy this situation. Fortunately, over the years, many articles have appeared that constitute self-contained, organized accounts of the historical development of some specific aspect or concept in mathematics. My own inquiries into the history of mathematics have led me to seek information from various sources. I have been fortunate enough to have been able to examine original inscriptions, manuscripts, correspondence, and artifacts as well as probe into the contents of learned reference books and other scholarly literature. Despite such research advantages, I have always been amazed and gratified at the information I have found in brief articles, written in a popular style and intended for large reading audiences. It is from this discovery that the concept for a book of readings on the history of mathematics first arose. A book of readings offers many learning advantages. Individual articles can be quickly read and usually provide a conceptual or thematic examination of a particular concept that can be readily appreciated. The research has already been done for the reader and a particular perspective developed or conclusion reached. In a collection of articles, a variety of perspectives is bound to exist, thus the history of mathematics is scrutinized and explained through the use of many viewpoints and a more comprehensive and perhaps penetrating understanding of this history emerges. It is with these advantages and benefits in mind that the following collection of readings on the history of mathematics has been compiled.

Several criteria were employed for the choice of articles included in this book. Primarily, articles had to convey something of the essence and evolutionary nature of mathematics, that is: how concepts evolved; how they were modified; and how human and environmental constraints influenced mathematical thinking. Of course, the articles had to be historically accurate and also correct according to present knowledge. Secondly, all articles had to be of value to a wide audience, not just to mathematics specialists. While some articles are quite mathematical in content, their level of understanding should be within the grasp of a secondary school graduate. Biographical material, per se, has been avoided except where its rendering also provides insights into the mathematical climate of the times. Obviously, appropriate articles on all aspects and events in the history of mathematics are not available or necessarily desirable. This collection presents a selected survey and, in a sense, a particular view of the growth of mathematical knowledge. It is not intended to be comprehensive in a global sense. In addition to the bibliographies and notes that accompany the individual essays, for the reader who requires a wider purview of historical examination a suggested reading list is included for each part and a general bibliography is given at the end if the book. Organizational format is basically chronological. Each group of articles represents a certain time period or aspect of mathematical development. In some instances, authors have chosen to extend their comments beyond the confines of a fixed time frame; a few such survey articles of this type are included as long as their main points pertain to the issues under consideration. Historical exhibits— one-page illustrations or informational digests—are included occasionally to illuminate a particular development or to round out a presentation. Each chapter is prefaced by remarks on the significance of the historical period in question and the mathematical themes about to be examined. Such remarks are intended to be thought-provoking.

Many people, either directly or indirectly, have contributed to the creation of this book. All deserve recognition and an expression of thanks for their efforts. First, a special "thank you" to all the authors whose works are represented in this collection. It is through their research and the articulation of that research in the following articles that an understanding of the history of mathematics emerges. To all the publishers and journal editors who permitted me access to their copyrighted materials, I thank you. More complete acknowledgments are provided with each article. I am especially grateful to the Publications Committee of the Mathematical Association of America for allowing me to reprint their materials free of cost.

A book, any book, cannot be all things to all people. This is especially true of a book surveying the history of mathematics. The history of mathematics can be likened to a work of impressionist art— certain objects may be discerned and moods conveyed; however, different people view it in different ways. While there are certain events in the history of mathematics that are *not* a matter of interpretation, there are also many events that *are*. Every effort has been made to check the facts and ascertain the accuracy of the information contained in this book. In this collection of readings, a variety of authors ranging from acknowledged and respected historians of mathematics (such as Howard Eves, Morris Kline, Phillip Jones, and Dirk Struik) to ordinary classroom teachers give their interpretations of certain aspects or events in the history of mathematics. Although widely divergent in their research approaches and writing styles, they all share in the consensus that the history of mathematics is important for understanding mathematics itself as well as being fascinating in its own right. Ultimately, it is with this conviction that this book is offered to the reader.

FRANK SWETZ
Harrisburg, Pennsylvania

PART I

WHY THE HISTORY OF MATHEMATICS?

*T*HERE IS A FAMOUS MATHEMATICAL TEXT that ends with the words:

> Catch the vermin and the mice, extinguish the noxious weeds; pray to the God Ra for heat, wind and high water.

Certainly these are strange words—"mice," "weeds," "wind," and "water"—to find associated with mathematics. If we review the writing that precedes these comments, we find a collection of 85 mathematical problems. These problems deal with the feeding of animals, the baking of bread, the brewing of beer, and the storage of grain. They are the common-day mathematical problems of an agricultural society and the comments at the end of this list of problems is a prayer asking for the conditions to ensure a good harvest. At the beginning of this problem list, we are met by an introduction that prepares its readers for the exercises they are about to encounter. Readers are assured that the text: undertakes "a study of all things," provides "insights into all that exists" and "knowledge of obscure secrets." Surely this mathematics must be wonderful! However, when we examine the content of the enclosed problems, we find that they are concerned with: basic computations involving whole numbers and fractions, the solution of simple linear equations, and the computation of area for several plane figures and the volumes of a few selected solids. To a modern reader judging by modern standards this mathematics might seem elementary, something easily undertaken by a schoolchild. However, the text we have been examining is the Rhind Papyrus compiled by the Egyptian scribe Ahmes in about 1650 B.C. It was retrieved from Egypt in 1858 by the Scottish antiquary, A. Henry Rhind and, thus, bears his name. The papyrus reveals the mathematics of an ancient people, a mathematics that met their needs and served them well.

In a sense, becoming involved with the history of mathematics is similar to the examination of the Rhind papyrus we have just undertaken. We work with mathematics, use mathematical ideas, and undertake mathematical processes. These concepts and processes that we work with are really end products or conclusions like the prayer at the end of the papyrus. Unconnected with their background, their history, they lose meaning and gain a certain mystique. To dispel this mystique, to understand the meaning of these conclusions, we must step back and read the text that precedes them, that is, the record of human involvement with mathematics. We must become familiar with the history of mathematics. This familiarity will lead to an understanding of the relationship of mathematics to society in general—how it evolved and is used in a societal context. It is to this end that the history of mathematics admirably lends itself by exposing how and why mathematics techniques came into being. The history of mathematics tells a story; it reveals the human connection between a problem recognized and a mathematical solution conceived. It illuminates mathematical activity as an ongoing human concern and interrelates that activity to the development of the parent society and culture. It is little wonder that one of the first historiographies of a science to appear (Eudemus, *History of Geometry*, ca. 335 B.C.) concerned mathematics.

The articles in this first section present broad perspectives on the historical relationships of mathematics to society. They survey the development of mathematics from several intellectual viewpoints and identify various influences—political, social, economic, and cultural—that have helped shape mathematics. In their discussions, the authors raise questions as to the nature of mathematics: Is it invented or discovered? What are its constraints and its potential? These articles begin to answer the question Why the History of Mathematics? and help to set the stage for further inquiries that can only be answered by undertaking a more complete journey through that history.

Mathematics in the
History of Civilization

NATHAN ALTSHILLER COURT

*T*HE DISCUSSION of the subject at hand would gain in clarity if I could ask the members of my invisible audience whether they can remember themselves far enough back in their childhood, when they were not able to count, that is when they could not recite the series of words one, two, three, four, and so on. Radio being what it is, the question must remain unanswered. Fortunately, I know what the answer would be, if it could be given. It is just like asking somebody whether he can recall the time when he could not walk or when he could not talk.

The same situation prevails with regard to the human race as a whole. We know quite definitely that there was a time when the notion of number was totally alien to mankind. Who was the genius who first asked the momentous question: "How many?" We will never know. At a certain stage of social development the need arises to determine how many objects constitute a given collection. The answer to the question becomes a social necessity. Contributions toward finding that answer are made by many individuals confronted with the same need, and the notion of number slowly emerges.

How slow and painful a process of creation this was may be judged from the fact that there are human tribes whose languages have no words for numbers greater than four, and even no greater than two. Beyond that any group consists of "many" objects.

Our numbers are applied to any kind of objects in the same way, without discrimination, they have a kind of "impersonality." That is not the attitude of the primitive man. With him the number applied to a group is modified in accordance with the nature of the group. The number characterizes the group in the same way as an adjective applied to a noun describes the object to which it is applied. The English language has preserved some traces of that attitude. A group of cattle is a *herd*, while a group of birds is a *flock;* a group of wolves is a *pack*, while a group of fish form a *school*. It would be shocking indeed to speak of a school of cows. Other languages offer much more striking proofs of such an attitude towards numbers in their relation towards the objects they are applied to. Thus in English we use the singular grammatical form when one object is involved, and we use the plural grammatical form for any number of objects larger than one. Some of the languages of the Western world, in their earlier stages of development, had a special grammatical form, a dual form, when two objects were spoken of. Some languages even had separate grammatical forms when reference was made to three objects, and still another form for four objects. An instructive example of the way the form of the same number may be modified to fit the group to which it is applied is furnished by the Polish language in its use of the number two. In

Reprinted from *Mathematics Teacher* 41 (Oct., 1948): 104–11; with permission of the National Council of Teachers of Mathematics.

that language a different form of "two" is used when applied to two men, to two women, to a man and a woman, and to inanimate objects or animals. These forms are, respectively: *dwaj, dwie, dwoje, dwa.*

The process of accumulating enough words to answer the question: how many? to satisfy the growing needs was slow and laborious. Man derived a great deal of help from the natural set of counters he always carries with him—his fingers. Of the many examples that could be cited to illustrate the use of fingers as counters let me quote a report of Father Gilij who described the arithmetic of the Indian tribe of the Tamanacas, on the Orinoco river.

The Tamanacas have words for the first four numbers. When they come to five they express it by a phrase which literally means "a whole hand"; the phrase for the number six means literally "one on the other hand," and similarly for seven, eight, and nine. When they come to ten, they use the phrase "both hands." To say eleven they stretch out both hands, and adding a foot, they say "one on the foot," and so on, up to 15, which is "a whole foot." The number 16 is "one on the other foot." For twenty they say "one Indian," and 21 is expressed by saying "one on the hands of the other Indian"; forty, sixty, . . . is "two Indians," "three Indians, " and so on.

When the question: how many? has once been raised, mere counting becomes insufficient. Further steps in civilization bring about the need of computation. The strongest single factor that stimulated the development of methods of computation was trade. According to the mythology of the ancient Egyptians, arithmetic was invented by their god of commerce. As with counting, the beginnings of reckoning were slow and laborious, awkward and painful. A trader in tropical South Africa during the last century has this to say about the members of the Dammara tribe. "When bartering is going on, each sheep must be paid for separately. Thus, suppose two sticks of tobacco to be the rate of exchange for one sheep; it would

sorely puzzle a Dammara to take two sheep and give him four sticks. I have done so, and seen a man put two of the sticks apart and take a sight over them at one of the sheep he was about to sell. Having satisfied himself that *that* one was honestly paid for, and finding to his surprise that exactly two sticks remained in his hand to settle the account for the other sheep, he would be afflicted with doubt; the transaction seemed to come out too "pat" to be correct, and he would refer back to the first couple of sticks; and then his mind got hazy and confused, and he wandered from one sheep to the other, and he broke off the transaction, until two sticks were put in his hand, and one sheep driven away, and then two other sticks given him and the second sheep driven away." It would seem that at least to this representative of humanity it was not obvious that two times two makes four.

The story illustrates the blundering beginnings of the art of reckoning. To relate the evolution of this art from its humble beginnings to the heights of power and perfection it has achieved in modern times, and how this art has followed and served the ever growing needs of mankind is to tell one of the most exciting sagas in the history of civilization. Only a mere outline can be attempted here.

Various human activities, and in particular commerce, require the keeping of some numerical records. Some kind of marks had to be invented for the purpose. The devices used through the ages were many and various. Among them were knots tied in a rope and notches cut in sticks. It may surprise some of my readers that such sticks, called *tallies*, were used as a method of bookkeeping by the Bank of England way into the nineteenth century.

The first written symbols for numbers were, naturally, sticks: *one stick, two sticks, three sticks*, and so on, to represent "one," "two," "three," etc. This worked fairly well as long as the numbers to be represented were small. For larger numbers the sticks occupy too much space, it becomes difficult to count them, and it takes too much time. The sticks had to be condensed into groups, thus representing larger units, and these new units in turn had to be condensed into larger units and thus a hierarchy of units had to be formed. The Greeks and the Hebrews used the letters of their alphabets as numerals. The Babylonians had special numerical symbols.

All these symbols or marks for numbers had one feature in common—they did not lend themselves to arithmetical computations. The art of reckoning had to be carried out with the help of different devices, the chief among them being the counting frame, or the *abacus*. This instrument most often consisted of a rectangular frame with bars parallel to one side. The operations were performed on the beads or counters strung on these bars. This instrument was widespread both in Asia and in Europe. When the Europeans arrived in America they found that the abacus was in use both in Mexico and in Peru.

The method of writing numbers and computing with them that we use now had its origin in India. The most original feature of that system, namely the Zero, the symbol for nothing, was known in Babylon and became common in India during the early centuries of the Christian era. Due to its flexibility and simplicity it gradually forced out the abacus. This system was brought to Europe by the Arabic and Jewish merchants during the twelfth century. The first printing presses set up in Europe, in the middle of the fifteenth century, rolled off a considerable number of commercial arithmetics. Two centuries later the abacus in Western Europe was little more than a relic of the past.

The very heavy demands that modern life in its various phases makes upon computation seem to be turning the tide against paper and pencil reckoning. We are about to enthrone the abacus back again, in a much improved form, to be sure, but nevertheless in the form of an instrument. In fact, we are using a considerable number of them, like the slide rule, the cash registers, the various electrically operated computers, to say nothing of the computing machines which operate on a much higher level, like those which give the solutions of differential equations. Such is the devious and puzzling road of human progress.

"How many?" This question is the origin of arithmetic and is responsible for much of its progress. But this question cannot claim all the credit. It must share the credit with another, a later arrival on the scene of civilization, but which is even more far reaching. This question is: "how much?" How much does this rock weigh? How much water is there in this barrel? How much time has passed between two given events? How long is the road from town A to town B? etc. The answers to these questions are numbers, like the answer to the question; "how many?" There is, however, a vast difference between the numbers which answer the two kinds of questions.

The answer to the question; "how many?" is obtained by counting discrete objects, like sheep, trees, stars, warriors, etc. Each of the objects counted is entirely separate from the others. These objects can be "stood up and be counted." Something vastly different is involved in the question: How much does this rock weigh? The answer can only be given by comparing the weight of the given rock to the weight of another rock, or to the weight of some other object taken for the unit of weight, say a pound or a ton. Obviously this is a much more involved process and implies a much more advanced social and intellectual level than the answer to the question: how many?

The question: "how many?" is always answered by an integer. Not so the question: "how much?" Given 17 trees, is it possible to plant them in five rows so that each row would have the same number of trees? The answer is: "No," and this is the end of the story. But given 17 pounds of salt in a container, it is possible to distribute this salt into five containers so that each of them will hold the same amount of salt. But the question: "how many pounds of salt does each container hold?" cannot be answered by an integer. Thus, the question:

"how much?" is responsible for the invention of fractions. It is also responsible for the introduction of irrational numbers. But about that we may have to say something later on.

The question: "how much?" that is, the introduction of measurements, has involved us in another kind of difficulty which did not bother us in connection with the question: "how many?" We can ascertain that the group at the picnic consisted of forty boys. But when we say that this table is forty inches long, we can only mean that it is closer to forty inches than it is either to 39 or 41 inches. We may, of course, use more precise instruments of measurement. That may narrow down the doubtful area, but it will not remove it. Results of measurements are necessarily only approximations. The degree of approximation to which we carry out these measurements depends upon the use we are to make of these measured things.

The herdsman is much concerned with the question: "how many?" The shepherd, in addition, is also interested in the question: "how much?" after he is through sheering his flock. When a human tribe turns to agriculture, the question: "how much?" imposes itself with increased insistence. Agriculture requires some methods of measuring land, of measuring the size of the crop, that is measuring areas and volumes as our school books call it. Furthermore, the agricultural stage of society implies already a considerable degree of social organization, and the tax collector appears on the scene. This official is vitally interested in the size of the crop. He also has to have some numerical records of the amount of taxes collected and of the amount of taxes due. Now you may not like the tax collector. Few people waste too much love on this maligned official. It is nevertheless quite obvious that no organized society is possible without the collection of taxes, that is without contributions from the individual members of that society towards the necessary enterprises that are of benefit to the members of the entire community. And such collections cannot be made in any orderly fashion, unless answers can be given to the two questions: "how much?" and "how many?"

The cultivation of the land faced the human race with problems of geometry. Egypt with its peculiar dependence upon the flood waters of the river Nile was confronted with extra difficulties of a geometrical nature. That is the reason why geometry found such fertile soil in the valley of the Nile.

Much geometry had to be discovered in order to construct human habitations. When civilization progresses beyond the cave dwelling stage, shelter becomes a problem of the first magnitude. The construction of dwellings involves in the first place the knowledge of the vertical direction, as given by the plumb line. It was observed very early that the plumb line or a pole having the same direction as the plumb line makes equal angles with all the lines passing through its foot and drawn on level ground. We have thus what we call a right angle, as well as the famous theorem of our textbooks that all right angles are equal.

However important the answers to the question: "How much?" may have been in the connections we just considered, the most important answer to this question is the one connected with the measuring of time. With the most rudimentary attempts at agricultural activity comes the realization that success is dependent upon the seasons; this dependence is even exaggerated. We still worry about the phases of the moon when we want to plant our potatoes.

Various tribes on the surface of the globe noticed that the shortest shadow cast by a vertical pole during the day always has the same direction. This is the north and south direction. The sun at that time occupies the highest point in the sky. It is essential to have a way of marking this direction. Here is how it can be done.

A circle is drawn on the ground having for center the foot of the pole used in the observation. The two positions of the shadow are marked, the tips of which just fall on the circumference. The north-south line sought is the line mid-way between the two lines marked, and that north-south line was found by many human tribes by bisecting this angle, and was done by the methods still in use in our textbooks.

Measurements connected with the sun, the moon, and the stars in general cannot be made directly. Some roundabout method must be used. Neither could the size of the earth be determined directly. On the elementary level such artifices are based on geometry and trigonometry. Two centuries B.C. Eratosthenes, the librarian of the famous Alexandrian Library, succeeded by the use of such methods to determine the length of the diameter of the earth with a surprising degree of accuracy. He thus made his contemporaries realize that the world they knew was only a very small part of the surface of the earth.

These very sketchy indications may have given you an idea of the role mathematics played in the development of mankind from the earliest times up to the time of the great civilizations of antiquity.

II

The Renaissance was the age of the revival of secular learning in Europe. It was also the age of the great voyages and of the discovery of America, the age of gunpowder and of mechanical clocks.

The new interest in seafaring has raised many pressing problems that had to be solved. The most obvious one was the need for a way of **determining** the position of a ship on the high seas, **that is**, the need of determining the longitude and the latitude of the ship at any time. This involved a great deal of laborious computation. The invention of logarithms reduced this labor to a fraction of the work it used to require. This accounts for the great success that the invention of logarithms enjoyed, as soon as it became available.

The process of finding the longitude requires an accurate clock which could be relied upon. I pointed out in Part I, above the important role the need of determining the seasons played in the

history of civilization, and the mathematical problems that had to be solved in this connection. The navigation of the Renaissance required the measuring of time with great precision. It was a question not of seasons and days but of minutes and seconds. The instrument that made such an accuracy possible was the mechanical clock moved by a pendulum, then by springs. This moving mechanism raised many problems of a mathematical nature that the available mathematical resources were insufficient to cope with. New mathematical methods were needed.

New mathematical problems were also raised by the cannon. It may be observed, in passing, that a cannon was just as much a necessary piece of equipment of a ship starting out on a long voyage towards unexplored shores as was a map, or a clock.

A gunner is frequently in need of determining the distance to some inaccessible objects. The information can be obtained by indirect measurements. This is a problem that was met with much earlier in the history of civilization and was solved in different ways. The cannon has stimulated further development in this connection, thus contributing to the progress of trigonometry.

But artillery presented problems of a new type. The cannon ball was an object which moved with a speed that was unprecedented in the experience of man. Motion took on a new significance and called for mathematical treatment and study. It required the study of the path that the projectile describes in the air, the distance it travels, the height it reaches at any given distance from the starting point, and so on. In short it required what we now call a graph.

The computation of the longitude of a ship at sea is based on astronomical observations and computations made in advance and published for that purpose. The greater the accuracy of these data, the more correctly can the position of the ship be determined. Thus navigation made necessary a more accurate knowledge of the motion of heavenly bodies.

The mathematics that the Renaissance inherited from preceding periods was inadequate for the study of motion. The new mathematical tools that were invented for the purpose of answering the new questions raised were: (1) Analytic Geometry, discovered by René Descartes (1637) and (2) the Infinitesimal Calculus, the contribution of Newton and Leibniz to the learning and technical proficiency of man.

I have already mentioned that the path of a cannon ball, or, for that matter, the motion of any object is most readily studied by a graphical presentation of that motion. Nowadays graphs are very common. We see them even in the newspapers when things like the fluctuation of the price, say, of wheat is discussed. But it took nothing less than the invention of Analytic Geometry to put this simple and powerful device at the service of man.

If a body travels along a curved path, it does so under the action of a force exerted upon it. If the force suddenly stops, the moving object continues nevertheless to move, not along the curve, however, but along the tangent to that curve at the point where the object was when the action of the force ceased. Thus, in the study of motion, it is important to be able to determine the tangent to the path at any point of that curve. The resources that mathematics had to offer up to the middle of the seventeenth century were insufficient to solve that apparently simple problem. The differential calculus provided the answer.

The calculus provides the tools necessary to cope with the questions involving the velocity of moving bodies and their acceleration, or pickup. The ancients had only very hazy notions about these concepts. The unaided imagination seems to find it very difficult to handle them successfully. The methods furnished by the calculus take all the sting and all the bitterness out of them. When velocity and acceleration are presented to students of mechanics who do not have the calculus at their disposal, these notions are still explained in terms of the calculus, in a roundabout fashion.

Our own age is confronted with technological problems of great difficulty. The mathematical tools they call for were not in existence at the time of Newton, two centuries ago. The airplane alone is sufficient to make one think what a variety of questions of an unprecedented kind had to be answered, what complicated problems had to be solved to enable the flier to accomplish all the wonders of which we are the surprised and admiring witnesses. The difficulties of constructing the airplane wings necessitated the concentration of mathematical talent and mathematical information that hardly has any parallel in history.

As has been pointed out, seafaring called for the solution of many problems. However, a ship sailing the high seas has one important feature in common with a vehicle traveling on land: both move on a surface. From a geometric point of view the problems related to their motion are two-dimensional. An airplane that roams in the air above is engaged in three-dimensional navigation. The geometrical aspect of flight belongs to the domain of Solid Geometry, and the problems connected with it are thus much more difficult, other things being equal.

Mathematics plays an enormous role in the field of social problems, through the use of statistics. I have already pointed out the value of mathematics in connection with the collecting of taxes, at earlier stages of civilization. The functions of a modern government are vastly more complex, more varied, and applied on an enormous scale. The variety and scope of problems modern government is interested in can be gleaned from the questions the citizen is asked when he receives the census blank, every ten years. To study the wealth of information that is thus gathered on millions of blanks is the function and the task of the census bureau. The inferences that can be drawn from these data are as involved as they are far-reaching in their applications. Such a statistical study requires a wide range of mathematical equipment, from the most elementary arithmetic to the most abstruse branches of mathematical analysis. If one thinks of the new functions of social welfare that the government has taken on, like social security or old age pensions, as well as of those that are in the offing, like health insurance, and the millions of individuals that these services cover, one is readily led to the realization that the intelligent dealing with these services sets before the government new statistical problems of vast magnitude.

The government is not the only social agency to use statistics. Far from it. Insurance companies have been using statistics for a long time. Banks and other organizations which study the trends of business arrive at their conclusions and their predictions by statistical analysis. The study of the weather raises many very difficult statistical problems. Statistics is used to determine the efficiency of the methods of instruction in our public schools. This list could be made much longer and become boring by its monotony. As it is it will suffice to convey the idea of the all-pervading role this branch of applied mathematics plays in our modern life.

I have alluded several times to the fact that during the course of the centuries mathematics was called upon to provide solutions for problems that have arisen in various human pursuits, for which no solution was known at the time. This, however, is not always the way things occur. In many cases the reverse is true. When the need arises and the question is asked, mathematics reaches out into its vast store of knowledge accumulated through the centuries and produces the answer. The astronomer Kepler had before him a vast number of observations concerning the motion of the planets. These figures were meaningless until he noticed that they would hang nicely together if the planets followed a path of the form which the Alexandrian Greek Apollonius called an ellipse. Another plaything of the same Apollonius, the hyperbola, came in very handy to locate enemy guns during World War I, when the flash of the gun could be observed twice.

This readiness of mathematics goes much further. Various branches of science, when they pass and outgrow the purely descriptive stage and are ready to enter the following, the quantitative stage, discover that the mathematical problems which these new studies present have already been solved and are ready for use. Thus, Biology has in the last decades raised many questions, answers for which were available in the storeroom of mathematics. At present the scope of mathematics used in "Mathematical Biology" exceeds by far the mathematical education which our best engineering schools equip their graduates with. A similar tale can be told of psychology, economics, and other sciences.

I have tried to point out the close relation of the mathematics of any period of civilization to the social and economic needs of that period. Mathematics is a tool in the work-a-day life of mankind. It is closely connected with the well-being of the race and has played an important role in the slow and painful march of mankind from savagery to civilization. Mathematics is proud of the material help it has rendered the human race, for the satisfaction of these needs is the first and indispensable step that must be taken before higher and nobler pursuits can be cultivated. The Hebrew sages of yore said: "Without bread there is no learning." The Russian fable writer Ivan Krylov put it in a more crude but striking way: "Who cares to sing on a hungry stomach?"

But mathematics does not limit its ambition to serving utilitarian purposes. "Man does not live by bread alone." It would be utterly erroneous to think that the development of mathematics is due solely to the stimulus it received from the outside, or was always guided by the needs of the moment, or of the period. Rather the contrary is true. To be sure, the mathematician, like any other scientist, is not unmindful of those needs and is not indifferent to the acclaim that would be his if he succeeded in supplying the answer to a pressing question of the day. But, by and large, the development of mathematics is mostly stimulated by the innate human curiosity and by the pleasure derived from the exercise of the human faculties, the only extraneous element is perhaps the approval of a small group of likeminded people. Beyond that the reward that the mathematician receives for his efforts consists in the pains of creation and the joys of discovery. If his labors can be useful now, so much the better. If they are not, they have a chance to be so in the future, perhaps in the distant future. "Without learning there is no bread," to quote the other half of the saying of the Hebrew sages. But to the mathematician his labor is primarily a labor of love, and the product of his labor—a further addition to the stately edifice that he calls the Science of Mathematics.

In the cultural history of mankind mathematics occupies a unique place. It is the model science. Other sciences try to approach its objectivity and its rigor. Mathematics was the first to point out that any branch of learning must start out with a certain fund of data that must be accepted as given and true, and which cannot be further analyzed.

When stated in this form, this idea seems so obvious that it borders on triviality. But it represents the product of a century of keen analysis and much labor. It led to the postulational approach to mathematics. The mathematician stopped worrying what a point is or what a straight line is. He simply says: "these are things that I accept as my original stock in trade." Neither is he trying to prove that two points determine a line, or that two lines can be parallel, or that through a point a line can be drawn parallel to a given line. This is simply so much more stock in trade. When he has accumulated enough of such, he is ready for action. He is ready to start building his science, in this case, of geometry. But how much is "enough"? Could there be too much of it? It is much easier to ask these rather obvious questions than to answer them. They were the object of much debate and are still one of the items of unfinished business in Mathematics.

11

Mathematics as a Cultural Heritage

WILLIAM L. SCHAAF

$\mathcal{B}$ROADLY SPEAKING, major cultures can be identified, at least in part, by certain outstanding characteristics. Thus, Babylonia and Egypt were steeped in mysticism and sensuality; the Greeks were preoccupied with ideas and ideals; the Romans, with politics, military prowess, and conquest. The culture of Western Europe from A.D. 600 to A.D. 1100 was expressed largely by its theology. From 1200 to 1800 it was the exploration of nature and the beginnings of science that marked the essence of the period. The spirit of the nineteenth and twentieth centuries, unless it is too early to judge in proper perspective, is typified by man's increasing mastery over his physical environment. This is evidenced not only by the general achievements of science and technology, but also by unprecedented industrial production, effective mass communication, and increasing automation. The creative language of the culture of today is science, and mathematics is the alphabet of science.

Contributions from the Age of Empiricism in Mathematics

In past ages, mathematics was very largely a tool that not only facilitated the development of a culture, but which was itself more or less shaped by the culture. The pre-Grecian period may be aptly characterized as an age of empiricism in mathematics. Babylonian and Egyptian mathematics were concerned chiefly with astronomical and cal-

endar questions, with the construction of tombs, temples, and other religious buildings, and with practical problems of land measurement, surveying, and primitive engineering. In fact, the knowledge of arithmetic and mensuration possessed by the Babylonians appears to have been derived from the earlier work of the Sumerians who preceded them. By 2500 B.C., merchants of Sumer were already thoroughly acquainted with weights and measures, and were using the arithmetic of simple and compound interest with more than ordinary zeal. The Babylonians were prolific in the creation of elaborate multiplication tables and tables of squares and square roots. They may even have used the zero, although this invention is generally attributed to the Hindus. In short, the Babylonians were very skillful computers.

The Egyptians, whose mathematics was similarly empirical, were likewise adept at calculation. As early as 3500 B.C., they had extended their use of numbers to include hundreds of thousands and millions. Since there was neither inflation nor a national debt, they presumably had little need for billions. The Egyptians revealed an amazing ingenuity in their use of unit fractions; they were aware of the value of checking computation; and there is reason to believe that they anticipated the generalized number concept by using negative numbers as numbers.

Despite the fact that the mathematics of Babylon and Egypt were basically empirical, these two cultures nevertheless left their stamp upon the future in several ways. The idea of number was pressed into service for the market place as well as for the contemplation of the heavens; the idea of

Reprinted from *Arithmetic Teacher* 8 (Jan., 1961): 5–9; with permission of the National Council of Teachers of Mathematics.

geometric form was embraced in practical measurement in surveying, engineering, and astronomy; a distinct if feeble beginning was made in the use of algebraic symbolism; the generalized extension of the system of natural numbers was at least anticipated if not consciously fashioned; and out of experiences with measurement, there grew some awareness of the notion of the mathematical infinite.

The Influence of Greek Contributions and Attitudes

With the ancient Greeks, some six centuries before the Christian era, mathematics came of age. For the next nine hundred years the contributions of Greek culture to mathematics were of the greatest significance, although oddly enough their influence upon arithmetic, as we know it at the present time, was little more than trivial.

It must be appreciated that the Greeks distinguished carefully between two aspects of knowledge about ordinary numbers: *logistiké*, the art of calculating, and *arithmetiké*, an abstract theory of numbers. *Logistica* (in the later Latin form) comprised the techniques of numerical computation in everyday trade and commerce, and in the arts and sciences, including geography and astronomy. *Arithmetica*, on the other hand, dealt with the properties of numbers as such, and in this sense was roughly, comparable to contemporary "higher arithmetic" or the elementary theory of numbers (primes, factorization, congruences, etc.). In the eyes of the Greeks, to be concerned with *logistica* was considered beneath the dignity of mathematicians and philosophers, who assigned this drudgery to lesser persons, while they devoted their attention to *arithmetica*, geometry, and philosophy. It should be added that Greek *logistica*, compared to modern methods of computation, was exceedingly cumbersome and crude, due chiefly to an inadequate system of numeration.

These two aspects of the study of numbers were regarded separately until about the time of the invention of printing, although from time to time the names were changed. In the Middle Ages, *logistica*, or "practical arithmetic," was referred to by Italian writers as *practica* or *pratiche*; Latin writers of the Renaissance spoke of the art of computation as *ars supputandi*; Dutch writers called it "ciphering." We still sometimes speak of computing as reckoning, figuring, or calculating; the Germans use the term *Rechnen* for computation, although their term *arithmetik* covers both aspects. Our modern word "arithmetic" began to be used for both branches in the early part of the sixteenth century.

Not much need be said here about Greek *arithmetiké*. To see it in appropriate perspective, it may be recalled that the Pythagoreans (500 B.C.) divided mathematical studies into four branches, the *quadrivium*: numbers absolute, or *arithmetic*; applied numbers, or *music*; magnitudes at rest, or *geometry*; and magnitudes in motion, or *astronomy*. The weakness of Greek arithmetic lay in the fact that almost from the very beginning, Greek mathematics succumbed to the number mysticism of the oriental cultures. As the centuries slipped by, the superstitious beliefs and esoteric lore associated with numbers increased to the detriment of Greek science and mathematics. Number "magic," *Gematria*, and other precursors of medieval and modern numerology were without doubt one of the facts which prevented the Greeks from embracing algebra and, possibly, even inventing the calculus.

Despite these shortcomings, however, during a period of nearly a thousand years, the Greeks did make several important contributions to arithmetic, or the theory of numbers; for example, (1) they arrived at certain basic theorems concerning the divisibility of numbers; (2) they discovered and proved that the number of primes is infinite; (3) they proved that $\sqrt{2}$ was irrational, and, in general, that the side of a square is incommensurable with its diagonal.

The greatest cultural contribution of the Greeks to mathematics was not their insight into number theory, which, after all, was essentially elementary, nor their skill in the art of computing, which was inconsequential. Their contribution had to do rather with two fundamental concepts or attitudes. One was their faith in the method of deductive reasoning as a sound basis upon which to build the structure of geometry. The other was the belief that our physical environment could be described in mathematical terms; that, in short, number is the language of science. Both legacies were destined to exert a profound influence upon Western civilization for the next two thousand years.

A Barren Period

After the fall of Rome, Western Europe entered upon a period of stagnation and groping for nearly a thousand years. From about A.D. 400 to 1500, mathematics reflected the general cultural condition of the times—meager and barren. Such as it was, this mathematics was kept alive by a handful of individual laymen and ecclesiastical scholars. One of the earliest influential laymen of this period was Boethius (ca. A.D. 500). Among the ecclesiastical mathematicians were Alcuin (ca. 775), and the French monk Gerbert, who later became pope under the name of Sylvester II (ca. 1000). Gerbert traveled extensively in Italy and Spain, becoming familiar with Arabic mathematics— especially the Hindu-Arabic numerals.

With the decline of feudalism during the twelfth, thirteenth, and fourteenth centuries there emerged such powerful commercial city-states as Milan, Venice, Florence, Pisa, and Genoa. At the same time, mathematics began to feel the impact of advances in technology and crafts, as well as the effect of a rising trade and money economy. In particular, mathematics was influenced by architecture, military engineering, navigation, and astronomy on the one hand, and, to a lesser extent, by trading, accounts, and commercial activities. The latter involved barter, exchange, customs, drafts, interest, discount, usury, rents, annuities,

insurance, partnerships, and stocks. An outstanding writer was Leonardo of Pisa, also known as Fibonacci, whose widely known book, *Liber Abaci* (1202), helped spread the Hindu-Arabic system of numeration in Western Europe, but not without considerable resistance.

Introduction of Hindu-Arabic Numerals and a New Interest in Arithmetic

The Hindu-Arabic Numerals had found their way into Europe originally through the contacts of traders with Moorish merchants, and through scholars who studied in Spanish universities. At first not well received, their adoption was greatly hindered by inertia and prejudice, and it was not until the sixteenth century that their numerals were in common use throughout Europe.

The fall of the Byzantine Empire, about 1450, saw a revival of interest in mathematics, particularly a study of the original Greek works. Scholarly activities had also moved northward into Central Europe. One of the most influential writers at this time, the German mathematician Regiomontanus, famous for his work in trigonometry, gave considerable impetus to the general interest in mathematics. During the two centuries following Fibonacci, the pace of progress had accelerated: printing had become a fact; the Hindu-Arabic numerals were beginning to take hold; and interest in mathematics was spreading beyond the Italian cities. Close upon the heels of the trigonometry of Regiomontanus came Luca Pacioli's *Summa de Arithmetica* (1494), one of the earliest and most-celebrated printed books on arithmetic. By now the Hindu-Arabic numerals were fairly well established.

Sixteenth-century arithmetic flourished, although new horizons in mathematics also began to dawn. More and more scholars became interested in mathematics; textbooks became more plentiful; interest in science and mechanics increased; and mathematics was being studied for its own sake.

During this fruitful period we find Robert Recorde's *The Ground of Artes* (London, 1540), one of the most popular arithmetics ever printed; the Dutch writer Gemma Frisius, author of a popular text on combined theoretical and commercial arithmetic; Simon Stevin, the Flemish mathematician who was instrumental in furthering the general acceptance of decimal fractions; and John Napier, the aristocratic Scotch baron whose invention of logarithms (1614) was epoch-making. From 1650 to 1850, the interest of mathematicians, having been touched off by the analytic geometry of Descartes and the calculus of Newton and Leibniz, was focused chiefly upon modern analysis, as well as upon algebra and geometry. As for elementary arithmetic, it was, comparatively speaking, neglected. With the universal acceptance of Hindu-Arabic notation and the recognition of the utility of the decimal notation, the "books were closed" for the time being.

The Dual Role of Arithmetic

Arithmetic from 1850 to the present time may be said to play a dual role in the cultural history of mankind. The more familiar role, and the more prosaic, is that of handmaiden to the arts and sciences, as well as to business. The extraordinary utility of arithmetical computation, as well as of elementary mathematical analysis, has been aptly described by L. Hogben, H. G. Wells, Herbert McKay, and many others. It is a well-known story which need not be reiterated here.

Perhaps the most spectacular development in the field of computation in the last decade or two has been the amazing development of electronic computers. To be sure, both in theory and in practice, these so-called "giant brains" involve far more mathematics than elementary arithmetic. The story of the development of computing machines is a fascinating one, too long to be told here in detail. But it is a far cry from Babbage's "analytic engine" of 1850 to the now famous Eniac of 1950. Ironically enough, our culture, insofar as science

and technology are concerned, is now at a point where progress, certainly in some areas, is no longer possible by individual effort alone, but requires the cooperative efforts of a group of related specialists. Thus in developing a typical I.B.M. machine, a fifty-man team may well be used: twenty mathematicians, twenty engineers, and ten technicians. That there is a leader of the team, a person who co-ordinates the work and directs the project, does not alter the sober fact that no one person can understand all of the theory and intricacies of the machine.

Notably less familiar, at least to the layman, is the second role played by arithmetic in the last hundred years. Yet in many ways it is more subtle and more profound. We refer to the role of arithmetic as catalyst to the comparatively modern examination of the logical foundations of all mathematics, the search for the "structure" of mathematics. This tremendously significant hallmark of twentieth-century mathematics was touched off first by making arithmetic more abstract and generalized, and then by subjecting algebra and analysis to severe "arithmetization." We shall try to make this clear in a few words.

The Search for the Structure of Mathematics

Pythagoras was convinced that all mathematics could be based on the ordinary numbers 1, 2, 3, ... Mathematicians of the eighteenth and early nineteenth centuries departed drastically from this naïve point of view by successive extensions of the concept of number from the ordinary whole numbers—extensions to the negative integers and zero, to fractions, to irrationals, to real and to complex numbers—concepts that would doubtless have bewildered Pythagoras. But the middle of the nineteenth century was to witness a revolution: mathematics won a freedom of imagination hitherto unknown. It was anticipated by the invention of non-Euclidean geometry by Lobachevski and the abstract approach to algebra initiated by Peacock,

Gregory, and De Morgan, all in the 1830s. The essence of their approach was to regard geometry and algebra, each respectively as an abstract hypothetico-deductive system in the manner of Euclid. A dozen years later, when Hamilton rejected the commutative law of multiplication (saying in effect, "Let us see what kind of an algebra or arithmetic we get if we assume that $a \times b$ does *not* equal $b \times a$"), the floodgates were opened. From that moment on, mathematicians devoted more and more attention to deliberate generalization and abstraction, exploring the full implications of postulates, and seeking an underlying structure of mathematics.

Contemporary Mathematics

The movement gathered momentum. About 1850, Boole expounded his *Laws of Thought*, foreshadowing modern symbolic logic; about 1875, the nature of the real number system was attacked in earnest by Cantor, Dedekind, Weierstrass, and others. In 1899 the die was cast: Hilbert's logical foundations of geometry sounded the keynote for postulational methods. Accordingly, geometric entities and numbers, as such, became pure abstractions, and the really important question for investigation was the nature and structure of the relations between these abstract concepts. Oddly enough, some ten years earlier, Peano had set forth his set of postulates for common arithmetic and deduced from them, by rigorous logic, the entire body of arithmetic based upon the ordinary, or natural, numbers. So the pendulum returned once more to Pythagoras.

The postulational technique thus initiated proved to be the most powerful single influence of twentieth-century mathematics. Contemporary mathematics is to be distinguished from all previous mathematics in two vital respects: (1) the intentional study of abstractness, where the important considerations are not the things related, but the relations themselves; and (2) the relentless examination of the very foundations—the fundamental ideas—upon which the elaborate superstructure of mathematics is based.

The validity of mathematical reasoning cannot be ascribed to the nature of things; it is due to the very nature of thinking. But the average man or woman does not customarily engage in a level of thinking that involves such abstraction and generalization. It is chiefly for this reason that mathematics repels so many people; the subject is too recondite. In this connection we recall the words of the late Professor C.H. Judd,* who reminds us that

> . . . children are not born with a number system as a part of their physical inheritance; they are not endowed at birth with number ideas in any form. The school puts them in contact with a system of number symbols which is one of the most perfect creations of the human mind. In the course of their acquisition of this system, they learn how to think in abstractions with precision. They learn how to use an intellectual device which no single individual, no single generation, could possibly have evolved. In the short span of a few years a child becomes expert in the use of a method of expressing ideas of quantity which cost the race centuries of time and effort to invent and perfect.

Mathematics is a linguistic activity; its ultimate aim is preciseness of communication. Second only to the mother tongue, the language of number is without doubt the greatest symbolic creation of man. And in some ways it is an even more effective agency of communication than the vernacular. In short, mathematics is a great cultural heritage, and although the beginnings have been lost in the mists of time, it is a heritage we should be proud to transmit to the world of tomorrow.

---

*Charles H. Judd, *Educational Psychology* (New York: Houghton Mifflin, 1939), p. 270.

Mathematics and World History

WILLIAM L. SCHAAF

A SUGGESTIVE if not unique interpretation of the history of mathematics is that proposed by Oswald Spengler in the first volume of his *Decline of the West*, in the chapter entitled "The Meaning of Numbers." This explanation, which might be termed a *morphological* interpretation, is based on the thesis that every outstanding culture is (or was) an organic structure; as such, every culture has its own mathematic, which is a morphological and distinctive characterization of that culture. Accordingly, the so-called growth of mathematics throughout the ages has *not* been a continuous, homogeneous and cumulative development. Furthermore, the particular "mathematic" of any given culture is inherently a necessary part of that culture, and expression of its world-being, just as its literature, music, art, and morals are also an intrinsic part of that culture. In short, the mathematic of a culture is the culmination of the symbolic expression of the soul and spirit of that culture.

The following quotation sets forth Spengler's contention succinctly. He says:

> If mathematics were merely a science like chemistry or astronomy, it would be a simple matter to define its object. But we know that this is not so. In short, there is no mathematics, only several mathematics. The so-called history of mathematics, implying the progressive actualizing of a single invariable ideal, is in reality, below the deceptive surface of history, a complex of self-contained and independent developments—an ever repeated process of bringing to

Reprinted from *Mathematics Teacher* 23 (Dec., 1930): 496–503; with permission of the National Council of Teachers of Mathematics.

birth new form-worlds, with the concomitant process of appropriating and transforming alien form-worlds. In other words, it is the typical organic process of blossoming, ripening, wilting and dying within a given period. We must not let ourselves be deceived. The mathematics of the western soul, while possessing the classical soul outwardly, though by no means inwardly, had to win its own self-realization by apparently altering and perfecting, but in reality destroying the Euclidean system.

The outstanding characteristic of Greek mathematics is the conviction that number is the essence of all things perceptible to the senses. The pervading spirit was that of measuring the immediate here-and-now. The outlook of the classical man in general was confined to the sensibly perceptible present. The dominant features of the nude statue are its dimensions, its surfaces, and the sensuous relation of the parts.

Similarly with the notion of space. The classical conception was to regard mathematics as an expression of magnitude and form through the medium of number. Number was to the ancient mind an "optical symbol" rather than the description of an abstract relation. Fundamentally it was simply stereometry, where numbers were merely units of lengths or surfaces or volumes, and no other extension was conceivable. Thus to Euclid, a triangle is the boundary of a fixed, rigid body, never a system of three intersecting variable lines or a group of three movable and related points. And again, to Euclid a "line is length without breadth"; to us such a statement is no definition at all, yet in the classical mathematics it was brilliant. In short, classic number is a thought process dealing not with spatial relations, but with visibly limited, tangible, and concrete entities.

Hence it is not surprising that the Greeks concerned themselves primarily with natural numbers. The Greek spirit could not appreciate the significance of the irrational number, the non-ending decimal of our present day notation. Euclid himself declared that "incommensurable lines are not related to one another like numbers." Here, according to Spengler, was a very real fear; the "unspeakable and formless must be left forever hidden." It was the same kind of fear that prevented the Greeks from expanding politically, confining themselves for the most part to the gates of their own cities; from entering boldly upon unknown paths in the Mediterranean, and from delving into Babylonian astronomy—a metaphysical fear that the sense-comprehended-world-of-the-present might otherwise precipitate itself into an unknown primitive abyss of darkness. Every product of the classical mind had to do with number. That which could not be drawn was not number. The Greeks used terms equivalent of "face number" and "volume number" for the second and third power. Hence the higher powers were quite unreal to them, and expressions like $x^{2/3}$, $x^{1/2}$, or e^{ix} would have been meaningless, and indeed, quite unthinkable. Similarly zero was not considered a number, since it could not be drawn.

When the mathematics of the seventeenth century was "revived," there was no concomitant revival of Greek mentality. Philosophy is very sensitive to differences in the state of mind and the culture of an age. Our mind—the mind of the modern Western world—is very differently constituted from the Classic mind. While we may produce a faithful replica of an ancient statue, we cannot similarly reproduce an ancient state of mind. We can sympathetically understand the past, but there is a profound contrast between modern and ancient reactions to the same stimuli. Hence there is apparently but little justification for regarding the classical mathematics as an early "stage" in the development of modern mathematics. For the

ancients, their mathematics was complete: for us, it is incomplete. Moreover, modern mathematics, while true for us, and complete so far as we are concerned, is for us a masterful creation of the Western spirit; yet to Plato it would have seemed meaningless, even ridiculous, since it would have been in utter discord with truth as he conceived it.

The Magian Mathematics

It is generally said that Diophantus extended the limits of the classical arithmetic. What he really did was to create not only a new symbolism, but an entirely new algebra. Properly speaking, Diophantus does not belong among the classical mathematicians at all. He is the spokesman of a new number-feeling which transcended the Hellenic feeling of sensuously present limits that had produced Euclidean geometry, the nude statue, and the coin.

The details of the origins of this new mathematic are not well known. In general, there is the presumption of an Indian influence. Diophantus lived about A.D. 250 (the third century of Arabian culture), and the influence of Persian and Arabian culture is plainly seen in this work. No Greek, for example, would have stated anything about an *undefined number a* or an *undenominated number six*—for these are neither magnitudes nor lines. Diophantus grasped and used a new number feeling, one which is distinct from the classic feeling of intention. Thus Diophantus did not *extend* the notion of number as magnitude—he *eliminated* it, and so unwittingly overthrew the classical meaning of number, supplanting it by an entirely new number spirit which Spengler designates the "Magian."

The beginnings of the Magian or Arabic culture, which are somewhat obscured by the last vestiges of the decaying classical civilization, had in all likelihood been germinating in the East since the time of Augustus in the regions between Armenia and Southern Arabia. The fact that Alexandria and Antioch still wrote in Greek is accidental, and of no more importance than the fact that Latin was the scientific language of the Western world up to the beginning of the eighteenth century.

With Diophantus number no longer represented measure and the essence of static, immovable things. While he did not recognize zero and the negative, he no longer regarded numbers in the old Pythagorean sense. On the other hand, the Arabic-Persian indeterminateness of number was something very different from the systematic variability of numbers which appeared later in the Western mathematics and which in fact is the beginning of the function concept. So, while the Magian mathematics is somewhat vague in detail, the trend is clear enough. Diophantus was indeed not its creator—it had begun before his time, and culminated in the ninth century with al-Khowarizmi and others. His importance lies in the fact that, so far as we know, he was the first mathematician in whom the new number feeling was unmistakably present.

Western Mathematics

In the light of the foregoing, let us reconsider the contribution of Descartes in 1637. It was evidently *not* the introduction of a new method or concept in the field of traditional geometry, i.e., the rigid system of Cartesian coordinates—this had already been used and understood by Oresme. Descartes neither improved nor extended the notion of coordinates—instead he *overcame* this semi-Euclidean concept by introducing an entirely *new number idea,* that of a point being a group of coordered pure numbers. The replacement of lengths by positions carried along with it a purely abstract, spatial conception of extension instead of a material one. Defining a point as a *manifold of numbers* is the beginning of analysis—and this is Descartes's contribution to the birth of Western mathematics.

As a striking illustration of the annihilation of the inherited "optical" geometry of extension, take the notion of angular functions. In the Indian mathematics these were numbers in the early sense; but with the beginning of the Western mathematics they pass into periodic functions, and thence, through the medium of analysis, into infinite series, where not the least vestige remains of the original Euclidean figure. Consider, for example, expressions like e , $\sqrt[n]{x}\, i^{i}$, or sin $x = e^{x}$, where no one thinks of drawing diagrams or of calculating the powers numerically. We recall the story concerning Benjamin Peirce, who, during one of his lectures, having derived the relation $e^{i\pi} = -1$, turned to his class and said: "Gentlemen: that is surely true; it is absolutely paradoxical, we cannot understand it, and we don't know what it means; but we have proved it, and therefore we *know* it must be the truth!"

In other words, as Spengler points out, the classical concept of number as pure magnitude was definitely supplanted by the concept of number as pure relation. Thus the keynote of Western culture is the function concept, a notion not even remotely hinted at by any earlier culture. And the function concept is anything but an extension or elaboration of previous number concepts—it is rather a complete emancipation from such notions. To quote from Spengler once more:

> In all history, there is no second example of one culture paying to another culture long extinguished, such reverence and submission in matters of science as ours has paid to the classical. It was very long before we found courage to think our proper thought. But though the wish to emulate the classical was constantly present, every step of the attempt took us further away from the imagined ideal. The history of Western knowledge is thus one of *progressive emancipation* from classical thought, an emancipation never willed, but enforced in the depths of the unconscious. And so the development of the new mathematics consists of a long, secret and finally victorious battle against the notion of magnitude.

For the classical mathematician, the guiding spirit was always an ordering of the become in terms of the visible present, the measurable and the numerable. For the mathematician of the West, whose symbol is imperceptible, unlimited space, the guiding principle is an unrestrained, far-reaching, form-feeling Gothic spirit. Originally, geometry meant the art of measuring; arithmetic, the art of numbering. Modern mathematics has long since ceased to deal with either of these as such, although it has not yet found a new name for its own entities—for the word analysis is hopelessly inadequate. What we understand by that term embraces among a host of related concepts, the notion of a universe of infinite space, a concept which simply did not exist for classical man—it is not even capable of being presented to him.

Several outstanding distinctions between classical and Western mathematics are given below in the form of a summary:

SIGNIFICANT CONTRASTS

Classical Mathematics	*Western Mathematics*
1. Accepts only what can be seen and grasped; where definitive and *physical visibility* ceases, there mathematics and logic also cease.	1. Abandons classical fetters, and becomes absorbed in highly *abstract n*-dimensional manifolds of space, spurning diagrams and other commonplace aids.
2. Concentrates on the consideration of the *small*, being handicapped by the principle of *visible limits;* hence the impossibility of conceiving non-Euclidean geometry.	2. Centers interest in the consideration of the *infinite* and *"ultra-visional,"* including the infinitely large as well as the infinitesimally small.
3. Conceives of a limit as an infinitely small quantity, yet *fixed or static* (Euclid).	3. Conceives of the lower limit of every possible finite magnitude; a *becoming and remaining smaller* than any previously assignable quantity, however small (Cauchy).
4. Interested primarily in *magnitude,* and hence in *proportion;* all proportion assumes the constancy of its elements. Statues and frescos admit of enlargements and reductions.	4. Interested primarily in *relationships,* and hence in *functions;* all transformation implies variability in its elements. Transformations are related to the theory of modern musical composition, but enlargements and reductions are meaningless here.
5. Interested chiefly in *particular* cases and individual instances, i.e., a singly visible figure, a once-and-for-all construction. *Geometric constructions affirm appearances.*	5. Interested chiefly in *generalizations,* i.e., operations not dealing with fixed visible figures, such as groups of relations, infinite curves, transformations, etc., where the process is of greater interest than the result. *Operations deny appearances.*
6. Gives artistic expression to its consciousness through the media of bronze and marble, where the human figure, whether dancer or gladiator, is given that *fixed form* in which contour, surface and texture are most expressive and effective.	6. Manifests its artistic feeling in *formless* music, where harmony and polyphony call forth feelings of an infinite beyondness anything but visible; or in a gloriously colored canvas, where *light and shade* alone suffice to mark the outline.

The annihilation of the bondage of geometry to the visual, and of algebra to the notion of magnitude, meant the substitution of the function concept. It meant replacing static and concrete forms by dynamic and abstract space relations. The elements of space became a group of numbers, i.e., images of the point, line, plane, etc. Spiritual perception instead of sensual actuality—this is the Spirit of the West. The culmination of this spirit is to be found in modern group theory; groups are mathematical images, such as the totality of differential equations of a certain type, various transformations, and the like. The Faustian spirit was fulfilled with the culmination of Western mathematics. Each of these two gigantic culture structures arose out of a wholly new number concept—in one case, that of Pythagoras, in the other, Descartes. Each flourished for 300 or 400 years, and then completed the structure of its ideas at the same moment that the respective cultures to which they belonged passed into the phase of megalopolitan civilization.

What remains for Western mathematics is the process of preserving, selecting and refining—in place of big dynamic creation, the same kind of "clever, subtle, patient detail work which characterized the Alexandrian mathematic of late Hellenism." And so the day of the *great* mathematician, according to Spengler, is past.

Conclusion

Many of Spengler's critics, of whom there are not a few, argue that he begins with a definite thesis or point of view, of which he is already convinced, and therefore selects only such historical data as will support his contention, carefully omitting any allusion to such evidence as would conflict with it. Be this as it may; but certain it is, that much can be said for this reinterpretation of the history of mathematics.

If the reader should feel that Spengler's interpretation, while perhaps interesting, nevertheless merely represents the opinion of an historian, he may be interested in the following extract from J. W. N. Sullivan's *Aspects of Science (First Series)*, in the chapter entitled "Mathematics as an Art":

The significance of mathematics is precisely that it is an art—it informs us of the nature of our own minds. It does not enable us to explore some remote region of the eternally existent; rather it helps to show us how far what exists depends upon the way in which we exist. We are the law givers of the universe—it is even possible that we can experience nothing but what we have created, and that the greatest of our material creations is the material universe itself.

According to the Classical conception, geometry was considered a necessity of thought. More than that, it was regarded as the necessary geometry of any space. Yet the recent development of non-Euclidean geometry and its application to physical phenomena by Einstein has shown us that Euclidean geometry is not only not a necessity of thought, but that it is not even the most convenient geometry to apply to existing space.

Thus the modern mathematician, in choosing his postulates, is free to imagine what he wishes; he constructs worlds of his own fancy and desire, limited only by the imagination. Hence he is not discovering fundamental laws of the Universe, nor is he becoming acquainted with God. If he finds a perceptual set of entities which follows the same logical scheme as his mathematics, he has applied mathematics to the external world. But why the external world should obey the laws of logic—why, indeed, science should be at all possible, is still a difficult question to answer. There are indications in the modern theories of physics which cause some scientists to doubt whether the universe will turn out to be rational after all.

Thus we return to a sort of inverted Pythagorean outlook. Mathematics is of profound significance in the universe, not because it exhibits principles that we *obey*, but because it exhibits principles that we *impose*. It shows us the laws of our own being, and the necessary conditions of experience. In short—we understand Nature so well because we have designed it.

HISTORICAL EXHIBIT I.I

The Growth of Mathematical Knowledge

Mathematical knowledge is cumulative, that is, people use and build upon the mathematics developed before them. If this fact seems reasonable then a model for the growth of mathematical knowledge can easily be devised by counting the number of recorded mathematicians that have lived up to any period of time and then plotting that number against the chronological year. Such a count was undertaken by noting the mathematicians listed in the comprehensive *Dictionary of Scientific Biography*, 16 vols. (New York: Scribner's, 1970–80). Cumulative sums were computed for the various years 600 B.C. to A.D. 1850 and plotted against their respective years. The following graph was produced:

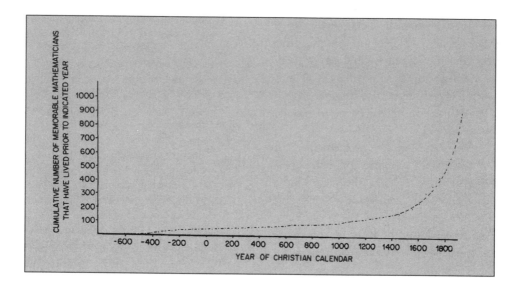

Several interesting trends can be noted from this graph, for example, the historical period A.D. 600 to 1200 is often termed the "Dark Ages," a time when little intellectual progress was taking place in Europe; yet the graph indicates a steady, albeit slow, growth rate of mathematical knowledge over this period. Around 1400, the rate of mathematical activity increased, due possibly to the introduction of printing and the writing of books in the common language. After A.D. 1500, the growth rate of mathematical knowledge increased exponentially.

The History of Mathematics
as a Part of the History of Mankind

ARTHUR O. GARDER

MANY CONCLUSIONS useful for the future of mathematics might result from a sustained investigation into the relationship between the history of mathematics and the history of the societies in which mathematics has developed. For example, is it possible to reach some conclusion about the nature of the social, economic, and political milieu which is optimal for the development of science in general and of mathematics in particular?

In raising such a question, we must remember that mathematics is unique among the sciences in the breadth of motivations which have generated it. Engineers, physicists, and philosophers have stimulated the growth of mathematics. The range of motivations lying between those of an Archimedes manufacturing clever devices of war for the tyrant of Syracuse to fight the Romans and those of a Bertrand Russell examining the foundations of logical thought spans the spectrum of intellectual purpose.

Thus we would expect the development of mathematics to be closely related to the level of tolerance and encouragement which a given society exhibits toward the development of the arts and sciences as a whole. To take a very well-known example: The Greek civilization, which first pursued mathematical ideas for their own sake, also excelled in almost every field of the human intellect.

A less well-known example, at the opposite extreme, is provided by the almost total lack of significant mathematics in European culture during the period which Will Durant [2]* has well named the "Age of Faith." It began with the accession of Constantine in A.D. 324 and ended with the death of Dante in A.D. 1321. The only significant creative mathematician to appear in Europe during this period was Leonardo Fibonacci of Pisa (1175–1250) [5]. It is surely no coincidence that Fibonacci was a contemporary and subject of the Emperor Frederick II of the House of Hohenstaufen (1194–1250).

Had Frederick not been an emperor, he would surely have been burned at the stake for his views. Had he not brought (if only for a historical moment) the precepts of the Age of Reason into the heart of the Age of Faith, there might well have been no Fibonacci sequence in our number-theory textbooks. Frederick spoke several languages, including Arabic. Recognizing the superiority of the contemporary Moslem culture, which was built upon the Hellenistic culture found by the Arabs in their conquered territories, Frederick abandoned the medieval Christian faith. Jewish scholars attended him at his court. In 1224, he founded the University of Naples—"a rare example of a medieval university established without ecclesiastical sanction" [2]. Rejecting generally accepted superstitions, Frederick asked: "How can a man believe

Reprinted from *Mathematics Teacher* 61 (May, 1968): 524–26; with permission of the National Council of Teachers of Mathematics.

* See the bibliography at the end of the article.

that the natural heat of glowing iron will turn cool without an adequate course or that, because of a seared conscience, the element of water will refuse to accept the accused?"

Fibonacci and Frederick died in the same year. No significant mathematics appeared again in Europe until the sixteenth century. We must, however, remark that except during the Dark Ages which spanned the early part of the Age of Faith until the First Crusade in 1098, art, music, architecture, literature, and philosophy were generated. But all of these bore the stamp of conformity to the prevailing theological view of life.

The Spanish civilization used the Inquisition to prolong the medieval view of life well into modern times with identical results, i.e., a complete extinction of scientific achievement combined with continued artistic creativity. Undoubtedly the urge to create science is strong enough to survive—and it may even be stimulated by—a certain level of persecution. The humiliation of Galileo before the Inquisition, for example, by no means ended mathematical achievement in Italy, as evidenced by the work of Torricelli, Viviani, and Cassini.

It is also clear that the attitudes of leaders of human thought and human affairs outside of mathematics may greatly affect the course of mathematics itself. The birth of the modern intellect began with the three almost simultaneous movements of the Renaissance, the Reformation, and the awakening of science. Sometimes these developments helped, and sometimes they hindered, each other. The battles of Galileo and Newton were in large part directed against the reverence for Greek views of the universe which the Renaissance had fostered [1].

Martin Luther, whose viewpoints were highly influential in Germany, shared the distrust of his theological contemporaries for the Copernican view of the universe. However, by supporting the mathematics textbook writer Adam Riese, Luther furthered mathematical development. In a widely circulated tract published in 1524, Luther wrote:

If I had children, they should not only study language and history, but they should also learn singing and music, together with the whole of mathematics. [7]

After his famous visit to England, Voltaire popularized Newton's work in France. He pointed out to the unwilling ears of the French that their patriotic addiction to the memory of Descartes was hindering their scientific progress.

II

What can we say of the effect of the modern societies' sponsorship of science and mathematics? Surely the bulk of the government support for science in Russia and the United States is not intended to serve the disinterested goals of free human inquiry. Has any study been made of such influences upon the course of modern mathematics?

Surely those influences upon the course of the nature of man himself of which Omar Khayyam and Galileo complained in such a similar fashion are with us still and are helping to shape the course of mathematical development. In the preface to his algebra, written about A.D. 1000, the Arab astronomer, poet, and mathematician wrote:

We have been suffering from a dearth of men of science, possessing only a group as few in numbers as its hardships have been many. Most of our contemporaries are pseudoscientists, who mingle truth with falsehood, who are not above deceit, and who use the little that they know of the sciences for base material purposes only. When they see a distinguished man intent on seeking the truth, one who prefers honesty and does his best to reject the falsehood and lies—avoiding hypocrisy and treachery—they despise him and make fun of him. [4]

Six hundred years later, Galileo wrote to Kepler:

I think, my Kepler, we will laugh at the extraordinary stupidity of the multitude. What do you say of the leading philosophers here to whom I have offered a thousand times of my own accord to show my studies but who have never consented to look at the planets, or moon, or telescope. [3]

The same theme was taken up again by Bertrand Russell in his speech accepting the Nobel Prize [6]. Russell catalogued the really influential motives in human politics as: (1) vanity, (2) rivalry, (3) acquisitiveness, and (4) love of power. Surely the development of mathematics and science has reached a stage where the creation of new scientific knowledge is vying for influence with these four motives. But if the developments of science are to be controlled by the motives listed by Russell, as indeed seems inevitable, what are the prospects for the future of man?

III

How shall we view the relationship between war and the development of science and mathematics? It is clear that war, and the fear of war, has vastly increased the sum of money spent upon scientific development. It is also clear that scientists have generally welcomed the opportunity to spend the money. The desirability of the accelerated development of science is seldom questioned.

It is, however, not unlikely that future wars may be even more destructive in their final effects upon mathematical and scientific development than was the Thirty Years' War in its destruction of German science and culture. In 1618, the year in which this war began, Johann Kepler published "The Harmony of the World" [3], which contained his third law of planetary motion. The world promptly became so disharmonious that Kepler managed to publish only one more significant work, at the cost of incredible personal hardship. In 1627 he published the Rudolphine tables, which contained the stellar observations of Tycho Brahe. These had been made before Tycho's death in 1601. The next significant German mathematician, Leibniz, was not born until 1646, two years before the war's end.

At the other end of the scale, the "Golden Age of Mathematics," which history may finally record as the Golden Age of Western civilization, very nearly coincided with the century between the end of the Napoleonic wars and the outbreak of World War I. Surely it is not an accident that this century marked a closer approximation to world peace than any other century since the Roman peace. If there were holocausts, at least they were localized.

If such considerations cannot be made as precise as the theorems of mathematics, they nonetheless contain the vital issues for the future of mathematics and of man. If mathematics is a science, the history of man and the history of mathematics are not sciences. There seems to be a tendency in all phases of history to pursue only questions of factual interpretation, because such limitations make the subject appear to be a science. This turns history into a pseudoscience and prevents its practitioners from coming to grips with its vital problems, which still, unfortunately can only be dealt with by the methods of philosophy.

Perhaps one conclusion we may draw is that a too-great pressure to conform, a too-high level of intolerance of a certain variety, can stifle mathematics and science. Does not this also suggest the possibility that too great a conformity to the goals of science might in modern societies stifle other components of the human spiritual and/or intellectual life?

BIBLIOGRAPHY

1. Bixby, William. *The Universe of Galileo and Newton.* New York: Harper & Row, 1964. Pp. 57, 62, 97.

2. Durant, Will. *The Age of Faith.* New York: Simon & Schuster, 1950.

3. Koestler, Arthur. *The Watershed.* New York: Doubleday & Co., 1960. P. 213.

4. Lamb, Harold. *Omar Khayyam.* New York: Doubleday & Co., 1936. P. 309.

5. Newman, James. *The World of Mathematics*, Vol. I. New York: Simon & Schuster, 1956. P. 117.

6. Russell, Bertrand. *Human Nature in Politics.* New York: Heritage Productions LPA–1202, 1953.

7. Smith, D. E. *History of Mathematics*, Vol. I. New York: Dover, 1958. P. 338.

The Dependence of Mathematics on Reality

DAVID W. HANSEN

$\mathcal{M}$OST MATHEMATICIANS today regard mathematics as an abstract structure built up from axioms by the use of logical principles. By altering these axioms, different types of mathematics can be developed; for example, Euclidean and non-Euclidean geometries differ by only one axiom—the parallel postulate—but this difference is quite sufficient to lead to very different conclusions in each system. This modern concept of mathematics stresses the idea that man *invents* his mathematics (and hence controls it to some extent) rather than *discovers* it (the assumption here being that mathematics exists apart from and independent of man). Mathematics becomes an intellectual game. You select your axioms according to taste and work out the corresponding mathematics, using logical principles. If the results are aesthetically pleasing or elegant, this new system becomes a part of mathematics. Notice here that for the modern mathematician there is no mention of applicability to reality. The system stands alone on its consistency and beauty. If, for some reason, the system can be used in explaining certain properties of the real world, this is of no import to the mathematician. G. H. Hardy, in his book *A Mathematician's Apology*, expresses this view quite forcefully. It has been said by Huxley that six monkeys pecking unintelligently on

typewriters would, after millions of millions of years, eventually write all the books in the British Museum (Sir James Jeans quotes Huxley's statement on page 4 of *The Mysterious Universe*). Similarly, mathematicians randomly selecting various axioms would eventually produce significant mathematical systems, but undoubtedly only after an enormous amount of time had been wasted.

Now I do not claim that this is really the way mathematics is produced. In fact, my thesis in this paper is that great mathematics is seldom done by playing these intellectual games; rather, the great mathematicians, in doing their mathematics, have had an eye out for something external to themselves—let us call it the real world—and their hope has been that their mathematics would lead not only to beauty and an internal truth but to external truths as well—namely, *significant knowledge of the real world*. Now, the present-day concern with and emphasis on axiomatics (to the exclusion of everything else) has been of fairly recent origin, getting its start in the late nineteenth century or very early twentieth century. Some observers feel that this emphasis has caused a split between mathematics and science. By tracing the development and interests of several mathematicians from the past who have without doubt significantly contributed to the advancement of mathematics, I hope to show that this modern trend of overemphasis on axiomatics should at least be modified to include consideration of the

Reprinted from *Mathematics Teacher* 64 (Dec., 1971): 715–19; with permission of the National Council of Teachers of Mathematics.

significance of reality in any development of new mathematics.

In *Men of Mathematics* (1937, p. 218), E. T. Bell says:

> Archimedes, Newton, and Gauss, these three are in a class by themselves among the great mathematicians, and it is not for ordinary mortals to attempt to range them in order of merit. All three started tidal waves in both pure and applied mathematics: Archimedes esteemed his pure mathematics more highly than its applications; Newton appears to have found the chief justification for his mathematical inventions in the scientific uses to which he put them, while Gauss declared that it was all one to him whether he worked on the pure or applied side.

Although practically all mathematicians before 1900 were concerned with the real world, because of space limitations I shall use the achievements of the three great mathematicians mentioned by Bell as being representative of the works of many others.

Let us begin with Archimedes (ca. 287–212 B.C.). Plutarch reported that Archimedes considered

> mechanical work and every art concerned with the necessities of life an ignoble and inferior form of labor and therefore exerted his best efforts only in seeking knowledge of those things in which the good and the beautiful were not mixed with the necessary.

And yet he has been called the "spiritual father of our modern institutes of technology" (Meschkowski 1964, p. 14). A brief list of his accomplishments would have to include his discovery of the laws of levers; his creation of the science of hydrostatics; and his determination of methods for calculating not only the areas and centers of gravity of flat surfaces and solids but also the areas of curvilinear plane figures and volumes bounded by curved surfaces, using the techniques of the integral calculus, thus anticipating Newton by more than 2,000 years! Now, these accomplishments were tremendous and clearly indicate that Archimedes was vitally concerned with the real world, Plutarch's remark notwithstanding. But, it is of some importance to consider his method for creating all this new knowl-

edge. His works, being expressed in severely economical and logical terms, do not contain any hints. However, in 1906, J. L. Heiberg discovered in Constantinople a previously unknown treatise, *On Mechanical Theorems, Method*, written by Archimedes to his friend Eratosthenes. In it he explains how, by weighing in his imagination a figure whose area or volume he wanted against a known one, he was able to determine the facts he sought. Archimedes says (Meschkowski 1964, p. 15):

> Certain theorems first became clear to me by means of a *mechanical* method. Then, however, they had to be proved geometrically since the method provided no real proof. *It is obviously easier to find a proof when we have already learned something about the question by means of the method* [through experience of the real world] *than it is to find one without such advance knowledge.* [Italics mine.]

Thus Archimedes, disdaining the practical side of knowledge, nevertheless used his mechanics as a means to discover his mathematics. Archimedes's development of his mathematics, by his own admission, was dependent on information supplied by other than pure axioms or postulates. His pure mathematics was based on mechanical conceptions suggested by the real world.

Let us now turn to Newton (1642–1727). What can be said or written of Newton that is not already familiar to all those in scientific professions? His creation of the fundamental laws of motion, his theory of universal gravitation, his work in optics would each have made him immeasurably famous; but in addition he created the differential and integral calculus, a subject that has been developing now for over 300 years and has led to modern function theory. But why was this imaginative and significant body of knowledge created? It was in order to make mathematically precise Newton's entire theory of celestial mechanics. For example, his solution to the problem of finding the rate of change of momentum, as stated in his second law of motion, gave a workable method for determining the velocity of a particle moving in a continuous manner and became a fundamental part of the differential calculus. A related problem—namely, that of determining the total distance traveled in a given time by a moving particle whose velocity is continuously varying—was solved by the use of the integral calculus he created. Margenau states (in Woodruff 1941, p. 100) that it is quite certain that Newton invented the calculus partly to prove that Kepler's laws were exact consequences of the inverse-square law of universal gravitation.

Although Newton first became aware of the consequences of the law of gravitation about 1664 or 1665, it was not until twenty years later that he published them. Why the long delay? According to E. T. Bell (1937, p. 106), this delay was caused by Newton's inability to solve the problem of finding the total attraction of a solid homogeneous sphere on any mass particle outside the sphere. It is a problem in integral calculus and was finally solved by Newton when he showed that the attraction is the same as it would be if the entire mass of the sphere were concentrated in a single point at its center. Surely this shows clearly that Newton's motive in creating and developing the calculus was not solely to deduce a pure, abstract system. It has taken mathematicians over 250 years to give rigor to Newton's basic ideas of the calculus, and in the process much that is interesting mathematically has been born; but I doubt that anyone would challenge the statement that it was the creation of the calculus under the impetus of elucidating a theory about the real world which gave birth to the developments of modern mathematics. This is not to say that *all* mathematics is based on a search for an explanation of the real. As Morris Kline (1959, p. 534) says:

> The history of such (mathematical) creations shows that while nature is the womb from which mathematical ideas are born, these ideas can be studied for themselves while nature is left behind.

It is simply my contention that the truly great developments in mathematics have been stimulated by considerations of physical reality.

Let us finally turn to Gauss (1777–1855), who has been aptly called the "prince of mathematicians." His works encompass pure number theory, algebra and analysis, applied mathematics, astronomy, and physics. His elegant, compact style of writing, like that of Archimedes, gives no inkling of his methods of discovering proofs. As Abel (Norwegian mathematician, 1802–1829) said of Gauss, "He is like the fox, who effaces his tracks in the sand with his tail" (Meschkowski 1964, p. 62). Gauss himself said that a cathedral is not a cathedral until the last scaffolding is down and out of sight. He also described himself as "all mathematician." This is simply not true unless one realizes that *mathematician* in his day included what we would now call a mathematical physicist (Bell 1937, p. 230). It is important here to recognize that much of his scientific work was given over to the study of astronomy. This great mathematician also founded the science of magnetism and has been honored by having one of its units of measurement—the gauss—named after him. His second motto (his first was *pauca sed matura*, "few but ripe"), taken from *King Lear*, was

> Thou, nature art my goddess; to thy laws
> My services are bound

Gauss anticipated the discovery of non-Euclidean geometry; gave the first wholly satisfactory proof of the Fundamental Theorem of Algebra (a polynomial equation with complex coefficients has at least one root); wrote *Disquisitiones Arithmeticae*, a work of fundamental importance to the modern theory of numbers; and contributed notably to geodesy, electromagnetism, terrestrial magnetism, differential geometry, and the method of least squares (Eves 1969, pp. 374–75). Now if Gauss, who developed much pure mathematics, was not concerned about explaining reality, it is hard to understand why: (1) he devoted valuable time to calculating the orbit of the newly discovered asteroid Ceres from meager data and in his work "Theory of the Motion of the Heavenly Bodies Revolving around the Sun in Conic Sections" expounded methods for determining planetary and cometary orbits from observational data which dominated astronomy for many years; (2) he contributed fundamental research to the mathematical theories of electromagnetism; (3) he attempted, along with Weber, to find a unifying theory for all electromagnetic phenomena; (4) he spent time in geodetical researches, out of which came the mathematics of differential geometry and conformal mapping; and (5) he created new knowledge concerning capillarity, gravitational attraction of ellipsoids, and systems of lenses. It seems quite clear that Gauss, prince of mathematicians, had strong desires to explain the real world and, in attempting to do so, created significant mathematics.

We have now briefly examined the three mathematicians called the greatest in the world and find that far from being oblivious to the real world, all three were influenced in their mathematical researches by considerations of reality: Archimedes, who used mechanical methods to suggest mathematical ideas; Newton, who developed his mathematics to aid in explaining his celestial mechanics and Kepler's laws; and Gauss, who received inspiration for his mathematics from his physical studies. I have not mentioned Archimedes's magnificent practical inventions, such as the catapult, nor Gauss's invention of the electric telegraph! Nor do I have space to go into Newton's intense interest in theological matters (he spent a large part of his life in studying theology) or Gauss's stated interest in philosophy (Bell 1937, p. 240):

> There are problems to whose solution I would attach an infinitely greater importance than to those of mathematics, for example touching ethics, or our relation to God, or concerning our destiny and our future; but their solution lies wholly beyond us and completely outside the province of science.

In conclusion, I wish to stress that these great mathematicians did not play intellectual games by weakening various axioms and deducing pretty baubles of mathematics with little significance. Rather, they let the influence of the real world guide them in their choice of mathematical axioms that led to far-reaching results. Of course, *when once started by ideas suggested by the real world*, mathematics can grow of its own accord, forming pictures of great beauty that are valuable in themselves, whether or not they have an application to physical reality. However, if left *completely* to its own devices, mathematics tends to become sterile and meaningless, pure symbol manipulation and nothing more.

Fortunately the great mathematicians have kept one eye on reality. As Morris Kline (1959, p. 537) puts it:

> Mathematics may be the queen of the sciences and therefore entitled to royal prerogatives, but the queen who loses touch with her subjects may lose support and even be deprived of her realm. Mathematicians may like to rise into the clouds of abstract thought, but they should, and indeed they must, return to earth for nourishing food or else die of mental starvation. They are on safer and saner ground when they stay close to nature. As Wordsworth put it, "Wisdom oft is nearer when we stoop than when we soar."

Today, as in the past, new mathematics must and is being developed to expound new theories of the real world. The quantum theory of Planck requires a calculus allied to that of finite differences, which is quite different from the calculus of Newton, which is dependent on continuous phenomena. Einstein's theory of relativity has produced differential equations that should supply mathematicians with many problems of great difficulty and interest for some time to come. It becomes readily apparent now, as in earlier years, that mathematics develops to its fullest when pursued with the hope of clearing away the fog and doubts surrounding our concepts of reality. This must be its primary goal.

REFERENCES

Bell, E. T. *Men of Mathematics.* New York: Simon & Schuster, 1937.

Bochner, Salomon. *The Role of Mathematics in the Rise of Science.* Princeton: Princeton University Press, 1966.

Eves, Howard. *An Introduction to the History of Mathematics.* 3d ed. New York: Holt, Rinehart & Winston, 1969.

Kline, Morris. *Mathematics and the Physical World.* Garden City, N. Y.: Doubleday & Co., Anchor Books, 1959.

Meschkowski, Herbert. *Ways of Thought of Great Mathematicians.* San Francisco: Holden-Day, 1964.

Woodruff, L. L., ed. *The Development of the Sciences.* New Haven: Yale University Press, 1923.

———. *The Development of the Sciences.* Second Series. New Haven: Yale University Press, 1941.

6

Seeking Relevance?
Try the History of Mathematics

FRANK J. SWETZ

$\mathcal{W}$HAT IS MATHEMATICS? This seems like a curious question to pose to a reading audience of mathematics educators, but I would wager that, on reflection, it is a difficult one to answer. Of course, a variety of quips and clichés can supply the answer: "It's an art," "A science—in fact, it's the queen and the servant of science," "It's what I use when I balance my checkbook," "A game that we play with rules we're not quite sure of," and the apologetic favorite, "Something I was never good at"; and so the list can go on and on. What these answers seem to avoid is the fact that mathematics is a necessary human activity, one that reflects a response to needs dictated by human existence itself. Mathematics is evident in all societies and cultures. The needs to which it responds are both material and intellectual, and as they change, so does the nature of the mathematics that serves them; thus, mathematics is a body of knowledge that is constantly evolving in response to societal conditions.

Just as each of the responses given in reply to the initial question is one dimensional, so, all too often, is our teaching of mathematics one dimensional. We frequently find ourselves concentrating on the teaching of "mathematics"—the symbols, the mechanics, the answer-resulting procedures—without really teaching what mathematics is "all about"—where it comes from, how it was labored on, how ideas were perceived, refined, and developed into useful theories—in brief, its social and human relevance. At best, this practice will produce knowledgeable technicians who can dispassionately use mathematics, but it will also produce students who perceive mathematics as an incomprehensible collection of rules and formulas that appear en masse and threateningly descend on them.

Unfortunately, in our schools, this latter group is in the majority. These students build psychological barriers to true mathematical understanding and develop anxieties about the learning and use of mathematics. Teachers can partially remedy this situation by incorporating a historical perspective into the teaching of mathematics. History is commonly taught in schools to initiate the young into a community—to give them an awareness of tradition, a feeling of belonging, and a sense of participation in an ongoing process or institution. Similar goals can be advocated for the teaching of the history of mathematics. By incorporating some history into teaching mathematics, teachers can lessen its stultifying mystique. Mathematics isn't something magic and forbiddingly alien, but rather it's a body of knowledge naturally developed by people over a 5000-year period—people who made mistakes and were often puzzled but who worked out solutions to their problems and left records of those solutions so that we can benefit from them.

Reprinted from *Mathematics Teacher* 77 (Jan., 1984): 54–62; with permission of the National Council of Teachers of Mathematics.

Its teaching should recognize and promote these people-centered facts.

Certainly, the chronicles of mathematics possess all the facets of human drama that capture the imagination and perpetuate interest: mystery, adventure, intrigue, and so on. High drama, as well as a nice introduction to the theory of mathematical modeling, can be found in a consideration of Galileo's challenge to the Ptolemic theory of planetary motion. Mystery abounds in numerology, the construction and purpose of the Egyptian pyramids, and the appearance of mathematical constants such as pi and *e* in diverse phenomena. The historical search for higher order constructible regular polygons provides adventure. Intrigue can be found in "Dungeon and Dragon"-type secret numerical codes that were popular in the Middle Ages or in the appearance of descriptive geometry in 1789 as a military weapon. The history of mathematics can convey the heights of human ingenuity and creative genius—for example, Cantor's diagonal proof of the uncountability of the real numbers—or demonstrate the foibles of human understanding as illustrated by the continued existence of proofs of the trisectibility of an angle. Such material can breathe life into mathematics lessons.

Incorporating History into Mathematics Teaching

Mathematics teaching can thus be humanized by the inclusion of historical perspectives in classroom discussions. This can, and should, be done unobtrusively by the use of historical anecdotes, films, projects, displays, and problems in teaching presentations as well as the replication of relevant historical experiments, e.g., the estimation of π by a method of exhaustion (NCTM 1969). It should be done in a natural way as an integral part of the lesson—no announcements to the effect that "now we're going to talk about the history of mathematics" should take place!

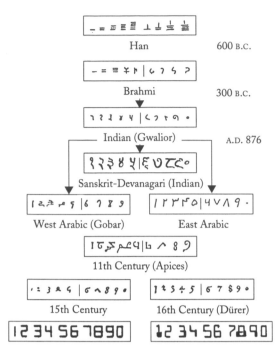

FIGURE I   *An evolution of numerals*

Consider, for example, how lessons on the development of a numeration system might be historically enhanced. A discussion can point out how our present system of numerals evolved from simple tally symbols such as $| = \equiv$ and how writing familiarity and a quest for speed transformed these symbols to $| \angle \xi$. The evolution of arabic numerals from such primitive tally strokes to the modern liquid-crystal displays of hand-held calculators or the numerals on bank checks can be demonstrated by the use of a simple chart (fig. 1). Are the black lines employed by optical scanners to read the price of a box of soap powder the modern equivalent of tally strokes?

Frequently in the teaching of number bases, references are made to Egyptian and Babylonian number systems. Instead of just talking about hieroglyphic and cuneiform numerals, one can generate much more excitement by giving students an opportunity actually to translate them from facsimiles of ancient records. A minimum student knowledge of hieroglyphic numerals and guided

FIGURE 2   *Egyptians knew formulas for the volume of pyramids*

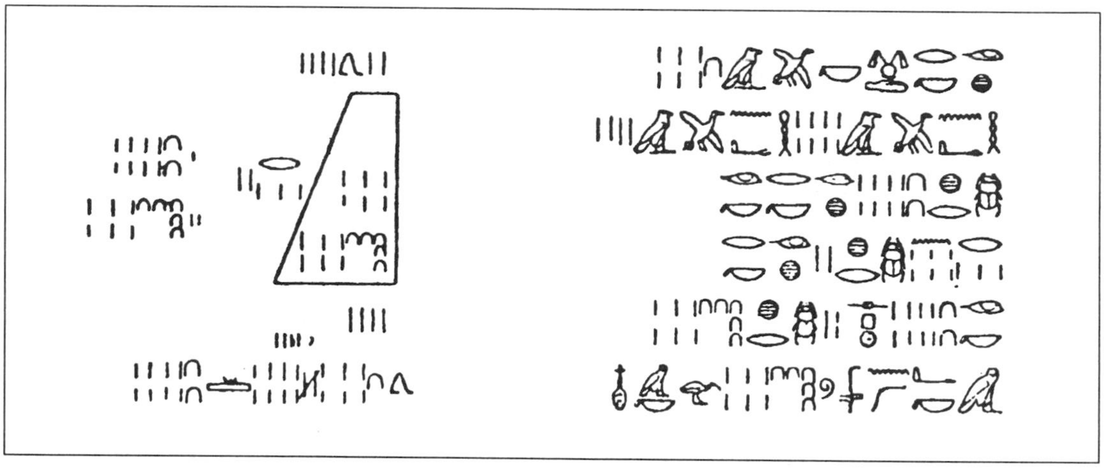

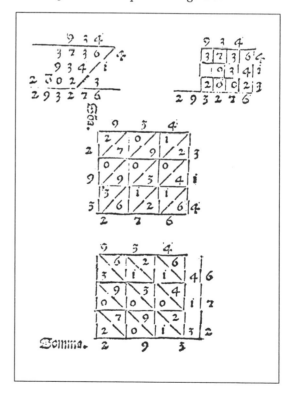

inspection of the contents of figure 2 could reveal the fact that the ancient Egyptians knew the formula for the computation of the volume of the frustum of a square pyramid, i.e.,

$$V = \frac{1}{3} \, h \, (a^2 + ab + b^2)$$

where *a* and *b* are measures of the sides of the bases and *h* a measure of the height.

As a result of such activities many questions arise. Each question, itself, offers new learning opportunities, and its answers can lead to profound insights into the growth process of mathematical ideas and techniques. As an illustration of such possibilities, consider the following sequence of questions and findings that might logically emanate from a discussion of numeration systems.

Why did the use of Roman numerals dominate European culture for such a long period of time? How did our modern algorithms for the basic operations evolve? For example, in fifteenth-century Italy eight algorithmic schemes were accepted for obtaining the product of two multidigit numbers. (Some are shown in fig. 3.) Even when attempts to standardize a particular method took place, the desired result was slow in coming, as demonstrated by the format of the same illustrative multiplication example appearing in different

FIGURE 3   *Some multiplication algorithms*

editions of Taglient's *Libro Dabaco* over a fifty-year period (fig. 4).

Was such confusion the result of the printer's incompetence, or does it reflect the author's con-

FIGURE 4 *Attempts to standardize an algorithm*

(1515)	(1520)	(1541)
456	456	456
23	23	23
1368	1368	136 8
912	912	912
10488	10488	1048 8

(1550)	(1567)
456	456
23	23
136 8	1368
912	912
104 88	10488

FIGURE 6 *Finding square roots*

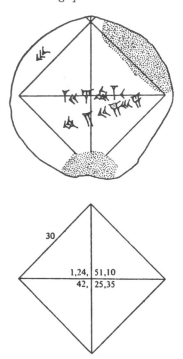

tinuing modification of the algorithmic form? What factors contributed to the popularization of an algorithm? Consider the case of the "galley," or "battelo," method of division that was favored among Renaissance mathematicians. Although it was mathematically efficient, it was not popular with printers who had the task of setting the type to reproduce it; thus, the algorithm fell into disuse (fig. 5).

FIGURE 5 *"Galley division"*

Examinations of the content of the history of mathematics frequently reveal striking similarities to present-day mathematical situations and procedures. How were square roots handled in antiquity? See figure 6. It is interesting to note that the Babylonians used a "divide and average" method of finding square roots: Given a number x, we wish to find the square root of x. Guess at the root and call it a. Then use the following procedure:

$$\frac{x}{a_1} = r_1, \quad \frac{r_1 + a_1}{2} = a_2,$$

$$\frac{x}{a_2} = r_2, \quad \frac{r_2 + a_2}{2} = a_3,$$

$$\frac{x}{a_3} = r_3, \quad \frac{r_3 + a_3}{2} = a_4,$$

Continue this process until the desired degree of accuracy is achieved. What makes the technique

34

even more interesting is that it is one of the earliest examples of an iterative process that we know, and iterative computing processes lie at the heart of modern computer operations. Our mathematical continuity with the past is ever present. In particular, iteration and approximation have always been and always will be integral components of computational processes. When discussing the history of mathematics, it is important to recognize the role of approximation in the efforts of human beings to quantify their world. "How good and useful were such approximations?" is a question that should frequently be brought before students. For example, Chinese mathematicians of the early Han dynasty (ca. 300 B.C.) used a trapezoidal approximation for the area A of the segment of a circle,

$$A = \frac{S}{2} \, (C + S),$$

where C is the measure of the length of the chord involved and S is the measure of the length of the sagitta of the segment (fig. 7). How good is this approximation?

Mathematics has sometimes been described as a "study of patterns." History would seem to affirm this description. Numbers and number patterns, especially those that demonstrated an emergence of regularity out of apparent disarray, fascinated early observers, and magical number meanings and

FIGURE 8

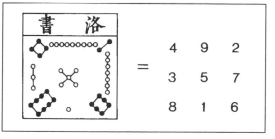

(a) Lo shu

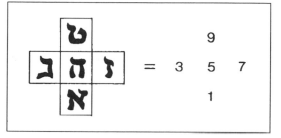

(b) Yahweh

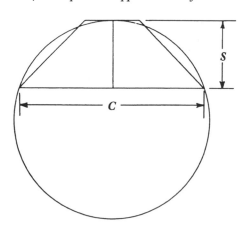

(c) "Clean plate" talisman

FIGURE 7 *A trapezoidal approximation for area*

configurations were created. The magic square exemplifies this facet of mathematical history. See figure 8. The Chinese saw in their *Lo shu* magic square (fig. 8a) the origins of science and mathematics, whereas the Hebrews used a serrated version of the same square to present the sacred name "Yahweh" in coded form (fig. 8b); in turn this square found its way into Islam, where it became a talisman painted on dinner plates that were sold to Europeans to ward off the Black Death (fig. 8c). Although magic squares and number configurations began as symbolic devices of supernatural

FIGURE 9

52	61	4	13	20	29	36	45
14	3	62	51	46	35	30	19
53	60	5	12	21	28	37	44
11	6	59	54	43	38	27	22
55	58	7	10	23	26	39	42
9	8	57	56	41	40	25	24
50	63	2	15	18	31	34	47
16	1	64	49	48	33	32	17

(a) Franklin 8 x 8 square

(b) Magic circles from Ming period

FIGURE 10

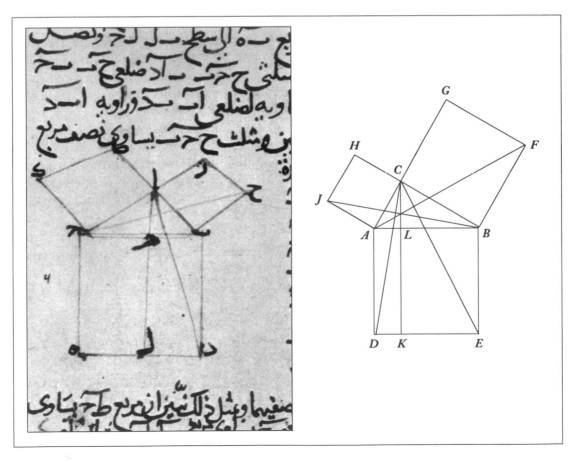

significance, they eventually evolved into intellectual challenges, intriguing such notable personages as Benjamin Franklin (fig. 9a), and were used as schemes to sharpen inductive reasoning powers and build problem-solving skills among Chinese mathematicians of the Ming dynasty (fig. 9b). Their appeal and computational attraction are equally relevant to our students, who can explore and unravel their mysteries with the assistance of hand-held calculators.

Frequently, historical material furnishes truly elegant examples and explanations that can be adopted for classroom use. Consider the traditional and universally popular "bride's chair" proof of the Pythagorean theorem shown in figure 10. Although this proof is mathematically appealing, its level of sophistication limits its pedagogical usefulness. At times it has been described as a "mouse trap proof" and "a proof walking on stilts, nay, a mean, underhanded proof" (Kline 1962, p. 50). An alternative proof, one that lends itself nicely to an overlay-overhead projector demonstration, is the "husan-thu" proof offered by Chinese mathematicians of the early Han period (ca. 300 B.C.). See figure 11. Throughout the history of mathematics, a variety of such dissection proofs have been offered to affirm the proposition that given a right triangle whose legs have measures of length a units and b units, respectively, and whose hypotenuse has a measure of length c units then $a^2 + b^2 = c^2$ (Loomis 1968). Over the years people with many backgrounds and interests shared a common obsession in seeking a solution to the Pythagorean theorem.

Illustrating the Growth and Evolution of Mathematical Ideas

The history of mathematics spans time in testifying to the similarity of human mathematical concerns and experiences, but it also transcends cultures and affirms the global nature of mathematical awareness and involvement. Frequently, the same mathematical situation can be examined

FIGURE II

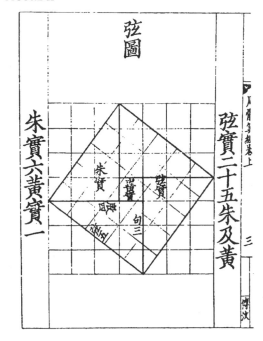

(a) Square of area c^2 composed of four right triangles whose legs measure a and b units, hypotenuse c units, plus a small square in the center

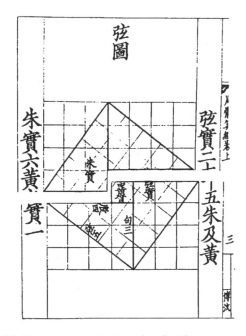

(b) Pieces reaaranged, thus $c^2 = a^2 + b^2$.

in different time periods and cultural milieus through the use of extant problems.

> When a ladder 25 feet long rests against the vertical wall of a building, its top is 17 feet farther from the base of the building than its foot. How high on the wall does the ladder reach? (Dolciani 1970)

> A spear 20 feet long leans against a tower. If its end is moved out 12 feet, how far up the tower does the spear reach? (Italy, A.D. 1300)

> The height of a wall is 10 feet. A pole of unknown length leans against the wall so that its top is even with the top of the wall. If the bottom of the pole is moved 1 foot farther from the wall, the pole will fall to the ground. What is the height of the pole? (China, 300 B.C.)

> A beam of length 30 feet stands aganst a wall. The upper end has slipped down a distance 6 feet. How far did the lower end move? (Babylonia, 1600–1800 B.C.)

It is interesting to note that as we progress back almost 4000 years through this sequence of right-triangle problems, their conceptual content does not become easier; in fact, it becomes more complex.

When Newton said that he could devise calculus because he "stood on the backs of giants," he was acknowledging the fact that mathematical discoveries are usually gradual in coming and are assisted in their birth by the efforts of many men and women working over a long period of time. The gradualness of this process should be made evident to students. For example, the phrase "Cartesian coordinates" would seem to credit the popularization of the use of rectangular coordinates to the seventeenth-century French mathematician and philosopher René Descartes. Was Descartes the originator of such an idea? No, it does not seem so, for systems of rectangular coordinates were employed by Renaissance artists as a technical aid in achieving perspective in their paintings; by early Greek, Roman, and Chinese cartographers who used accurate rectangular grid systems in thier map designs; and by Egyptian tomb painters who as early as the XVIII dynasty (1552–1306 B.C.) constructed graphical grids to assist in the transfer

FIGURE 12   *Grids help transfer sketches to tomb walls*

of sketches from their working tablets to tomb walls. See figure 12.

Time and time again in the history of mathematics, a similar scenario unfolds, a scenario that shows how mathematics evolved as a science: a mathematical idea or concept is pursued and studied for its utilitarian or sociological significance, but gradually its considerations become abstract and divorced from empirical reality. An example of this phenomenon is illustrated by the three classical problems of Greek antiquity:

1. The duplication of the cube
2. The trisection of an angle
3. The quadrature of the circle

These problems were to be solved with the use of a straightedge and compass alone. It was over two thousand years before this feat was proved impossible, but yet, in the interim search for solutions, many useful mathematical discoveries were made, including a theory of conic sections and the development of cubic, quadratic, and transcendental curves. Out of this particular legacy emerged a series of geometric problems that can challenge and fascinate modern students. In the search to achieve a quadrature of the circle, a theory of lunes developed and problems like those that follow resulted.

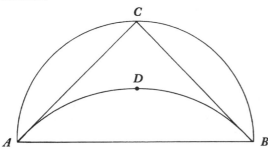

Given a semicircle with diameter *AB*, arc *ADB* is inscribed in the semicircle. The region bounded between the semicircle and the arc is called a lune. Show that the area of the lune *ACBD* is equal to the area of the inscribed triangle *ACB* where $\overline{AC} \cong \overline{CB}$.

Problems become more complex and further removed from reality, as shown by a consideration of the arbelos and its properties (Raphael 1973):

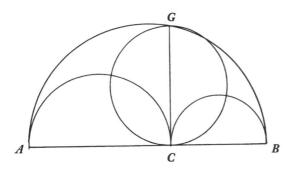

Let *A*, *C*, and *B* be three points on a straight line. Construct semicircles on the same side of the line with *AB*, *AC*, and *CB* as diameters. The region bounded by these three semicircles is called an *arbelos*. At *C* construct a perpendicular line to *AB* intersecting the largest semicircle at point *G*. Show that the area of the circle constructed with *CG* as a diameter equals the area of the arbelos.

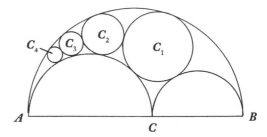

Given an arbelos packed with circles $C_1, C_2, C_3, \ldots$, as indicated, show that the perpendicular distance from the center of the nth circle to the line *ACB* is n times the diameter of the nth circle.

In tracing the cultural, temporal, and intellectual migration of mathematical concepts and techniques, it is important that the social, political, and economic forces that have influenced the growth of mathematics be recognized and acknowledged. Just why is it that the rise of deductive mathematics appeared in classical Greece? Why did the emergence of mercantile capitalism in fourteenth-century Italy spawn a mathematical Renaissance? How did the invention of printing affect the growth and spread of mathematical ideas? Is a nation's mathematical competency contingent on its wealth? Does war change the nature of a country's mathematical activity? It is from the contemplation of such questions and their answers that the dynamic interrelationship of mathematics and society emerges.

Conclusion

A recognition of the pedagogical importance of the history of science and mathematics in teaching is certainly not new. George Sarton, the noted historian of science, frequently pleaded this cause. In addressing a general audience in 1953, Sarton delivered a message that is quite appropriate today:

I said that if you do not love and know science, one cannot expect you to be interested in its history; on the other hand, the teaching of the humanities of science would create the love of science as well as a deeper understanding of it. Too many of our scientists (even the most distinguished ones) are technicians and nothing more. Our aim is to humanize science, and the best way of doing that is to tell and discuss the history of science. If we succeed, men of science will cease to be mere technicians, and will become educated men (Sarton 1958).

Unfortunately, in recent years the history of mathematics has been experiencing neglect. Around the turn of the century, the importance of the history of mathematics in relation to the teaching of mathematics was widely recognized and promoted. Almost all universities and teacher-training colleges offered studies in the subject, but gradually its status diminished as mathematical content, itself, became the sole focus of mathematics teaching. A survey of the offerings of American colleges and universities in 1980–81 revealed that only 30 percent offered a course on the history of mathematics (ISGHPM 1982). In light of this revelation, it is rather ironic to note that during this same period many professional schools (law, medicine, architecture, and so on) are requiring their graduates to take courses in the history of their discipline. This rather recent curriculum innovation is intended to refocus attention and concern on the people-centered origins and applications of the disciplines in question. Perhaps it is also time to attempt such a reorientation in our teaching of mathematics.

REFERENCES

Dolciani, Mary P., Simon L. Berman, and William Wooten. *Modern Algebra and Trigonometry, Book 2.* Boston: Houghton Mifflin Co., 1970.

International Study Group on the Relations Between the History and Pedagogy of Mathematics (ISGHPM). *Newsletter* (February 1982): 3.

Kline, Morris. *Mathematics: A Cultural Approach.* Reading, Mass.: Addison-Wesley Publishing Co., 1962.

Loomis, Elisha Scott. *The Pythagorean Proposition.* Washington, D.C.: NCTM, 1968.

National Council of Teachers of Mathematics. *Historical Topics for the Mathematics Classroom.* Thirty-first Yearbook. Washington, D.C.: The Council, 1969.

Raphael, L. "The Shoemaker's Knife." *Mathematics Teacher* 67 (April 1973): 319–23.

Sarton, George. *Ancient and Medieval Science during the Renaissance.* New York: A. S. Barnes & Co., 1958.

A SUGGESTED BIBLIOGRAPHY
FOR SELF-STUDY

Aaboe, A. *Episodes from the Early History of Mathematics.* New York: Random House, 1964.

Al-Daffa, Ali A. *The Muslim Contributions to Mathematics.* Atlantic Heights, N.J.: Humanities Press, 1977.

Boyer, Carl B. *A History of Mathematics.* New York: John Wiley & Sons, 1968.

Datta, B., and A. N. Singh. *History of Hindu Mathematics.* Bombay: Asia Publishing House, 1962.

Eves, Howard. *An Introduction to the History of Mathematics.* New York: Holt, Rinehart & Winston, 1969.

————. *Great Moments in Mathematics,* Vol. I (before 1650), Vol. II (after 1650). Washington, D.C.: Mathematical Association of America, 1980, 1981.

Gillings, Richard. *Mathematics in the Time of the Pharaohs.* Cambridge, Mass.: M.I.T. Press, 1972.

Kline, Morris. *Mathematics: A Cultural Approach.* Reading, Mass.: Addison-Wesley Publishing Co., 1962.

Kramer, Edna. *The Mainstream of Mathematics.* New York: Fawcett World Library, 1961.

May, Kenneth. *Bibliography and Research Manual of the History of Mathematics.* Toronto: University of Toronto Press, 1973.

Medonick, Henrietta, ed. *The Treasury of Mathematics: A Collection of Source Material.* New York: Philosophical Library, 1965.

Menninger, Karl. *Number Words and Number Symbols.* Cambridge, Mass.: M.I.T. Press, 1969.

Moffatt, Michael, ed. *The Ages of Mathematics,* 4 vols. New York: Doubleday, 1977.

National Council of Teachers of Mathematics. *Historical Topics for the Mathematics Classroom.* Thirty-first Yearbook. Washington, D.C.: The Council, 1969.

Needham, Joseph. "Mathematics." In *Science and Civilization in China.* pp. 1–168. Cambridge: At the University Press, 1959.

Osen, Lynn. *Women in Mathematics.* Cambridge, Mass.: M.I.T. Press, 1974.

Popp, Walter. *History of Mathematics: Topics for Schools.* Milton Keynes, U.K.: Open University Press, 1975.

Read, Cecil. "Periodical Articles Dealing with the History of Mathematics: A Bibliography of Articles in English Appearing in Nine Periodicals." *School Science and Mathematics* 68 (1970): 415–53.

Smith, David Eugene. *History of Mathematics,* 2 vols. New York: Dover Publications, 1958.

Struik, D. J. *A Concise History of Mathematics.* New York: Dover Publications, 1967.

Swetz, Frank J., and T. I. Kao. *Was Pythagoras Chinese? An Examination of Right Triangle Theory in Ancient China.* University Park, Pa.: Pennsylvania State University Press; Reston, Va.: National Council of Teachers of Mathematics, 1977.

Zaslavsky, Claudia. *Africa Counts: Number and Pattern in African Culture.* Boston: Prindle, Weber & Schmidt, 1973.

SUGGESTED READINGS FOR PART I

Bergamini, David. *Mathematics.* Life Science Library. New York: Time-Life Inc., 1972.

Davis, Philip and Hersh, Reuben. *The Mathematical Experience.* Boston: Berkhauser, 1981.

Kramer, Edna E. *The Main Stream of Mathematics.* New York: Oxford University Press, 1951.

Rényi, Alfréd. *Dialogues on Mathematics.* San Francisco: Holden Day, 1967.

Resnikoff, H. L. and Wells, R. O. *Mathematics in Civilization.* New York: Holt, Rinehart and Winston, 1973.

Schaaf, William L. *Our Mathematical Heritage.* New York: Collier Books, 1966.

Wilder, Raymond. *Mathematics as a Cultural System.* London: Pergamon, 1981.

PART II

IN
THE
BEGINNING . . .

HISTORICAL EVIDENCE for the existence of human mathematical activity extends back in time at least thirty thousand years. Mathematically speaking, in the beginning there was number and from number mathematics proceeded. Old Stone Age hunters quantified elements in their environment simply by making scratches or tally marks on a hard, convenient surface such as a stone, a piece of wood, or a bone. Extant bone specimens of these early tally records have been found in Africa and Central Europe. Their markings are quite extensive and suggest attempts at chronological record keeping with each notch representing the passing of a day. Perhaps these bones were part of a ritual timekeeping process. For their food supply, hunting societies frequently depended on migratory prey and the seasonal appearance of certain food plants. They had to be attuned to the cycles of nature and some omens, numerically devised or otherwise, were needed to forecast hunting and gathering seasons. The number knowledge of such societies was rudimentary. Other than for the purpose of timekeeping, extensive quantification was not needed. Early humans lived in close contact with their environment; they knew objects around them concretely and had little need to abstract their properties. A large number vocabulary is only needed when information must be transmitted at a remote distance from the objects in question, a situation indicative of technological societies.

The ritual origins of mathematics also extended to the use of simple geometry in the construction of megalithic ceremonial altars and observatories. These observatories, the most famous of which is at Stonehenge, England, allowed our ancestors to undertake simple observations of the heavens and determine noteworthy events such as the yearly solstices or eclipses of the sun or moon. Later uses of geometric form and principles in the decoration of pottery were contingent upon the existence of technological bases and work specialization and mark a distinct societal advance for the status of mathematics in relation to its users.

In the following selection of articles in Part Two, the authors examine and discuss the place of mathematics in early and traditional societies including those of the Americas. Archeological evidence is presented that would seem to support the intriguing premise that numeration actually preceded formal writing. Counting procedures and the development of numeration techniques in several societies are also discussed in fascinating detail. Finally, the *magical* properties of number are considered. For our ancestors, and many people living today, numbers remain powerful, mystical devices which control the destinies of individuals, groups of people, and even nations.

7

Oneness, Twoness, Threeness

How Ancient Accountants Invented Numbers

DENISE SCHMANDT-BESSERAT

*P*EOPLE appear to be born to compute. The numerical skills of children develop so early and so inexorably that it is easy to imagine an internal clock of mathematical maturity guiding their growth. Not long after learning to walk and talk, they can set the table with impressive accuracy—one plate, one knife, one spoon, one fork, for each of the five chairs. Soon they are capable of *noting* that they have placed five knives, spoons, and forks on the table—and, a bit later, that this amounts to fifteen pieces of silverware. Having thus mastered addition, they move on to subtraction. It seems almost reasonable to expect that if a child were secluded on a desert island at birth and retrieved seven years later, he could enter a second-grade mathematics class without any serious problems of intellectual adjustment.

Of course, the truth is not so simple. This century, the work of cognitive psychologists, notably Jean Piaget, has illuminated the subtle forms of daily learning on which intellectual progress depends. Piaget observed children as they slowly grasped—or, as the case might be, bumped into—concepts that adults take for granted, as they refused, for instance, to concede that quantity is unchanged as water pours from a short stout glass into a tall thin one. Psychologists have since demonstrated that young children, asked to count the pencils in a pile, readily report the number of blue or red pencils, but must be coaxed into finding the total. Such studies have suggested not only that the rudiments of mathematics are mastered gradually, and with effort, but that the very concept of abstract numbers—the idea of a oneness, a twoness, a threeness that applies to any class of objects and is a prerequisite for doing anything more mathematically demanding than setting a table—is itself far from innate.

This observation draws support from linguistics and anthropology, in particular from the study of cultures that have evolved in isolation from modern society. Anthropologists have found that when a Vedda tribesman, of Sri Lanka, wanted to count coconuts, he would collect a heap of sticks and assign one to each coconut. Every time he added a new stick he said, "That is one." But if asked how many coconuts he possessed, he could only point to his pile of sticks and say, "That many," for the Vedda had no words devoted to expressing quantities. Thus, while capable of a kind of counting—counting in one-to-one correspondence, rather like the child setting the table—the Vedda apparently had no conception of numbers that exist independently of sticks and coconuts and can be applied to either without reference to the other.

Indeed, even languages with numerous words for enumeration sometimes do not reflect a clear understanding of *abstract* numbers; they lack words, such as the English *two,* that are applicable to a

Reprinted from *The Sciences* 27 (1987): 44–48. Photos supplied through the courtesy of Denise Schmandt-Besserat and respective institutions.

wide range of objects. Franz Boas, studying the Tsimshians, of British Columbia, in the late nineteenth century, found that they denoted one, two, or three canoes, respectively, with the words *k'amaet, g'alpeeltk,* and *galtskantk,* whereas one, two, and three men were *k'al, t'epqadal,* and *gulal.* In short—to use a loose but instructive metaphor—societies appear to develop in the fashion of children; whereas one-to-one counting may come early, almost innately, abstract numbers do not.

In the case of children, it has been known for some time when the concept of abstract numbers typically arrives: the average six-year-old applies such words as *five* with considerable confidence and generality. But in Western culture, evidence bearing on the origin of abstract numbers has emerged only in recent years. It takes the form of clay tokens first used in the Middle East ten thousand years ago, five thousand years before the pictographic clay tablets that are generally associated with the invention of writing. Careful study of these tokens has radically improved our understanding not only of when abstract numbers arose but of how they arose. And it has shown with new clarity how writing itself originated and how these two seminal developments of Western culture were intertwined from the start.

II

The earliest Middle Eastern artifacts connected with counting are notched bones found in Israel and Jordan, in caves inhabited between 15,000 and 10,000 B.C., during the Paleolithic period. The bones seem to have served as lunar calendars, each notch representing one sighting of the moon. And such instances of counting by one-to-one correspondence appear again and again in preliterate societies: pebbles and shells aided the census in early African kingdoms, and, in the New World, cacao beans and kernels of maize, wheat, and rice were used as counters.

All such systems suffer from a lack of specificity. A collection of pebbles, shells, or notches indicates a quantity but not the items being quantified, and, thus, it cannot serve as a way of storing detailed information for long periods. The first means of enumeration known to have solved this problem was devised in the Fertile Crescent, the rich lowlands stretching from Syria to Iran. Clay tokens, usually an inch or less across, were used to represent, by virtue of their shapes, specific commodities: a cylinder stood for an animal; cones and spheres referred to two common measures of grain (approximately equivalent to a peck and a bushel), which the Sumerians would later call the *ban* and

FIGURE I

Plain tokens from Susa, Iran: from left to right, sphere, flat disk, 2 tetrahedrons, cone, cylinder. The plain tokens are among the earliest. The cone, sphere and disk probably represent various units of grain. The tetrahedrons may stand for units of work (a day's work?). Courtesy Musée du Louvre, Département des Antiquités Orientales.

FIGURE 2

Complex tokens from Susa, Iran: from left to right a parabola (garment?), triangle (metal?), ovoid (oil?), disk (sheep), biconoid (honey?), rectangle (?), parabola (garment?). Courtesy Musée du Louvre, Département des Antiquités Orientales.

the *bariga*. These tokens, like the notches of the Paleolithic period and the shells of African kings, did not reflect a clear conception of abstract numbers; two animals could be recorded only with two cylinders, and two bushels of grain with two spheres. Nonetheless, in their specificity, the tokens possessed a huge advantage over past technologies of information storage; they could be kept for years without any loss of meaning. This fusion of simple one-to-one counting with specific symbolic identification of the objects being counted might be called *concrete* counting.

It is no coincidence that this relatively sophisticated accounting practice appeared in the Fertile Crescent around 8000 B.C. For there, at that time, plants and animals were first domesticated. Indeed, the tokens are often found at sites where rectangular storage silos had been erected amid the round huts of villages, and the soil at these sites has an unnatural density of cereal pollen. It is only logical that agriculture, ushering in an economy dependent on the redistribution of foodstuffs and the planning of subsistence over many seasons, would require a reliable system of record keeping, and that the tokens were a device for organizing and storing economic data—an extrasomatic brain not liable to human memory failure.

By 6000 B.C., clay tokens had spread throughout the Middle East. Every major archaeological site of that era in Iraq, Iran, Syria, Turkey, and Israel yields somewhere between a handful of tokens and hundreds. Their nature and essential function seem to have changed little, if at all, over the millennia. As late as 3300 B.C., the same cones, spheres, and cylinders that had appeared five thousand years earlier were still in use. Then, near the end of the fourth millennium B.C., came a second, more complex type of token, one that bears varied, sometimes elaborate markings and assumes a wide assortment of shapes. There are ovoids, rhomboids, biconoids, bent coils, parabolas, quadrangles, and miniature representations of tools and animals.

Why the sudden profusion of complex tokens? The answer lies largely in the increasingly complex structure of the Sumerian society in which these tokens were used. By 3200 B.C., Sumer was seeing the birth of the first great cities, and complex tokens appear to represent the increasing number of finished goods that are the hallmark of an urban economy. Whereas plain tokens belonged to the pens and granaries, and stood for the several staple foodstuffs, complex tokens were used in the workplace and stood for finished merchandise—

garments, metalworks, jars of oil, loaves of bread.

Another clue to the origin of this new information technology lies in its association with the rise of the great temples. In the Sumerian metropolis of Uruk, for example, the earliest group of complex tokens was found in the ruins of Eanna, the temple precinct dedicated to Inanna, the goddess of love. These temples—monumental, ornately decorated public buildings—signal the advent of a strong, economically powerful state government, and perhaps the first regime of coercive taxation and redistribution. Sumerian reliefs from this period depict processions of citizens delivering their dues to the temple in the form of foodstuffs—which, presumably, were then distributed among the public servants who maintained the temples and the laborers who built walls around the cities. There is even a hint that tribute was not always willingly paid; some of the carvings show scenes of punishment, perhaps being inflicted on the first tax delinquents. This possibility underscores the fact that taxation required tight administrative control, complete with a precise system of reckoning and record keeping, such as the tokens provided.

Even as complex tokens proliferated, the relatively plain tokens from an earlier age remained in use and kept their essential form and function, referring to agricultural commodities. And, though both plain and complex tokens are found in temple archives, they evolved separately, into two of the greatest inventions of the human species—abstract numbers and phonetic writing.

III

One catalyst in both of these inventions was the habit of storing plain tokens in clay envelopes. About two hundred of these envelopes have been recovered, most still intact, at half a dozen sites. They typically bear the impression of one or more seals—the small, usually cylindrical stones engraved with a design that served as a sort of signature. (Just about all Sumerians—even slaves—appear to

have possessed personal seals.) These imprints refer, presumably, to the person or government agency whose account was recorded within a clay envelope or to the parties who had agreed on a transaction that might have lain in the future or been recently completed. In other words, the sealed envelopes could have served as contracts or receipts.

FIGURE 3

Unbroken envelope filled with tokens from Susa, Iran. Note the traces of seals showing a line of prisoners with their hands tied at the back. The seals identified the parties involved in the transaction. *Courtesy Musée du Louvre, Département des Antiquités Orientales.*

A clay envelope has one obvious drawback as a means of storing information: it is not transparent; if you forget what is inside, the only way to find out is to break open the seal, which amounts to symbolically breaking the deal. That, presumably, is why Sumerian accountants began impressing the tokens on the soft exteriors of the envelopes before enclosing them, thus leaving a visible record of the number and shape of tokens held inside. At some point, accountants must have realized that

FIGURE 4

Broken envelope with its token contents from Uruk, Iraq. The incised ovoid tokens enclosed were units of oil. Courtesy Deutsches Archaeologisches Institut, Abteilung Baghdad.

the markings on the envelope—reflecting, as they did, everything significant about its contents—rendered the tokens superfluous. Thus were the first written tablets created, as two-dimensional representations of three-dimensional symbols; the circle replaced the sphere, and the wedge the cone.

Complex tokens, unlike plain ones, were not stored in envelopes. Rather, they were perforated and laced on a string that was then attached to a *bulla*, an oblong piece of clay. These strings of tokens were supplanted by two-dimensional symbols at roughly the same time that plain tokens suffered this fate, and this is almost surely no coincidence; once the economy of two-dimensional information storage had been demonstrated via plain tokens, accountants must have hastened to apply the lesson to complex ones. The means of rendering the new symbols differed, however, in part, perhaps, because it would have been difficult to impress the finely marked complex tokens on a clay tablet and get a sharp, legible image. Instead, these markings, along with the outline of the token, were replicated on tablets with a stylus—a technique that is scarcely surprising, since the stylus had been used to mark the tokens in the first place. Thus, an ovoid with an incision—the symbol for a jar of oil—became a neatly inscribed oval with a slash across it.

FIGURE 5

Envelope showing impressed markings corresponding to the tokens enclosed from Susa, Iran. The large and small cones probably stood for measures of grain and the lenticular disks seem to have been units of animal count (10 sheep?). Courtesy Musée du Louvre, Département des Antiquités Orientales.

Two-dimensional renderings of plain and complex tokens were in a sense pictographs—pictures not of the items ultimately being represented but of the symbolic objects (cones, spheres, ovoids, quadrangles) that had in turn represented those items. Pictographic writing was thus a distinct step toward abstraction. The original symbols, the clay tokens, had been fundamentally concrete things; like the items they stood for, they could be grasped. The pictographs were much further removed—by an entire dimension, to be precise—from these items. This switch from three- to two-dimensional representation may seem an obvious step to the modern, who is steeped in the notion of abstraction, but the very fact that the step was not immediately taken—that for some time the unambiguously marked clay envelopes redundantly carried tokens—suggests that two-dimensional representation was a cognitive leap of some distance.

In the course of this transition, a second momentous step was taken. For a time, the images of tokens on clay tablets had maintained a kind of concreteness in their expression of plurality; a set of three jars of oil had been depicted with three identical slashed ovals in a row. But with the use of clay tablets, another kind of abstraction was attained; plurality was no longer expressed by repeating a pictograph in one-to-one correspondence. Instead, pictographs were preceded by numerals applicable to all kinds of objects.

In selecting symbols to stand for numbers, the Sumerians pressed into service shapes that stood for measures of grains. A wedge, which originally meant a small quantity of grain, now meant 1. A circle, which had represented a larger quantity, stood for 10. The two could be combined, somewhat as Roman numerals later would be: 23 was two circles and three wedges. (The same symbols continued to refer to grain as well, but differences of context were sufficient to avoid confusion.) A large wedge came to signify 60, and a large circle 360, giving the Sumerians a strangely hybrid system of numerical notation—elements of a base-six system mixed in with the decimal notation.

FIGURE 6

Tablets showing impressed markings of tokens from Susa, Iran. The tablet is a record of quantities of grain. Courtesy Musée du Louvre, Département des Antiquités Orientales.

Why the symbols for grain were selected as abstract numerals can only be guessed, but there is no shortage of possible reasons. First, they were signs in common use, since grain was the staple crop in the ancient Middle East. Second, the circle and the wedge already referred to two different magnitudes, so their use to signify 1 and 10 was conceptually natural. Finally, since the circle and the wedge were derived from plain tokens, they were *impressed* signs, easily differentiated from the *incised* signs with which they were typically paired, so the numerals stood out clearly from a body of writing, much as they do today. This contrast underscores the remarkable feat the Sumerians had accomplished: with the invention of numerals, they developed a discrete category of signs, used exclusively to indicate quantity and capable of combination with any member of a second set of symbols representing tangible items. The Sumerians had invented abstract numbers—the concept of oneness, twoness, threeness.

50

IV

Not long after the evolution of certain plain tokens into numerals, the pictographs derived from complex tokens crossed a second great threshold. By about 2900 B.C., these symbols could function phonetically, representing not objects but, in selected cases, sounds. Thus did the Sumerians invent the precursor of the modern alphabet—a syllabary, which represents each syllable with a single symbol.

The exact route of invention can only be guessed. As Sumerian society evolved, and the business of taxation and redistribution grew more complicated, the requisite systems of accounting became more elaborate. Accountants may have faced a problem: since a person's identity could be recorded only with his personal seal, his accounts could not be duplicated, nor substantially amended, in his absence. At some point—perhaps as a solution to a problem such as this—accountants began referring to individuals with separate sequences of symbols. This technique was facilitated by the fact that personal names were often built from concepts for which symbols existed. For instance, one name consisted of the word for *god* and the word for *life* and meant, literally, "The god that gives life." It was easy to represent this name with the symbols for *god* (a star) and *life* (an arrow). The crucial step came when names that resisted such easy depiction—names not built of concepts for which symbols already existed—were nonetheless recorded in comparable fashion. Thus, the slashed oval might be employed if a syllable in a name sounded like the Sumerian word for *oil*—even if the idea of oil had played no role in the name's origin. This was the beginning of phonetic writing—the use of symbols to stand not for objects or concepts but simply for the sounds they brought to mind.

The resultant Sumerian script, which led to the cuneiform writing that prevailed in the Middle East until the Christian era, was based on three major elements. Numerals were ideographs; each symbol stood for a distinct concept. Some words were pictographs, resembling the tokens they had replaced. And proper names usually were depicted phonetically, albeit with symbols that once had been strictly pictographic. In time, other words would also be phonetically encoded. But practicality eventually demanded changes in the nature of this coding. After all, it takes a cumbersomely large number of symbols to fill a syllabary, since any language has thousands of syllables. But if symbols are made to represent smaller units of sound—the sound of *o* or of *l* but not of *oil*—then a relatively small number of symbols can, in diverse combination, form all syllables and hence all words. The Phoenicians, by inventing the first truly phonetic alphabet, around 1500 B.C., fully exploited this fact (though their alphabet lacked vowels, which the Greeks added later).

Thus abstract numbers, sometimes thought of as a mere subset of the alphabet, actually ushered it in. Indeed, during the first three hundred years of writing, there were no literary, religious, or historical texts. Except for a small number of lexical texts—catalogues of signs, used to train scribes—the only written works produced between 3100 and 2900 B.C. were the ledgers of accountants: lists of the names and quantities of goods. There were no symbols for verbs, adverbs, or prepositions, which would have allowed the expression of entire sentences. The human species had to build its abstractions one block at a time.

HISTORICAL EXHIBIT 2.1

Bodily Mathematics

When man first began counting and communicating number information, he used his fingers to establish a 1:1 correspondence with the objects under concern. Readily available, the human digits, fingers and toes, have always lent themselves to the tasks of counting. The vestiges of this practice can be traced through the etymology of current number terms: in Sanskrit, *pancha* means 'five,' this word is related to the Persian *pentcha*, 'hand;' the Russian word for 'five,' *piat*, is related to *piast*, 'outstretched hand.' Among traditional peoples of the past as well as today, hand-related words or phrases are also number identifiers. The Jibaros Indians of the Amazon rain forest express 'five' by the phrase *wehe amukei*, 'I have finished one hand' and 'ten' by *mai wehe amukei*, 'I have finished both hands.' In North America, the Takelma of southwestern Oregon used the word *ixdil*, 'hands,' for 'ten' and *yapanis*, 'one man,' for 'twenty' indicating they continued their counting practices beyond fingers to include toes. The present day Kewa people of Papua New Guinea speak of 'hand' for 'five,' however, their upper body count which begins with the fingers of the right hand proceeds across the upper body to the left hand to total sixty-eight as illustrated below.

Human reliance on digit counting has resulted in the popular adoption of quinary, decimal, and vigesimal based number systems.

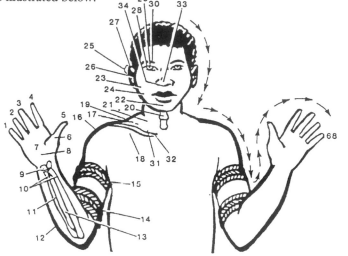

Body parts have also been used for physical measurement. Among traditional Malay people of Southeast Asia, the width of an object may be designated by a finger nail, *sebesar kuku*, 'the circumference of a forearm,' *selengan* or 'a human embrace,' *sepemeluk*. Their rice is measured by the handful, *gengam*. Hands, arms, and feet in various postures are employed for the measure of length. The modern measurement words 'foot,' 'cubit,' 'span,' and 'fathom' all derive their origins from the use of body parts. A study of the history of numeration and measurement affirms the adage that "Man is the measure of all things."

8

The Evolution of Modern Numerals from Ancient Tally Marks

CHARLES E. WOODRUFF

*N*O ACCEPTABLE EXPLANATION has ever been given as to the origin of our numerals though the literature on the subject is enormous. The multiplicity of the theories is evidence of the inadequacy of each. Of course it is known that the Arabs obtained the signs from Southern Asia or India before the ninth century A.D. and brought them to Europe in the tenth, but the remote origin has never been discovered. The later Greeks and Hebrews used the first letters of their respective alphabets to represent units, the second groups to represent tens and the third for the hundreds. Sometimes the initial letter of the word for the number was used as a symbol for that number, as in the early Greek, and possibly the Roman C and M, though the latter may have been evolved from earlier symbols. For these reasons quite a number of scholars have sought for the original forms of our numerals in the letters of some alphabet. There is certainly quite a remarkable resemblance between some of the old numerals and some of the letters of old alphabets, but that is no proof of common origin. Indeed it would be strange if we could not find many such coincidences in the innumerable forms which both letters and numerals have taken. Indeed numerals were used where there were no alphabets or before alphabets were evolved as in modern China and ancient Egypt.

In addition, Sir E. Clive Bagley and George Bühler point out the fact (*Journal Royal Asiatic Society*, p. 335, 1882) that there is no known reason why certain letters should have been selected to represent the numerals four to nine, which the former accepts as being derived from letters from several alphabets widely separated. Bagley curiously enough also states that all numerals at first were merely shorthand ways of expressing their names, which wholly contradicts the idea of derivation from letters, and ignores the fact that primitive tallies or numerals must have existed before they had names.

It has also been frequently asserted that the numerals were inventions which sprang up in a very short time, although such a phenomenon is contrary to human experience. Sudden appearance generally means borrowing, for all written symbols were slow in their evolution.

Reasoning from analogy, one would assume that if letters had their origin in the first crude attempts to represent things and ideas, the numerals must necessarily have their origin in the first crude attempts to record numbers.

The natural way of recording numbers is by tally marks, and it is the universal custom of mankind, at least of all who were intelligent enough to count. Historians of mathematics use the term tally-mark to refer to the notched stick (French, *tailler*, 'to cut'), but it is here meant to refer to any simple marks or scores. The Maya numerals are described by George B. Gordon (*The Century,*

Reprinted from *American Mathematical Monthly* 16 (Aug., 1909): 125–33; with permission of the Mathematics Association of America.

January, 1898), as follows: "The numbers from one to four are represented by dots; a bar signifies five; a bar and a dot six; ten is written by two bars; and so on up to nineteen, the sign of which is three bars and four dots; after this number the signs employed are in doubt." The Egyptians used tally marks up to nine but ten is an inverted U. Cuneiform numerals are also tally marks. Even as late as the third century B.C. in India the Asoka edicts record numbers up to five as vertical tally marks.

In the accompanying chart, there is some evidence that, as a rule, tally marks originally went only to five after which the one or five was repeated with extra markings. Professor Edwin S. Crawley (*Popular Science Monthly,* August, 1897), says that all systems of numbers were originally quinary from the use of the fingers of one hand in counting and became decimal as soon as two hands were used, but others state that there is no foundation for such a generalization. A few people have used both hands and feet, a vigesimal system, whose

In order to equalize the plates it was necessary to include one Indian column on the first and two on the third plate.

CHINESE

Probable Origin. (Chalfant.)	Evolutionary Forms.	Seal Form A.D. 100.	Modern Form.	Running Hand.	Commercial Form.	Nepal Mss.
1. One line or one weapon.						
2. Two lines or two weapons.						
3. Three lines.						
4. Four lines gradually connected in cursory style.						
5. Five lines variously indicated.						
6. Probably six lines united.						
7. Probably seven lines united.						
8. Eight lines united.						
9. Probably nine lines united.						
10. Two contracted signs for five united.						

remnants are seen in the "quatrevingts" of the French from 81 to 99 and our "score" of one finished tally group. Duodecimal systems were sometimes grafted on as in Babylonia, making also a sexagesimal system, side by side with the decimal. The Chinese numerals for 11 to 20 are merely those from one to ten, with a ten (X) over or before them, but the commercial forms are quinary, the six to eight being based on the one to five with extra tally marks beneath.

The numerals in the accompanying chart are copied from Isaac Taylor's "The Origin of the Alphabet," from the pamphlet on the subject by Professor John C. C. Clark, and from a paper by Karl Mischke (*Deutsche Japan Post*, Yokahama, Japan, December 29, 1906). Edward Clodd, "The Story of the Alphabet," derived his from Taylor, who in 1882, described the changes in Arabic numerals. The Chinese forms are copied mostly from the monograph on early Chinese writing by Rev. Frank H. Chalfant, a missionary in China (*Memoirs of the Carnegie Museum*, Pittsburg, Vol. IV, No. 1), to which have been added a few old forms obtained through the courtesy of Mr. L. C.

INDIAN

Jaina Mss.	Jaina Mss.	Nepal Mss.	"Bower" Mss.	Old Nepal Handwriting.	"Old Forms."	Eastern Caves.	Western Caves.	Bactrian.	"Very Old."	First Century.	Fifth Century.

Solyom of the Congressional Library, who copied them from "Luh-shoo-fun-luy," a dictionary of ancient forms and characters, by Foo Lwan-tseang. Others were obtained through the kindness of Professor Friedrich Hirth, Professor of Chinese in Columbia University in New York, who searched through Chinese archeological inscriptions for the purpose. Two columns of Nepal and Jaina manuscript and the "Bower" manuscript are copied from George Bühler's "Indische Paleographie," in which there are also a large number of variants, so closely resembling those copied elsewhere that they have been omitted here. Identical forms are also found in Burnell's "South Indian Paleography." There are also an enormous number of Medieval European forms in Cappelli's "Dizionario di Abbreviature Latine ed Italiane" and in Chassant's "Dictionaire des Abréviations Latines et Française" but they are almost identical with the ones here given. This infinite variety is partly due to an occasional scribe's fondness for decorating with flourishes, as well as due to the gradual evolution of simple forms from more complex ones. In all of them there must also be considerable allowance

ARABIC						EUROPEAN				
Tenth Century.	Modern Print.	Old Gobar.	Gobar.	Tenth Century Persian.	Common Modern.	Medieval Latin.	Twelfth Century.	Thirteenth Century French.	Fourteenth Century.	Twelfth Century English.

made for incorrect transcribing in the numerous times they have been copied from one book to another since the original was taken from some manuscript, coin, monument, or vase, and some were probably erroneously translated.

Tally marks must be so distinctive that they can be read, no matter how the coin or stick is held, vertically, horizontally, or inverted. The modern six and nine would have been useless, for they would have been indistinguishable unless the top was known. Consequently we find many instances in which the numerals have been inverted or inclined at various angles as it was of no practical importance at first whether they were upside down or not. Our two, for instance, as well as the four,

five, six, or seven have all been inverted and reversed at times. There may also have been reversals of the forms due to the fact that the people from whom the Arabs obtained the numerals, probably wrote from left to right while the Arabs wrote from right to left, and the Europeans from left to right. In addition to this, one Sanskrit language was written from right to left, and another from left to right. It is remarkable that the old Arabic forms have changed less in Europe than in Arabia. The final European forms of 6, 7, 8, 9, and 10 after many vicissitudes of change are nearly identical with old Arabic forms and show strong relationship to the early Indian. The vast majority of the old forms are like fossils of extinct species which

PROBABLE EVOLUTION OF THE NUMERALS

57

have left no descendants; but by picking out forms here and there, it is possible to secure a gradation of changes from the first to the last, the zero, by the way, being a very late development, possibly from the position dot of the Arabic, which in turn was probably a modification of the Indian O or ०. The use of position of some sort to indicate multiple value is a recent invention, G. R. Kaye stating that there is no evidence of position value prior to A.D. 1000 (*Journal Royal Asiatic Society*, July, 1907).

The numeral one, is always a single tally mark, straight or curved, horizontal, vertical, or at an angle. Two was evidently two tally marks and cursively became the modern 2 or the reversed or inverted form. Three in all its forms is evidently derived from three parallels. But the higher numerals have not so clear an origin. It is quite difficult to distinguish groups of parallel lines of more than three. We use a cross tally for every fifth stroke making groups of five, and probably we once grouped a "score" together also to indicate each time all the fingers and toes were counted. The Egyptians used groups of three strokes up to nine. Consequently some other system than parallel groups must have been used for the original numerals of four to ten. The modern Arabic has even returned to figures of four and five strokes for these two numerals, after having evolved other forms which they later abandoned, and there seems often to be an effort in ancient Indian forms of four and five to make them of four and five strokes, respectively.

The evolution of the Chinese numerals from tally marks is evident from a mere glance at the chart, for in every case there is a gradation to the modern. There are several facts which must be kept in mind in interpreting these changes. In the first place there are two evolutions recorded side by side—the oldest or quinary system which developed into the modern commercial, and the later decimal which formed the common numerals. Both of these are exceedingly ancient in origin, possibly a thousand years before the Christian era or even much earlier still for they had developed into the present decimal forms by A.D. 200. To represent six, seven, eight, and nine, the Chinese evidently first used five vertical tally marks under which they placed one to four horizontals, respectively, but the five verticals were early contracted into one. The commercial forms for six, seven, and eight are thus exceedingly old. The nine was evidently abandoned for the more convenient decimal system. The X which represents the commercial four is a direct descendant of the four vertical tally marks, the horizontal marks were used for the numeral four as late as the Han dynasty, A.D. 206–220. After this time only the new form appears though its origin was probably very much more remote. The commercial five which looks like our eight, is also a direct descendant of a very old form of five strokes which early replaced the five parallel ones— the transition forms given by Chalfant being quite conclusive as to this point.

When a decimal system became necessary, the Chinese, or those from whom they borrowed, had to invent other forms of tally marks for the numerals from five to nine, as it was clearly impracticable to continue the use of parallel strokes, but why they should have hit upon the curious forms recorded is a mystery.

Another fact must be kept in mind in explaining some of these changes. The use of a brush for a pen made it impossible to write the old curved forms which were carved on stone or written with a solid pencil. Chinese writing thus very early abandoned curves for straight lines—the squares and angles of the modern writing. This fact explains the change from the rounded forms of A.D. 100 into the modern numerals at about that date. The modern four, for instance, is sometimes thought to have been four vertical strokes enclosed in a square and that it changed into the modern form of a square containing two short strokes. The process was the opposite, for the square is the manner in which the oval of the older form had to be written with a brush, and the oval form is evidently a descendant of the earlier vertical four strokes. The Chinese "running" characters for two and three are the same as our own. The standard numerals (Chêng-tzu) are not given on the chart as they are complicated late inventions to prevent swindling, the commercial and common numerals being very easily changed to higher ones by the addition of a stroke or two. The human figure or a weapon of some sort seems to have been used in addition to the tally marks in the original decimal numerals. Bühler (*Journal Royal Asiatic Society,* 1882) says that there is no doubt that the numerals were introduced into India from without, as they appear rather suddenly and in a well-evolved form. The original forms, then, are not found in India and the oldest known Indian must be evolutionary forms of some nearby Mongolian system, though Bühler and many others are inclined to believe that they came from Egypt. Prior to their introduction the Hindus used quite a variety of methods of expressing numbers which do not concern us here. Though letters and syllables were used, there are others having no such derivation. Indeed one writer, Kern, is quoted by Bühler as believing the old Indian four and five to be four and five strokes, respectively. The identity of the first three numerals in Chinese and Indian would lead us to suspect identity of origin of the two systems, or more probably that the Chinese forms were carried by traders into India. Indeed Dr. Fritz Hommel ("The Civilization of the East") states that Indian "Culture is in the main an offshoot of the Chinese." The Chinese would also be suspected on account of the identity of the old Nepal nine and an evolutionary Chinese nine and there is a perfect gradation of forms from the oldest Chinese nine tally marks to our present nine.

Doubtless there was a considerable interval between the date of introduction of the numerals into India and the date of the oldest surviving forms—an interval in which their origin was forgotten if it was ever known. This fully accounts for the fact that in some cases considerable change had taken place. Probably many of the ancient Indian evolutionary forms will never be found, nevertheless those now known are within the bounds of ordinary variation of writing. They were undoubtedly introduced while some still retained an evident form of tally marks, which led to the frequent writing of four and five with four and five strokes, respectively. Some of the old Indian fives are almost identical with the Chinese original form, and in the six there are undoubted affinities, the sevens are closer still and nines identical. Moreover these resemblances are between forms of practically the same period. The well evolved "western cave" forms are ascribed to the first and second century A.D., and the older Indian must be set at least two centuries earlier which makes them contemporaneous with the later Chinese evolutionary forms which they resemble. Indian dates prior to the Christian era are very uncertain, for it is said the Brahma laws were not written until about 325 B.C. As the Asoka edicts did not have the numerals in 250 B.C., and as the cave numerals were well evolved forms in the first century A.D., we have a fairly definite period in which they were introduced and evolved.

Among the old Indian forms given by Kaye (*Journal Asiatic Society of Bengal,* July, 1907) there is an eight which is evidently an attempt to write the Chinese eight cursively, and it makes this evolutionary series almost as complete as the nine. The interesting thing about our eight and two, is that each seems to be the result of an attempt to write two parallel strokes cursively, the two was two horizontal parallel bars and the eight two curved vertical lines, and some of the old forms of eight are practically the same as the two turned over 90°.

It is quite likely also that when forms of numerals are evidently tally marks, the ancient tribes would not stick to any particular arrangement, but form new ones provided they indicated numbers. This is the most reasonable explanation of the very evident tally-mark nature of the numerals in the Jaina manuscript. Ten is a nine with an extra stroke, and the eights are sevens with an extra stroke. The Jaina four, five, and six are also clearly derived from groups of marks. In course of time, by slurring, omission of strokes and adding embellishing flourishes, the Nepal and Bower manuscript forms arose. Indeed in the seven there is a perfect gradation of evolutionary forms to our present seven. In the four the resemblance is seen by making an assumption. In the five there is more evidence of an attempt to write cursively one of the X forms of the Chinese, but the six is not so evident without making two assumptions.

The supremacy of the Chinese numerals is explained by the fact that they were the first ideographs in the field. Egyptian pictographs evolved in the direction of representing sounds and, besides, their tally marks elsewhere were in groups of parallels, and not the fortunate Chinese groupings which lent themselves to change into ideographs. The invention of position value of course killed all the numerals above nine.

9

Ancient Mathematics and the Development of Primitive Culture

WILLIAM A. CORDREY

THE HISTORIAN has been unable to ascertain whether the number sense was a part of the original endowment of the *genus homo* or not. It is known, however, that this innate or acquired ability possessed by man in the early periods of his development permitted him to recognize that a change had been made in a small group of objects when one had been deleted or added.

Counting came much later than the number sense and was developed much more rapidly by some people than by others. This evolution of counting was followed by number words and number symbols. Like counting, number systems evolved much earlier in some places than in others. An examination of the histories of uncivilized peoples reveals the absence of a workable number system. By a usable or workable number system is meant a system of number words and number symbols sufficient to permit the development of elementary mathematics. Anthropology cites no instance of an advanced social culture preceding a workable number system.

The Ahmes Papyrus evidences the fact that the Egyptians had a well developed number system nearly two millennia before the beginning of our era. This number system made possible the development of units of weights and measures which must of necessity precede trade and indus-

try. With commerce came the propagation of ideas and learning. Without an increase and dissemination of knowledge a civilization could not have evolved in the fertile valley of the life-giving Nile.

It was natural for the measurement of area to follow on the heels of linear measurement. Likewise, a study of three dimensional geometry was naturally begun before the elements of plane geometry were well developed. Prior to the inception of the Ahmes Papyrus, Egyptian mathematicians were able to determine the areas of rectangles, triangles, and circles, and the volumes of cylinders and prisms, and they knew that in a right triangle the relation of the length of two sides determines one of the angles. The author of the papyrus took up problems of volume, then problems of area, and finally problems involving relative length of the sides of a triangle.[1] The crowning achievement of ancient Egyptian mathematicians is the so-called formula for the volume of the frustum of a square pyramid.[2] Thus nearly four thousand years ago the development of Egyptian geometry had been sufficient to permit the measurement and division of grain and other produce. It made possible surveying and the division of land among the people and a redivision after the swollen Nile had receded. Aristotle said that mathematics had its birth in Egypt, because there the priestly class had the leisure needful for its study.[3] Regardless of what claims may truthfully be made for the antiquity of mathematics in various other countries, claims of

Reprinted from *Mathematics Teacher* 32 (Feb., 1939): 51–60; with permission of the National Council of Teachers of Mathematics.

even greater validity can justly be made for the science in the Nile valley.[4]

Here mathematics made it practicable for people to dwell near each other in peace and harmony. An elementary knowledge of this useful science made it possible for the Egyptian farmer to return after the annual flood to an equal amount of more fertile land. Had a knowledge of surveying not been developed, the return following the recession of the overflow would probably have been accompanied by strife and bloodshed.

Northeastward from the Nile valley, at a distance of more than a thousand kilometers, was located ancient Babylon. In and around this city lived the Sumerians, who evolved one of the earliest known civilizations. Like the Egyptians they inhabited a fertile valley where food could be grown in abundance. They developed a number system and the elements of mathematics four thousand years ago and with this came the evolution of commerce and industry.

The Babylonians knew that the area of a rectangle was equal to the product of two adjacent sides and that the area of a right triangle could be found by taking one half the product of the length of the sides about the right angle. They were acquainted with the so-called Pythagorean theorem and also with the fact that the angle in a semicircle is a right angle.[5] These early Sumerians had tables of square root and tables of cube root, and they understood equations as well as arithmetic and geometric progression.[6] Numerous problems prove that the Babylonian mathematicians were familiar with our formula for the solution of a quadratic equation twenty centuries before the dawn of the Christian era.[7] They knew how to dig canals, measure grain, calculate interest for loans of silver, and numerous other problems pertaining to business and industry.

The people of the Tigris-Euphrates valley had a system of numbers that made possible the development of simple arithmetic operations. These simple operations were followed by more advanced and more useful mathematics—making trade and industry practicable. Their knowledge of geometry made feasible the division of real estate which was necessary to peace and harmony at home. The arithmetic of ancient Babylon gives evidence to the fact that these people traded and exchanged goods. Industry at home was followed by commerce and foreign trade. Local trade stimulated the dissemination of knowledge at home and the exchange of goods with foreign people was accompanied by an exchange of ideas and learning. This would have been impossible without a knowledge of mathematics which had been preceded by usable number symbols and number words.

A brief examination of the Egyptian and Babylonian number systems and a comparison of their symbols with our modern symbols should be of interest. [*Editor's note:* In William Cordrey's original discussion of Babylonian numeration, he conveyed the erroneous impression that it was a base 10 system and discussed it accordingly. This confusion remains unexplained as his notes indicate that he was familiar with the writing of Archibald (1936) which clearly points out the sexagesimal nature of Babylonian mathematics (see Archibald's contribution, chapter 17). Thus for the benefit of general readers, the following passages have been partially edited to correct this situation. The spirit and intent of Cordrey's original discussion have been maintained; however, erroneous and ambiguous statements have been corrected. This particular editing ends at the asterisk*]. The basic figure of the Babylonian system which was impressed on a moist clay tablet was the wedge. A lone wedge standing on its point, ▼, represented 1. Nine wedges properly arranged and of the right magni-

tude, ▼▼▼ ▼▼▼ ▼▼▼ represented 9. A double wedge or wing-shaped mark, ◀, represented 10. Such double wedges arranged in a group could represent integer values of 10 up to 50, ◀◀◀ ◀◀ . A number such as 36 would be represented in the Babylonian cuneiform markings as ⟨⟨ ▼▼▼ ⟨ ▼▼▼ , that is, 30 + 6, thus an additive principle was used in composing numerals. However, the Babylonians also employed the concept of a base in their numeral system; within a Babylonian numeral, each grouping of single and double wedges represented a sexigesimal magnitude: 60^0, 60^1, 60^2, ... Accordingly 1000 was written as ⟨▼▼ ⟨⟨ ▼▼ ⟨⟨ , $(16 \times 60^1) + (40 \times 60^0)$. In determining the magnitude of each grouping of symbols, multiplication is used and the relative position of the symbols has a significance. Thus the sexigesimal Babylonian numeral system is positional in nature and employs both additive and multiplicative principles. Note the quantity of individual symbols used to represent a number in the Babylonian manner as compared to that required in the Hindu-Arabic system. To represent 8976, the Babylonian scribe wrote symbols designating the following numbers in the indicated order: 2, 29, 36, that is, $(2 \times 60^2) + (29 \times 60^1) + (36 \times 60^0)$. If each cuneiform mark is considered a character, the number of characters used was 2 + 11 + 9 = 22. If the scribe wrote three characters per second, he could write approximately three such numbers in one minute. At the same rate of writing characters, forty-five such numbers can be written by using the Hindu-Arabic symbols. For this specific example, our modern symbols require approximately one-fifteenth as much time for writing.

The Egyptian system resembled the Babylonian system in that it employed the additive principle and differed in that it did not employ the multiplicative principle. Nine was represented by nine strokes, | | | | | | | | |, each stroke representing 1. Ahmes used nine inverted U's, ∩∩∩ ∩∩∩ ∩∩∩, in writing 90, each character representing 10. Ten to any integral positive power was represented by a distinct symbol unless the number was beyond the system.[8]

One hundred was represented by the coil of a rope, ℮ ; 1,000 by a lotus flower, ⚘ ; 10,000 by an upright bent finger, ℔ ; 100,000 by a tadpole, ∽; 1,000,000 by a god with uplifted hands, 𓁨 .[9]

Therefore 1,984,567 would have been written: a god, then nine tadpoles, followed by eight bent fingers, followed by four lotus flowers, followed by five rope coils, followed by six inverted U's, followed by seven strokes. Thus the number of Egyptian characters necessary to represent the above number was 1 + 9 + 8 + 4 + 5 + 6 + 7 = 40. The number of characters necessary to represent 8976 was 8 + 9 + 7 + 6 = 30. In representing any number by Egyptian symbols, the actual number of characters required is equal to the sum of the individual figures. Thus if 8976 is again used as a time comparison number, nearly eight times as many symbols are required as is the case when the Hindu-Arabic notation is used. Assuming that Ahmes could make an Egyptian symbol in the same time required for a modern mathematician to write a modern symbol, his speed in writing numbers was approximately one-eighth that of the modern mathematician.

The absence of a zero was another deficiency in the early number systems. In writing 4230, ⚘⚘⚘⚘ ℮℮ ∩∩∩ , the Egyptians used four symbols, each representing 1,000, two symbols, each representing 100, and three symbols, each representing 10. In ancient Babylon 4230 was written thus: ▼ ⟨ ⟨⟨ ⟨⟨. The Babylonian method differed from the Egyptian in that a multiplicative principle was used: $(1 \times 60^2) + (10 \times 60^1) + (30 \times 60^0)$. If, in our Hindu-Arabic numeral rendering, the zero be placed between the two and four, the same symbols are still used; but not so with Ahmes and the Babylonian writers. Following the representation of 4,000, Ahmes would have used two symbols each representing 10 and then three symbols, each equivalent to 1, $\text{ϡϡϡϡ} \cap \cap | | |$.

In representing 4023, the Babylonian scribe wrote $\blacktriangledown \; \blacktriangledown\blacktriangledown\blacktriangledown \; \blacktriangledown\blacktriangledown\blacktriangledown$, that is, $(1 \times 60^2) + (7 \times 60^1) + (3 \times 60^0)$, which for an uninitiated reader could be interpreted several ways, for example $(8 \times 60^1) + (3 \times 60^0) = 483$. A similar ambiguity is also present in the rendering of 4230 above. Positional notation is completely lacking in the Egyptian system and "zero" as a place holder is absent from both systems. [*Editor's note:* Cordrey was considering the early Babylonia system. By about 300 B.C., the Babylonians had devised a "zero" symbol to act as a place holder in their positional numeral system].* This made it impossible for the early mathematicians to use our modern method of multiplication or of long division, since both presuppose place value. These were both slow processes with the Egyptians and Babylonians.

The world was yet to wait nearly three millennia for a zero and place value. With this "Columbus egg" and the nine Hindu-Arabic number symbols, the development of mathematics was greatly enhanced. Had the early inhabitants of the Tigris-Euphrates and Nile valleys possessed our number symbols, science, as well as mathematics, would no doubt be far in advance of its present development.

Despite the fact that the Egyptians and Babylonians were handicapped by the absence of zero and local value, various tribes and social groups were far inferior to them in the development of number. In fact, languages have existed in which there were no pure number words. The Tacanas of Bolivia have no numerals except those borrowed from other tribes or from the Spanish.[10] The Chiquitos of the same country had no real numbers before they contacted the white man, but expressed their idea for "one" by a word meaning "alone."[11] A few other South American languages are almost as destitute of numeral words. The entire number system of the forest tribes of Brazil appears to be limited to three.[12] The Encabellada of the Rio Napo have only two distinct numbers; meaning "one" and "two"; the number words of the Puris are three, meaning "one," "two," and "many,"[13] and the Botocudos count "one" and "many." The Campas[14] of Peru possess separate words for "one," "two," and "three"; but when they count beyond three they proceed by combinations, as one and three for four, one and one and three for five—expressing any number beyond ten as "many." The Conibos[15] had only two number words when the Spanish invaded South America. In the languages of the extinct or rapidly disappearing tribes of Australia, we find a remarkable absence of numeral expressions. In the Gudang[16] dialect only two numbers are found, meaning "one" and "two." Among the natives of Port Essington[17] and in a number of other instances the numerals stop at three.

Not one of the tribes referred to above developed a civilization. The transaction of business, excepting in an extremely crude way, was impossible. A member of a forest tribe of Brazil could not ask another tribesman to exchange four fish for a fowl unless he used the sign language. A Puri could not demand of a compatriot five of one kind of fruit for three of another. A transfer of goods only on a small scale was possible because of the absence of a number system. Trade among the different tribes was next to impracticable. Hence the transfer of knowledge from one tribe to another could have taken place only by an extremely slow process.

These people represented a point near the beginning of the curve of educational development. Their knowledge of mathematics and other academic fields placed them on a rung near the bottom of the ladder. With reference to the Egyptians and Babylonians, these people represented the antipodes in mathematics and general learning as much as in actual geographic location.

Let us now observe some of the mathematics and general knowledge of a people who had made a number of steps toward the development of a civilization, but still far from an advanced social culture. These people inhabited the vast area north of our southern border long before Columbus discovered America. A study of the habits and customs of the American Indians north of Mexico before the intervention of the white man brings to light many interesting examples of their use of various arithmetical and geometrical concepts and principles, although the absence of an adequate symbolism very greatly limited their mathematical progress.[18]

Counting on the fingers was practiced universally by the Indian tribes of this region. In fact, man's ten fingers have left their permanent imprint everywhere.[19] None of the American Indians north of Mexico had an advanced system of number symbols. Other than the spoken word only a few symbols have been found, and these are on buffalo robes, grave markers, tattooing, and other mnemonic pictographs. The most frequently used method for representing numbers was by the repetition of single strokes, that is, the additive principle in its most elementary form.[20] The Dakotas represented a count of one by a vertical line and a count of one hundred by one hundred such marks.[21] The use of notches cut in wood was frequent in the middle west, in California, and on Puget Sound. At the San Gabriel mission in California every tenth notch extended entirely across the stick. The Creeks used nine vertical marks to indicate nine, but represented ten by a cross. A number of regular systems extended to include one thousand, but in many cases larger numbers were not introduced until after the advent of the white man.

In the Yucatan Peninsula was a tribe of Indians which possessed, long before the birth of Christ, a number system that extended beyond one thousand. These people used two different methods of writing numbers. In their codices are found symbols for one to nineteen expressed by bars and dots: each dot represented one unit and each bar represented five. The additive principle was used in forming these symbols, and the principle of local value entered in writing twenty, which was expressed by a dot placed over the symbol for zero. The numbers were written vertically, the lowest order being placed in the lowest position. Accordingly, thirty-six was expressed by the symbol for sixteen (three bars and one dot) in the *kin* (lowest order) place, and one dot representing twenty placed above sixteen. The highest number found in the codices is, in our notation, nearly twelve and a half millions.

The second number system employed the zero, but not the principle of local value. In this system each of the numerals from zero to thirteen is expressed by a distinct type of head. From fourteen through nineteen the head numerals are expressed by the application of the essentials of the head for ten to the head for four to nine inclusive.[22] Both of the Maya number systems were essentially vigesimal. They had a unique word for each positive integral power of twenty up to and including the fourth power.[23]

Tribes employing the decimal system expressed "one hundred" by a unique word or by such forms as "completed," "stock of tens," and "ten times ten"; and "thousand" by "ten times ten times ten," "ten times one hundred," "big hundred," "old man hundred," and "large stock of tens."

The Greenland Eskimo uses "other hand two" for seven, "first foot one" for eleven, "other foot three" for eighteen, and similar combinations up to twenty, which is "man ended." The Unalit employs the same base to twenty; but forty is given by "two sets of animals' paws," sixty "three sets of animals' paws," and so on to four hundred, where there is an interesting change in the formation of this primary base (twenty times twenty) from animals back to man, four hundred being "twenty sets of man's paws."[24] Wintum alternates the decimal and vigesimal bases; twenty, forty, and sixty being "one Indian," "two Indians," "three Indians," while thirty and fifty are "three times ten" and "five times ten" respectively.[25] Kopiagmiut is quinary to ten, decimal for the formation of the odd tens and vigesimal for integral multiples of twenty.[26] The

Haida number system is alternately quinary-decimal and quinary-vigesimal to one hundred, then vigesimal, "four-hundred" being "twenty times twenty times one" and "eight hundred" being "twenty times twenty times two."[27] The Santa Barbara Indians employed the quarternary system reaching to eight and in forming some numbers above eight, "twelve" being "three times four" and "fifteen" being "twelve plus three."[28] The Cuchan words for three, six, and nine are *hamook, humhook,* and *hum-hamook*.[29]

The Coahuiltecan of Texas is a very interesting example of the use of a number of bases. It exhibits binary, ternary, quarternary, quinary, decimal, and vigesimal bases.[30] Represented in Hindu-Arabic some of its numbers are $1, 2, 3 = 2 + 1, 6 = 3 \times 2, 7 = 4 + 3, 8 = 4 \times 2, 9 = 4 + 5, 11 = 10 + 1, 12 = 4 \times 3, 13 = 12 + 1, 15 = 5 \times 3, 17 = 15 + 2, 30 = 20 + 10, 40 = 2 \times 20, 50 = 40 + 10$.[31]

The additive and multiplicative principles existed in the number systems of many North American tribes. Cantor states that addition and multiplication are two methods of counting as old as the formation of number words.[32] The languages of these American tribes verify the statement by Fink that subtraction is very rarely used in the verbal formation of number words.[33]

The number words of the North American Indians were often cumbersome and difficult to use. Examples of an unwieldy system can be seen in the combination *lagwau swaunshung wok swaunshung,* which was used by the Haidas to represent twenty-one, and in *wich a chimen ne nompah sam pah nep a chu wink ah,* which was used by the Sioux to express twenty-nine. The Tshimshians of British Columbia used seven sets of numerals. Each set was used for one of their seven distinct classifications of objects.[34] For example, one set of numbers was used in abstract counting, another set for round objects and division of time, etc. The Takulli dialect of the Athapascan language presents another example of the modification of number words.[35] For example, *tha* was used to represent three things; *that,*

three times; *thauh,* in three ways; *thane,* three persons; *thatser,* in three places; and *thailtoh,* all three things.

These Indian tribes north of Mexico give evidence to the fact that some of the elements of mathematics can be developed with a crude number system. Also their arithmetic and geometry substantiate the statement that mathematics cannot evolve rapidly where a people depend on a cumbersome and unwieldy system of number words and number symbols. It is interesting to note that these Indians were further advanced in the development of a civilization than were those heretofore mentioned of South America and the Australian aborigines who had only the rudiments of a number system. The use of one hundred strokes to represent one hundred is an advancement far beyond the absence of number symbols. Such representation of one hundred is, however, not only much inferior to the Babylonian and Egyptian symbols for the same number, but is indicative of less cultural development.

The arithmetic of these Indian tribes appears to have been confined to addition, subtraction, and in a few cases simple multiplication with integers. Only a relatively small number of tribes had any conception of fractions. Their failure to evolve an adequate system of number words and number symbols prevented their advancing far beyond the simple fundamentals. Their knowledge of geometry was also meager. The geometric forms employed in the building of their houses were principally rectangular and circular. McGee and Thomas believe that the Indians had some method of drawing one line perpendicular to another.[36]

All the essentials of our modern elementary arithmetic, including numeration by position and a symbol representing zero, had been devised by the ancient Mayas two millennia ago, and at least five centuries before the Hindus had evolved the fundamentals of the so-called Arabic notation in India.[37] Little progress has been made in deciphering the Maya glyphs excepting those pertaining to their calendar and chronology.[38]

Goods were exchanged by members of the same tribe and business and social communion took place between members of different tribes. These North American aborigines transacted business on a much smaller scale than did the citizens of ancient Babylon. To a limited extent, there existed a division of labor which is vitally essential to business intercourse. This division of labor was insufficient to necessitate the exchange of large quantities of goods.

Unlike the forest tribes of Brazil, the progress of learning among the Indians north of Mexico was not prevented by a lack of energy. Their warlike nature precluded peace and harmony, which are always necessary to the production of leisure. Without leisure, the existing knowledge can increase only by a very slow process. Had peace and harmony existed among the early Americans, as it did in ancient Babylon and Egypt, the American aborigines would have had a much different general history, as well as a much different history of mathematics.

The preceding discussion evidences the fact that mathematics had its inception at a very remote date. The study of astronomy was also begun early and in this subject laws of nature were open to discovery by observers, endowed with clear vision and persistence of purpose. This science started with simple and familiar concepts which resulted from inductive thought. Its advancement was stimulated by clarity and definiteness of ideas of space and figure, time and number, and motion and reoccurrence.

Like primitive men everywhere, the Egyptian peasants began at a very remote date to scan the heavens and to observe the stars and the moon. There is probably no other region in which Sirius is such a bright and noticeable spectacle in the evening sky. Aristotle related that the Egyptians attempted scientific observations of the heavens about 2200 B.C. Long before this, however, they observed the periodic changes of the moon, but the lunar month contributed nothing to the determination of the length of the year. The total incommensurability

of the stellar or solar year and the so-called lunar year could not be discerned until the approximate length of the solar year had been determined and the fact that there is no such thing as a lunar year recognized. The length of the stellar year as measured by successive sunrise reappearances of Sirius was first roughly estimated by the Egyptians as 360 days more than six millennia ago.[39] As their observations of the heliacal rising of Sirius accumulated, they discovered that the solar year was longer than 360 days. In 4241 B.C. an unknown ruler of prehistoric Egypt added five feast days at the end of the existing 360-day unit, making a convenient calendar of 365 days. This new unit was divided into eleven months of thirty days each and one month of thirty-five days, the months being grouped into three seasons of four months each.

A calendar dominated by the lunar month existed in Babylon as early as the fourth millennium B.C. Instead of five days each year, however, the Babylonians intercalated an entire month every few years. The insertion of an extra month was originally the result of a royal command. Whenever the king noticed that "the year hath a deficiency," he ordered the insertion of an intercalary month.

Babylonian tablets of some five millennia ago contain astronomical reports and astrological deductions based upon them. These tablets contain many lists of stars, and observations of the moon, planets, and comets. The early Babylonians made angular measurements of position, determined the period of revolution of the planets, and predicted successfully eclipses of the sun and moon.

The week of seven days and the day of twenty-four hours had their origin in Babylon. The inhabitants of the Tigris and Euphrates valleys divided the hour into sixty minutes and the minutes into sixty seconds. They divided the circle into three hundred and sixty degrees, the degree into sixty minutes of arc, and the minute into sixty seconds. The savants of today measure time and angles in the same units as did the Babylonians and Sumerians of some five millennia ago. These units enabled the scientists of ancient Babylon to record and accumulate astronomical data.

In contrast to the Egyptians and Babylonians, there are a number of South American and Australian tribes who have not used the periodic changes of the moon in measuring time, neither have they used the heliacal rising of any star in determining the length of the year. In fact, there are several primitive tribes who have no conception of a year, but who recognize half-year periods, corresponding to the wet and dry seasons, or to the monsoon winds, which reverse their directions semi-annually.[40] The Indians of the Orinoco anticipate the rains when the monkeys scream at midnight or at the break of day, when the yams which have lost their leaves suddenly grow green again, and when the Pleiades first appear in the eastern sky before sunrise. The Chinhwan of Formosa know that a new year has dawned when a certain flower blooms again. The Tupi of Southern Brazil call the year "cashew-tree," which blooms once a year and bears a fruit used in the making of wine.

The most important bases for calculating time, used by the Indians north of Mexico, were the changes of the moon and the seasons, and the succession of days and nights. Four seasons were generally recognized, although some tribes distinguished more and some less.[41] One tribe counted twelve moons, with some extra days, as a year. Another tribe used thirteen. The Creeks intercalated an extra moon every second year, thus averaging twelve and one half moons in each year. The Sarcees advanced from moon counts by stages of five to six and then to twelve months of thirty days each long before the coming of the white man to the North American continent.[42] Some tribes divided the day into three parts, others used four, and one tribe had a day of nine divisions. The Choctaws expressed a fraction of a day as the time required for the sun to traverse the distance between two parallel lines drawn on the ground.[43] The Indians of South Carolina used the number of handbreadths the sun was above the horizon in determining the time of day.[44]

The methods of reckoning time used by these Indian tribes of the United States and Canada, like their number systems, are indicative of a people low in the evolution of a civilization; much lower than the Babylonians of three millennia before Columbus but far superior to the Australian aborigines of about the same periods.

The Maya made the day, which he called *kin*, the primary unit of his calendar. There were twenty such units following each other in a definite order. When the last day in the list, *Ahau*, had been reached, the count began anew with *Imix*, the first day. This twenty-day sequence continued in endless repetition. Calling the day by one of the twenty names, however, did not sufficiently describe it according to the Maya notation. In order to complete the name of a day it was necessary to prefix to it a number ranging from one to thirteen, inclusive, as 10 *Ahau* or 3 *Imix*. By this method the Mayan day received its complete designation and found its appropriate place in the calendar. Since there were thirteen numbers and twenty days, any given day could not reappear in the sequence until the two hundred and fifty-nine days immediately following it had elapsed. This method of numbering days was doubtless introduced long after the calendar had assumed a regular form, and probably by the priests, for the purpose of complicating it and rendering it as far as possible unintelligible to the people.[45] This group of two hundred and sixty differently named days was called *tonalamatl*, or "book of days," by the Aztecs. The Maya name for this unit is unknown. The *haab* or solar year of the Mayas consisted of three hundred and sixty-five days. It was divided into eighteen "months" of twenty days each with a closing period of five unlucky days.[46] The least common multiple of the number of days in the solar year and the *tonalamatl* was another unit of time. This period was called the "Calendar Round" and consisted of fifty-two times 365 days.

Bishop Landa stated that the Mayas knew their solar year to be nearly six hours short of the true length of the year.[47] The Mayan calendar had no elastic period corresponding to our month of February. It is probable that the priests corrected the calendar by adding calculations which showed the exact number of days the recorded year was ahead of the true year at any given time. Thus the Mayan calendar year consisted of three hundred and sixty-five days.

The Mayas in their knowledge of the apparent movements of the heavenly bodies—the sun, moon, Venus, and probably other planets as well—far exceeded both the ancient Egyptian and Babylonian, yet their greatest intellectual achievement was the invention of a chronology exact to a day within a period of more than two millennia, which is as accurate as our own Gregorian calendar.

As we press our way into the meaning of the still undeciphered hieroglyphs, it becomes increasingly apparent that they deal more and more with the subject matter of astronomy.[48] This science, never wholly wanting in the culture of the most lowly aborigines and often occupying a large amount of attention among nations, might easily be employed to furnish an index of the intellectual power of group minds. Thus an intelligence test of tribes, nations, and other units of people might be prepared on the true knowledge possessed by these social groups of the movements of the celestial bodies.[49] On the basis of such a test, the Mayas, Egyptians, and Babylonians, the people who early developed mathematics, would surely be certified as a people blessed with creative minds.

NOTES

1. Arnold Chase and Henry Manning, *The Rhind Mathematical Papyrus* (Oberlin, Ohio: The Mathematical Association of America, 1927), I, 35.

2. G. A. Miller, "A Wide-Spread Error Relating to Egyptian Mathematics," *Science* LXXXI (1935): 152.

3. Florian Cajori, *A History of Mathematics* (New York: The Macmillan Co., 1931), p. 9.

4. David Eugene Smith, *History of Mathematics* (New York: Ginn and Company, 1923), I, 41.

5. R. C. Archibald, "Babylonian Mathematics," *Science* LXXX (1929): 67.

6. "Two Babylonian Multiplication Tables in Ontario," *Art and Archaeology* XXVI (1928): 145.

7. R. C. Archibald, "Babylonian Mathematics with Special Reference on Recent Discoveries," *The Mathematics Teacher* XXIX (1936): 217.

8. Moritz Cantor, *Vorlesungen über Geschichte der Mathematik* (Leipzig: Verlag und Druck von B. F. Teubner, 1922), I, 82.

9. R. C. Archibald, "Mathematics Before the Greeks," *Science* LXXI (1930): 110.

10. L. L. Conant, *The Number Concept* (New York: Macmillan and Co., 1923), p. 2.

11. D. G. Brinton, *American Race* (New York: N. D. C. Hodges, Publisher, 1891), pp. 296, 359.

12. L. L. Conant, *op. cit.*, p. 7.

13. Edward B. Taylor, *Primitive Culture* (New York: Brentano's, 1924) I, 243.

14. L. L. Conant, *op. cit.*, p. 22.

15. P. Marcoy, *Travels in South America* (London: Blackie, 1874), II, 47.

16. E. M. Curr, *Australian Race: Its Origin, Language, and Customs* (London: Trubner; 1888), I, 282.

17. J. Bonwick, *The Daily Life and Origin of the Tasmanians* (London: Low, 1870), p. 144.

18. F. L. Wren and Ruby Rossman, "Mathematics Used by American Indians North of Mexico," *School Science and Mathematics* XXXIII (1933): 363.

19. Tobias Dantzig, *Numbers, the Language of Science* (New York: The Macmillan Co., 1933), p. 9.

20. W. C. Eells, "Number Systems of North American Indians," *American Mathematical Monthly* XX (1913): 271–272.

21. H. R. Schoolcraft, *Historical and Statistical Information Respecting the History, Condition, and Prospects of the Indian Tribes of the United States* (Philadelphia: J. B. Lippincott, 1852), II, 178.

22. S. G. Morley, "An Introduction to the Study of the Maya Hieroglyphs," *Bureau of American Ethnology Bulletin* 57 (Washington: Government Printing Office, 1915), p. 96.

23. Moritz Cantor, *op. cit.*, p. 9.

24. E. W. Nelson, "Eskimo About Bering Strait," *Eighteenth Annual Report of the Bureau of Ethnology,* 1896–1897, p. 238.

25. R. B. Dixon and A. L. Kroeber, "Numeral Systems of Languages of California," *American Anthropologist* IX, (1896): 675.

26. E. Petitot, *Vocabulair Française-Esquimau* (Paris: Maisonneuve Freres, 1876).

27. C. Harrison, "Haida Grammar," *Royal Society of Canada Proceedings and Transactions,* I (1895): 142–143.

28. W. C. Eells, *op. cit.,* p. 296.

29. J. H. Trumbull, "On Numerals in American Indian Languages," *Transactions of the American Philological Association,* 1874, p. 74.

30. A. Gallatin, *Transactions of the American Ethnological Society,* 1845, I, 1.

31. W. C. Eells, *op. cit.,* p. 297.

32. Moritz Cantor, *op. cit.,* p. 8.

33. Karl Fink, *A Brief History of Mathematics* (Chicago: The Open Court Publishing Co., 1903), p. 8.

34. Franz Boaz, "First General Report of the Indians of British Columbia," *Report of the British Association for the Advancement of Science,* 1889, pp. 880–881.

35. F. W. Hodge, "Handbook of American Indians North of Mexico," Bureau of American Ethnology, *Bulletin* 30, Part I (Washington: Government Printing Office, 1907), p. 354.

36. W. J. McGee and Cyrus Thomas, *Prehistoric North America, the History of North America* (Philadelphia: George Barrie and Sons, 1905), XIX, 376.

37. S. G. Morley, "Yucatan, Home of the Gifted Maya," *National Geographic Magazine* LXX (1936): 595, 597.

38. Florian Cajori, "The Zero and the Principle of Local Value Used by the Maya of Central America," *Science* XLIV (1916): 715.

39. James H. Breasted, "The Beginning of Time-Measurement and the Origin of Our Calendar," *Scientific Monthly* XLI (1935): 293.

40. Charles F. Talman, "Backgrounds of the Calendar," *Nature Magazine* XIV (1929): 22.

41. H. H. Bancroft, *The Native Races, the Works of H. H. Bancroft* (San Francisco: The History Co., I, 1886), 564.

42. Moses B. Cotsworth, "The Evolution of Calendars and How to Improve Them," *Pan American Union* LIV (1922): 544.

43. H. B. Cushman, *History of the Choctaw, Chickasaw, and Natchez Indians* (Greenville, Texas: Headlight Printing House, 1889), p. 249.

44. E. L. Green, *The Indians of South Carolina* (Columbia, South Carolina: The University Press, 1904), p. 45.

45. Cyrus Thomas, "A Study of the Manuscript Troano," *Contributions to North American Ethnology* (Washington: Government Printing Office, 1882), p. 7.

46. John Kinsella, "The Mayan Calendar," *The Mathematics Teacher* XXVII (1934): 341.

47. Diego De Landa, *Relacion De Las Cosas De Yucatan* (Paris: Brasseur de Bourbourg, 1864).

48. S. G. Morley, "The Foremost Intellectual Achievement of Ancient America," *National Geographic Magazine* XLI (1922): 129.

49. Herbert J. Spinden, "Ancient Mayan Astronomy," *Scientific American* CXXXVIII (1928): 9.

10

Mayan Arithmetic

JAMES K. BIDWELL

*I*T IS NOT UNUSUAL IN THE SCHOOLS to study numeration systems other than our own decimal system. A good unit of this type can stimulate the students' interest in arithmetic and can be helpful in providing for practice in computation. Usually little is done with systems employing strange numerals besides learning how to write names for numbers. This article will discuss the Mayan numeration system and suggest computational schemes for the system. If pursued with some effort, these schemes can produce a better appreciation of the essential features of a numeration system, such as its additive nature, place value, base, and need of special symbols. It would be possible to combine this numeration work with a more comprehensive unit on the Mayan civilization, which was one of the great ancient cultures. Several popular books are now available on the Mayans, and some of these are mentioned in the bibliography.

A Brief History of the Mayans

The span of the Mayan civilization is generally placed from 1500 B.C. to about A.D. 1700. The period of greatest cultural growth was from A.D. 300 to A.D. 900. Knowledge is sketchy of the civilization previous to this time, and archaeologists have as yet little evidence to offer. Obviously, a civilization of the extent of the Mayan did not spring up "overnight," so it is safe to assume that the Yucatan Peninsula, in Mexico, was the place of a developing Mayan culture for many centuries before Christ.

The religious rituals of the people became calendric at a very early period. The priestly class was totally absorbed in time and developed an all-embracing philosophy of time as divine and eternal. Days and months were considered sacred and were personified as gods who "carried" time with them. This early and continuing development about the concept of time gives us almost all of our present knowledge about the Mayan culture, since to a large extent the calendar records and astronomical data are the only translated artifacts. At the present time the written language has not been translated.

During the classic period the populated region covered approximately 100,000 square miles, with no large settlements. The population density was about 30 people per square mile. The permanent buildings were small stone temples. Emphasis was placed on skilled workmanship by stone masons, rather than on masses of slave labor common to other ancient civilizations. The main labor of the people was devoted to the cultivation of corn. The priests, the presumed governing group, succeeded in avoiding war for centuries. Hence, a flourishing culture was able to develop.

The chief source of information on the Mayans is the writings of Father Diego de Landa (1524–1579), who came to Mexico as a missionary in 1549. His chief work was *Relación de las cosas de Yucatán*, which is the authority for the customs, history, and intellectual development of the culture. A few original manuscript copies exist, be-

Reprinted from *Mathematics Teacher* 74 (Nov., 1967): 762–68; with permission of the National Council of Teachers of Mathematics.

sides the many stone stelae, or date-stones, that remain. Father Landa burned many of the existing manuscripts of the classic period because he considered them to be heretical. Consequently, much remains unknown about the culture of the classic period.

Mayan Numeration Systems

The Mayans developed two numeration systems, one for the priestly class and one for the common people. These systems were radically different in terms of the base value and of the symbols used to denote numbers. These facts are in opposition to the popular belief that only one numeration was used, namely the calendric numeration system.

The numeration system of the priestly class was dominated by religious ritual. Since the days of the year were thought of as gods, the formal symbols for day names and numerals were decorated heads. Artistry made copies of these symbols even more complex and difficult to read. The Mayan numeration system employed in recording time was based on 360 days of the year, and hence it was a mixed-base system using multiples of both 20 and 360 for place values. Because of these complexities, this article will be concerned only with the much simpler common form of expressing numerals. For those interested in pursuing the more complex system, the bibliography supplies several excellent sources.

The system used by the common people has not survived in written form. Father Landa indicated the common system of numeration that was used:

> They count by fives up to twenty, by twenties to a hundred and by hundreds to 4 hundred, 400's up to 8000. This count is much used in merchandizing the cacao. They have other very long counts, extending to infinity, counting twenty times 8000 to 160,000; then they multiply this 160,000 again by twenty and so on until they reach an uncountable figure. They do all this counting on the ground or a flat surface.[1]

This indicated a pure system of position values of units 20, 400, 8,000, 160,000, and so on. The lack of manuscripts using this base leaves open to conjecture how they may have computed in this base and how numerals were written.

However, the manuscripts do use some simple symbols for the numerals from zero to nineteen in the mixed-base system. A vertical system is used to record higher numbers. We will adopt both of these methods of notation. The lowest position will be units, the next higher position will represent twenties, the third position will represent four hundreds, etc. Table 1 indicates this vertical place value and also gives the position names in the native language.

TABLE I *Vertical Place Value and Place Names of the Mayan Common Counting System*

12,800,000,000	=	20^7	. . . *hablat*
64,000,000	=	20^6	. . . *alau*
3,200,000	=	20^5	. . . *kinchil*
160,000	=	20^4	. . . *cabal*
8,000	=	20^3	. . . *pic*
400	=	20^2	. . . *bak*
20			. . . *kal*
1			. . . *hun*

To write the common numerals, a *normal form* was employed, using combinations of dots and bars, where each dot represented one unit, and a bar represented five units. These numerals are illustrated in Table 2.

We observe that these numerals use only three symbols: a dot, a bar, and a special symbol for zero. This Mayan system may have been the first to make use of zero as a place holder, for this system was used by the Mayans many centuries ago. This system is repetitive and semiadditive, and with the vertical place-value scheme, allows very simple representation of large numbers.

Figure 1 illustrates five numerals using the base-twenty system of the Mayans. We will agree to read the numeral from the top down. So Figure

TABLE 2

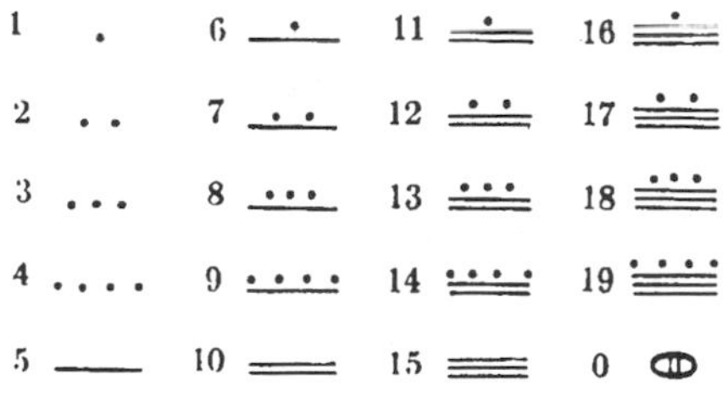

1	.	6	⎯•⎯	11	⩵•	16	⩶•
2	..	7	⎯••⎯	12	⩵••	17	⩶••
3	...	8	⎯•••⎯	13	⩵•••	18	⩶•••
4	....	9	⎯••••⎯	14	⩵••••	19	⩶••••
5	⎯⎯	10	⩵	15	⩶	0	⊕

1(a) is a two-place numeral. It represents 15 *kal* 6 *hun*, or 15 (20) + 6 = 306 in our decimal system. Figure 1(b) is a three-place numeral: 17 *bak* 6 *kal* 7 *hun*, or 17 (400) + 6 (20) + 7 = 6,927. Figure 1(c) illustrates the use of the zero symbol as a place

FIGURE I

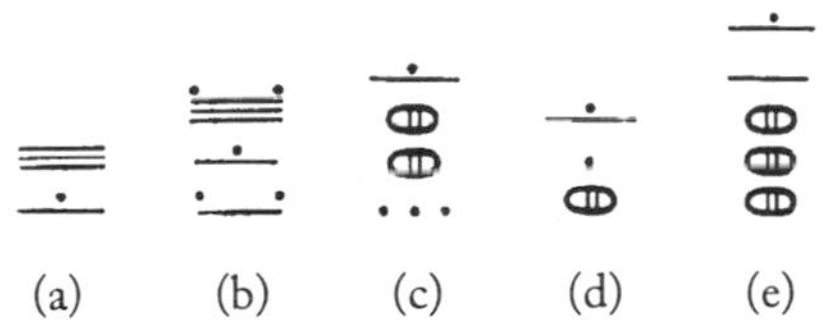

(a) (b) (c) (d) (e)

holder indicating completion of that place count. Figure 1(c) represents 6 *pic* 3 *hun*, or 6 (8,000) + 3 = 48,003. Figure 1(d) shows 6 *bak* 1 *kal*, or 6 (400) + 1(20) = 2,420. The reader will notice a saving in the number of places required for large numbers. For example, one million requires seven places in our decimal system but only five places in the Mayan system. Figure 1(e) represents one million, that is, 6 *cabal* 5 *pic*.

Arithmetic

In some of the manuscripts, tables of repeated addition are found which might be called multiplication tables, but only in a limited sense. The Mayan calendar development and the knowledge of planetary and lunar cycles indicate some acquaintance with common multiples. Addition and subtraction are indicated in the codices; but no multiplication or division, as we think of the operations, can be found. Hence, no fractions or

decimals are found. Calculations could have been carried out, however, and not recorded. Actually, most of the extant writings are either dates or tables of data in condensed form, where calculations have been eliminated. The remainder of this article will explore the possibilities of written calculations using the base-twenty system.

1. *Addition.*—Addition is easily handled in this vertical notation by simply combining bars and dots and carrying to the next higher place. After the combining of dots and bars, the second step is to exchange every five dots for one bar in the same position. Then, lastly, every four bars in one position are exchanged for one dot in the next higher position. As an example consider the addition of 89 + 124, that is, 4 *kal* 9 *hun* + 6 *kal* 4 *hun*, in Figure 2.

FIGURE 2

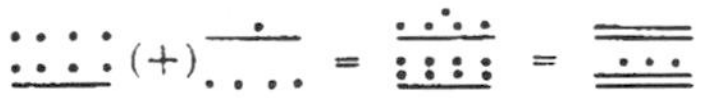

Addition could be performed with a minimum of writing by mental computation. That is, the partial sum in the middle of Figure 2 could be eliminated. As another example consider 6 *bak* 12 *kal* 16 *hun* + 5 *bak* 10 *kal* 6 *hun*, as shown in Figure 3.

FIGURE 3

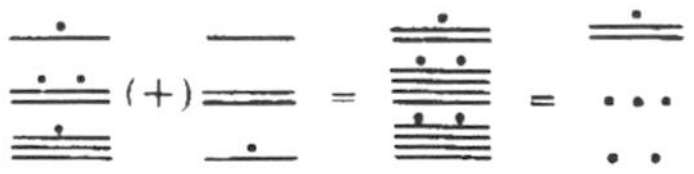

2. *Subtraction.*—Subtracting two numbers can be performed with Mayan numerals by actually taking away or by cancelling symbols from the numeral for the larger number. Consider 94 - 53, or 4 *kal* 14 *hun* - 2 *kal* 13 *hun*. Figure 4 shows the result of 2 *kal* 1 *hun*, or 41. The calculator can actually mark out the symbols in the numeral of the larger number and be left with the correct answer (Fig. 5).

FIGURE 4

$$\begin{matrix} \bullet\bullet\bullet\bullet \\ \overline{} \\ \overline{} \\ \bullet\bullet\bullet\bullet\bullet \end{matrix} (-) \begin{matrix} \bullet\bullet \\ \overline{} \\ \bullet\bullet\bullet \end{matrix} = \begin{matrix} \bullet\bullet \\ \bullet \end{matrix}$$

FIGURE 5

$$\begin{matrix} \prime\prime\bullet\bullet \\ \overline{} \\ \bullet\bullet\bullet\bullet \end{matrix} (-) \begin{matrix} \bullet\bullet \\ \overline{} \\ \bullet\bullet\bullet \end{matrix} = \begin{matrix} \bullet\bullet \\ \bullet \end{matrix}$$

In any numeration system using place value, borrowing becomes necessary. From the earlier quotation from De Landa's book, we note that the natives counted "on the ground." This suggests that the Mayans used sticks and stones with which to calculate. If this is true, then "borrowing" could have been an actual process of removing one stone and replacing it with four sticks in the next lower position. Also, five stones could replace one stick. Using actual sticks and stones would make addition and subtraction a combining and a taking away in reality.

FIGURE 6

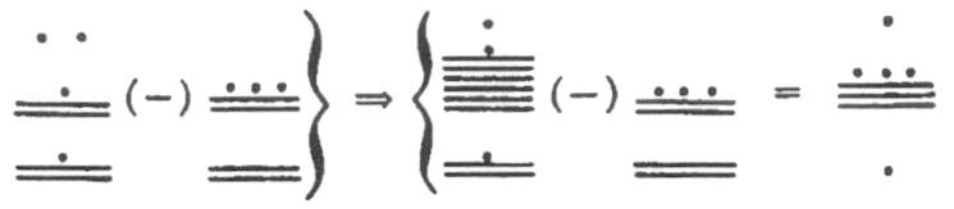

$$\left. \begin{matrix} \bullet\bullet \\ \overline{} \\ \overline{} \end{matrix} (-) \begin{matrix} \bullet\bullet\bullet \\ \overline{} \\ \overline{} \end{matrix} \right\} \Rightarrow \left\{ \begin{matrix} \overline{} \\ \overline{} \\ \overline{} \end{matrix} (-) \begin{matrix} \bullet\bullet\bullet \\ \overline{} \end{matrix} = \begin{matrix} \bullet\bullet\bullet \\ \overline{} \\ \bullet \end{matrix} \right.$$

Figure 6 shows a problem requiring borrowing. In the *kal* position the problem requires subtracting 13 from 11. Hence we borrow 20 from the higher position by removing one dot and adding four bars. The problem then is finished as before.

3. *Multiplication.*—No records indicate that the Mayans had any sophisticated multiplication scheme. Lacking any factual knowledge, we rely on ingenuity and invent a scheme which utilizes properties of whole numbers and the particular nature of the normal-form notation. Since any base numeral is composed of up to three bars and four dots, we need only consider a basic multiplication table of forty-nine entries. These entries involve only the products of dots and bars separately. The multiplication facts are shown in Table 3. We note that there are only about 60 percent of the number of facts in our decimal scheme of products from 1 × 1 to 9× 9. Hence, the memorization of the Mayan table is simpler by far.

TABLE 3

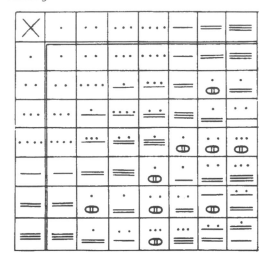

Since any number is a combination of bars and dots, Table 3 is sufficient. This is true because of the distributive law for whole numbers. Thus, any product with factors using both bars and dots can be split into four products with these separate products added to obtain the required numeral. Thus a product like ($\overline{\bullet\bullet}$) ($\overline{\bullet\bullet\bullet\bullet}$) is considered as (• •) (• • • •) + (• •) ($\overline{}$) + ($\overline{}$) (• • • •) + ($\overline{}$) ($\overline{}$). Using Table 3, one can write down the partial products and add. (See Fig. 7.)

The calculation of Figure 7 can be checked by using base-ten numerals: 12 × 14 = 168 = 8(20) + 8. Since the bars and dots in the partial products act as tallies, they can be combined as one works to write the final answer immediately. Figure 8 shows 11 × 8 = 88, using condensed partial products. Now this multiplication scheme offers an excellent opportunity for review of multiplying by multiples of the base. That is, in base ten, 10 × 10 = 100, so 30 × 20 = 600; or 10 × 100 = 1,000, so 20 × 400 =

FIGURE 7

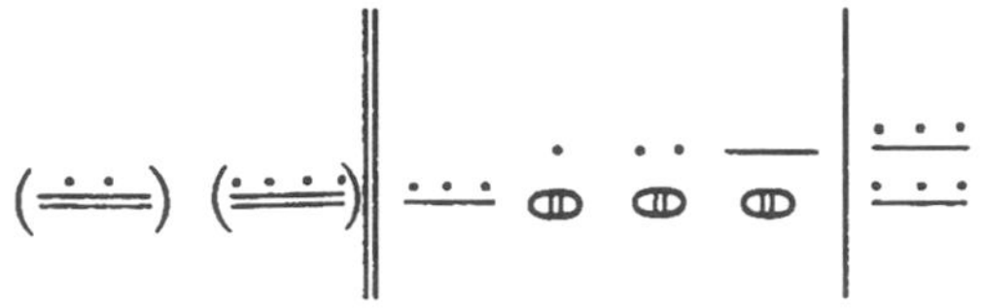

$$(\overline{\bullet\bullet}) \; (\overline{\bullet\bullet\bullet\bullet}) \left\| \begin{matrix} \bullet \\ \bullet\bullet\bullet \; \oplus \end{matrix} \begin{matrix} \bullet\bullet \\ \oplus \end{matrix} \begin{matrix} \overline{} \\ \oplus \end{matrix} \right\| \begin{matrix} \bullet\bullet\bullet \\ \overline{} \\ \bullet\bullet\bullet \end{matrix}$$

FIGURE 8

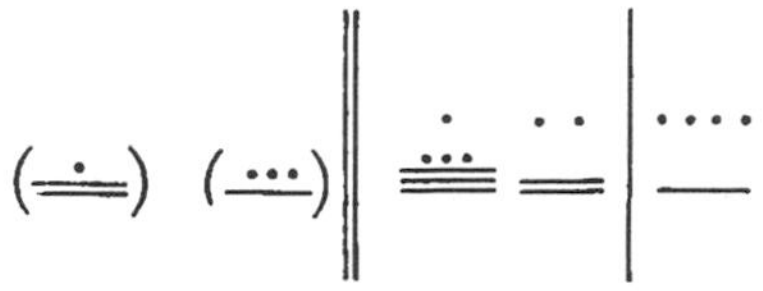

$$(\doteq) \; (\cdots) \parallel$$

8,000 and $30 \times 500 = 15,000$. Similarly, in Mayan numerals one can do the same work (Fig. 9).

FIGURE 9

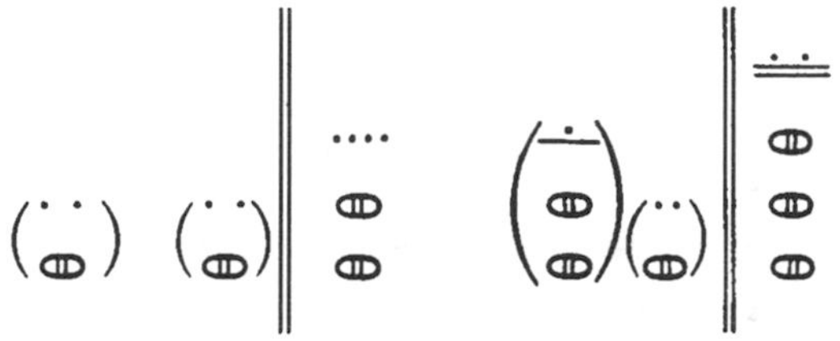

In multiplying numbers of two or more places, we have to consider further partial products of this type. These may be eliminated with practice, as in our normal algorithm for multiplication. Figure 10 shows that the product of 2 *kal* and 6 is 12 *kal*, and the 12 is written in the second place. With practice, the placement of the symbol $\overset{\bullet\bullet}{=\!=}$ would be automatic.

Figure 11 illustrates a more complicated product, with the partial products written. Checking in base ten: $606 \times 103 = 62,418 = 7(8,000) + 16(400) + 0\,(20) + 18$.

As with our regular algorithm, the zero symbols may be omitted with practice. Figure 12 can be studied as an example without the symbol for zero. The base-ten check shows that $[12(400) + 11] \times [15(400) + 9(20) + 5] = 4,811 \times 6,185 = 29,756,035 = 9(3,200,000) + 5(160,000) + 19(8,000) + 10(400) + 1(20) + 15$. With practice and memorization of the 7×7 multiplication table, computation in this system can be rapid and accurate.

Computation can be even more rapid than our own system, which requires a larger number of basic addition and multiplication facts and ten different symbols to express numerals. The dot and bar notation is easy to use on slate or on clay, or with sticks and stones for small children. Since we have twenty fingers and toes, grouped by fives, it is easy to teach young children to record by fives, using both feet and one hand, and by units, using the other hand.

4. *Division.*—In the same manner that multiplication can be developed, division can also be developed. That is, since the Mayan system is a positional system, a method similar to our long-division algorithm can be utilized. We can begin by considering a relatively simple case: $561 \div 33 = 17$. We establish a framework for our calculation by using double vertical bars to separate quotient, divisor, dividend, and partial dividends, in that order (Fig. 13).

FIGURE 10

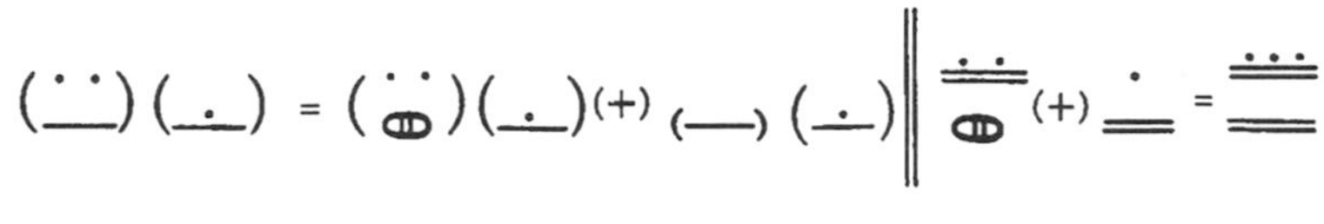

FIGURE 11

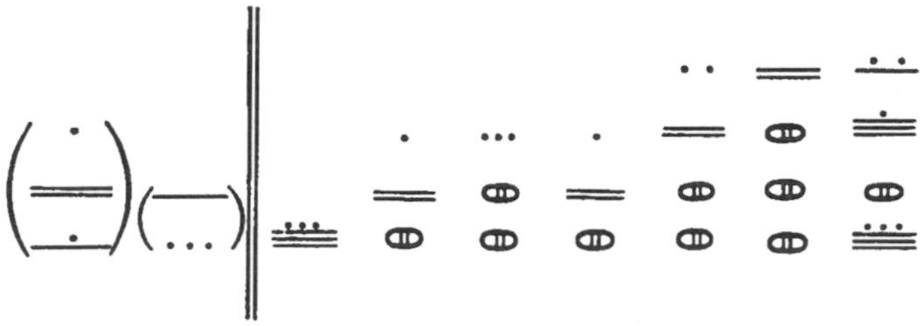

FIGURE I2

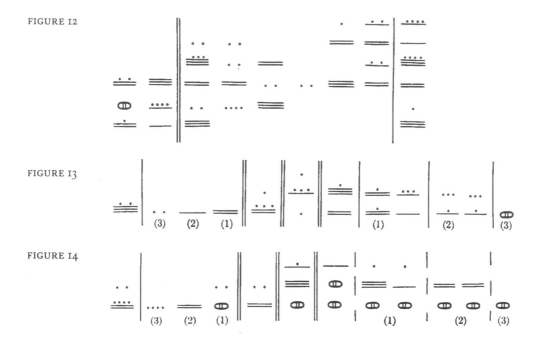

FIGURE I3

FIGURE I4

First we observe that 1 *kal* × 1 *kal* 13 *hun* = 1 *bak* 13 *kal* and hence is too large, so that our quotient is a number of *huns*. First the maximum number of bars is found. We use 2 *kal* as approximately 1 *kal* 13 *hun* and estimate that at least 10 *hun* is a partial quotient. This leads to the first product and subtraction. The subtraction leaves a remainder of 11 *kal* 11 *hun*. We see that 5 *hun* will be the next partial quotient. That is, we could have used 15 *hun* (or three bars) originally. Subtracting the product we have 3 *kal* 6 *hun* as a remainder. We now try 2 *hun* as our last partial quotient and obtain a remainder of zero. Hence the quotient is 17 *hun*.

Now the preceding exactly parallels the development of our base-ten division algorithm. One proceeds by estimating maximum partial quotients and obtaining remainders until we have a remainder less than the divisor. In Mayan work we first estimate the maximum number of bars (fives) in a position and then the number of dots (ones). It requires years to become proficient enough in our number system to perform efficient long division. Consequently, the student of Mayan numeration should expect slow progress in division. However,

with practice one can become quite skilled in division, particularly after the multiplication tables have been memorized.

Figure 14 shows the division problem: 2,700 ÷ 50 = 54. The work shows three trial divisions are utilized and hence three partial quotients were subtracted. These have been labelled in order.

Conclusions

In summary, the following features of the Mayan numeration system can be emphasized:

1. The system uses a base of twenty.

2. Only three symbols in repetition are needed to represent any number.

3. Only forty-nine basic multiplication facts are needed.

4. Addition and subtraction involve simple manipulation of dots and bars.

5. Decimal-system calculations can be mimicked in the system.

6. Visual aids, such as sticks and stones, can easily be employed both to count and to express numbers.

These positive facts about Mayan arithmetic make a study of this system a worthwhile addition to the topics which can be offered in the schools as special projects, change-of-pace topics, and the like. With a little practice, teachers can accustom themselves to the system and find it most helpful to add to the repertoire of teaching units.

The bibliography below offers a source for further reading on the Mayan culture. These references are nontechnical and are accessible in most libraries. The reader will find a definite emphasis on the calendric system in most of these publications.

NOTES

1. See page 40 of the work by W. E. Gates cited in the bibliography.

BIBLIOGRAPHY

Gallenkamp, Charles. *Maya*. New York: McKay Co., 1959.

Gates, W. E. *Yucatan Before and After the Conquest, by Friar Diego de Landa, etc.* Translated, with notes. Maya Society Publication No. 20. Baltimore: The Society, 1937.

Morely, S. G. *The Ancient Maya*. Stanford, Calif.: Stanford University Press, 1956.

Thompson, J. Eric S. *Maya Arithmetic*. ("Contributions to American Anthropology and History," Vol. VII, No. 36.) Washington, D.C.: Carnegie Institution of Washington, 1942.

Thompson, J. Eric S. *The Rise and Fall of Maya Civilization*. Norman, Okla.: University of Oklahoma Press, 1954.

Von Hagen, Victor W. *Maya, Land of the Turkey and the Deer*. Cleveland: World Publishing Co. 1960.

HISTORICAL EXHIBIT 2.2

Mayan Head Variant Numerals

Perhaps the most exotic system of numerals devised by man are the head variant hieroglyphic numerals used by the ancient Maya. Each face depicted below signifies a numerical value from 0 to 19 and identifies a god within the Mayan Pantheon.

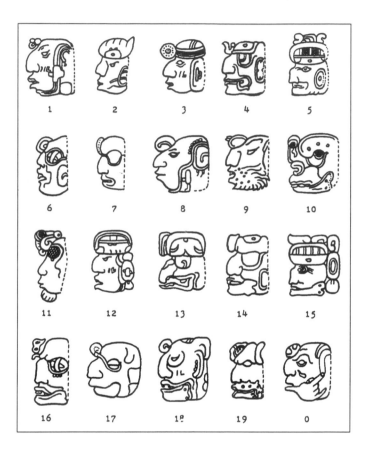

Identities of the Gods

1. Young earth goddess
2. God of sacrifice [?]
3. God of wind and rain
4. Sun god
5. Aged god of the underworld
6. Rain god [?]
7. Jaguar god of the underworld
8. Maize god
9. Serpent god [?]
10. Death god
11. Earth god
12. Unknown
13. Reptilian monster

Although apparently imaginatively conceived, these numerals were formed according to an established system of rules. For example, the hieroglyphic form for 16 is composed of the fleshless jaws of the character for 10 combined with the "hatchet" eye from the character for 6. The primal elements for the numerals for 0 to 19 are as follows:

Number	Numeral Characteristics
0	Clasped hands across lower part of face.
1	Forehead ornament composed of more than one part.
2	Undetermined.
3	Banded head dress.
4	Bulging eye with square iris, snag tooth, curling fangs from back of mouth.
5	"Tun" sign for head dress.
6	Hatchet eye.
7	Large scroll passing under eye and curling under forehead.
8	Forehead ornament composed of one part.
9	Dots on lower cheek or around mouth.
10	Fleshless lower and upper jaws.
11	Undetermined.
12	Undetermined—type of head known.
13 to 19	Head for 3, 4, 5, 6, 7, 8, 9 with fleshless lower jaw for 10.

Reprinted from "The Mathematical Notation of the Ancient Maya" in *Native American Mathematics* by Michael P. Closs© by the University of Texas Press. By permission of the publisher.

The Peruvian Quipu

LIND MAE DIANA

$\mathcal{M}$AN, having curiosity as one of his basic characteristics, has long desired to unlock the secrets of those who have gone before him in time. Thus, he has embarked on archeological expeditions, has studied history books, has consulted ancient sources of information, and has generally set about looking for the answers to his questions.

Mathematicians, possessing that same curiosity, have been no different from their fellow beings. They, too, have searched and traveled, and from their efforts have emerged hundreds of books written on the historical foundations of modern mathematics. Especially has the greatest effort been concentrated in the eastern hemisphere (i.e., Europe, Africa, and Asia). The logical question, then, concerns the apparent lack of books containing information as to the nature of the mathematics of the Americas—that is, of the original inhabitants of North and South America. The answer is twofold. First, there really was a lack of specialized mathematics in these geographical areas; and secondly, many of the secrets of ancient American mathematics have been, and are being, carefully guarded.

Such is apparently the situation when one inquires into the nature of the mathematical system of the Incas of Peru. Many allusions to the subject have been made by a great number of the Spanish chroniclers whose works are still preserved today. Yet we find that no distinct literary work has been completely dedicated to the subjects of ancient

Peruvian mathematics or the place of the Incan quipu (a mnemonic device used to keep record of computations) within that system. All that the inquirer finds are generalities, which do not pinpoint specific information.

However, using some of the hints given to us by the aforementioned chroniclers, research studies have been carried out by several men in an effort to discover what mathematical knowledge was known to the Incas and what importance the Peruvian quipu enjoyed in that mathematics. Two such scientists are Leland Locke and Erland Nordenskiöld.

The remainder of this discussion will be centered around the results of these studies. The reader, however, should note that many of the conclusions reached by these men (especially by Nordenskiöld) are nothing more than beliefs. There is no definite way to prove the truth of them, even though they are based on long and rigorous studies.

Before discussing the intricacies of the Incan mathematical system, it is necessary to note that, in comparison with other cultural achievements of the civilization, the field of mathematics was relatively barren. For example, the Incas lacked the complex computational system of the Mayans of Mexico; being somewhat like the ancient Babylonians, they were primarily interested in keeping records, rather than developing elaborate methods of computation.

Thus, we arrive at the subject of our discussion: the Peruvian quipu. The quipu was the device used by the Incas to record results of various kinds of mathematical problems. We have proof that quipus were used not only in Peru but also in

Reprinted from *Mathematics Teacher* 60 (Oct., 1967): 623–28; with permission of the National Council of Teachers of Mathematics.

other areas of South America.[1] Most of those that we now use for study were found in dry graves along the coastal areas of Peru. This is especially true of those used by Mr. Locke and Mr. Nordenskiöld.

The Incas were not especially adroit in advanced mathematical computation. They were much more interested in recording the results of the yearly census and in keeping an account of the number of sheep they had in a herd. Their mathematical achievements were not great, but their method of keeping these statistics, which they apparently valued so highly, was ingenious and still intrigues many twentieth-century scholars.

But exactly how did they compute the results they recorded on the quipus? An abacus was used, but not of the same type as came into such widespread use, during the time of the Roman Empire, in the Mediterranean region of the world. Instead of the wires and beads of the more conventional one, the Incan abacus consisted of a rectangular slab of stone into which were cut a number of rectangular and square compartments so that a free octagonal space was left in the middle, and two opposite corner rectangles were raised. Another two sections were mounted on the originally raised portion such that there were now three levels represented. An overhead view would thus be depicted by Figure 1[2] with the darkened areas denoting raised portions.

Pebbles were used to keep the accounts, and their positions within the various rectangles and squares that composed the instrument yielded the total. For example, a pebble in a small square was one unit; when it was put into a rectangular one, its value was doubled; and when it was put into the central octagonal area, its value was tripled. The value of the pebble would be multiplied by six if put on the first of the upper levels; if it were placed on the uppermost level, its value would be multiplied by twelve. The color of the pebbles used indicated the nature of the objects being counted. Since this method of computation was efficient in arriving at results but did not yield a permanent record, the Incas (who most likely borrowed this abacus from the Canaris of Ecuador) felt a need for another instrument, which could be used not as a calculating device but as a numerical record keeper.[3]

The quipu fulfilled this need. A first look at a quipu might yield the impression that it is nothing more than a knotted cord with no special significance, but this is not the case. It is true that a quipu is a collection of cords in which knots have been tied, but these are definitely and systematically arranged to have special significance to whoever embarks on the task of deciphering it.

The quipu consists of a main cord to which branches (other cords) are attached like a fringe.

FIGURE 2 *Quipu*

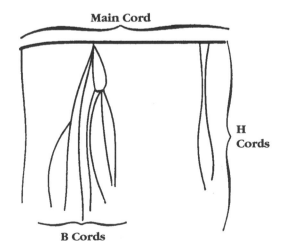

FIGURE I *Incan Abacus*

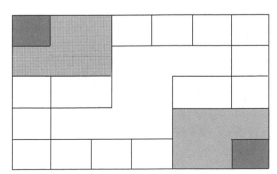

These branches have been named H cords by Locke. They are fastened to the main cord in groups, and many times these groups contain equal numbers of H cords. To the H cords are attached still other cords, these being called B cords. Any additional ones are labeled B cords of the second order (see Fig. 2).

The knots themselves are seldom present on the main cord. Instead, they are concentrated on the H and B cords. As to the nature of the knots, it has been noted by Mr. Locke that there are three main types of knots. These represent different values, depending on the particular knot used in a particular position on the cord in question. The Incas used a decimal arithmetical system much like ours of today. It is interesting to note that, like ours, their system was based on ten and was a place system. Thus, on the quipu each type of knot that was used had a specific decimal value. There were so-called overhand knots, simple single knots in the cord. These represented tens, hundreds, thousands, and ten thousands. The Incas rarely used numbers greater than ten thousand. Then there were "Flemish knots," S-shaped configurations used to denote "one." All other integers were depicted by "long knots" in which the specific value could be determined by counting the number of times the cord was wound before tying the knots. "Long knots" were also used at times to represent tens and hundreds. In such cases the position of the knots on the cord would denote the value. The units were placed nearest the bottom, the tens being placed immediately above them, then the hundreds, thousands, and ten thousands.

Let us illustrate exactly how the Incas used the quipu by the following example. A native desires to keep an account of the number of animals in his herds. He uses his abacus and arrives at the fact that he has 10 sheep, 30 llamas, 3 dogs, and 100 goats, giving him a total of 143 animals. To construct his quipu he would begin with a horizontal cord (main cord) from which he would suspend five H cords (see Fig. 3). On the first (left) cord he would place one overhand knot for the 10 sheep he

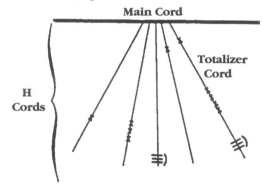

FIGURE 3 *Quipu*

has; on the second, three overhand knots would tally the number of llamas; the third would boast a long knot with three twists for the 3 dogs; and the fourth cord would have one overhand knot tied in it and placed in the appropriate spot to denote 100 goats. The fifth cord is the totalizer cord, and it would contain knots representing the total number of animals in the herd, 143.[4]

The colors of the cords are significant and indicate that the 10 registered on the quipu stands for sheep and not llamas. Thus, each thing to be counted was assigned a color to help avoid confusion. But what was done when the Inca ran out of colors? It is evident that colors had to be reused but with different significance. As a result, the deciphering of the quipu became an intricate art, and a specially trained individual was needed to do the job. He was the Quipucamayoc (keeper of the quipus), and his job was the construction, preservation, and decoding of the quipu.[5]

The purposes of the quipu were numerous, with authorities agreeing and disagreeing vehemently concerning specific uses.

Mr. Mason asserts that the Incas used the quipu to keep an account of their history and traditions. In short, he says, ". . . they never developed any system of writing—pictographic, ideographic, hieroglyphic, or alphabetic."[6] Their history was not written down; it was tied into the many knots that composed the quipu.

In accord with this view of the purpose of the quipu, we find an interesting reference in the

Comentarios Reales written by El Inca Garcilaso de la Vega. He suggested that the quipu was indeed used as a replacement for writing because it formed a basic part of the Incan postal system. He says,

> *Otros recaudos llevaban, no de palabra, sino por escrito, digámoslo así, aunque hemos dicho que no tuvieron letras, las cuales eran ñudos dados en diferentes hilos de diversos colores. . . . Los ñudos y los colores de los hilos significaban el número de gente, armas, o vestidos, o bastimiento, o cualquiera otra cosa que se hubiese de hacer, enviar o aprestar. A estos hilos añudados llamaban los indios quipu . . .* [7]

[Other messages were carried not by word but in writing. We say this although we have said that they (the Incas) had no writing. This writing was composed of knots of different colors on different cords. . . . The knots and the colors of the cords signified the number of men, arms, dresses, supplies, or whatever that they had to make, send, or prepare. These knotted cords were called quipu . . .]

Another interesting use of the quipu was proposed by Herrera. He says that after the conquest of the Incas by the Spaniards and the conversion to Catholicism, there still remained a communication problem between the two groups because of the difference in language. As a result, when an Inca wanted to confess to a Spanish priest, he did so by means of a quipu on which were written all of his infringements of church law.[8]

It has also been suggested that one of the purposes of the quipu was the registering of astronomical and magic numbers. Erland Nordenskiöld's extensive research has resulted in the promulgation of this theory. The quipus he used for his research were solely those that came from dry coastal graves; and all were in excellent condition, the majority being perfect specimens. It should be noted before discussing the issue any further that Nordenskiöld's opinions are not wholly accepted by modern authorities, but they do plant seeds of speculation in the reader's mind.

Nordenskiöld believed, even before consulting the quipus he used for his study, that they were not accounts of historical events or compilations of figures that corresponded to the yearly census. He based this presumption on the fact that primitive peoples usually believe that entombing anything containing information about the living is likened to entombing the living themselves. As a result, he conjectured that the numbers recorded on the quipu might be connected with magic of some sort. This conviction was further strengthened when he stumbled upon Bertonio's statement of the preoccupation of the Incas with magic and counting.

> *Adevinar tomando un puñado de algo y contando después los granos. Contar una almoçada de papas para saber si el año será buena. Es gran superstición.* [9]

[Foretell future events by taking a handful of something and counting the grains. Count a sack of potatoes to know if it will be a good year or not.]

Upon examination of a group of quipus, he did discover a predominance of the number 7, not as a number present on the cords themselves, but as a divisor of the totals registered on single cords and as the last digit of a number that was a divisor of the totals registered on neighboring cords. Since this number appeared so often in these divisors and since Bertonio stated that the Incas connected magic with counting, Nordenskiöld concluded that the number 7 must have been considered as possessing mystical value.[10]

The quipus studied also were shown to contain astronomical numbers. An example is the first quipu Nordenskiöld examined. It, like the others, showed a predominance of numbers that contained 7 or numbers whose divisors ended in 7. But it also contained a total equal to 1,172, which in turn equals 2×586. Nordenskiöld reasoned that these numbers represented days and that this total was to denote the period of two synodic revolutions of Venus, which is equal to 2×584 days. The four-day discrepancy was explained away as being added so that the difference between three of our solar years (3×365) and two synodical revolutions of Venus would be a magical number—in this case 77.

This particular example is one of the reasons why experts believe Nordenskiöld has looked for relationships and therefore forced a discovery of what he desired. His findings cannot be completely refuted, though, because other quipus he studied did show forth a prevalence of astronomical numbers.

The Spanish chroniclers themselves have stated the relationship between the quipus and the Incas' interest in heavenly bodies.

More exactly, Molina states:

No obstante que ussavan de una quenta muy subtil de unas ebras de lana de dos ñudos y puesta lana de colores en los ñudos los quales llaman quipos entendíanse y entiéndense tanto por esta quenta que dan razón de más de quinientos años de todas las cossas que en esta tierra en este tiempo an passado: tenían yndios yndustriados y maestros de los dichos quipos y quentas y estos yban de generación en generación mostrando lo pasado y empapándolo en la memoria a los que avían de entrar, que por maravilla se olvidaban cosa por pequeña que fuese tenían en estos quipos que cassi son a modo de pavilos con que las biejas recan en nuestra España salvo ser ramales tenian tanta quenta en los años, messes, y luna de tal suerte que no avia lunar, luna, año, ni mes aunque no con tanta pulcia como después que Ynga Yupanqui empeço a deñorear y conquistar esta tierra; porque hasta entonces los yngas no avian salido de los alrededores del Cuzco. Como por la relactión que Va. Sa. Rma. tiene, parece este ynga fue el primero que empeçó a poner quenta y razon en todas las cossas y el que quito cultos y discultos y cerimonias que en cada uno dellos hacen, porque no obstante que antes que reynasen sus antecesores, tenían meses y años por sus quipos, no se regían con tanto concierto como despues que este fue señor que se regían por los ynviernos y veranos. [11]

[Notwithstanding the fact that they used in a very subtle manner some strings of wool with two knots and that they put colored wool in the knots that they called quipus, having used them and using them now in acknowledging for more than 500 years all of the things that have happened in this land, some industrious and wise Indians had some quipus that they passed from generation to generation. These contained every little thing that had happened and were such good accounts of the years, months, and moons that there was not a single moon, year or month that was not included. But after Ynga Yupanqui began to conquer the land, he began to construct accounts of everything. And even though before he became ruler, they had recorded months and years on their quipus, now they included winters and summers too.]

Armed with the above information, it is no wonder that Nordenskiöld felt certain that some of the numbers registered on the quipus he studied must have been astronomical—especially when he did encounter them himself, too.

One quipu studied by Nordenskiöld was of an odd circular shape. It yielded numbers having to do with the types of months the Incas used. He notes that the sum of the numbers depicted on this specimen equals 3,481 or 118×29.5. Also, the total of the numbers on all the cords of one color results in a number divisible by 30. The total on the remaining colors is divisible by 29.5. Over and over again, examination of various other specimens yields the same results: years of 365 days divided into months of either 29.5 or 30 days each. [12]

Therefore, based on the previous discussion, Nordenskiöld concluded that the quipus used by the Incas and examined by him contained: (1) the number 7 combined in various forms and often represented, (2) numbers that expressed days, and (3) astronomical numbers.

In conclusion, one should fully be aware of the fact that the secrets of the Peruvian quipus have not yet been completely unlocked. This is clearly demonstrated when we note that Nordenskiöld's study occurred in 1925, and little progress has been made since then. In summation, we do know that the Incas counted in groups of ten, used a decimal placement system, and figured on the abacus. The results of these computations were then transferred to the quipu by the Quipucamayoc, who was a specialist in his field. The purposes of this mnemonic device were varied: to record historical events, to keep statistical records, to aid in the confessions of Incan converts, to record magic numbers, and perhaps even to record astronomical data.

Much additional research needs to be undertaken. The possibilities of further discovery are still present for those who are intrigued by the mysteries that are tied into the quipu.

NOTES

1. Louis Baudin, *A Socialist Empire: The Incas of Peru* (New York: D. Van Nostrand Co., 1961), p. 124.

2. *Ibid.*, p. 125.

3. *Ibid.*, pp. 124–25.

4. *Ibid.*, p. 126.

5. J. Alden Mason, *The Ancient Civilizations of Peru* (Baltimore: Penguin Books, 1957), p. 228.

6. *Ibid.*, p. 226.

7. *Comentarios Reales* (Buenos Aires: Espasa-Calpe S.A., 1961), pp. 65–69.

8. "Quipos," *Enciclopedia Universal Ilustrada Europeo Americana* (Madrid: Espasa-Calpe S.A., 1985), pp. 1417–18.

9. Nordenskiöld, *The Secret of the Peruvian Quipus* (Sweden: Comparative Ethnographical Studies, 1925), VI, 9.

10. *Ibid.*, pp. 5–37.

11. *Ibid.*, pp. 10–11.

12. Nordenskiöld, *Calculations with Years and Months in the Peruvian Quipus* (Sweden: Comparative Ethnographical Studies, 1925), II, Part 2, 5–34.

ADDITIONAL REFERENCES

Bushnell, G. H. S. *Peru.* New York: Frederick A. Praeger, 1960, pp. 125–28.

Cieza de Leon, Pedro De. *The Incas.* Norman, Okla.: University of Oklahoma Press, 1959, pp. 39–40, 105, 163, 172–75, 177, 187.

Prescott, William H. *History of the Conquest of Peru.* Philadelphia: J. B. Lippincott Co., 1847, pp. 122–26.

12

Mathematics
Used by American Indians
North of Mexico

F. L. WREN AND RUBY ROSSMANN

*T*HE MATHEMATICAL CONCEPTS of primitive races are of interest to both the mathematician and the historian; the mathematician accepts them as indicators of the natural order of development of his subject while the historian studies them for an insight into the cultural development of the race and of mankind. A study of the habits and customs of the American Indians north of Mexico, before the intervention of the white race, brings to light many interesting instances of their use of various arithmetical and geometrical concepts and principles. Although their bonds of superstition and lack of an adequate symbolism very greatly limited their mathematical progress we find that number played a very important role in their various religious beliefs and that they made rather elaborate use of many geometric figures in ornamentation and construction.

I. Arithmetical Concepts and Principles

Sacred Numbers. Most of the Indians of North America attributed mystical properties to certain numbers. Specific reference may be found to the use of three, four, five, seven, and thirteen in the performance of religious ceremonies. Among all the different numbers so employed four seemed to be the most popular, probably due to the recognition of the four cardinal points of the compass. Five was the mystical number of some of the Pacific Coast Indians; seven was reverenced by the Zuñi, Cherokee, Creeks, and most of the Plains tribes; and thirteen, which was quite prominent in Central America, was adopted by the Hopi, Pawnee, and Zuñi.[1]

Among the *four* cults ceremonial acts were repeated in sets of four and certain rites lasted four days and nights. In the Pueblo Snake Dance the Snake Men prepared eight days for the ceremony; the snakes used were of four kinds obtained from a four days' hunt in the four directions.[2] An Apache prayed to his gods at least once every four days and if it were expedient he might pray every day, four, or any multiple of four, times a day.[3] Medicine-men of the Apache tribe often used this sacred number in their remedies; in a prescription the roots of four varieties of herbs or four roots of one herb were used. When singing, four songs were sung or one song was sung four times or any multiple of four.[4] To a person destined to become a medicine-man in the Papago tribes there came four dreams on four successive nights.[5] On the four following nights an animal appeared before him, and at the fourth appearance led the spirit of the man through

Reprinted from *School Science and Mathematics* 33 (Apr., 1933): 363–72; with permission of the School Science and Mathematics Association.

four mountains, showing him the medicine to be found.

If a member of the Potawatomi tribe were accused of murder and the tribal chief thought he was not guilty, a pipe bearer with flint and steel attempted to light the chief's pipe; if he were successful in four strokes of the steel the man went free, but if he failed the accused was executed. An influential man might have escaped punishment for three murders, but if he killed four people there was nothing that could save him.[6]

Three and five were sacred numbers to the Iroquois.[7] They would take three puffs from a pipe when smoking according to ceremonial custom and only three trials were allowed in contests of skill or strength. It was necessary for five days or some multiple thereof to elapse between the announcement and the beginning of a celebration, and a grain of corn was sent as an invitation because corn sprouted five days after sowing. In the Ghost Dance of the Plains tribes the sacred number was seven and the dancers formed in groups of seven or fourteen.[8]

Many other examples of the use of mystical numbers are to be found in the legends and creation myths. These, in general, have reference to the number of repetitions of events or to intervals of a certain number of days.

Methods of Counting. The system of counting which was practically universal among the Indian tribes was that of counting on the fingers. The count began with the little finger of the left hand and proceeded to the thumb which was five. The method of numerating the second five varied, but usually the thumb of the right hand was six and the little finger ten. Often the fingers were bent inward as they were counted, but some of the western tribes began with a clenched fist and opened the fingers outward. Some tribes used the toes in counting the second ten, but others continued with the fingers, designating the completion of the first ten by a clap of the hands or a wave of the right hand. The Zuñi counted the second ten on the knuckles.[9]

The Dakotas made a vertical line for a count of one and represented one hundred by one hundred such marks.[10] The Creeks also counted each vertical mark as one unit but represented ten by the sign of the cross.[11] Another use of tally marks was found in the census roll of a Mille Lac band of the Chippewa; the thirty-four families were symbolized by pictographs under which was indicated the number of persons in each family.[12]

Evidences of subtraction were found in the formation of some number words; for example, in the Bellacoola language of British Columbia:

16 = one man less four,
18 = one man less two,
26 = one man and two hands less four,
36 = two men less four.[13]

In a letter of November 17, 1893, Dr. J. Owen Dorsey called to Conant's attention a gesture of throwing the hand to the left used by some of the prairie Indians to signify multiplication.[14] A count of five followed by a wave of the hand to the left meant fifty. Traces of multiplication were found also in the construction of number words, as is exemplified by the following numbers found in the Zuñi scale:

10 = all the fingers,
20 = two times all the fingers,
100 = the fingers all the fingers,
1000 = the fingers all the fingers times all the fingers.[15]

Number Systems. Although a number system with five as base might be considered the most natural development of the finger method of counting, it was quite universally the practice among these North American Indians to make use of a system of numbers whose unities grouped themselves into tens. In fact among some of the tribes, particularly those north of the Columbia River and along the Pacific Coast, the most generally employed scale was the vigesimal. Scales even more primitive than the quinary have been found and vestiges of them appear in more advanced systems. Traces of a binary scale have been discovered in the Siouan and Algonquian languages,[16] as well as evidences of a quaternary scale among the Indian dialects of British Columbia.[17]

Many of the regular systems extended to include one thousand. Larger numbers existed in the languages but it was quite often the case that they were not introduced until after contact with white people. Schoolcraft[18] illustrates the difference between a native combination and that made under outside influence by contrasting the Winnebago expression for one billion, *ho ke he hhuta hhu cheu a ho ke he ka ra pa ne za*, with that of the Choctaw *bil yan chuffa*. There were some tribes whose numerals were extremely limited, and quite frequently it was the fundamental formula of word combination that caused this limitation. The cumbersome nature of such a system can be seen in the combination *wick a chimen ne nompah sam pah nep e chu wink ah* which was used by the Sioux to express twenty-nine.[19] Bancroft says that the Lower Californians were unable to count more than five, though a few understood that two hands signified ten.[20] Numbers above this level were expressed by terms as "many" or "much."

An unusual number system was that of the Micmacs, in which the numerals were verbs and they were conjugated through all forms of mood, tense, person, and number.[21] In the Tshimshian language of British Columbia, there were seven sets of numerals, one set used for counting under each of the following distinct classifications of objects: (1) with reference to no definite object or merely abstract counting, (2) flat objects and animals, (3) round objects and divisions of time, (4) men, (5) long objects, (6) canoes, and (7) measures.[22] The Takulli dialect of the Athapascan language furnishes another interesting example of the modification of the number name according to the thing counted, for example,[23]

tha = three things,
thane = three persons,
that = three times,
thatser = in three places,
thauh = in three ways,
thailtoh = all three things.

Very few instances of the use of a system of ordinal numbers have been found. Boas[24] found that the Tlingit tribe had an ordinal scale from *first* through *eighth*, and that they also used the numeral adverbs;

tlehaden = once,
daqdahen = twice,
natsk dahen = three times,
dak'on dahen = four times.

He also states that the Haida used ordinal numbers from *first* through *fifth*, and that the Kutonaqa had a system of numeral adverbs somewhat similar in meaning to those of the Tlingit tribe. The only reference to the use of fractions were the Kutonaqa expressions for one-half and one-third.

Methods of Reckoning Time. The two most important bases for reckoning time were the changes of the moon and the seasons, and the succession of days and nights. The moons and seasons were named for some natural phenomenon which occurred at that time. The Indians of Virginia divided the year into five seasons: *The budding of spring; the earing of corn*, or *the roasting-ear time; summer*, or *the highest sun; corn-gathering*, or *the fall of the leaf*; and *winter*.[25] Four seasons were usually recognized, although some tribes distinguished five, some two and the Lower Californians six.[26] The Teton Sioux, in general, acknowledged only two seasons.[27] They counted twelve moons, of twenty-seven days each, in a year, and three days had to elapse after each moon before the count was resumed. In order to complete the solar year the Cree counted thirteen moons; the Haida inserted a moon between their two seasons; and the Creeks added a moon at the end of every second year, thus resulting in twelve and one-half moons in each year.[28]

Some tribes divided the day into four parts—sunrise, noon, sunset, and midnight; others, omitting midnight, had only three divisions. The Salish divided the day into nine parts according to the position of the sun.[29] the Indians of South Carolina determined the time of day by the number of handbreadths the sun was above the horizon.[30] The Choctaws expressed a period shorter than a day as the length of time it took the sun to travel the distance between two parallel lines drawn on the ground.[31]

Several calendars have been found, one of the first being *Lone Dog's Winter Count* which was a product of the Dakota Indians.[32] Each figure of this calendar, which was painted on a buffalo hide, represented an outstanding event in one year. The pictographs were arranged in an outward spiral starting from a central point, and covered a period of seventy-one years. The discovery of other Dakota calendars reveals that the year-characters were arranged in straight lines and serpentine curves as well as in spirals. They counted years by winters and a man's age was recorded as "so many snows old."

Among other tribes calendars were found to employ tally marks, notches, and circles along with the pictograph for recording and reckoning time.

The arithmetic of these Indian tribes seems to have been limited to additional simple subtraction, and in a few cases, simple multiplication with integers. Their bonds of superstition and failure to develop an adequate system of number words and symbols prevented any great amount of progress beyond the simple fundamentals.

II. Geometrical Concepts and Principles

Indian Houses. Geometric forms were used by the Indians in the construction of their houses which were principally rectangular or circular. No definite indication as to the method of constructing right angles and circles has been found. The drawing of a circle was probably accomplished by using two stakes and a cord or strip of rawhide as a pair of compasses. MacLean found what he believed to be an incomplete circle with points on the circumference marked by small heaps of earth,[33] thus indicating that in cases where the complete circle could not be drawn, points on the circumference were marked by special devices.

The typical dwelling of the Eastern Indians and the Plains tribes was the conical tent. The Timuqua lived in circular houses, but generally the Indian home of the southeastern section was rectangular with a curved roof.[34] A drawing by Le Moyne, 1563, portrays the Seminole Indians carrying their crops to a cylindrical storehouse which has a conical top.[35] Cyrus Thomas states that, according to Le Moyne's drawings, most of the Indian houses of Florida were circular while those farther north were oblong.[36]

The Menomini, Winnebago, Sauk, and Fox tribes occupied dome-shaped lodges in the winter and rectangular bark houses in the summer.[37] The Mandan huts were circular in form and were from forty to sixty feet in diameter; the roof was conical and the interior was divided into triangular compartments.[38] The Northern Californians built either conical or rectangular gabled houses, but their doors were always circular.[39] The Indians of Central and Southern California as well as those of the entire Southwest except for the Pueblos, built conical or dome-shaped huts. An outline of the ground plan of a Navajo hogan, given by Mindeleff, shows the circular form of the hut and within the sacred circular path which the Indians followed in their ceremonials.[40] The Pueblos lived in rectangular stone houses of the communal type. This style of architecture was preceded by the circular and it is thought by some that the rectangular form of construction used by the Pueblos was a result of crowding a large number of circular houses into a small space.[41] Another interesting rectangular variation was found among the Tlingit who built their houses upon a base which was in the shape either of a square or a parallelogram.[42]

The snow houses of some of the Eskimo bands were quite distinctive in their structure. The blocks of snow were arranged in spiral courses which produced a hemispherical effect. Boas gives a sketch of the ground plan of a snow house found on Davis Strait, which shows a small dome and a small elliptical vault as an entrance to the main section of the house.[43]

Mounds and Other Earthworks. Mounds have been discovered principally in the eastern section of the United States. Most of the burial mounds were conical with a circular or oval base. The typical pyramidal mound was a truncated quadrangular pyramid, but a few irregular pentagular ones have been found. The diameters of the bases of the conical mounds varied from six feet to three hundred feet, and the pyramidal mounds were usually larger. The largest mound of the entire section, located in Illinois, was one of the pyramidal type one hundred feet high with a base covering about seven hundred square feet.[44]

One group of earthworks in Ohio consisted of circles, squares, and octagons. The circles were quite unusual in their close resemblance to true circles and the octagon, while not a regular polygon, was a symmetric eight-sided figure with remarkable approximations to the properties of regularity. The sides of the square were 928, 926, 939, and 951 feet in length and the greatest variation of the angles from a true right angle was fifty-seven minutes. McGee and Thomas think that it is improbable that such accuracy could have been attained even by a well-trained eye and that the Indians must have had some method of drawing one line perpendicular to another.[45] Brownell states that two earthen enclosures were found at Circleville, Ohio, of which one was an exact circle and the other an accurate square whose corners corresponded to the cardinal points of the compass.[46] The town itself seems to have been built in the shape of a circle and it is from this fact that it derives its name.

Ornamentation. East of the coast mountains and south of Vancouver Island the majority of the ornamental designs were distinctly geometric in form. The technique of right angle and radiate weaving in basketry tended to produce various geometric patterns. Quite frequently the pottery designs were borrowed from the textile arts and basketry. A rather interesting principle found in weaving was that no line could intersect itself, which idea was carried over into pottery decoration and it is said that nowhere in native American art have loops in the curved forms or intersections in the angular forms been found.[47] In the development of designs rectangular figures are regarded as preceding the circular ones.[48] In the Navajo blanket designs all figures consisted of straight lines and angles, and Reagan lists the following patterns as those found on Navajo pottery: opposed sets of isosceles triangles, line bordering dots, hooked spirals, double spirals, vertical and horizontal lines, and stepped figures.[49]

The Sioux separated designs into their structural elements and named them accordingly. Some of these divisions as listed by Parker are:

(1) Rectangles in a stepped position,
(2) A cross with two bars,
(3) Two opposite rows of right-angled triangles,
(4) Parallel lines,
(5) Inverted equilateral triangles,
(6) Long triangles standing upon others of equal size,
(7) Lozenges,
(8) A stepped pyramid,
(9) Parallel lines with small squares between,
(10) A rectangle divided into nine rectangles of contrasting colors.[50]

The patterns used by the eastern Indians consisted chiefly of circles, opposed curves, parallel lines, dots and dashes. The Indians of the Mississippi Valley made little progress in the textile arts and their designs were, to a great extent, curvilinear. Their motifs included meanders, scrolls, circles, and various combinations of curved lines. The principle rectilinear figures were lozenges, zigzags, and checkers.[51] One of the designs on pottery from

Florida consisted of semicircles of six concentric circles separated at regular distances by five parallel lines.[52]

Other interesting applications of geometric figures in artistic design are: the use of the trapezoid by the Apache, the hexagon by the Mojave, and the combinations of lines, points, lozenges, and circles in the carvings of the Eskimo.

The Indian people as a race generally attached great importance to the practice of ornamenting the body. One such form of personal ornamentation was that of tattooing which was indulged in more extensively among the tribes of the west. Some designs were totemic while others were merely simple combinations of dots and lines. The California women tattooed their chins in three blue perpendicular lines drawn downward from the center and corners of the mouth; the Mojaves used similar vertical lines but drew them closer together.[53] Among certain groups of Eskimos the manner of tattooing was dictated by social rank; a plebeian woman had one vertical line in the center of her chin and one parallel to it on each side, while a woman of the nobility had two vertical lines from each corner of her mouth.[54] The Kiowa women frequently had small circles tattooed on their foreheads.

Paint was also used for personal ornamentation, especially during ceremonies. One of the favorite designs used in the dances of the Central Californians was made up of broad stripes painted up and down, across, or spirally around the body.[55] In the Sun Dance of the Teton Sioux the Chief Dancer was painted with a black semicircle from the forehead down each cheek, others at his shoulder joints, and complete circles about his elbows and wrists. In a ceremony of the Arikara medicine-men the leader painted a black circle around his face and others about his wrists and ankles.[56]

Many other instances of the use of geometric figures in artistic design and ornamentation may be found in the legends of Indian lore. In most cases the basic configuration was quite simple though the resultant design was rather intricate. The variations of form ranged from the simple combinations of point and line used in carving and tattooing to the spiral of the Eskimo hut and the truncated cone found among the hat designs of the Tlingit. The fact that the circle was the most popular of all geometric figures was probably due to its symmetry of form and facility of construction. It is probably the most natural of all geometric configurations.

NOTES

1. F. W. Hodge, "*Handbook of American Indians North of Mexico,*" Bulletin No. 30, Bureau of American Ethnology, Part I (Washington, D.C.: Government Printing Office, 1907), p. 354.

2. A. C. Parker, *The Indian How Book* (Garden City, N.Y.: Doubleday, Doran, and Co., 1927), p. 271.

3. E. S. Curtis, *The North American Indian*, Vol. I (Cambridge: University Press, 1907), p. 42.

4. *Ibid.*, pp. 36, 41.

5. *Ibid.*, Vol. II, (1908), p. 34.

6. W. D. Strong, The Indian Tribes of the Chicago Region, Anthropology Leaflet, 24, Field Museum of Natural History, Chicago, 1926, p. 22.

7. J. N. B. Hewitt, "Sacred Numbers Among the Iroquois," *The American Anthropologist*, (1889) II, p. 165–166.

8. A. C. Parker, *op. cit.*, p. 266.

9. F. W. Hodge, *op. cit.*, pp. 353–354.

10. H. R. Schoolcraft, Historical and Statistical Information Respecting the History, Condition and Prospects of the Indian Tribes of the United States, J. B. Lippincott, Philadelphia, Vol. II, (1852), p. 178.

11. *Ibid.*, Vol. I, (1851), p. 273.

12. *Ibid.*, Vol. II, p. 222.

13. L. L. Conant, *The Number Concept, Its Origin and Development* (New York: The Macmillan Co., 1896), p. 46.

14. *Ibid.*, p. 59.

15. *Ibid.*, p. 49.

16. F. W. Hodge, *op. cit.*, p. 353.

17. L. L. Conant, *op. cit.*, p. 113.

18. H. R. Schoolcraft, *op. cit.*, Vol. II, pp. 206, 216.

19. *Ibid.*, p. 207.

20. H. H. Bancroft, *The Native Races.* The Works of H. H. Bancroft, I (San Francisco: The History Co., 1886), p. 564.

21. H. R. Schoolcraft, *op. cit.*, Vol. V, p. 587.

22. Franz Boas, "First General Report on the Indians of British Columbia," Report of the British Association for the Advancement of Science, (1889), pp. 880–881.

23. F. W. Hodge, *op. cit.*, p. 354.

24. Franz Boas, *op. cit.*, pp. 857–858, 869, 890.

25. F. W. Hodge, *op. cit.*, p. 189.

26. H. H. Bancroft, *op,. cit.*, p. 564.

27. E. S. Curtis, *op. cit.*, Vol. III, p. 30.

28. F. W. Hodge, *op. cit.*, p. 189.

29. H. H. Bancroft, *op. cit.*, p. 275.

30. E. L. Green, *The Indians of South Carolina*, (Columbia: The University Press, 1904), p. 45.

31. H. B. Cushman, *History of the Choctaw, Chickasaw, and Natchez Indians* (Greenville, Texas: Headlight Printing House, 1899), p. 249.

32. Garrick Mallery, "Pictographs of the North American Indians," Fourth Report of the Bureau of Ethnology (Washington, D. C.: Government Printing Office, 1886), p. 92.

33. W. J. McGee and Cyrus Thomas, *Prehistoric North America, The History of North America*, Vol. XIX, (Philadelphia: George Barrie & Sons, 1905), p. 376.

34. Clark Wissler, *The American Indian* (New York: Douglass C. McMurtrie, 1917), pp. 222–223.

35. Minnie Moore-Willson, *The Seminoles of Florida*, Moffatt (New York: Yard and Co., 1911), p. 6.

36. Cyrus Thomas, *The Indians of North America in Historic Times, The History of North America*, Vol. II, (Philadelphia: George Barrie & Sons, 1903), p. 60.

37. Clark Wissler, *op. cit.*, p. 221.

38. George Catlin, *Letters and Notes on the Manners, Customs and Condition of the North American Indians*, Vol. I (New York: Wiley and Putnam, 1842), p. 81. Livingston Farrand, *Basis of American History, 1500-1900. The American Nation; A History*, (New York: Harper and Brothers, 1904), p. 136.

39. Clark Wissler, *op. cit.*, p. 213.

40. Cosmos Mindeleff, "Navajo Houses," Seventeenth Report of the Bureau of American Ethnology, Part II (Washington, D. C.: Government Printing Office, 1898), p. 517.

41. F. H. Cusling, "A Study of Pueblo Pottery as Illustrative of Zuñi Culture Growth," Fourth Report of the Bureau of Ethnology, (1882–83), (Washington, D. C.: Government Printing Office, 1886), p. 476.

42. H. H. Bancroft, *op. cit.*, p. 102.

43. Franz Boas, "The Central Eskimo," Sixth Report of the Bureau of Ethnology, (1884–85) (Washington, D. C.: Government Printing Office, 1888), pp. 541–42.

44. W. J. McGee and Cyrus Thomas, *op. cit.*, p. 307.

45. *Ibid.*, pp. 373–377.

46. C. D. Brownell, *The Indian Races of America*, (Boston: Dayton and Wentworth 1885), p. 41.

47. W. H. Holmes, "Pottery of the Ancient Pueblos," Fourth Report of the Bureau of Ethnology; (1882–83) (Washington, D. C.: Government Printing Office, 1886), p. 359.

48. J. W. Fewkes, "Archeological Expedition to Arizona in 1895," Seventeenth Report of the Bureau of American Ethnology, (1895–96), Part II (Washington, D. C.: Government Printing Office, 1898), p. 702.

49. A. B. Reagan, "Some Notes on the Archaeology of the Navajo Country," *El Palacio*, Vol. XXIV, (1928), pp. 339–342.

50. A. C. Parker, *op. cit.*, p. 95.

51. W. H. Holmes, "Ancient Pottery of the Mississippi Valley," Fourth Report of the Bureau of Ethnology, (1882–83) (Washington, D. C.: Government Printing Office, 1886), pp. 374–75.

52. H. R. Schoolcraft, *op. cit.*, Vol. III, p. 80.

53. H. H. Bancroft, *op. cit.*, Vol. I, pp. 332, 369, 480.

54. *Ibid.*, p. 48.

55. *Ibid.*, p. 393.

56. E. S. Curtis, *op. cit.*, Vol. III, p. 95, and Vol. V, p. 67.

13

Hawaiian Number Systems

BARNABAS HUGHES

ENGLISH-SPEAKING VISITORS to the Hawaiian Islands delight in interspersing Hawaiian words with their own language. The islands are perhaps the only place in the United States where the English-speaking *haole* (Caucasian) can mix a different language with one's own and be understood. The tourist easily learns many Hawaiian words and phrases from pamphlets available in stores and hotels. For instance, in Honolulu, directions are *mauka* ("toward the mountain"), *makai* ("toward the sea"), *ewa* ("toward the west") or *Diamond Head* ("toward the east") rather than compass oriented. Everywhere in the islands, *aloha* means "love," "greetings," "welcome," or "farewell," depending on circumstances of time, place, and persons.

The mathematics teacher from the mainland soon discovers the Hawaiian words for counting numbers (Table 1). The pronunciation of several words is very similar to the English pronunciation: *tausani* sounds like "thousand," *miliona* like "million." Moreover, *haneri* is phonetically close to "hundred." These Hawaiian words seem to echo American words; or, are they really American words made more liquid under languid, tropical skies?

The list also betrays an obvious decimal formation—is that the way it was? In fact, what is

known about Hawaiian number words? What was their origin? What influenced their development? Answers to these questions form an interesting chapter in the history of number systems.

The history of Hawaiian number systems follows closely the history of the people of the islands. The latter is customarily divided into two phases, one before and the other after the arrival of Captain James Cook in 1778. The period before witnessed the Polynesians sweeping across the South Pacific in huge sailing canoes from Asia by way of the Malay Peninsula and Java. Their easterly migration took them through Tahiti and the Marquesas until they reached Hawaii during the fifth century A.D. Bringing families and food, livestock and languages, the various Polynesians merged their cultures into what is now recognized as the Hawaiian culture.

A certain affinity among the number words of the inhabitants of Polynesia was found and reported by John Davis in the *Hawaiian Spectator*, January 1839. He listed the number words from one to ten as used by the inhabitants of Tahiti, the Marquesas, Rapa, Rarotongan, New Zealand, and the Easter and Hawaiian islands. There is no significant difference among these number words, regardless of their place of use, except for ten: the Hawaiians use *'umi*, the others use a variant of *angauru*. The latter word, however, is known to the Hawaiians in the form *anahulu*, which means a "decade" or "ten-day week."

The references report older counting words than those listed in tourist books. The evidence hinges on the word for *four* and multiples of it by

Reprinted from *Mathematics Teacher* 75 (Mar., 1982): 253–56; with permission of the National Council of Teachers of Mathematics.

Author's note: I want to thank Rubellite Kawena Johnson (University of Hawaii) for her critical assistance in completing this manuscript.

TABLE I

kahi	1	ono	6
lua	2	hiku	7
kolu	3	walu	8
hā	4	iwa	9
limā	5	ʻumi	10

ʻumi-kumana-kahi	11
ʻumi-kumana-lua	12
"	"
"	"
"	"
iwakalua	20
iwakalua-kumana-kahi	21
"	"
"	"
"	"

kanakolu	30
kanahā	40
kanalima	50
kanaono	60
kanahiku	70
kanawalu	80
kanaiwa	90
haneri	100
tausani	1,000
miiiona	1,000,000

ten. *Kauna* is the earlier term for four; *kaʻau* antedated *kanaha* for forty. In groups, therefore, the number words are as follows:

ahā kahi				
(four ones)	=	kauna	(4)	
ʻumi kauna				
(ten fours)	=	kaʻau	(40)	
ʻumi kaʻau				
(ten forties)	=	lau	(400)	
ʻumi lau	=	mano	(4,000)	
ʻumi mano	=	kini	(40,000)	
ʻumi kini	=	lehu	(400,000)	
ʻumi lehu	=	nalowale	(4,000,000)	

For instance, the following are found:

12	=	ekolu kauna (three fours)
20	=	elima kauna (five fours)
760	=	ekahi lau me iwa kaʻau (one four hundred and nine forties)

Numbers too large to count were often denoted by *kinikini* or *lehulehu*. In fact, the modern mathematical term "infinity" is represented by *nalowale*, which also means "out of sight."

Why the emphasis on "four"? The answer is difficult to come by. Alexander (1968 reprint) remarks that the significance of four arose from the custom of counting fish, coconuts, taro, and such by taking a couple in each hand or by tying them in bundles of four. The Hawaiian philologist Rubellite Kawena Johnson thinks that the importance of the number four may have come from basket weaving and astronomy. To begin a basket, two pieces of *peʻa* (i.e., "cross") were placed at right angles to one another; hence, there are four times two strands (or 8). This figure is found likewise in the constellation *Hānai-a-ka-mālama*, Southern Cross, called *Peka* (i.e., cross, "8") in Taumotuan.

Prior to the introduction of writing by the missionaries, computations were performed mentally or by counting on fingers. The number words simply indicated the results of computations. Hawaiians who were particularly skilled in computing were much in demand by the *aliʻi*, local chief. Their job was to keep an account of tapas, mats, fish, and other property that the chief would distribute to dependents. Thus, before the arrival of missionaries, there was little, if any, interest in an academic appreciation of arithmetic.

The first missionaries arrived from New England on 31 March 1820. Within two years they had learned the Hawaiian language, reduced it to literal form based on Latin pronunciation, and begun printing books. The native Hawaiians were eager to learn and flocked to the missionary schools. Education brings change. Among the first things that changed were the Hawaiian number words.

The Hawaiian alphabet has only twelve letters: seven consonants (*h, k, l, m, n, p, w*) and five vowels (*a, e, i, o, u*). The consonants are pronounced as in English, except for *w*. When *w* follows *e* or *i*, it is given a *v* sound. Otherwise *w* is generally pronounced as in English. The vowels are pronounced thus:

a . . . uh (in unaccented syllables)
      ah (in accented syllables)
e . . . eh
      ah (when marked ē or E)
i . . . ee
o . . . oh
u . . . oo.

If a bar is over a vowel, the vowel sound is held just a bit longer. If an inverted comma, as in *aliʻi*, precedes a vowel, the breath is stopped momentarily (glottal stop). Syllables consist of (1) a consonant and vowel—in that order or (2) a single vowel. There are no diphthongs. The accent ordinarily falls on the next to last syllable, unless the final vowel has a bar. In this case, it gets the accent. For instance, *kanahā* (40) is pronounced *káh-nuh-hÁH*, and *kanakolu* (30) is pronounced *káh-nuh-kÓH-loo*. Spoken properly, the Hawaiian language seems like gentle waves; the syllables rise and fall.

Using the vocabulary at hand, the missionaries reformed the number system from a mixed base-four-base-ten to a strict base-ten structure. The method was easy because Hawaiian terms were at hand. The multiple prefix *kana* meaning "tens of" was placed before *kolu* to form *kanakolu* (30), be-

fore *hā* to make *kanahā* (40) and so on up to *kanaiwa* (90). This was an easy change, for the Hawaiians were used to thinking in terms of "tens." Their year was based on ten-day periods, *kana ʻekā* was ten bunches of bananas, and *kana ko ʻoluna mai* was ten two-man canoes. Larger numbers, of course, were adopted from English: *haneri* for hundred, *tausani* for thousand, and *miliona* for million. Thus arose the counting number words found in tourist books today.

There is little, if any, evidence that words for fractions existed in any Polynesian language before the arrival of missionaries. They supplied the gap. In Tahiti the missionaries introduced the word *afa* (half) and *tuata* (quarter); in Hawaii they created *hapa* for half. In time this became a general word for *part*, which is conjoined with the counting numbers to form whichever fraction one desires. It must be noted that these words for fractions are conceptually different from *pukahi, paluu,* and so on. The latter signify a separating of a quantity into groups or a counting by ones, by twos, and so on.

Changes in number words were not restricted to counting. The missionaries also changed the names of the days of the week. Before the arrival of the missionaries, the calendar was a lunar calendar consisting of twelve or thirteen months of twenty-nine or thirty days, depending on the year. Each day had its own name taken from the number order of the night before—names that suggested deities. The missionaries adapted Latin terminology for the days of the week. where Monday is *feria prima* in Latin, Tuesday *feria secunda*, and so on; in Hawaiian, Monday became *Pō ʻakahi*, night the first; Tuesday was *Pō ʻalua*, night the second; and so on. Sunday received a special name, *Lāpule*, day of prayer. Nor did the ancient month-names escape the changes wrought by the missionaries. Indeed, the missionaries brought not only a new religion but also the trappings of a new culture.

There is no evidence that the change was immediate or universal. Chamisso (1837, pp. 55–57) noted that an early translation of the Bible into

Hawaiian employed the modified base-four system. For example, the passage from Exod. 7:7 was "elua kanaha makahiki o Mose a elua kanaha makahiki o Aarona a me kumamakolu" (Moses was twice forty years old and Aaron twice forty and three). Kanepuu (1867) wrote of conflict between older and younger Hawaiians. Some of the latter knew only the decimal system, whereas the older men and women who sold fish and produce in the markets reckoned in the old way with specific terminology. *Iako* (40) was used only for counting tapas and canoes; *ka'au* (40) was used with fish. But change did come.

Clark (1839) justified the change to the decimal system with the remark that the new method of computation is better adapted (!) to mathematical calculations. Could the reason for his judgment have been his own experience with problems employing numbers expressed only in base ten? Nonetheless, he recognized the natural ability of the Hawaiians for mathematics. For he wrote that they were fond of arithmetic, both mental and written, and were capable of making good progress not only in common arithmetic but in the higher branches of mathematics.

Opportunities to learn and use the "new math" of the early 1800s were at hand. Primary arithmetic books had been published for children as early as 1833 (*He helu kamalii*, Oahu, 2d ed.), and in the same year and place a text on mental arithmetic was published (*He helunaau*). A more challenging work was available at least by 1870 when C. J. Lyons translated and published James B. Thomson's *Higher Arithmetic (Ka Hainahelu Hou; Oia Hoi Ka Arimatika Kulanui)*. At the beginning of this last text, the translator placed mathematical terms in Hawaiian and English, in parallel columns. The Hawaiianization of certain English terms is obvious: *akioma* ("axiom"), *avakupoi* ("avoirdupois"), *bila* ("bill of goods"), *teroe weta* ("troyweight"). Mathematical terminology had arrived.

BIBLIOGRAPHY

Alexander, W. D. *A Short Synopsis of the Most Essential Points in Hawaiian Grammar*. Rutland, Vt.: Charles E. Tuttle, 1968, reprint of 1864 edition.

Chamisso, Adelbert. "Uber die hawaiische Sprache." *Vorgelegt der König*. Leipzig: Akademie der Wissenschaften zu Berlin am 12 Januar 1837.

Clark, E. W. "Hawaiian Method of Computation." *Hawaiian Spectator* 2 (January 1839): 91–94.

Dixon, R. B., and Alfred Kroeber. "Numeral Systems of the Languages of California." *American Anthropologist* 9 (1907), (new series) pp. 663–90.

Kanepuu, J. J. "Ka Helu Hawaii." *Ke Au Okoa*, 21 January 1867. Related to the subject of this essay and based on Kanepuu's letter is the article by Russell and Peg Apple, "Hawaiian Math," which appeared in the *Honolulu Star Bulletin* 11 April 1970.

14

The Lore of Number

JULIAN BARIT

*T*HE ORIGINS OF COUNTING AND NUMBERS are shadows of prehistory. Certainly they must have sprung from the needs of the earliest herdkeeping beings who, finding themselves with possessions, had to keep track of them. It is equally certain that number became far more than a tool of convenience and need—it penetrated inexorably into men's beliefs, mores, and very lives. Evidence of number's pervasive function may be found in every oral and written record left by the human race. Through the years a huge quantity of fact and fancy, science and symbolism, sense and superstition has grown up. It is the purpose of this paper to discuss and collate some of this lore of number.

It could not have been long after man began to find usefulness in numbers that he must have seen something of the *wonder* of numbers. His attempts to explain and rationalize some of the coincidences and oddities apparent to the mathematician have created a large body of both real science and pseudoscience. Many of his questions he solved with superstitious reasoning, some of which found its way into his religions and persists in his beliefs to this day. Other questions, eventually attacked by a more objective and scientific method made a base for present-day mathematics. Probably no other notion, save possibly that of a deity, has penetrated into so much of mankind's life as has that of number. References to it are to be

found in man's religion, his laws, his folk stories, his songs, and his nursery rhymes.

In the following, a brief listing of number lore is attempted. This is a mixture of superstition, symbolism, and the occult. It is to be noted that there is far more of the latter two categories than there is of the first one. The most bizarre and esoteric reasoning has been used by man in order to give numbers some of their attributes.

Superstition, Symbolism, the Occult

Zero symbolizes nonbeing, the latent and potential, death. Because of its circular shape, it is sometimes construed as eternity.

One symbolizes being, revelation, unity, divinity, and light. All peoples used the idea of oneness to set their chief deity apart from all else. The Hebrew word for one is *echod*, and was sometimes used to connote God, as in "*Adenoi Echod*," "The Lord is One." *One*, with the numeral "1" being a phallic symbol, stands also for male procreativity. It is the father of all odd numbers, because if one is added to any even number, an odd number is produced. Odd numbers have always been considered better than even ones; the Greeks assigned only odd numbers to their major gods, relegating even numbers to the minor deities. Even Shakespeare said, "There is divinity in odd numbers."[1]

Two was taken by the mystics to mean echo, reflection, and conflict. It was the number of contrast, an ominous number because of its ambiva-

Reprinted from *Mathematics Teacher* 61 (Dec., 1968): 779–83; with permission of the National Council of Teachers of Mathematics.

lence in posing evil against good, death against life. It was a female number, contrasted with the male "1." The malefic effect of *two* has resulted in the bad luck always attributed to it. All kings of England who were the second of the name met with disaster—William II, Henry II, Edward II, Richard II, and James II. The deuce in cards is the lowest and is a bad one to have "in the hole" in stud poker or blackjack. The bad luck of the two-dollar bill can be averted only if a small triangle is torn off a corner of the bill. The triangle, being three-sided, opposes the evil of the *two* with the power for good of the *three*, more about which in the next paragraph.

Three symbolically unites the *one* of the spiritual with the *two* of the mundane, so it is called the perfect number. Its triplicity denotes birth, existence, and then descent [1].[2] *Three* has been one of the most-used occult numbers in all civilizations; references to it may be found everywhere [6]. The Egyptians referred to the basic family unit of *three*—the father, the mother, and the child. The Druids used *three* as the number of "the unknown god." Among the Brahmins, there is a trinity of gods—Brahma, Vishnu, and Siva. The Romans considered the world as being under the rule of three gods—Jupiter (of Heaven), Neptune (of the Sea), and Pluto (of Hades). Jupiter holds triple-forked lightning, Neptune a trident, and Pluto has a three-headed dog. The Greeks were very partial to *three*. There were three Fates and three Graces; man has body, mind, spirit; the world has earth, sea, and air; nature is animal, vegetable, and mineral. References to *three* are endless, right down to the present day when the Christian church has made the Trinity its own.

Three is inextricably woven into folklore and custom. Axel Olrik has postulated the Law of Three (*das Geset der Dreizahl*) [8, pp. 139 ff.]. This stipulates that repetition in the European folk tales is almost always tied with the number *three*—to wit, three sons, three giants, three wishes, three tests.

Present-day superstition and custom also pay attention to *three*. Why do we eat three meals a day? The doctor tells us to take his medicine three times a day; is there a magic in this? Troubles come in threes, as do events, and we mustn't light three cigarettes on one match for fear of bad luck.

Four pictures the arms of the cross; therefore it becomes a symbol of Earth because of the cross formed by the north-south line and the east-west line. As man himself, with arms outstretched, can be likened to a cross, *four* also represents the human condition. Rational organization and tangible achievement are likewise found here [11]. The ancient Hebrews used the four-letter configuration IHVH to stand for the awesome, unmentionable name of God. The tetragrammaton became vowelized to the modern word Jehovah. *Four* was much used by the Hebrews—four winds, four rivers, four quarters of heaven. Forty days and forty nights of rain produced Noah's flood; Jesus, Moses, and Elijah fasted for forty days. The Israelites went from Egypt and spent forty years in the wilderness; David and Solomon both reigned for forty years.

Followers of Pythagoras of Greece almost made a cult of *four*. In the Christian era of the Middle Ages, scholars were equally partial to *four*. There were four elements, four ages, four seasons, four humors, four temperaments, four cardinal points, four winds, four senses, and four Gospels. And strangely enough, the Indic folk tales show the same preoccupation with the repetition of things *four* times as the Europeans do with *three*, as shown in an earlier paragraph.

Five, being likened to man's four limbs topped by his thinking mind, then symbolizes completeness. *Five* is also coupled with man's five-fingered hand, or with man's five senses—sight, touch, taste, hearing, and smell. This idea then expanded to imply health and well-being. With the Greeks the five-pointed star, or pentagram, was the symbol of Hygeia, or health. It was the "Seal of Solomon" among the Hebrews, and in the Middle Ages it became the "Wizards' Star." Because *five* is the union of the first odd and even numbers after the unit, it had a great appeal as a charm or amulet; a numeral "five" painted on the door in ancient Egypt or Greece sufficed to ward off evil spirits. *Five* was considered circular because whenever it is multiplied by an odd number it restores itself by appearing in the unit's place of the product.

Six, to the followers of the occult, was also a number of perfection and completeness. It was pictured as a double triangle, bases together; the triangle with the apex upward was the male symbol, that with apex down, the female. Together, then, they were union or marriage [11]. This idea of union follows also from the joining together of two and three to produce *six* by multiplication, two and three being the first numbers after unity. *Six* also perpetuates itself, because when multiplied to itself any number of times, the number *six* is always found in the unit's place of the answer, as in 6, 36, 216, 1,296, 7,776, ... The Druids favored *six* over all other numbers—their principal ceremonies took place on the sixth day of the moon, and they went in groups of six people to gather the sacred mistletoe. *Six* has its malefic qualities, too, for it is the number of sin. On the sixth hour of the sixth day of creation, sin came into the world. And 666 is the emblem of the Evil One, the "beast."

Seven is the combination of three, the triangle representing man's mind, and four, the square representing man's earthly house. *Seven,* therefore, is perfection and virtue. It corresponds to the seven basic musical notes, the seven basic colors, the seven planets, the seven gods. Power has been ascribed to the number *seven* from time immemorial; probably no other number has had so many things attributed to it [11]. Some 2,500 years before Christ, the Babylonian story of creation mentions the seven winds, the seven spirits of storm, the seven evil diseases, and the seven divisions of the underworld closed by seven doors. They said "Remember the seventh day to keep it holy." The Hebrews, too, thought much of *seven*—the six days of creation with the seventh day for the Lord, the seven lean years in Egypt following seven plentiful ones. The fall of Jericho was full of sevens—on the seventh day the city was circled seven times by seven priests bearing seven trumpets. Jesus cast out seven demons from Mary, spoke seven words from the cross, and commanded his followers to forgive their enemies seventy times seven times [11]. The great seven-branched candelabrum, the menorah, was and is proof of the meaning of *seven* to the Hebrews. The Greeks noted that there were seven planets (wanderers) in the sky and that there are seven stars in some of the most important constellations, like the Pleiades and the Great Bear. The concern shown for *seven* has come down through the ages to the present day, when people exhort the dice, "Come on, *seven!*"

Eight is symbolically related to regeneration and baptism, to a balancing out of opposing forces. Eight is the first cube after one, hence as a cube it represents the planet Earth. Among the Greeks, *eight* was sacred to Dionysoe, who was born an eight-month baby. The Pythagoreans, however, suspected there was something wrong with *eight*; they made it the symbol of death [3, p. 140]. In numerology 888 is the special number of Jesus Christ, the opponent of 666, "the Beast."

Nine is the triplication of the triple in symbolic language; therefore it is a complete image of the three worlds: corporal, intellectual, and spiritual. With the Hebrews, *nine* was the symbol of truth because when used as a multiplier it is never destroyed. They used this kind of illustration:

$$37,643 \times 9 = 338,787;$$
$$3 + 3 + 8 + 7 + 8 + 7 = 36,$$
$$3 + 6 = 9.$$

Children of today, learning from their teacher how to check multiplication by "casting out 9's," recognize the same phenomenon. More sophisticated mathematicians recognize modulo-nine arithmetic in this. In John Heydon's *Holy Guide,* 1662, we find that he asserts the number nine to have curious properties: "If writ or engraved on silver, or Sardis, and carried with one, the wearer becomes invisible. . . . It prevails against plagues and fevers, it causes long life and health, and by it Plato so ordered events that he died at the age of nine times nine." [11, p. 89.]

Ten, according to the Cabalists, is composed of the first four extra special numbers:

$$1 + 2 + 3 + 4 = 10.$$

Being the recipient of all those numbers, ten was called the "decad," from the Greek *dechomai,* to receive. This idea of "to receive," therefore, extends to mean heaven, ordained to receive all men. Ten was important to the Hebrews as in the Ten Commandments, the ten plagues which afflicted the Egyptians before the exodus, the ten men needed for a minyan, the legally convened body of worshippers.

Eleven seems to have an evil reputation among most peoples. It is more than ten, the number of perfection, so it stands for excess, peril, and conflict. It is called the number of "sins" because it exceeds the number of the Commandments and is less than twelve, which is the number of grace and perfection [11, p. 100].

Twelve undoubtedly owes its popularity over the years to its relationship to the zodiacal configurations which have been noted by all ages and all civilizations. There were twelve Tribes of Israel; the Greek and Roman mythology had twelve major gods or goddesses; practically every people had some kind of twelve-part theology, all due to the influence of the zodiac.

For sheer persistence, no number can compare with *thirteen.* The fear of *thirteen* is at least two thousand years old. The common theory explaining the origin of the superstition is that there were thirteen present at the Last Supper, Judas being the thirteenth. Much evidence exists, however, which suggests that the fear and dislike of *thirteen* go back much further than the Christian era. Here is a disturbing thought for the superstitious: on the Great Seal of the United States one will find thirteen stars and thirteen bars, an eagle with thirteen feathers in its tail holding in its left claw thirteen darts and in its right claw an olive branch bearing thirteen leaves and thirteen olives, and the motto *E Pluribus Unum* containing thirteen letters [3, p. 144].

This compendium could go on and on; there seem to be very few numbers that have not had attention paid them by mankind. Sometimes the reasoning seems forced and weak to the sophisticated modern, but certainly faith in the occult has prevailed for centuries. The subject has been only lightly touched upon here; if the reader is interested in knowing more of the esoteric and mystic functions of number, he is urged to read more by Westcott, cited several times in this paper.

NOTES

1. *The Merry Wives of Windsor,* Act V, Scene 1, line 2.

2. Numerals in brackets refer to the Bibliography appearing at the end of the article.

BIBLIOGRAPHY

1. Bailey, Henry Turner, and Pool, Ethel. *Symbolism for Artists.* Worcester, Mass.: The Davis Press, 1925.

2. Baring-Gould, William S., and Baring-Gould, Ceil. *The Annotated Mother Goose.* New York: Clarkson N. Potter, 1962.

3. Berry, Brewton. *You and Your Superstitions.* Columbia, Mo.: Lucas Brothers, 1940.

4. Boni, Margaret Bradford. *Fireside Book of Folk Songs.* New York: Simon & Schuster, 1947.

5. *The Union Hagaddah.*

6. Cirlot, J. E. *A Dictionary of Symbols.* New York: Philosophical Library, 1962.

7. DeLys, Claudia. *A Treasury of Superstitions.* New York: Philosophical Library, 1957.

8. Olrik, Axel, "Epic Laws of Folk Narrative," in *The Study of Folklore,* ed. Alan Dundos. Englewood Cliffs, N. J.: Prentice-Hall, 1965.

9. Smith, David Eugene, *History of Mathematics,* Vol. II. Boston: Ginn & Co., 1925.

10. Thompson, C. J. S. *The Handbook of Destiny.* London: Rider & Co., 1932.

11. Westcott, W. Wynn. *Numbers, Their Occult Powers and Mystic Virtues.* London: Theosophical Publishing Society, 1911.

SUGGESTED READINGS FOR PART II

Ascher, Marcia and Ascher, Robert. *Code of the Quipu: A Study in Media, Mathematics and Culture.* Ann Arbor: University of Michigan Press, 1980.

Closs, Michael P., ed. *Native American Mathematics.* Austin, TX: University of Texas Press, 1986.

Dantzig, Tobias. *Number: The Language of Science.* New York: Macmillan, 1939.

Flegg, Graham. *Numbers, their History and Meaning.* New York: Schocken Books, 1983.

Ifrah, Georges. *From One to Zero.* New York: Viking Penguin Inc., 1985.

Menninger, Karl. *Number Words and Number Symbols: A Cultural History of Numbers.* Cambridge, MA: MIT Press, 1969.

Seidenberg, A. "The Ritual Origin of Counting." *Archive for History of Exact Sciences* 2 (1962): 1–40.

———. "The Origins of Mathematics." *Archive for History of Exact Sciences* 18 (1978): 301–42.

Zaslavsky, Claudia. *Africa Counts: Number Patterns in African Culture.* Boston: Prindle, Weber & Schmidt, 1973.

PART III

HUMAN IMPACT AND THE SOCIETAL STRUCTURING OF MATHEMATICS

*A*BOUT TEN THOUSAND YEARS ago, a new plant hybrid appeared in the Middle East. It was a cross between two grasses that grew abundantly on the plains of that region. Endowed with a generous supply of large seeds, the new plant soon became an attractive food source for animals and humans. This new food was wheat. Wheat could be cultivated, processed, and stored for future consumption. It served as a nutritionally satisfying food source and when made into a gruel and allowed to ferment, it produced a stimulating and intoxicating beverage. Attracted to the properties of this new food, humans turned from survival strategies based on hunting-gathering to those centered around agriculture. The resulting agricultural revolution altered the status of individuals and communities. Whereas hunting-gathering was a collective societal endeavor, farming became an individual-family responsibility. People still banded together into communities for security; however, interrelations between members of the community changed radically. The cultivation of wheat is a sedentary activity. Between preparing the soil for planting and the harvesting and storing of wheat a considerable period of time is consumed. Young shoots of grain must be protected and fields tended. Farmers had to live near their fields and develop and maintain implements for farming, harvesting, and storing grain. Thus simple technologies appeared and labor and craft specialization developed. Hunter-gatherers, due to their migratory nature, had small families and accumulated few possessions. Their possessions were usually personally crafted and maintained—each hunter-gatherer was self-sufficient. However, settled farmers needed more children to work the land and accumulated varied possessions, some of which were beyond their personal capacity to produce. They had to rely on the manufacturing skills of others. Patterns of trade and systems of values and exchange developed. For example, one flint sickle blade might be worth three small baskets of grain. To accommodate these new social interactions, mathematical knowledge and techniques had to be broadened.

During this period of time, the emerging environmental demands of agriculture limited human settlement to certain attractive geographical regions, namely, fertile river valleys. River flood plains allow for a yearly revitalization of soil. At such locations, water is available for crop irrigation and the river itself provides both a supplementary food source and an avenue of transportation. Anthropologists have termed such societies, "hydraulic societies." Due to their dependence on water resources, hydraulic societies usually engaged in extensive engineering projects such as the construction of dikes, reservoirs, and irrigation channels. Control of water became a central societal concern. In order to facilitate the realization of hydraulic projects, a strong, central authority such as a chief or king had to be designated. This leader would then develop a supporting bureaucracy to directly administer the projects. These bureaucracies were frequently composed of priests and scribes who developed and amassed the scientific and mathematical knowledge necessary for the undertaking of the tasks at hand. The mathematical knowledge possessed by hydraulic societies had distinctive characteristics. It was empirical in nature and the computational techniques and problems were focused on specific situations, usually related to hydraulics. In a sense, there was no *theory* of mathematics, merely a collection of prescribed computational schemes for designated situations. In such settings, mathematics did not evolve into an abstract science.

Early hydraulic societies existed worldwide, notably: in the mid-East in the Tigris-Euphrates flood plain; along the banks of the lower Nile River; on the Indian subcontinent along the Indus and Ganges rivers; and in China along the Huang Ho and Yang-tse rivers.

In ancient times, the region bounding the confluence of the Euphrates and Tigris rivers, in what is present-day Iraq, was known by many names including Mesopotamia, Assyria, and Babylonia. This region has been identified as the "cradle of Western civilization" for it was here that the first evidence of wheat cultivation and corresponding human settlements was found. In fact, the early Babylonians denigrated their neighbors "who did not live in houses and plant wheat" as barbarians. Extant systems of irrigation channels, dikes, and earthworks testify to the fact that the Babylonian people belonged to a hydraulic society. But their society prospered and advanced from purely agricultural pursuits to ones based on trade and commerce. Eventually, they evolved a monied economy and such social institutions as taxes. For many years, historians believed that the principal source of Western mathematics originated in the Hellenistic world of the Greek peoples but twentieth-century research focusing on the deciphering of cuneiform tablets reveals that the Babylonians had achieved a high level of mathematical sophistication centuries before the Greek work existed. Today, we know that the early Greeks and Egyptians borrowed heavily from Babylonian sources. The scope of Babylonian mathematics is surveyed in the readings in Part Three.

Herodotus, the Greek historian, described Egypt as the "gift of the Nile." This description holds true today as almost all life and activity in Egypt centers around the Nile River. Ancient Egypt was a hydraulic society whose mathematical needs were carried out by a priest bureaucracy. Whereas numerous cuneiform clay tablets supply evidence about Babylonia's mathematical achievements, few such sources exist to testify for Egyptian accomplishments. Egyptians wrote on scrolls made of papyrus, a paper-like substance that survived poorly over the centuries. Our knowledge of Egyptian mathematics has been gleaned from about half a dozen papyrus sources, scrolls, and scroll fragments. Principal among these sources is the Rhind Papyrus, a mathematical manual compiled in about 1650 B.C. An examination of this source and its contents can provide a comprehensive picture of Egyptian life and the uses of mathematics in that ancient society.

The rise of the Greek empire along the shores of the eastern Mediterranean marks a transitional point in the development of mathematical reasoning. The Greeks used a deductive approach to mathematics. They organized mathematics into a systemized abstract science where facts could be related through chains of reasoning. Apparently, they were the first people to formally ask the "why?" of a mathematical happening. The principles of mathematical proof that are commonly used originated with the Greeks. Why was it that among the Greeks a new mathematical world outlook was formed? Firstly, it would be noted that ancient Greece was not a hydraulic society and was not subject to many of the intellectual restrictions of a hydraulic way of life. For example, the Greeks were not subservient to a strong central authority and were not bound to prohibiting religious constraints. A spirit of individualism was evident in the Greek character. Further, the harsh realities of an agricultural survival based on water control was absent from Greek life. Trade became the backbone of the Greek empire and the resulting accumulation of wealth allowed for leisure time. A contributing factor to the appearance of mathematical introspection in Greece was the simple fact that the Greeks had unencumbered time to think.

A more recent theory of why Western science and philosophy began with the Greek people is that they were among the first societies to develop and use a phonetic alphabet. The use of an alphabet requires abstraction—a symbol is associated with a sound, a sound with a word, a word with a meaning. Written words convey meaning and imagery upon which concepts are formed. Such a process encourages deductive thought which can readily be transferred to mathematical and scientific problems.

There exists an extensive literature on the mathematical accomplishments of ancient Greece; therefore, the following considerations of Greek mathematical accomplishment are rather brief. This selection of articles is intended to convey the spirit and power of the Greek approach to mathematical reasoning, an approach that has endured through the present day.

Prime Numbers

JAMES RITTER

$\mathcal{M}$ATHEMATICS and writing have a close, symbiotic relationship. Indeed, recent archaeological discoveries have shown that it was the need to measure, divide, and distribute the material wealth of societies that gave birth to the first writing systems.

For a society to develop a mathematics that goes beyond simple counting, a material support of some kind is essential. Without writing, the limitations of human memory are such that only a certain degree of numerical sophistication can be achieved.

By allowing us to follow the development of two writing systems, one in southern Mesopotamia towards the middle of the fourth millennium B.C., and the other in the area around Susa in Iran, slightly later, archaeological discoveries in the last few decades have shown that the inverse is equally true. For a society to develop writing, material needs, and, in particular, the need for record-keeping, are central.

In these societies the material support was clay, which is virtually indestructible, and the first documents are accounts. Cuneiform (wedge-shaped) Mesopotamian writing in particular was to know great success over the succeeding 3,000 years. Used to write not only the original Sumerian and Akkadian, but also Hittite, Elamite, Hurrian, and many other languages of the ancient Near East, it died out only at the beginning of our era.

Meanwhile, an independent civilization developed rapidly in Egypt towards the end of the fourth millennium. Here the situation concerning writing is less clear. First of all, the material support, for other than monumental inscriptions, was principally papyrus, the reed-like plant growing along the Nile and in the Delta, and, to a lesser extent, other perishable materials. Egypt has thus yielded fewer documents than Mesopotamia by a factor of thousands.

Number Systems

The third millennium marks, for both the Mesopotamian and the Egyptian civilizations, the gradual emergence of an abstract concept of number. Originally each number is attached to a given system of units. The "four" of "four sheep" and "four measures of grain" are not, for example, written with the same symbol.

Equally, the different systems of units are not connected among themselves. Measures of area, for instance, have no simple relationship to measures of length since the connection between the two—that area could be calculated from the product of a length and a width—was not yet operational.

But the very practice of writing things down, yielding permanent records of measures, opens up the possibility of observing regularities and patterns. This possibility was taken advantage of in these two societies over a period of about 1,000 years so that by the end of the third millennium Egyptian and Sumerian scribes had learned how to calculate areas and volumes from lengths, how to divide rations among workers, how to calculate the

Edited from *UNESCO Courier* (Nov., 1989): 12–17.

time necessary for a given job of work from volumes, numbers of men, and work rates. Next our evidence shows how a new level of abstraction was reached in which the concept of number became more and more detached from its metrological context.

By the beginning of the second millennium both civilizations had succeeded in developing equally abstract systems of numeration, though they had chosen different paths to represent their numbers. The Egyptians, like the vast majority of modern societies, had a written number system based on ten; that is, one counts up to nine of each unit before moving up to the next higher unit—after nine "ones" comes a "ten", after nine "tens" comes a "hundred", etc. But, unlike modern systems, their writing of numbers is "additive", that is there are separate signs for units, tens, and hundreds, and these are repeated as necessary.

The Mesopotamians used a base of sixty for their mathematical calculations, and they developed the first known system of position. The signs for numbers repeat after fifty-nine, the actual value being determined by the position of the digit in the number as a whole.

A Scribe's Training

Learning how to manipulate such number systems requires a specialized training and the creation of schools can be traced back to the same time as the invention of writing. Furthermore we know that the learning of arithmetic began early in the schoolchild's career, along with reading and writing, and that mathematics, then as now, was considered one of the "hardest" subjects.

Around 2000 B.C., Shulgi, one of the kings of the Ur III empire in Mesopotamia, was the subject of a literary hymn which became a model text used as a school exercise during the first half of the second millennium. In this document he boasts of his academic achievements and proudly states: "I am perfectly able to subtract and add, [clever in] counting and accounting."

Over a thousand years later the Assyrian king Assurbanipal was to repeat much the same thing in one of his hymns: "I can find the difficult reciprocals and products which are not in the tables."

What would be the history of a young scribe destined to become a "mathematician" in ancient Egypt or Mesopotamia? He (and it is almost certain that it would be a boy; girls, while not forbidden a scribal formation, are almost entirely absent from the records) would of course first go to school. The sons of the rich and powerful rubbed shoulders with less fortunate young people, who found in education one of the rare means of social ascension.

What would the scribe learn during his schooling, which lasted at least ten years? We possess, from both civilizations, examples of school exercises including mathematical texts, and can catch a glimpse of student life through the so-called "scribal disputes". An example from Mesopotamia has one scribe boasting to another of his achievements:

"I want to write tablets:
tables [of measures] from 1 *gur* of barley to 600 *gur;*
tables [of weights] from 1 shekel to 20 minas of silver,
with the contracts of marriage that may be brought me,
the commercial contracts . . .
the sale of houses, fields, slaves,
pledges in silver, contracts for the rental of fields,
contracts for raising palm trees . . .
even the tablets of adoption; I know how to write all that."

In this typical text from Egypt one scribe taunts another as follows:

"You come here and fill me up with your office. I will show up your boastful behaviour when a mission is given to you. I will show up your arrogance when you say: 'I am the scribe, commander of the work gang.'. . .

"A ramp, 730 cubits [long] and 55 cubits wide, must be built, with 120 compartments filled with reeds and beams; with a height of 60 cubits at its peak, 30 cubits in the middle; with a slope of 15 cubits; with a base of 5 cubits. The quantity of bricks is demanded of the troop commander.

"The scribes are all assembled but no one knows how to do it. They put their faith in you and say: 'You are a clever scribe, my friend! Decide quickly for us, for your name is famous.' . . . Let it not be said: 'There is something he does not know.' Give us its quantity in bricks. Behold, its measurements are before you; each of its compartments is 30 cubits [long] and 7 cubits [wide.]"

But these are literary and not real mathematical texts. In fact, we possess some mathematical school texts from each civilization, almost all of them from two distinct periods: the first half of the second millennium and the period of Greek and Roman domination at the end of the first millennium. They are of two kinds, table texts and problem texts.

A typical example of a table text is a square root table from early second millennium Babylonia. The systematic and organized writing of a table shows the high degree of abstraction already attained. What does the scribe do if he needs a square root? If it is not on the table, it can easily be calculated from the surrounding values that do appear. The procedure was standard in almost all cultures including Western ones until a quite recent date, and this is how the Babylonians and the Egyptians used their tables of multiplication, square roots, and sums of fractions. This is also the way in which they used the problem texts. A typical example, from an Egyptian papyrus of the middle of the second millennium, begins with a statement of the problem to be resolved:

> "A pyramid. Its side is 140 [cubits] and its slope is 5 palms 1 finger [per cubit]. What is its height?"

Here the data are presented in the form of concrete numbers rather than abstract variables, followed by a step-by-step solution with the answer at the end. Each step makes use of either the result of a preceding step or of one of the data given at the beginning of the problem.

No argument is given to justify the procedure nor is any explanation offered for its form. But even with the numerical values included, the nature of this form is quite clear. Thus the student would be able to solve any other problem of the same type. Moreover these problems are often grouped in such a way that the techniques learned can be immediately applied in other cases. The above problem, for example, follows a problem concerned with the determination of the slope of a pyramid given its length and height, and is followed in its turn by a problem involving the calculation of the slope of a cone.

However, not all mathematical problems were as practically oriented as this one. The main purpose of mathematical school exercises was to give the student scribe practice in the mathematical techniques used in solving problems. Technical drill, not direct application, was the main point. For this reason many of the apparently "practical" problems in these texts are far removed from real life: a Babylonian tablet proposes a problem in which a broken measuring rod is used to measure a field; an Egyptian document asks a scribe to calculate the original size of a herd of cattle based on the number of cows used to pay a cowherd's taxes…

The pedagogical purpose of all this is clear. Moreover, the structure of the problem and table texts permits an alternative approach to abstraction and generalization in mathematics. Rather than take the path of increasing symbolization marked by a hierarchy of "levels of generality", the Egyptian and Babylonian approach is to create a network of typical examples in which a new problem can be related—by a form of interpolation—to those already known.

Exactly the same approach can be seen at work in other areas of ancient thought such as medicine, divination, and astrology. All these subjects were treated by the Egyptians and Mesopotamians as constituting a special domain of "rational practice."

No Word for 'Mathematician'

Ample documentary sources—accounting texts, lists of professions, references in literary and historical texts, even painting and sculpture—enable us to follow the professional life of the aspiring scribe once his education is complete. We will look in vain, though, for any mention of a 'mathematician' in the modern sense of a person working within a recognized community of researchers on the properties of number and geometric figure. No word for 'mathematician' exists in any of the ancient languages of Egypt and Mesopotamia.

There were two paths that a young scribe could follow. A few would become teachers of mathematics themselves, perhaps exploring further problems that could be presented to a new generation of schoolboys. In the course of time this did lead to a deepening and broadening of the mathematical techniques available to the two societies.

The graduate could also become an accountant—a calculator of work, rations, land, and grain. Scribes are omnipresent and their industrious labourings are carefully depicted on Egyptian wall frescos and Assyrian palace reliefs, always hard at work on the business of their master, who might be either the state or a private landowner. They seem to have held privileged though subservient positions. Like their schoolteaching colleagues, they were not the movers of ancient society but they served those who were—and well enough to be immortalized on their patrons' walls as the visible symbols of that concentration of power and wealth they laboured so hard to calculate.

110

16

Further Evidences of Primeval Mathematics

DANIEL B. LLOYD

IN AN EARLIER DISCUSSION on archaeological discoveries of a mathematical nature in Iraq [3],* the author described a 1958 digging operation near Baghdad. The present article describes another, more recent, digging in 1962 at Tel Dhibayi, a tell some four kilometers east of the Tigris and within the area of one of Baghdad's new housing projects. The two diggings were a part of the same Iraqi operation, but were separated by several years because of unfavorable climatic conditions of summer heat and winter rains.

Tel Dhibayi is an Arabic name meaning "Mounds of the Hyenas." The city underneath this tell, discovered by Iraqi archaeologists, flourished during the early Babylonian era in the reign of Ibalpiel II, king of Eshnunna (ca. 1000 B.C.). Brought to light was an interesting array of public buildings and private homes, as well as a large Babylonian temple.

The tablets recently unearthed were of baked clay, which served for centuries as paper for the ancient Mesopotamians. Altogether, there were found some five hundred such tablets containing texts of letters, business and economic transactions, contracts, and other administrative records. The complexity and variety of the records kept indicate the high level of culture and civilization which these people had attained.

Reprinted from *Mathematics Teacher* 57 (Nov., 1966): 668-70; with permission of the National Council of Teachers of Mathematics.

* Numerals in brackets indicate the references at the end of the article.

Among the tablets found, the one of greatest interest to mathematicians was one measuring approximately $4\frac{1}{2}$ by $2\frac{3}{4}$ by $1\frac{1}{4}$ inches and containing on its face the inscription of a mathematical problem. It was a geometric problem concerning a rectangle (see Figure 1). Apparently, it was to find the dimensions, given the diagonal and area.

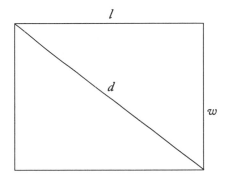

FIGURE I

Accompanying the diagram shown, the following steps were inscribed on the tablet:

$$\text{Let } d = 5/4 \text{ and } A = 3/4.$$

$$2A = 3/2, \quad d^2 - 2A = 1\tfrac{9}{16} - 1\tfrac{1}{2} = \tfrac{1}{16},$$

$$\sqrt{d^2 - 2A} = \tfrac{1}{4}, \quad \tfrac{1}{2}\sqrt{d^2 - 2A} = \tfrac{1}{8},$$

$$\frac{d^2 - 2A}{4} = 1/64, \quad \frac{d^2 - 2A}{4} + A = \frac{d^2 + 2A}{4},$$

$$\tfrac{1}{2}\sqrt{d^2 + 2A} + \tfrac{1}{2}\sqrt{d^2 - 2A} = l = 1,$$

$$\tfrac{1}{2}\sqrt{d^2 + 2A} - \tfrac{1}{2}\sqrt{d^2 - 2A} = w = 3/4.$$

The following steps were omitted, but apparently used.

$$A = \text{area} = lw$$
$$d^2 = l^2 + w^2,$$
$$d^2 - 2A = d^2 - 2lw,$$
$$d^2 - 2A = l^2 - 2lw + w^2 = (l - w)^2,$$
$$\sqrt{d^2 - 2A} = l - w.$$

Similarly,

$$\sqrt{d^2 + 2A} = l + w.$$

Therefore, their result.

Thus they seemed to know the Pythagorean relationship and the area formula for the rectangle.

A natural question arises here regarding the method of solution used by these ancient people. The archaeologist, or even the psychologist, might be better prepared to answer this question than the mathematician, to whom it is quite baffling. Their complex approach implies a knowledge of far simpler relations which could solve their problem more directly. Whether prompted by pedantry, ostentation, or merely a penchant for "being different," their precise motive remains obscure. At least, it is an ingenious method, avoiding the formal solution of a quadratic equation, which they may not have had at their disposal.

Such recent discoveries at these Iraqi sites serve to strengthen the evidence already existing [4] that the Greeks leaned heavily on these earlier sources of mathematical knowledge. Much which had formerly been attributed to the Egyptians was much earlier accessible from the Sumerians and Babylonians. Thus the Greek contributions were not as original as formerly believed, but were confined to elaboration and organization of the acquired knowledge. For instance, the Pythagorean theorem, a masterpiece of Euclidean geometry and proved so elegantly by the Greeks, was known to the Babylonians, either with or without a general proof, around 1800 B.C.—some fourteen centuries earlier. This was revealed by the discovery and translation in 1945 [1] of the Babylonian tablet known as Plimpton 322, catalogued as item 322 in the G.A. Plimpton Collection at Columbia University, New York City. This tablet is inscribed with a large number of Pythagorean triplets, representing the sides of right triangles. In contrast to this, there is no evidence that the Egyptians were aware of even a single case of this right-triangle theorem.

Archaeology, even with its modern advanced techniques, can present nothing new in terms of modern mathematical knowledge. But in the history of early mathematical developments it serves as a vital source of enlightenment. Just as new mathematics is growing today, likewise our knowledge of early history is expanding through these archaeological revelations. In very recent times, as illustrated in the above case, this information has been highly revealing.

REFERENCES

1. Eves, Howard, *An Introduction to the History of Mathematics*. New York: Holt, Rinehart & Winston, 1964.

2. Kramer, Samuel N. *From the Tablets of Sumer*. Indian Hills, Colorado: Falcon's Wing Press, 1956.

3. Lloyd, Daniel B. "Recent Evidences of Primeval Mathematics," *The Mathematics Teacher*, December 1965.

4. Lloyd, Seton, *Foundations in the Dust*. New York: Oxford University Press, 1945.

5. Neugebauer, O., and Sachs, A.J. *Mathematical Cuneiform Texts*, American Oriental Series, Vol. XXIX. New Haven: American Oriental Society, 1946.

6. *Sumer*, Vol. XVI, 1964. Published quarterly by the Iraqi Directorate of Antiquities, Baghdad, Iraq.

17

Babylonian Mathematics
with Special Reference to Recent Discoveries*

RAYMOND CLARE ARCHIBALD

*I*N A VICE-PRESIDENTIAL ADDRESS before Section A of the American Association for the Advancement of Science just six years ago, I made a somewhat detailed survey[1] of our knowledge of Egyptian and Babylonian Mathematics before the Greeks. This survey set forth considerable material not then found in any general history of mathematics. During the six years since that time announcements of new discoveries in connection with Egyptian mathematics have been comparatively insignificant, and all known documents have probably been more or less definitely studied and interpreted. But the case of Babylonian mathematics is entirely different; most extraordinary discoveries have been made concerning their knowledge and use of algebra four thousand years ago. So far as anything in print is concerned, nothing of the kind was suspected even as late at 1928. Most of these recent discoveries have been due to the brilliant and able young Austrian scholar Otto Neugebauer who now at the age of 36 has a truly remarkable record of achievement during the past decade. It was only in 1926 that he received his doctor's degree in mathematics at Göttingen, for an interesting piece of research in Egyptian mathematics; but very soon he had taken up the study of Babylonian cuneiform writing. He acquired a mastery of book and periodical literature of the past

fifty years, dealing with Sumerian, Akkadian, Babylonian, and Assyrian grammar, literature, metrology, and inscriptions; he discovered mathematical terminology, and translations the accuracy of which he thoroughly proved. He scoured museums of Europe and America for all possible mathematical texts, and translated and interpreted them. By 1929 he had founded periodicals called *Quellen und Studien zur Geschichte der Mathematik*[2] and from the first, the latter contained remarkable new information concerning Babylonian mathematics. A trip to Russia resulted in securing for the *Quellen* section, Struve's edition of the first complete publication of the Golenishchev mathematical papyrus of about 1850 B.C. The third and latest volume of the *Quellen*, appearing only about three months ago, is a monumental work by Neugebauer himself, the first part containing over five hundred pages of text, and the second part in large quarto format, with over 60 pages of text and about 70 plates. This work was designed to discuss most known texts in mathematics and mathematical astronomy in cuneiform writing. And thus we find that by far the largest number of such tablets is in the Museum of Antiquities at Istanbul, that the State Museum in Berlin made the next larger contribution, Yale University next, then the British Museum, and the University of Jena, followed by the University of Pennsylvania, where Hilprecht, some thirty years ago, published a work containing some mathematical tables. In the Museum of the Louvre are 16 tablets; and then there are less than

---

Reprinted from *Mathematics Teacher* 29 (May, 1936): 209–19; with permission of the National Council of Teachers of Mathematics.

8 in each of the following: the Strasbourg University and Library, the Musée Royaux du Cinquanteraire in Brussels, the J. Pierpont Morgan Library Collection (temporarily deposited at Yale) the Royal Ontario Museum of Archaeology at Toronto, the Ashmolean Museum at Oxford, and the Böhl collection at Leyden. Most of the tablets thus referred to date from the period 2000 to 1200 B.C. It is a satisfaction to us to know that the composition of this wonderful reference work was in part made possible by The Rockefeller Foundation. Some two years ago it cooperated in enabling Neugebauer to transfer his work to the Mathematical Institute of the University of Copenhagen, after Nazi intolerance had rendered it impossible to preserve his self respect while pursuing the intellectual life. This new position offered the opportunity for lecturing on the History of Ancient Mathematical Science. The first volume of these lectures,[3] on "Mathematics before the Greeks," was published last year, and in it are many references to results, the exact setting of which are only found in his great source work referred to a moment ago. In these two works, then, we find not only a summing up of Neugebauer's wholly original work, but also a critical summary of the work of other scholars such as Frank, Gadd, Genouillac, Hilprecht, Lenormant, Rawlinson, Thureau-Dangin, Weidner, Zimmern, and many others.[4] Hence my selection of material to be presented to you tonight will be mainly from these two works. Before turning to this it may not be wholly inappropriate to interpolate one remark regarding Neugebauer's service to mathematics in general. Since 1931 his notable organizing ability has been partially occupied in editing and directing two other periodicals, (1) *Zentralblatt für Mathematik* (of which 11 volumes have already appeared), and (2) *Zentralblatt für Mechanik,* (3 volumes)—a job which of itself would keep many a person fully employed. *Mais, revenons a nos moutons!*

From about 3500 to 2500 years before Christ, in the country north of the Persian Gulf between the Tigris and Euphrates Rivers, the non-semitic Sumerians, south of the semitic Akkadians, were generally predominant in Babylonia. By 2000 B.C. they were absorbed in a larger political group. One of the greatest of the Sumerian inventions was the adoption of cuneiform script; notable engineering works of the Babylonians, by means of which marshes were drained and the overflow of the rivers regulated by canals, went back to Sumerian times, like also a considerable part of their religion and law, and their system of mathematics, except, possibly, for certain details. As to mathematical transactions we find that long before coins were in use the custom of paying interest for the loan of produce, or of a certain weight of a precious metal, was common. Sumerian tablets indicate that the rate of interest varied from 20 per cent to 30 per cent, the higher rate being charged for produce. At a later period the rate was $5^{1}/_{2}$ per cent to $33^{1}/_{3}$ per cent for produce.[5] An extraordinary number of tablets show that the Sumerian merchant of 2500 B.C. was familiar with such things as weights and measures, bills, receipts, notes, and accounts.

Sumerian mathematics was essentially sexagesimal and while a special symbol for 10 was constantly used it occupied a subordinate position; there were no special symbols for 100 or for 1000. One hundred was thought of as $60 + 40$ and 1000 as $16 \cdot 60 + 40$. But in these cases the Sumerian would write simply 1,40 and 16,40;

$= 12 \times 60^2 + 25 \times 60 + 33 = 44733$. Hence the Sumerians had a relative positional notation for the numbers. The word cuneiform means wedgeshaped and the numbers from one to nine were denoted by the corresponding number of wedges, where the Egyptian simply employed strokes. For 10, as we have seen, an angle-shaped sign was used. Practically all other integers were made up of combinations of these in various ways. There is great ambiguity because, for example, a single upright wedge may stand for 1 or 60 or any positive or negative integral multiple of 60. Hence there was a special sign D for 60; Ð or Ð or ⅊ for 600, the last of which suggests 60×10; O for 3600; and ◎ for 36000, again suggesting a product. No special sign for zero in Sumerian times, other than an empty space, has yet been discovered. But by the time of the Greeks

$$\langle\text{YY} \; \overset{\triangle}{\triangle} \; \langle\langle\langle \; \text{YYY} = 12 \times 60^2$$

$+0+33 = 43233$, the sign $\overset{\triangle}{\triangle}$ being for zero. But to matters of numeral notation we shall make no further reference, except to remark that the Babylonians thought of any positive integer $a = \sum c_n 60^n$, and in the form

$$a = \cdots c_2 c_1 c_0 c_{-1} c_{-2} \cdots .$$

This may not, of course, correspond to what we call integers. By means of negative values of n, fractions were introduced.

Babylonian multiplication tables are very numerous and are often the products of a certain number, successively, by 1, 2, 3 ··· 20, then 30, 40, and 50. For example, on tablets of about 1500 B.C. at Brussels are tables of 7, 10, $12\frac{1}{2}$, 16, 24, each multiplied into such a series of numbers. There are various tablets giving the squares of numbers from 1 to 50, and also the cubes, square roots, and cube roots of numbers. But we must be careful not to assume too much from this statement; the tables of square roots and cube roots were really exactly the same as tables of squares and cubes, but differently expressed. In the period we are considering the Egyptian really had nothing to correspond to any of these tables, nor do we know that even the conception of cube root was within his ken. Until two years ago it was a complete mystery why the Babylonians had tables of cubes and cube roots, but finally a tablet in the Berlin Museum gave a clue. This is a table of $n^3 + n^2$, for $n = 1$ to 30. Certain problems on British Museum tablets were found to lead to cubic equations of the form $(ux)^3 + (ux)^2 = 252$. Hence Neugebauer reasoned in his article of 1933 in the Göttingen *Nachrichten* that the purpose of the tablet in question was to solve cubic equations in this "normal form." He contended that it was within the power of the Babylonians, by a linear transformation $z = x + c$, to reduce a four-term cubic equation $x^3 + a_1 x^2 + a_2 x + a_3 = 0$ to $z^3 + b_1 z^2 + b_2 = 0$. Multiplying this equation by $1/b_1^3$ we have at once (on setting $z = b_1 w$, and $a = -b_2/b_1^3$) the normal form

$$w^3 + w^2 = a.$$

Neugebauer's theory as to the possibility of such a reduction is in part based on problems to which I shall later refer. Up to the present, however, Neugebauer has found no four-term cubic equation solved in this way. And indeed in these same British Museum tablets are two problems which lead naturally to such equations but are solved by a different method.[6]

Neugebauer feels that *tables are the foundation of all discussion of Babylonian mathematics,* that more tables, such as the one to which we have just referred, are likely to be discovered, and to illuminate other mathematical operations. There are many tables of parallel columns of integers such as

2	30
3	20
4	15
5	12
6	10
8	7, 30
9	6, 40

which is nothing but a table of reciprocals $1/n = \bar{n}$ in the sexagesimal system. $n \cdot \bar{n}$ is always equal to 60 raised to 0, or some positive or negative integral power. It is notable that in the succession of numbers chosen, the divisor $n = 7$ does not appear, the reason being that there is no integer $\bar{n}$ such that the product is the power of 60 indicated. Hence every divisor, a, with a corresponding $\bar{n}$ must be of the form $a = 2^a \cdot 3^\beta \cdot 5^\gamma$. All such reciprocals are called regular; and such reciprocals as of 7 and 11 irregular.[7] When irregular numbers appear in tables the statement is made that they do not divide.

Some of these tables are extraordinary in their complexity and extent. One tablet in the Louvre, dating from about the time of Archimedes, has nearly 250 reciprocals of numbers many of them six-place, and some seven. For example, here is the second last entry for a six-place number:[8] 2, 59, 21, 40, 48, 54, 20, 4, 16, 22, 28, 44, 14, 57, 40, 4, 56, 17, 46, 40 that is, the product of $(2 \times 60^5 + 59 \times 60^4 + \cdots + 54) \times (20 \times 60^{13} + 4 \times 60^{12} + \cdots + 40) = 60^{19}$.

The object of a table of reciprocals is to reduce division to multiplication since b/a equals b multiplied by the reciprocal of a.

I referred a few moments ago to one-figure tables of squares (that is, the squares of numbers from 1 to 60). In a tablet of the Ashmolean Museum at Oxford is the only example at present known of a two-figure table of squares.[9] This dates from about 500 B.C. The tablet is of further interest from the fact that on it are several examples of the sign for zero, e.g.

$$(15, 30)^2 = 4, \cdot, 15$$
$$(39, 30)^2 = 26, \cdot, 15.$$

The latter is equivalent to $2370^2 = 5,616,900$.

Among table-texts are also certain ones involving exponentials. From Neugebauer's volume of Lectures we may easily gain the impression[10] that these are tables for $c^n, n = 1$ to 10, for $c = 9$, $c = 16$, $c = 100$, and $c = 225$. On turning, however, to his work published three months ago we find that the tables in question are on Istanbul tablets, which are in very bad condition, so that for $c = 16$ there is not a single complete result; for $c = 9$ there are only three complete results, and similarly for the others. Enough is present however to show that the original was probably at one time as described.

One use of such tablets is in solving problems of compound interest. For example in a Louvre text dating back to 2000 B.C. is a question as to how long it would take for a certain sum of money to double itself at 20 per cent interest.[11] The problem here, then, is to find x in the equation

$$(1; 12)^x = 2.$$

The answer given is $4 - 0; 2, 33, 20 = 3; 57, 26, 40$ years, not so very different from the more accurate result 3; 48. That is, from $(1; 12)^4 = 2; 4, 24, 57, 36$, 4 was found too large, giving a quantity greater that 2. How the amount to subtract was discovered is not indicated in the text, and can not now be surmised. This is a conspicuous example of a solution by the Babylonians of an equation of the type $a^x = b$ where x was not integral.

Both in the Berlin Museum and in Yale University are tablets with other problems in compound interest. If for no other reason than to point out that five-year plans are not wholly a modern invention I may refer to a problem in a Berlin papyrus, the transcription and discussion of which occupies 16 pages of Neugebauer's new book.[12] As yet I have not mastered all the discussion of this problem, but certain facts can be stated with assurance. There is a very curious combination of simple and compound interest which is naturally suggestive of what may have been customary in old Babylonia. If P is an amount of principal, r is the rate of interest per year (here 20 per cent), and we suppose that through a five-year period P accumulates at simple interest it will amount to $2P$ at the end of the first five-year period. This amount $2P$ is then put at interest in the same way for a second five-year period and the principal is again doubled to 2^2P. The amount of capital at the end of any year is therefore given by the formula

$$A = 2^n P(1 + rm)$$

where $0 \leq m < 5$, and n is the number of five-year periods. One of the problems is: How many five-year periods will it take for a given principal P to became a given sum A? The particular case when $m = 0$ gives us the equation $A = 2^n P$. In modern notation $n = \log_2 A/P$. Now Neugebauer suggests as a possible theory in explanation of the text that something equivalent to logarithms to the base 2 was here used. In the problem $P = 1$, $A = 1,4$ whence $n = 6$.

Two other suggestive problems of the Babylonians involving powers of numbers are in a

Louvre tablet of about the time of Archimedes.[13] We have here 10 terms of a geometric series in which the first term is 1, and 2 the constant multiplier; the sum is given correctly,

$$\sum_{i=0}^{9} 2^i = 1 + 2 + 2^2 + \cdots + 2^9 = 1023.$$

But what is of special interest is the apparent suggestion as to how this number 1023 was obtained. On the tablet it is stated that it is the sum of $511 = 2^9 - 1$, and $512 = 2^9$. That is

$$1023 = 2^9 + 2^9 - 1 = 2 \cdot 2^9 - 1 = 2^{10} - 1.$$

Does this imply a knowledge of Euclid's formula leading to the sum of the ten terms of the geometric progression as, $(2^{10} - 1)/(2 - 1)$?

On the same tablet is the following

$$1 \cdot 1 + 2 \cdot 2 + \cdots + 10 \cdot 10$$
$$= (1 \cdot 1/3 + 10 \cdot 2/3) \cdot 55 = 385.$$

That is, we have the sum of the squares of the first 10 integers, and this sum is the product of two integers, one of which is 55, the sum of the first 10 integers. In general terms this relation may be stated

$$\sum_{i=1}^{n} i^2 = (1 \cdot 1/3 + n \cdot 2/3) \sum_{i=1}^{n} i.$$

Now if we set $\sum_{i=1}^{n} i = \frac{1}{2} n(n+1)$, a formula known to the Pythagoreans, we have $\sum_{i=1}^{n} i^2 = \frac{1}{6} n(n+1)(2n+1)$.

This formula is practically equivalent to one known to Archimedes.

Turning back to tables for a moment one finds a word for subtraction, *lal;* 19 is 20 *lal* 1, 37 is 40 *lal* 3; *a lal b = a − b*. Neugebauer refers to a late astronomical text in which *before* each of 12 numbers the words *tab* and *lal* (plus and minus)[14] are placed, suggesting the arrangements of points above and below a line which lie on a wave-shaped curve. This seems extraordinary. Neugebauer promised

more about the matter in the third volume of his Lectures which is to deal with mathematical astronomy.

It is also a matter of great historical interest that, in at least three different problems in simultaneous equations in two unknowns, *negative numbers occur as members*. These examples are in Yale University texts,[15] and it is very noteworthy that such conceptions were not current in Europe, even 2500 years later.

As a point of departure for certain other things let us now consider some geometrical results known to the Babylonians. There will of course be no misunderstanding when I state general results. These simply indicate operations used in many *numerical* problems of the Babylonians.

1. The area of a *rectangle* is the product of the lengths of two adjacent sides.

2. The area of a *right triangle* is equal to one half the product of the lengths of the sides about the right angle.

3. The sides about corresponding angles of *two similar right triangles* are proportional.

4. The area of a *trapezoid* with one side perpendicular to the parallel sides is one-half the product of the length of this perpendicular and the sum of the lengths of the parallel sides.

5. The perpendicular from the vertex of an *isosceles triangle* on the base, bisects the base. The area of the triangle is the product of the lengths of the altitude and half the base.[16] Indeed the Babylonians would probably think of the area of a triangle, other than right or isoseles, as the product of the lengths of its base and altitude—an easy deduction from two adjacent, or overlapping, right triangles. A large rectilineal area portrayed in a Tello table, in the Museum at Istanbul,[17] was calculated by dividing it up into 15 parts: 7 right triangles, 4 rectangles (approximately), and 4 trapezoids.

6. The *angle in a semicircle* is a right angle, a result till recently first attributed to Thales of Miletus, who flourished 1500 years later.

7. $\pi = 3$, and the *area of a circle* equals one twelfth of the square of the length of its circumfer-

ence (which is correct if $\pi = 3$). $A = \pi r^2 = (2\pi r)^2/4\pi$.

8. The *Pythagorean theorem,* a result entirely unknown to the Egyptian.[18]

9. The volume of a *rectangular parallelopiped* is the product of the lengths of its three dimensions, and the volume of a *right prism* with a trapezoidal base is equal to the area of the base times the altitude of the prism. Such a volume as the latter would be considered in estimating the amount of earth dug in a section of a canal. In a British Museum tablet the volume of a solid equivalent to that cut off from a rectangular parallelopiped by a plane through a pair of opposite edges is given correctly as half that of the parallelopiped.[19]

10. The volume of a *right circular cylinder* is the area of its base times its altitude.

11. The volume of the *frustrum of a cone* is equal to its altitude multiplied by the area of its median cross-section.[20]

12. The volume of the *frustrum of a cone,* or of a square pyramid, is equal to one-half its altitude multiplied by the sums of the areas of its bases. [Contrast this approximation to the volume of a frustrum of a square pyramid with the exact formula known to the Egyptians of 1850 B.C., $V = \frac{1}{3}h(a_1{}^2 + a_1a_2 + a_2{}^2)$, where a_1, a_2 are the lengths of sides of the square bases, and h the distance between them.] On the other hand Neugebauer believes that the Babylonians also had an exact value for the volume of the frustrum of a square pyramid, namely[21]

$$V = h\left[\left(\frac{a_1+a_2}{2}\right)^2 + \frac{1}{3}\left(\frac{a_1-a_2}{2}\right)^2\right];$$

concerning the second term there has been more than one discussion.

Practically all of these results are in British Museum texts of 2000 B.C.

That the Pythagorean theorem was known to the Babylonians of 2000 B.C. is certain from the following problems of a British Museum text:[22] (1) To calculate the length of a chord of a circle from its *sagitta* and the circumference of the circle;

and (2) To calculate the length of the *sagitta* from the chord of a circle, and its circumference. If c be the length of the chord, a of its *sagitta* and d of the diameter (one-third of the circumference) of the circle, the formulae used are evidently

$$c = \sqrt{[d^2 - (d-2a)^2]}$$

$$a = \frac{1}{2}[d - \sqrt{(d^2-c^2)}].$$

Now every step of the numerical work is equivalent to substitution in these formulae.

The same is true of the following problem in another British Museum tablet.[23] A beam of given length l was originally upright against a vertical wall but the upper end has slipped down a given distance h, what is the distance d of the other end from the wall? Each step is equivalent to substitution in the formula

$$d = \sqrt{[l^2 - (l-h)^2]}$$

and then follows the converse problem, given l and d to find h,

$$h = l - \sqrt{l^2 - d^2}.$$

In these problems a, c, d, h, and l are integers.

A third problem involving the use of the Pythagorean theorem is one on a Louvre tablet of the Alexandrine period:[24] Given in a rectangle that the sum of two adjacent sides and the diagonal is 40 and that the product of the sides is 120. The sides are found to be 15 and 8 and the diagonal 17.

There are, however, various problems in Babylonian mathematics where square roots of non-square numbers, such as 1700, are discussed. In this particular case the problem, on an Akkadian tablet of about 2000 B.C., is to find the length of the diagonal of a rectangle whose sides are ;40 and ;10. It is worked out twice, as if by two approximation formulae.[25] If the lengths of the diagonal and sides of a rectangle are respectively d, a, and b, $d = \sqrt{(a^2 + b^2)}$ and the approximation formulae are:

(1) $$d = a + \frac{b^2}{2a};$$

(2) $$d = a + 2ab^2.$$

The first of these is equivalent to the one employed several times, two thousand years later, by Heron of Alexandria, in his *Metrica*.

On the other hand the dimensions of the second formula are incorrect. After considerable calculation Neugebauer shows that a correct and good approximation to d is given by the following

(3) $$d = a + \frac{2ab^2}{2a^2 + b^2}.$$

Since, in the particular problem in question, $1/(2a^2 + b^2) = 12/11$ that is, almost unity, Neugebauer has surmised that the equivalent of this third formula may have been used.

The Heron approximation formula was also used by the Babylonians to find[26] $1^5/_{12}$ for $\sqrt{2}$ and $17/24$ for $1/\sqrt{2}$.

In a British Museum text of about 2000 B.C. there is an interesting attempt to approximate $\sqrt{2^1/_2}$ by a step equivalent to that of seeking the solutions of the Diophantine equation[27] $y^2 + 22^1/_2 = x^2$. When $x = 5$, $y = \sqrt{2^1/_2}$. The values $y = 1^1/_2$, $x = 5^1/_4$ are found in the text. One readily finds also $y = 1^1/_8$, $x = 4^7/_8$ so that the required y is between $1^1/_8$ and $1^1/_2$. Babylonian tables of squares might well give much closer approximations.

Two of the geometrical theorems referred to, a few moments ago, are employed in the solution of the following problem of a Strasbourg tablet.[28]

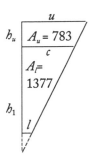

Consider two adjacent trapezoids, sections of the same right triangle and with a common side of length c as in the figure. The upper area of height h_u (between u and c) is given as 783; the lower area of height h_l(between c and l) is 1377. It is further given that

(1) $$h_l = 3h_u$$

(2) $$u - c = 36$$

Then by applying the theorems mentioned

(3) $$h_u \cdot \frac{u+c}{2} = 783$$

(4) $$h_l \cdot \frac{c+l}{2} + 1377$$

(5) $$u - c = (1/3)(c - l)$$

five equations from which the five unknown quantities are found.

There are many similar problems, one, of a group, leading to ten equations in ten unknowns. This is in connection with the division of a right triangle (by lines parallel to a side) into six areas of equal altitudes, while their areas are in arithmetic progression.[29] This problem seems to show mathematics studied for its own sake, just as problem 40 of the Rhind papyrus suggested a similar thought there.

Consider now another Strasbourg problem, of a different type, leading to a quadratic equation:[30] The sum of the areas of two squares is a given area. The length (y) of the side of one square exceeds a given ratio (α/β) of the length (x) of the side of the other square, by a quantity d. The problem is to find x and y. Here

$$x^2 + y^2 = A,$$

$$y = \frac{\alpha}{\beta} x - d.$$

If we set $x = X\beta$ it may be readily shown that we are led to the equation

$$X^2 - \frac{2d\alpha}{\alpha^2 + \beta^2} X - \frac{A - d^2}{\alpha^2 + \beta^2} = 0$$

whence

$$X = \frac{1}{\alpha^2 + \beta^2} \{ d\alpha + \sqrt{d^2\alpha^2 + (\alpha^2 + \beta^2)(A - d^2)} \}$$

Now every step of the solution of this problem is equivalent to substitution in this formula.

There are scores of problems which prove this amazing fact, that the Babylonians of 2000 B.C. were familiar with our formula for the solution of a quadratic equation. Until 1929 no one suspected that such a result was known before the time of Heron of Alexandria, two thousand years later.

In general only the positive sign before the radical in the solution of a quadratic equation is to be considered; but in the following problem[31] (because of its nature) both roots are called for. The problem on a Berlin tablet deals with the dimensions of a brick structure of given height h, of length l, of width w, and of given volume v. The exact nature of the structure is not clear but it is given that $v/a = hlm$, where $1/a$ is a given numerical factor. $l + m$ is also a given quantity S; it is required to find l and m. They are evidently roots of the quadratic equation

$$X^2 - SX + \frac{v}{ah} = 0 ;$$

when l and m are given by

$$\frac{S}{2} \pm \sqrt{\left(\left(\frac{S}{2} \right)^2 - \frac{v}{ah} \right)}.$$

The upper sign gives the required value for l and the lower for m. Of course both roots are positive.

On another Berlin tablet[32] is a problem divorced from geometrical connections but which may possibly illustrate another point of interest. Two unknowns y_1, y_2 are connected by relations

(1) $$y_1 - \frac{\alpha}{\beta} (y_1 + y_2) = D$$

(2) $$y_1 y_2 = 1$$

where α, β, D are given, $\beta > \alpha$. New variables are then introduced,

(3) $$x_1 = (\beta - \alpha) y_1, \quad x_2 = \alpha y_2$$

whence $x_1 - x_2 = \beta D$, $x_1 x_2 = \alpha (\beta - \alpha)$. From the resulting quadratic equation

$$X^2 - \beta DX - \alpha (\beta - \alpha) = 0$$

x_1 and $- x_2$ are found to be,

$$\pm \frac{\beta D}{2} + \sqrt{\left[\left(\frac{\beta D^2}{2} \right) + \alpha(\beta - \alpha) \right]},$$

y_1 and y_2 are then found from (3). Neugebauer emphasizes that here, and in other texts we have a transformation of a quadratic equation to a normal form with unity as coefficient of the squared term. And also we have another example of an equation in which both roots are positive and the double sign before the radical is taken in solving the question.

We have now considered Babylonian solutions of simultaneous equations, exponential equations, quadratic equations, and cubic equations. Before giving examples leading to equations of higher degree some general remarks may be made about 17 of the 35 mathematical tablets at Yale.[33] In size they are from 9.5×6.5 cm. to 11.5×8.5 cm. They belong to series and contain the enunciation of problems systematically arranged. No solutions are given. On one tablet there are 200 problems and on the seventeen over 900. Since only a few tablets have been preserved there must have been thousands of problems in the original series.

To give an idea of what is meant by problems being arranged in a series it may be noted that on one tablet are 55 problems of the type to find x and y, given[34]

$$\begin{cases} xy = 600 \\ (ax + by)^2 + cx^2 + dy^2 = B \end{cases}$$

where some coefficients can be zero. The first equation is the same for all of these problems. The second equations for the first seven problems are as follows:

1. $(3x)^2 + y^2 = 8500$
2. $+ 2y^2 = 8900$
3. $- y^2 = 7700$

4. $(3x + 2y)^2 + x^2 = 17800$
5. $+ 2x^2 = 18700$
6. $- x^2 = 16000$
7. $- 2x^2 = 15100$

Problems 48 and 49 are

$$[3x + 5y - 2(x - y)]^2 - 2y^2 = 28100$$
$$+ x^2 + y^2 = 30200.$$

The solution of all the equations leads at once to a biquadratic equation which is a quadratic equation in x^2.

On another tablet, however, are problems of the type[35]

$$xy = A$$
$$a(x + y)^2 + b(x - y) = C = 0$$

which leads to the most general form of biquadrdratic equation (if $d = c + 2aA$)

$$x^4 + \frac{b}{a} x^3 + \frac{d}{a} x^2 - \frac{bA}{a} x + A^2 = 0.$$

The second equation of one of the problems in this group is

$$\tfrac{1}{5}(x + y)^2 - 60(x - y) = -100,$$

one of the extraordinary examples of a negative number in the right hand member, to which we have already referred.

Problems on another tablet[36] lead to the most general cubic equation. How the Babylonians found the solution of such equations is unknown. It is true that $x = 30$, $y = 20$ gives the solution of every one of these, and of hundreds of problems in other series; Neugebauer believes, however, that it is nonsensical to imagine that such values were merely to be guessed (*Quellen*, v. 3, part 1, p. 456).

There are problems about measurement of corn and grain, workers digging a canal, interest for loan of silver, and more problems like the Strasbourg texts where algebraic questions are derived from consideration of sections of a triangle. Neugebauer concludes the second part of his great work with an italicized statement to the effect that the Strasbourg and Yale texts prove that the chief importance of Babylonian mathematics lies in algebraic relations—not geometric.

In his work of thirty years[37] ago Hilprecht was guilty of more than one disservice to truth. One such was his great emphasis on mysticism in Babylonian mathematics. Its association with what he called "Plato's number," $60^4 = 12,960,000$. In spite of the protests of contemporary scholars such ideas were widely disseminated. We have noted enough to realize that such an idea is purest bunk—rather freely to translate Neugebauer's expression.[38]

While it has been possible for me to draw your attention to only a few somewhat isolated facts, I trust that you have received the impression that in Babylonian algebra of 4000 years ago we have something wonderful, *real* algebra without any algebraic notation or any actual setting forth of general theory. And if all this was known in 2000 B.C., how far back must we go for the beginnings of the Sumerian mathematics, simple arithmetic operations? Probably back to 3000 B.C. at least.

One thing which is of great interest in the study of Egyptian and Babylonian mathematics is that we handle and study the actual documents which go back to those days, four thousand years ago. Contrast with this the way in which we learn of Greek mathematics. Even in the case of such a widely used work as Euclid's Elements, there is not a single manuscript which is older than 1200 years after Euclid lived, that is about a thousand years ago.

Eight years of work by a young genius standing on the shoulders of great pioneering scholars, have in extraordinary fashion greatly advanced the frontiers of our knowledge of Babylonian mathematics. One can not help feeling that the inspiration of such achievement will cause more than one man to shout, "Let knowledge grow from more to more," as *he* too joins in the endless torch race to

"pass on the deathless brand
From man to man."

NOTES

* Delivered at a joint meeting of The National Council of Teachers of Mathematics, the American Mathematical Society, and The Mathematical Association of America, at St. Louis, Mo., on January 1, 1936.

1. "Mathematics before the Greeks," *Science*, n.s. v. 71, 31 Jan. 1930, p. 109–121.

2. I shall later refer to the two periodicals simply by the words *Quellen*, and *Studien*.

3. *Vorlesungen über Geschichte der antiken mathematischen Wissenschaften*, v. 1, *Vorgriechische Mathematik* (Die Grundlehren der mathematischen Wissenschaften, v. 43), Berlin, 1934. Reference to this work will later be made simply by the word *Vorlesungen*.

4. For the literature of Babylonian mathematics prior to 1929, see my *Bibliography* in the Chace-Manning-Bull edition of the *Rhind Mathematical Papyrus*, v. 2; for later items see K. Vogel's bibliography in *Bayer, Blätter f. d. Gymnasialschulwesen*, v. 71, 1935, p. 16–29.

5. M. Jastrow, Jr., *The Civilization of Babylonia and Assyria*, Philadelphia, 1915, p. 326, 338; C. H. W. Johns, *Babylonian and Assyrian Laws, Contracts and Letters*, New York, 1904, p. 251, 255–256. See also D. E. Smith, *History of Mathematics*, vol. 2, 1925, p. 560.

6. Compare Göttingen, *Nachrichten, Math.-phys. Kl.*, 1933, p. 319; also *K. Danske Videnskabernes Selskab, Mathem.-fysiske Meddelelser*, v. 12, no. 13, p. 9. Also *Quellen*, v. 3, part 1, p. 200–201, 210–211.

7. It is easy to approximate to 1/7, e.g. 7/28 =; 8, 45; 13/90 =; 8, 40, etc., but there is no case known where this was done.

8. *Quellen*, v. 3, part 1, p. 22.

9. *Quellen*, v. 3, part 1, p. 72–73 and part 2, plate 34.

10. *Vorlesungen*, p. 201; *Quellen*, v. 3, part 1, p. 77–79 and part 2, plate 42.

11. *Quellen*, v. 3, part 2, p. 37–38, 40–41.

12. *Quellen*, v. 3, part 1, p. 351-367, and part 2, plates 29, 32, 54, 56, 57; *Vorlesungen*, p. 197–199.

13. *Quellen*, v. 3; part 1, p. 96–97, 102–103 and part 2, plate 1; *Studien*, v. 2, 1932, p. 302–303.

14. *Vorlesungen*, p. 18.

15. *Quellen*, v. 3, part 1, p. 387, 440, 447, 455, 456, 463, 470, 474; and part 2, plates 23, 48, 59.

16. *Quellen*, v. 3, part 2, p. 43, 46–47, 50–51 and part 1, p. 97, 104.

17. A. Eisenlohr, *Ein altbabylonischer Feldplan*, Leipzig, 1896. J. Oppert, Académie d. Inscriptions et Belles-Lettres, *Comptes Rendus* s. 4, v. 24, 1896, p. 331–348; also in *Revue d'Assyriologie et d'Archéologie Orientale*, v. 4, 1897, p. 28–33. F. Thureau-Dangin, *Revue d' Assyriologie*, v. 4, 1897, p. 13-27.

18. After many misstatements by mathematical historians it was an Egyptologist, the late T. E. Peet, in his *Rhind Mathematical Papyrus* (London, 1923, p. 31–32) who brought out the fact that there is not one scrap of evidence that the Egyptians knew the Pythagorean theorem, even in the simple 3–4–5 case. He gave also interesting new information about the *harpedonaptai*, or rope stretchers, referred to by Democritus. It is of course true that there are problems involving the relations of such numbers as 8, 6, and 10, as in Berlin Papyrus 6619 (about 1850 B.C.): Distribute 100 square ells between two squares whose sides are in the ratio 1 to $^3/_4$. The same equations arise in problem 6 of the Golenishev papyrus: Given that the area of a rectangle is 12 arurae and the ratio of the lengths of the sides 1:$^1/_2$ $^1/_4$, find the sides; see also the Kahun papri, ed. by Griffith (1898).

19. *Quellen*, v. 3, part 2, p. 43, 47, 52.

20. *Studien*, v. 1, 1929, p. 86–87; *Vorlesungen*, p. 171; *Quellen*, v. 3, part 1, p. 176.

21. *Studien*, v. 2, 1933, p. 348–350; *Vorlesungen*, p. 171; *Quellen*, v. 3, part 1, p. 150, 162, 187–188. Heron of Alexandria (second century A.D.?) found the volume of such a pyramid, for which $a_1 = 10$, $a_2 = 2$, $h = 7$ (*Heronis Alexandrini Opera quae supersunt omnia*, Leipzig, v. 5, 1914, p. 30–35), every step being equivalent to substituting in this formula.

22. *Studien*, v. 1, 1929, p. 90–92.

23. *Quellen*, v. 3, part 2, p. 53.

24. *Studien*, v. 2, 1932, p. 294; *Quellen*, v. 3, part 1, p. 104.

25. *Studien*, v. 2, 1932, p. 291–294; *Vorlesungen*, p. 33–36; *Quellen*, v. 3, part 1, p. 279–280, 282, 286–287, and part 2, plates 17, 44.

26. *Studien*, v. 2, 1932, p. 294–295; *Vorlesungen*, p. 37; *Quellen*, v. 3, part 1, p. 100, 104, and part 2, plate 1.

27. *Studien*, v. 2, 1932, p. 295–297, 309; *Quellen*, v. 3, part 1, p. 172.

28. *Studien*, v. 1, 1929, p. 67–74; *Quellen*, v. 3, part 1, p. 259–263.

29. *Quellen*, v. 3, part 1, p. 253; *Studien*, v. 1, 1929, p. 75–78; *Vorlesungen*, p. 180–181.

30. *Studien*, v. 1, 1930, p. 124–126; *Quellen*, v. 3, part 1, p. 246–248.

31. *Quellen*, v. 3, part 1, p. 280–281, 283–285.

32. *Quellen*, v. 3, part 1, p. 350–351; *Vorlesungen*, p. 186–187.

33. *Quellen*, v. 3, part 1, p. 381–516, and part 2, plates 36, 37, 57–59, p. 60-64; *Studien*, v. 3, 1934, p. 1–10.

34. *Quellen*, v. 3, part 1, p. 418-420.

35. *Quellen*, v. 3, part 1, p. 455–456.

36. *Quellen*, v. 3, part 1, p. 402.

37. *Mathematical, Metrological and Chronological Tablets, from the Temple of Nippur* (Univ. Pennsylvania), s.A, Cuneiform Texts, v. 20, part 1, Philadelphia, 1906.

38. *K. Danske Videnskabernes Selskab, Matem.-fysiske Meddelelser*, v. 12, no. 13, 1934, p. 5–6.

Recent Discoveries in
Babylonian Mathematics I

Zero, Pi, and Polygons

PHILLIP S. JONES

ONE OF THE BEST SUMMARIES of Babylonian mathematical achievements is to be found in the late Raymond Clare Archibald's article on this topic in the *Mathematics Teacher* for May 1936. In this note he paid particular honor to Otto Neugebauer, the young Austrian scholar who, possessing the rare combination of a fine knowledge of mathematics, mechanics, and astronomy with the ability to read Babylonian cuneiform symbols, had been responsible for many of the discoveries listed there. Archibald's concluding statement was,

> Eight years of work by a young genius standing on the shoulders of great pioneering scholars [such as Frenchman F. Thureau-Dangin and H.V. Hilprecht of the University of Pennsylvania] have in extraordinary fashion greatly advanced the frontiers of our knowledge of Babylonian mathematics. One can not help feeling that the inspiration of such achievement will cause more than one man to shout, "Let knowledge grow from more to more," as *he* too joins in the endless torch race to
> > "pass on the deathless brand
> > From man to man."[1]

Since 1936 a war has intervened, but Otto Neugebauer has continued to unveil new aspects of Babylonian mathematics and astronomy even while moving from Germany to Denmark to the United States and here taking the time to serve as the first editor of *Mathematical Reviews*. The United States was fortunate indeed to have available a man who had previously founded and edited two German abstracting journals at a time when it became apparent that although we were not yet in the war, we were being cut off from those journals which were so important to a critically needed acceleration in the development of American mathematics, especially applied mathematics. We in America owe much to Otto Neugebauer who is now chairman of the Department of the History of Mathematics at Brown University, for his services in setting up and for many years editing the now well-established and flourishing *Reviews*. However, all mathematical historians owe him a debt for his continued research and the postwar publication of two new books on Babylonian mathematics.[2]

The first of these contains several remarkable new discoveries, among which are the use by the Babylonians of an averaging or division method for approximating square roots and a table of Pythagorean triples which have been discussed in earlier numbers of this journal.[3]

The purpose of this note and its successors is to report several additional recent discoveries due to Neugebauer, and his collaborator A. Sachs, the Netherlander; M.E.M. Bruins, who has been studying materials located at Susa, the capital of An-

Reprinted from *Mathematics Teacher* 50 (Feb., 1957): 162–65; with permission of the National Council of Teachers of Mathematics.

cient Elam, by a French archaeological expedition in 1936; and the Arab Taha Baqir, who has been studying tablets recently found at Tell Harmal in Iraq. In this note we shall turn our attention to three topics only: *zero, π,* and *polygons.* For persons who would like a broader survey of Babylonian writings and Babylonian and Egyptian mathematics, we recommend in addition to Archibald's article and the more technical references cited here, two recent "pocket books," V. Gordon Childe's *Man Makes Himself* [4] and Edward Chiera's *They Wrote on Clay.* [5]

It is rather remarkable that although the importance of zero is recognized the world around, its history is still clouded with doubts and uncertainties, and the concept of zero and the operations with it remain pedagogical stumbling blocks through the grades. Van der Waerden writes, "The zero is the most important digit. It is a stroke of genius, to make something out of nothing by giving it a name and inventing a symbol for it." He quotes Halsted as saying "It is like coining the Nirvana into dynamos," and elaborates Freudenthal's theory that the Hindus developed if not their whole decimal system, at least their zero concept and a round symbol for it out of the Greeks' round symbol for zero. This concept the Greeks, in turn, had obtained from the Babylonians, he believes. [6]

Neugebauer agrees with this when he writes, "It seems to me rather plausible to explain the decimal place value notation [of the Hindus] as a modification of the sexagesimal place value notation with which the Hindus became familiar through Hellenistic Astronomy." [7] He cites further evidence for this line of communication from the history of astronomy.

The facts about zero at the moment seem to be these: There was no zero in use in old Babylonian times (1800–1600 B.C.). One had to determine the sexagesimal places where we would have written a zero either by the context of the problem or, occasionally, by the spacing of the symbols. In the later Babylonian or Seleucid period (300 B.C.—0), a special symbol, which was also used as a separation mark between sentences, came into use for a zero in both mathematical and astronomical texts. Further, there is a definite possibility that the Babylonians used this mark for a zero within a number, as early as the end of the eighth century B.C. [8]

The Greeks took over Babylonian data and sexagesimal fractions when Hipparchus began to study and write about astronomy. In fact, we today get our minutes and seconds via the Arabs from this Greek use of Babylonian sexagesimal fractions, into the writing of which the Greeks introduced their own symbols for zero. [9] One of these was a round symbol which van der Waerden and others have regarded as coming from the first letter of the Greek word for *nothing.* Neugebauer doubts this and cites many other symbols used by early Greeks for representing a zero within a number, but agrees that the later Greek astronomer Ptolemy (ca. A.D. 150) was the first to write a zero at the *end* of a number. For this he used a circular symbol. [10]

There seems to be no evidence that the Babylonians ever regarded zero as a number by itself which could enter into operations with other numbers. Boyer presents good evidence that this distinction belongs to Aristotle, who discussed division by zero in connection with speed through a vacuum as compared to speed through a resisting medium. Brahmagupta (Hindu, ca. A.D. 600), who has often been given credit for the first discussion of division by zero, did write of it more explicitly and was probably the first to discuss the division of zero by zero. [11]

On page 59 of his *Mathematical Cuneiform Texts,* Neugebauer, after offering a rather involved and tentative explanation of the multiplier used on a tablet dealing with the volume and circumference of a log, remarks that the result would have been obtained directly if one had assumed π = 3;7,30. This is his notation for the sexagesimal number

$$3 + \frac{7}{60} + \frac{30}{60^2},$$

which would have been, decimally, 3.125 or 3¹/₈. He then continues "This assumption, however, seems unwarranted because this approximation of π is not attested elsewhere in Babylonian mathematics."

This is not the first time that Neugebauer has anticipated new discoveries. In fact, he had speculated that the Babylonians had probably made advances in number theory some time before the discovery of the Plimpton tablet containing a table of Pythagorean triples. In the case of π, Neugebauer was able to add a note after the completion of the manuscript of *The Exact Sciences in Antiquity* pointing to a tablet newly published by Bruins. Here he found a correction factor which the Babylonians inserted into their older processes when a more accurate determination of area was needed.

The Babylonians in general determined the circumference of a circle by multiplying its diameter by 3, this is equivalent to saying $C = \pi d$, with $\pi = 3$. They found the area as one twelfth of the circumference squared. In modern symbols, this amounts to saying $A = {}^1/_{12}C^2$ which, if π were 3, would have been equivalent to $A = C^2/(4 \cdot \pi)$, a correct formula if the correct value of π were used. Bruins reported in 1950 a tablet excavated by French archaeologists in 1936 on which it was directed that for more accurate results the ¹/₁₂ should be multiplied by ;57,36. This means

$$\frac{57}{60} + \frac{36}{60^2}$$

which is ²⁴/₂₅. Since the correct multiplier would be 1/4π, and since the Babylonian corrected value would have been

$$\left(\frac{24}{25}\right)\left(\frac{1}{12}\right),$$

by equating these two values and solving for π, we find that this Babylonian process for determining the area of a circle was equivalent to using $\pi = 3^1/_8$!

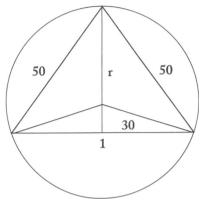

FIGURE I

These tablets discussed by Bruins also contain a problem in which the radius of the circle circumscribed about an isosceles triangle with sides 50, 50, and 1, was determined.[12] These peculiar dimensions need to be interpreted in terms of the sexagesimal system in which one half of 1 is 30, i.e., if, lacking a terminal zero, 1 really represented 1, 0 in this case, i.e., one sixty and no units, then one half of it is 30. Or if 1 really means 1, then one half of it is ³⁰/₆₀, which the Babylonian would write as ;30 (using the corresponding cuneiform symbols of course).

Figure 1 shows the diagram. The altitude is 40, and the Babylonians found r by the process which we would symbolize as solving the equation $(40 - r)^2 + 30^2 = r^2$.

This and some problems on regular polygons indicate that the Babylonians went farther into more theoretical geometry that has been previously realized. For example, a table lists good values for the coefficients which multiplied times the square of a side give the area of a regular pentagon, hexagon, and heptagon.

These coefficients are in agreement with the ones given by Heron (ca. A.D. 75) in his *Metrica* for determining the sides of the regular inscribed pentagon, hexagon, and heptagon, or nonagon, by multiplying the radius of the circle by 6/5, 6/6, 7/8, and 2/3 respectively. Bruins suggests that the first two and last of these might have been derived by the Babylonians by processes equivalent to the formula $S_n = 6/nr$. This could have been derived by assuming that the perimeter p_n of the $n$sided polygon was nearly equal to the circumference of a circle. Thus $p_n = S_n \cdot n \approx 2\pi r$, and then using the Babylonian approximation, 3, in place of π. Heron derived the ratio 7/8 for the heptagon by using for the side of the heptagon the altitude of the central triangle associated with a hexagon and by using 7/4 as an approximation to the $\sqrt{3}$. This still gives a good approximate construction for a heptagon which, of course, we now know to be nonconstructible with ruler and compasses.

These recent discoveries show, then, that the Babylonians had gone farther in geometry than has been previously known. However, it is still true that their most remarkable and extensive achievements were algebraic.

NOTES

1. "Babylonian Mathematics with Special Reference to Recent Discoveries," *The Mathematics Teacher*, XXIX (May 1936), 219.

2. O. Neugebauer and A. Sachs, *Mathematical Cuneiform Texts* (New Haven: American Oriental Society, 1945); O. Neugebauer, *The Exact Sciences in Antiquity* (Princeton: Princeton University Press, 1952).

3. P.S. Jones, "$\sqrt{2}$ in Babylonia and America," *The Mathematics Teacher*, XLII (October 1949), 307, and "The Pythagorean Theorem," *The Mathematics Teacher*, XLII (April 1950), 162.

4. (New York: New American Library of World Literature, 1951.) This paper-backed Mentor Book was originally published in England in 1936 and revised in 1941 and 1951. Chapter VIII, "The Revolution of Human Knowledge," deals fairly extensively with Egyptian and Babylonian mathematics.

5. George G. Cameron (ed.), *They Wrote on Clay* (Chicago: University of Chicago Press, 1938); reprinted in 1956 as a paper-backed Phoenix Book. Little space is devoted to Babylonian mathematics (pp. 154–157), but among its profuse illustrations are good pictures of tablets containing maps (p. 160) and "blueprints" (p. 163).

6. B.L. van der Waerden, *Science Awakening* (Gronin- gen, Holland: P. Noordhoff Ltd., 1954), p. 56.

7. *The Exact Sciences*, p. 180.

8. Neugebauer, *The Exact Sciences*, pp. 16, 20, 26, and *Mathematical Cuneiform Texts*, p. 34.

9. P.S. Jones, "Angular Measure—Enough of Its History to Improve Its Teaching," *The Mathematics Teacher*, XLVI (October 1953), 419.

10. *The Exact Sciences*, pp. 13–14.

11. C.B. Boyer, "An Early Reference to Division by Zero," *The American Mathematical Monthly*, VI (October 1943), 487–91.

12. M.E.M. Bruins, *Nouvelles Découvertes sur les Mathématiques Babyloniennes* (Paris: Université de Paris, 1952), pp. 18–21.

Recent Discoveries in Babylonian Mathematics II

The Earliest Known Problem Text[1]

PHILLIP S. JONES

*I*T IS INTERESTING and perhaps remarkable that, along with the current rapid development of mathematics, our knowledge of the history of mathematics is also expanding. Some of this most recent expansion has been in our knowledge of some of the most remote periods. A current newspaper report tells of evidences of a prehistoric arithmetic recently found in archaeological excavations in a site in the Belgian Congo dating back 8,000 to 9,000 years ago.[2] It will be interesting to learn more of this.

Added data on the earliest historical mathematics has been accumulating as the result of excavations begun in 1945 at Tell Harmal in Iraq. This note is based on the first discussions of one of these clay tablets published by Taha Baqir, Curator of the Iraq Museum in 1950.[3]

This tablet is 9.5 × 6 × 3 cm. and is thought to date back to about 2000 B.C. This dating makes it the earliest problem text known (Babylonian mathematical tablets are classified as problem texts and table texts). It not only is the earliest text but it also implies a breadth of understanding that, though conjectured, had not been so well evidenced previously.

The problem essentially was to find lengths *BD, AE, AD, DF* given the sides of triangle *ABC*

as shown, *AB* = 45, *AC* = 60, *BC* = 75. (We write them here in decimal notation. Originally, written with cuneiform symbols and using 60 as a base, they were 45; 1, 0; and 1, 15.) The areas of the successive triangles *BAD* = 486, *ADE, DEF*, and *EFC* were also given.

The original lengths assure us that the given triangle is a right triangle, and the diagram and procedures used by the Babylonian indicate that lines *AD, DE,* and *EF* are perpendicular to *BC* and *AC*. If this is true, we have the beginning of an infinite series of similar triangles. This similarity and perhaps the series concept seem to have been recognized by the Babylonians who, however, did not write general statements, theorems, or outlines of procedures, but typically went directly to computations. We shall outline their processes, then

FIGURE I

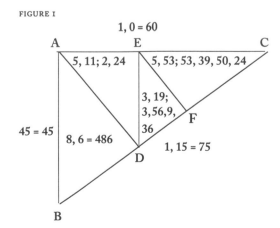

Reprinted from *Mathematics Teacher* 50 (Oct., 1957): 442–44; with permission of the National Council of Teachers of Mathematics.

interpret them in terms of general or literal expressions and the principles on which the computations may have been based.

The Babylonian steps were:

(1) Take the reciprocal of 60 and multiply it by 45, $1/60 \cdot 45 = {}^3/_4$

(2) Multiply this by 2, $2 \cdot {}^3/_4 = {}^3/_2$

(3) Multiply this by 486, $486 \cdot {}^3/_2 = 729$

(4) $\sqrt{729} = 27 = BD$

By this process the leg BD of the right triangle ABD was found from a knowledge of its hypotenuse and area. The same procedures were applied to find AD and AE and could have been reapplied ad infinitum. In fact, computation for ED was begun on this tablet.

There are two things which are remarkable about this. We will comment on them before explaining the process itself. In the first place, this is a sort of "trumped-up" or "textbookish" problem. It is difficult to conceive of a natural problem situation in which the area and one side of a triangle would be known and the other sides unknown. There is nothing in the context which would give rise to such a problem. It seems quite probable that this was a problem "dreamed up" for schoolboys, or perhaps as a puzzle for colleagues or a demonstration of the mathematical power of a writer.

Further, the problem as stated—find a leg given a leg and the area—is indeterminate. Something more has to be known or assumed in order to solve it. We will see in a moment that the procedures used repeatedly here are justified by the fact that all the triangles are similar. The Babylonian never stated such a generalization as: *if a perpendicular is drawn from the right angle to the hypotenuse of a right triangle, it divides the triangle into two triangles similar to the original and to each other.* However, he seems to have used this fact consistently.

If we analyze his steps in modern symbols and ideas, we have:

(1) $1/AC \cdot AB = {}^3/_4$

(2) $2 \cdot \dfrac{AB}{AC} = {}^3/_2$

(3) $2 \cdot \dfrac{AB}{AC} \cdot (\text{area of } \triangle ABD) = 729$

Steps (1) and (2) seem unmotivated, and step (3) seems almost absurd until we look a little deeper at this problem. The area of ABD would have been $^1/_2 (BD \cdot AD)$. Further, if we note that $\triangle ABD \sim \triangle CBA$, then

$$\frac{AB}{AC} = \frac{BD}{AD}.$$

Substituting these facts into step (3) we now have

$$2 \cdot \frac{BD}{AD} \cdot {}^1/_2 \cdot BD \cdot AD = \overline{BD^2} = 729$$

This does explain why the Babylonian's final step

(4) $\sqrt{729} = 27$

gave him the correct value for BD.

In addition to confirming that the Babylonians had some idea of similar triangles and showing some knowledge of the theorem about the perpendicular to the hypotenuse of a right triangle, this problem from 2000 B.C. also illustrates their use of multiplication by a reciprocal to do division, their knowledge of the area formula for triangles, the notion of square root, and perhaps of the solution of simultaneous equations. This latter may be more hindsight to us that a real understanding by the Babylonians, at least as exhibited in this problem. We could interpret steps (1) – (4) as the simultaneous solution of equations derived from the condition determined by the given area and the condition determined by the similarity of the triangles. It may be a little unwarranted to attribute this type of thinking to the Babylonians. Their thinking is sufficiently remarkable without adding this!

NOTES

1. The first note in this series, "Recent Discoveries in Babylonian Mathematics I: Zero, Pi, and Polygons," appeared in *The Mathematics Teacher*, L (Feb., 1957), 162 ff.

2. John Hillaby, "Ancient Africans Knew Arithmetic," *New York Times* (June 9, 1957).

3. Taha Baqir, "An Important Mathematical Text from Sumer," *Sumer*, Vol. VI (1950), 39–54. We have corrected several errors noted later by Taha Baqir. These grew out of the fact that at this period the Babylonians used no symbol for zero and hence when one finds the cuneiform equivalent of 1, or of 1, 30 one must determine from the context whether these represent 1 = 1 or

1, 0 = 60 in the former case, or

1, 30 = 60 + 30 = 90 or

$$1;30 = 1 + \frac{30}{60} = \frac{3}{2}$$

in the latter case.

Further discussions by Taha Baqir, Drenckhaha, Goetze, and Bruins of this and other texts from Tell Harmal appear in volumes VI, VII, VIII, IX, X, XI (1950–1956) of *Sumer*. I owe thanks to Dr. Wassel M. Al Dahir of the College of Arts and Sciences of Bhaghdad for a reprint of those in volume VI.

20

Recent Discoveries in Babylonian Mathematics III
Trapezoids and Quadratics

PHILLIP S. JONES

*I*N THE FIRST TWO NOTES in this series[1] we discussed several really new extensions of our knowledge of Babylonian mathematics. In this note we will discuss a newly discovered tablet which gives another example of mathematical skills which, although previously known to have been a part of the Babylonian culture, may not be familiar to all of our readers—and which, in any case, would provide an interesting enrichment problem for our students.[2]

This problem associates the area of a trapezoid with simultaneous equations and quadratic equations. It occurs on a clay tablet $15.9 \times 9.7 \times 3.4$ cms. along with two other problems and a list of coefficients which seem to relate to engineering

FIGURE I

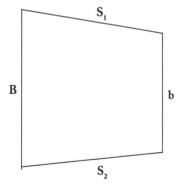

problems associated with the use of mortar, bricks, and rammed earth in construction.

The problem is to find the length of the sides of an isosceles trapezoid (see Figure 1) given that the area is 150, that the difference of its bases is 5, i.e., $B - b = 5$, and that its equal sides are 10 greater than two-thirds the sum of its bases, i.e., $S_1 = S_2 = \frac{2}{3}(B + b) + 10$.

We might solve this problem by using the Pythagorean theorem to find the height of the trapezoid, and then using the area formula to get a quadratic equation in B and b from which we could eliminate B by letting $B = b + 5$.

The Babylonians followed essentially this same procedure, *except* that they used an incorrect formula for the area of a trapezoid. They found the area as the product of the averages of the two pairs of opposite sides. (This formula is correct for some quadrilaterals. Can you find for which ones?) Thus they set the problem up as follows. (Of course, they used sexagesimal cuneiform numerals instead of Hindu-Arabic decimal-based numerals, and they used words instead of symbols or letters, and they actually wrote no formula but merely said "do this—do that" in such a way as to be equivalent to what we are writing.)

$$A = \frac{B + b}{2} \cdot \frac{S_1 + S_2}{2}$$

(This formula is false, but is a fair approximation.)

Reprinted from *Mathematics Teacher* 50 (Dec., 1957): 570–71; with permission of the National Council of Teachers of Mathematics.

(1) $$150 = \frac{2b+5}{2} \cdot \frac{2}{2} \left[\frac{2}{3}(2b+5) + 10 \right]$$

or, if we let $x = B + b + 2b + 5$

(2) $$150 = \frac{x}{2} \cdot \left[\frac{2}{3}x + 10 \right].$$

We would now clear of fractions and expand to get

(3) $$150 \cdot \frac{3}{2} \cdot 2 = x^2 + \frac{3}{2} \cdot 10 \cdot x.$$

Completing the square we would have

(4) $$450 + \left[\frac{1}{2} \cdot \frac{3}{2} \cdot 10 \right]^2 = x^2 + \frac{3}{2} \cdot 10x$$
$$+ \left[\frac{1}{2} \cdot \frac{3}{2} \cdot 10 \right]^2$$

(5) $$\pm\sqrt{506\tfrac{1}{4}} = \pm 22\tfrac{1}{2} = x + 7\tfrac{1}{2}$$

(6) $$x = -7\tfrac{1}{2} \pm 22\tfrac{1}{2}.$$

The Babylonian followed identically these same steps except that he had no idea of negative numbers or two square roots. Hence, he found

(7) $$x = 15 \quad \text{or} \quad B + b = 15.$$

From this he said

(8) $$\frac{B+b}{2} = \frac{15}{2} \quad \text{and} \quad \frac{B-b}{2} = \frac{5}{2}$$

and hence, by adding, $B = 10$, and by subtracting, $b = 5$.

Substituting in the given formula he completed the solution by finding

$$S_1 = S_2 = \frac{2}{3} \cdot 15 + 10 = 20.$$

It is interesting to note that he checked his work by verifying that these values of B, b, S_1, S_2 did give, by his formula, an area of 150.

In order that you may see a little more clearly exactly how the Babylonian proceeded without symbols or formulas, we'll quote what he wrote. After stating the problem he gave the following instructions:

(a) "Take the reciprocal of $2/3$; you get $3/2$."
(b) "Halve $3/2$; you get $3/4$."
(c) "Multiply $3/4$ by 150, the area, you get 225."

[Step (b) was a mistake. He did not need to halve $3/4$. The result of step (c), 225, is correct but would be obtained as $150 \cdot 3/2$ as in our step (3).]

(d) "Double 225; get 450. Keep it in your head."

[This completes the calculation of the left member of our equation (3).]

(e) "Take the reciprocal of $2/3$; it is $3/2$."
(f) "Halve $3/2$, get $3/4$. Multiply it by the 10 which you added, and get $7\tfrac{1}{2}$. Keep in your head."
(g) "Lay down (another) $7\tfrac{1}{2}$ and square it. You get $56\tfrac{1}{4}$."
(h) "Add $56\tfrac{1}{4}$ to 450 which your head has kept, and you get $506\tfrac{1}{4}$."
(i) "Extract the square root of $506\tfrac{1}{4}$. Its square root is $22\tfrac{1}{2}$."
(j) From the square root, subtract $7\tfrac{1}{2}$ and 15 is the value of the unknown."
(k) "Halve 15, get $7\tfrac{1}{2}$."
(l) "Halve 5 by which one width exceeds the other, and you get $2\tfrac{1}{2}$."
(m) "Add $2\tfrac{1}{2}$ and $7\tfrac{1}{2}$; 10 you get. Subtract it from the second, 10 is the upper width and 5 is the lower width."
(n) [Check]
(o) "Such is the procedure."

It is on the basis of a number of such problems that we conclude that the Babylonians could solve quadratic equations with positive rational roots by processes which could have been extended to all quadratics even though they never did the extension nor used symbols nor stated general theorems.

NOTES

1. *The Mathematics Teacher*, vol. L (Feb., Oct., 1957), p. 162 ff., p. 442 ff.

2. This is based on Taha Baqir, "Another Important Mathematical Text from Tell Harmal," *Sumer*, vol. VI (1950), pp. 130–148.

HISTORICAL EXHIBIT 3.1

Sources of Information on Ancient Egyptian Mathematics

Frank Swetz examines Rhind papyrus at the British Museum

While some information concerning ancient Egyptian mathematics can be gleaned from inscriptions on stone tombs and monuments, the majority of our information on this subject is found on less durable material, papyrus and leather scrolls. Papyrus is a form of paper made from a reed that grows in abundance along the banks of the Nile river. Egyptian scribes using ink wrote information in hieroglyphics, either in pictogram form or as a hieratic cursive script, onto such scrolls. Both the ravages of time and human conquests have left few early Egyptian documents from which a mathematical testimony can be secured. The principal sources of information available to researchers come from:

1. The Moscow papyrus, a collection of 25 problems dating from 1850 B.C. This scroll was purchased in Egypt in 1893 and is now housed at the Museum of Fine Arts, Moscow. It is about 18 feet long and 3 inches wide. A translation of its contents was published in 1930.

2. The Rhind or Ahmes papyrus, a collection of 85 problems compiled in approximately 1650 B.C. Written in hieratic, the scroll is 18 feet long and 13 inches wide. It was acquired by the Scottish antiquarian A. Henry Rhind in 1858. Upon his death, the scroll passed to the British Museum where it can be viewed today.

3. The Rollin papyrus from 1350 B.C. supplies examples of Egyptian bread accounts. It is housed in the Louvre in Paris.

4. The Harris papyrus was prepared by Rameses IV to document the accomplishments of his father, Rameses III. This papyrus dates from 1167 B.C.

By far, the most mathematically informative document in this collection is the Rhind papyrus.

This scroll was compiled by the scribes Ahmes under the patronage of King Hykos who ruled Egypt sometime within the period 1788-1580 B.C. It is a practical "handbook" of applied mathematics claiming to supply "a thorough study of all things, insight into all that exists, knowledge of all obscure secrets." Obviously, for the Egyptians, mathematics was felt to be obscure and mysterious. The contents of the Rhind papyrus reveal that its users employed an additive arithmetic, they multiplied by doubling and divided by halving, and depended on a system of unit fractions. They could solve simple linear equations and sum finite numerical progressions. In geometry, they could compute the areas of triangles, rectangles, trapezoids, and circles as well as find the volumes of cylinders and selected prisms. Their "formula" for the area of a circle with diameter, d, was $(8/9\ d)^2$ which supplies a value $\pi = 3.16$. Clearly, the contents of the Rhind papyrus reflect the administrative needs of an agriculturally based society. In fact, the scroll closes with a farmer's prayer "Catch the vermin and the mice, extinguish the noxious weeds; pray to the god Ra for heat, wind and high water."

21

Problems 1 to 6 of the Rhind Mathematical Papyrus

R. J. GILLINGS

*W*HEN THE RHIND MATHEMATICAL PAPYRUS was received by the British Museum in 1864, it was broken in some places, and portions of the brittle fragments were missing. Subsequently the papyrus roll was unrolled and mounted in two glass frames, the first portion, called the Recto (reading from right to left), having a jagged break because of these missing fragments. The whole papyrus was originally 18 feet long and a little over a foot in width, the second half, called the Verso, having a blank space about 10 feet long towards the left-hand end, upon which nothing is written.

Section I of the papyrus [1]* consists of an introduction, followed by the division of the number 2 by all the odd numbers from 3 to 101, 50 separate divisions in all, with the answers expressed as unit fractions except for the quite common use of the special fraction 2/3 [2] for which the scribe had a particular liking, so that he used it whenever it was possible. This takes up about one-third of the Recto.

Section II, which is relatively much smaller— occupying only about nine inches of the papyrus— consists of a short table giving the answers in unit fractions of the divisions of the numbers 1 to 9 by 10, followed by Problems 1 to 6 [1] on the division of certain numbers of loaves of bread equally among ten men, in which problems the scribe uses, for his answers, the values given in his table of reference,

and then proves arithmetically that they are correct. (This table and an explanatory version are shown on the following page.)

One observes that as soon as it is possible to introduce the fraction 2/3 into his table, the scribe does so, even though for 7 divided by 10 he could have used the simpler value 1/2 + 1/5, and for 8 divided by 10, 1/2 + 1/5 + 1/10, and for 9 divided by 10, 1/2 + 1/3 + 1/15 [3].

The problems on the division of loaves are available to us for examination and discussion as a result of the restoration of the missing fragments of the papyrus, which, while in the possession of the New York Historical Society, were recognized as such by the English archaeologist Professor Newberry in 1922, and were brought to the British Museum by Edwin Smith to allow them to be put in their appropriate places on the Recto.

In the text of the R. M. P. itself, no statement occurs giving the title, or describing the purpose, of the table. The scribe clearly regarded these as self-evident. The table occurs as if (for example) a student about to solve a group of problems which would require multiplications and divisions of numbers by 7 (let us say) were to write down first of all the 7 times table, for ready reference in the operations which were to follow.

The scribe now sets down the divisions of 1, 2, 6, 7, 8, and 9 loaves among 10 men, using the values he has already given in his table. These constitute Problems 1 to 6. He omits the division of 3, 4, and 5 loaves among 10 men, but in this

---

* Numbers in brackets refer to the notes at the end of the article.
Reprinted from *Mathematics Teacher* 56 (Jan., 1962): 61–69; with permission of the National Council of Teachers of Mathematics.

discussion of the Egyptian division of loaves we shall include them, following exactly the methods he uses for the others.

1 divided by 10 is		1/10			
2	"	" 10 "	1/5		
3	"	" 10 "	1/5	1/10	
4	"	" 10 "	1/3	1/15	
5	"	" 10 "	1/2		
6	"	" 10 "	1/2	1/10	
7	"	" 10 "	2/3	1/30	
8	"	" 10 "	2/3	1/10	1/30
9	"	" 10 "	2/3	1/5	1/30

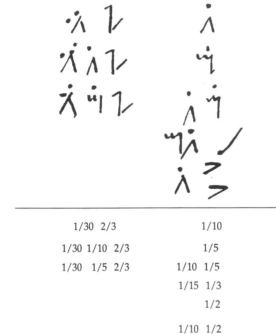

1/30 2/3	1/10
1/30 1/10 2/3	1/5
1/30 1/5 2/3	1/10 1/5
	1/15 1/3
	1/2
	1/10 1/2

Table of division of 1 to 9 by 10

Problem 1
Division of 1 loaf among 10 men.
[Each man receives 1/10.] (From his reference table.)
For proof multiply 1/10 by 10.
Do it thus:

[If]		1 [part is]	1/10
[then]	√	2 [parts are]	1/5
"		4 "	1/3 + 1/15

(From the Recto, 2 ÷ 5=   1/3+1/15.)
"   √ 8 "   2/3+1/10+1/30

(From the Recto, 2 ÷ 15 = 1/10 + 30.)
[Add the fractions on the lines with check marks (since 2 + 8 = 10): 1/5 + 2/3 + 1/10 + 1/30 = 1.]

Total 1 loaf which is correct.

Those portions enclosed within square brackets are not given by the scribe, but are included here to explain his methods. He does not prove that 1 divided by 10 is 1/10. His readers are supposed to know this by referring to his table which precedes his problems. Further, the steps in his proof that 1/10 multiplied by 10 is in fact one loaf may be verified by referring to his more elaborate table of "2 divided by all the odd numbers from 3 to 101," with which the R.M.P. begins [4]. Thus when it is required to multiply the fraction 1/5 by 2, it is only necessary to look in his table, for 2 divided by 5, to find the answer 1/3 + 1/15. And similarly for 1/15 multiplied by 2, he finds from his table that 2 divided by 15 is 1/10 + 1/30.

For the addition of the fractions, the standard method, used elsewhere in the R.M.P., was analogous to our modern concept of Least Common Multiple. Taken as parts of (say) the abundant number 30, 1/5 is 6, 2/3 is 20, 1/10 is 3, and 1/30 is 1, so that 6 + 20 + 3 + 1 = 30, and thus the whole.

We may fairly assume that the Egyptian loaves were long and regular in cross-section, either cylindrical or rectangular, so that the actual cutting up of a loaf into unit fractions (and of course 2/3) did not present real difficulties. The sign in hieroglyphics for a loaf is

which permits us to make this assumption. The division of one loaf among ten men would be done, therefore, as in Figure 1.

Problem 2
Division of 2 loaves among 10 men.
[Each man receives 1/5.] (From his reference table.)

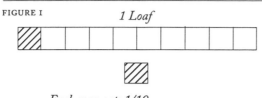

FIGURE 1   *1 Loaf*

Each man gets 1/10

For proof multiply 1/5 by 10.

Do it thus:

| | | [If] | 1 [part is] | 1/5 |
| | | [then] √ | 2 [parts are] | 1/3 + 1/15 |

(From the Recto, 2 ÷ 5 =   1/3 + 1/15.)

       "   4   "   2/3 + 1/10 + 1/30

(From the Recto, 2 ÷ 15 =   1/10 + 1/30.)

       "   √   8   "   1 1/3 + 1/5 + 1/15

[Add the fractions on the lines with check marks (since 2 + 8 = 10):

1/3 + 1/15 + 1 1/3 + 1/5 + 1/15 = 2.]

Total 2 loaves which is correct.

The steps in Problem 2 are practically the same as those for Problem 1, and the division of two loaves among ten men would be done therefore as in Figure 2.

If the scribe had included the division of three loaves among ten men as his next problem, he would have set it down as follows (if we adopt the same procedure shown in the other problems).

Division of 3 loaves among 10 men.
Each man receives 1/5 + 1/10. (From his reference table.)
For proof multiply 1/5 + 1/10 by 10.
Do it thus:

     If   1 part is   1/5 + 1/10
       then   √   2 parts are   1/3 + 1/15 + 1/5

(From the Recto, 2 ÷ 5 =   1/3 + 1/15.)

       then   4 parts are
2/3 + 1/10 + 1/30 + 1/3 + 1/15

(From the Recto, 2 ÷ 15 = 1/10 + 1/30.)

       then   √   8 parts are
1 1/3 + 1/5 + 1/15 + 2/3 + 1/10 + 1/30

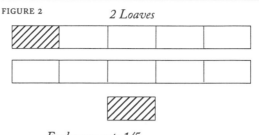

FIGURE 2   *2 Loaves*

Each man gets 1/5

[Add the fractions on the lines with the check marks (since 2 + 8 = 10):

1/3 + 1/15 + 1/5 + 1 1/3 + 1/5
+ 1/15 + 2/3 + 1/10 + 1/30 = 3.]

Total 3 loaves which is correct.

The addition of the row of fractions is merely (1 + 2), from the additions in Problems 1 and 2. The division of three loaves among ten men would therefore be as shown in Figure 3.

It now becomes clear to us why the scribe omitted this division. It is merely a repetition of his first two examples, whose answers were 1/10 and 1/5, giving his answer 1/10 + 1/5. A glance at Figures 1, 2, and 3 also shows this. Now in the case of the division of one and two loaves equally among ten men, there is no difference whatever between what an Egyptian would have done following the R.M.P. and what would be done in the present day. But in the case of the division of three loaves,

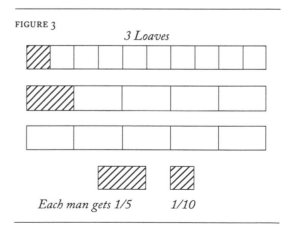

FIGURE 3

3 Loaves

Each man gets 1/5   *1/10*

FIGURE 4

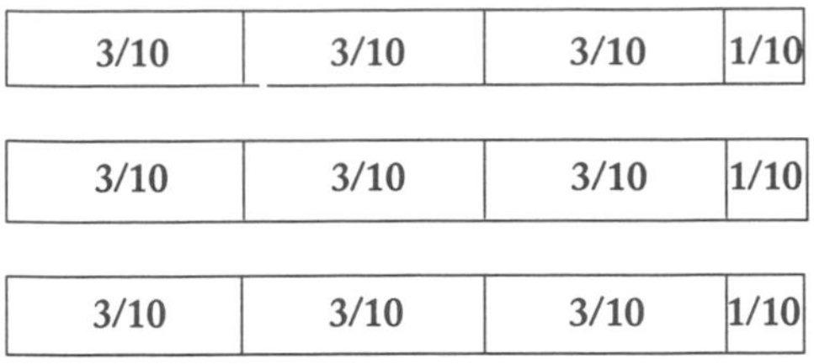

a modern division would result in nine men getting 3/10 of a loaf each in one whole piece, and the tenth man getting three smaller pieces each 1/10 of a loaf, as in Figure 4 .

The Egyptian foreman charged with the duty of sharing the food fairly and equally among his gang of ten men, would give each man exactly the same share both in quantity (i.e., size) and number of pieces, and thus to the uneducated and ignorant laborers, not only is justice done, but justice also *appears* to have been done. In our modern division it could be supposed that one man had received more than his share, because he had three pieces, to the others' one piece.

We proceed now to the next case, the division of 4 loaves among 10 men. It might have appeared as follows.

Division of 4 loaves among 10 men.
Each man receives 1/3 + 1/15. (From his reference table.)
For proof multiply 1/3 + 1/15 by 10.
Do it thus:
    If   1 part is 1/3 + 1/5
    then √ 2 parts are 2/3 + 1/10 + 1/30

(From the Recto, 2 ÷ 15 = 1/10 + 1/30.)

    then   4 parts are 1 1/3 + 1/5 + 1/15
    then √ 8 parts are 2 2/3 + 1/3 + 1/15 + 1/10
+ 1/30

(From the Recto, 2 ÷ 5 = 1/3 + 1/15.)
[Add the fractions on the lines with the check marks (since 2 + 8 = 10):

2/3 + 1/10 + 1/30 + 2 2/3 + 1/3
+ 1/5 + 1/10 + 1/30 = 4.]

    Total 4 loaves which is correct.

The addition of the long row of fractions to give 4 is much simpler than it appears at first glance. The scribe would observe that, fraction by fraction, the array is exactly twice the summation he has already made in Problem 2. Thus,

$$2 \times 1/3 = 2/3$$
$$2 \times 1/15 = 1/10 + 1/30$$
$$2 \times 1\ 1/3 = 2\ 2/3$$
$$2 \times 1/5 = 1/3 + 1/15$$
$$2 \times 1/15 = 1/10 + 1/30.$$

The division of four loaves among ten men is therefore as shown in Figure 5.

One again notices that not only is justice done in the Egyptian system, but justice also *appears* to be done. A modern division would result in eight men getting each one portion, namely 2/5 of a loaf, and two men getting each two smaller pieces of 1/5 of a loaf, an apparent inequity perhaps to the ignorant members of the working gang. This is shown in Figure 6.

Continuing we have:

Division of 5 loaves among 10 men.
Each man receives 1/2. (From his reference table.)
For proof multiply 1/2 by 10.
Do it thus:

FIGURE 5

4 Loaves

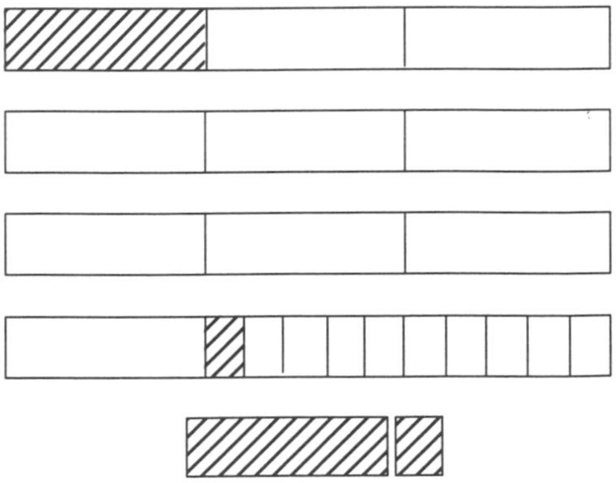

Each man gets 1/3   *1/15*

FIGURE 6

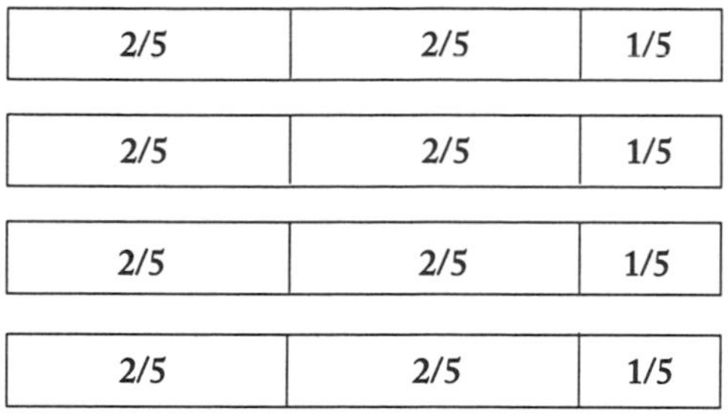

2/5	2/5	1/5
2/5	2/5	1/5
2/5	2/5	1/5
2/5	2/5	1/5

If		1 part is	1/2
then	√	2 parts are	1
then		4 parts are	2
then	√	8 parts are	4

[Add the numbers on the lines with check marks (since 2 + 8 = 10): 1 + 4 = 5.]

Total 5 loaves which is correct.

This is the simplest of all the divisions, and it was omitted by the scribe as were the divisions of three and four loaves. The actual cutting up of the five loaves is shown in Figure 7 and, of course, is the same it would be done today.

Problem 3

Division of 6 loaves among 10 men.
[Each man receives 1/2 + 1/10.] (From his reference table.)
For proof multiply 1/2 + 1/10 by 10.
Do it thus:

[If]		1 [part is]	1/2 + 1/10
[then]	√	2 [parts are]	1 + 1/5
"		4 "	2 + 1/3 + 1/15

(From the Recto, 2 ÷ 5 = 1/3 + 1/15.)

| " | √ | 8 " | 4 + 2/3 + 1/10 + 1/30 |

(From the Recto, 2 ÷ 15 = 1/10 + 1/30.)
[Add the fractions on the lines with check marks (since 2 + 8 = 10): 1 + 1/5 + 4 + 2/3 + 1/10 + 1 + 1/30 = 6.]

Total 6 loaves which is correct.

One notes, *en passant*, that the addition of the row of fractions in this problem is, except for the integers, exactly the same as that for Problem 1. The actual cutting up is shown in Figure 8.

This is perhaps the best example to show the efficiency of the Egyptian unit fraction system in the cutting up of loaves, for here again each man

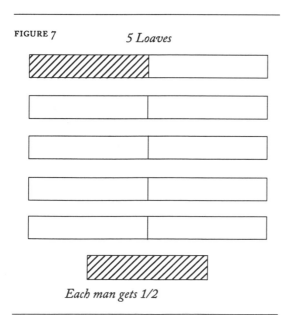

FIGURE 7   *5 Loaves*

Each man gets 1/2

FIGURE 8   *6 Loaves*

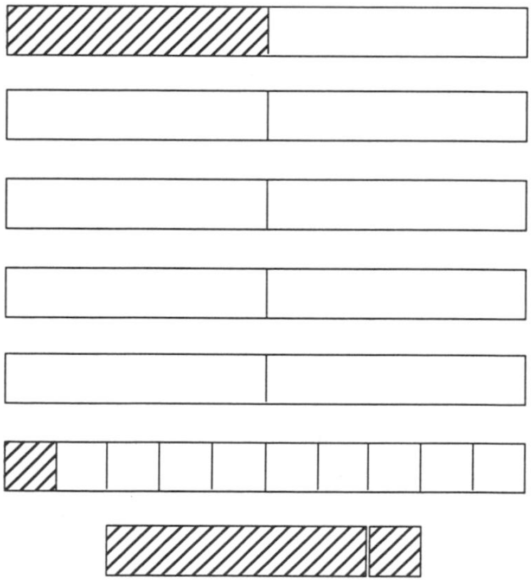

Each man gets 1/2   1/10

FIGURE 9

3/5	2/5	

3/5	2/5	

3/5	1/5	1/5

3/5	2/5	

3/5	1/5	1/5

3/5	2/5	

gets exactly the same portions, both in quantity and numbers of pieces of bread, whereas a modern division would very likely be as shown in Figure 9.

In this case an apparent inequity is even more apparent, since six men each receive a single piece which is 3/5 of a loaf, and four men receive two unequal pieces each, one 1/5 of a loaf and the other 2/5 of a loaf.

Problem 4
Division of 7 loaves among 10 men.
[Each man receives 2/3+1/30.] (From his reference table.)
For proof multiply 2/3+1/30 by 10.
Do it thus:

    [If] 1 [part is]   2/3 + 1/30
    [then] √ 2 [parts are]1 1/3 + 1/15
      "   4   "   2 2/3 + 1/10 + 1/30

(From the Recto, 2 ÷ 15 = 1/10 + 1/30.)

      "   √ 8   "   5-1/3 + 1/5 + 1/15

[Add the fractions on the lines with check marks.]

    Total 7 loaves which is correct.

On the last line, in the R.M.P., the scribe has written instead of the simple 5 1/3 + 1/5 + 1/15 (which together with the fractions 1 1/3 + 1/15 is exactly the same as the fractions he has already added in Problem 2) the values 5 1/2 + 1/10. It is not clear at this stage why he should have done this, for while 1/2 + 1/10 is equivalent to 1/3 + 1/5 + 1/15, the substitution does not make the summation to a total of 7 any easier, nor in fact does it make it any more difficult [5]. We shall see later why the scribe preferred to use this alternative form, but the change here made does not in any way alter the general thread of the working of the problem. In the division of seven loaves among ten men, the scribe comes to his first chance to use the fraction 2/3 for which he has such a great propensity [2]. And here it gets him into trouble, if we may use the phrase. His division results in seven men getting two pieces, 2/3 and 1/30, while three men get three pieces, two equal pieces 1/3 each and one piece 1/30. This is shown in Figure 10.

FIGURE 10

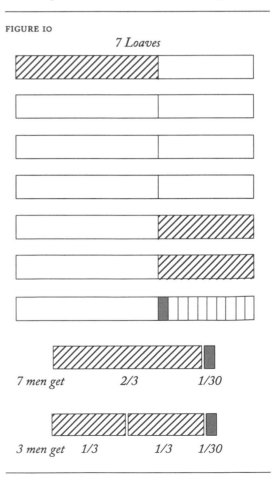

7 Loaves

7 men get   *2/3*   *1/30*

3 men get *1/3*   *1/3*   *1/30*

The scribe had at his disposal the simple division of each man getting 1/2 + 1/5, and the seven loaves could have been equally distributed with only 13 cuts instead of the 16 cuts required for the division which he recommends. It is just another illustration of the Egyptian's desire to use the nonunit fraction 2/3 wherever he possibly could. A modern division, the simplest I can think of, would give seven men each one portion of 7/10 of a loaf, and the remaining three men would each receive three pieces, 3/10 of a loaf twice, and a smaller piece, 1/10 of a loaf. This is just about as complicated as the scribe's division, but would require only nine cuts.

Problem 5
Division of 8 loaves among 10 men.
[Each man receives 2/3 + 1/10 + 1/30.] (From his reference table.)
For proof multiply 2/3 + 1/10 + 1/30 by 10.
Do it thus:

    [If]   1 [part is]   2/3 + 1/10 + 1/30

    [then] √ 2 [parts are] 1 + 1/2 + 1/10 (As in Problem 4 [6].)

      "   4   "   3 + 1/5
      "   √ 8   "   6 + 1/3 + 1/15

(From the Recto, 2 ÷ 5 = 1/3 + 1/15.)
[Add the fractions on the lines with check marks.]

    Total 8 loaves which is correct.

One notes that the summation of the fractions, here 1/2 + 1/10 + 1/3 + 1/15 = 1, is exactly the same as the summation of the fractions 1/3 + 1/15 + 1/2 + 1/10 = 1 in Problem 4, as the scribe wrote it. And now the efficacy of using this alternative form 1/2 + 1/10 instead of the conventional 1/3 + 1/5 + 1/15 becomes quite clear to us, and in the subsequent doubling, 3 + 1/5 is now much simpler that 2 + 2/3 + 1/3 + 1/15 + 1/10 + 1/30, and the next doubling is simplified as well. The division of eight loaves among ten men as the scribe directs is shown in Figure 11.

Here seven men get three distinct portions, 2/3, 1/10, 1/30, and three men get four distinct portions, 1/3 and 1/3, 1/10, 1/30, which appear to

FIGURE II

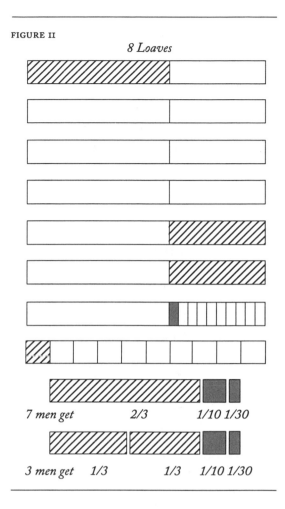

8 Loaves

7 men get   *2/3*   *1/10 1/30*

3 men get *1/3*   *1/3*   *1/10 1/30*

be more or less identical if the two separate 1/3 pieces are kept together, and so again justice *appears* to have been done. A modern cutting up of loaves would result in eight men getting each 4/5 of a loaf in one piece, while the remaining two men get each four pieces, each 1/5 of a loaf, which might be construed as being inequitable.

Problem 6
Division of 9 loaves among 10 men.
[Each man receives 2/3 + 1/5 + 1/30.] (From his reference table.)
For proof multiply 2/3 + 1/5 + 1/30 by 10.
Do it thus:

    [If]   1 [part is] 2/3 + 1/5 + 1/30
    [then] √ 2 [parts are]1 2/3 + 1/10 + 1/30

(i.e., 1 1/3 + 1/3 + 1/15 + 1/15 as before,
= 1 2/3 + 1/10 + 1/30 from 2 ÷ 5 in the Recto.)

"   4   "   3 + 1/2 + 1/10 [6]

"   √   8   "   7 1/5

[Add the fractions on the lines with check marks.]

    Total 9 loaves which is correct.

Again the use of the equivalent 1/2 + 1/10 instead of 1/3 + 1/5 + 1/15 leads to simpler answers, for 2/3 + 1/10 + 1/30 + 1/5 = 1, exactly as in Problem 1, and the scribe is to be commended on an elegant arithmetical solution.

The division would be performed as in Figure 12, and the apparent equity of the shares is observable if the two separate 1/3 pieces are placed together to make 2/3.

A modern distribution of nine loaves among ten men would result either in nine men receiving 2/10 of a loaf and the tenth man getting nine small pieces each 1/10 of a loaf, which would be clearly inequitable to the uneducated laborer, or in the ten men getting each 1/2 a loaf, and then eight of them receiving in addition 2/5 of a loaf while the remaining two men would get four small pieces each 1/10 of a loaf, which is even worse.

As in Problem 4, the scribe had at his disposal simpler divisions in terms of unit fractions, as for example 1/2 + 1/5 + 1/10 for 8 ÷ 10, and 1/2 + 1/3 + 1/15 for 9 ÷ 10, but speculations on these dissections would not serve us. The simple fact is that the Egyptian scribe always preferred to use 2/3 if he could, and then by halving his answer to get 1/3 of a quantity, "an arrangement," as Neugebauer remarks, which is "standard even if it seems perfectly absurd to us [7]."

Problems 1 to 6 of the R.M.P. have had considerable interest for students of Egyptian mathematics, although I know of none who has looked at the practical aspect of the actual cutting up of the loaves, at least in publication. I quote from *The Scientific American* of 1952, "The Rhind Papyrus," by James R. Newman, reprinted in *The World of Mathematics* (New York: Simon & Schuster, Inc., 1956), I, 173. Discussing the division of nine loaves among ten men, that is, Problem 6, Newman says:

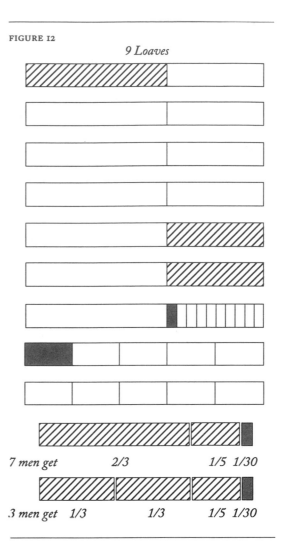

FIGURE 12

9 Loaves

7 men get   2/3   1/5   1/30

3 men get   1/3   1/3   1/5   1/30

The actual working of the problem is not given. If 10 men are to share 9 loaves, each man, says A'h-mosè, is to get 2/3 + 1/5 + 1/30 (i.e., 27/30) times 10 loaves (*sic.* [8]); but we have no idea how the figure for each share was arrived at. The answer to the problem (27/30, or 9/10) is given first and then verified, not explained. It may be, in truth, that the author had nothing to explain, that the problem was solved by trial and error—as, it has been suggested, the Egyptians solved all their mathematical problems.

But of course we know how he would have obtained the values he gives in his table! The first part of the Recto of the R.M.P. is filled with examples of how he would do it. Chace (R.M.P., Vol. 1, p. 23), for example, gives his methods for 2 and 7 divided by 10, but they are so simple from the scribe's point of view that he merely sets the answers down in order, without even a heading let alone any explanations. A'h-mosè neither here nor elsewhere indulged in unnecessary detail or verbosity. He was writing a mathematical text! Having "no idea how the figure for each share was arrived at," as Newman puts it, is so far from the truth that I will set down here how the scribe could have performed the divisions of the numbers 1 to 9 by 10. All that is necessary is to note that he performed division by continually multiplying the divisor until the dividend was reached, usually by doubling and halving, and of course taking 2/3 wherever possible [2].

I follow these by his abbreviated proofs as given in the R.M.P. (except, of course, those for 3, 4, and 5, which I reconstruct) so that one can see the order and regularity with which the Egyptian scribe sets down his work. It is a matter of wonderment to me that an Egyptian scribe could perform such arithmetical operations with the limited tools at his disposal.

1	10
√1/10	1
Ans. 1/10	1

1	10
√1/2	5
√1/10	1
Ans. 1/2 1/10	6

1	10
1/10	1
√1/5	2
Ans. 1/5	2

1	10
√2/3	6 2/3
√1/30	1/3
Ans. 2/3 1/30	7

1	10
√1/10	1
√1/5	2
Ans. 1/10 1/5	3

1	10
√ 2/3	6 2/3
√1/10	1
√1/30	1/3
Ans. 2/3 1/10 1/308	

1	10
2/3	6 2/3
√ 1/3	3 1/3
√1/15	2/3**
Ans. 1/3 1/15	4

1	10
√ 2/3	6 2/3
√1/5	2
√ 1/30	1/3
Ans. 2/3 1/5 1/30	9

1	10
√1/2	5
Ans. 1/2	5

1	1/10
√2	1/5
4	1/3 1/15
√8	2/3 1/10 1/30
	Total 1

1	1/2 1/10
√2	1 1/5
4	2 1/3 1/15
√8	4 2/3 1/10 1/30
	Total 6

2	1/5
2	1/3 1/15
4	2/3 1/10 1/30
√8	1 1/3 1/5 1/15
	Total 2

1	2/3 1/30
√2	1 1/3 1/15
4	2 2/3 1/10 1/30
√8	5 1/2 1/10
	Total 7

1	1/5 1/10
√2	1/2 1/10
4	1 1/5
√8	2 1/3 1/15
	Total 3

1	2/3 1/10 1/30
√2	1 1/2 1/10
4	3 1/5
√8	6 1/3 1/15
	Total 8

1	1/3 1/15
√2	2/3 1/10 1/30
4	1 1/2 1/10
√8	3 1/5
	Total 4

1	2/3 1/5 1/30
√2	1 2/3 1/10 1/30
4	3 1/2 1/10
√8	7 1/5
	Total 9

1	1/2
√2	1
4	2
√8	4
Total	5

SUMMARY OF THE SCRIBE'S ADDITION OF
UNIT FRACTIONS
From the Recto of the R.M.P.

2 ÷ 5 = 1/3 1/15
2 ÷ 15 = 1/10 1/30
2/3 1/5 1/10 1/30 = 1

**Written probably as 3 30 1/3, meaning that, since 3 × 10 = 30, then 1/30 × 10 = 1/3, and hence 1/15 × 10 = 2/3.

Problem 1
1 ÷ 10 = (2/3 1/5 1/10 1/30) = 1

Problem 2

$2 \div 10$	= 1 1/3 1/3 1/5 1/15 1/15
	= 1 2/3 1/5 (2÷15)
	= 1 (2/3 1/5 1/10 1/30) = 2
$3 \div 10$	= 1 1/3 2/3 1/3 1/5 1/5 1/10 1/15 1/15 1/30
	= 2 1/3 (2÷5)1/10 1/15 1/15 1/30
	= 2 1/3 (1/3 1/15) 1/10 1/15 1/15 1/30
	= 2 2/3 (1/15 1/15 1/15) 1/10 1/30
	= 2 (2/3 1/5 1/10 1/30) = 3
$4 \div 10$	= 2/2/3 2/3 1/3 (1/10 1/10) 1/15 (1/30 1/30)
	= (2 2/3 1/3) 2/3 1/5 1/15 1/15
	= 3 2/3 1/5 (2÷15)
	= 3 (2/3 1/5 1/10 1/30) = 4
$5 \div 10$ = 1 4 = 5	

Problem 3

$6 \div 10$ = 4 1 (2/3 1/5 1/10 1/30) = 6

Problem 4

$7 \div 10$	= 5 1 (1/3 1/3) 1/5 (1/15 1/15)
	= 6 2/3 1/5 (2 ÷ 15)
	= 6 (2/3 1/5 1/10 1/30) = 7

Problem 5

$8 \div 10$	= (6 1) 1/2 (1/3 1/10 1/15)
	= 7 1/2 1/2 = 8

Problem 6

$9 \div 10$	= 7 1 (2/3 1/5 1/10 1/30) = 9

A dispassionate view of this array of unit fractions, with its order and symmetry, and indeed its elegance and simplicity (if one considers the mathematical issues involved), leaves one with a feeling of hopeless admiration and wonderment at what was achieved by the Egyptyian scribe with the rudimentary arithmetical tools at his disposal.

NOTES

1. Following the nomenclature of Chace, Bull, Manning, *The Rhind Mathematical Papyrus* (Oberlin, Ohio: Mathematical Association of America, 1929).

2. See Gillings, "The Egyptian 2/3 table for fractions," *Australian Journal of Science*, XXII (December, 1959), 247.

3. It is convenient to use the plus sign at this stage. The Egyptian scribe had no such notation; mere juxtaposition indicated addition. I also use the signs = and ÷ which also do not have equivalents in the R.M.P., although in the Egyptian Leather Roll

❗❗❗

meaning "this is," is repeatedly used.

4. See Gillings, "The division of 2 by the odd numbers 3 to 101 from the Recto of the R.M.P. (B.M. 10058)," *Australian Journal of Science*, XVIII (October, 1955), 43-49.

5. The scribe is familiar with many alternatives evident throughout the R.M.P., and he uses the values $7 \div 10 = 1/2 + 1/5$ in Problem 54, while in his table, and in Problem 4, he uses $7 \div 10 = 2/3 + 1/30$.

6. Here again the scribe writes $1/2 + 1/10$ as equivalent to $1/3 + 1/5 + 1/15$, both arising from the multiplication of $2/3 + 1/10 + 1/30$ by 2. In Problem 4, the scribe does not have any further doubling, so that there is little advantage in using the shorter form there. Thus $2(1 \ 2/3 + 1/10 + 1/30) = 2 + 1 \ 1/3 + 1/5 + 1/15 = 3 + (1/3 + 1/5 + 1/15) = 3 + 1/2 + 1/10$.

7. O. Neugebauer, *The Exact Sciences in Antiquity* (Princeton, N.J.: Princeton University Press, 1952), p. 76.

8. An error. "Times 10 loaves," should read "of one loaf."

22

The Volume of a Truncated Pyramid in Ancient Egyptian Papyri

R. J. GILLINGS

$\mathcal{T}$HE MOSCOW MATHEMATICAL PAPYRUS, which was acquired by the Moscow Museum of Fine Arts from the Egyptologist Golenischeff in 1912, contains 25 problems of varying interest and importance. Problem No. 14 shows quite clearly that the Egyptian scribes were familiar with the formula,

$$V = \frac{h}{3}(a^2 + ab + b^2)$$

for the volume of a truncated pyramid (or frustrum), where h is the height and a and b are the edges of the square base and the square top, respectively.

While it has been generally accepted that the Egyptians were well acquainted with the formula for the volume of the complete square pyramid,

$$V = \frac{h}{3}a^2,$$

it has not been easy to establish how they were able to deduce the formula for the truncated pyramid, with the mathematics at their disposal, in its most elegant and far from obvious form, which, in the words of Gunn and Peet, "has not been improved on in 4000 years."

It would have been a simple enough operation to determine that a square pyramid made hollow had a capacity exactly one-third that of a cubical box of the same height, by merely pouring sand or water into them. By making them solid, using

Nile mud or clay, the same result could have been achieved by weighing. Not so simple is the method of dissection—that of cutting up, say, a cube—to obtain three equal pyramids with the same square base and height. The only convincing dissection method known to me is to make six congruent "Juel" pyramids which fit together to form a cube. These would have a vertical height exactly half of the edge of the resulting cube, for a "Juel" pyramid has its sides sloping at 45°. The dissectionist Harry Lindgren, of Canberra, Australia, has communicated to me a method by which three models of a truncated "Juel" pyramid can be dissected into three slabs, $a \times a$, $b \times b$, $a \times b$, each of a thickness h, which would establish the formula.

Earlier attempts to establish the formula as an Egyptian might have done it have been made by Gunn and Peet (with R. Engelbach), *Jr. Egypt. Arch.*, 1927; by Kurt Vogel, *Jr. Egypt. Arch.*, 1930; by P. Luckey, *Z. für Math. und Naturwiss.*, LXI, 1930; by W. R. Thomas, *Jr. Egypt.. Arch.*, 1931; and more recently by Van der Waerden, *Science Awakening*, 1954.

In the specific case which we are considering, No. 14 of the Moscow Papyrus, the scribe sets the problem of finding the volume of a truncated square pyramid of height 6, base edge 4, and top edge 2 (see Fig. 1). The solution is correctly given as,

$$V = \frac{6}{3}[4^2 + 4 \cdot 2 + 2^2]$$
$$= 2[16 + 8 + 4]$$
$$= 2 \cdot 28$$
$$= 56.$$

Reprinted from *Mathematics Teacher* 57 (Dec., 1964): 552–55; with permission of the National Council of Teachers of Mathematics.

FIGURE I

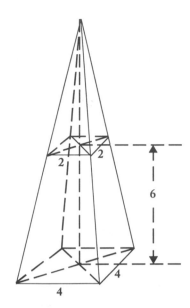

FIGURE 2

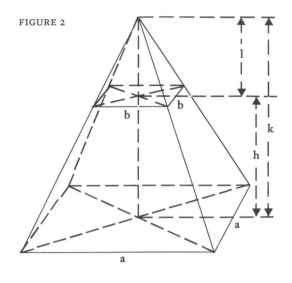

Assuming a knowledge of the formula for the volume of a pyramid, as one-third of the area of the base times the height, the question is, how may the scribe have derived the formula,

$$V = \frac{h}{3}\,[a^2 + ab + b^2],$$

keeping in mind that a knowledge of the elementary algebra involved in the identities

$$a^3 - b^3 = (a - b)(a^2 + ab + b^2),$$
$$a^2 - b^2 = (a - b)(a + b),$$

is nowhere attested in the extant Egyptian papyri; therefore, they may not be utilized.

The volume of the frustrum is the difference between the volumes of the original pyramid and the smaller pyramid cut off from the top. Then (see Fig. 2)

$$V = \frac{1}{3}\,a^2k - \frac{1}{3}\,b^2l$$
$$= \frac{1}{3}\,a^2(h + l) - \frac{1}{3}\,b^2l$$
$$= \frac{1}{3}\,a^2h + \frac{1}{3}\,a^2l - \frac{1}{3}\,b^2l. \quad (1)$$

We take first the special case which the scribe gives in No. 14 of the Moscow Papyrus, where $a = 2b$ and hence, by simple geometry, $l = h$. We then have, from (1),

$$V = \frac{1}{3}\,a^2h + \frac{1}{3}\,a^2h - \frac{1}{3}\,b^2h$$
$$= \frac{h}{3}\,[a^2 + a^2 - b^2]. \quad (2)$$

We evaluate the last two terms in the brackets, $a^2 - b^2$, which are the square bases of the original pyramid and the smaller pyramid cut from the top, by cutting the smaller square from the larger and examining the area left (see Fig. 3). Then the area

$$a^2 - b^2 = ab + b^2,$$

so that, from (2),

$$V = \frac{h}{3}\,[a^2 + ab + b^2].$$

Now consider the case in which $a = 3b$ and, again by simple geometry, $l = {}^1\!/_2h$. We then have, from (1),

$$V = \frac{1}{3}\,a^2h + \frac{1}{3}\,a^2\,\frac{1}{2}\,h - \frac{1}{3}\,b^2\,\frac{1}{2}\,h$$

FIGURE 3

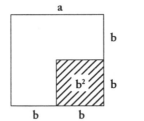
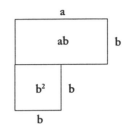

$$= \frac{h}{3} \left[a^2 + \frac{1}{2} (a^2 - b^2) \right]. \qquad (3)$$

We evaluate the last two terms in the brackets, which equal half the difference of the top and bottom of the truncated pyramid, by cutting the smaller square from the larger, and again examining the remaining area (see Fig. 4). Then the area

$$a^2 - b^2 = 2ab + 2b^2,$$

so that

$$\frac{1}{2} (a^2 - b^2) = ab + b^2,$$

and, from (3),

$$V = \frac{h}{3} \left[a^2 + ab + b^2 \right]$$

as before.

Consider further the case in which $a = 4b$ and hence, as before, $l = \frac{1}{3}h$.

We have, from (1),

$$V = \frac{1}{3} a^2 h + \frac{1}{3} a^2 \frac{1}{3} h - \frac{1}{3} b^2 \frac{1}{3} h$$

$$= \frac{h}{3} \left[a^2 + \frac{1}{3} (a^2 - b^2) \right]. \qquad (4)$$

We again evaluate the last two terms in the brackets, which equal one-third of the difference of the top and bottom of the truncated pyramid. We do this again by cutting the smaller square from the larger and examining the remaining area by elementary geometry (see Fig. 5). Then the area

$$a^2 - b^2 = 3ab + 3b^2,$$

so that

$$\frac{1}{3} (a^2 - b^2) = ab + b^2,$$

and, from (4),

$$V = \frac{h}{3} \left[a^2 + ab + b^2 \right].$$

Clearly, this process can be continued for all cases where $a = nb$, where n has the integral values $2, 3, 4, 5, \cdots$. Let us therefore consider a fractional value for n and assume $a = \frac{3}{2}b$ and hence, as before, $l = 2h$. We have, from (1),

$$V = \frac{1}{3} a^2 h + \frac{1}{3} a^2 2h - \frac{1}{3} b^2 2h$$

$$= \frac{h}{3} \left[a^2 + 2a^2 - 2b^2 \right]$$

FIGURE 4

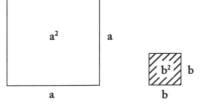
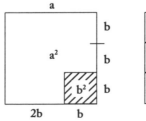
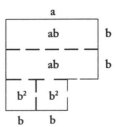

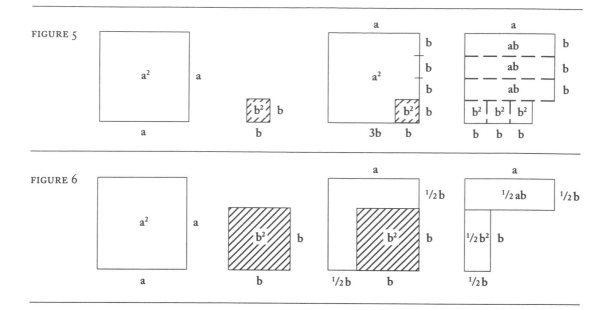

FIGURE 5

FIGURE 6

$$= \frac{h}{3} \left[a^2 + 2 \left(a^2 - b^2 \right) \right] . \qquad (5)$$

We evaluate the last two terms in the brackets, $2(a^2 - b^2)$, which in this case is twice the difference of the areas of the top and bottom of the truncated pyramid. By cutting the smaller square from the larger, we evaluate this by simple geometry (see Fig. 6). Then the area

$$a^2 - b^2 = \frac{1}{2} ab + \frac{1}{2} b^2$$

so that

$$2(a^2 - b^2) = ab + b^2,$$

and, from (5),

$$V = \frac{h}{3} \left[a^2 + ab + b^2 \right].$$

The Egyptian scribe is now entitled to conclude inductively that, since the formula holds for the fractions $1/2$, $1/3$, $1/4$, $\cdots$, and also for his familiar $2/3$ fraction, it therefore holds in all possible cases.

23

On the Practicality of the Rule of False Position

HOWARD EVES

IN THE RHIND (or Ahmes) papyrus, which dates back to at least 1650 B.C., we find the following problem: "A quantity, its $2/3$, its $1/2$, and its $1/7$, added together, become 33. What is the quantity?" Today, with our algebraic symbolism, we would probably solve this problem by letting x denote the sought quantity. Then we have

$$x + 2x/3 + x/2 + x/7 = 33$$

which, upon solving, yields $x = 1386/97$.

But how would an Egyptian of 1650 B.C. solve this problem? The procedure for solving linear equations employed in those early days was a form of the method later to become known as the *rule of false position*. The method, in brief, is to assume any convenient value for the desired quantity, calculate the sum of the required parts, and then "blow up" the assumed value in the ratio that will convert the calculated sum into the given sum. In the problem of the Rhind papyrus, for example, we might assume the desired quantity to be 42 (this is convenient because 42 is divisible by 3, 2, and 7). Then we find $42 + (2/3)42 + (1/2)42 + (1/7)42 = 42 + 28 + 21 + 6 = 97$, instead of 33. Since 97 must be multiplied by $33/97$ to give the required 33, the correct value of the quantity must be $42(33/97)$, or 1386/97. It is stimulating to contemplate that from a pure guess of the answer we are able to obtain the true answer.

Reprinted from *Mathematics Teacher* 51 (Dec., 1958): 606–8; with permission of the National Council of Teachers of Mathematics.

It must be confessed that there is a certain element of beauty and simplicity in the primitive rule of false position, and it is no wonder that this rule, or some modification of it, was used even as late as the nineteenth century. Without a suitable algebraic symbolism, what else could one do? Actually, the method has both pedagogical and practical value, for not only do students find it interesting, but there are instances where the procedure has decided advantages. Let us consider the following examples, the first three of which are taken from the *Greek Anthology* (ca. A.D. 500).

1. After staining the holy chaplet of faireyed Justice that I might see thee, all-subduing gold, grow so much, I have nothing; for I gave 40 talents under evil auspices to friends in vain, while O ye varied mischances of men, I see my enemies in possession of the half, the third, and the eighth of my fortune (How many "talents" did the unfortunate man once possess?)

Suppose the man's original fortune had been 24 talents. Then $24 - 24/2 - 24/3 - 24/8 = 1$, instead of 40. Therefore the man once possessed $(24)(40)$, or 960, talents.

2. Brickmaker, I am in a hurry to erect this house. Today is cloudless, and I do not require many more bricks, for I have all I want but three hundred. Thou alone in one day couldst make as many, but thy son left off working when he had finished two hundred, and thy son-in-law when he had made two hundred and fifty. Working all together in how many days can you make them?

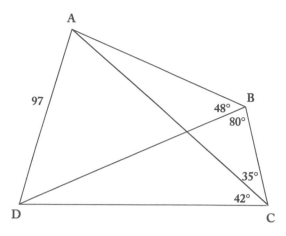

FIGURE 1

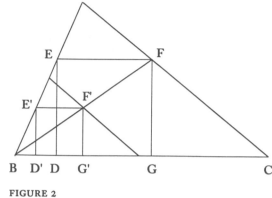

FIGURE 2

Suppose the required time is ⅕ of a day. Then, in this time, the three men can make 60 + 40 + 50, or 150 bricks. Since 300 bricks are needed, the answer is ⅖ of a day.

3.   I am a brazen lion; my spouts are my two eyes, my mouth, and the flat of my right foot. My right eye fills a jar in two days [1 day = 12 hours], my left eye in three, and my foot in four. My mouth is capable of filling it in six hours. Tell me how long all four together will take to fill it.

If the time were 12 days, then the right eye would fill 6 jars, the left eye 4 jars, the foot 3 jars, and the mouth 24 jars, whence, in 12 days, all spouts together would fill 37 jars. Therefore, it takes ¹²/₃₇ of a day, or ¹⁴⁴/₃₇ hours, for all spouts together to fill one jar.

4.   Find the length of BC in the quadrilateral pictured in Figure 1.

Apply the rule of false position by taking BC to be any convenient length, say 1 unit. Then, by the

law of sines, we find CD = 2.52 and AC = 2.69. Next, by the law of cosines, we find AD = 1.88, instead of the given 97. It follows that BC = 1($^{97}/_{1.88}$) = 51.7. This is a highly practicable procedure for solving this problem, and is actually the method a surveyor would probably use.

5.   In the study of geometrical constructions there is a counterpart of the rule of false position, generally known as the *method of similitude*. The method lies in constructing a figure similar to the one desired, and then, by the use of proportion, "blowing it up" to proper size. Suppose, for example, we wish to inscribe a square $DEFG$ in a given triangle ABC so that side DG of the square lies along the base BC of the triangle. (See Figure 2.)

Choose any point D' on base BC and draw the square $D'E'F'G'$, where E' is on BA and G' is on BC. Now, using B as a center of similitude, blow up the square $D'E'F'G'$ to the proper size by projecting F' into F on AC, etc.

24

The Dawn of Demonstrative Geometry

NATHAN ALTSHILLER-COURT

WHEN THE GREEKS APPEARED in the history of culture, ca. 1000 B.C., Egypt, Babylon, and some other neighboring countries had already behind them more than a millennium of civilized existence. During that period of time, the Egyptians and the Babylonians succeeded in accumulating a considerable amount of learning, including mathematics. This knowledge was taken over by the first, largely mythical, Greek mathematicians, Thales of Miletus and Pythagoras of Croton (ca. 600 B.C.).

That both Thales and Pythagoras learned their mathematics from the Egyptian priests is explicitly and unequivocally attested to by Greek authors, including Aristotle. On the other hand, it is beyond a doubt that the geometry the priests had to offer to those eager foreigners was of an empirical and practical nature, for such was the status of geometrical learning in Egypt at that time—the same as it had been already for about a thousand years.

Thales is credited with having been the first mathematician to use logical proofs in support of his geometrical propositions. He is also supposed to have arranged his propositions in a systematic, rational order. These assertions are based on a passage which the philosopher Proclus (A.D. 410–485) included in his commentary upon the first book of Euclid's *Elements*. Proclus states that he

Reprinted from *Mathematics Teacher* 57 (Mar., 1964): 163–66; with permission of the National Council of Teachers of Mathematics.

copied the passage in question from the *History of Mathematics* by Eudemus (third century B.C.), a work that has not come down to us. Van der Waerden vigorously supports the credibility of Eudemus' statement [3; pp. 87–90].

We are thus led to the conclusion that the Egyptians, Babylonians, and some others, under the pressure of practical needs, accumulated and made use of a wealth of empirical propositions, some of them correct, others wrong. It was left to the Greeks to invent demonstrative geometry and transform this subject into a science—the first deductive science.

The cleavage between the two stages in the history of geometry is thus so sharp and so abrupt that it seems to invite further scrutiny. Is it credible that the Egyptian geometers had no incentive in their work to have recourse to some logical thinking? True it is that the most valuable and the most extensive Egyptian mathematical writings now available, namely the Moscow and the London (Ahmes) papyri, do not include any attempts at proofs. However, it should be noted that both Egyptian documents happen to be schoolbooks containing exercises for pupils. The task before those apprentices was to follow procedures by which to obtain a particular result. By their nature those textbooks might not have been the place for theories or cogitation.

We have thus no documentary information regarding the question. We are reduced to our own resources, if any light is to be shed upon the subject. For want of anything better, we may attempt to imagine an Egyptian geometer at work.

Let us consider a concrete example. Suppose we have two rulers of equal length covering each other. If we slide one of the rulers along the other so as to uncover a part of each ruler, will those two uncovered parts be equal [1; p. 78]? For the sake of argument, let us say that a pupil in an Egyptian school of scribes would have to resort to the empirical method of direct measurement in order to answer our question. But it would be fair to grant that a teacher in such a place of learning may have had enough experience with this kind of situation to know the answer sought without any further experimental help. He may not have formulated his experimental knowledge into the succinct form of Euclid's third axiom, "If equals are taken from equals the remainders are equal," but he knows its content and notices that the case under consideration comes under this "principle." But that is reasoning, is it not? To be sure, it is very simple reasoning. It is important to notice, in passing, that our hypothetical teacher is well rewarded for this trick of his. This indirect method saved him all the time and all the effort his pupil had to spend to find the same result directly.

The temptation to get something done with the least expenditure of time and effort may have led our teacher to resort to the same trick on other occasions.

In his profession, opportunities of this kind were of frequent occurrence. The teacher knew the isosceles triangle. He came across them, say, when drawing the diagonals in a rectangle—an important figure when dealing with areas. Now our teacher, the experienced rope-stretcher that he was, must have encountered many situations where he had to split an isosceles triangle into two pieces along its middle line, that is, along the bisector of the angle at the vertex opposite the base. He may have been surprised when measuring the areas of the two triangles, by whatever methods he had at his disposal, to find that the areas were very nearly equal. He may have been impressed by this, and perhaps a little puzzled as well. If he were, could it

not have occurred to him, say, on a bright morning, after a good night's rest, that one half of his isosceles triangle when swung around that bisector would just exactly cover the other side? He certainly was very close to this idea, and if it did not occur to him, it may have occurred to one of his colleagues, or to one of his successors.

And when that event did happen, when this intellectual "breakthrough" did take place, the perspicacious and lucky teacher of the school of scribes may also have noticed that the coincidence of the two parts of the isosceles triangle show that: "The bisector of the angle at the vertex of an isosceles triangle also bisects the base of the triangle."

This "windfall" should have been to the teacher's liking. It should not be surprising to us if it whetted his appetite for more of the same. And he was close on the trail of another such discovery. Very little effort and not much astuteness on his part would be needed for him to notice that the figure before him shows that: "The base angles of an isosceles triangle are equal."

The French have a very clever saying, *ce n'est que le premier pas qui coûte* (only the first step is costly). If the teacher was in possession of the secret regarding the base angles of an isosceles triangle, what did he do when he was confronted with an equilateral triangle *ABC?* Did he try to ascertain whether the angles of that triangle are equal using the "shortcut method," or did he resort to his empirical way of direct measurement? Even if you are not willing to grant that he could have used the "shortcut method," you will concede that after having arrived at his result empirically, he may have noticed that by considering two isosceles triangles, each having the same angle *A* for one of the base angles, it became at once apparent that *B* and *C* are equal to *A* and therefore *B = C*, by an implicit use of Euclid's first axiom. Moreover, that experience would once more warn our Egyptian teacher that it is better to look for a shortcut *before* he jumps, not *after*.

Our Egyptian geometer would find it quite advantageous to heed that warning. The axioms of Euclid are generalizations of simple concrete facts.

The geometer had ample opportunities to meet with particular cases of each axiom in his agricultural, architectural, and many other undertakings. Repeated occurrence of such cases of the same axiom would make him aware, however dimly, of the general principle. This "insight" would enable him to find the axiom in places familiar to him, but where he had not noticed it before. For instance, our Egyptian had known empirically that if

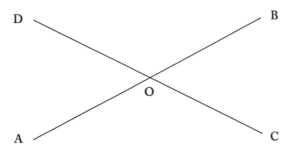

two lines *AOB* and *COD* cross at *O*, the two angles *AOC* and *BOD* are equal. Now when he reached the point that enabled him to answer the question regarding the two rulers we considered above, the Egyptian may, perhaps inadvertently, have noticed that the angle *BOC* is a common part of the two straight angles *AOB* and *COD*, hence the remaining parts of those two straight angles, namely *AOC, BOD,* are equal.

Encouraged by discoveries of this kind, our geometer would then be looking out for occasions to find answers to new questions by the application of the axiom.

As has already been pointed out, we find no logical arguments in the mathematical writings of the Egyptians which have come down to us, but this cannot be taken as conclusive proof that geometrical demonstrations were unknown to the builders of the pyramids. In Egypt, geometrical learning, and, for that matter, all other learning, was a monopoly of the priesthood and was jealously guarded from all outsiders. Practical geometry, by the very nature of things, had to be taught to a relatively large number of scribes. But theoretical geometry, as much or as little as there was

of it, was considered a privilege of the "inner sanctum." If any of this "higher" learning was ever committed to writing, copies of that papyrus were very few in number. The chance that such a work could come down to us would be very slim indeed.

On the other hand, there are some indications that geometrical reasoning, or at least some rudimentary beginning of it, was not foreign to the priests of Egypt. In the first place, some of the results obtained by them could not possibly have been arrived at by the empirical method alone. The most striking and shining example of this kind is the formula,

$$V = \frac{h\,(a^2 + ab + b^2)}{3},$$

for the volume of a truncated pyramid. Moreover, the greatest of the Greek physical philosophers, Democritus (ca. 410 B.C.), relates that he had traveled in many lands, talked to many learned men, but with regard to combining lines with demonstrations, no one of them could best him, "not even those who in Egypt are known as rope-stretchers." Kolman quite judiciously points out that this is a direct testimony to the effect that the Egyptians used logic in connection with "lines" [2; pp. 67–70].

Van der Waerden remarks that the Egyptian priests could decipher for Thales the rules of computing, but the train of thought underlying them was no longer known [3; p. 89].

After the practitioners of empirical geometry reached a certain level of erudition, the desire to make their work more efficient and to save themselves time and effort prompted them to have recourse to logical reasoning. They were amply rewarded for their efforts. After having acquired some skill in handling this new "shortcut method" they discovered, to their own amazement, that this new trick not only simplified their methods of solving their old problem, but enabled them to solve problems that their empirical procedures could not tackle at all, like, say, the determination of the distance from the shore of a ship out at sea.

If the considerations presented above be accepted as plausible, then the traditional view of the abrupt passage from the empirical to the demonstrative stage in the history of geometry becomes untenable. Rather, one assumes there was a gradual transition from one to the other. When enough skill and erudition had been acquired in handling logical proofs, geometry became an object of study in its own right, and the first deductive science was created. This view does not in any way belittle the great achievement of ancient Greece.

REFERENCES

1. Court, N. A., *Mathematics in Fun and in Earnest.* New York: The Dial Press, 1958.

2. Kolman, E., *History of Mathematics in Antiquity.* Moscow: 1961 (in Russian).

3. Van der Waerden, B. L., *Science Awakening.* Groningen, Holland: P. Noordhoff, 1954.

HISTORICAL EXHIBIT 3.2

A Chronology of Greek Mathematicians

Mathematical traditions in early Greece can be traced to particular mathematicians. A listing of some of these mathematicians and their principal work is given below:

Period	Mathematician	Principal Accomplishment
622 – 547 B.C.	Thales of Miletus	Introduced deductive geometry into Greece
585 – 501 B.C.	Pythagoras of Samos	Advocated a philosophy based on geometric and numerical harmony; discovery of incommensurable magnitudes
ca. 470 B.C.	Theodorus of Cyrene	Proved irrationality of square roots of nonsquare integers from 3 to 17
ca. 450 B.C.	Zeno of Elea	Revealed paradoxes concerning infinite processes
460 – 380 B.C.	Hippocrates of Chios	Involved with quadrature of circle; theory of lunes
ca. 420 B.C.	Hippias of Elis	Worked on the "Three Classical Problems"; developed the quadratrix
408 - 355 B.C.	Eudoxus of Enidos	Developed a theory of proportion to accommodate dilemma of incommensurables
323 – 285 B.C.	Euclid	Compiled the *Elements;* organized mathematics as a deductive system
287 – 212 B.C.	Archimedes	Employed a method of exhaustion to approximate infinite limiting processes
262 – 190 B.C.	Apollonius of Perga	Formulated a theory of conic sections
ca. 230 B.C.	Eratosthenes of Cyrene	Calculated the circumference of the earth
ca. A.D. 75	Heron of Alexandria	Developed formula for area of a triangle based on the measure of its sides
A.D. 85 – 160	Claudius Ptolemy	Devised epicyclic model for planetary motion
ca. A.D. 250	Diophantus	Worked with indeterminant equations

The Odyssey of Reason

BERNARD VITRAC

*I*T IS SCARCELY necessary to recall the importance of the role played by the mathematics of ancient Greece in the development of this discipline in the West. The very words "mathematics" and "mathematician," or their equivalents in most European languages, are derived from the Greek word meaning "to know" or "to learn." Before the classical era, however, when it took on the specialized meaning that it has today, the Greek word *mathema* meant "that which is taught," in other words all branches of knowledge.

The Teaching of Mathematics

We know very little about the teaching of mathematics in ancient Greece. It seems that some philosophical schools played a major part in the training of mathematicians, at a time when intellectual specialization was still the exception.

During the classical era there is reference to the existence of "scientific" schools, such as those at Chios and Cyzicus. However, we do not know whether these schools offered general or specialized education, or whether they were anything more than groups of scholars gathered around a famous teacher.

As with medicine—for the existence of schools of which there is earlier and more solid evidence—it seems that family background may have had an influence on professional training. We have few biographical details about the early mathematicians, but we do know that Archimedes was the son of an astronomer, that Hypsicles' father was a mathematician, that the geometers Menaechmus and Dinostratus were brothers and that Hypatia, the only Greek woman mathematician about whom we know anything much, was the daughter of the mathematician Theon of Alexandria.

In the centralized states of the Near East of ancient times, the need to train a class of scribes and functionaries was recognized at an early stage. The Greece of the classical era, however, consisted of small, independent city-states, continually warring among themselves, or of loose, clan-like associations, and had no need of organized educational systems such as those of Egypt, Babylon, or China.

Although trade, land surveying, and navigation demanded a certain minimum knowledge of mathematics, and although elementary calculation was taught in schools, the Greek city-state paid scant attention to the intellectual and technological education of children and youths. Schools, some of which were to achieve considerable renown, were the fruit of private initiatives. The two great Athenian masters of the beginning of the fourth century B.C., Isocrates and Plato, each founded his own educational establishment—Isocrates a school of rhetoric and Plato a school of philosophy.

Both of them regarded mathematics as an indispensable tool for intellectual development and valued the "mental gymnastics" and concentration this discipline required. But their approaches to mathematics differed. Isocrates thought that mathematics, like the adversarial debates that young people were so fond of, should develop "well-

Edited from *UNESCO Courier* (Nov., 1989): 29–35.

formed minds," even if the subject matter was of little value to the citizen, who, ideally, would devote himself to political life. Plato, however, whilst recognizing its propaedeutic value, saw mathematics as a preliminary to the study of philosophy—in other words, of Platonic idealism—and also as a method of selection, since the mathematics and the philosophy he taught together constituted a form of intellectual asceticism essential to his project for political reform.

The third and second centuries B.C. witnessed a considerable development in the mathematical sciences. Most of the works that have come down from this period are those of mathematicians more or less closely linked to Alexandria, the capital of the Greek Ptolemaic dynasty which ruled Egypt from 306 to 31 B.C. It is known that the Ptolemies adopted a broader policy of state patronage—previously restricted to a few individuals, often poets—by creating a number of institutions, the best known of which were the Alexandria Library and Museum. These new institutions undoubtedly gave an impetus to literary studies. That they had an effect on scientific development is probable but less certain, but the favourable climate they created could only be beneficial to it.

However, we do not even know if the great scholars of the time, whose presence in Alexandria is well documented—Herophilus of Chalcedony, Euclid, Strato of Lampsacus, Aristarchus of Samos, Eratosthenes of Cyrene, Apollonius of Perga—taught their disciples or gave lessons or even lectures, whether under the auspices of the Museum or privately. Indeed, it was only in the time of the Roman Empire that the Museum was to operate as a university, to be copied in this at Ephesus, Athens, Smyrna, and Aegina.

Alongside pure mathematics, in the Greek tradition, there was also a corpus of mathematical texts that might be termed "calculative"—similar to those to be found in Egyptian, Babylonian, or Chinese mathematical writings. For example, a body of comparatively late mathematical texts, attributed to Hero of Alexandria, was compiled and used right up to the Byzantine era, probably for the training of technicians. As in Babylonian and Egyptian texts, the problems set refer explicitly to a factual situation, even if this is merely a teaching device.

Nothing like this is to be found in the classic treatises of Euclid, Archimedes, or Apollonius, who showed scant regard for practical applications. Euclid's exposition of the theory of numbers even manages to avoid reference to numerical examples. The works that have survived seem to indicate a strict separation between pure research and practical applications. Yet even though the two are so clearly distinguished, the same authors gave equal attention to both "pure" and "applied" mathematics.

That section of Greek mathematics which, by convention, we have designated as "pure" had four basic characteristics:

Deductive presentation: the classic treatises, such as Euclid's *Elements,* are organized on deductive lines. The results are proved by demonstration, either from results previously achieved, or on the basis of principles established from the start. This might be described as a semi-axiomatic approach, which emphasized the logical, ineluctable aspect of mathematics. It should, however, be noted that this is often difficult to dissociate from the rhetorical aspect, which had the advantage of compelling the attention of the student, aimed at psychological and pedagogical effectiveness, and concentrated on the necessary objective structure of reasoning.

Geometrical orientation: even when they were concerned with the theory of numbers, statics or astronomy, those treatises that were demonstrative in style were basically geometric in their approach. Ancient mathematicians had introduced various symbols to designate numbers and fractions and also employed certain abbreviations, although these did not constitute a complete set of algebraic notations such as we use today.

However, it was in the use of geometrical figures that the Greeks went farthest in their experimentation with the use of "representative" symbols. The possibility of breaking figures up into elements, of establishing rules of construction that could be quoted, and the revelation of properties which seemed already to be "present" in the figures, all these features were perfectly suited to the method of deductive exposition.

The ideal of disinterested science: the love of knowledge itself was the driving force behind the study of mathematics.

Mathematics and philosophy: the development of mathematics was contemporaneous with the development of philosophy.

Philosophers and Mathematics

These developments in mathematics were paralleled by the opening of a methodological and ideological debate on the sciences. The classification of the sciences proposed by the Greek astronomer-mathematician Geminus (see box on next page) affords a good example. His classification assumed the existence of a body of scientific knowledge already fairly well developed and very varied. It adopted a differentiation between the sciences which favoured abstract study, which it detached from possible practical applications.

According to Aristotle, mathematics was the study of properties that could be "abstracted" from the objects of the physical world. Furthermore, like all the demonstrative sciences, it was based on principles, so that one science could presuppose another and, in Aristotle's words: "be subordinate to another." Optics for example, was "subordinate" to geometry. Thus there was a logical hierarchy of the sciences. This had to be distinguished from the division, commonly stressed by Greek authors, between "practical" and "disinterested" mathematics. Only the latter, according to Aristotle, was worthy of inclusion in a liberal education. "To be free" was an end in itself.

The art of embellishment was superior to practical technology, but disinterested science, an end in itself, was the supreme activity. For Plato, the mathematics developed by the Barbarians, prestigious though their civilizations might have been, ranked as mere arts, since they were encumbered by the constraints of necessity. Greek philosophers thus based their argument on a mixture of methodological and ideological considerations.

Though treatises on optics and astronomy adopted the style of geometrical exposition, the deductive method being particularly suited to the elimination of anything that smacked of the "tangible" and the "practical" (see box on next page), it is not easy to see how Greek mathematicians managed to identify themselves with this way of looking at their activities. Moreover, though it is tempting to do so, we should refrain from confusing the modern distinction between "pure" and "applied" mathematics with the ancient division between "intelligible" and "tangible," since the concepts do not coincide.

Evocation of the ideal of "disinterested science" was for the advancement of mathematics. A distinction has to be made between external forces and what might be described as internal influences. Among the former, an important role was played by optics and astronomy, which we classify as physics but which the ancients included in mathematics. To these should be added statics, the science of equilibria.

What is known of the internal influences? Perhaps they are to be found in the prefaces with which, from Archimedes onwards, author/mathematicians seem to have introduced their works. What emerges from these prefaces is that far from being the product of psychological characteristics peculiar to the Greek mind, disinterested research presupposes the existence of a community of mathematicians all abiding by certain standards.

First and foremost, these mathematicians felt under no obligation to justify the fact that they were practitioners of science for its own sake; that went without saying. At the most they might acknowledge that they had opted for mathematics rather than for "physics" or "theology" because it was a more certain and rigorous discipline that was "stable" (unlike physics), yet nevertheless "accessible" (unlike theology).

During the Hellenistic era, mathematicians formed an "international" community whose members were scattered around the shores of the Mediterranean (Greece, Asia Minor, Egypt, and Sicily). They maintained personal contacts, either by paying each other visits or by circulating their latest works. Above all, they would set each other problems, solve those that were sent to them, or criticize solutions submitted by others which they considered to be flawed. By this means some of them acquired a commonly recognized status of authority. Publications were submitted to their judgement which they in turn circulated to others considered worthy to receive them. Among them also were a few impostors who would be unmasked by the expedient of setting them impossible problems which they would claim to have solved. These contacts, of course, remained personal and were not made within the framework of any form of institution in the modern sense of the term.

The ideal of "disinterested mathematics" seems, in short, to have been linked to the existence of a group within which problems of rivalry and competition were not altogether unlike those characteristic of the modern scientific community. The comparison should not be pressed too far. There is, for example, a marked difference in scale between the ancient and the modern communities. The number of scientists, and in particular of mathematicians, in the Hellenistic era certainly never exceeded a few hundred. Likewise, there can be no doubt that, in the absence of any true form of institution, the community they formed was a very precarious one. From the Roman era onwards, the best authors (Ptolemy, Pappus) were concerned, so it would seem, only with perfecting results already achieved. The rivalry and the search for new theories that marked the preceding era had apparently disappeared.

Geminus' Classification of *Mathemata*

Others, among them Geminus, claim that mathematics should be sub-divided in another fashion. On one side they place tangible things, or anything connected with them. No doubt they would call intelligible the subjects of contemplation that the soul elaborates within itself, cutting itself off from things material. They rank arithmetic and geometry as the two first and most important sectors of that form of mathematics that deals with that which is intelligible. With regard to the mathematics which covers things tangible, they designate six sectors: mechanics, astronomy, optics, geodesy, canonics, and logistics.

On the other hand, for them, unlike others, tactics is not worthy to be considered a part of mathematics, even though at times it involves logistics, as in the enumeration of troops, and geodesy, as in the surveying and the division of land. Even less do they consider either history or medicine to be part of mathematics, despite the fact that the authors of historical works often refer to mathematical theorems to indicate climatic conditions or to calculate the size of a town, its diameter and its perimeter, and that physicians use these procedures to throw light on many matters within their competence. The value of astronomy to medicine has been made perfectly clear by Hippocrates and all those who studied the seasons and regions.

Proclus, Commentary on Book I of Euclid's *Elements*, 38.

The Evolution of Geometry [1]

BRUCE E. MESERVE

THE CONSTANTLY CHANGING nature of our concept of geometry has intrigued me for many years. As a student I first visualized geometry, and indeed all of mathematics, as a static body of knowledge which could be mastered if one had sufficient perseverance. As a college student I began to recognize that mathematics was different in the eighteenth century from what it was in the nineteenth or twentieth centuries. When I first became aware of these changes, the elusive nature of mathematics disturbed me. Indeed, I am still a bit disturbed. I shall try to describe to you the high points of the historical evolution of geometry so far as I understand the geometry of our times.

I hope that the brevity imposed upon me will also keep me from becoming lost in a maze of details. I shall try to present the growth or evolution of our concept of geometry in such a way that we shall not only be aware of the elusive nature of the subject but shall also gain sufficient knowledge of the past to enable us to understand the changes that are in process in our own time. I shall consider primarily the evolution of Euclidean geometry and shall mention other geometries only as they arise through the application of new principles to certain fundamental concepts of Euclidean geometry.

The word "geometry" is derived from the Greek words for "earth measure." This derivation of the word also connotes the origin of the science of geometry. Early geometry was a practical science and an empirical science, that is, a science based

upon man's experiences and observations. General theories, postulates, and proofs came much later. Thus our geometry has evolved from a few practical procedures to a deductive science based upon undefined terms, postulates, and the logical deduction of theorems.

We do not know the complete history of geometry. However, we can see several major influences that have contributed to the evolution of geometry. I shall consider very briefly ten of these influences.

1. The empirical procedures of the early Babylonians and Egyptians

2. The Greeks' love of knowledge for its own sake and their use of classical constructions

3. The organization of early geometry by Euclid

4. The embellishment of Euclid's work during the Golden Age of Greece

5. The contributions of Hindu, Arabian, and Persian mathematicians during the Dark Ages in Europe

6. The reawakening in Europe with the growth of the universities, the printing press, and the flowering of all branches of knowledge

7. The introduction of coordinate systems and the recognition of the relationship between differentiation and integration, giving birth to calculus in the seventeenth century

8. The application of algebra and calculus to geometry in the eighteenth century

9. The recognition of abstract points, giving rise to many different geometries in the nineteenth century

Reprinted from *Mathematics Teacher* 49 (May, 1956): 372–82; with permission of the National Council of Teachers of Mathematics.

10. The emphasis upon generalizations, arithmetization, and axiomatic foundations during the past fifty years

As a conclusion I shall endeavor to show how the above influences may be used to predict the influence of our most recent innovation, the machine calculators, and to indicate an appropriate emphasis in our classrooms. Students of history may also observe that the development of geometry is closely related to the development of other areas of mathematics and, indeed, to the development of our culture.

Early Measurements

We have evidences of the geometry of the early Babylonians and Egyptians. Records from pictorial tablets indicate that the Babylonians of 4000 B.C. used the product of the length and the width of a rectangular field as a measure of the field, probably for taxation purposes. The pyramids of Egypt provide striking evidence of early engineering accomplishments that probably required the use of many geometric concepts. For example, the granite roof members over the chambers of a pyramid built about 3000 B.C. are 200 feet above the ground level, weigh about 50 tons each, and were probably brought from a quarry over 600 miles away.

The geometry of the early Babylonians and Egyptians was concerned with areas and volumes. Many of our elementary formulas were known in both cultures by 1500 B.C. Undoubtedly some incorrect formulas were used in both cultures. For example, Coolidge[2] states that in both cultures the area of a quadrilateral with sides a, b, a', and b' was expressed as the product of the averages of the pairs of opposite sides $(a + a')(b + b')/4$. We now know that this formula is valid only for rectangles. We must recognize that each of the early formulas was an empirical result to be considered solely on its own merits. There did not exist any underlying body of theoretical geometric knowledge. In general, the Babylonian and Egyptian concepts of

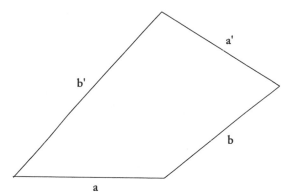

FIGURE I

geometry appear to have remained at a utilitarian and empirical level until about 600 B.C. when the influence of the Greeks began to have an effect.

The Greeks

The Greeks with their love of reason and knowledge left a profound impression upon geometry. They encouraged the study of geometry as a science independent of its practical applications. They enlarged the scope of geometry to include not only empirical formulas for areas and volumes but also (1) the use of line segments to represent numbers; (2) the study of properties of polygons and parallel lines; (3) properties of circles and other conic sections; (4) classical constructions with straightedge and compasses; (5) ratios and proportions arising from a study of similar polygons; and (6) proofs of consequences of a set of postulates.

Most of the intellectual progress of the Greeks came from schools centered around such outstanding scholars and groups of scholars as

Thales: about 600 B.C., one of the Seven Wise Men of early times and often called the creator of the geometry of lines.

Pythagoras: about 540 B.C., a brilliant scholar and mystic known for a famous theorem which probably provided the basis for the discovery of the irrationality of $\sqrt{2}$ and the corresponding incommensurability of the side and diagonal of a square.

The Sophists: a group of teachers in Athens who were more concerned with theory than practice, who introduced the ideas of the Pythagoreans into Athens, and, among other things, considered the classical construction problems. In particular, they proposed the construction of the trisection of any given angle, the construction of a cube with volume double that of a given cube, and the construction of a square with area equal to that of a given circle.

It is interesting to note here that although these three problems were solved by other methods within a few years, it was over two thousand years before geometry, the theory of equations, the theory of numbers, and the use of algebra in solving geometric problems had developed sufficiently to enable anyone to prove that all three of these constructions were impossible, using only straightedge and compasses.

Finally we consider the school of

Plato: about 400 B.C., who emphasized the recognition of geometry as a part of a liberal education. His school used analysis with reversible steps to discover geometric proofs. His pupil Eudoxus developed the "method of exhaustion" for determining ratios of areas (a forerunner of integration in calculus). Another pupil discovered the parabola, ellipse, and hyperbola as conic sections. A few years later Aristotle's work as a systematizer of logic prepared the way for Euclid's organization of the geometry of his time.

Thus we find the influence of the Greeks in the recognition of geometry as a science independent of its practical applications, the broadened scope of geometry, the recognition of a study of geometry as a part of a liberal education, and the study of geometry as a logical system.

Euclid

The center of mathematical activity shifted from Egypt and Babylonia to the Ionian Islands with Thales, to southern Italy with Pythagoras, and then to Athens. About 300 B.C. we find the center of activity shifting back to Egypt to the newly established university of Alexandria. Euclid was a professor of mathematics at Alexandria. He had probably studied in Athens. He wrote at least ten treatises covering the "mathematics" of his time.

His most famous work is called Euclid's *Elements* and contains thirteen books in which he presents an elegant organization of

(1) plane geometry (Books I to IV),
(2) the theory of proportions (Books V and VI),
(3) the theory of numbers (Books VII to IX),
(4) the theory of incommensurables (Book X) and,
(5) solid geometry (Books XI to XIII).

These topics were more closely related than we now consider them. Proportions were based upon similar polygons, the theory of numbers upon the lengths of line segments, and incommensurables upon proportions and the construction of line segments. The books on geometry included nearly all of the concepts that are now considered in a high school geometry course. They also included geometric proofs of algebraic identities and geometric solutions of quadratic equations.

The logical structure of Euclid's proofs was excellent. They included

(1) a statement of the proposition,
(2) a statement of the given data (usually with a diagram),
(3) an indication of the use that is to be made of the data,
(4) a construction of any needed additional lines or figures,
(5) a synthetic proof, and
(6) a conclusion stating what has been done.

We cannot be sure whether this logical structure is due primarily to Euclid or to his training. In either case it is a consequence of the trend in early Greek philosophy.

Lest we idealize Euclid, we should recognize that he adopted many Greek ideas including Aristotle's distinction between postulates and common notions; that he often used definitions as descriptions of terms; and that he tacitly assumed the existence of points and lines, order relations on a line, continuity, and the infinite extent of a line. However, the results of two thousand years of experience should not seriously detract from the significance of Euclid's *Elements*.

Euclid's first three postulates are probably Plato's assumptions for classical constructions. However, Euclid's geometry included much more than classical constructions. He simply used the constructions to demonstrate the existence of the points and lines under consideration. The fourth postulate concerned the equality of right angles; the fifth postulate was the parallel postulate. Apparently Euclid felt uneasy about the fifth postulate since he avoided using it as long as possible. We shall find that this skepticism was justified and was shared by other mathematicians. In general, Euclid's *Elements* rendered geometry a tremendous service as an organization of geometry and indeed of all the mathematical knowledge of that time. We do not know how much of this material was original with Euclid. We do know that the *Elements* represents a logical outgrowth of the geometry of the early Greeks.

Early Euclidean Geometry

The word "mathematics" is derived from the Greek word meaning "subject of instruction," and at the time of the Pythagoreans referred to geometry, arithmetic, music, and astronomy. As knowledge increased, the concepts of mathematics and geometry were restricted. For example, optics, surveying, music, astronomy, and mechanics gradually became recognized as separate bodies of knowledge. Algebra, trigonometry, and the theory of numbers became separate branches of mathematics. Today such specialization is decreasing since it is not unusual for a modern research problem to arise in classical algebra, have significance in geometry, and be solved using the theories of analysis. From this point of view, our concern with geometric concepts (in a narrow sense) is against the modern trend.

Since Euclid's *Elements* contained a logical organization of nearly all the mathematical concepts of his time, it provides a basis for a consideration of the mathematical achievements of the next thousand years. The first half of this period is characterized by a continued development of mathematical concepts, the second half has decreasing mathematical significance. At the end of this period Alexandria was destroyed by the Arabs, and Europe was entering its Dark Ages.

The prestige of Alexandria lasted for many years after Euclid. Archimedes (about 250 B.C.) studied there and, in addition to discoveries in other fields, made many contributions to geometry. For example, his work on areas and volumes of revolution provided another forerunner of integration. Apollonius (about 225 B.C.) studied at Alexandria and wrote an extensive treatise on conics. His descriptions of figures in terms of a diameter and a tangent line were equivalent to a coordinate system. He recognized that the distances of a point from the two foci have a constant sum for the points on an ellipse, a constant difference for the points on a hyperbola. Later, about A.D. 300, Pappus enhanced the prestige of Alexandria through his work on volumes of revolution, the construction of conics through given sets of points, and the study of special curves including spirals.

The remaining mathematicians of note prior to the destruction of Alexandria were primarily commentators upon the works of their predecessors. Thus by A.D. 700 the geometry of measurements of the Babylonians and Egyptians had been molded in accordance with the Greek love of knowledge and reason and was in need of new influences. This is not intended to minimize the Greek influence but rather to indicate that it had run its course.

Hindu, Arabian and Persian Influences

Each culture that passes along a body of knowledge makes a contribution to that knowledge. Thus, especially during the Dark Ages in Europe, the mathematical achievements of the Greeks were modified by Hindu, Arabian, and Persian mathematicians. Most of these influences were of a practical and utilitarian nature. There were noteworthy advances in number notation, areas, volumes, classical constructions, astronomy, and trigonometry. Euclid's parallel postulate was seriously questioned. Omar Khayyám, the author of the *Rubáiyát*, wrote a treatise on algebra and determined roots of some cubic equations as intersections of conics. Like their Greek predecessors, these mathematicians used line segments to represent numbers and therefore did not recognize "negative" roots. Also like their predecessors, they proposed replacements for Euclid's parallel postulate, but failed to prove either the postulate or their replacement as a theorem. We now know that the parallel postulate cannot be proved as a theorem unless an equivalent statement is taken as a postulate.

Throughout the development of Euclidean geometry we shall find that the attempts to prove the fifth postulate were generally of one of the following types: (1) direct proof from Euclid's other postulates; (2) replacement of the fifth postulate by a more "self-evident" postulate (either explicitly or tacitly and unknowingly) and a proof of the fifth postulate as a consequence of the new assumption and Euclid's other postulates; or (3) indirect proof by showing that the fifth postulate cannot fail to hold. All of these approaches were doomed to failure as "proofs" of the fifth postulate. They did, however, encourage the study of several possible geometries using primarily (1) synthetic methods; (2) algebraic concepts; (3) curvature and differential geometry; (4) distance relationships; and (5) groups of transformations. As a result of these studies, mathematicians have considered the axiomatic basis for Euclidean geometry in detail, and now recognize the consistency of the non-Euclidean geometries.

The Reawakening in Europe

In mathematics as in trade and art, the first signs of European awakening were in Italy. There were

some evidences of mathematical life in England in the eighth century and in France in the tenth century. However, progress was very slow. About the end of the twelfth century Leonardo returned to Italy after traveling extensively, and soon published several treatises making available a vast amount of information regarding previous achievements in number notation, arithmetic, algebra, geometry, and trigonometry. These ideas were soon picked up in the new European universities where groups of scholars provided the stimulus to each other that is necessary for great achievements. Then in the fifteenth century the printing press provided a new means of disseminating knowledge, and intellectual activity began to spread rapidly throughout Europe. At first there was an intellectual smoldering while the scholars acquired additional knowledge of past achievements. Soon Alberti and the Italian artists developed some of the principles of descriptive geometry. A treatise on plane and spherical trigonometry was prepared in Germany. Letters were introduced for numbers in France. Gradually mathematical activity acquired momentum and by the end of the seventeenth century was well under way. During the last three centuries there has burst forth such an avalanche of activity, constantly expanding without any definite signs of spending its force, that we shall be hard pressed to assess the evolution of geometric concepts. However, I shall make the attempt and let you be the judge of the success of the venture.

The Seventeenth Century

Kepler and Galileo made their contributions to geometry and astronomy at the beginning of the seventeenth century. One of Galileo's pupils, Cavalieri, assumed that a line could be generated by a moving point, a plane by a moving line, a solid by a moving area. Many of you have taught *Cavalieri's principle* regarding the equality of the volumes of any two solids of the same height and of the same cross-sectional areas at corresponding heights.

Our Cartesian coordinate systems are based upon the work of René Descartes who applied algebraic notation to the analysis of conics by Apollonius, visualized all algebraic expressions as numbers instead of geometric objects, and found equations representing several curves (considered as loci). His interpretation of such symbols as x^2 and x^3 as numbers and, therefore, as lengths of line segments was very important. Previously, linear terms such as x or $2y$ had been considered as line segments, quadratic terms such as x^2 or xy had been considered as areas, and cubic terms such as x^3 or x^2y had been considered as volume. The old interpretations were restrictive in the sense that only like quantities could be added. For example, it was permissible to add x^2 and xy (areas), but it was not permissible to add x^2 and x (i.e., an area and a line segment). Descartes' interpretation of all algebraic expressions as numbers and, therefore, as line segments made it possible to consider sums such as $x^2 + x$. This new point of view provided a basis for the representation of curves by equations.

The study of geometric figures as loci corresponding to equations introduced a new era and an entirely new point of view into the study of geometry. Since we shall find enthusiastic supporters of both the new and the old points of view, we shall endeavor to distinguish between the two as follows: The study of figures in terms of their algebraic representation by equations will be called *analytic geometry;* the study of figures directly without using their algebraic representations will be called *synthetic geometry.* Since many geometers will use both algebraic and synthetic methods, the above distinction is at best a relative one. Modern geometers consider the two geometries as two equivalent points of view of the same body of knowledge.

Fermat was a contemporary of Descartes who also worked in analytic geometry. Desargues and Pascal visualized the circle, ellipse, parabola, and hyperbola as projections of circles, discovered other properties of conics, and prepared the foundations for synthetic projective geometry.

The last half of the seventeenth century is marked by the independent discovery by Newton and Leibniz of the relationship between differentiation (often visualized as rate of change) and integration (often visualized as summation). Newton's discovery of calculus was based upon a study of figures in motion, Leibniz' upon a study of static figures. Our present notation is largely that of Leibniz. Newton also made contributions to analytic geometry through his classification of cubic curves and to synthetic geometry.

The application of new ideas to geometry typifies the seventeenth century. There was the application of algebra initiated by Descartes and Fermat; the applications of projections initiated by Desargues and Pascal; and the applications of calculus initiated by Newton and Leibniz. These new approaches were the beginnings of three major phases of geometry—analytic geometry, synthetic geometry, and differential geometry.

The Eighteenth Century

In the eighteenth century we find a broadening of the applications of the new ideas started in the seventeenth century and a new development in relation to Euclid's parallel postulate.

The work of Descartes and Fermat was extended to three dimensions. Previously known results and new results were restated in algebraic notations. Transformations of coordinates were considered.

Newton's work was continued by Maclaurin who discovered many special curves, including the cissoid, cardioid, and lemniscate. The techniques of differential geometry were used to develop a theory of curvature and to study quadric surfaces.

In synthetic geometry Euler proved that for any given triangle the point of intersection of the altitudes, the point of intersection of the medians, and the point of intersection of the perpendicular bisectors of the sides are collinear. He also considered networks of arcs and developed a general theory of traversability which is often considered as the starting point of topology.

The work on Euclid's parallel postulate was another development of synthetic geometry. Early in the eighteenth century Saccheri tried to prove the postulate by denying it and showing that the results so obtained could not hold. He succeeded in proving that exactly one of three situations must consistently hold:

1. the sum of the angles of a triangle is always equal to two right angles,
2. the sum of the angles of a triangle is always greater than two right angles, or
3. the sum of the angles of a triangle is always less than two right angles.

Under each hypothesis he could prove several theorems. He disposed of the last two hypotheses by accepting Euclid's tacit assumption that lines are not reentrant and by assuming that two lines cannot merge into a single line at infinity. We now recognize these assumptions as postulates leading to independent geometries. We know that the first hypothesis holds for plane triangles, the second for spherical triangles. Lambert (still in the eighteenth century) suggested that the third hypothesis might hold for triangles on a sphere with an imaginary number as its radius. This suggestion is now recognized as mathematically sound. Lambert based it upon his study of the geometry on the surface of a sphere. As a part of this study he proved that the area of a spherical triangle was a function of the excess of its angle sum relative to two right angles.

Saccheri's approach was also adopted by Legendre (1752–1833). However, he too was trying to prove the parallel postulate. Legendre is well known for his work in the theory of numbers, theory of functions, and calculus. He has had a great influence upon American high school geometry texts through his publication of a geometry text rearranging and modifying Euclid's *Elements*.

During the last half of the eighteenth century we find increasing evidence of the forthcoming crescendo of mathematical activity that characterizes the nineteenth and twentieth centuries. Because of the increased specialization of terminology and the detailed and abstract character of many of the contributions, our treatment of the evolution of geometry will, of necessity, become more expository.

The Nineteenth Century

In the early nineteenth century Gauss apparently visualized the possibility of geometries in which Euclid's fifth or parallel postulate did not hold. However, he did not publish his results. Probably he observed that Euclid's fifth postulate is quite different from his first four postulates. The first four postulates are concerned with finite segments of lines and the possibility of extending a finite segment to form a line. The fifth postulate asserts a property of lines in their full extent. Thus, since all measurements and constructions are finite (even the most distant star visible in the most powerful telescope is only a finite distance from the telescope), the fifth postulate is based upon a faith or conviction that if two lines are cut by a transversal such that the sum of the interior angles on one side of the transversal is less than two right angles, then the lines will intersect if they are extended sufficiently far. When the sum of the interior angles differs very slightly from two right angles, the assumption that the lines will intersect cannot be definitely established.

Given a line m and a point P that is not on m, we may draw a circle with center P and any given radius d intersecting line m in points R and S. Then it is clear that there are infinitely many lines through P that do not intersect m inside the circle, i.e., within the distance d from P. When d is the radius of the circle that I have drawn, these nonintersecting lines are expected. Why shouldn't there also be infinitely many lines through P and not intersecting m when d is a mile, the distance to the horizon, the distance to the moon, the distance to the most distant visible star? This question could not be answered 150 years ago. It is now known that one may obtain consistent geometries by assuming either

1. there are infinitely many lines through P that do not intersect m, or
2. there is exactly one line through P that does not intersect m.

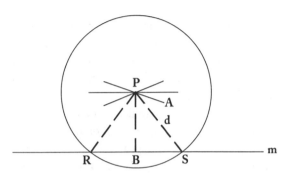

FIGURE 2

If the line PB is perpendicular to m, then Euclid's fifth postulate asserts that any line PA such that angle BPA is less than a right angle must intersect the line m if the lines PA and m are extended sufficiently far.

Bolyai and Lobachevski independently developed theorems in a geometry based upon the assumption that there are infinitely many lines through P that do not intersect m. Later this geometry was called *hyperbolic geometry* in recognition of the two distinct lines, PR and PS, "parallel" to the line m. Hyperbolic geometry corresponds to Saccheri's third hypothesis that the angle sum of a triangle is less than two right angles.

Riemann developed a geometry based upon Saccheri's hypothesis that the angle sum of a triangle is greater than two right angles. He considered space as a set of undefined objects, called "points," where each point was determined by its coordinates. He assumed that there existed a distance function and that the square of the differential of the distance function was homogeneous and of the second degree in the differentials of the

coordinates. For example, in Euclidean plane geometry the distance function

$$s^2 = (x_1 - x_2)^2 + (y_1 - y_2)^2$$

may be expressed in the form

$$ds^2 = dx^2 + dy^2$$

in terms of the differentials of the distance and the coordinates. Then just as we consider transformations (rotations and translations) in Euclidean geometry that do not change the Euclidean distance function, Riemann considered transformations that do not change his generalized distance function. Thus, as in Euclidean geometry, Riemann assumed that the measurements of figures and objects do not depend upon their position. Accordingly, on any surface the figures must be freely movable as on a plane (on which figures may slide or rotate) or on the surface of a sphere.

The distance functions or metrics were developed further by Cayley and Klein. The new theories were extended to three dimensions and then to n-dimensions. Klein also classified various geometries by considering them as studies of properties that are invariant (unchanged) under groups of transformations. Lie considered groups of transformations leaving distance functions invariant and proved that if figures are to be freely movable (slide in any direction and rotate about any point), there are exactly four possible types of geometry in three-dimensional space. These are: (1) Euclidean geometry; (2) the geometry on a sphere; (3) the geometry of Riemann; and (4) the geometry of Bolyai and Lobachevski. The last two are often called the *non-Euclidean geometries.*

The use of algebra (especially the theory of groups) and differentials in the development of the above geometries provides another illustration of the interdependence of the various branches of mathematics. This interdependence is also seen in the search for proofs of the consistency of geometries. It can now be proved that Euclidean geometry is consistent if the real number system is consistent. (This dependence of geometry upon properties of numbers is an example of the arithmetization of modern geometry.) The non-Euclidean geometries are also consistent if the real number system is consistent. Accordingly, even though several prominent philosophers have based their beliefs in natural logical systems upon the existence of a natural geometry—Euclidean geometry—we must now recognize that Euclidean geometry may not be the inherent geometry of our universe. It is probable that the development of philosophy as well as mathematics would have been noticeably changed if this had been recognized several centuries ago.

Let us now return to the early nineteenth century. At the end of the eighteenth century Monge was active in France. He organized descriptive geometry and applied the new techniques of algebra and calculus to curves and surfaces. One of his pupils, Poncelet, wrote the first text on projective geometry and developed the concept of duality. Von Staudt considered a geometry of positions independent of all measurements and proved that geometry did not need the techniques of algebra and calculus. Gauss' ideas on curvature were extended by Riemann and eventually provided a basis for relativity theory. Riemann also used n sheets (in a sense, planes) to render n-valued functions single-valued on these new Riemann surfaces. This concept had important implications in both the theory of functions and geometry. Plücker, Cayley, and Grassman developed new coordinate systems and extended their results to n-dimensional geometrics. Moebius introduced new coordinates and discovered a new type of surface, a one-sided surface.

As indicated by the introduction of several types of coordinates, by Riemann's concept of a point as an undefined entity determined by its coordinates, and by the introduction of n-dimensional spaces, the concept of a point was undergoing a change in the middle of the nineteenth century. The complete abandonment of all visual intuitive concepts of a point was gradually accepted by theoretical mathematicians. The new concepts of a point were used in the non-Euclidean geometries and as a basis for abstract geometry. Thus near the end of the nineteenth century we find geometry extending its break with the physical world. This break was completed with the axiomatic developments of the twentieth century.

The Twentieth Century

About the end of the nineteenth century the Italian geometers enjoyed a period of active leadership. In recent years this leadership has probably shifted to the United States with the influx of scientific immigrants and the development of topology and abstract algebraic geometry. Topology is a very general and important geometry. In many ways it represents the peak of our geometric achievements to date. It also involves several elementary concepts that could be useful in secondary schools. Some of these concepts are described in the November, 1953 issue of *The Mathematics Teacher.*[3]

The shift in leadership in algebraic geometry appears to be due to the introduction of the arithmetic ideas of Dedekind and Weber and the modern algebraic concepts of group, ring, field, and ideal. Since the trend in algebraic geometry is indicative of trends in other branches of geometry, I shall present the following introductory remarks of Oscar Zariski in a paper presented at the 1950 International Congress of Mathematicians in order to give the viewpoint of one of the leaders in this area.

The past 25 years have witnessed a remarkable change in the field of algebraic geometry, a change due to the impact of the ideas and methods of modern algebra. What has happened is that this old and venerable sector of pure geometry underwent (and is still undergoing) a process of arithmetization. This new trend has caused consternation in some quarters. It was criticized either as a desertion of geometry or as a subordination of discovery to rigor. I submit that this criticism is unjustified and arises from some misunderstanding of the object of modern algebraic geometry. This object is not to banish geometry or geometric intuition, but to equip the geometer with the sharpest possible tools and effective controls. It is true that the lack of rigor in algebraic geometry has created a state of affairs that could not be tolerated indefinitely. Effective controls over the free flight of geometric imagination were badly needed, and a complete overhauling and arithmetization of the foundations of algebraic geometry was the only possible solution. This preliminary foundational task of modern algebraic geometry can now be regarded as accomplished in all its essentials.

But there was, and still is, something else more important to be accomplished. It is a fact that the synthetic geometric methods of classical algebraic geometry, operating from a narrow and meager algebraic basis and faced by the extreme complexity of the problems of the theory of higher varieties, were gradually losing their power and in the end became victims to the law of diminishing returns, as witnessed by the relative standstill to which algebraic geometry came in the beginning of this century. I am speaking now not of the foundations but of the superstructure which rests on these foundations. It is here that there was a distinct need of sharper and more powerful tools. Modern algebra, with its precise formalism and abstract concepts, provided these tools.

An arithmetic approach to the geometric theories which we were fortunate to inherit from the Italian school could not be undertaken without a simultaneous process of generalization; for an arithmetic theory of algebraic varieties cannot but be a theory over arbitrary ground fields, and not merely over the field of complex numbers. For this reason, the modern developments in algebraic geometry are characterized by great generality. They mark the transition from classical algebraic geometry, rooted in the complex domain, to what we may now properly designate as *abstract algebraic geometry*, where the emphasis is on abstract ground fields.[4]

We have seen in our previous discussion and in the above quotation that the trend of geometry in the first half of the twentieth century has been toward generalization, arithmetization (i.e., the use of properties of numbers), and the axiomatic foundations of geometry. Pasch visualized geometry as a deductive science based upon a set of postulates. Hilbert and Veblen were especially active in this

formalized concept of geometry. Most of us slyly appreciate Hilbert's concept of mathematics as a game played according to certain rules with meaningless marks on paper.

Conclusion

This paper has been concerned with the origins of geometry. What about the future? Are we steadily pushing back the horizons of mathematical knowledge? Are we gaining an ever firmer grasp upon this body of knowledge, its underlying principles, and its implications? Such appears to be the case. Moreover, let us not lose sight of the manner in which this progress is being made—new mathematical tools are being combined with an emphasis upon generalizations and fundamental operations. The secondary schools cannot hope to introduce all the new tools; the schools could and, in my opinion should, place a major emphasis upon the importance and use of generalizations and the fundamental operations.

In geometry the influence of intellectual curiosity has been evident since the time of the early Greeks. Many phases of geometry have been primarily intellectual achievements and have not found a practical application until they were well developed. In other cases utility came earlier in the development. Recently, technological advances have provided a wide range of activity for the mathematically curious as well as those who are trying to accomplish specific tasks. The electronic and mechanical computers provide a striking example of the influence of technological advances. Within the past decade these new mathematical tools have gained recognition for the ways in which they may remove most of the drudgery from mathematical computations in both applied and theoretical branches of mathematics. Within another decade it seems reasonable to predict that these machines will revolutionize many of our approaches to mathematical problems. This new technique appears destined to go through the same stages that we have observed in the use of the techniques of algebra and calculus. At present we are in the period of the discovery of the technique and the initial explorations. Problems of rigor are arising. We may expect a rash of exploitations of the new technique comparable to the exploitations of calculus in the eighteenth century. As in the case of algebraic and synthetic methods in geometry, there will undoubtedly be strong proponents of old and new techniques. Also as in the case of algebraic and synthetic methods in geometry, we may look forward to the eventual absorption of the new technique into accepted mathematical procedures with a resulting unification of viewpoints and increase in our knowledge of mathematics.

Scientific advances in many phases of our society are creating a very fertile source of problems for calculating machines. The greatest problem at present is the preparation of the problems for the machines. As the president of a nationally known research corporation said recently: "We have not run out of problems, we have run out of mathematicians." Here again the secondary schools can serve their students most effectively by emphasizing fundamental operations and basic procedures. Throughout the evolution of geometry and, indeed, all of mathematics we find progress dependent upon people who understand basic operations and have the courage to try new ideas. Geometry today represents but a fleeting instant of an ever-changing body of knowledge. Even in our minute personal worlds, we may have an influence upon the evolution of geometry by keeping ourselves aware of the changes that are taking place and by transmitting this awareness to our students through our teaching of mathematics as a living

subject, based upon generalizations of the fundamental operations and procedures that we develop in our classrooms.

Thus you see that I do not consider our hasty journey through the history of geometry as a momentary intellectual pastime. Rather I find in such journeys both inspiration and guidance for my daily classroom teaching. I find a living subject, a subject of great importance to our national and cultural welfare, a subject whose growth depends upon the efforts of thousands of people like you and me who must keep trying to teach the underlying principles behind the formal details in our textbooks and to keep our students active in the formulation and testing of their own ideas and generalizations.

NOTES

1. Based upon a chapter of the same title in *Fundamental Concepts of Geometry* (Addison-Wesley, 1955) and presented April 15, 1955, at the Thirty-third Annual Meeting of the National Council of Teachers of Mathematics, Boston, Massachusetts.

2. J. L. Coolidge, *A History of Geometrical Methods*, (Oxford: Cambridge University Press, 1940) p. 5.

3. B. E. Meserve, "Topology for Secondary Schools," *The Mathematics Teacher*, Vol. XLVI (November 1953), pp. 465–474.

4. Oscar Zariski, "The Fundamental Ideas of Abstract Algebraic Geometry," *Proceedings of the International Congress of Mathematicians*, Vol. II, 1950, p. 77.

The Platonic Solids

A three dimensional solid whose sides are formed by regular, congruent polygons is called a regular polyhedron. There are only five such polyhedrons: the tetrahedron formed from equilateral triangles; the hexahedron, or cube, made from squares; the octahedron also comprised of equilateral triangles; the dodechahedron formed from pentagons and, lastly, the twenty-sided icosahedron composed of equilateral triangles. The uniqueness of this set, that is the fact that there were only five such solids, fascinated the Pythagoreans. They assigned symbolic representations to each solid: the tetrahedron designated fire; the cube, earth; the octahedron, air; and the icosahedron water. In the numerology of the Pythagoreans, these four "elements" were called the tetractys or "holy fourfoldness" and their number representations were 1, 2, 3, 4. Holy 10, representing the universe, was derived from the sum of the tetractys, i.e. 1 + 2 + 3 + 4 = 10; thus in Pythagorean mysticism the dodecahedron symbolized the universe. The tradition of these solids was passed down to Plato (ca. 380 B.C.), who incorporated discussions of them in his *Timaeus,* and they popularly became known as the "Platonic Solids." Mathematical treatments of these solids were also rendered by Euclid in his *Elements*, Book XIII (300 B.C.) and later appeared in the works of Pappus (A.D. 300).

| TETRAHEDRON | CUBE | OCTAHEDRON | ICOSAHEDRON | DODECAHEDRON |
| Fire | Earth | Air | Water | the Universe |

It was known to the ancient Greeks that the regular polyhedron could be inscribed in a sphere.

In the sixteenth century, the German mathematician, astronomer, and numerologist Johannes Kepler (1571–1630) sought a Divine, harmonious order for the solar system. Kepler sought a more perfect model for the orbits of the known planets and, initially, found it in the radii of circumscribing spheres for the Platonic solids. As he explained in his *Mysterium cosmographticum* written in 1596:

> The orbit of the Earth is a circle: round the sphere to which this circle belongs, describe a dodecahedron; the sphere including this will give the orbit of Mars. Round Mars describe a tetrahedron; the circle including this will be the orbit of Jupiter. Describe a cube round Jupiter's orbit; the circle including this will be the orbit of Saturn. Now inscribe in the Earth's orbit an icosahedron; the circle inscribed in it will be the orbit of Venus. Inscribe an octahedron in the orbit of Venus; the circle inscribed in it will be Mercury's orbit. This is the reason of the number of the planets.

Thus, Pythagorean mysticism survived in the theories of Kepler but these speculations were soon replaced by more rational estimates based on observation and calculation.

27

Irrationals or Incommensurables I: Their Discovery, and a "Logical Scandal"

PHILLIP S. JONES

*T*HE STORY OF THE EARLIEST BEGINNINGS of the idea of incommensurable quantities is so uncertain as to be still the basis of lengthy scholarly discussions airing somewhat inconsistent scholarly opinions. No one *knows* exactly who first discovered two incommensurable quantities, when this was done, how the discovery was made, or what was the motive or interest which led to the discovery.[1] However, in spite of the lack of agreement on details, enough of the story is known and enough agreement does exist to reveal rather clearly the development of mathematical thought and Greek motives at an interesting period in the history of the world as well as to demonstrate still usable approaches to the fascinating and conceptually difficult topic of irrational or incommensurable numbers.

To put the Greek work on incommensurables into proper perspective, one should first note that two problems which we today recognize as involving irrationals had occupied important positions in Egyptian and Babylonian mathematics without apparently, however, stimulating in their minds any thought of the concepts which are our topic. These are the problems of the circumference and area of circles in connection with which we today think of the number π which is not only irrational

but also transcendental, and the problems involving the Pythagorean theorem and Pythagorean triples.[2] The circle problems do not seem to have brought either Egyptians or Babylonians at all close to the concept of irrationals. However, the Babylonian use of the Pythagorean theorem in situations where the hypotenuse was irrational and their approximations to the square roots involved in these cases may be regarded as steps toward a discovery which they never made. Further, the recently discovered Babylonian tables of Pythagorean triples preceded and could have stimulated the later Greek study of these triples and of figurate numbers. Both of these topics can be associated with the Greek discovery of irrationals.

Although modern scholars varyingly locate the discovery of the incommensurables from 450 B.C. to 375 B.C., they all agree that it took place after Pythagoras (540 B.C.) and before Plato (430?–349? B.C.) in spite of Proclus (A.D. 450) who wrote, "Pythagoras transformed this science [geometry] into a free form of education; he examined this discipline from its first principles and he endeavored to study the propositions, without concrete representation, by purely logical thinking. He also discovered the theory of irrationals (or of proportions) and the constructions of the cosmic solids (regular polyhedra)."[3] Fritz's recent study supports the ancient legends that it was Hippasus of Metapontum (ca. 500 B.C.) who first *discovered* the existence of incommensurables.

Reprinted from *Mathematics Teacher* 49 (Feb., 1956): 123–27; with permission of the National Council of Teachers of Mathematics.

The legends also relate that he was later lost at sea as punishment presumably for having revealed his discovery.

To understand either the ancient meaning of this legend or more recent interpretations, one must recall that the Pythagorean brotherhood was actually a somewhat aristocratic group of people carrying on political as well as philosophical-mathematical activities in Italy where Pythagoras himself had lived at Crotona. These followers of Pythagoras, whose recognition symbol was the mystic pentagram formed by extending the sides of a regular pentagon, were finally broken up and driven out of Italy after the rise of Athenian democracy which followed the defeat of the Persians by Athens and Sparta. Pythagorean philosophy had two interrelated aspects, one mystical-religious, and the other scientific-mathematical. B. L. van der Waerden's summary of this is:

> Their doctrine proclaims that God has ordered this universe by means of numbers. God is unity, and the world is plurality, and it consists of contrasting elements. It is harmony which restores unity to the contrasting parts and moulds them into a cosmos. Harmony is divine, it consists of numerical ratios. Whosoever acquires full understanding of this number-harmony, he becomes himself divine and immortal. Music, harmony, and numbers—these three are indissolubly united according to the doctrine of the Pythagoreans.[4]

From this you may see why anything which would upset their theory of numbers would be most distressing to the Pythagoreans. Since their theory of numbers was essentially the theory of integers and the ratios of integers, the discovery of incommensurable geometric magnitudes, that is, magnitudes whose ratio could not be represented by pairs of integers, was to them what one modern writer has called a "logical scandal." Further, in its earliest days Pythagorean doctrine was to be transmitted only by word of mouth and the works of his followers were credited to the master; hence, when Hippasus not only revealed a Pythagorean discovery, but also revealed such an upsetting one, he

certainly made himself a fair target for the wrath of his fellows if not of the gods.

Van der Waerden reads a much more modern situation into the Hippasus legend. He sees the existence of a conflict between those followers who after his death wished to regard Pythagoras' teachings as the source of all knowledge and those who felt one must rely on one's own thoughts in the search for truth. He believes that science cannot advance if it is a secret known only to initiates, that scientists will inevitably come into conflict with a pledge of silence, and that Hippasus' expulsion from the Pythagoreans came because he deliberately both added to and revealed Pythagorean doctrines.

So far we have discussed some of the historical arguments and legends relating to the *discovery* of the existence of incommensurables. A related question is: How were given pairs of quantities first *proved* to be incommensurable? Although one could argue that nothing, especially in mathematics, is discovered until it has been proven it is also important for teachers and students to recognize that the discovery or formulation of a conjecture is often as important, and sometimes more inspired, than the proof which must be worked out before the conjecture is fully established as a valid theorem.

The oldest definitely known *proof* associated with incommensurables corresponds in its essential elements to the modern proof that $\sqrt{2}$ is irrational. This is the proof that the diagonal and side of a square are incommensurable and is to be found as an addition, probably by Theaetetus, at the end of Euclid's Book X. In spite of its geometric formulation the proof there depends not only on the Pythagorean theorem (which probably was not proven until after Pythagoras!), but also on the Pythagorean theory of even and odd numbers which does go back to the time of Pythagoras. Euclid expounded this latter theory in Book IX of the *Elements*. The essential facts of number theory needed are that an odd number times an odd number is odd and an even times an even is even. From this, since all integers are even or odd, it

FIGURE 1

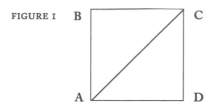

FIGURE 2

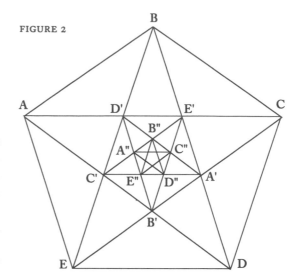

follows that if a perfect square is even, its square root is also even. The indirect method of proof was used. The negative of the desired conclusion was assumed and a contradiction was derived from this assumption. In condensed form the proof was: Assume that in square $ABCD$ in Figure 1, AC and AB are commensurable, i.e., $AC{:}AB = m{:}n$ where m and n are integers with no common factor, that is, not both m and n are even. By the Pythagorean theorem $AC^2 = 2AB^2$, whence $AC^2{:}AB^2 = 2 = m^2{:}n^2$. From this $m^2 = 2n^2$. Thus since m^2 is even, m is also even, or $m = 2p$. Thus, since m is even, n is odd. Substituting for m and simplifying we have $2p^2 = n^2$ which shows n also to be even. Thus, we have shown the same number, n, to be both even and odd, which is impossible. Hence the original assumption that AC and AB are commensurable is false.

The earlier study of Pythagorean triples and of figurate or polygonal numbers may well have paved the way for such a proof by associating the theory of integers with geometric figures and by stimulating mathematicians to search for common elements, such as ratios, possessed by a class of figures such as similar triangles or right triangles. However, though this may have been the first proof that a pair of lines were incommensurable, Kurt von Fritz believes that the original *discovery* of the existence of incommensurables was due to Hippasus and grew out of a study of the pentagon rather than the square. This conjecture is based on the historical facts that the pentagon and pentagram were important Pythagorean symbols, that Hippasus was also concerned with the pentagon when he solved the problem of inscribing a regular dodecahedron in a sphere, and the psychological fact that the incommensurability of the side and diagonal of a pentagon is almost visually evident as shown in Figure 2.

Further, the chief theorem needed to construct a proof of incommensurability in this case is the isosceles triangle theorem associated with Thales, the first known Greek geometer. The only other concepts needed are that angles inscribed in equal arcs are equal and the method of successive subtractions which was also a common tool in Greek mathematics. The gist of the proof is to show that in the successive smaller and smaller pentagons formed by repeatedly drawing diagonals as shown in Figure 2 a common measure of the side and diagonal of one pentagon would also be a common measure of the side and diagonal of each smaller pentagon. Since these approach zero as a limit, the common measure is zero or does not exist.

The proof for all of this follows from observing that all the angles at each vertex of $ABCDE$ are equal and that $\angle BAC'$ is equal to $\angle BC'A$. From this, $AB = BC'$ and

(1) $BE - BA = C'E = EB' = B'D' = BD'$.
(2) $AB - BD' = BC' - BD' = C'D'$.
(3) $B'D' - C'D' = E'B' = A'D''$.

If we symbolize successive sides and diagonals as follows: $AB = s_1$, $C'D' = s_2$, $A'E' = s_3$, and $BE = d_1$, $B'D' = d_2$, $A'D'' = d_3$, we then have

(1) $d_1 - s_1 = d_2$ (3) $d_2 - s_2 = d_3$
(2) $s_1 - d_2 = s_2$ (4) $s_2 - d_3 = s_3$, etc.

From (1) we conclude that a common measure (divisor) of d_1 and s_1 will also measure d_2. From (2) and (3) we see that such a common measure will also measure s_2 and d_3, etc. This intuitively completes the proof that there is no common measure, or that the side and diagonal of the pentagon are incommensurable, because we see that the assumption that there is such a (finite) common measure leads to the contradiction that it must be zero. A rigorous proof requires either a more modern idea of limit or the use of some form of the axiom of Archimedes. This axiom, which really antedates Archimedes, says that if from any quantity we take away its half or more, and then from the remainder we take away its half or more, and so on, we can ultimately reach a remainder which will be less than any previously given or assumed fixed quantity.[5] This axiom is, of course, closely allied to the limit concept, which is fundamental in any modern treatment of irrationals.

Whether or not Hippasus *discovered* the existence of irrationals, and if so whether via studying the mystic pentagon or pentagram or a square, whether or not he *proved* this existence and if so whether via successive subtractions or the Pythagorean theory of even and odd numbers may never actually be known. However, these conclusions all seem valid:

1. The *discovery* and *proof* both took place in the period 450 B.C. to 375 B.C.

2. Both the pentagon and the square were the subject of study at that time.

3. Both the methods of successive subtractions and of even and odd numbers were known at about this period, were being used, and were applicable to this problem, although the latter is the method which appears in the earliest discussions of incommensurables which we now have.

4. All of these are reasonably understandable, concrete, and elementary approaches to a problem which is still of importance in mathematics.

In our next installment of this story of irrational numbers, we will see how the method of successive subtractions could be applied to the side and diagonal of a square and how this associates with a method of finding rational approximations to the $\sqrt[2]{2}$. After this we will present the Greek solution to the dilemma of the incommensurable, namely, Eudoxus' method of exhaustion.[6]

NOTES

1. Some of the differences in opinion are revealed in the content and footnotes of the most extensive recent discussion by Kurt von Fritz, "The Discovery of Incommensurability by Hippasus of Metapontum," *Annals of Mathematics*, Vol. 46 (April 1945), pp. 242–64. Other basic sources which give further data and references are Sir T. L. Heath, *History of Greek Mathematics* (Oxford: The Clarendon Press, 1921) and Sir T. L. Heath, *The Thirteen Books of Euclid's Elements* (Cambridge: The University Press, 1908).

2. For discussions of the Egyptians' procedure for finding the area of a circle and more recent facts about π see *The Mathematics Teacher*, XLIII (March 1950), pp. 120–21, and (May 1950), p. 208; N. T. Gridgeman, "Circumetrics," *The Scientific Monthly*, LXXVII (July 1953), p. 31 ff. as well as the standard histories of mathematics, of course.

For discussions of the Pythagorean theorem, Pythagorean triples, and Babylonian extractions of square root, see especially *The Mathematics Teacher*, XLII (October 1949), pp. 307–8, and XLIII (April 1950), pp. 162–63; as well as XLIII (October 1950) p. 278; XLIII (November 1950), p. 352; XLIV (April 1951), p. 264; XLIV (October 1951), p. 396; XLV (April 1952), p. 268; XLVI (March 1953), p. 188; XLVI (April 1954), p. 269.

3. B. L. van der Waerden, *Science Awakening* (Groningen, Holland: P. Noordhoff Ltd., 1954), p. 90, which quotes from Proclus' commentary on Euclid's *Elements* as reported by Friedlein.

4. van der Waerden, *op. cit.*, p. 93.

5. This statement is actually Euclid X.1. which he proved by use of the more generally recognized form of Archimedes' Axiom. See Sir T. L. Heath, *Manual of Greek Mathematics* (Oxford: The Clarendon Press, 1931).

6. *Editor's note:* This second installment by P. S. Jones can be found in *Mathematics Teacher* 49 (March 1956), pp. 187–91. The third installment appears in the present volume as chapter 28 and immediately follows this chapter.

28

Irrationals or Incommensurables III: The Greek Solution

PHILLIP S. JONES

$\mathcal{B}$OTH THE PYTHAGOREAN number-theoretic proof of the irrationality of $\sqrt{2}$ (or of the incommensurability of the side and diagonal of a square) and the successive subtraction process which we applied to the square and the pentagon in earlier articles[1] required generalization before one could claim that mathematicians had a theory of irrationals. Some generalizations developed quite rapidly. We shall merely sketch in the steps in this development and explain the finished product as it appeared in Euclid.

The chief contributors to this development were Theodorus of Cyrene (ca. 400 B.C.), Theaetetus (ca. 375 B.C.), and Eudoxus of Cnidos (ca. 400–347 B.C.). Theodorus showed the irrationality of the non-square integers from 3 to 17. Exactly how he did this is not certain, nor why he stopped

Reprinted from *Mathematics Teacher* 49 (Apr., 1956): 282–85; with permission of the National Council of Teachers of Mathematics.

FIGURE I

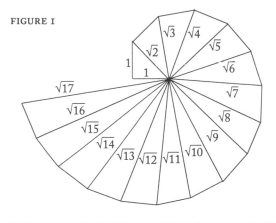

with 17; but the fact that he treated only these cases indicates that he still did not have a general process.[2] Theaetetus, on the other hand, is supposed to have been the original source for such general theorems as Euclid, Book X, Proposition 8: *If two magnitudes have not to one another the ratio which a number has to a number, the magnitudes will be incommensurable;* and Book X, 9: *The squares on straight lines commensurable in length have to one another the ratio which a square number has to a square number; . . . ; and Squares which have not to one another the ratio which a square number has to a square number will not have their sides commensurable in length either.*[3]

The use in these propositions of the words "magnitude," "linc," and "squares on lines" as different from "numbers" indicates that "number" was still "integer" to the Greeks. It still had not occurred to them to extend the number system to include new numbers—irrational numbers—in order that the ratios of all magnitudes could be represented by numbers. However, there is at this time explicit recognition of that which was a "logical scandal" to the Pythagoreans—the existence of magnitudes, areas or lines, which did not have ratios representable by integers.

Eudoxus provided the definition of equal ratios or proportion and the method for using the definition which for centuries and until relatively recent times was the tool for dealing with the ratios of incommensurable quantities without the use of irrational numbers. This definition as stated in Euclid's Book V, Definition 5, was: *Magnitudes are said to be in the same ratio, the first to the second and the third to the fourth, when, if any equimultiples whatever be taken of the first and third, and any equal multiples whatever of the second and fourth, the former equimultiples alike exceed, are alike equal to, or alike fall short of, the latter equimultiples respectively taken in corresponding order.*[4] This says that $a:b = c:d$ if and only if

$$ka > mb \text{ implies } kc > md$$
$$ka = nb \text{ implies } kc = nd$$
$$ka < pb \text{ implies } kc < pd.$$

An illustration of the use of this definition is to be found in the proof of the theorem: *Triangles and parallelograms which are under the same height are to one another as their bases.* In Figure 2, *ABC* and *ACD* are two triangles with the same height. Produce *BD* in both directions laying off any number of segments $BG = GH = BC$ and $DK = KL = CD$. Complete the triangles with these segments as bases and with vertex at *A*. Therefore triangle *AHC* is the same multiple of *ABC* that base *HC* is of *BC*.

Similarly triangle *ADL* is the same multiple of *ACD* as *DL* is of *CD*.

"Thus there being four magnitudes, two bases *BC, CD,* and two triangles *ABC* and *ACD,* equimultiples have been taken of the base *BC* and the triangle *ABC,* namely the base *HC* and the triangle *AHC;* and of the base *CD* and the triangle *ADC* other, chance, equimultiples, namely, the base *LC* and the triangle *ALC;* and it has been proved that if the base *HC* is in excess of the base *CL,* the triangle *AHC* is also in excess of the triangle *ALC;* if equal, equal; and if less, less. Therefore as the base *BC* is to the base *CD,* so is the triangle *ABC* to the triangle *ACD.*"[5]

The proof for parallelograms was similar. Today, as then, logically some recognition of the possible irrationality of the dimensions of geometric figures must be made at the time of defining their areas or the ratios of areas. This may be done in several ways. One procedure is to postulate a correspondence between the "lengths" of lines or between points on lines and the real numbers (which of course include irrationals) and to assume the ability to operate with irrationals. A second procedure is to use a limit idea explicitly in getting the ratio of a figure, e.g., a rectangle, to a figure defined to have a unit area. Of course and unfortunately, there has also been a tendency to ignore the whole question in secondary school instruction.

Both Eudoxus' definition of proportion and his "method of exhaustion" are illustrated by Euclid XII, 2, which says: *Circles are to one another as squares on the diameters.* The proof is an indirect proof in which the desired conclusion is reached by showing that assuming the ratio of the circles to be different from the ratio of the squares of the diameters leads to a contradiction.

The essential steps in the proof are shown on the next page with Figure 3 to illustrate them.

(1) If
$$\frac{\text{circle } BXARD}{\text{circle } FKENH} \neq \frac{BD^2}{FH^2}$$
then
$$\frac{BD^2}{FH^2} = \frac{\text{circle } BXARD}{S}$$

where *S* is some circle other than *FKENH*.

(2) *S* is either greater or less than circle *FKENH.* Assume it is less. (Here another type of indirect proof is being used. We will follow through only one-half of it, the "less than" part.)

(3) If *EFGH* is a square inscribed in *FKENH* it is more than one-half of *FKENH* because the in-

FIGURE 2

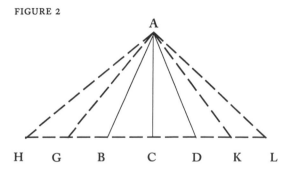

177

FIGURE 3

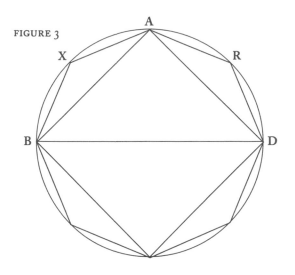

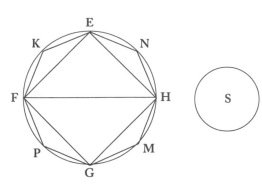

scribed square is one-half of a circumscribed square and the circle is less than its circumscribed square.

(4) Use the midpoints *KNMP* of the arcs as vertices of triangles based on the sides of the inscribed square. These triangles are together more than one-half of the area between the inscribed square and the circle.

(5) By repeatedly inscribing triangles in smaller and smaller circular arcs one can eventually arrive at an inscribed polygon which differs in area from circle *FKENH* by an amount less than the difference between circle *FKENH* and *S*. In other words, there is a polygon in *FKENH* which is greater than *S*.

(6) Now inscribe a similar polygon in *BXARD*. Then:

$$\frac{\text{area of polygon in } BXARD}{\text{area of polygon in } FKENH}$$

$$= \frac{BD^2}{FH^2} = \frac{\text{circle } BXARD}{S}$$

(7) But circle *BXARD* is greater than the polygon in *BXARD*, hence the proportion shows that *S* is greater than the polygon in *FKENH*.

But this conclusion contradicts the conclusion of step (5), and hence the assumption of step (2) that *S* is less than circle *FKENH* is false.

A similar procedure would show that *S* cannot be larger than circle *FKENH* and thus the original theorem is established.

Note that Eudoxus' "method of exhaustion" makes use of Euclid X, 1: *Two unequal magnitudes being set out, if from the greater there be subtracted a magnitude greater than its half, and from that which is left a magnitude greater than its half, and if this process be repeated continually, there will be left some magnitude which will be less than the lesser magnitude set out.*[6] In XII, 2 the lesser magnitude was the difference between circle *FKENH* and *S*.

This proposition was also used to prove X, 2: *If when the less of two unequal magnitudes is continually subtracted in turn from the greater, that which is left over never measures the one before it, the magnitudes will be incommensurable.* This is a *general* theorem on incommensurables, and like XII, 2, also has embodied in its proof the subtractive approach which we discussed earlier.

The subtractive approach coupled with X, 1 (which can be derived from what is today called the Axiom of Archimedes) comes close to modern limit ideas. In Greek times they were closely related to the philosophical-physical problem of whether matter or magnitude is infinitely divisible. Were we to tell the entire story of incommensurables and all the related questions of Greek mathematics, we would have to discuss Democritus (470?–370? B.C.), the first atomist who wrote two books on irrational lines, Zeno (ca. 450 B.C.) and his paradoxes, the ideas of Anaxagoras (ca. 440

B.C.) on infinite divisibility, as well as Antiphon (ca. 430 B.C.), Bryson (ca. 450 B.C.), and Archytas (ca. 400 B.C.), who had some conception of the method of exhaustion before Eudoxus.

However, we will close our discussion of incommensurables through the Greek period by quoting B. L. van der Waerden on a topic also related to some studies of Greek art. Some persons have termed the Greeks "tactile" minded, emphasizing sculpture more than painting, for example, and have cited the geometric nature of Greek algebra as further evidence of this. Van der Waerden writes, however, "The Greeks knew irrational ratios very well—that they did not consider $\sqrt{2}$ a number was not ignorance, but a strict adherence to the definition of number . . . . In the domain of numbers [as viewed by the Greeks] $x^2 = 2$ cannot be solved, not even in the ratios of numbers. But it is solvable in the domain of segments; indeed the diagonal of the unit square is a solution . . . . It is

therefore logical necessity, not the mere delight in the visible, which compelled the Pythagoreans to transmute their algebra into a geometric form."[7]

NOTES

1. P. S. Jones, "Irrationals or Incommensurables I, II," *The Mathematics Teacher*, vol. XLIX (February 1956), pp. 123–127, (March 1956), pp. 187–91.

2. B. L. van der Waerden, in *Science Awakening* (Groningen, Holland: P. Noordhoff, Ltd., 1954), pp. 145–46, summarizes several interesting theories which explain why Theodorus stopped with $\sqrt{17}$, one reason being that the special diagram of Figure 1 laps itself at 17.

3. T. L. Heath, *The Thirteen Books of Euclid's Elements* (Cambridge: The University Press, 1908), vol. III, p. 28.

4. *Ibid.*, vol. II, p. 114.

5. *Ibid.*, vol. II, p. 191.

6. *Ibid.*, vol. III, p. 14.

7. Van der Waerden, *op. cit.*, p. 125 ff.

HISTORICAL EXHIBIT 3.4

The Method of Archimedes

Archimedes of Syracuse (ca. 287–212 B.C.) was perhaps the greatest mathematician of the ancient world. His creativity was expressed in the production of numerous inventions, all marvels of his time, the clarification of scientific principles and applications of mathematics. Archimedes was particularly proud of the fact he discovered that if a cone was contained in a hemisphere which in turn was enclosed in a right circular cylinder as shown below, the volumes of the three solids are in the ratio 1:2:3.

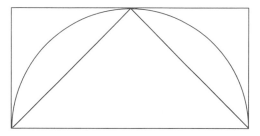

To arrive at this conclusion, Archimedes had to have known the formula for the volume of a sphere. Since it was unknown at this time, he must have derived this correct formula himself—how did he do it? This mystery remained unsolved until 1906 when the Danish philologist Johan Ludvig Heiberg, examining ancient manuscripts in a monastery library in Constantinople, came across a tenth-century palimpsest on which mathematical text could be discerned. This text proved to be a lost work of Archimedes called *The Method* in which he explained his technique for discovering the formula for the volume of a sphere. Combining principles of mathematics and physics, Archimedes proceeded as follows.

Consider a cone, a sphere, and a right circular cylinder situated along a radial axis *POQ* as shown below.

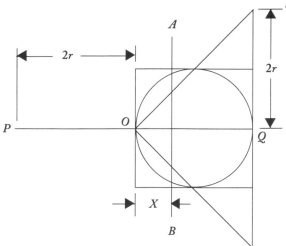

The sphere has radii of length r; the radii of the cone's base is $2r$. A plane, *AB*, slices the configuration a distance X from point O. This plane then slices discs of thickness Δx, from the cone, sphere, and cylinder respectively. Let the volumes of these discs be represented by v_k, v_s, v_c, then

$$v_k = x^2 \pi \Delta x$$

$$v_s = x \pi (2r - x) \Delta x$$

$$v_c = r^2 \pi \Delta x$$

If the discs for the cone and sphere are hung from point P, they will cause a counter-clockwise turning moment about point O of magnitudes

$$2r \left[x\pi (2r - x) + x^2\pi \right] \Delta x = 4r^2 x \pi \Delta x$$

but this is equal to $4v_c$. Assume that the volume of cone, V_k, sphere, V_s, and cylinder, V_c, can be considered as a sum of such thin discs, then:

$$2r \left[V_s + V_k \right] = 4rV_c \text{ or}$$

$$2r \left[V_s + \frac{8\pi r^3}{3} \right] = 8\pi r^4 \text{ and}$$

$$V_s = \frac{4}{3} \pi r^3$$

29

Diophantus of Alexandria

J. D. SWIFT

1. Introduction

The name of Diophantus of Alexandria is immortalized in the designation of indeterminate equations and the theory of approximation. As is perhaps more often the rule than the exception in such cases, the attribution of the name may readily be questioned. Diophantus certainly did not invent indeterminate equations. Pythagoras was credited with the solution $(2n + 1, 2n^2 + 2n, 2n^2 + 2n + 1)$ of the equation $x^2 + y^2 = z^2$; the famous Cattle Problem of Archimedes is far more difficult than anything in Diophantus, and a large number of other ancient indeterminate problems are known. Further, Diophantus did not even consider the most common type of problem called by his name, the linear equation or system of equations, to be solved in integers.

Nevertheless, on at least three grounds the place of Diophantus in the development of mathematics is secure. On all the available data he was the first to introduce systematic algebraic procedures to the solution of non-linear indeterminate equations and the first to introduce extensive and consistent algebraic notation representing a tremendous improvement over the purely verbal styles of his predecessors (and many successors). Finally, the rediscovery of the book through Byzantine sources greatly aided the renaissance of mathematics in western Europe and stimulated many mathematicians, of whom the greatest was Fermat [1].

Reprinted from *American Mathematical Monthly* 63 (Mar., 1956): 163–70; with permission of the Mathematical Association of America.

(Much of Fermat's work is known from notes written in his copy of Diophantus.)

Of Diophantus as an individual we have essentially no information. A famous problem in the Greek Anthology indicates that he died at the age of 84, but in what year or even in which century we have no definite knowledge. He quotes Hypsicles and is quoted by Theon, the father of Hypatia. Now Hypsicles, in the introduction to his book, the so-called Book XIV of Euclid, places himself within a generation or so of Apollonius of Perga whose time is definitely established by the rulers to whom he dedicates his works. Thus we may put Hypsicles in the early or middle part of the second century B.C. with reasonable accuracy [17]. Theon, on the other hand, definitely saw the eclipse of A.D. 364 [10]. Within this gap of five hundred years, historians are at liberty to place Diophantus wherever he best fits their theories of historical development [10, 14]. The majority follow [2] and, on the basis of a dubious reference by the Byzantine Psellus (ca. 1050), assign him to the third century A.D.

2. The Arithmetic

The surviving work of Diophantus consists of six books (sometimes divided into seven) of the *Arithmetic* and a fragment of a work on polygonal numbers. The introduction to the *Arithmetic* promises thirteen books. The position and content of the missing six or seven books is a matter of conjecture. (The reader is reminded that a "book" is a single scroll and represents the material contained in twenty to fifty pages of ordinary type.)

These books may be summarized as follows: Book I—Determinate systems of equations involving linear or quadratic methods. Books II–V—Equations and systems of equations, the majority of which are quadratic indeterminate, although Books IV and V contain a selection of cubic equations, determinate and indeterminate. Book VI—Equations involving right triangles. All books consist of individual problems and their solutions in positive rationals. In the ordering of the problems some consideration has been given to relative difficulty and interrelation of material, but the over-all impression is of a disconnected assortment.

3. Notation

The numerical notation used by Diophantus is, of course, the standard Hellenistic notation which uses the letters of the Greek alphabet with three archaic letters added to give 27 different symbols; [6, 7] the first nine stand for units, the second for tens and the last for hundreds. Thus, for any further notation, either non-alphabetic symbols or monogrammatic characters were required.

There is a single symbol for the unknown quantity. This may be a monogram for $\alpha\rho\iota\theta\mu\sigma\varsigma$. The symbol for "minus" is apparently a monogram for the root of $\lambda\epsilon\iota\psi\iota\varsigma$ [5]. Addition is indicated by juxtaposition. The powers of the unknown are designated by easily recognizable monograms, the square by Δ^{υ} for $\delta\upsilon\nu\alpha\mu\iota\varsigma$, the cube by K^{υ} for $\kappa\upsilon\beta\sigma\varsigma$. Higher powers are formed from these by addition, i.e., the fifth power is considered as square-cube. To avoid ambiguity it is necessary to have a special symbol for the zero-order terms also, a monogram for $\mu\sigma\nu\alpha\delta\sigma\varsigma$, and to write all the negative terms together. Thus, if we adopt an equivalent set of conventions in English, retaining Arabic numerals and the letter x for the indeterminate, the expression: $6x^4 + 23x^2 - 2x^2 + x - 5$ would appear as $SS^q 6 C^u 23 X1 MS^q 2 U^n 5$.

Fractions were represented either in the inverse position to the present day or by inserting the word for "divided by" between the numerical expressions on the same line. Reciprocals of integers and negative powers of the unknown are designated by a special symbol placed after the number or power.

The most important limitation of this notation is the restriction to one unknown. Since practically all the problems require the determination of several quantities, a considerable part of Diophantus' work lies in the reduction to a single quantity. Further, no general solution in expressed parameters is possible. Even if a general method is indicated, it must be restricted in its presentation to a specific numerical case.

A particular problem will illustrate the situation. In problem 1, Book IV, it is desired, in modern terms, to solve the system: $x^3 + y^3 = a$, $x + y = b$. Essentially the method is to let $x = z + b/2$, $y = b/2 - z$. Substitution in the first equation now yields a binomial quadratic. Let us look at this problem in a translation as bald as possible:

To partition a given number into two cubes of which the sum of the sides is given: Let the number to be partitioned be 370 and the sum of the sides $U^n 10$. Let the side of the first cube be $x1 U^n 5$, the latter term of which is half the sum of the sides. Therefore, subtracting, the side of the other cube is $U^n 5 Mx$. Then the sum of the cubes will be $S^q 30 U^n 250$. This is equal to $U^n 370$ as is given and x becomes $U^n 2$. As to the original numbers, the first side will be $U^n 7$ and the second, $U^n 3$. The first cube, 343; the second, 27.

4. Diophantine Algebra

With this problem in mind let us turn to some aspects of Greek and Babylonian mathematics. A number of tablets [15, 16], both old Babylonian (1800–1600 B.C.) and Seleucid (300 B.C. and later), exist which teach the solution of equations which can be reduced to the forms $x + y = a$; $xy = b$ [10, 12, 13]. Again Euclid's Elements II, 5, 6 can best be viewed as giving solutions to these problems [12]. In modern notation, the procedure in both cases is

183

to write $x = a/2 + z$, $y = \pm (a/2 - z)$; $xy = b = \pm (a^2/4 - z^2)$; $z = \sqrt{a^2/4 \pm b}$. Now Diophantus in I (27, 30) considers the same equations, solves them the same way and applies the basic idea repeatedly as in the quoted problem. Other examples can be followed in a similar way, e.g., $x^2 + y^2 = a$, $xy = b$. (See [13] for a complete discussion of quadratic equations in antiquity.)

Let us now compare the treatment in the three cases: The tablets consist of lists of problems of varying complexity each framed in specific numbers and quantities. The problems are not "practical" nor in any sense rigorously geometric; men are added to days, lengths to areas, areas are multiplied, etc. It is clear that the basic thought is purely algebraic. The problems are so set that the solutions are positive integers or terminating sexagesimal fractions such that the roots can be obtained from tables of squares, but from other tablets we learn of approximations to non-terminating rationals like $1/7$ or irrationals like $\sqrt{2}$.

The Euclidean problems are cast in the form of propositions about line segments, squares, and rectangles. Their generalizations in II, 28, 29, concern parallelograms. The propositions are general and the result is deduced rigorously from the postulational basis. The results are line segments which may well be incommensurable with the original segments; i.e., "irrational" answers are acceptable.

In Diophantus the problems are formulated in terms of abstract numbers but a "number" is always positive rational. The solutions are worked out in terms of particular numerical examples. This procedure may be considered analogous to carrying out a geometrical construction in terms of particular line segments and, indeed, Diophantus probably intended that his problems should be read in this manner. There is, however, no pretense at postulational development. No general propositions are stated even where the solution implies them. Restrictions on the choice of initial values are not always given; in the case of I, 27, we are informed that $a^2/4 - b$ must be a square but in the problem in the previous section no restriction is

mentioned. The most reasonable conclusion is that he did not know the form of the restriction or did not know how to express numbers that satisfied the restriction. The authors of [5] and [4] disagree but on the naive ground that since he did come up with workable numbers 370 and 10, he must have had some way of generating them. The answer is obvious; he generated them from the answer.

Like the Babylonians, Diophantus had no qualms about adding areas and lengths (see VI, 19 in 5) although, to be precise, he says that he adds "the number in the area" to "the number in the length." His algebraic technique is tremendously advanced beyond anything we possess of the Babylonians. The complicated cubic and higher degree equations and the indefinite equations are not even suggested in Babylonian algebra. The latter had examples of binary cubics and a few other higher degree equations soluble by tables; they also knew general forms for Pythagorean numbers and obtained solutions of $x^2 - 2y^2 = 1$ but this is as far as our present evidence takes them. Even in the quadratic case there may be a difference [13]. When a quadratic is to be solved, Diophantus makes some effort to choose the variable so that a binomial equation results, but if this is not practicable, the general quadratic formula (positive sign before the radical) is used without further comment. The question is still at issue whether the Babylonians ever solve a quadratic without bringing it into some normal form involving a known sum or difference and product.

It is useless even to try to guess what proportion of the advanced problems and methods are Diophantus' own. Most modern historians postulate a continuous underlying tradition of oriental algebraic methods in Greek mathematics rather than a sudden invasion in the Roman period. If this be so, texts and problem lists would certainly have existed. It is probable that the *Arithmetic* was in good part a compilation of such a quality that the predecessors were no longer held in repute. There are traces of the Diophantine notation elsewhere; Heron (A.D. 60) used the same minus sign, for example, but no evidence exists that the semi-algebraic notation or the general methods it permitted were used before the publication of the *Arithmetic.*

To sum up, the basic algebraic approach in Diophantus is Babylonian. The generality and abstraction is Greek. The work may be viewed as an episode in the decline of Greek mathematics [12] or as the finest flowering of Babylonian algebra [10].

5. *Indeterminate Problems*

In giving translations of several illustrative problems, I have avoided the usual practice of direct translation. Instead, I adhere carefully to the method of the original while replacing the particular numbers used by parameters. The rationale may thus be conveyed with less verbal explanation than if the presentation were given in its original special form. At the same time, the full power of the method is apparent.

II. 9: If $n = a^2 + b^2$, find other representations of n as the sum of two squares: Modify a to $x + a$. The corresponding modification of b may be written $(rx - b)$. Here x is the unknown, a, b and r were assigned specific values.

$$n = a^2 + b^2 = (x + a)^2 + (rx - b)^2$$
$$= (r^2 + 1)x^2 + (2a - 2br)x + a^2 + b^2.$$

Thus $x = (2br - 2a)/(r^2 + 1)$ where r may be any rational such that the required quantities are positive.

Note the clever choice of the unknown; fixing x and solving for r would leave a condition on x still to be met; $b^2 - x^2 - 2ax$ would have to be made a square. Again, if a is increased by a fixed amount and the unknown is taken as the corresponding decrease in b, the result does not come out at once. The choice of $rx - b$ instead of $b - rx$ was dictated solely by the numerical values selected which happened to make $b - rx$ negative. (But see V, 24, below.) Euler wrote $a^2 + b^2 = (a + rx)^2 + (b - qy)^2$

which results in a more symmetric solution but this concept is foreign to Diophantus' notation and the solution above is quite general.

Here would be a perfect opportunity to state a proposition instead of a problem. A proof of the theorem: "Any number which is the sum of two squares can be represented as such in an infinite number of ways," is contained in the solution above.

III. 6: Find three numbers whose sum is a square and such that the sum of any two is a square:

$$x + y + z = t^2$$
$$x + y = u^2$$
$$y + z = v^2$$
$$x + z = w^2.$$

Here Diophantus assigns a definite value to w, or, in modern notation, lets it play the role of the parameter. He then chooses an unknown, r, restricting it as follows: Let $t = r + 1$, $u = r$, $v = r - 1$. Then $z = 2r + 1$, $y = r^2 - 4r$, $x = 4r$ and $w^2 = 6r + 1$. Thus $r = (w^2 - 1)/6$ where w is an arbitrary rational exceeding 5 (so that y is positive). So $x = (2w^2 - 2)/3$; $y = (w^2 - 1)(w^2 - 25)/36$; $z = (w^2 + 2)/3$.

This problem was chosen to illustrate two points. First Diophantus is not interested in generality except as an incidental by-product. A considerable increase in generality can be obtained merely by replacing $r \pm 1$ by $r \pm s$ in the solution and this possibility could easily have been indicated by the addition of a single phrase. Second, the choice of wording of the problems is often peculiar from a modern viewpoint. This problem is clearly equivalent to the single equation: $u^2 + v^2 + w^2 = 2t^2$.

Incidentally, using methods available to Diophantus but probably exceeding his control of notation, a much more general solution of this equation is available than the system $(u, v, w, t) = ((w^2 - 1)/6, (w^2 - 7)/6, w, (w^2 + t)/6)$ given above. The equation being homogeneous, it will be more convenient to solve in integers. Let $w = rs$, $u = s^2 - p$, $v = s^2 - q$, then

$$s^4 + (r^2/2 - p - q)s^2 + (p^2 + q^2)/2 = t^2.$$

The left hand side is a perfect square if $p^2 + q^2 = 2k^2$, $r^2/2 - p - q = 2k$. The first of these is the problem of finding three squares in arithmetic progression. It does not occur specifically in the *Arithmetic*, probably because it is too simple in the rational case, reducing essentially to $a = b = 1$ in the problem above. It will be more convenient to take a solution derived from the solution given to the Pythagorean equation by Euclid. If $X^2 + Y^2 = Z^2$, $(X + Y)^2 + (X - Y)^2 = 2Z^2$. Thus $p = -m^2 + 2mn + n^2$, $q = m^2 + 2mn - n^2$, $k = m^2 + n^2$; $r^2 = 4(m + n)^2$ and $r = 2(m + n)$. Thus $(u, v, w, t) = (s^2 + m^2 - 2mn - n^2, s^2 - m^2 - 2mn + n^2, 2(m + n)s, s^2 + m^2 + n^2)$. The previous solution is obtained by setting $m = 2$, $n = 1$ and dividing by 6.

V. 24: Find a solution of $x^4 + y^4 + z^4 = t^2$.

If $t^2 = (x^2 - m)^2$, $x^2 = (m^2 - y^4 - z^4)/2m$. Thus an integer m must be found so that $(m^2 - y^4 - z^4)/2m$ is a square. Let $m = y^2 + z^2$ so $x^2 = y^2z^2/(y^2 + z^2)$. Thus $y^2 + z^2$ must be a square, say $(y + r)^2$. Then $y = (z^2 - r^2)/2r$. Thus

$$(x, y, z, t) = \frac{z^3 - r^2z}{z^2 + r^2}, \quad \frac{z^2 - r^2}{2r}, \quad z, \quad \frac{z^8 + 14z^4r^4 + r^8}{4r^2(z^2 + r^2)^2}.$$

This example has been chosen for three reasons. First, it is of great historical interest. To this problem Fermat appended a note: "Why does Diophantus not ask for the sum of *two* biquadrates to be a square? This is, indeed, impossible . . . ." Later, Euler conjectured that it was also impossible to find three fourth powers whose sum was a fourth power; i.e., to replace t^2 by t^4. This question remains unsolved.

Second, the problem indicates what happens when the notation is insufficient. First, the chosen unknown is x; m, y, z are assigned specific values to indicate that they play the role of parameters. But the problem cannot be completed, so the author turns to a sub-problem in which y is the unknown.

Finally, the problem contains a curious case of indifference to sign. The quantity $x^2 - m$ is, in fact, the negative root of $t^2 = (x^2 - m)^2$. Since only the square is used, no harm is done but we must remember that, to Diophantus, the quantity $x^2 - m$, which he used, did not exist. The reader may find it interesting to see why $x^2 + m$ was not used by trying it. Why $m - x^2$ which is positive and produces the same result was not preferred is a matter for conjecture.

VI. 19: Find a right triangle such that its area added to one of its legs is a square while the perimeter is a cube.

First form the triangle $(2x + 1, 2x^2 + 2x, 2x^2 + 2x + 1)$. The perimeter is $4x^2 + 6x + 2 = (4x + 2)(x + 1)$. Since it is difficult to make a quadratic a cube, consider in turn the triangle $(2x + 1)/(x + 1), 2x, 2x + 1/(x + 1)$ obtained by dividing through by $x + 1$. The perimeter is $4x + 2$ and the area is $(2x^2 + x)/(x + 1)$. Adding $(2x + 1)/(x + 1)$ to the latter we have $2x + 1$. Thus $4x + 2$ is required to be a cube and $2x + 1$ a square. The obvious value for $2x + 1$ is 4. Thus $x = \frac{3}{2}$ and the triangle is $\frac{8}{5}$, 3, $\frac{17}{5}$.

It is not clear whether or not Diophantus implies the more general solution $2x + 1 = 4r^6$, $x = (4r^6 - 1)/2$; probably not.

This problem is illustrative of the rather peculiar problems considered throughout Book VI and of the complete freedom from geometrical considerations. To Euclid such phrases as "the sum of one side and the area" would have been shocking nonsense.

6. *An Approximation Problem*

In V. 9 it is required as a sub-problem to find two squares, both exceeding 6, whose sum is 13. Since we have, in the first example of the preceding section, a general method of partitioning a number into two squares when one such partition is given, it is merely necessary to set the two values equal, solve for the parameter and approximate this solution in rationals. If this is done with $a = 2$ and $b = 3$, we find $r = 5 + \sqrt{26}$. Approximating by $r = 10$, $13 = (258/101)^2 + (257/101)^2$ and it is readily seen that the conditions are met.

Of course Diophantus could not do this since the parameters were not expressed. He first finds a number slightly greater than $\sqrt{13/2}$. $13/2 = 26/4$; if $\sqrt{26} < 5 + 1/x$, $x^2 < 10x + 1$; let $x = 10$, then $\sqrt{13/2} \sim 51/20$. Now $51/20 = 3 - 9/20 = 2 + 11/20$. Thus we wish to find a number near $1/20$ such that $(3 - 9y)^2 + (2 + 11y)^2 = 13$. Then $y = 5/101$ and the squares are precisely those obtained above.

The problem is typical of the approximate methods used. To approximate the nth root of a rational, first write it in the form p/q^n by multiplying numerator and denominator by the necessary integer to make the denominator a perfect nth power. Then multiply p by the nth powers of successive integers until pa^n is sufficiently close to a perfect nth power, say b^n. The approximation is then b/aq. To improve an approximation a_1 to $\sqrt{a}$, set $(a_1 + 1/x)^2 = a$ and approximate x.

7. Transmission of Diophantus

When the Arabs overran the Southeastern Mediterranean in the seventh century, they came into possession of manuscripts of works which had been published in sufficiently large editions to survive the wars attendant on the breakup of the Roman Empire and the lack of interest in learning of the early Christians. Among these was the *Arithmetic* or at least a portion of it. Translations and commentaries were published in Arabic. These have all been lost; their only trace is in bibliographers' references. When the Arabs formulated their own algebra, they apparently appealed directly to the basic Oriental tradition previously cited. The beginnings of an algebraic notation and the abstract numbers are nowhere to be seen. With the sole exception of the problems mentioned in section 3 as common to the whole ancient world (Diophantus—I, 27–30) not one problem from the *Arithmetic* is found in the algebra of Al-Khwarizmi or, as far as is known, in any other basic Oriental text [13]. Probably the Arabs found Diophantus too impractical for their utilitarian mathematics and the Hindus, if they ever saw the *Arithmetic*, were interested in other problems such as the theory of linear indeterminate problems.

In the other reservoir of learning, Byzantium, the manuscripts of Diophantus lay almost unnoticed for eight centuries. We do not know when the missing books were lost but the part which we now possess escaped the sack of Constantinople by the Crusaders in 1204 and later in the same century M. Planudes and G. Pachymeres wrote commentaries on the first part of the *Arithmetic*. At some time, probably in the course of the emigration of the Byzantine scholars during the Turkish conquests, copies were brought to Italy and Regiomontanus saw one there between 1461 and 1464.

The first translation to Latin was made by W. Holzmann who wrote under the Greek version of his name, Xylander. This translation was published in 1575. Meanwhile, Bombelli, in 1572, distributed all the problems in the first four books among problems of his own in a text on algebra. Bachet, borrowing liberally from Bombelli and Holzmann, made another translation in 1621 and a second edition was published in 1670 including Fermat's marginal notes. In the next two centuries various translations were made into modern languages which were based primarily on the editions just mentioned by Holzmann and Bachet. Finally in 1890, P. Tannery prepared a definitive edition of the Greek text with a translation into simple mathematical Latin using modern numerical and algebraic notation. From this work the three excellent translations listed in the bibliography have been prepared. The references to the last two paragraphs are the commentaries in [2] and [6], particularly the latter which has been followed rather closely.

BIBLIOGRAPHY

A. Greek Text with Latin Translation:

[1] C. G. Bachet: Diophanti Alexandrini Arithmeticorum libri sex, *etc.* Paris, second edition, 1670.

[2] P. Tannery: Diophanti Alexandrini opera omnia cum Graecis commentariis, Teubner, vol. 1, 1893, vol. ii. 1895.

B. Modern translations based on [2]:

[3] A. Czwalina: Arithmetik des Diophantos aus Alexandria, Göttingen, Vandenhoek, 1952.

[4] P. ver Eeke: Diophante d' Alexandrie. Les six livres arithmetiques et le livre des nombres polygones. Bruges, Descée, 1926.

[5] T. L. Heath: Diophantus of Alexandria, Cambridge, 1910.

C. Histories and Compilations of Ancient Mathematics:

[6] T. L. Heath: History of Greek Mathematics (Two volumes), Oxford, 1921.

[7] ———A Manual of Greek Mathematics, Oxford, 1931.

[8] G. Loria: Le Scienze esatte nell' antica Grecia, 2nd edition, Milan, Hoepli, 1914.

[9] G. H. F. Nesslemann: Die Algebra der Griechen, Reimer, Berlin, 1842.

[10] O. Neugebauer: The Exact Sciences in Antiquity, Princeton, 1952.

[11] I. Thomas: Selections Illustrating the History of Greek Mathematics, Harvard and Cambridge (Loeb Classics) 1939.

[12] B. L. van der Waerden: Science Awakening, P. Noordhoff, Groningen, 1954 (English translation by A. Dresden).

D. Miscellaneous References:

[13] S. Gandz: The Origin and Development of the Quadratic Equations in Babylonian, Greek and Early Arabic Algebra, Osiris, vol. 3 (1938) pp. 405—557.

[14] J. Klein: Die griechische Logistik und die Entstehung der Algebra, Quellen and Studien zur Ges. der Math. Abt. B, vol. 3, 1934–6, pp. 18–105; 122–235.

[15] O. Neugebauer: Mathematische Keilschrift-Texte, 3 Vols., Springer, 1935–7 = Quellen und Studien zur Ges. der Math. Abt. A, vol. 2.

[16] ———and A. Sachs: Mathematical Cuneiform Texts, American Oriental Society, New Haven, 1945 = Am. Oriental Series, vol. 29.

[17] Paulys Real Encyclopädie der Classischen Altertums Wissenshaft, Stuttgart, 1893.

30

How Ptolemy Constructed Trigonometry Tables

BROTHER T. BRENDAN

Introduction

Historians point out that, while in economics Gresham's Law holds ("bad currency drives out good currency"), in mathematics a kind of reverse law holds: Good books drive out bad or inferior books. Thus, historians can offer evidence that Euclid's *Elements* drove out of circulation the geometry texts of his predecessors, and, in a less well-known instance, they can show that Ptolemy's *Almagest* made the trigonometric writings of Archimedes, Hipparchus, and perhaps others disappear. And in Ptolemy, just as in the case of Euclid, there are some beautiful gems of mathematical literature available—even if his system of astronomy has been superseded.

Little is known about the author save what can be inferred from the treatise itself, *Mathematike Syntaxis*. (The modern title, *Almagest*, has come down to us by way of the Arabs.) In this work he refers to stellar observations made personally near Alexandria between the years A.D. 127 and A.D. 151. Speculation about his name, Claudius Ptolemaeus, has been widespread, but nothing historically certain can be inferred from it. What is of interest is that in Book I he supplies, among extant works, the first genuinely mathematical treatment of the essentials of plane trigonometry. (He does

much the same for spherical trigonometry in Book IX.) Book I served as the basis for all the work in plane trigonometry among the Arabs and in the West until about the year 1650. At that time, Wallis, Gregory, Mercator, Newton, Leibniz, and others began to apply the new techniques of infinite series to the subject [1].*

The viability of Ptolemy's trigonometry is partly explained by the fact that with it one could solve any of our usual "triangle problems." The chief instrument of this was his "table of chords," a brilliantly constructed equivalent of our sine tables and of amazing accuracy. It offers, for every 15 minutes between 0° and 90°, values of the sine correct to four decimal places—as well as careful means for interpolation at every $1/2$ minute. Augustus De Morgan is credited [2] with saying that this construction is "one of the most beautiful in the Greek writers."

The purpose of the present paper [3] is to outline the elegant mathematics to be found in Book I of the *Almagest* and at the same time to call attention to its many items of enrichment for the student; applications of Euclid, uses for the sexagesimal number system, employment of inequalities and approximations to the beginning of real number theory, practice in interpolation, insights into some trigonometric identities, and historical oddities such as the possible first use of "zero."

Reprinted from *Mathematics Teacher* 58 (Feb., 1965): 141–49; with permission of the National Council of Teachers of Mathematics.

* Numerals in brackets refer to the Notes at the end of this article.

Preliminaries

In constructing his table, Ptolemy used the traditional mixed sexagesimal-decimal number system, assigning degrees, minutes, seconds, and thirds to the angles, and corresponding measures (here called "parts" instead of degrees) to the chords subtending those angles. Instead of a unit circle as his basic circle of reference, he chose to use a circle of radius = 60 parts. This choice kept his table entries all of a reasonable order of magnitude. He could almost immediately, then, write down Table 1. For an angle of 60° has a chord equal to a radius, and from elementary geometry we readily find that an angle of 90° has a chord of 84 parts, 51 min, and 10 sec (approximately); also, the chord of 120° has a chord of 103 parts, 55 min, and 23 sec (approximately); and, of course, the chord of 180° is exactly 120 parts. The original tables of Ptolemy were constructed with the Greek system of numeration, that is, they used the letters of the alphabet for numbers according to the well-known schemes described in references [4] and [5]. What is notable is that we have here the use of the Greek letter omicron, which looks like our zero, to indicate vacancies in the table. We do not have evidence, however, to show that Ptolemy actually calculated with this symbol in the way that we calculate with zero, for, unfortunately, he does not display any of his actual computations. But it seems likely, to some observers, that he did more than use "zero" for marking vacancies in his tables.

The problem now is to expand Table 1 into a complete "table of chords." In order to do this systematically, Ptolemy used a set of theorems and corollaries.

TABLE I

Angle (in degrees)	Chord (in parts, minutes, seconds)		
60	60	0	0
90	84	51	10
120	103	55	23
180	120	0	0

Ptolemy's First Theorem

The simplest chords are sides of regular polygons, so Ptolemy starts with a neat construction and proof relative to the sides of regular inscribed pentagons, hexagons, and decagons.

Given (Fig. 1): A circle with center at D and with diameter ADC; DB perpendicular at D; DC bisected at E; EF constructed equal to EB.

Prove: FD is equal to a side of a regular inscribed decagon; BF is equal to a side of a regular inscribed pentagon.

The proof depends on readily established converses of theorems in Book XIII of Euclid's *Elements*, to which Ptolemy appeals. They state: (1) If $CF/DC = DC/FD$, then FD is the side of a decagon; (2) if $BF^2 = DB^2 + FD^2$, then BF is the side of a pentagon. A proof of these converses and a check on the validity of this construction (using similar triangles) is a good exercise for students.

Note that this first theorem puts Ptolemy in a position to find BF, or chord 72°, and FD, or chord 36°, since

$$FD = EF - ED = [(30^p)^2 + (60^p)^2]^{1/2} - 30^p$$
$$\doteq 37^p\ 4'\ 55'',$$
$$BF = (FD^2 + DB^2)^{1/2} \doteq 70^p\ 32'3''.$$

FIGURE I

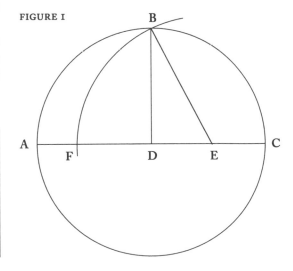

And, of course, as in computing Table 1, we can apply the Pythagorean theorem to right triangles inscribed in semicircles of diameter 120^p to get what we might call "supplementary chords":

$$\text{crd } 144° = [(120^p)^2 - (\text{crd } 36°)^2]^{1/2}$$
$$\doteq 114^p\ 7'\ 37",$$
$$\text{crd } 108° \doteq 97^p\ 4'\ 56".$$

Since, as remarked above, Ptolemy does not exhibit his actual calculations, we can try to perform them ourselves. Another good exercise for the student!

Ptolemy's Second Theorem

Ptolemy's second theorem is well known as "Ptolemy's lemma."

Given (Fig. 2): An inscribed cyclic quadrilateral *ABCD*.
Prove: $(AC)(BD) = (AB)(DC) + (BC)(AD)$.

Select *E* on *AC* so that $\angle ABE = \angle DBC$. This assures similar triangles *ABD* and *EBC* and similar triangles *ABE* and *DBC*. From the proportions *BC/EC = BD/AD* and *AB/AE = DB/DC* we get the desired result $(AE + EC)BD = (AB)(DC) + (BC)(AD)$.

The power of this second theorem for our purpose lies in the cases where one side of the quadrilateral is a diameter of 120^p.

FIGURE 2

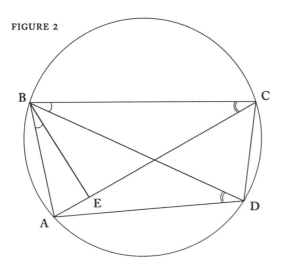

FIGURE 3

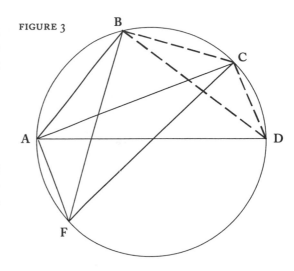

A. Thus we get first a "difference corollary." (See Fig. 3.)

Given: Chords *AB* and *AC*.
Find: Chord *BC*

By "supplementation" we calculate *BD* and *DC*, and with $AD = 120^p$ we have only one unknown in the equation of Ptolemy's lemma. That is, we can calculate *BC* from the equation $(BC)(AD) + (AB)(DC) = (AC)(BD)$.

For example, we get crd 12° from crd 72° and crd 60°, by this "difference corollary," to be $12^p\ 32'\ 36".$

B. Similarly we have a "summation corollary" (same figure).

Given: Chords *BC* and *CD*.
Find: Chord *BD*.

Supply the additional diameter *CF* as shown and also chords *AF* and *BF* so that by "supplementation" we can find *AC* and *BF*. Then, by using the "difference corollary" in quadrilateral *ABCF*, we can get *AB*. But *AB* is the "supplement" of the required *BD*.

C. Another result of Ptolemy's lemma is a "half-angle corollary" (see Fig. 4).

Given: Chord *BC* and the midpoint *D* of arc *BC*.
Find: Chord *CD*.

191

FIGURE 4

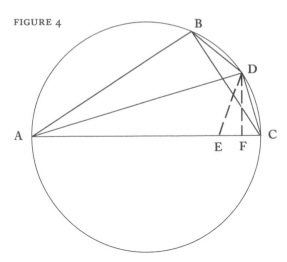

FIGURE 5

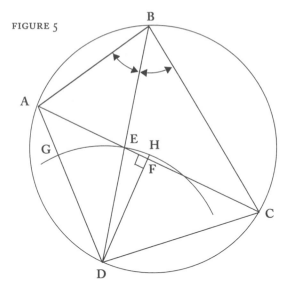

AB is easily found as the "supplement" of *BC*, and, if we mark off *AE* = *AB*, we can show that *BD* = *DE* = *CD*. Further, if *F* is chosen so that

$$EF = CF = \frac{(AC - AB)}{2},$$

we have right triangle *ADC* similar to right triangle *DFC*, and thus $(AC)(CF) = CD^2$. Therefore *CD* can be calculated.

For example, knowing crd 12°, we can get crd 6° $\doteq$ 6^p 16' 49", and then crd 3° $\doteq$ 3^p 8' 28", and then crd (³/₂)° $\doteq$ 1^p 34' 15", and crd (³/₄)° $\doteq$ 0^p 47' 8".

At this juncture, however, Ptolemy notes that we are not able to get crd (¹/₂)° or crd 1° for our table. The reason he offers, though he does not prove it, is that, "The chord of a third of the arc is in no way geometrically given." Which is to say that even if we know the chord of (³/₂)° we have no geometrical constructions to give us the chord of (¹/₂)°. He therefore has to resort to another idea (which he calls "a little lemma," but which commentators regard as one of his greatest achievements).

Ptolemy's Third Theorem

Ptolemy's "little lemma," or third theorem, as we shall call it, establishes an inequality.

Given (see Fig. 5): Chord *BC* > chord *AB*.

Prove: $\dfrac{\text{chord } BC}{\text{chord } AB} < \dfrac{\text{arc } BC}{\text{arc } AB}$.

Bisect $\angle B$, making crd *AD* = crd *CD*, and drop perpendicular *DF* on *AC*. Then *CE* > *AE* and *AD* > *DE* > *DF*, so that the constructed arc *GEH* must fall as shown. Now sector *DEH* > $\triangle DEF$, and $\triangle DEA$ > sector *DEG*, so that, by Euclid V–8, we have

$$\frac{\triangle DEF}{\triangle DEA} < \frac{\text{sector } DEH}{\text{sector } DEG},$$

where

$$\frac{\triangle DEF}{\triangle DEA} = \frac{EF}{EA}$$

and

$$\frac{\text{sector } DEH}{\text{sector } DEG} < \frac{\angle FDE}{\angle EDA}.$$

It follows that

$$\frac{EF}{EA} < \frac{\angle FDE}{\angle EDA}.$$

Hence we have

$$\frac{EF + EA}{EA} < \frac{\angle FDE + \angle EDA}{\angle EDA}$$

or

$$\frac{AF}{EA} < \frac{\angle FDA}{\angle EDA}.$$

And since $\triangle CAD$ is isosceles, this means that

$$\frac{CA}{EA} < \frac{\angle CDA}{\angle EDA}.$$

But then

$$\frac{CA - EA}{EA} < \frac{\angle CDA - \angle EDA}{\angle EDA}$$

or

$$\frac{CE}{EA} < \frac{\angle CDB}{\angle BDA}.$$

The proof is completed by substituting the following values into the last inequality:

$$\frac{\text{crd } BC}{\text{crd } AB} = \frac{CE}{EA} \quad \text{and} \quad \frac{\text{arc } BC}{\text{arc } AB} = \frac{\angle CDB}{\angle BDA}.$$

Ptolemy uses his third theorem as follows. We have

$$\frac{\text{crd } 1°}{\text{crd } (3/4)°} < \frac{1}{3/4} \quad \text{and crd } (3/4)° \doteq 0^p \, 47' \, 8''$$

imply that crd $1° < 1^p \, 2' \, 50''$. But

$$\frac{\text{crd } (3/2)°}{\text{crd } 1°} < \frac{3/2}{1} \quad \text{and crd } (3/2)° \doteq 1^p \, 34' \, 15''$$

imply that crd $1° > 1^p \, 2' \, 50''$. Ptolemy concludes from these two inequalities, "And so, since it has been proved that the chord of an arc of 1° is both greater than and less than the same number of parts, clearly we shall have crd $1° = 1^p \, 2' \, 50''$; and by means of earlier proofs we get crd $(1/2)° = 0^p \, 31' \, 25''$." Toeplitz remarks [6] that Ptolemy's bland assertion of the absurdity "both greater than and less than" expresses a very modern viewpoint.

Completion of the Table of Chords

Table 2 indicates some of the entries which Ptolemy is now in a position to make for his complete table of chords. The third column is labeled "sixtieths" and is carried out to "thirds" (an extra place beyond "seconds") to help in interpolation in steps of $1/60$ of a degree. It represents sixtieths of a whole de-

TABLE 2

ARCS	CHORDS			SIXTIETHS			
$1/2$	0	31	25	0	1	2	50
1	1	2	50	0	1	2	50
$1^{1}/_{2}$	1	34	15	0	1	2	50
2	2	5	40	0	1	2	50
$2^{1}/_{2}$	2	37	4	0	1	2	48
3	3	8	28	0	1	2	48
5	5	14	4	0	1	2	46
10	10	27	32	0	1	2	35
20	20	50	16	0	1	1	51
30	31	3	30	0	1	0	35
40	41	2	33	0	0	59	0
60	60	0	0	0	0	54	21
70	68	49	45	0	0	51	23
90	84	51	10	0	0	44	20
100	91	55	32	0	0	40	17
120	103	55	23	0	0	31	18
135	110	51	57	0	0	23	55
140	112	45	28	0	0	21	22
160	118	10	37	0	0	10	47
180	120	0	0	0	0	0	0

gree or thirtieths of a half-degree. In calculating sine values (instead of chords) this would lead, of course, to values in steps of $1/2$ minute [7].

It is apparent that by using such a table, most ordinary triangle problems found in modern trigonometry texts were solvable in Ptolemy's time. Students can check this assertion, and they can also be given other exercises derived from the table, such as to find the values of $\sqrt{2}$ and $\sqrt{3}$. In some of these problems, Figure 6 is useful. In that figure we note the following relationships:

$$\sin A = (\text{crd } 2A)/120^p$$

and

$$\cos A = [\text{crd } (180° - 2A)]/120^p$$
$$= [(120^p)^2 - (\text{crd } 2A)^2]^{1/2}/120^p.$$

Of somewhat more interest is the fact that several familiar analytical expressions are derivable from the corollaries given above. For example, the

FIGURE 6

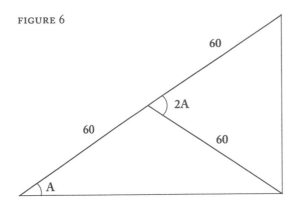

"difference corollary" mentioned above involves a relation

$$BC = [(AC)(BD) - (AB)(DC)]/120^p,$$

and from this we get, since $\sin A = (\text{crd } 2A)/120^p$,

$$\sin(\alpha - \beta) = \sin \alpha \cos \beta - \sin \beta \cos \alpha.$$

Other such standard identities can be derived from the "summation corollary" and the "half-angle corollary." These derivations make good exercises.

As a concluding remark, I want to express my thanks to Professor Philip S. Jones of the University of Michigan for encouragement in writing this paper and for assistance in researching it.

BIBLIOGRAPHY AND NOTES

1. "Trigonometry," article in *Encyclopaedia Britannica*, 11th ed.

2. "Ptolemy," *ibid.*

3. The present paper is based on a talk given at the Monterey conference of the California Mathematics Council, Dec. 8, 1963.

4. O. Neugebauer, *The Exact Sciences in Antiquity.* (New York: Harper and Brothers, 1962.) This is a Harper Torchbook, "Science Library/TB 552." See especially pp. 10 and 209ff.

5. B. L. Van der Waerden, *Science Awakening.* (Groningen, Netherlands: P. Noordhoff, Ltd., 1954.) See especially pp. 206ff. and 271ff.

6. O. Toeplitz, *Calculus, a Genetic Approach.* (Chicago: University of Chicago Press. 1963.) This is a paperback, part of the "Phoenix Science Series," No. 520, and highly recommended. See especially pp. 17–22.

7. The complete text of Ptolemy's *Almagest* is available with the tables in the following:

a) R. Hutchins (ed.), *Great Books of the Western World*, Vol. 16. (Chicago: Encyclopaedia Britannica, 1952.) This is the only source in English.

b) M. Halma, *Composition Mathématique de Claude Ptolémée*, Vols. 1 and 2. (Paris: Henri Grand, 1813; reprinted by J. Helmann, 1927.) This contains both Greek and French.

c) J. L. Heiberg, *Almagest.* (Leipzig: 1898.) This is the critical Greek text. (Translated by Manitius into German in 1912.)

8. Additional help in understanding and appreciating the role of Ptolemy can be found in:

a) A. Aaboe, *Episodes from the Early History of Mathematics.* (New York: Random House, 1964.) This is Vol. 13 of the New Mathematical Library. See Chapter 4.

b) T. Dantzig, *The Bequest of the Greeks.* (New York: Charles Scribner's Sons, 1955.) See especially Chapter 15.

c) Dedron and Itard, *Mathématiques et Mathématiciens.* (Paris: Magnard, 1959.) A copiously illustrated history.

d) H. Eves, *An Introduction to the History of Mathematics.* (2nd ed.; New York: Holt, Rinehart & Winston, Inc., 1964.) See pp. 154–56 and Problem Study 6.9 (Ptolemy's Table of Chords), pp. 171–72.

e) J. Gow, *Short History of Greek Mathematics.* (New York: Stechert, 1923.)

f) T. Heath, *A History of Greek Mathematics*, Vol. 2. (Oxford: The Clarendon Press, 1921; reprinted, 1960.)

g) I. Thomas, *Greek Mathematical Works*, Vol. 2. (Cambridge, Mass.: Harvard University Press, 1941.)

SUGGESTED READINGS FOR PART III

Aaboe, Asgar. *Episodes from the Early History of Mathematics.* New York: Random House, 1964.

Chace, A. B., et al., eds. *The Rhind Mathematical Papyrus.* 2 vols. Oberlin, OH: Mathematical Association of America, 1927–1929.

Dantzig, Tobias. *The Bequest of the Greeks.* New York: Charles Scribner's, 1955.

Gillings, Richard J. *Mathematics in the Time of the Pharaohs.* Cambridge, MA: MIT Press, 1972.

Heath, T. L. *History of Greek Mathematics.* 2 vols. New York: Dover, 1981 reprint.

Knorr, Wilbur. *The Ancient Tradition of Geometric Problems.* Boston: Birkhauser, 1986.

Neugebauer, O. *The Exact Sciences in Antiquity.* 2nd ed. New York: Dover Publications, 1969.

Sarton, George. *A History of Science: Hellenistic Science and Culture in the Last Three Centuries B.C.* New York: W. W. Norton, 1970.

van der Waerden, B. L. *Science Awakening.* Groningen: Noordhoff, 1954.

———. *Geometry and Algebra in Ancient Civilizations.* Berlin: Springer-Verlag, 1983.

PART IV

EUROPEAN MATHEMATICS DURING THE "DARK AGES"

THE HISTORICAL PERIOD from the close of the sixth century to approximately the beginning of the twelfth has frequently been referred to as the "Dark Ages" of European intellectual and cultural achievement. Such a designation results from a relative judgment based on retrospective comparison between the preceding accomplishments of the Greek world and those to come later during the Renaissance. True, chronologically, the "Dark Ages" existed between these two periods of illustrious human achievement, but, in itself, it was not a period of barren stagnation as has too easily been inferred. The centuries in question mark a period of strife, social experimentation, and societal and intellectual transition. The political void left by the fall of the Roman Empire was filled by the institutional authority of the Catholic Church. Its priorities were spiritual not secular. Revelation was frequently placed in a position of official dominance over reason. While some priorities changed, the spirit of human ingenuity still thrived, and cultural and technological innovations abounded. In the period from the sixth century to the twelfth, the uses of wind, water, and animal power for agricultural and industrial purposes were greatly enhanced, freeing man from much tedious labor. Technologies of mineral mining and processing were perfected. The iron plow and the concept of contour planting and crop rotation were introduced into agriculture, revolutionizing food production. In architecture, the use of vaulted arches allowed man to capture space and often that space, as in the case of the gothic cathedral, was filled with ecclesiastic art and music. Many of the inventions of this time might be viewed as mundane by today's standards, for example, the use of an iron stirrup for horseback riding or the construction of a chimney on a building, but such innovations drastically improved the quality and potential of daily life in the Middle Ages. It was this general improvement of living standards that helped to develop a climate of self-sufficiency in which creativity could eventually be expressed in the flowering of a humanistic renaissance.

Mathematical practices of the Middle Ages were inherited from the Romans. The Romans were builders and engineers, pragmatic users of mathematics who demonstrated little interest in the subject's theoretical aspects. Their geometry and trigonometry, while founded on solid principles, were used for engineering works and were mainly descriptive in nature. Calculation was accomplished on a counter or slab abacus and the results recorded in Roman numerals. Some scholars of the time attempted to preserve classical mathematical traditions. Nicomachus of Gerasa, who lived around A.D. 100, was a neo-Pythagorean who presented a Pythagorean view of arithmetic in his *Introductio Arithmetica*. This work became a model for later medieval writers such as Boethius (ca. 480–524). Pythagorean numerology and mysticism as evident in these early works affirmed the Church's suspicions of the diabolical nature of mathematics as St. Augustine warned in A.D. 400, "The good Christian should beware of mathematicians and all those who make empty prophecies. The danger already exists that mathematicians have made a covenant with the devil to darken the spirit and confine man in the bonds of hell."

Despite such opposition, mathematical activities continued. Mathematical manuscripts of this period were of three types: Computi which taught mathematical methods for computing the determination of Church holy days, particularly Easter; abacus arithmetics intended to convey traditional mathematics using Roman numerals and a counter; and algorisms, texts on new computational techniques based on the use of Hindu-Arabic numerals and written algorithms. It was this latter group of writings that would irrevocably change the character of Western mathematics. Perhaps the most influential of these algorisms was *Liber abaci* published by Leonardo of Pisa (Fibonacci) in 1202. In his book, Leonardo introduced European audiences to the new numerals, their algorithms, and their capacity for problem solving.

The following inclusions were selected to reflect on the climate for mathematics and its learning during the "Dark Ages."

31

Practical Mathematics
of Roman Times

MARGARET FIELDS

*F*EW PROFESSIONAL WRITINGS have exerted so great an influence on the practitioners of art as the *De Architectura* of Vitruvius (ca. 20 B.C.). Upon the invention of printing it was one of the first books to be published, three editions being included among the incunabula, and during the Renaissance it was the handbook of all the great architects and engineers.

The *De Architectura* becomes immediately of interest to the mathematician when, in his introduction, Vitruvius writes as follows: "Geometry, also, is of much assistance in architecture, and in particular it teaches us the use of the rule and compasses, by which especially we acquire readiness in making plans for buildings in their grounds, and rightly apply the square, the level and the plummet. By means of optics, again, the light in buildings can be drawn from fixed quarters of the sky. It is true that it is by arithmetic that the total cost of buildings is calculated and measurements are computed, but difficult questions involving symmetry are solved by means of geometrical theories and methods."

It being impossible to give here an exhaustive account of the many topics treated by Vitruvius, only those which seem to be of greatest interest are presented.

Requirements of Architecture

Architecture depends on Order, Arrangement, Eurythmy, Symmetry, Propriety, and Economy.

Of these, Order concerns symmetrical agreement of the proportions of the whole. *Arrangement* has to do with forms of expression such as groundplan, elevation, and perspective. *Symmetry* considers the proper agreement between different parts and the whole general scheme. In the human body there is symmetrical harmony between the forearm, foot, palm, and finger—"so it is with perfect buildings."

On Symmetry: In Temples and in the Human Body

The design of a temple depends on symmetry which is due to proportion. Proportion is a correspondence among the measures of the members of an entire work to a certain part selected as standard. From this result the principles of symmetry. Without symmetry there can be no principles in the design of any temple; that is, if there is no precise relation between its members as in the case of those of a well-shaped man. The human body is so designed that the face from the chin to the top of the forehead is a tenth part of the whole height; the open hand from the wrist to the tip of the middle finger is just as the same, etc. Other members, too, have symmetrical proportions and it was by employing them that the famous painters and sculptors of antiquity attained to great and endless renown. Thus there is good reason for the rule that the different members of a perfect building must be in exact symmetrical relation to the whole general scheme.

Reprinted from *Mathematics Teacher* 26 (Feb., 1933): 77–84; with permission of the National Council of Teachers of Mathematics.

In the actual building of the temple the columns should be placed so that there are twice as many intercolumniations on the sides as in front, thus the length will be twice the width. The steps in front should be arranged so that there is always an odd number; thus the right foot with which one mounts the first will be the first to reach the level of the temple. The columns at the corners should be made thicker than the others by one fiftieth of their own diameter because they are sharply outlined by unobstructed air and seem to be more slender than they are. Hence, we must counteract the ocular deception by an adjustment of proportions.

Ionic Architecture

All parts above the capitals of the columns should be inclined to the front a twelfth part of their own height for when we stand in front of them, if two lines are drawn from the eye, one reaching to the top of the building and the other to the bottom, the one from the upper part will be longer and make it look as if it were leaning back, but when inclined to the front, they will appear to be plumb.

Doric Columns

The Doric columns should be fluted with twenty flutes. If these are to be channeled out, the contour of the channeling may be determined thus: Draw a square with sides equal in length to the breadth of the fluting, and center a pair of compasses in the middle of this square. Then describe a circle with a circumference touching the angles of the square, and let the channelings have the contour of the segment formed by the circumference and the side of the square.

Directions of the Streets

In laying out a town, streets must be laid with due regard to climatic conditions; cold winds are disagreeable, hot winds are enervating, and moist winds are unhealthful. Therefore, by shutting out winds from our dwellings, we make them health-

ful. The method of procedure is as follows: In the middle of the city let a spot be made "true" by means of the rule and level. In the center of this spot set up a bronze gnomon or "shadow tracker." At about the fifth hour of the morning mark the end of the shadow cast by the gnomon. Then opening your compasses to this point—B—describe a circle from the center. In the afternoon when the shadow once more touches the circumference of this circle, mark it with a point C. From these two points describe intersecting arcs, and through their intersection and the center let a line—DE—be drawn to the circumference of the circle to give us the quarters of north and south. Then using a sixteenth part of the circumference of the circle as a diameter, describe a circle with its center on the line to the south, at the point where it crosses the circumference, and put points to the right and left on the circumference on the south side, repeating the process on the north side. From the four points thus obtained draw lines intersecting the center and from one side of the circumference to the other. Thus we shall have an eighth part of the circumference. The rest of the circumference is then to be divided into three equal parts on each side, and thus we have designed a figure equally apportioned among the eight winds. Then let the direction of the streets and alleys be laid down on the lines of division between the quarters of two winds.

Leveling and Leveling Instruments

Vitruvius considers three types of leveling instruments—dioptrae, water levels, and chorobates. However, he considers the first two as being very deceptive and describes the chorobate only. It is a straight-edge about twenty feet long. At the extremities it has legs, made exactly alike and jointed on perpendicularly to the extremities of the straightedge, and also crosspieces fastened by tenons connecting the straightedge and the legs. These crosspieces have vertical lines drawn upon them, and there are plumblines hanging from the straightedge over each of the lines. When the straightedge is in position and the plumblines strike both the lines alike and at the same time, the instrument is level. If the wind interposes, have a groove on the upper side 5 feet long and one digit wide and a digit and a half deep and pour water into it. If the water comes up uniformly to the rims of the groove, the instrument is level.

Pythagoras's 3-4-5 Plan

Vitruvius gives an interesting application of this plan in the building of staircases so that the steps may be at proper levels. Suppose the height of the story to be divided into three parts, five of these will give the right length for the stringers of the stairway. Let four parts, each equal to one of the three, be set off from the perpendicular, and there fix the lower ends of the stringers.

Hodometer

This instrument, for "telling the number of miles while sitting in a carriage or sailing by sea," is one of the most interesting. The wheels of the carriage were made 4 feet in diameter, so that if a wheel has a mark made upon it and begins to move forward, it will have covered exactly 12 and one-half feet after one revolution. Let a drum with a single tooth projecting beyond the face of its circumference be firmly fastened to the inner side of the hub of the wheel. Then above this, let a case be firmly fastened to the body of the carriage, containing a revolving drum set on edge and mounted on an axle; on the face of the drum there are 400 teeth engaging the tooth of the drum below. The upper drum has, moreover, one tooth standing out farther than the others. Then above, a horizontal drum is placed in another case with its teeth engaging the tooth fixed to the side of the second drum and let as many holes be made in this third drum as will correspond approximately to the number of miles in a day's journey. Let a small round stone be placed in every one of these holes, and in

the case containing that drum let one hole be made with a small pipe attached through which when they reach the point, the stones fall one by one into a bronze vessel underneath in the body of the carriage. The result is that the upper drum is carried around once for every 400 revolutions of the lowest and the tooth fixed to its side pushes forward one tooth of the horizontal drum. Since with 400 revolutions of the lowest drum, the upper will revolve once, the progress will be a distance of $400 \times 12\frac{1}{2}$ or 5000 feet or one Roman mile.

The same principle was used on ships. The axle passed through the sides of the ship with its ends projecting and four-foot wheels were mounted on them, with projecting flatboards fastened round their faces and striking the water. The middle of the axle in the middle of the ship carried a drum with one tooth projecting and to this the other drums were attached just as in the land hodometer. When the ship was making headway, the flatboards on the wheel struck against the water and were driven violently back, thus turning the wheels which moved the axle and the axle the drum, etc.

Aqueducts

With the exception of the *De Architectura* the only other extant work on Roman engineering is the *De Aquis Urbis Romae* of Frontinus. Frontinus was "curator aquarum" or superintendent of the water supply of the city of Rome appointed under Nerva in A.D. 97. He describes in much detail the materials used and the dimensions of the various aqueducts. The most interesting thing mathematically is the method used by Frontinus for determining the amount of water brought to Rome by the aqueducts, as it affords an interesting illustration of the difficulties of Roman arithmetic. He assumed that the discharge of an aqueduct was equal to the total discharge of a large number of small pipes whose combined cross-sectional areas were equal to the cross-sectional areas of the aqueduct, which of course is far from correct. He also failed to take into account the effect of the velocity of the flowing stream on the rate of discharge. Consequently his calculations were of little value. It is estimated that only one half as much water actually was brought to Rome as was stated by Frontinus. Regardless of this, when we read Frontinus's account of the Roman aqueducts we cannot but agree with Pliny: "fatebitur nihil magis mirandum fuisse in toto orbe terrarum."

Bridges

Throughout the Roman dominions, especially during the Imperial period, stone bridges with wide spanning arches of the most massive kind were erected in great numbers. Many were of remarkable size—one bridge over the Acheron being 1000 feet in length—and show in a striking way the skill of the Roman engineers. Many of these bridges still exist in various states of preservation.

One of the most interesting bridges was that described by Julius Caesar (*Bell. Gall.* IV, 17) which was a temporary wooden bridge constructed over the Rhine in the incredibly short space of ten days. It was supported on a series of double piles, formed by two baulks of timber, each 18 inches square in section, pointed at one end and driven into the bed of the river by machines called *fistucae.* These fistucae, which were also used for ramming down pavements and threshing floors, are supposed to have been very similar to our pile driving machines, which lifted a heavy log of wood shod with iron to a considerable height and then let it fall on the head of the pile. The piles were set in a sloping direction to resist the force of the current and a corresponding row of piles was driven in at a distance of 40 feet, thus forming a wide roadway for the Roman army. The cross-pieces were 2 feet thick and were supported by cross-struts so as to diminish the bearing.

A more permanent military bridge constructed by Trajan across the Danube and designed by the celebrated engineer Apollodorus shows great skill and ingenuity in the way in which he spanned wide spaces with short pieces of timber. According to Gest "it showed the greatness of invincible courage." The piers of this bridge rested upon the foundations made by sinking barges loaded with concrete, foreshadowing rather rudely the modern caisson. There is an interesting story, probably untrue, to the effect that though Hadrian demolished Apollodorus' bridge on the pretense that it might facilitate incursions of barbarians into the Roman provinces, he did it really from jealousy at the success of so great an undertaking.

Tunnels

The Roman engineers, as stated above, were very ingenious. They fitted the lines of their aqueducts to the contour of the ground and selected favorable points for crossing valleys. In building bridges they made up for a lack of theoretical knowledge by an excessive use of materials, and when they drove tunnels they expected the drifts to meet with a fair amount of precision. However, occasionally they did make a mistake. One such mistake is recorded by Gest. Nonius Datus, an engineer of the Third Legion, surveyed and marked the line for a tunnel and then was called away on military duty. During his absence the tunnels were dug, passed the expected meeting point and the drifts were found to have missed each other entirely. Nonius was summoned post haste. When he arrived, he observed in true modern style that as usual they put all the blame on the engineer when it was really all the fault of the contractor, who, in producing the alignment had in both cases deviated to the right. Fortunately the tunnel was for an aqueduct and not a railroad so Nonius was able to locate a traverse tunnel connecting the two.

Clocks

According to Pliny there was no sundial in Rome until eleven years before the war with Pyrrhus (about 460 A.U.C.), the first being placed by Cursor on the temple of Quirinus. Later the sundials became of very general use and in various forms. Of the numerous kinds, two have been preserved—the hollow hemispherical one and the flat one. The hour lines are in almost every instance engraved in the same manner. They are mostly bounded by the segments of the circle. A mid-day line *m* is cut by another line running from east to west upon the intersection of which with the hour lines the shadow of the gnomon *g* must fall at fixed times. On these intersecting points the hours are marked. In order to know the hour without giving themselves any trouble, slaves were kept on purpose to watch the "solarium" and "clepsydra" and report each time that an hour expired.

Mart. VIII. 67: Horas quinque puer nondum tibi nunciat, et tu Iam conviva mihi, Caeciliane, venis.

JUVEN. *X. 216:* . . . clamore opus est, ut sentiat auris, Quem dicat venisse puer, quot nunciet horas.

Agrimensores or Gromatici

The Roman surveyors were called Agrimensores—Land Measurers or Gromatici, after the name of their most important instrument. Their business was to measure unassigned lands for the state and ordinary lands for the proprietors and to fix and maintain boundaries. In the case of land surveying the augur looked to the south; for the gods were supposed to be in the north and the augur was considered as looking at the earth in the same way as the gods. Hence the main line in land surveying was drawn from north to south and was called *cardo*, as corresponding to the axis of the world; the line which cut it was termed *decumanus*, because it made the figure of a cross. These two lines were produced to the extremity of the ground which was to be laid out, and parallel to these were drawn other lines, according to the size of the quadrangle required.

The most important instrument of the surveyors was the *groma*, which was the prototype of the modern surveyors' cross. The groma is represented on the gravestone of a gromaticus found some years ago at Irvea. Two small planks crossing one

another at right angles are supported on a column or post (ferramentum). Plummets are suspended from the planks to guide the operator in securing a vertical position of the column. They were used to guide a surveyor in drawing real or imaginary lines at right angles to one another. They were finally superseded by the "dioptra" which closely resembles the modern theodolite. A treatise on the dioptra by Heron contains a number of propositions illustrative of the use of this instrument. Some of them sound very modern; e.g.,

1. To cut a straight tunnel through a hill from one given point to another.

2. To find the height of an inaccessible point.

3. To sink a shaft which shall meet a horizontal tunnel. (The measurement of angles in degrees was never used by the Greeks or Romans for any but astronomical purposes.)

The most common problem was that of finding the width of a river—"fluminis varatio." One interesting solution proposed by Heron is as follows: *ee* is the accessible shore of a river and *a* is a point on the opposite side. The dioptra is set up at *b* which is a point farther from the bank than the breadth of the river. The sights are directed so that line *ba* shall cut the river at right angles in *h*. The dioptra is then turned through a right angle and the point *c* is taken on *bc* at right angles to *ab*. Then *bc* is bisected at *g* and *gf* is drawn parallel to *bc* meeting *ab* in *d*. A preceding proposition has shown that the sides *ab, ac* of the triangle *abc* are bisected in *d* and *f*. Therefore *ad* is equal to *db* and it remains only to measure *db* and *df* and to deduct the measure of *df* from that of *db*. The remainder is the measure of *ah*—the width of the river.

Much more might be said of the practical mathematics of Roman times. Vitruvius, especially, cites many other illustrations of the skill and knowledge of the Romans in architecture and engineering, yet no more need be said to show that "they were resolute in what they undertook, they were good thinkers, they were born engineers and builders."

BIBLIOGRAPHY

W. A. Becker. *Gallus* or *Roman Scenes of the Time of Augustus*.

Frontinus. *De Aquis Urbis Romæ*, translation by Herschel (1913).

A. P. Gest. *Engineering*.

Sanford. *A Short History of Mathematics*.

Smith. *Dictionary of Greek and Roman Antiquities*.

Vitruvius. *De Architectura*, translation by M. H. Morgan (1914). *Dictionary of Classical Antiquities*.

32

De arithmetica, Book I, of Boethius

DOROTHY V. SCHRADER

The Works of Boethius

IN THE YEAR A.D. 525, Theodoric the Ostrogoth commanded the execution of a political prisoner. When his order was carried out, the active work of Ancius Manlius Torquatus Severinus Boetius, otherwise known as Boethius, came to an end. In a relatively short time, for he was but forty-three when he died,[1] Boethius had made a tremendous contribution to the intellectual world by his original writings and by his translations and paraphrases of the works of earlier scholars. He had written on philosophy, theology, and music; he had translated and written commentaries on works of Aristotle, Porphyry, and Cicero. Although his death prevented the carrying out of his plan to translate into Latin and write a commentary on every one of the works of Aristotle,[2] he did translate the entire *Organon*, the logical writings of Aristotle. He translated and commented on Porphyry's *Introduction to the "Categories"* and wrote a commentary on Cicero's *Topica*. He also composed several theological works and made translations, with added explanations, of Nicomachus' *Arithmetica*, two books of Euclid's *Elements*, a work on music, and probably a work on astronomy, although this last is lost to us. His best-known work is *The Consolation of Philosophy*, written when he was in prison awaiting execution.

It is with his mathematical work that we are here concerned, particularly with Book I of *De arithmetica*. Boethius presented his mathematics in four parts—arithmetic, music, geometry, and astronomy—the four subjects of the medieval quadrivium. Indeed, Boethius may have been the first to use the term "quadrivium" in this sense.[3] Arithmetic is the basic subject of the quadrivium, the other three subjects depending upon it for their very existence.

Boethius quite frankly borrowed his arithmetic from Nicomachus of Gerasa, a neo-Pythagorean Greek arithmetician who flourished about A.D. 100. The *De arithmetica* of Boethius is very largely a translation of this earlier work, although in the Preface he informed his reader that he was going to feel free to condense the original in some places and expand it in others. This he did, although most of the expansions are merely explanations, additional illustrative examples, and tables rather than extensions or refinements of Nicomachus' work. By some scholars, Boethius is censured for not holding strictly to his source, although he stated clearly that he intended "to roam freely in the path, not in the actual footprints" of Nicomachus.[4] By others he is denounced as a "mathematical writer who . . . displays too little originality, independence, and progressiveness and too much prolixity."[5]

One indisputable fact remains. Boethius' *De institutione arithmetica libri duo* was the source of

Reprinted from *Mathematics Teacher* 61 (Oct., 1968): 615–28; with permission of the National Council of Teachers of Mathematics.

all arithmetic taught in the schools and universities for over a thousand years. It was a standard work even after the Hindu-Arabic system of notation and computation was introduced to the Western world.[6] Written at the very beginning of the sixth century, when the author was only about twenty years old,[7] the book was copied, summarized, rephrased, and recopied, serving as the basis for the arithmetical work of Martianus Capella, Cassiodorus, Isidore, and other writers as late as Jordanus Nemorarius. The latest known edition of Boethian *Arithmetic* was printed in Paris in 1521, a complete edition of the two books with commentary. There were at least four other earlier editions in the sixteenth century: the *Arithmetic* of John de Muris, taken directly from Boethius, undated but judged to be from the early sixteenth century; *The Arithmetica of Boethius* explained and amended, with rules for operation, Paris, 1503; the *Arithmetic* of Boethius and Jordanus Nemorarius, Paris, 1514; and *Boethius* by John de Muris, Vienna, 1515.[8]

Boethius was at once the last of the ancients and the first of the scholastics, holding an intermediate position between classical antiquity and the Middle Ages.[9] "It is not easy to exaggerate the influence of Boethius in late mediaeval thought."[10] His work was one of the major channels of transmission of ancient learning. The great importance of his *De arithmetica* lies in the fact that it was "the chief medium through which the Roman world and the Middle Ages learned the principles of formal Greek arithmetic."[11]

De arithmetica, Book I

In Book I of his *De arithmetica*, Boethius gave various methods of classifying numbers and the properties which could be observed in the several classifications. In the following discussion, the principles enunciated and the examples illustrating them are taken directly from Boethius' text,[12] although the ideas are set forth in modern terminology whenever possible.[13]

MAGNITUDE AND MULTITUDE

Things which are unified and continuous, as a tree or a stone, have the quality of magnitude, while those which are disjoint and discontinuous, as a flock or a chorus, have the quality of multitude. Magnitude involves size, and multitude involves quantity. Two disciplines deal with multitude: arithmetic, which is concerned with absolute quantity, and music, which is concerned with relative quantity. Likewise, magnitude involves two disciplines: geometry, which deals with size at rest, and astronomy, which deals with size in motion. These four disciplines make up the science of mathematics.[14]

Arithmetic is a complete and perfect science. The other three branches of mathematics depend upon arithmetic for their perfection and even for their very existence. Thus music cannot exist without arithmetic, but arithmetic is in no way dependent on music for its existence or perfection. Therefore, arithmetic is said to have priority over the other branches of mathematics.[15]

ODD AND EVEN NUMBERS IN GENERAL

Number is defined as a collection of units, or as a flow of quantity made up of units. Numbers can be divided into classes in many ways; the primary division is into even and odd. Even is that which can be separated into two equal parts without a unit remaining in the middle, while odd is that which cannot be so divided.[16] This is the ordinary definition.[17]

According to the Pythagoreans, an even number is one which can, in the same division, be separated into the greatest and the smallest parts, greatest in quantity and smallest in number, while an uneven number can be separated only into two unequal parts. Perhaps an example will make the definition clearer. The number 8 can be separated into two parts, 4 and 4. These are the largest parts that can be obtained by separating 8 into the smallest possible number of parts, two. Should 8 be divided any other way, the number of parts would be greater and the size of the parts would be smaller. Thus a number is even if it contains parts which are greatest in size when least in number.[18]

An ancient definition says that an even number is one which can be separated into two equal and into two unequal parts and the parts are always alike. For example: 10 = 5 + 5, both odd, and 10 = 7 + 3, both odd; 8 = 4 + 4, both even, and 8 = 5 + 3, both odd. An odd number can be divided only into unlike parts: 7 = 3 + 4, one odd and one even.[19]

Lastly, each of the divisions of odd and even may be defined in terms of the other. Thus, that number is called odd which differs from an even number by 1, either greater or less. If 1 is added to or taken from an even number, an odd results, or if the same is done to an odd number, an even number results.[20]

Every number is equal to one-half the sum of the numbers next above and next below it, the next one but one above and below, the next one but two, and so on. Thus, 5 = ½ (4 + 6); 5 = ½ (3 + 7); and 5 = ½ (2 + 8). This is true of every number but unity, which has nothing below it and so is equal to one-half the number above it. Thus, 1 = ½ (2). Hence, unity is the natural base of numbers.[21]

Even Numbers

Even numbers may be classified into three types: evenly-even, evenly-odd, and oddly-even. Evenly-even and evenly-odd numbers have opposite properties, while oddly-even numbers share some of the properties of each of the other types and so may be considered as the means between the extremes.[22]

When an evenly-even number is divided into two equal parts, the parts are always evenly-even and can again be separated into two evenly-even parts. This division can be carried right down to unity.[23] As an example, consider the evenly-even number 64.

$$64 = 32 + 32$$
$$32 = 16 + 16$$
$$16 = 8 + 8$$
$$8 = 4 + 4$$
$$4 = 2 + 2$$
$$2 = 1 + 1$$

Such numbers are obtained by successive doubling:

$$1, 2, 4, 8, 16, 32, 64, 128, 256, 512, \ldots^{24}$$

Each number so produced is evenly-even. If there is an even number of terms in the double ratio, there are two means, while if there is an odd number of terms, there is but one mean.[25] The parts correspond in pairs as factors of the whole. Hence, for an even number of terms such as 1, 2, 4, 8, 16, 32, 64, 128:

$1 \times 128 = 128$	and	$1/128 \times 128 = 1$
$1/2 \times 128 = 64$	and	$1/64 \times 128 = 2$
$1/4 \times 128 = 32$	and	$1/32 \times 128 = 4$
$1/8 \times 128 = 16$	and	$1/16 \times 128 = 8$

If the double ratio has an odd number of terms, such as 1, 2, 4, 8, 16, 32, 64, then the mean term, in this case 8, has no term to which it corresponds.[26]

$1 \times 64 = 64$	and	$1/64 \times 64 = 1$
$1/2 \times 64 = 32$	and	$1/32 \times 64 = 2$
$1/4 \times 64 = 16$	and	$1/16 \times 64 = 4$
$1/8 \times 64 = 8$		

When the terms of a double ratio are added together, each sum is one less than the next term. Thus 1 + 2 = 3, which is 1 less than the next term, 4; 1 + 2 + 4 = 7, which is 1 less than the next term, 8; 1 + 2 + 4 + 8 = 15 which is 1 less than the next term, 16; etc. This may be continued as far as one wishes to go. These sums are always odd, since each is just 1 less than an even number. Moreover, in a double ratio series, if the number of terms is even, the product of any pair of extreme terms is equal to the product of the two mean terms.

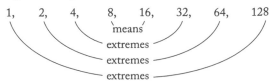

1,   2,   4,   8,   16,   32,   64,   128

means
extremes
extremes
extremes

If the number of terms is odd, the product of any pair of extreme terms is equal to the square of the mean term.

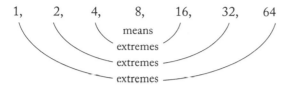

Evenly-odd numbers are those which can be divided into equal parts, but the halves cannot again be separated into equal parts: 6, 10, 14, 18, and 22 are examples of such numbers. Their factors have names opposite in meaning to their true value, as far as even and odd are concerned. Thus, $^{1}/_{2}$ of 18 is 9, yet $^{1}/_{2}$ is an even number, while 9 is odd. Likewise, $^{1}/_{3}$ of 18 is 6, yet $^{1}/_{3}$ is an odd number, while 6 is even. That which is odd in name is even in value, and that which is even in name is odd in value.

Evenly-odd numbers are generated by multiplying odd numbers by 2.[27] The successive numbers in the evenly-odd sequence differ by 4.

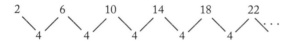

These numbers are said to be opposite in property to the evenly-even numbers because, while evenly-even numbers can be divided until the very last value is unity, that alone being indivisible, evenly-odd numbers can be divided only once.

Moreover, in evenly-even numbers, the product of the extremes is equal to the product of the means for an even number of terms and to the square of the mean for an odd number of terms. In evenly-odd numbers the sum of the extremes is equal to the sum of the means for an even number of terms and to twice the mean for an odd number of terms.

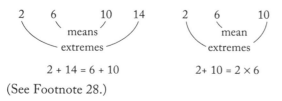

$$2 + 14 = 6 + 10 \qquad 2 + 10 = 2 \times 6$$

(See Footnote 28.)

Oddly-even numbers have some of the properties of evenly-even numbers and some of those of evenly-odds. An oddly-even number is one which can be divided into two equal parts; the parts can again be divided, even several times, but not to unity.

$$24 = 12 + 12$$
$$12 = 6 + 6$$
$$6 = 3 + 3,$$

and the number cannot be further reduced. In admitting of more than one division, the oddly-even number resembles the evenly-even; in not being divided to unity, it resembles the evenly-odd.

In some cases, factors of oddly-even numbers have names opposite in meaning to their true value, as do the evenly-odd. Thus, $^{1}/_{3}$ of 24 is 8, yet $^{1}/_{3}$ is odd while 8 is even. In other cases, however, the name of the factor is of the same meaning as the name of the true value; for example, $^{1}/_{6}$ of 24 is 4, both $^{1}/_{6}$ and 4 being even.

The very way the oddly-even numbers are produced shows that they are of a mixed type. To generate the oddly-evens, list the sequence of odd numbers beginning with 3, and below them, list the sequence of evenly-even numbers, beginning with 4; multiply the first number of the first sequence by each term of the second sequence; multiply the second term of the first sequence by each term of the second sequence; continue the process indefinitely, as far as is desired. Each number so produced is oddly-even.

3	5	7	9	11	13 . . .
4	8	16	32	64	128 . . .

$3 \times 4 = 12$, $3 \times 8 = 24$, $3 \times 16 = 48$, $3 \times 32 = 96$, $5 \times 4 = 20$, $5 \times 8 = 40$, $5 \times 16 = 80$, $5 \times 32 = 160$, etc. These, and other numbers similarly obtained, are all oddly even.

Many interesting properties and relationships of these numbers can be seen from the diagram in Figure 1 (see Footnote 29). Generate sixteen oddly-even numbers as explained above. Place them in

FIGURE I

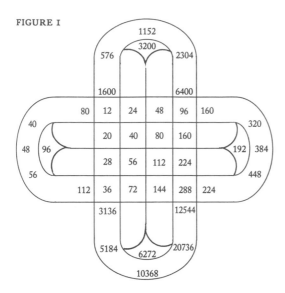

order in a square, the multiples of 3 in the first row, of 5 in the second row, etc. The multiples of 4 will be in the first column, of 8 in the second, etc.

Each vertical column shows the evenly-odd property that the sum of the extremes is equal to the sum of the means. Thus:

$$
\begin{aligned}
12 + 36 &= 48 = 20 + 28 \\
24 + 72 &= 96 = 40 + 56 \\
48 + 144 &= 192 = 80 + 112 \\
96 + 288 &= 384 = 160 + 224
\end{aligned}
$$

Each horizontal row shows the evenly-even property that the product of the extremes is equal to the product of the means. Thus:

$$
\begin{aligned}
12 \times 96 &= 1{,}152 = 24 \times 48 \\
20 \times 160 &= 3{,}200 = 40 \times 80 \\
28 \times 224 &= 6{,}272 = 56 \times 112 \\
36 \times 288 &= 10{,}368 = 72 \times 144
\end{aligned}
$$

(See Footnote 30.)

Moreover, one can see that if the vertical columns are considered three terms at a time, the sum of the extremes is twice the mean, an evenly-odd property. Thus:

$$
\begin{aligned}
12 + 28 &= 40 = 2 \times 20 \\
20 + 36 &= 56 = 2 \times 28 \\
24 + 56 &= 80 = 2 \times 40 \\
40 + 72 &= 112 = 2 \times 56
\end{aligned}
$$

$$
\begin{aligned}
48 + 112 &= 160 = 2 \times 80 \\
80 + 144 &= 224 = 2 \times 112 \\
96 + 224 &= 320 = 2 \times 160 \\
160 + 288 &= 448 = 2 \times 224
\end{aligned}
$$

If the horizontal rows are considered only three terms at a time, the product of the extremes is equal to the square of the mean, an evenly-even property. Thus:

$$
\begin{aligned}
12 \times 48 &= 576 = 24 \times 24 \\
24 \times 96 &= 2{,}304 = 48 \times 48
\end{aligned}
$$

$$
\begin{aligned}
20 \times 80 &= 1{,}600 = 40 \times 40 \\
40 \times 160 &= 6{,}400 = 80 \times 80
\end{aligned}
$$

$$
\begin{aligned}
28 \times 112 &= 3{,}136 = 56 \times 56 \\
56 \times 224 &= 12{,}544 = 112 \times 112 \\
36 \times 144 &= 5{,}184 = 72 \times 72 \\
72 \times 288 &= 20{,}736 = 144 \times 144
\end{aligned}
$$

(See Footnote 31.)

UNEVEN NUMBERS

There are three classes of uneven numbers: the prime and incomposite, the secondary and composite, and those which are secondary and composite in themselves but relative to another number are prime and incomposite. The first and second classes are opposites, while the third class is midway between the other two as a means between extremes.[32]

Prime numbers are those which have no part except a fraction with the number itself as denominator; this part is really unity. Thus 3 is prime; its only part is $1/3$ of 3 or 1. The number 5 is prime; its only part is $1/5$ of 5, or 1. And 7 is prime; its only part is $1/7$ of 7, or 1. A prime number can be

FIGURE 2

1, 3, 5, 7, 9, 11, 13, 15, 17, 19, 21 , 23, 25, 27, 29, 31, 33, 35, . . .

produced only by the product of the number itself and 1.[33]

Secondary, composite numbers are odd and are obtained from the product of odd numbers. They have parts other than fractions with themselves as denominators; they have other factors than themselves and 1. The number 15 is a secondary number; $1/15$ of 15 is 1; $1/3$ of 15 is 5; $1/5$ of 15 is 3. Such numbers are called secondary because they are measured by some value other than themselves; they are called composite because they are made up of other numbers than themselves and 1.[34]

The third type of odd number is a secondary and composite number; but, when it is compared to another number, there is no common measure except unity. As an example, consider 9 and 25. Both are secondary numbers, 9 having 3 as a factor and 25 having 5. But 3 is not a factor of 25 nor is 5 a factor of 9. Therefore the two numbers are to one another as if they were prime and incomposite; they have no common factor except unity.[35]

There is a method of producing prime and secondary numbers, the sieve of Eratosthenes, so-called because one takes all the odd numbers and "sieves" out the secondary ones, leaving the primes. All the odd numbers are listed in order; starting with 3, every third number is marked; starting again with 5, every fifth number is marked; starting again with 7, every seventh number is marked; this is repeated as far as one wishes to go. (See Fig. 2.) Every number not marked is prime and incomposite, while every number marked is secondary and composite.[36]

Given two odd numbers, the following method will determine whether they are prime or secondary in relation to one another—in other words, whether or not they have a common factor other than unity. Subtract the smaller from the larger as many times as possible.[37] Subtract the final difference from the number that was being subtracted, as many times as possible. Subtract that final difference from the number that was being subtracted, as many times as possible. Carry on that process until no further subtractions can be carried out.

If the final difference is 1, the numbers are relatively prime; if it is greater than 1, the numbers are not relatively prime and the difference is the common factor.

Compare 21 and 9.

$$
\begin{array}{rr}
21 & 9 \\
-9 & -3 \\
\hline
12 & 6 \\
-9 & -3 \\
\hline
3 & 3 \\
\end{array}
$$

Since 3 cannot be subtracted from 3 (see Footnote 37), the numbers are not relatively prime but have a common factor of 3.

Compare 29 and 9, in the following way:

$$
\begin{array}{rrr}
29 & 9 & \\
-9 & -2 & \\
\hline
20 & 7 & \\
-9 & -2 & \\
\hline
11 & 5 & 3 \\
-9 & -2 & -2 \\
\hline
2 & 3 & 1 \\
\end{array}
$$

Since the final difference is 1, 29 and 9 are relatively prime.[38]

A SECOND CLASSIFICATION OF EVEN NUMBERS

There is a second type of classification of even numbers: superabundant, deficient, and perfect. Superabundant numbers are those the sum of whose divisors is greater than the number. The number 12 is divisible by 6, 4, 3, 2, 1.

$$6 + 4 + 3 + 2 + 1 = 16,$$

which is greater than 12. Therefore, 12 is superabundant. The number 24 is divisible by 12, 8, 6, 4, 3, 2, 1.

$$12 + 8 + 6 + 4 + 3 + 2 + 1 = 36,$$

which is greater than 24. Therefore, 24 is superabundant.

Deficient numbers are those the sum of whose divisors is less than the number. The number 8 is divisible by 4, 2, 1.

$$4 + 2 + 1 = 7,$$

which is less than 8. Therefore, 8 is deficient.

The number 14 is divisible by 7, 2, 1.

$$7 + 2 + 1 = 10,$$

which is less than 14. Therefore, 14 is deficient.

Perfect numbers are those the sum of whose divisors is equal to the number. The number 6 is divisible by 3, 2, 1.

$$3 + 2 + 1 = 6.$$

Therefore 6 is a perfect number. The number 28 is divisible by 14, 7, 4, 2, 1.

$$14 + 7 + 4 + 2 + 1 = 28.$$

Therefore 28, also, is a perfect number.[39]

There are many superabundant and many deficient numbers, but few perfect numbers. In fact, in each power of ten, there is but one perfect number. In the units (10^0), there is only the perfect number 6. In the tens (10^1), the perfect number is 28. In the hundreds (10^2), it is 496. In the thousands (10^3), it is 8,128.[40]

To generate perfect numbers, list the evenly-even numbers, as far as you choose to go: 1, 2, 4, 8, 16, 32, 64, 128, . . . Add these numbers in order. Whenever the sum is prime and incomposite, multiply it by the last number used. This product is a perfect number, as in $1 + 2 = 3$, which is prime, and $3 \times 2 = 6$, which is perfect; $1 + 2 + 4 = 7$, which is prime, and $7 \times 4 = 28$, which is perfect; $1 + 2 + 4 + 8 + 16 = 31$, which is prime, and $31 \times 16 = 496$, which is perfect.[41]

INEQUALITY

From absolute quantity, let us now consider relative quantity. The first division of quantities considered relatively is equality and inequality. Equality exists when one thing is neither greater nor less than the thing to which it is compared. Thus a denarius is equal to a denarius, a cubit to a cubit, and a foot to a foot. All things which are equal measure alike. Equality is of only one kind and is naturally indivisible. Equals are called by the same name—as friend, neighbor, etc.

Unequals are divided. One is greater than the other; the other is less. The greater is greater than the less; the less is less than the greater. Unequals are not signified by the same name but by various terms like teacher and pupil, conqueror and conquered.[42]

The greater in inequality is divided into five classes:

Multiple

Superparticular

Superpartient

Multiple superparticular

Multiple superpartient

The lesser is similarly divided:

Submultiple

Subsuperparticular

Subsuperpartient

Multiple subsuperparticular

Multiple subsuperpartient [43]

The multiple is the first subdivision of the greater. When compared with a number, a multiple contains the number more than once. All natural numbers are multiples of unity. Thus, 1 is unity; 2 is double unity; 3 is triple unity; and this continues to infinity.

Opposite to this is the submultiple, the first of the subdivisions of the lesser. It is the number which, in comparison to the larger, measures it more than once. If it measures it twice, it is called the subdouble; three times, the subtriple, etc. There are infinitely many multiples and submultiples. Thus both the primary set and all the multiple sets make infinite sequences.

Primary	1	2	3	4	5	6	7	8	...
Double	2	4	6	8	10	12	14	16	...
Triple	3	6	9	12	15	18	21	24	...
Quadruple	4	8	12	16	20	24	28	32	...

Double and quadruple sequences are all even, while primary, triple, and quintuple sequences are alternately odd and even.[44]

The superparticular is the second subdivision of the greater. The superparticular contains the whole of the number to which it is compared, plus a unit factor.[45] If the factor is one-half, the greater term is called the sesquialter and the lesser term the subsesquialter; if the factor is one third, the terms are sesquitertian and subsesquitertian; one-fourth, sesquiquartan and subsesquiquartan; and so on to infinity.

The sesquialter ratio has as its antecedents (larger terms) the triples of the natural numbers and as its consequents (smaller terms) the doubles.

1	2	3	4	5	6	7	8	9	10
3	6	9	12	15	18	21	24	27	30
2	4	6	8	10	12	14	16	18	20

$$3 = 1\tfrac{1}{2} \times 2$$
$$6 = 1\tfrac{1}{2} \times 4$$
$$9 = 1\tfrac{1}{2} \times 6$$

or, expressed differently, $3/2 = 6/4 = 9/6 = \ldots$

In the sesquitertian ratio, the antecedents are the quadruples and the consequents are the triples of the natural numbers.

1	2	3	4	5	6	7	8	9	10
4	8	12	16	20	24	28	32	36	40
3	6	9	12	15	18	21	24	27	30

$$4 = 1\tfrac{1}{3} \times 3$$
$$8 = 1\tfrac{1}{3} \times 6$$
$$12 = 1\tfrac{1}{3} \times 9$$

or expressed differently, $4/3 = 8/6 = 12/9 \ldots$

The antecedent term is the superparticular, and the consequent term is the subsuperparticular. Thus, 9 is the superparticular in relation to 6, and 6 is the subsuperparticular in relation to 9.[46]

We should observe the intervals of numbers in the superparticular pairs. Consider the sesquialter ratio. The first sesquialter is $^4/_3$. There is no number between 3 and 4. The second sesquialter is $^8/_6$. There is one number between 6 and 8. The third sesquialter is $^{12}/_9$. There are two numbers between 9 and 12. Similar observations can be made in the other ratios.[47]

Let us construct a diagram of multiples in order: doubles, triples, quadruples, etc., in line as far as tenfold multiples, as in Figure 3. We may observe many useful facts and principles from such a diagram.[48]

Consider the first and second rows or columns. (The relationship is the same whether one reads vertically or horizontally.) The second row is double the first and differs from the first row by a primary sequence.

The third row is a triple of the first and differs from it by the second.

The fourth row is a quadruple of the first and differs from it by the third row. Each row will be seen to differ from the first row by quantities equal to the row directly above itself.

The third row is a superparticular, the sesquialter, of the second: 3 is the sesquialter of 2; 6 is the sesquialter of 4; 9 is the sesquialter of 6, etc. The fourth row is the sesquitertian of the third: 4 is the sesquitertian of 3; 8 is the sesquitertian of 6; 12 is the sesquitertian of 9. The fifth row is the sesquiquartan of the fourth, and so on.

Note the numbers in the corners:

They are the unit forms of the various classes of numbers, powers of ten: units, tens, and hundreds. The diagonal products are equal: $1 \times 100 = 10 \times 10$. Moreover, all the terms on the diagonal from 1 to 100 are squares.

The numbers on either side of the squares are heteromecic.[49]

The sum of two squares plus twice the number between is a square.[50]

FIGURE 3

1	2	3	4	5	6	7	8	9	10
2	4	6	8	10	12	14	16	18	20
3	6	9	12	15	18	21	24	27	30
4	8	12	16	20	24	28	32	36	40
5	10	15	20	25	30	35	40	45	50
6	12	18	24	30	36	42	48	54	60
7	14	21	28	35	42	49	56	63	70
8	16	24	32	40	48	56	64	72	80
9	18	27	36	45	54	63	72	81	90
10	20	30	40	50	60	70	80	90	100

First row	1	2	3	4	5	6	7	8	9	10
Second row	2	4	6	8	10	12	14	16	18	20
Difference	1	2	3	4	5	6	7	8	9	10

First row	1	2	3	4	5	6	7	8	9	10
Third row	3	6	9	12	15	18	21	24	27	30
Difference	2	4	6	8	10	12	14	16	18	20

The sum of two heteromecic numbers plus twice the square between them is a square.[51]

There are many other interesting relationships which those who are interested may derive from this diagram.[52]

The superpartient is the third type of subdivision of the greater. The superpartient contains the whole of the number to which it is compared plus more than one part of it.[53] The whole plus two parts (two-thirds) is the superbipartient; the whole plus three parts (three-fourths) is the super-tripartient; and so on, as far as one chooses to go.

The subsuperpartient is the reverse of the superpartient. The number 9 is the superpartient of 5; $9 = 1^4/_5 \times 5$. The number 5 is the subsuperpartient of 9; $5 = 9 \div 1^4/_5$.

To generate the superpartient, one lists the primary sequence beginning with 3 and below it the odd numbers beginning with 5.

3	4	5	6	7	8	9	10
5	7	9	11	13	15	17	19

The first pair have a superbipartient ratio, the second a supertripartient ratio, the third a superquadripartient ratio, the third a superquadripartient ratio, etc.[54]

$$5 = 1^2/_3 \times 3,$$
$$7 = 1^3/_4 \times 4,$$
$$9 = 1^4/_5 \times 5,$$

etc. The same ratio holds for the various multiples of the ratios.

5	2	7	4	9	5	11	6
10	6	14	8	18	10	22	12
15	9	21	12	27	15	33	18
20	12	28	16	36	20	44	24
25	15	35	20	45	25	55	30
30	18	42	24	54	30	66	36

(See Footnote 55.)

The fourth subdivision of the greater is the multiple superparticular. In the multiple superparticular ratio, the greater term contains the lesser term more than once plus some one part of it.[56] The multiple superparticular has a double name; the first part is double, triple, quadruple, etc., according as the lesser number is contained two, three, four or more times in the larger number; the second part is called sesquialter, sesquitertian, sesquiquartan, etc., according to the superparticular terminology. Thus 5 is the double sesquialter of 2, that is, $5 = 2^{1}/_2 \times 2$; 9 is the double sesquiquartan of 4, that is $9 = 2^{1}/_4 \times 4$; 11 is the double sesquiquintan of 5, that is, $11 = 2^{1}/_5 \times 5$.

The multiple superparticulars are produced by comparison of pairs of sequences, chosen according to the ratio desired. To generate double superparticulars, compare the primary sequence beginning with 2 to the sequence of odd numbers beginning with 5. This gives the basic double superparticular pairs.

$$2 \quad 3 \quad 4 \quad 5 \quad 6 \ldots n$$

$$5 \quad 7 \quad 9 \quad 11 \quad 13 \ldots 2n+1$$

(See Footnote 57.)

Reading from the above, one sees that 5 is the double sesquialter of 2, 7 is the double sesquitertian of 3, 9 is the double sesquiquartan of 4, etc. To find the double sesquialters in order, compare the sequence of multiples of 2 with the sequence of multiples of 5.

$$2 \quad 4 \quad 6 \quad 8 \quad 10 \quad 12 \quad 14 \ldots 2n$$

$$5 \quad 10 \quad 15 \quad 20 \quad 25 \quad 30 \quad 35 \ldots 5n$$

Each pair will be a double sesquialter ratio. To find the double sesquitertians in order, compare the sequence of multiples of 3 with the sequence of multiples of 7.

$$3 \quad 6 \quad 9 \quad 12 \quad 15 \quad 18 \quad 21 \ldots 3n$$

$$7 \quad 14 \quad 21 \quad 28 \quad 35 \quad 42 \quad 49 \ldots 7n$$

Each pair will be a double sesquitertian ratio. Similarly, any double superparticulars desired can be found by comparing the two sequences composed of multiples of the proper pair of numbers.

The basic triple superparticular pairs can be found by listing the primary sequence beginning with 2 and beneath it a sequence beginning with 7 and using every third number.

$$2 \quad 3 \quad 4 \quad 5 \quad 6 \quad 7 \quad 8 \ldots n$$

$$7 \quad 10 \quad 13 \quad 16 \quad 19 \quad 22 \quad 25 \ldots 3n+1$$

(See Footnote 58.)

Reading from the above, one sees that 7 is the triple sesquialter of 2, that is $7 = 3^{1}/_2 \times 2$; 10 is the triple sesquitertian of 3, that is, $10 = 3^{1}/_3 \times 3$; 13 is the triple sesquiquartan of 4, that is, $13 = 3^{1}/_4 \times 4$; etc.

To find the triple sesquialters in order, compare the sequence of multiples of 2 with the sequence of multiples of 7.

$$2 \quad 4 \quad 6 \quad 8 \quad 10 \quad 12 \quad 14 \ldots 2n$$

$$7 \quad 14 \quad 21 \quad 28 \quad 35 \quad 42 \quad 49 \ldots 7n$$

Each pair is a triple sesquialter ratio.

Other multiple superparticulars can be found in the same fashion.[59]

Referring to the table in Figure 3, one can see that multiples, superparticulars, superpartients, and multiple superparticulars may be read from the table by comparison of the proper rows (or columns). Any row compared to the first is a multiple. Each row, from the second on, compared to the one above itself shows the superparticular. The odd-numbered rows, from the third on, compared in order to the rows from the fifth on will give superpartient ratios.[60] The multiple superparticulars will be seen by comparing the odd rows from the fifth on to the rows from the second on.[61] It is to be noted that the smaller terms have names comparable to the larger terms with the prefix "sub" added.[62]

The fifth subdivision of the greater is the multiple superpartient. In the multiple superpartient ratio, the greater number contains the smaller term more than once plus several parts of the number, as explained in the superpartient.[63] The number 8 is the double superbipartient of 3, that is, $8 = 2\frac{2}{3} \times 3$; and 16 is the double superbipartient of 6; that is, $16 = 2\frac{2}{3} \times 6$. For the multiple superpartients, tables similar to those for the multiple superparticular can be devised. The prefix "sub" indicates the smaller of the two numbers in the multiple superpartient ratio.[64]

All forms of inequality can be derived from equality. This can be shown by the construction of a series of tables. The first is made as follows: Make a table of three columns beginning with 1, 1, 1 such that the numbers in the first column are the same as in the row above, the numbers in the second column are the sums of the first and second numbers in the row above, and the numbers in the third column are the sums of the first, twice the second, and the third numbers in the row above.

1	1	1
1	2	4
1	3	9
1	4	16

The resulting rows will be doubles, triples, quadruples, etc. Reverse the rows:

4	2	1
9	3	1
16	4	1

Make a new table in the following way: The first number in each row will be the same as the number in the above table. The second number in each row will be the sum of the first two numbers in the corresponding rows in the above table. The third number in each row will be the sum of the first number, twice the second, and the third.

4	6	9
9	12	16
16	20	25

The rows are superparticular ratios: $\frac{6}{4} = \frac{9}{6}$, sesquialter; $\frac{12}{9} = \frac{16}{12}$, sesquitertian; $\frac{20}{16} = \frac{25}{20}$, sesquiquartan.

Reverse the table above and repeat the above calculations.

9	6	4		9	15	25
16	12	9		16	28	49
25	20	16		25	45	81

This table is made up of rows of superpartients: $\frac{15}{9} = \frac{25}{15} = 1\frac{2}{3}$, superbipartient; $\frac{28}{16} = \frac{49}{28} = 1\frac{3}{4}$, supertripartient; $\frac{45}{25} = \frac{81}{45} = 1\frac{4}{5}$, superquadripartient.

If the superparticular table were not reversed but the calculations repeated, the double superparticulars are obtained.

4	6	9		4	10	25
9	12	16		9	21	49
16	20	25		16	36	81

The results: $\frac{10}{4} = \frac{25}{10} = 2\frac{1}{2}$, double sesquialter; $\frac{21}{9} = \frac{49}{21} = 2\frac{1}{3}$, double sesquitertian; $\frac{36}{16} = \frac{81}{36} = 2\frac{1}{4}$, double sesquiquartans.

If the superpartient table is not reversed but the calculations repeated, the double superpartient is obtained.

9	15	25		9	24	64
16	28	49		16	44	121
25	45	81		25	770	196

The results: $\frac{24}{7} = \frac{64}{24} = 2\frac{2}{3}$, double superpartient; $\frac{44}{16} = \frac{121}{44} = 2\frac{3}{4}$, double supertripartient; $\frac{70}{25} = \frac{196}{70} = 2\frac{4}{5}$, double superquadripartient.

These tables can be extended and the process repeated as far as is desired.[65]

Conclusions

Book I of *De arithmetica* is largely devoted to a discussion of the classification of numbers, their properties and relationships. It gives no rules of computation, no practical application of arithmetic to daily-life problems, nothing useful or practical in the ordinary sense of those terms. In this, the work shows very clearly its Greek antecedents. Here is the Greek love of learning, of speculative knowledge, for its own sake.

Boethius is often criticized for his impracticality, for not giving at least the elementary rules of computation. It must be remembered, however, that *arithmetica* was not similar to our present-day arithmetic but rather to number theory, with which science it still has many features in common. The Greek and Roman equivalent of today's arithmetic, which is really the art of reckoning, was *logistica*.

It is doubtful if Boethius could have done much, even had he so desired, in teaching the art of reckoning. He had to express his ideas in the currently-used Roman-number system, which employed cumbersome additive numerals; these numerals do not lend themselves to even elementary calculations. The simple figuring necessary for daily intercourse in home and marketplace was done by a system of finger-reckoning and was not the concern of philosophers and scholars. They were interested in the theoretical aspect of numbers: how they combined, the patterns they formed, and the laws that governed their functioning. It was of this phase of number that Boethius treated. It was not until the introduction of the Hindu-Arabic number system, in which digits have place value in powers of ten, and the consequent extension of the number concept to include zero and negative numbers, that it was possible clearly and unambiguously to define the various number relations which had attracted Boethius' attention. Indeed, it was not until the advent of the electronic computer, almost fourteen and a half centuries after Boethius wrote, that some number-theory problems, notably those concerned with perfect numbers, could be handled adequately. Some problems, such as many of those dealing with prime numbers, have not yet been solved.

It is true that Boethius' ideas were borrowed from Nicomachus; it is also true that some of them were sterile, leading into intellectual blind alleys and blank walls, and that his arithmetic was not a series of logical cause-and-effect relationships. He seems to have been trying to discover all he could about these mysterious entities called numbers, hoping that, in a sea of accumulated facts, he could detect a current of relationships which would indicate the mainstream of arithmetic. His method was an ancient one—and a very modern one. Is this not also a method that has been used most fruitfully in the discovery, development, and organization of several branches of modern mathematics?

NOTES

1. Henry Osborn Taylor, *The Mediaeval Mind* (Cambridge, Mass.: Harvard University Press, 1951), p. 89.

2. *Ibid.*, p. 91.

3. *Ibid.*, p. 90.

4. Eleanor Shipley Duckett, *The Gateway to the Middle Ages* (New York: Macmillan Co., 1938), p. 154.

5. Martin Luther D'Ooge (trans.), Nicomachus of Gerasa's *Introduction to Arithmetic* (New York: Macmillan Co., 1926), p. 133.

6. Paul Abelson, *The Seven Liberal Arts* (New York: Columbia University, 1906), p. 96.

7. Taylor, *op. cit.*, p. 90.

8. Augustus De Morgan, *Arithmetical Books from the Invention of Printing to the Present Time* (London: Taylor & Walton, 1847), pp. 3, 4, 10, 11, 13.

9. Sir John Edwin Sandys, *A History of Classical Scholarship* (Cambridge, England: Cambridge University Press, 1921), p. 253.

10. D'Ooge, *op. cit.*, p. 137.

11. *Ibid.*, p. 135

12. *Boetii opera omnia (Patrologiae cursus completus*, ed. Jacques Paul Migne, Vol. LXIII [Paris, 1847]), cols. 1079–1114.

13. In some cases there is no modern equivalent for Boethius' term.

14. *Boetii opera omnia*, cols. 1082–83, chap. i, "The Division of Mathematics."

15. *Ibid.*, col. 1083, chap. ii, "The Essence of Number."

16. *E.g.*, 8 = 4 + 4: even, while 9 = 4 + 1 + 4: odd.

17. *Boetii opera omnia*, col. 1083, chap. iii, "Definition and Division of Numbers and of Even and Odd."

18. *Ibid.*, cols. 1083–84, chap. iv, "Definition of Even and Odd According to the Pythagoreans."

19. *Ibid.*, col. 1084, chap. v, "Ancient Definition of Even and Odd."

20. *Ibid.*, col. 1085, chap. vi, "Definition of Even and Odd in Terms of One Another."

21. *Ibid.*, col. 1085, chap. vii, "The Superiority of Unity."

22. *Ibid.*, chap. viii, "The Classes of Even Numbers,"

23. The general expression for evenly-even numbers is 2^n where n is a positive integer.

24. This type of sequence, a geometric progression with the common ratio of 2, Boethius referred to as a double ratio.

25. D'Ooge, *op. cit.*, p. 194, note.

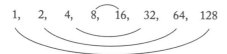

Even number of terms, two means

Odd number of terms, one mean

26. *Boetii opera omnia*, cols. 1085–87, chap. ix, "Evenly-Even Numbers and Their Properties."

27. The general expression for an evenly-odd number is $2(2n - 1)$ where n is a positive integer.

28. *Boetii opera omnia*, cols. 1087–89, chap. x, "Evenly-Odd Numbers, and Their Properties."

29. *Ibid.*, col. 1091, has an error in this figure. It is given there as 2726 instead of 6272.

30. *Ibid.*, cols. 1089–91, chap xi, "Oddly-Even Numbers and Their Properties."

31. *Ibid.*, cols. 1091–92, chap xii, "Description of Oddly-Even Numbers in Comparison to the Other Two Types."

32. *Ibid.*, col. 1092, chap. xiii, "The Division of Unequal Numbers."

33. *Ibid.*, cols. 1092–93, chap xiv, "Prime, Incomposite Numbers."

34. *Ibid.*, col. 1093, chap xv, "Secondary and Composite Numbers."

35. *Ibid.*, chap. xvi, "Numbers Which Are, in Themselves, Secondary and Composite but, Relative to Others, Are Prime and Incomposite."

36. *Ibid.*, cols. 1094–96, chap. xvii, "On the Production of Prime and Secondary Numbers."

37. Negative numbers and zero are not considered as numbers at all.

38. *Boetii opera omnia*, col. 1096, chap xviii, "A Method of Determining Whether Numbers Are Prime to One Another."

39. *Ibid.*, col. 1097, chap. xix, "Perfect, Deficient, and Superabundant Numbers."

40. Perfect numbers may be written in the form $2^n (2^{n+1} - 1)$, but all such numbers do not yield perfect numbers although all even numbers which are perfect are of this form. The first such eleven perfect numbers are

$$2 (2^2 - 1) = 6$$
$$2^2 (2^3 - 1) = 28$$
$$2^4 (2^5 - 1) = 496$$
$$2^6 (2^7 - 1) = 8,128$$
$$2^{12} (2^{13} - 1) = 33,550,336$$
$$2^{16} (2^{17} - 1) = 8,568,910,416$$
$$2^{18} (2^{19} - 1)$$
$$2^{30} (2^{31} - 1)$$
$$2^{60} (2^{61} - 1)$$
$$2^{88} (2^{89} - 1)$$
$$2^{126} (2^{127} - 1)$$

Modern use of electronic computers has made possible the determination of perfect numbers to any number of digits, an impossible task for anyone in Boethius' time, due to the physical limitation of the human lifespan.

41. *Boetii opera omnia*, cols. 1098–99, chap. xx, "The Generation of Perfect Numbers."

42. *Ibid.*, chap. xxi. "Relative Quantity."

43. *Ibid.*, col. 1100, chap. xxii, "The Kinds of Greater and Less in Inequalities."

44. *Ibid.*, cols. 1100–1101, chap. xxiii, "Multiples, Their Kinds and Generation."

45. By a unit factor is meant a fraction whose numerator is 1. The superparticular can be expressed in general terms as $a + (1/n)a$ where n = 2, 3, 4, 5, ... and both a and $(1/n)a$ are integers.

46. *Boetii opera omnia*, cols. 1101–2, chap. xxiv, "Superparticulars, Their Kinds and Generation."

47. *Ibid.*, cols. 1102–3, chap. xxv, "An Observation Concerning Superparticulars."

48. *Ibid.*, cols. 1103–4, chap. xxvi, "A Description of a Diagram of Multiples."

49. Heteronecic numbers are those which are formed from the product of two consecutive numbers. They are of the form $n(n + 1)$.

50. For example, $9 + 16 + 2 \cdot 12 = 49 = 7^2$.

Algebraically, this can be expressed as $n^2 + (n + 1)^2 + 2n(n + 1) = 4n^2 + 4n + 1 = (2n + 1)^2$.

51. For example, $12 + 20 + 2 \cdot 16 = 64 = 8^2$.

Algebraically, this can be expressed as $(n-1)n - n(n + 1) + 2n^2 = 4n^2 = (2n)^2$.

52. *Boetii opera omnia*, cols. 1104–6, chap. xxvii, "Relationships Observed from the Diagram."

53. The "more than one part" must be of a definite order. Always the parts are of the form $(n - 1)/n$ so that the superpartient is $1^2/_3$, $1^3/_4$, $1^4/_5$, $1^5/_6$, ... $1 + (n - 1)n$ of the number to which it is compared.

54. The general term of the ratio would be $(2n + 3)/(n + 2)$, which reduces to $(n - 1)/n + 1$.

55. *Boetii opera omnia*, cols. 1106–7, chap. xxviii, "Superpartients, Their Kinds and Generation."

56. The multiple superparticular is a number of the form $xa + (1/n)a$ where x, a, n, and $(1/n)a$ are all integers.

57. $2n + 1$ is the double superparticular of n. $(2n + 1)/n = 2 + 1/n$, which is the basic form of any double superparticular ratio.

58. The general term for quadruple superparticulars would be $n/(4n + 1)$, for quintuple superparticulars would be $n/(5n +1)$, etc.

59. *Boetii opera omnia*, cols. 1107–9, chap. xxix, "Multiple Superparticulars."

60. That is, the 5th row is compared to the 3rd, the 7th to the 4th, the 9th to the 5th, etc.

61. That is, the 5th row is compared to the 2nd, the 7th to the 3rd, the 9th to the 4th, etc.

62. *Boetii opera omnia*, col. 1109, chap. xxx, "Examples Set Forth in the Diagram."

63. The multiple superpartient is of the form $xa + [n/(n + 1)]a$ where x, a, n, and $[n/(n + 1)]a$ are all integers.

64. *Boetii opera omnia*, cols. 1109–10, chap. xxxi, "Multiple Superpartient."

65. *Ibid.*, cols. 1110–14, chap. xxxii, "Demonstration that All Inequality Comes from Equality."

BIBLIOGRAPHY

Abelson, Paul. *The Seven Liberal Arts.* New York: Columbia University, 1906.

Boetii opera omnia. (Patrologiae cursus completus, ed. Jacques Paul Migne, Vol. LXIII.) Paris, 1847.

Bolgar, R. R. *The Classical Heritage and Its Beneficiaries.* Cambridge, England: Cambridge University Press, 1954.

De Morgan, Augustus. *Arithmetical Books from the Invention of Printing to the Present Time.* London: Taylor & Walton, 1847.

Duckett, Eleanor Shipley. *The Gateway to the Middle Ages.* New York: Macmillan Co., 1938.

Haskins, Charles Homer. *The Rise of Universities.* New York: Henry Holt Co., 1923.

———. *Studies in the History of Mediaeval Science.* Cambridge, Mass.: Harvard University Press, 1927.

Laistner, M. L. W. *Thought and Letters in Western Europe.* London: Methuen & Co., 1931.

Nicomachus of Gerasa. *Introduction to Arithmetic.* Translated by Martin Luther D'Ooge. New York: Macmillan Co., 1926.

Sandys, Sir John Edwin. *A History of Classical Scholarship.* Cambridge, England: Cambridge University Press, 1921.

Taylor, Henry Osborn. *The Classical History of the Middle Ages.* New York: Columbia University Press, 1903.

———. *The Mediaeval Mind*, Vol. I. Cambridge, Mass.: Harvard University Press, 1951.

HISTORICAL EXHIBIT 4.1

Bede's Finger Mathematics

Bede was a British monk who lived in the seventh century (ca. 673–735) and who was considered to be one of the greatest of the medieval Church scholars. Among his works were mathematical tracts on the Church calendar, i.e. computi, ancient number theory, and finger mathematics or numeration which was actually a method for designating numbers by the use of finger gestures. The following diagram, published a thousand years after Bede's death, illustrates some of Bede's medieval number postures.

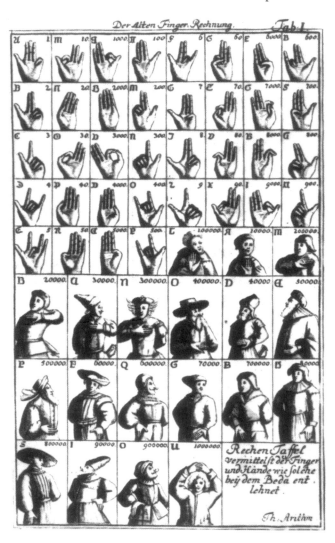

Counters: Computing if You Can Count to Five

VERA SANFORD

*O*RDINARY computation can be accomplished with a minimum of learning by using the loose counter abacus and the counting board. The counting board is a flat surface marked with a series of parallel lines whose values are 1, 10, 100, 1000. The counters are small, easily handled objects,— pebbles, metal disks about the size of a penny, or even, for present-day experimentation, paper clips. The position of a counter shows its value. The number 1432 is indicated by placing one counter on the thousands line, four on the hundreds line, three on the tens, and two on the units line.

Reckoning with counters has left its mark in such words as "calculater" and "calculus" from the Latin *calculus* (a small stone), the "counter" in a store which originally was a counting board, to "cast up accounts" from throwing the counters on the board, and the terms "carry" and "borrow" which described the actual process.

The origin of the counting board and much of its history are not known. Herodotus (ca. 425 B.C.) notes that, "In writing and in reckoning with pebbles, the Greeks move the hand from left to right, but the Egyptians from right to left." This indicates that the reader was familiar with the process and would be interested in the difference between the Greek practice and that of the Egyptians and it also indicates that the lines of the

abacus were vertical. In both Greek and Latin literature, references to the counters appear in context such as the following,—"He (Solon) used to say that men who surrounded tyrants were like the pebbles used in calculations; for just as each pebble stood now for more, now for less, so the tyrants would treat each of their courtiers now as great and famous, now as of no account." We also know that the equipment of a Roman school boy included a bag of counters as well as a wax tablet. And there are still extant Roman abacuses of the type in which beads slide on rods or where studs move in grooves. Details as to the use of the abacus are lacking.

There are references to the use of counter casting in the fourteenth and fifteenth centuries, but there appears to be no actual description of its operation.

In the sixteenth century, a considerable number of arithmetics appearing in northern Europe, especially in Germany, included accounts of computation with the loose counter abacus which seems to have become a well established mercantile practice. German arithmetics were outstanding in this regard. In England, the earliest known book in English on arithmetic (1539) has the following descriptive title:

AN INTRODUCTION

for to lerne to recken with the pen, or with the counters accordynge to the trewe cast of Algorysme, in hole nombers or in broken/newly corrected. And certayne notable and goodlye rules of false posytions

Reprinted from *Mathematics Teacher* 43 (Nov., 1950): 368–70; with permission of the National Council of Teachers of Mathematics.

therevnto added, not before sene in oure Englyshe tonge, by whiche all maner of difficyle questionyons may easily be dissolved and assoylyd.

This was quickly followed by Robert Recorde's *Ground of Artes* (ca. 1542) which had a section on the use of counters. The subject seemed to have been neglected in Italy and in France, the text-books keeping to the arithmetic with the pen. Strangely enough, although the subject was omitted from the earlier editions of a popular arithmetic in France, a book first printed in 1656, a section on counters was introduced in the edition of 1705 and appeared in at least three editions including that of 1781. It is a bit surprising that the later editors chose to put this topic in, but the explanation is as follows: "This arithmetic is quite as useful as that which is done with the pen since by counters, one can perform every calculation of which he has need in business. This way of computing is more practiced by women than by men; nevertheless many people who are employed in the Treasury and in all the government departments make use of this with great success."

The loose counter abacus of the sixteenth century differed from the Greek one in that the lines were marked from right to left instead of up and down. The line nearest to the computer had the lowest value. The counting board might be of stone with the lines cut on it. It might be of wood with lines drawn in chalk for temporary use or painted on permanently. It might be a table cloth with the lines embroidered on it. The lines had the values 1, 10, 100, etc., and the spaces between had the values 5, 50, 500. There is a close correspondence between these markings and Roman numerals. A number is indicated by placing counters on the lines and spaces as the case requires. The accompanying figure shows how the numbers 1285 and 431 would appear.

In addition, the addends are indicated on the counting board. Then whenever a line has five counters on it, as is the case with the tens line and the hundreds line in the case given here, the five

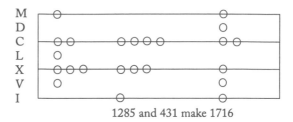

1285 and 431 make 1716

counters are picked up and one is "carried" to the next space. In the example under consideration, there now are two counters in the fifties space. But two fifties make one hundred, so these two counters are picked up and a counter is laid on the hundreds line. The process is repeated until no line has more than four counters and no space more than one.

In subtraction, minuend and subtrahend are entered on the counting board. Then the counters of the subtrahend are matched with those of the minuend and each pair is removed from the board. In some cases it is necessary to "borrow" a counter of higher value from the minuend and to replace it by the equivalent value of counters in the next lower space or line. The process continues until no counters are left in the subtrahend.

To multiply a number by 10, the counters are laid out as if the tens line were actually the units line. To multiply by 100, the hundreds line represents the units. To multiply by 200, you multiply by 100 twice. Since 50 is half of 100, multiplying by 50 is accomplished by taking half of the number of counters on each line or space in 100 times the number. In the following example the number 284 is multiplied by 153. (See solution below.)

Division is difficult and a number of different methods are used. The computer is expected to know the multiplication facts. He decides on the proper quotient figure, and subtracts the partial product from the dividend, using the various lines as the units line as was done in multiplication.

Except for the process of division, computation with the loose counter abacus made no demands on the learner beyond learning how to enter the counters, how to read a number represented by counters, and how to count to five.

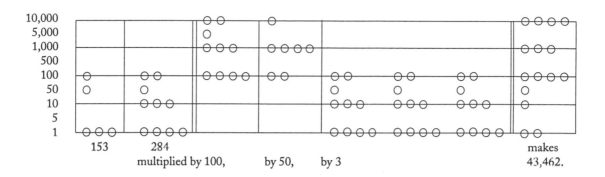

153   284
multiplied by 100,   by 50,   by 3   makes
43,462.

Multiplication was clumsy. Division demanded a knowledge of multiplication combinations unless the computer avoided this issue by using repeated subtraction.

The loose counter abacus is simpler and slower than is the Chinese or the Japanese abacus which requires more mental work. On the other hand, it is a simpler device and one which is easier to master.

REFERENCES

Barnard, F. P. *The Casting Counter and the Counting-Board*, Oxford, 1916. This is technical and exhaustive.

Smith, D. E. *History of Mathematics* II, Boston, 1925, pp. 156–192.

Yeldham, F. A., *The Story of Reckoning in the Middle Ages*, London, 1926.

34

Gerbert's Letter to Adelbold

G. A. MILLER

ONE OF THE MOST astounding comedies of errors in the history of elementary mathematics relates to Gerbert's letter to Adelbold, as may be inferred from the brief articles by the present writer published in the May, 1921, number of the *Scientific Monthly* and in the June, 1921, number of the *American Mathematical Monthly*. Neither of these two articles furnishes the full data upon which the conclusions were based and hence they serve mainly to direct the attention of the careful reader to a few sources of interesting information and to present to the indifferent reader statements which he may accept or reject as his confidence, or lack of confidence, in the judgment of the present writer on such questions may dictate.

Teachers of mathematics should be especially interested in Gerbert, who died in 1003 as Pope Sylvester II, since he is the only man who rose to the position of Pope of the Catholic Church from that of being the most influential teacher of mathematics and other subjects in his generation. The mathematics which Gerbert taught would now be regarded as very elementary, and hence he should be classed with the teachers of secondary mathematics rather than with those of college grade. He lived at a time when even the leading mathematicians of the world used little of their mathematical heritage and added practically nothing thereto for the good of future generations.

Statements found in various well known modern histories of mathematics might lead one to infer that the letter under consideration had been a real contribution towards the increase of mathematical knowledge. Among these statements are the following: It is "the first mathematical paper of the Middle Ages which deserves this name." It explains "the reason why the area of a triangle, obtained by taking the product of the base by half its altitude, differs from the area calculated according to the formula $\frac{1}{2} a (a + 1)$." It "gives the correct explanation that in the latter formula all the small squares, into which the triangle is supposed to be divided, are counted in wholly, even though parts of them project beyond it." To provide a solid basis for further observations and on account of the intrinsic value of this letter for teachers we quote in full the extant part thereof, as follows[1]:

"In these geometric figures which you have taken from us there was an equilateral triangle whose side was 30 feet, altitude 26, and, according to multiplication of side and altitude, the area 390. If you measure the same triangle without paying attention to the altitude, viz., that one side be multiplied into itself and to this product the numerical measure of one side be added and from this sum half be taken, the area is 465. Do you not see how these two rules disagree? But also the geometric rule, which, by taking into consideration the altitude makes the area 390 feet, has been minutely discussed by me, and I grant that its altitude is only $25\frac{5}{7}$ feet and its area $385\frac{5}{7}$ feet. And let this be a universal rule for you for finding the altitude in every equilateral triangle; from the side always take away a seventh and assign the six remaining parts to the altitude."

Reprinted from *School Science and Mathematics* 21 (Oct., 1921): 649–53; with permission of the School Science and Mathematics Association.

"To make what has been said more intelligible permit me to exemplify with smaller numbers. I give you a triangle whose side is 7 feet long. This by the geometric rule I measure thus. Take away a seventh of the side and the six sevenths which are left I give to the altitude. I multiply the side by this and say 6 times 7, that makes 42, the half of this, 21, is the area of said triangle. If you measure the same triangle by the arithmetic rule and say: 7 times 7, that makes 49, and you add the side, making 56, and divide so that you may find the area, you will obtain 28. Behold thus in a triangle of one magnitude there are different areas, which is impossible."

"But that you may not wonder longer I shall explain to you the cause of the diversity. I believe it is known to you what feet are said to be linear, what square, and what cubic, and that in measuring areas we take only square feet. However small a part of these the triangle touches the arithmetic rule computes them as integral. Allow me to give a diagram[2] that what is said may be more clear."

"Behold in this little diagram there are included 28 feet, not all of which are integral. Whence the arithmetic rule, taking the part for the whole, counts the halves with the integers. But the skill of the geometric discipline throwing away the small parts extending beyond the sides and putting together the remaining halves cut in two on the inside of the lines considers that only which is enclosed by the lines. For in this little diagram of which the sides are each 7 feet long if you seek the altitude it is 6 feet. Multiplying this number 6 by 7 you complete, as it were, a rectangle of which the front is 6 feet, the side 7 and the area of it you thus determine to be 42 feet. If you take half of this you leave a triangle of 21 feet. To understand more clearly use your eyes and always remember me."

It may be assumed that all modern teachers of mathematics agree that if any of their students would present the vague arguments found in this letter to explain why the formula $\frac{1}{2}a\,(a + 1)$ does not give the same result for the area of an equilateral triangle of side a as the formula $(a/2)^2\,\sqrt{3}$,

which is equivalent to the product of one-half the base into the altitude, they would reply that these arguments failed to explain the difference in question. Hence the statement that Gerbert gave a *correct* explanation of this difference, which appears in various well-known modern histories of mathematics, including that of M. Cantor, volume I (1907), page 865, is inaccurate.

It may be of interest to consider here briefly a possible explanation of the fact that the Roman surveyors assumed that the formula $\frac{1}{2}a\,(a + 1)$ represents the area of an equilateral triangle whose side is a. The ancient Greeks used the term *triangular numbers* for the positive integers of the form $\frac{1}{2}a\,(a + 1)$. From the facts that the number of equal tangent circles arranged

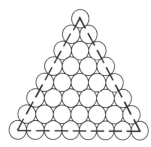

as in the adjoining figure is of this form and that the centers of the outside circles in this figure determine an equilateral triangle, it seems natural to assume that the term triangular numbers originated in connection with some such figure.

If, in the given figure, the number of circles is increased indefinitely without changing the centers of the corner circles, the number of these equal tangent circles will always be expressed by the formula $\frac{1}{2}a\,(a + 1)$ where a represents the number of those circles whose centers determine a side of the given equilateral triangle. As the lower limit of each of these circles is a point it would appear that the number of points in the surface bounded by the given equilateral triangle should be $\frac{1}{2}a\,(a + 1)$, where a represents the number of points in a side. On the other hand, if the area of the given triangle is expressed in smaller and smaller square units, such that the length of the base is equal to a of the corresponding linear units, the area of this triangle is always $(a/2)^2\sqrt{3}$. In fact, $\frac{1}{2}a\,(a + 1)$ has for its limit as a is indefinitely increased the area of an isosceles triangle whose base is equal to the altitude as may be seen directly by means of figure below.

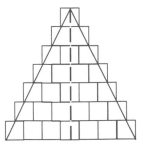

It would thus seem that the assumption that a surface is composed of points which are the limits of equal tangent circles leads to a contradiction. At any rate, if our hypothesis in regard to the origin of the formula $\frac{1}{2}a(a + 1)$ for the area of an equilateral triangle whose side is a is correct, it is interesting to note the dilemma into which the ancient Roman surveyors were led by assuming that a surface is made up of points which are the limits of circles. While these surveyors were not good mathematicians they had sufficient mathematical instinct to recognize that if two assumptions lead to contradictory results they cannot both be true—an instinct which is sometimes disregarded by eminent modern mathematicians.

Returning to Gerbert's letter it may first be noted that the rule for finding the altitude of an equilateral triangle which is found therein is inaccurate since this altitude is equal to a side multiplied by $\frac{1}{2}\sqrt{3} = .8660+$, while $\frac{6}{7} = .8571+$. It is true that the latter is a fairly close approximation but the language of the letter implies that the writer does not realize that his rule gives a result which is only approximately correct.

The main claims for the mathematical merits of the letter in question relate to the so-called explanation of the reasons why the "geometric rule" and the "arithmetic rule" give different re-

sults. Judging from his language the diagram to which Gerbert refers may have been like the adjoining figure which he probably assumed to represent an equilateral triangle. At any rate, if the number of squares on the base line had been 30, in accord with the first triangle to which Gerbert refers in this letter, the vertex of the equilateral triangle constructed on this base would have been below several of the topmost squares, and hence a number of the squares of the corresponding figure would not have been touched by this triangle, lying entirely outside of it.

The method of counting as wholes also those squares which lie only partly within the triangle would therefore not have yielded the number 465, required by the formula $\frac{1}{2}a\,(a + 1)$. Even for the special case when $a = 7$ the explanation is too vague to be called correct. Hence it seems clear that the statement that Gerbert gives a *correct* explanation[3] of the fact that different results are obtained by using two different methods for finding the area of an equilateral triangle is very misleading.

The claim that the letter in question is "the first mathematical paper of the Middle Ages which deserves this name" is even more misleading. It is true that not much mathematical progress was made from the beginning of the Middle Ages to the time of Gerbert, but various papers which were produced during this period are much more meritorious than this letter by Gerbert. In support of this assertion it is only necessary to recall that much of the work of the Hindu and the Arabian mathematicians belongs to this period. In particular, the work from which our term for algebra is derived appeared therein, as well as the dissertation on amicable numbers by Tabit ibn Korra, which has been called "the first known specimen of original work in mathematics on Arabian soil."[4]

Teachers of mathematics should be interested in the letter of Gerbert quoted above not so much on account of the many historical errors relating thereto as on account of the fact that it exhibits a type of the best mathematical thinking among the

Romans during the tenth century. From this standpoint the letter is very instructive and presents to us a picture of weak mathematical thinking on the part of a gifted man. We are thus reminded of the need of carefully graded mathematical aids for our students in order that they may reach even at an early age a mathematical maturity which excels that of the most gifted among the Romans who had to find their way without such aids. Gerbert's letter can profitably be read repeatedly by teachers of mathematics, since it illustrates a type of mathematical thinking which is not only interesting historically but is represented by many students at the present time.

NOTES

1. Gerberti, *Opera Mathematica,* by N. Bubnov, 1899, p. 43. Professors H. J. Barton and W. A. Oldfather assisted me in the translation of this letter.

2. Cf. Fig. 2.

3. This claim seems to have been made first by H. Hankel, *Zur Geschichte der Mathematik,* 1874, p. 314.

4. F. Cajori, *History of Mathematics,* 1919, p. 104.

HISTORICAL EXHIBIT 4.2

The Geometry of Gothic Church Windows

One of the outstanding architectural features of European Gothic churches is tracery, ornamental stone work of interlacing or branching arcs. Tracery depends on the ingenious and creative use of circular arcs and attests to the skill of medieval stonemasons as both craftsmen and users of geometry. Individual designs were quite simple but combined they resulted in structures of striking symmetry and beauty. Two tracery patterns used in constructing vaulted windows are given below.

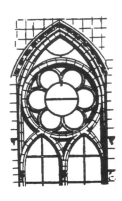
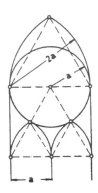

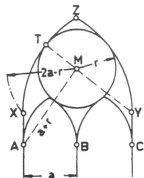

Some other designs are:

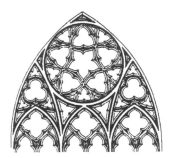

Illustrations adapted from Renno Artmann, "The Cloisters of Hauterine," *The Mathematical Intelligencer* 13 (Spring 1991): 44–49; with the permission of Springer-Verlag and the cooperation of the author.

35

The Arithmetic of the Medieval Universities

DOROTHY V. SCHRADER

The Seven Liberal Arts in Antiquity

The seven liberal arts—the trivium composed of grammar, logic, and rhetoric and the quadrivium made up of arithmetic, music, geometry, and astronomy—were the basis of the curricula in the medieval universities. Whence came these liberal arts? They were not the creation of the Christian Middle Ages but rather a heritage from pagan antiquity. The very word "liberal" implies that these arts belonged to the education of free men, not to the technological training of slaves. The educational system of the medieval universities was an outgrowth, modification, and development of the ancient Greek and Roman educational patterns, adapted and oriented to the Christian ideal.

The Greeks were concerned with the education of free men as future citzens. Plato, whose plan was a theoretical one probably never put into actual practice but nevertheless reflecting the spirit and ideal of his period, conceived of such education as the sole occupation of the first thirty-five years of a man's life. He would have the first twenty years spent on gymnastics, music, and grammar, the next ten on arithmetic, geometry, astronomy, and harmony, and the next five on philosophy.[1] Only then would a man be equipped to take his rightful place as a useful member of society. Aristotle advanced a similar plan in which the elementary training consisted of reading, writing, gymnastics, and music; and the advanced studies included arithmetic, geometry, and astronomy, with the emphasis on the natural sciences.[2] The Sophists, who asserted that rhetoric rather than natural science was the essential study for higher education, advocated gymastics, music, and drawing in the early years and arithmetic and geometry as advanced studies.[3] Philo Judaeus, about A.D. 30, suggested grammar, music, arithmetic, geometry, astronomy, dialectic, and rhetoric as elementary studies, and philosophy as the one higher study.[4] Sextus Empiricus, in the first half of the third century, mentioned grammar, rhetoric, geometry, astronomy, music, and arithmetic as elementary subjects, reserving dialectic for advanced work.[5] Thus we see that there was a tendency among the Greeks to consider six or eight different types of learning as essential to the education of free men, with some difference of opinion as to the order in which the various subjects should be studied.

The Roman writers have mentioned similar subjects of study in their educational plans. Varro (116–27 B.C.) tried to institute a Roman educational system based on that of the Greeks but with reference to Roman rather than Greek literature. He mentioned grammar, literature, arithmetic, geometry, and music as subjects of study.[6] Seneca (2 B.C.–A.D. 65) was less definite in his plans, sometimes placing medicine, rhetoric, and dialectic as basic liberal arts and sometimes considering them as higher studies.[7] Quintillian (A.D. 35–95) suggested that a boy study grammar, music, geometry,

Reprinted from *Mathematics Teacher* 60 (Mar., 1967): 264–75; with permission of the National Council of Teachers of Mathematics.

and astronomy until he was sixteen, after which age he might advance to higher studies.[8] Thus the Romans, like the Greeks, seem to have accepted a few basic subjects as worthy of the efforts of free men and to have expected the young citizens to concentrate on such subjects. However, there is some difference. The Greeks, with characteristic interest in the speculative, placed a greater emphasis on mathematical theory than did the Romans, who rejected much of the Greek mathematics as impractical and therefore unimportant. Moreover, the Romans expected a boy to be ready for advanced work by the time he was sixteen; the Greeks considered him a beginner until he was twenty or older.

It was Martianus Capella in his *De nuptiis philologiae et mercurii*, written about A.D. 330, who set the number of liberal arts at seven and named them: grammar, dialectic, rhetoric, arithmetic, music, geometry, and astronomy. Capella rejected medicine and architecture as purely technical subjects, pursued only for practical and not speculative ends and so unworthy of free men.[9] By the fourth century, this curriculum of the seven liberal arts, as Capella named them, was well established in the pagan schools.

Christianity and the Liberal Arts

In the first ages of the Church, when Christianity was struggling for its existence in a pagan world, the seven liberal arts were denounced by such Christian writers as Origen, Tertullian, and St. Jerome, not because the arts were evil in themselves, but because they were the basis of the pagan educational system which was a threat to the infant Christian Church. Later, when Christianity gained the ascendency over paganism and the pagan schools were no longer a danger, the pagan educational methods were reexamined and were eventually adopted by the Christians. In fact, the study of the seven liberal arts became prerequisite to the study of theology.[10] St. Augustine wrote on all the liberal arts except astronomy and "although not

the originator of the curriculum of the seven liberal arts, he, more than anyone else, made possible its general adoption by the Christian world of the west."[11]

St. Augustine was thoroughly educated in both pagan and Christian learning. He could see much that was good, true, and irrefutable in the work of the pagan scholars and accepted truth where he found it. "We may well say that St. Augustine defends . . . the principles of logic as the inviolable foundations of knowledge. . . . Side by side with logic we find the truths of mathematics . . . all these truths are necessarily and unconditionally true; they cannot be contested."[12] Augustine himself said (A.D. 386), in illustration of the reality of absolute truth: "However, if there are one world and six worlds, it is clear that there are seven worlds, no matter how I may be effected . . . for even if the whole human race were fast asleep, it would still be necessarily true that three times three are nine and that this is the square of intelligible numbers."[13]

Cassiodorus Senator (A.D. 490–585) was probably the first to use the term "seven liberal arts." With his writing, these particular seven subjects became fixed in Christian education and remained so for the next nine hundred years. Seven, as the number of subjects to be studied, was considered to be sanctioned by Holy Scripture itself (Proverbs 9:1), "See where wisdom has built herself a house, carved out for herself those seven pillars of hers,"[14] and continued to hold full sway in educational circles throughout the Middle Ages.

Arithmetic in the Quadrivium

Of the seven liberal arts, the quadrivium comprised what we might call the scientific studies of the day. It has been frequently stated and more frequently implied that the trivium-quadrivium division of studies existed in theory but that in actuality only the trivium was studied. This is not entirely true. It is true that there was, especially in the early years of the Middle Ages, little creative work in the quadrivium and so little of interest for later generations, but instruction was given and the knowledge possessed was imparted to those students who were interested. That there was much teaching in quadrivial subjects is attested by the many manuscripts extant today in European libraries. Most of these manuscripts have not been published because, being mere copies, translations, or commentaries, they contain little original work and so do not evoke scientific interest. It is probably true that there were fewer students of the quadrivium than of the trivium, just as there are fewer graduate students than undergraduates in the average modern university, and for the same reasons.

Considering specifically the field of mathematics, it is known that Church councils from the time of Charlemagne demanded that the clergy have a knowledge of music and be able to compute the date of Easter. To fulfill these two apparently simple requirements, a knowledge of three subjects of the quadrivium was involved: music, arithmetic, and astronomy. In England, from the eighth to the twelfth centuries it was forbidden to the bishop to ordain a priest who could not compute the date of Easter and teach the method to others.[15] Therefore, it is evident that the quadrivium must have been studied.

All the subjects of the quadrivium were based on the learning of antiquity, but perhaps none shows its Greek origins better than arithmetic. Among the Greeks computation or reckoning, the arithmetic of business, was called logistic and was considered to be entirely different from the study of number as such, which philosophical study was called arithmetic. "Logistic is the theory which deals with numerable objects and not with numbers. It does not consider number in the proper sense of the term . . ."[16] but rather the counting of flocks, addition, subtraction, multiplication, and division, always dealing with sensible objects. Arithmetic corresponded roughly to present-day number theory, being a philosophical approach to what is implied in number; it was a mathematical discussion of properties of numbers, proof, and for-

mal demonstration, a mixture of mathematical rigor and pseudo-scientific, semi-magical mysticism. Arithmetic was a study of the universities; logistic was not.

There seems to be some confusion among modern writers concerning the distinction between logic and logistic. Some have asserted or implied that logistic, not logic, was studied in the trivium and then was followed by the advanced mathematical study, arithmetic. However, it should be clear from the Aristotelian treatises used as texts that logic and not logistic was studied in the trivium of the medieval university. Logistic was practical and utilitarian, a study for children and slaves; logic was a liberal art, a study for free men.

The arithmetical knowledge of the Middle Ages can be divided roughly into three periods, and the type of teaching in the schools and universities falls into similar divisions. The first period extended to the end of the tenth century. Arithmetical knowledge was based on the writing of Nicomachus of Gerasa and his interpreters and commentators. The abacus was used for calculations and Roman letter-numerals for recording the results. The most complicated practical problems attempted had to do with determining the date of Easter, and theoretical arithmetic was confined to the classification of numbers by varied and sometimes fantastic properties.

The secomnd period covered approximately the eleventh and twelfth centuries although the termination of this period varied in different localities. It was a period of little creative activity. However, these years did see an extension of the use of the abacus and the introduction of columnar computations.

The third period, from the end of the twelfth century to the end of the Middle Ages, was one of great activity and great change, one might almost say of intellectual revolution, in Europe. The Hindu-Arabic number system was introduced: zero became a familiar concept in a hitherto zeroless world; through translation into Latin from Arabic and Syriac, the ancient Greek mathematical writings were made available to the medieval world. Much of this intellectual activity was centered in the Islamic universities of Spain, and from there learning spread throughout Europe.

In each period, the arithmetic taught in the schools and universities paralleled the general development of arithmetical learning.

Stress on Number Mysticism and Calendric Reckoning

The first period of mathematical development in Europe closed at the end of the tenth century, before the rise of the great universities. However, arithmetic was considered to be an essential part of the curriculum in the cathedral schools and was also taught in all the monastery schools. The emphasis was on the art of computation, especially the method of establishing the date of Easter. In fact, "computus," which originally meant merely computation, soon came to be associated exclusively with the technical study of Easter reckoning.[17] The exclusive use of Roman numerals made computations with large numbers cumbersome, difficult, and close to impossible. Generally, the actual figuring was done on the abacus and then the results recorded in the Roman letter-numerals. There was much study of the mystical and symbolic number properties and relations, based on the work of Nicomachus.

Nicomachus of Gerasa (ca. A.D. 200) wrote much on number mysticism, a sort of theology of numbers. Numbers were identified with the various gods. He considered the odd numbers to be male and the even ones to be female. He made a strange distinction between the "divine number," a sort of general concept of number which existed only in the mind of the creator-god, and scientific numbers, which were the common numbers known to men on earth. Nicomachus' *Introduction to Arithmetic* is a restatement of commonly known facts, not an original treatise, and is largely Pythagorean. Theon of Smyrna wrote a book dealing with those mathematical matters which were essential for one

who was to read Plato; there are so many similarities between the work of Nicomachus and that of Theon that it is difficult to say which wrote first.[18] Nicomachus' book was the source for Martianus Capella, Boethius, Cassiodorus Senator, and Isidore of Seville.

Martianus Capella, who flourished from about 410 to 429, wrote *The Marriage of Philology and Mercury*, which is actually little more than a textbook, an allegorical setting for a book on the seven liberal arts. The first two chapters establish the setting; the rest of the book is devoted to instruction. It is "strictly instructive, as sapless as the rods of the mediaeval schoolmasters."[19] In it, "the most bizarrre imagination was allied with the most arid intellect."[20] The general level of the mathematical instruction contained in it is shown in the scene in which Philology, fearing to marry Mercury because he is a god and she cannot claim divinity, realizes, after much calculation, that the numbers forming her name and his indicate that this is a propitious match, and so she enters into the marriage happily and without fear.[21] Besides this numerology, Capella discusses proportions and multiples of proportions,[22] prime numbers, perfect numbers, perfect triads (cubes), and similar mystic properties of numbers, stressing the relationship of numbers with the planets.[23]

Martianus Capella's forty-seven pages of number mysticism pale to insignificance in the light of Boethius' hundred or more pages of mathematical work. In his *De institutione arithmetica libri duo*,[24] which was the direct or indirect source of arithmetical knowledge for close to a thousand years, Boethius (ca. 470–525) has not merely translated Nicomachus' treatise. True, he did write after the manner of the Greek, borrowing freely and at times obviously copying, but he also considerably augmented the other's work. In a sense he "baptized" Nicomachus, making his work acceptable to the Christian schoolmen. Boethius' work has the same sort of classification of number properties as does Nicomachus', with many extended or original interpretations and examples; he also includes a mystical interpretation of scriptural numbers. Like Nicomachus, Boethius gives no rules of operation.

The *De arithmetica* of Cassiodorus Senator (490–585) is almost a condensation of Boethius' arithmetical work. Cassiodorus quotes Scripture to show that God planned the universe on a basis of number, weight, and measure, and he advocated this as a reason for the study of arithmetic. At the end of his treatise on arithmetic, after a recapitulation of the importance of number, he pays tribute to Nicomachus, to one Apuleius of Madura, and to "Boethius, a man of distinction." He ends with a naively pious interpretation of the first seven digits: one God, two Testaments, three Persons in the Trinity, four Gospels, five books of Moses, six days of Creation, seven Gifts of the Holy Spirit.[25]

Isidore of Seville (560–636) also wrote a condensation of Boethius in his *Epistemologies*. There is little that is original in his encyclopedic work except his advanced number mysticism, which is almost ridiculous to the modern reader. He finds a mystic signification in all the numbers from one to 20 except 17; also 24, 30, 40, 46, 50, and 60.[26] He makes only the vaguest references to rules of operation, simply listing instances where like numbers occur and assuming from the similarity that there is an explanation and interrelation. To him, a pound is a perfect weight because it has the same number of ounces as there are months in a year.[27] He is deeply religious and finds mystic and scriptural significance in all numbers. The fact that God performed 22 works in creation is adequate explanation for Isidore as to why there are 22 generations from Adam to Jacob, 22 books of the Old Teatament to Esther, and 22 sextarii in a bushel.[28] He is Pythagorean in his stress on the vital importance of number. "Take number from all things and all things perish."[29] In spite of this phase of his work, Isidore has written a fairly good condensation of Boethius' *Arithmetica*.

Venerable Bede, in his *De tempore ratione*, wrote what was perhaps the first treatise on practical computation, a method of determining the date of Easter.[30] Rabanus Maurus, in *Liber de computo*, wrote a most complete textbook, giving a method of finding the date of Easter, instructions for finger reckoning, instructions for using Roman numerals, and some astronomical material.[31]

The method for fixing the date of Easter occupied a very important place in the arithmetic of the Christian era. Almost the entire liturgical year depends upon the date of Easter. The beginning of the Church year (the first Sunday of Advent) is determined by the fixed date of Christmas, but nearly all the other great feasts and important days—Septuagesima, Lent, Maundy Thursday, Good Friday, Ascension, Pentecost, the number of weeks after Epiphany and after Pentecost—all depend on the date of Easter. In the sixth century there was a serious schism threatened about this very topic; the Irish monks, calculating one way, arrived at a date a week different from the one determined by the continental Benedictines, who figured the date by another method. The controversy was settled by the Synod of Wisby.

The problem of setting the date of Easter is essentially a problem of determining the date on which will fall the first Sunday after the first full moon after the vernal equinox. To do this, one must know the vernal equinox, the day of the following full moon, and the method of correcting the error in the metonic cycle. (The metonic cycle is a period of nineteen years, after which the full moon falls on the same day of the year.) The astronomical problems were settled as early as 525. Tables of Dionysius, Isidore, and Bede were available: Bede's tables went as far as 1063. To use the tables, one had to be able to find the "Golden number" and the "Dominical letter" for the year involved. The "Golden number" was found by adding one to the number of the year, dividing by nineteen: the remainder is the "Golden number." The "Dominical letter" is found by dividing the number of the year by four, adding the quotient to the number of the year, adding four, dividing by seven, and subtracting the quotient from seven; the remainder determined the place of the "Dominical letter." By referring to the table, the date of Easter was easily determined, using these two answers.[32] After the Carolingian revival, every priest was required to understand the method and principles of computation.

This period was, then, one of slow growth and dissemination of arithmetical learning. At a time when many of the educated men were clerics, the major practical application of arithmetic was related to the Church. Nevertheless, there was an interest in every phase of arithmetic, which interest would grow during the second period and burst forth into the intellectual revolution of the third period.

The Rise of the Universities and the Influence of the Arabic Texts

The second period of medieval mathematical development extended from the end of the tenth to the end of the twelfth century. The outstanding mathematical genius of the period was Gerbert, who taught the quadrivium with marked success in the cathedral school at Rheims from 972 to 982. Gerbert improved the abacus by placing symbols at the top of each column, and so extended its use. He developed a method of division, making possible all four fundamental operations on the abacus. He introduced columnar computation but did not use zero. he wrote a treatise, *Regulae de abaci numerorum,* on the use of the abacus, perhaps the first of its kind.[33]

Towards the end of this period, the Hindu-Arabic number system was beginning to be known in Europe. Those who adopted the new system were known as algorists as distinct from the abacists, who used the older methods. The abacists used the abacus, the old Roman notation, no zero, and the Roman duodecimal fractions, and had no method for finding square root. The algorists used written calculations, the Hindu-Arabic numbers, zero, decimal fractions, and the Arabic method for finding square roots. Addition, subtraction, and multiplication were comparable in both systems, but division was very different, so much so that the complicated abacist division was called *divisio ferrea* while the simpler algorist division was called *divisio aurea.* Nevertheless, the algorists did not supplant the abacists at this time; in fact there was little controversy between the two groups.

Hermanus Contractus, in the first half of the eleventh century, wrote a *Liber abaci,* based on Gerbert's work, as did several other writers of the period.

With the rise of the universities, with their curricula based on the trivium and the quadrivium there came an opportunity for a more rapid spread of arithmetical knowledge. In the twelfth century and the first half of the thirteenth, the arithmetic taught at Oxford was largely derived from Boethius, Cassiodorus, and Isidore; it consisted of the study of the properties of numbers: ratio, proportion, fractions, and polygonal numbers. No practical calculations were done so no abacus was necessary.[34] An almost identical situation existed at Cambridge.[35] This was probably the situation in the continental universities, also. By the mid-thirteenth century, there was a wider range of mathematics available for those students who were interested but there seems to be no evidence that many took advantage of the opportunity.

The Arabic texts were being translated into Latin and so made available to the scholars of Europe. Gerard of Cremona is perhaps typical of the translators. Born in Cremona in 1114, he studied all the arts but was especially interested in astronomy. As Ptolemy's works were not available to him, he went to Toledo, where he learned Arabic and translated the *Almagest.*[36] In 1187 he translated Al-Khowarizmi's works on calculation and on algebra into Latin. Later he wrote a text on calculation, *Algorismus.* Various other translators did likewise: Adelard of Bath, Abraham ben Ezra, John of Seville, and others, whose names have not survived. These may not be considered typical works of this period but rather the foundation of the university texts which were to follow; these twelfth-century works formed a link between the second and third periods of intellectual development.

Arithmetic as a Science

The third period of mathematical development extended from the thirteenth century until the end of the Middle Ages. One might almost say that it was a period of intellectual revolution. Algorist arithmetic replaced the use of the abacus. There was a new emphasis on the practical and scientific aspects of arithmetic, although the mystical side was not neglected.

Leonard of Pisa, in his *Liber abaci* of 1202, exhibited an amazing theoretical and practical knowledge. He explained the Arabic numeration, using nine digits and zero, proved elementary formulae, and even solved some elementary algebraic equations.[37] Strangely, his work had little effect on the universities, although it seems to have had a great influence in the world of commerce. Until his time, the abacus and finger-reckoning had been the principal means of computation. Now the new symbols became popular, and algorism, as distinct from Boethian arithmetic, spread throughout Europe. By 1300, there were few almanacs which did not include an explanation of it. By 1350, the almanacs had added explanations of the four fundamental processes and the more important ones included rules of proportion, examples, and formulae for the more common commercial transactions.[38] In spite of the popularity of the Hindu-Arabic numbers for actual computation, the old Roman numerals were still used in keeping records by merchants until 1550, in colleges and monasteries until 1650, and for parish registers as late as 1700.

With the intellectual revolution of the twelfth and thirteenth centuries and the rise of the universities, brought about by a renewed interest in Roman and canon law, the systematization of theology, and a new interest in dialectic, there was a decided change in the status of the seven liberal arts. "Dialectic or logic became so important that it tended to obscure all the other arts."[39] There was a revival of interest in science, and for a while all the subjects of the quadrivium were carried along on the tide of the new intellectualism. This interest was short-lived, however, and the quadrivial subjects declined rapidly. In Paris and the other French universities, the quadrivium was almost entirely obscured during the latter part of the thirteenth and during the fourteenth centuries. "As for the quadrivium, as the sciences are called, since they have little to attract in themselves and produce only a meager profit, most of the students neglect them or else omit them entirely."[40] A utilitarian spirit grew up among the students. To obtain a prebend or a prelature, all that was necessary was to have studied the liberal arts. So, after the trivium, many of the students left the universities and received benefices. Those who did go on into law or medicine did so for profit. "Only in theology was the pure love of learning retained, and that was guarded jealously."[41]

In Oxford, the quadrivium was a little, but not much, more in vogue than in Paris. The quadrivium was neglected in Paris, Oxford, and Cambridge ostensibly because quadrivial studies had practical application and Paris, Oxford, and Cambridge "systematically discouraged all technical instruction, holding that a university education should be general and not technical."[42] The real reason, though, seems to have been that distinction could be more easily attained in theology and philosophy than in the sciences.

Nevertheless, by the mid-thirteenth century, arithmetic was being taught as a science in the universities. Algorism and arithmetic flourished together for a while, but gradually the mystical and over-refined classifications of Boethius gave place to the Arabic algorism and to algebra. Perhaps the most influential mathematician of this time was Jordanus Nemorarius. Manuscripts of Jordan's work have been found in Basle, Cambridge, Dresden, Erfurt, Munich, Oxford, Paris, Rome, Thorn, Venice, Vienna, and many places in southern Germany.[43] The three treatises with which we are most directly concerned are *Algorithmus demonstratus*, *Arithmetica demonstrata*, and *De numeris datis*, although Jordan is perhaps best known for his *De ponderibus*, a treatise on weights. *Algorithmus demonstratus* is an elementary treatment of practical computations, including Arabic notation, fundamental operations, and common and astronomical fractions. Jordan treats of nine operations: numeration, addition, subtraction, duplation, multiplication, mediation (finding the mean), division, progression, and extraction of roots.[44] About two-fifths of the work is devoted to fractions. *Arithmetica demonstrata*, in ten books, is in the Boethian tradition, treating of properties and relations of numbers, prime, perfect, polygonal, and solid numbers, and the Greek theory of ratio. Jordan made a notable advance over earlier mathematicians and imitators of Boethius in that he used letters as general symbols instead of only concrete numbers. That the *Arithmetica* was popular with generations of university students is shown by the number of printed editions which are extant. D. E. Smith has identified five authentic editions, those of 1496, 1503, 1507, 1510, and 1514, and one probable one, 1480.[45] *De numeris datis* is a treatise on arithmetic and algebra, four books of problems with linear and quadratic equations in one or more unknowns. All Jordan's books are purely academic, abstract and scientific, containing no business methods, and so they were eminently fitted for use as texts in the medieval universities.

The arithmetic of the schools did not receive the wholehearted approbation of all the people. This is readily seen in Henri d'Andeli's *Battle of the Seven Arts*. The medieval author, a cleric of the University of Paris, portrays a battle among the liberal arts, with logic and grammar striving for the ascendency. Arithmetic is portrayed as a maiden sitting under a tree, placidly counting and figuring, entirely unmoved by the practical problems of warfare surrounding her. In the lines, "And three times twenty by themselves make sixty," and "Five twenties make hundred . . . ," d'Andeli is ridiculing the affectation of the schoolmen who coin the terms "soixante" and "cent" instead of using the "trois-vingts" and "cinq-vingts" of the common people.

Arithmetic sat in the shade,

Where she says, and where she figures,

That ten and two and one make thirteen,

And three more make sixteen;

Four and three and nine to boot

Again make sixteen in their way;

Thirteen and twenty-seven make forty,

And three times twenty by themselves make sixty;

Five twenties make hundred and ten hundreds a thousand.

Does counting involve anything further? No.

One can easily count a thousand thousands

In the foregoing manner,

From the number which increases and diminishes

And which in counting goes from one to hundred.

The dame makes from this her tale,

That usurer, prince, and count

Today love the countess better

Than the chanting of High Mass.

Arithmetic then mounted

Her horse and proceeded to count

All the knights of the army; . . . [46]

Throughout the Middle Ages, university instruction was based on a lecture-disputation method. The students were obliged to "hear" certain books, that is, to attend lectures in which the text, glosses, and commentaries were discussed by the professors. Sometimes there were after-class discussions, reviews, and recapitulations of the lectures by the young bachelors. There were no examinations in the modern sense of the term. The student had simply to swear that he had read the books prescribed and attended the lectures. To qualify for a degree, he was required to participate in public disputations, either defending a proposition or opposing one defended by another student.

Detailed regulations and requirements varied in different universities, of course, and even at different periods in the same university, but the general practice was to follow this lecture-disputation method.

University Curricula

There was a rather wide range in the mathematical curricula offered by the various universities; this is to be expected, as some placed greater emphasis than others on arithmetical and scientific studies.

At Bologna, there was a chair of arithmetic in the faculty of arts, and a course on Jordan's Algorismus was prescribed for all students. Algorismi de minutes et integris, a text based on Jordan's book, was used and, in 1405, was the only arithmetical text prescribed. It was to be studied, along with the first book of Euclid's Elements, by the first-year students.[47]

In Paris, in 1215, no special books were prescribed for any subjects of the quadrivium. In 1366, in an effort to stimulate interest in mathematics, a university statute required Master of Arts candidates to attend lectures on "some mathematical books." This is generally interpreted to mean Sacrobosco's lectures on the sphere and one other book.[48]

Johannes Sacrobosco, also known as John of Holywood and John of Halifax, was a leading figure in the mathematical world of his day (d. 1244). Born in Yorkshire, he received his degree as Master at Oxford and then went to Paris, where he taught until his death. He was the first to give public lectures on algebra and algorism at any university. He wrote a work on algorism, based on Jordan's work, containing the rules of arithmetic, omitting fractions, and giving the rules of multiplication in verse. He also wrote on the sphere and the astrolabe. He was apparently a popular teacher and an influential member of the Paris faculty.

At Oxford, in the fourteenth century, mathematics flourished, especially at Merton College. Richard of Wallingford, Maudith, Simon Bredon,

John Ashender, William Rede all were capable mathematicians. Their works survive in manuscript but are little known. "Thus it is that we have no cognizance of work produced during a century in which Oxford could boast more mathematicians than any country in Europe."[49] Perhaps the outstanding Oxford mathematician of the century was Thomas Bradwardine, a Franciscan friar who was born near Chichester about 1290. Called "Doctor Profundis" because of his learning, he became proctor of Merton College in 1325. His mathematical writings include *De tractatus de proportionibus*, *Arithmetica speculativa*, *Geometria seculativa*, and *De quadratu circuli*. The tract on proportions was printed in Paris in 1495 and the arithmetic in 1502, some 160 years or more after they were written.[50] The *Tractatus de proportionibus* is included in the 1389 list of required texts at the university at Vienna.[51] Bradwardine left Oxford in 1335 to become Archbishop of Canterbury, which post he held until his death in 1349.

In the fifteenth century, Oxford mathematics declined so that between 1449 and 1463, all the mathematics required for the Master's degree was the first two books of Euclid and Ptolemy's astronomy, either in the original or in commentary.[52]

The last of the great Oxford mathematicians was Robert Recorde (1510?–1558). He must have been a fine teacher as well as an eminent mathematician, for it is said that the students used to applaud his lectures and comment that they learned much of general interest in science as well as the rules of arithmetic. The earliest use of the word "algebra" in English occurs in his *Pathway of Knowledge*, in 1551. He wrote an arithmetic, *The Grounde of Arts*; a geometry, *The Pathway of Knowledge*; and an algebra, *The Whetstone of Witte*. Particularly interesting are the devices Recorde proposes for the painless use of Arabic numerals. For example, one need learn multiplication tables only as far as five times five if he chooses to multiply by Recorde's method. Using 8×7 as an example:

$$8 \diagdown \quad 2 \qquad (10 - 8 = 2)$$
$$7 \diagup \diagdown \quad 3 \qquad (10 - 7 = 3)$$
$$\overline{5 \qquad 6} \qquad (2 \times 3 = 6)$$

$$(7 - 2 = 5)$$

$$\text{or} \ (8 - 3 = 5)$$

Therefore,

$$8 \times 7 = 56.^{53}$$

Might Recorde's little diagram be the origin of ×
as a symbol of multiplication? Recorde's works,
written in dialogue form as a discussion between
master and disciple, enjoyed great popularity in
university and in social circles. Both *The Grounde
of Arts* and *The Whetstone of Witte* went into several
editions.

The state of mathematics at Cambridge was
such that in the fourteenth century, only
Sacrobosco's *De sphaera* was required for a bachelor; six books of Euclid and a book on optics were
required for a master. There is record of lectures
being given in arithmetic, finger reckoning, and
the algorism of the integers, but none of these
seems to have been required.[54]

There was a somewhat different situation at
the University of Vienna, where there was a particular emphasis on science and mathematics. "Extraordinary" or "cursive"[55] lectures on mathematics
were considered to be especially appropriate diver-
sions for holiday afternoons. Also, disputations
were held on mathematical subjects, a practice
which was almost unknown elsewhere. In 1365,
Vienna prescribed the first book of Euclid and
some book on algorism for bachelors but required
the first five books of Euclid, Bradwardine's *Tract
on Proportion*, and two books on arithmetic for a
license to teach. In 1389, five books of Euclid,
treatises on proportional parts, perspective, and
measurement of areas were required for a master.[56]
In 1391–92, all of Jordan's works on integers,
common fractions, astronomical fractions, proportions, and *arithmetica* were required for a master's
degree. There was a distinction made between

algorism (practical) and arithmetic (theoretical);
the fees for arithmetic were twice the fees for
algorism.[57]

Most of the German universities seem to have
followed the lead of Vienna in respect to their
mathematical curricula. Prague in 1366 required
algorism and six books of Euclid;[58] Erfurt in 1449
demanded algorism, computus, and six books of
Euclid; Ingolstadt in 1472 prescribed the first book
of Euclid, algorism, and common arithmetic.[59]
Heidelberg in 1443 relegated algorism and proportion to the "brush up" category, charged extra
fees, and gave no master's credit for such courses.[60]

A general view of the medieval universities
would seem to indicate that, at any given time,
whatever arithmetic was known was taught, being
either required of degree candidates or made available to those students who were interested in reading beyond the narrow confines of course requirements. The teachers were often famous mathematicians, translators, commentators, and authors of
texts.

Present-Day Judgments

Although it is true that much of what was, in the
medieval university, course material for a master's
degree is today common knowledge for third-grade
school children, and although some of the more
profound medieval processes of ratio and proportion are today taught in eighth-grade arithmetic
classes, medieval arithmetic must not be regarded
as superficial or merely elementary. Many of the
concepts are as challenging to modern graduate
students of number theory as they were to medieval students of *arithmeitca*. Moreover, the acceptance and spread of the Hindu-Arabic number
system—a place system instead of an additive one,
the use of zero as a number, number symbols
instead of letter numerals, decimal instead of duodecimal fractions—with all of its implications and
ramifications, required the reeducation of the entire population of Europe, no small task to accomplish, even over two centuries.

It is interesting to conjecture whether or not we are today facing a similar upheaval in theoretical mathematics. Is it not possible that someday a high school student may laugh condescendingly and say, "And they got *graduate* credit for that!"

NOTES

1. Paul Abelson, *The Seven Liberal Arts, A Study in Mediaeval Education* (New York: 1906), p. 2.

2. *Ibid.*, p. 2.

3. *Ibid.*, p. 3.

4. *Ibid.*, p. 3.

5. *Ibid.*, p. 4.

6. *Ibid.*, p. 4.

7. *Ibid.*, p. 5.

8. *Ibid.*, p. 5.

9. *Ibid.*, p., 7.

10. *Ibid.*, p. 8.

11. *Ibid.*, pp. 8–9.

12. Denis J. Davanagh, O.S.A., *Answer to Skeptics* (New York: 1943), Introduction, p. xv.

13. *Ibid.*, p. 183.

14. R. A. Knox translation.

15. Abelson, *op. cit.*, p. 91.

16. Nicomachus, *Introduction to Arithmetic*, tr., Martin L. D'Ooge (New York: 1926), pp. 3–4.

17. See later discussion in this article.

18. Nicomachus, *op. cit.*, Notes, pp. 37 ff.

19. Percival R. Cole, *Later Roman Education* (New York: 1909), p. 19.

20. *Ibid.*, p. 22.

21. *Ibid.*, p. 21.

22. E.g., 6:8::9:12—

$$6 \times 72 = 432$$

$$8 \times 72 = 576$$

$$9 \times 72 = 648$$

$$12 \times 72 = 864$$

Therefore,

$$432:576::649:864$$

23. Adolfus Dick (ed.), *Martianus capella* (Leipzig: 1925), VII, 735–802.

24. J. P. Migne, *Patrologiae cursus completus*, Tomus LXIII (Paris: 1847), columns 1079–1168.

25. Cassiodorus Senator, *An Introduction to Divine and Human Readings* (New York: 1946), pp. 180–89.

26. Ernest Brehaut, *An Encyclopedist of the Dark Ages* (London: 1912), p. 29.

27. *Ibid.*, p. 65.

28. *Ibid.*, p. 65.

29. *Ibid.*, p. 65.

30. Ableson, *op. cit.*, p. 97.

31. *Ibid.*, p. 97.

32. *Ibid.*, p. 98.

33. *Ibid.*, pp. 100–102.

34. R. T. Gunther, *Early Science in Oxford* (London: 1922), I, 94.

35. W. W. R. Ball, *A History of the Study of Mathematics at Cambridge* (Cambridge: 1889), p. 2.

36. Moritz Cantor, *Vorlesungen über Geschichte der Mathematik* (Leipzig: 1894), pp. 853–54.

37. Ball, *op. cit.*, p. 4.

38. *Ibid.*, p. 7.

39. Louis J. Paetow, *The Arts Course in Medieval Universities* (Champaign-Urbana: 1910), p. 7

40. Achille Luchaire, *L'Université de Paris sous Philippe-Auguste* (Paris: 1899), p. 25.

41. *Ibid.*, p. 25.

42. Ball, *op. cit.*, p. 152.

43. Abelson, *op. cit.*, p. 105.

44. *Ibid.*, p. 106.

45. David Eugene Smith, *Rara arithmetica* (Boston: 1908), p. 62.

46. Henri d'Andeli, *La Bataille des VII Arts* (Berkeley: 1927), pp. 48–49.

47. Hastings Rashdall, *The Universities of Europe in the Middle Ages* (Oxford: 1942), p. 248.

48. *Ibid.*, p. 249.

49. Gunther, *op. cit.*, p. 96.

50. *Ibid.*, p. 93.

51. Karl von Raumer, *German Universities* (New York: 1859), p. 159.

52. Gunther, *op. cit.*, p. 98.

53. The algebraic justification for Recorde's method is quite obvious:

$$10[x - (10 - y)] + (10 - x)(10 - y)$$
$$= 10(x - 10 + y) + (10 - x)(10 - y)$$
$$= 10x - 100 + 10y + 100 - 10x - 10y + xy$$
$$= xy$$

54. Ball, *op. cit.*, p. 5 ff.

55. These lectures covered merely the text, without glosses and commentaries.

56. Ball, *op. cit.*, p. 8.

57. Abelson, *op. cit.*, p. 107.

58. von Raumer, *op. cit.*, p. 159.

59. *Ibid.*, p. 159.

60. Abelson, *op. cit.*, p. 111.

BIBLIOGRAPHY

Abelson, Paul. *The Seven Liberal Arts: A Study in Mediaeval Education.* New York: 1906.

Aron, Marguerite. *Saint Dominic's Successor.* St. Louis: 1955.

Augustine. *Answer to Skeptics.* A translation of St. Augustine's *Contra academicos* by Denis Kavanagh, O.S.A., S.T.M. Introduction by Rudolph Arbemann, O.S.A. New York: 1943.

Ball, W. W. Rouse. *Cambridge Papers.* London: 1918.

————. *A History of the Study of Mathematics at Cambridge.* Cambridge: 1889.

Boethius. *Patrologiae cursus completus* lxiii, Ser. 1 (*Opera omnia Boetii,* ed. J. P. Migne). Paris: 1847.

Brehaut, Ernest. *An Encyclopedist of the Dark Ages, Isidore of Seville.* London: 1912.

Cantor, Moritz. *Vorlesungen über Geschichte der Mathematik.* Leipzig: 1894.

Cassiodorus Senator. *An Introduction to Divine and Human Readings.* Translated by Leslie Webber Jones. New York: 1946.

Cole, Percival R. *Later Roman Education in Ausonius, Capella, and the Theodosian Code.* New York: 1909.

D'Andeli, Henri. *La Bataille des VII Arts,* ed. Louis J. Paetow, in *Memoirs of the University of California,* Vol. IV. Berkeley: 1927.

Dick, Adolphus (ed.). *Martianus capella.* Leipzig: 1925.

Gunther, R. T. *Early Science in Oxford,* Vol. VI, Part II. Oxford: 1922.

Isidore Hispalensis Episcopi. *Etymologiarum sive originium.* Oxford: 1911.

Luchaire, Achille. *L'Université de Paris sous Philippe-Auguste.* Paris: 1899.

Moody, Ernest A., and Clagett, Marshall. *The Medieval Science of Weights.* Madison: 1952.

Mullinger, James Bass. *The University of Cambridge from the Earliest Times to the Royal Injunctions of 1535.* Cambridge: 1873.

Nicomachus of Gerasa. *Introduction to Arithmetic.* Translated by Martin Luther D'Ooge ("Studies in Greek Arithmetic," ed. Frank E. Robbins and Louis C. Karpinski). New York: 1926.

36

The Craft of Nombrynge

E. R. SLEIGHT

*V*ERY few arithmetics appeared in the English language before the sixteenth century. As late as the middle of the fifteenth century every person using even the simplest operations could read Latin, in which language such information was written. Until modern commerce was established, arithmetic was required only for addition and subtraction, and as late as the thirteenth century a student well advanced in science generally knew nothing of division.

The Earliest Arithmetics in English, edited by Robert Steele, and published for the English Text Society, lists only five arithmetics appearing in the English language before the beginning of the sixteenth century. All of these are of the fifteenth century as is shown by the style of English used. Three of them are mere fragments and have received very little attention. The other two, *The Crafte of Nombrynge* and *The Art of Nombrynge,* are more extensive in scope, and it is the purpose of this paper to review the first of these with special emphasis on the methods of operation.

The Crafte of Nombrynge is an interpretation of the *Canto de Algorismo* of Alexander de Villa Dei (1220), and is bound up, together with other scientific treatises, in the British Museum Library. It deals with the science of arithmetic rather than with the art. Each separate idea in the English translation is introduced by one or more lines of the *Canto*, the Latin form being retained. There are only thirty-two pages in which are discussed certain definitions,

Reprinted from *Mathematics Teacher* 32 (Oct., 1939): 243–48; with permission of the National Council of Teachers of Mathematics.

notation, meaning and use of zero, how to read and write numbers, and the seven rules.

Algorism is the first definition introduced by

HEC ALGORISMUS ARS PRESENS DICITUR; IN QUA TALIBUS INDORUM FRUIMUR BIS QUINQUE FIGURIS.[1]

Then follows in fifteenth century English—*This book is called the book of Algorism, and the book treats the Craft of Numbering, the which Craft is called also Algorism. There was a King of India, the name of whom was Algor, and he made this his craft, and after his name he called it Algorism.*

Very frequently the question and answer method is used. *This present craft is called Algorism, in which we use ten signs of India. Questio. Why ten figures of India? Solucio. For as I have said before they were found first in India by a King of that country that was called Algor.*

It becomes necessary to define certain terms frequently used in this treatise. *Some numbers are digitus in Latin and digits in English. Some numbers are called articulus in Latin and articles in English. Some numbers are called composite in English.*

SUNT DIGITI NUMERI QUI CITRA DENARIUM SUNT.

Here he[2] tells what a digit is. A digit is a number that is within (or less than) ten, such as 9. 8. 7. 6. 5. 4. 3. 2. 1.[3]

ARTICUPLI DECUPLI DEGITORUM; COMPOSITI SUNT ILLI QUI CONSTANT EX ARTICULIS DEGITISQUE.

Here he tells what a composite number is, and what are articles. Articles are those which may be divided into numbers of ten and nothing left over, such as twenty, thirty, one hundred, one thousand, etc. A composite number is one that is composed of a digit and an article such as fourteen, fifteen, twenty-five, etc. And so every number that begins (at right) with a digit and ends with an article is a composite number. Thus twenty-five begins with the digit five and ends with the article twenty.

The meaning and place of these numbers is then given, followed by an explanation of the principles of notation with great emphasis on the use of the cypher.

NIL CIFRA SIGNIFICAT SED DAT SIGNARE SEQUENII

Explain this verse. A cypher means nothing but he[4] makes the figure that comes after him to mean more than he should if he were absent, as thus 10. Here the number means ten, and if the cypher were away and no figure in front of him it would mean only one, for then he would stand in the first place. And another cypher would mean nothing more unless it keeps the order of the place. A cypher is not a figure significant.

QUAM PRECEDENTES PLUS ULTIMA SIGNIFICABIT.

The last figure shall mean more than all the others though there were a hundred thousand before it. Thus 17689. The last figure, that is 1, means ten thousand, and all the other figures mean only seven thousand six hundred eighty and nine. And ten thousand is more than all that number. Ergo, the last figure means more than all the number before.

SEPTEM SUNT PARTES, NON PLURES ISTIUS ARTIS;
ADDERE, SUBTRAHERE, DUPLARE, DIMIDIARE,
SEXTAQUE, DIVIDERE, SED QUINTA MULTIPLICARE;
RADICEM EXTRAHERE PARS SEPTIMA DICITUR ESSE.

From this quotation the writer discovers that there are seven operations to be considered: addition, subtraction, duplication, mediation, multiplication, division, and extraction of roots. The remainder of the thirty-two pages of the treatise are devoted to the explanation and use of five of these seven operations, no mention being made of extraction of roots, and division is used only in mediation.

ADDERE SI NUMERO NUMERUM VIS, ORDINE TALI
INCIPE; SCRIBE DUAS PRIMO SERIES NUMERORUM
PRIMAM SUB PRIMA RECTE PONENDO FIGURAM, ET SIC
DE RELIQUIS FACIAS, SI SINT TIBI PLURES.

Here begins the craft of addition. In this craft thou must know four things. First thou must know what is addition. Next thou must know how many rows of figures thou must have. Next thou must know how many diverse cases happen in this craft. And next what is the profit of this craft. As for the first thou must know that addition is a casting together of two numbers into one number. As for the second thou must know that thou shalt have two rows of figures, one under the other

1234

as here you may see: 2168. As for the third thou must know that there are four different cases. As for the fourth thou must know that the profit of this craft is to tell what is the whole number that comes of different numbers.

The four cases are:

1. No partial sum being greater than 9.
2. At least one partial sum being greater than 9.
3. The case in which at least one partial sum is 10 or a multiple of 10.
4. The case in which there is a cypher in the upper row.

The method of procedure is discussed at length, but it resolves itself into the method used today, the operations being performed from right to left as at present.

A NUMERO NUMERUM SI SIT TIBI DEMERE CURA SCRIBE
FIGURARUM SERIES, VT IN ADDICIONE.

This is the chapter on subtraction in which thou must know four things.[5] These four things are identical in all operations, but the definitions are different. As for the first thou must know that subtraction is the withdrawing of one number from another. As for the second thou must know that there shall be two numbers. As for the third thou must know that there are four different cases. When all digits in the upper row are larger than the corresponding digit in the lower. When at least one digit in the lower row is larger than the corresponding digit of the upper. When the digit in the lower is larger than the corresponding digit in the upper, and the next figures to the left, both above and below are zeros.

Here again the method is entirely like the method of today, "borrowing" and all. A paragraph is given to *teaching the Craft how thou shalt know, when thou hast subtracted whether thou hast well done or no.* The method is the one now in use, adding the remainder to the subtrahend to give the minuend.

SEQUITUR DE MEDIACION. INCIPE SIC, SI VIS ALIQUEM NUMERUM MEDIARE SCRIBE FIGURARUM SERIEM SOLAM VELUT ANTE.

In this chapter is taught the Craft of mediation, in the which craft you must know four things. As for the first you shall understand that mediation is a taking out of a half of a number out of a whole number, as you would take 3 out of 6. As for the second thou shalt know that thou shalt have one row of figures and no more. As for the third thou must understand that five cases may happen in this craft, and as for the fourth, thou shalt know that the profit of this craft is when thou hast taken away a half of a number to tell what shall be left. If thou wouldst mediate, that is to say take a half of a number, thou must begin thus,—write one row of figures of what number you wish.

POSTEA PROCEDAS MEDIANS, SI PRIMA FIGURA SI PAR AUT IMPAR VIDEAS.

Here he says when thou hast written a row of figures, thou shalt take heed whether the first figure be even or odd in number, and understand that he speaks of the first figure in the right side. And in the right side thou shalt begin in this craft.

QUIA SI FUERIT PAR, DIMIDIABIS EAM, SCBIBENS QUIQUID REMANEBIT:

Here is the first case of this craft which is this, if the figure be even then thou shalt take away from the even figure a half and do away with that figure and set the

half in its place. Thus 4. *Take a half out of four and that leaves 2. Do away with the 4 and set the 2 in its place.*

IMPAR SI FVERIT VNUM DEMAS MEDIARE QUOD NON PRESUMAS, SED QUOD SUPEREST MEDIABIS INDE SUPER TRACTUM FAC DEMPTUM QUOD NOTAT VNUM.

Here is the second case of this craft, the which is this. If the first figure is a number that is odd, the odd number shall not be mediated, but thou shalt mediate that number less one, and write the result as in the first part of this craft. Where thou hast written that, then write such a mark as is here[w]. Lo an example, 245. The 5 is odd. Then mediate 4 and replace the 5 by 2. That is to say replace 5 by 2 and make such a mark as w upon his head as thus 242[w]. Then mediate the 4 and the 2. The half is written 121[w].

If the first figure is 1, *Thou shalt do away with the 1 and set there a cypher and a mark over his head.* The number 241 is used as an example which by mediation yields 120[w]. This is merely a special form of the second case.

The third case arises when the second digit is 1. But this again is merely a special form of the case in which the second digit is odd, which is listed as the fourth case and treated thus:

POSTEA PROCEDAS HAC CONDICIONE SECUNDA: IMPAR SI FUERIT HINC VNUM DEME PRIORI, INSCRIBENS QUINQUE, NAM DENOS SIGNIFICABIT MONOS PREDICTAM.

Here he puts forth the fourth case, the which is this. If it happens the second figure is an odd number, thou shalt take one away from the odd number, the which shall be reckoned as 10. The new number is then mediated, and a 5 is placed over the head of the second digit. As an example the mediated form of 4678 is 233[5]4.

SI MEDIACIO SIT BENE FACTA PROBARE VALEBIS DUPLANDO NUMERUM QUEM PRIMO DIMEDIASTI.

This couplet explains how the operation of mediation may be proved. *The second example was this, 245. When thou hast mediated this number, if thou hast done well, thou shalt have as the mediation this number, 122[w]. Now double this number, and begin with the left side. Double 1, that shall be 2. Do away with the 1 and set there the 2. Then double the 2 and set there the 4. Then double the other 2, that will be 4. Then double the mark (w) which stands for a half and that shall be 1. Cast that on to 4, and it shall be 5. Do away with the 2 and the mark and you shall have 5, then thou shalt have the number 245, and this was the number when thou began to mediate, as thou mayest see if thou takest heed.*

The same four things are to be known about multiplication as in the operations discussed. Multiplication is defined as *the bringing together of 2 numbers into one number.* The manner in which the two are united is illustrated by an example, thus: *Twice 4 is 8, and this number 8 contains as many times 4 as there are unities in the other number, the which is 2, for in 2 there are 2 unities and so 4 times 2 is 8, as thou knowest well.* The process is thus based on the number of "unities" involved.

In the "craft" of multiplication there are eight operations or types, all based on the product of a digit by a digit. After an elaborate description of this process of multiplying a digit by a digit the following consolation appears. *But nevertheless if thou hast haste to work thou shalt have here a table of figures whereby thou shalt see at once correctly what is the number that comes of the multiplication of two digits. Thus thou shalt work with this figure. As for example if 5 is to be multiplied by 3 look for the 5 in the left side of the triangle, look where the 3 sits in the lower-most row of the triangle. Now go from him upwards to the same row in which the 5 sits. And that number, the which you find here, is the number that comes from the multiplication of the 2 digits.*

1								
2	*4*							
3	*6*	*9*						
4	*8*	*12*	*16*					
5	*10*	*15*	*20*	*25*				
6	*12*	*18*	*24*	*30*	*36*			
7	*14*	*21*	*28*	*35*	*42*	*49*		
8	*16*	*24*	*32*	*40*	*48*	*56*	*64*	
9	*18*	*27*	*36*	*45*	*54*	*63*	*72*	*81*
1	*2*	*3*	*4*	*5*	*6*	*7*	*8*	*9*

A quotation of eight lines from the Latin shows how to work in this craft. An example will illustrate. It is desired to multiply 2465 by 232. The problem is thus written

2465
232 .

Then follows *Thou shalt begin to multiply from the left side. Multiply 2 by 2 and set the result over the head of 2, then multiply the same upper 2 by 3 of the lower number, thus thrice 2 shall be 6. Set the 6 over the head of 3, then multiply the same upper 2 by the 2 that stands under it, that will be four. Do away with the upper 2 and replace it by 4. The upper row will then be 464465, the 465 remaining unchanged as no operation was performed upon these digits.* Then follows a process called antery which means that the problem is now written thus: *464465*

232

The antery refers to the change produced on the lower line, the whole being moved one place to the right, the 2 which formerly appeared under the fourth digit from the right now appears under the third. Then as before the 4 of the upper line multiplies the digits of the lower line in succession, beginning with the left. The product is now placed above the lower digit, as in the case of the product of the 4 and 2, the 8 is placed above the 6. In case the product is a composite number, such as the product of 4 and 3 the units digit is placed above the corresponding digit of the lower line, while the tenths digit is placed above the topmost digit in the previous column; in this case above the 8. In the last multiplication in each antery, the units digit replaces the number above it in the upper line, while the tenths digit takes its position in the previous column. The result of this multiplication may thus be written:

	1				
	8	*2*			
4	*6*	*4*	*8*	*6*	*5*
	2	*3*	*2*		

Another antery and the problem becomes

1
82
464865, which on multiplication yields
   232

> *11*
>
> *121*
>
> *828*
>
> *464865*
>
> *232*

Another antery followed by the necessary multiplication and our problem takes this form, in which the multiplier is omitted.

> *11*
>
> *110*
>
> *1211*
>
> *8285*
>
> *464820*

The sum of these numbers yields the product. The addition may be performed in the usual manner, but we are told to: *begin with the left,—now draw all these figures down together, as thus 6.8.1.[6] and 1; that whole is 16. Do away all this number save 6. Let him stand alone and set 1 over the head of 4 toward the left side, then draw onto 4, that will be 5. Do away with that 4 and 1 and set there the 5. Then draw 4. 2. 2. 1 and 1 together, that will be 10. Do away all that and write 0 and set the 1 over the next figure to the left side, the which is 6. Then draw that 6 and 1 together and that will be 7. Now do away with the 6 and set there the 7. Then draw that 8. 8. 1. and 1 together and that will be 18; do away all the figures that stand over the head of the 8, and let 8 stand still, and write the 1 over the next figures head to the left, which is 0. Then do away with the 0 and set there the 1, the which stands over the head of 0. Then draw the 2, 5, and 1[7] together, that will be 8. Then do away all that and write the 8. And then thou shalt have this number, 571880.*

The above process might be thus indicated:

> *4*
>
> *16*
>
> *10*
>
> *18*
>
> *80*
> _____
>
> *571880*

It will be noted that addition here is performed from left to right, and not from right to left as defined in the previous discussion concerning addition. It leads one to wonder why the two methods are used. It is quite probable that this second plan is introduced to be consistent with the order of multiplication as here used.

Eight cases in multiplication are recognized, depending upon the types of products, or upon the form of the problem.

1. When the product is an article.
2. When the product is a composite number.
3. When the product is a digit.
4. When the lower digit multiplies the upper digit directly above it.
5. When the lower digit multiplies the upper digit, and that upper digit is not directly above it.
6. If it happens that a zero stands right over the figure by which you multiply.
7. If it happens that the figure by which you multiply is a zero.
8. If there are several zeros in the upper row.

Each of these cases is treated at length, but all of them reduce to the method outlined in the given example.

Although division and extracting roots are listed, among the seven operations, no mention is made of the latter, and division is mentioned only in the process of mediation. One is led to wonder why this omission occurs, since the original poem discusses these operations in detail. It is highly probable that some of the original manuscript has been lost, which would account for the omission of these topics.

NOTES

1. Throughout the rest of the article, the Latin quotations will be in capitals and the translations from the fifteenth-century English in italic type.

2. "He" refers to the Latin author.

3. Note the use of the dot.

4. Note the personal pronoun.

5. See addition.

6. Here written without the dot. But I have used it as was the case in naming the digits.

7. Here the comma is used to separate the digits.

BIBLIOGRAPHY

Athenae Cantabrigienses.

Ball, W. W. R.: *History of the Study of Mathematics at Cambridge.*

De Morgan, Augustus: *List of Books in Arithmetic.*

Halliwell, J.: *Rara Mathematica.*

Leslie, John: *Philosophy of Arithmetic.*

Smith, David Eugene: *Rara Arithmetica.*

Steele, Robert (Editor): *The Earliest Arithmetics in English.*

Wilson, Duncan: *History of Mathematical Teaching in Scotland.*

37

The Art of Nombryng

E. R. SLEIGHT

THE *Art of Nombryng* is a translation into fifteenth-century English of *de Arte Numerandi*, which is attributed to John of Holywood,[1] better known by the name of Sacrobosco. In this treatise nine operations are recognized—numeration, addition, subtraction, mediation, duplation, multiplication, division, progressions and extracting square and cube roots. The definitions of algorism, digits, articles, and composite numbers appear here in the same form and with the same meaning as in the *Craft of Nombrynge*. The method of writing and reading numbers is thoroughly discussed. *Understand that there are [2] .9. characters or figures that represent the .9. digits; they are .9.8.7.6.5.4.3.2.1. The .10.[3] is called a cypher. It gives significance to the others; for without a cypher a pure article could not be written. Thus with these nine characters and the cypher all numbers may be written. Note well that all digits shall be written with one figure alone, and every article with a cypher, for every article is named for one of the digits as .10. for .1., .20. for .2. and so for the others. All digits shall be set in the first place and all articles in the second. Also all numbers from .10. to .100., which is excluded, must be written with .2. figures. And if it be an article, by a cypher first, and the figure written towards the left, that signifies the digit by which the article is named; if it be a compound number, first write the digit that is a part of the compound then write at the left side the article. All numbers that are from one hundred to one thousand*

which is excluded, shall be written with .3. figures, and all numbers from one thousand to ten thousand[4] must be written by .4. figures; and so forth. And understand that every figure in the first place signifies a digit, in the second place .10. times his digit, in the third place one hundred times, etc.

The methods used in addition and subtraction are the same as those used in the *Craft*, but with insistence upon form, as follows:

Addition—

Result	25
To whom it shall be added	13
The number to be added	12

Subtraction—

The remainder	12
Wherefrom we shall withdraw	25
The number to be withdrawn	13

The plan of procedure in multiplication is here indicated. In the first place we are told that *there are 6 rules of multiplication. The first is a rule for multiplying a digit by a digit. As an example, to multiply 4 by 8. Take the difference between 8 and 10 which is 2. Now write the article corresponding to 4 which is 40, and subtract 4 × 2. The result follows.* The second case arises when a digit is multiplied by an article. Here we *must multiply the digit by the digit for which the article is named. Then every unity shall stand for .10. and every article for .100. For*

Reprinted from *Mathematics Teacher* 35 (Mar., 1942): 112–16; with permission of the National Council of Teachers of Mathematics.

example, to multiply 6 × 40. The product of 6 and 4 is first taken, in which product 2 is the article, and 4 the unit. According to the rule, the result may be written 2 × 100 + 4 × 10 = 240. The remaining cases are multiplying a composite by a digit, article by an article, composite by an article, and composite by a composite. As in the *Craft of Nombryng*, the process of antry is used, which means that the operations proceed from left to right, except in case the multiplier is a digit. The form of the results would indicate that the order in case of a digit is reversed, as is shown by the following illustrations:

Resultans	1	1	0		1	3	2		7	3	6
Multiplicandus			5			4				3	2
Multiplicans		2	2		3	3			2	3	

The 5 standing above the last 2 indicates that the operation is from right to left, while in the last illustration, the form shows that the process is reversed.

Division appears in the *Art of Nombryng. In order to divide one number by another there are 2 numbers proposed. It is used to separate the larger number into as many parts as there are unities in the lesser number. Note that in the operation of division, there are .3. numbers, the number to be divided, the number dividing, and the number resulting, or quotient. Therefore if thou wilt any number divide, write the number to be divided over the divisor so that the last[5] of the divisor be under the last of the number to be divided, the next last under the next last, and so with the others, as here indicated.*

To be dyvydede	3	4	2
The dyvser	2	1	

But there are cases in which the last figure of the divisor may not be set under the last figure of the dividend. This has reference to the case in which the last digit of the divisor is larger than the last digit of the dividend. When this is true then there is an antry of the digits of the divisor as is indicated in the first illustration below.

The Dyvydede	3	5	5	1	2	2		8	8	6	3	7	0	4
The Dyvyser			5	4	3					4	4	2	3	

The order being observed we must work from the last figure of the divisor and see how oft it may be withdrawn from the figure (or figures) above its head. Then remove the result this often. This being done we must set forward (antry) *the figures of the divisor by one difference towards the right as before obtaining this result:*

Dyvydede	2	9	3	2	2
Dyvyser		5	4	3	

Now performing the operation of removing .5. times the divisor from the dividend, and setting forward the digits in the divisor, .4. is obtained as the final digit in the quotient. The entire process is thus indicated in the text.

Quocient				6	5	4
Dyvydede	3	5	5	1	2	2
Dyvyser		5	4	3		

And if it happens that after setting forward the figures of the divisor the result may not be withdrawn from the dividend then we must set a cypher for the number in the quotient, then set the figures of the divisor forward as before, and so we shall do in all numbers to be divided, for where the divisor may not be withdrawn we must set there a cypher.

One paragraph is devoted to progressions. *A progression is an aggregation of numbers of equal excesses, beginning with one or two. Of progressions one* is natural or continuous, and the other broken or discontinuous. Natural is when we begin with one and increase by one, as .1.2.3.4.5.6., etc. So that the number following passeth the other before by .1. broken it is when we leap from one number to another and keep not the continuous order as .1.3.5.7., etc. Also we may begin with 2, thus, .2.4.6.8., etc., and the number in each case following passeth the other by .2. And note well that natural progressions always begin with one, but broken progressions begin with .1. or with .2. Of the natural progressions there are[6] two rules of which the first is when the progression ends in an even number, by the half[7] thereof multiply the next larger number. Example: .1.2.3.4. Here we multiply .5. by .2. and so .10. cometh, that is the total result. The second rule is when a natural progression ends in an odd number. Take the middle number and multiply this by the last. Example .1.2.3.4.5.6.7. Here multiply .4. by .7. and this shall be the result. Of broken progressions there are also two rules; and the first is this: when the broken progression ends in an even number by the half thereof multiply the next number to that half as .2.4.6.8. Here multiply .4. by .5. and .20. is the sum of all the progression. The second rule is this: when the broken progression ends in an odd number take the total (number) of all the odd numbers and multiply by himself as .1.3.5.7. Multiply .4. by himself and the sum of all these numbers will be .16.

Here followeth the extraction of roots, the first being square roots. Wherefore we shall see what is a square number and what is the root of a square number, and what it is to draw out the root of a number. And above all note this division: of numbers some are lineal, others superficial, others quadrates, and others cubic or solids. Here is brought out clearly that the lineal has one dimension, the superficial two while the solid has three. Because a number may be multiplied by another in .2. ways, that is by itself or by another, understand that if it is by itself the result is a quadrate. Wherefore it is seen that all quadrate numbers are superficial, the converse is not true. The root of a quadrate is that number which forms the number by himself. Thus .4. is the first quadrate number and .2. is its root. A solid number is one that comes by double

multiplication, and it is clear that it has .3. dimensions, length, breadth, and thickness. But a number may be multiplied twice in two ways, either by itself or by others. If the number be multiplied by itself twice, the first result is a quadrate, the other a cubic, and is a solid. If the numbers are not alike the solid is elliptic and not cubic. Wherefore all cubic numbers are solids but the reverse is not true. It is hereby clear that no number is the root of a quadrate and a cubic. Therefore since multiplying unity by himself once or twice gives unity, then unity is not a number.

Now understand that between every .2. quadrates there is a mean proportional. It is obtained thus: multiply the root of the one quadrate into the root of the other and the result will show the mean.

To extract the root of a quadrate number means to see whether the proposed number be a quadrate. And if it is not a quadrate, then find the root of the largest quadrate less than the number proposed. Therefore if thou wilt find the root of any quadrate write the number and count the number of figures, to determine if it (the number of figures) be even or odd. If it is even, then the work must be begun with the last save one, but if odd then with the last. Or to say it a shorter way always begin with the last[8] odd digit. We must then find a digit, which squared is equal to this last, or as near equal as possible. In case the number of digits is even, this means that the square must be equal to or less than the number represented by the last two digits. If the number is odd, then it must be equal to or less than the last digit only. We are then instructed to place the square in its proper place, and withdraw it from that which is over it. We must then double the digit and set the double under the next figure toward the right hand. The second digit in the answer is found by division, following which we multiply the double by the second figure and add after it the square of the second figure, then subtract.

As an illustration 15139 is given. This is not a perfect square, but the solution leaves no residue so I assume that an error has been made, all of the work indicating that the number should have been 15129. The process step by step is as follows: The number of digits being odd, the process first considers the digit 1. Obviously the first digit in the answer is 1. Squaring this and removing the result the number reduces to 5129. Now take two times the 1, and place the result under the next digit or 5, and the problem takes this form: $\frac{5^{129}}{2}$. By division 2 is found to be the second digit in the answer. Multiplying the double (2) by the second figure (2) and adding the square of the second (4) we must subtract 44 from 51 and there remains 739, and the first two digits in the answer are 1 and 2. Doubling this answer and placing it according to instructions there results $\frac{729}{24}$. Division gives 3 for the final digit in the answer. If to 3 times 26, 9 is annexed, the result is 729, showing that 15129 is a quadrate number. The form used for indicating the above process is here given.

Number	1	5	1	2	9
The double		2		4	
The result	1		2		3

It will be observed that in rather an indirect way the method now in use is here outlined. However, it was not until the middle of the sixteenth century when Robert Record's Arithmetic was published did the method of today appear in its entirety, and to him is given the credit for having first used it. The author completes the explanation concerning square roots thus: *that done, either ought or nought will remain. If nought, then it showeth that the proposed number was a quadrate. And if ought remaineth that showeth that the number proposed was not a quadrate. If thou wilt prove the work, multiply the last result by itself and thou shalt find the same number that thou haddest before. But if thou hast a residue, then with the addition thereof of this residue, thou shalt find the first number as before.*

Here followeth the finding of roots of cubic numbers. Wherefore we must see what a cubic number is, what is its root, and what is meant by the extraction of a root. As before said, a cubic number is one that comes by multiplying a number twice by itself, or once by its

quadrate. The root of a cubic number is a number that, multiplied twice by itself or once by its quadrate, will give the number. To find the root of a cubic, it is first necessary to find if the proposed number is a cubic. If not then one must extract the root of the largest cubic less than the proposed number whose root is to be found. First mark off the places in threes. And with the last three places, a digit which cubed must do away with that place, or as near it as possible. Here as in square root the process is very similar to the process now used—first subtract the proposed cube, then find the trial divisor, the two corrections, etc. The writer finishes the discussion of cube roots by calling attention to the fact that if nothing remains in the process, then the number proposed is a perfect cube but if a residue remains, then it is not.

To aid in determining the squares and cubes of numbers tables were formed as follows: in each case, the numbers in the principal diagonal represent the required results.

1	2	3	4	5	6	7	8	9
2	8	12	16	20	24	28	32	36
3	18	27	36	45	54	63	72	81
4	32	48	64	80	96	112	128	144
5	50	75	100	125	150	175	200	225
6	72	102	144	180	216	252	288	324
7	92	147	196	245	294	343	392	441
8	128	192	256	320	384	448	512	576
9	162	243	324	405	486	576	648	729

1	2	3	4	5	6	7	8	9
2	4	6	8	10	12	14	16	18
3	6	9	12	15	18	21	24	27
4	8	12	16	20	24	28	32	36
5	10	15	20	25	30	35	40	45
6	12	18	24	30	36	42	48	54
7	14	21	28	35	42	49	56	63
8	16	24	32	40	48	56	64	72
9	18	27	36	45	54	63	72	81

NOTES

1. Died 1244 or 1246.

2. Note the use of the dot.

3. Meaning the 10th character.

* *Mathematics Teacher*, vol. XXXII, no. 6. [Reprinted as chapter 36 in the present volume.]

4. Written .X.M.

5. "Last" to the left, as is true in all translated parts of this article.

6. Rules for Finding Sum.

7. Half Sum.

8. This refers to position, not to value.

BIBLIOGRAPHY

Ball: *History of Study of Mathematics at Cambridge.*

Halliwell: *Rara Mathematica.*

Gerhardt: *Manuscripts.*

De Morgan: *Arithmetical Books.*

Robert Steele: *Earliest Arithmetics in English.*

Cunningham: *Story of Arithmetic.*

Manuscript: *Johannes de Sacro-Basco de Arithmetica.*

38

Leonardo Fibonacci

CHARLES KING

THE *Fibonacci Quarterly* receives its name from Leonardo of Pisa (or Leonardo Pisano), better known as Leonardo Fibonacci (Fibonacci is a contraction of *Filius Bonacci,* son of Bonacci). Leonardo was born about 1175 in the commercial center of Pisa. This was a time of great interest and importance in the history of Western civilization. One finds the influence of the crusades stirring and awakening the people of Europe by bringing them in contact with the more advanced intellect of the East. During this time the Universities of Naples, Padua, Paris, Oxford, and Cambridge were established, the Magna Carta signed in England, and the long struggle between the Papacy and the Empire was culminated. Commerce was flourishing in the Mediterranean world and adventurous travelers such as Marco Polo were penetrating far beyond the borders of the known world.

It is in this growing commercial activity that we find the young Leonardo at Bugia on the northern coast of Africa. Here the merchants of Pisa and other commercial cities of Italy had large warehouses for the storage of their goods. Actually very little is known about the life of this great mathematician. No contemporary historian makes mention of him, and one must look to his writings to find information about him. In the preface of his first and most important work, *Liber abaci* (I), Leonardo tells us that his father, the head of one of

the warehouses of Bugia, instructed him to study arithmetic. In Bugia, he received his early education from a Moorish schoolmaster.

Leonardo then traveled about the Mediterranean visiting Egypt, Syria, Greece, Sicily, southern France, and Constantinople. He met with scholars and studied the various systems of arithmetic then in use. Leonardo was persuaded that the Hindu-Arabic system was superior to the methods then adopted in the different countries he had visited and that it was even superior to the Algorithma and the method of Pythagoras. He busied himself with the subject and carried on his own research, intent upon bringing the Hindu-Arabic system to his Italian countrymen. The study and research in mathematics so absorbed him that he seems to have devoted his life to this pursuit and spent little time in commerce which was flourishing at that time and was the favorite occupation of his fellow citizens. Yet most of the applications Leonardo makes in his works are in the field of commerce. In one place, he gives a careful evaluation of the money systems of the countries of his travels.

Leonardo returned to Italy about 1200 and in 1202 wrote *Liber abaci* (I), in which he gave a thorough treatment of arithmetic and algebra, the first that had been written by a Christian. The work is divided into fifteen chapters. The chapter contents are given here to indicate the scope of the work: (1) Reading and writing numbers in the Hindu-Arabic system; (2) Multiplication of inte-

Reprinted from *Fibonacci Quarterly* 1 (Dec., 1963): 15–19; with permission of the Fibonacci Association.

gers; (3) Addition of integers; (4) Subtraction of integers; (5) Division of integers; (6) Multiplication of integers by fractions; (7) Additional work with fractions; (8) Prices of goods; (9) Barter; (10) Partnership; (11) Alligation; (12) Solutions of problems; (13) Rule of false position; (14) Square and cube roots; (15) Proportions, and Geometry and algebra.

The last and most important chapter is divided into three parts; the first relates to proportions, the second to geometry and the third, to algebra. each of the three parts begins with definitions and demonstrations credited to the Arabs, then Leonardo considers six questions, three simple and three complex, giving solutions for them.

Leonardo, in 1228, gave a second edition of the *Liber abaci* which he dedicated to Michel Scott, astrologer to the Emperor Frederick II and author of many scientific works. Copies of this edition exist today. Leonardo profusely illustrated and strongly advocated the Hindu-Arabic system in this work. He gave an extensive discussion of the Rule of False Position and the Rule of Three. Leonardo did not use a general method in problem solving; each problem was solved independently of the others. In the solution of a problem he not only considered the problem as it might occur, but considered all of the variations of the question, even those that were not reasonable.

In the *Liber abaci*, Leonardo states and gives the solution to the famous Rabbit Problem [1, Vol. 1, p. 285]. A pair of rabbits is placed in a pen to find out how many offspring will be produced by this pair in one year if each pair of rabbits gives birth to a new pair of rabbits each month starting with the second month of its life; it is assured that deaths do not occur.

Leonardo traces the progress of the rabbits: The first pair has offspring in the first month: thus two pair. The second month there are three pair, the first reproducing in this month. In the third month there are five pair. Continuing in this manner through the twelve months, Leonardo gives the following table:

0	Sixth Month
Pairs	21
1	Seventh Month
First Month	34
2	Eighth Month
Second Month	55
3	Ninth Month
Third Month	89
5	Tenth Month
Fourth Month	144
8	Eleventh Month
Fifth Month	233
13	Twelfth Month
	377

It is this sequence of numbers, 1, 2, 3, 5, 8, 13, . . ., that gives rise to the Fibonacci Sequence.

Of the many problems of elementary nature in the *Liber abaci*, the following are given as examples.

Seven old women are traveling to Rome and each has seven mules. On each mule there are seven sacks; in each sack there are seven loaves of bread: in each loaf there are seven knives; and each knife has seven sheaths. How many in all are going to Rome?

A man went into an orchard which had seven gates; and there took a certain number of apples. When he left the orchard he gave the first guard half the apples he had and one apple more. To the second he gave half the remaining apples and one apple more. He did the same in the case of each of the remaining five guards, and left the orchard with one apple. How many apples did he gather in the orchard?

A certain man puts one denarius at such a rate that in five years he has two denarii and in every five years thereafter the money doubles. How many denarii would he gain from this one denarius in 100 years?

A certain king sent thirty men into his orchard to plant trees. If they could set out a thousand trees in nine days, in how many days would thirty-six men set out four thousand four hundred trees?

Many readers will recognize these problems.

In 1220, Leonardo wrote *Practica geometriae*, which he dedicated to Master Dominique, a person of whom there is no record. In this work Leonardo systematized the subject matter of practical geometry with a specialization in measurements of bodies. He included some algebra and trigonometry, square and cube roots, proportions and indeterminate problems. The use of a surveying instrument called the quadrans is included. The work is skillfully done with Euclidean rigor and some originality.

Leonardo's reputation grew and from his writings it can be seen that he had a vast range of knowledge concerning Arabian mathematics and mathematics of antiquity, especially Greek. His treatment shows much originality, completeness, and rigor. It is especially noted that his writings did not contain the mysticism of numerology and astrology that were so prevalent in the writing of his day.

Because of Leonardo's great reputation, the Emperor Frederick II, when in Pisa (1225), held a sort of mathematical tournament to test Leonardo's skill. The competitors were informed beforehand of the questions to be asked, some or all of which were composed by Johannes of Palermo [1, Vol. II, p. 227], who was one of Frederick's staff. This is the first case in the history of mathematics that one meets with an instance of these challenges to solve particular problems which were so common in the sixteenth and seventeenth centuries.

The first question propounded was to find a number of which the square when decreased or increased by 5 would remain a square. The correct answer given by Leonardo was $41/12$. The next question was to find by the methods used in the tenth book of Euclid a line whose length x should satisfy the equation $x^3 + 2x^2 + 10x - 20 = 0$. Leonardo showed by geometry that the problem was impossible, but gave an approximation of the root $1.3688081075 \ldots$, which is correct to nine places.

The third question was:

Three men possess a certain sum of money, their shares in the ratio 3:2:1. While making the division, they were surprised by a thief and each took what he could and fled. Later the first man gave up half of what he had, the second gave up one-third, and the third, one-sixth. The money given up was divided equally among them and then each man had the share to which he was entitled. What was the total sum? Leonardo showed that the problem was indeterminate and gave as one solution 47 which is the smallest sum.

The other competitors failed to solve any of these questions. Through the consideration of these problems and others similar to them, Leonardo was led to write his *Liber quadratorum* (1225) [1, Vol. II, p. 253], a brilliant and original work containing a well arranged collection of theorems from indeterminate analysis involving equations of the second degree such as $x^2 + 5 = y^2$, $x^2 - 5 = z^2$. This work has marked him as the outstanding mathematician between Diophantus and Fermat in this field.

Two or three works of Leonardo that are known are the *Flos* [1, Vol. II, p. 227] (blossom or flower), which contains the last two problems of the tournament; the first problem is found in the *Liber quadratorum*, and a *Letter to Magister Theodoris* [1, Vol. II, p. 247], philosopher to Frederick II, relating to indeterminate analysis and to geometry. The last three works show clearly the genius and brilliance of Leonardo as a mathematician and were beyond the abilities of most contemporary scholars.

The works of Leonardo Fibonacci are available in some universities in the United States through B. Boncompagni, *Scritti di Leonardo Pisano*, Rome, (1857–1862). The first volume contains the *Liber abaci* and the second volume contains *Practica geometriae*, the *Flos, Letter to Magestrum Theodorum*, and *Liber quadratorum*. A treatment of square numbers composed by Leonardo and addressed to the Emperor Frederick II seems to have been lost.

REFERENCE

Boncompagni, Baldassarre. *Scritti di Leonardo Pisano:* 2 vols. Rome, 1857.

39

Leonardo of Pisa
and His *Liber quadratorum*

R. B. McCLENON

*T*HE THIRTEENTH CENTURY is a period of great fascination for the historian, whether his chief interest is in political, social, or intellectual movements. During this century great and far-reaching changes were taking place in all lines of human activity. It was the century in which culminated the long struggle between the Papacy and the Empire; it brought the beginnings of civil liberty in England; it saw the building of the great Gothic cathedrals, and the establishment and rapid growth of universities in Paris, Bologna, Naples, Oxford, and many other centers. The crusades had awakened the European peoples out of their lethargy of previous centuries, and had brought them face to face with the more advanced intellectual development of the East. Countless travelers passed back and forth between Italy and Egypt, Asia Minor, Syria, and Bagdad; and not a few adventurous and enterprising spirits dared to penetrate as far as India and China. The name of Marco Polo will occur to everyone, and he is only the most famous among many who in those stirring days truly discovered new worlds.

Among the many valuable gifts which the Orient transmitted to the Occident at this time, undoubtedly the most precious was its scientific knowledge, and in particular the Arabian and Hindu mathematics. The transfer of knowledge and ideas from East to West is one of the most interesting phenomena of this interesting period, and accordingly it is worth while to consider the work of one of the pioneers in this movement.

Leonardo of Pisa, known also as Fibonacci,[1] in the last years of the twelfth century made a tour of the East, saw the great markets of Egypt and Asia Minor, went as far as Syria, and returned through Constantinople and Greece.[2] Unlike most travelers, Leonardo was not content with giving a mere glance at the strange and new sights that met him, but he studied carefully the customs of the people, and especially sought instruction in the arithmetic system that was being found so advantageous by the Oriental merchants. He recognized its superiority over the clumsy Roman numeral system which was used in the West, and accordingly decided to study the Hindu-Arabic system thoroughly and to write a book which should explain to the Italians its use and applications. Thus the result of Leonardo's travels was the monumental *Liber abaci* (1202), the greatest arithmetic of the middle ages, and the first one to show by examples from every field the great superiority of the Hindu-Arabic numeral system over the Roman system exemplified by Boethius.[3] It is true that Leonardo's *Liber abaci* was not the first book written in Italy in which the Hindu-Arabic numerals were used and explained,[4] but no work had been previously produced which in either the extent or the value of its contents could for a moment be compared with this. Even today it would be thoroughly worth while for any teacher of mathematics to become

Reprinted from *American Mathematical Monthly* 26 (Jan., 1919): 1–8; with permission of the Mathematical Association of America.

familiar with many portions of this great work. It is valuable reading both on account of the mathematical insight and originality of the author, which constantly awaken our admiration, and also on account of the concrete problems, which often give much interesting and significant information about commercial customs and economic conditions in the early thirteenth century.

Besides the *Liber abaci*, Leonardo of Pisa wrote an extensive work on geometry, which he called *Practica geometriae*. This contains a wide variety of interesting theorems, and while it shows no such originality as to enable us to rank Leonardo among the great geometers of history, it is excellently written, and the rigor and elegance of the proofs are deserving of high praise. A good idea of a small portion of the *Practica geometriae* can be obtained from Archibald's very successful restoration of Euclid's *Divisions of Figures*.[5]

The other works of Leonardo of Pisa that are known are *Flos*, a *Letter to Magister Theodorus*, and the *Liber quadratorum*. These three works are so original and instructive, and show so well the remarkable genius of this brilliant mathematician of the thirteenth century, that it is highly desirable that they be made available in English translation. It is my intention to publish such a translation when conditions are more favorable, but in the meantime a short account of the *Liber quadratorum* will bring to those whose attention has not yet been called to it some idea of the interesting and valuable character of the book.

The *Liber quadratorum* is dedicated to the Emperor Frederick II, who throughout his whole career showed a lively and intelligent interest in art and science, and who had taken favorable notice of Leonardo's *Liber abaci*. In the dedication, dated in 1225, Leonardo relates that he had been presented to the Emperor at court in Pisa, and that Magister Johannes of Palermo had there proposed a problem[6] as a test of Leonardo's mathematical power. The problem was, to find a square number which when either increased or diminished by 5 should still give a square number as result. Leonardo gave

a correct answer, $11^{97}/_{144}$. For $11^{97}/_{144} = (3^5/_{12})^2$, $6^{97}/_{144} = (2^7/_{12})^2$, and $16^{97}/_{144} = (4^1/_{12})^2$. Through considering this problem and others allied to it, Leonardo was led to write the *Liber quadratorum*.[7] It should be said that this problem had been considered by Arab writers with whose works Leonardo was unquestionably familiar; but his methods are original, and our admiration for them is not diminished by careful study of what had been done by his Arabian predecessors.[8]

In the *Liber quadratorum*, Leonardo has given us a well-arranged, brilliantly-written collection of theorems from indeterminate analysis involving equations of the second degree. Many of the theorems themselves are original, and in the case of many others the proofs are so. The usual method of proof employed is to reason upon general numbers, which Leonardo represents by line segments. He has, it is scarcely necessary to say, no algebraic symbolism, so that each result of a new operation (unless it be a simple addition or subtraction) has to be represented by a new line. But for one who had studied the "geometric algebra" of the Greeks, as Leonardo had, in the form in which the Arabs used it,[4] this method offered some of the advantages of our symbolism; and at any rate it is marvelous with what ease Leonardo keeps in his mind the relation between two lines and with what skill he chooses the right road to bring him to the goal he is seeking.

To give some idea of the contents of this remarkable work, there follows a list of the most important results it contains. The numbering of the propositions is not found in the original.

PROPOSITION I. THEOREM. Every square number[10] can be formed as a sum of successive odd numbers beginning with unity. That is,

$$1 + 3 + 5 + \cdots + (2n - 1) = n^2.$$

PROPOSITION II. PROBLEM. To find two square numbers whose sum is a square number. "I take any odd square I please, . . . and find the other from the sum of all the odd numbers from unity up to that odd square itself."[11] Thus, if $2n + 1$ is a square (= x^2) then

$$1 + 3 + 5 + \cdots + (2n - 1) + x^2 = n^2 + (2n + 1)$$
$$= \text{a sum of two squares} = (n + 1)^2$$

This is equivalent to Pythagoras's rule for obtaining rational right triangles, as stated by Proclus,[12] viz.,

$$\left(\frac{x^2 - 1}{2}\right)^2 + x^2 = \left(\frac{x^2 + 1}{2}\right)^2.$$

For, inasmuch as $2n + 1 = x^2$, we have

$$n = \frac{x^2 - 1}{2} \text{ and } n + 1 = \frac{x^2 + 1}{2}.$$

PROPOSITION III. THEOREM.

$$\left(\frac{n^2}{4} - 1\right)^2 + n^2 = \left(\frac{n^2}{4} + 1\right)^2.$$

This enables us to obtain rational right triangles in which the hypotenuse exceeds one of the legs by 2. It is attributed by Proclus to Plato. Leonardo also gives the rule in case the hypotenuse is to exceed one leg by 3, and indicates what the result would be if the hypotenuse exceeds one leg by any number whatever.

PROPOSITION IV. THEOREM. "Any square exceeds the square which immediately precedes it by the amount of the sum of their roots." That is, $n^2 - (n-1)^2 = n + (n-1)$. It follows from this that when the sum of two consecutive numbers is a square number, then the square of the greater will equal the sum of two squares. For, if $n + (n - 1) = u^2$, then $n^2 - (n - 1)^2 = u^2$ or $n^2 = u^2 + (n - 1)^2$.

PROPOSITION V. PROBLEM. Given $a^2 + b^2 = c^2$, to find two integral or fractional numbers x, y, such that $x^2 + y^2 = c^2$. Solution: Find two other numbers m and n such that[13] $m^2 + n^2 = q^2$. If $q^2 \neq c^2$, multiply the preceding equation by c^2/q^2, obtaining

$$\left(\frac{c}{q} \cdot m\right)^2 + \left(\frac{c}{q} \cdot n\right)^2 = c^2$$

so that $x = c/q \cdot m$, $y = c/q \cdot n$ is a solution.

PROPOSITION VI. THEOREM. "If four numbers not in proportion are given, the first being less than the second, and the third less than the fourth, and if the sum of the squares of the first and second is multiplied by the sum of the squares of the third and fourth, there will result a number which will be equal in two ways to the sum of two square numbers." That is,

$$(a^2 + b^2)(c^2 + d^2) = (ac + bd)^2 + (ad - bc)^2 = (ad + bc)^2 + (ac - bd)^2.$$

This very important theorem should be called Leonardo's Theorem, for it is not found definitely stated, to say nothing of being proved, in any earlier work. Leonardo considers also the case where a, b, c, and d are in proportion, and shows that then $(a^2 + b^2) \cdot (c^2 + d^2)$ is equal to a square and the sum of two squares. This gives him still another way of finding rational right triangles.[14]

PROPOSITION VII. THEOREM. $(x^2 - y^2)^2 + (2xy)^2 = (x^2 + y^2)^2$.[15] Leonardo proves this very simply as a corollary of Proposition VI.

PROPOSITION VIII. PROBLEM. "To find two numbers the sum of whose squares is a number, not a square, formed from the addition of two given squares." That is, to find x and y such that $x^2 + y^2 = a^2 + b^2$. Choose any two numbers c and d, such that $c^2 + d^2$ is a square, and write $(a^2 + b^2)(c^2 + d^2)$ as a sum of two squares, let us say $p^2 + q^2$; this we can do by Proposition VI. Construct the right triangle whose legs are p and q; then the similar triangle whose hypotenuse is equal to $\sqrt{c^2 + d^2}$ will have as its legs the two required numbers x and y.

PROPOSITION IX. THEOREM.

$$6(1^2 + 2^2 + 3^2 + \cdots + n^2) = n(n + 1)(2n + 1).$$

The proof of this is strikingly original, and proceeds from the identity

$$n(n + 1)(2n + 1) = n(n - 1)(2n - 1) + 6n^2.$$

Hence

$$n(n - 1)(2n - 1) = (n - 1)(n - 2)(2n - 3) + 6(n - 1)^2,$$
$$\cdots\cdots\cdots\cdots\cdots\cdots\cdots\cdots\cdots$$
$$2 \cdot 3 \cdot (2 + 3) = 1 \cdot 2 \cdot (1 + 2) + 6 \cdot 2^2,$$
$$1 \cdot 2 \cdot (1 + 2) = \qquad\qquad 6 \cdot 1^2.$$

It follows by addition that

$$n(n + 1)(2n + 1) = 6(1^2 + 2^2 + 3^2 + \cdots + (n-1)^2 + n^2).$$

PROPOSITION X. THEOREM.

$$12[1^2 + 3^2 + 5^2 + \cdots + (2n-1)^2] = (2n-1)(2n+1)4n.$$

Leonardo gives a proof very similar to that of Proposition IX.

PROPOSITION XI. THEOREM.

$$12[2^2 + 4^2 + 6^2 + \cdots + (2n)^2] = 2n(2n+2)(4n+2),$$

and likewise

$$18[3^2 + 6^2 + 9^2 + \cdots + (3n)^2] = 3n(3n+3)(6n+3),$$

and

$$24[4^2 + 8^2 + 12^2 + \cdots + (4n)^2] = 4n(4n+4)(8n+4),$$

and in general

$$6a[a^2 + (2a)^2 + (3a)^2 + \cdots + (na)^2] = na(na+a)(2na+a).$$

Here Leonardo has almost discovered the general result

$$a^2 + (a+d)^2 + (a+2d)^2 + \cdots + [a+(n-1)d]^2$$
$$= \frac{6na^2 + 6n(n-1)ad + n(n-1)(2n-1)d^2}{6}.$$

His method needed no change at all, in fact.

PROPOSITION XII. THEOREM. If $x + y$ is even, $xy(x + y)(x - y)$ is divisible by 24; and in any case $4xy(x + y)(x - y)$ is divisible by 24. A number of this form is called by Leonardo a *congruum*, and he proceeds to show that it furnishes the solution to a problem proposed by Johannes of Palermo.

PROPOSITION XIII. PROBLEM. "To find a number which, being added to, or subtracted from, a square number, leaves in either case a square number." Leonardo's solution of this, the problem which had stimulated him to write the *Liber quadratorum*, is so very ingenious and original that it is a matter of regret that its length prevents its inclusion here. It is not too much to say that this is the finest piece of reasoning in number theory of which we have any record, before the time of Fermat. Leonardo obtains his solution by establishing the identities

$$(x^2 + y^2)^2 - 4xy(x^2 - y^2) = (y^2 + 2xy - x^2)^2$$

and

$$(x^2 + y^2)^2 + 4xy(x^2 - y^2) = (x^2 + 2xy - y^2)^2.$$

PROPOSITION XIV. PROBLEM. To find a number of the form $4xy(x + y)(x - y)$ which is divisible by 5, the quotient being a square. Take $x = 5$, and y equal to a square such that $x + y$ and $x - y$ are also squares. The least possible value for y is 4, in which case

$$4xy(x + y)(x - y) = 4 \cdot 5 \cdot 4 \cdot 9 \cdot 1 = 720.$$

PROPOSITION XV. PROBLEM. "To find a square number which, being increased or diminished by 5, gives a square number. Let a congruum be taken whose fifth part is a square, such as 720, whose fifth part is 144; divide by this the squares congruent to 720,[16] the first of which is 961, the second 1681, and the third 2401. The root of the first square is 31, of the second is 41, and of the third is 49. Thus there results for the first square $6^{97}/_{144}$, whose root is $2^7/_{12}$, which results from the division of 31 by the root of 144, that is, by 12; and for the second, that is, for the required square, there will result $11^{97}/_{144}$, whose root is $3^5/_{12}$, which results from the division of 41 by 12; and for the last square there will result $16^{97}/_{144}$, whose root is $4^1/_{12}$."

PROPOSITION XVI. THEOREM. When $x > y$, $(x + y)/(x - y) \neq x/y$. It follows that $x(x - y)$ is not equal to $y(x + y)$, and "from this," Leonardo says, "it may be shown that no square number can be a congruum." For if $xy(x + y)(x - y)$ could be a square, either $x(x - y)$ must be equal to $y(x + y)$, which this proposition proves to be impossible, or else the four factors must severally be squares, which is also impossible. Leonardo to be sure overlooked the necessity of proving this last assertion, which remained unproved until the time of Fermat.[17]

PROPOSITION XVII. PROBLEM. To solve in rational numbers the pair of equations

$$x^2 + x = u^2,$$

$$x^2 - x = v^2.$$

The solution is obtained by means of any set of three squares in arithmetic progression, that is, by means of Proposition XIII. Let us take x_1^2, x_2^2, and x_3^2 for the three squares, and let the common difference, that is, the congruum, be d. Leonardo says that the solution of the problem is obtained by giving x the value x_2^2/d. For then

$$x^2 + x = \frac{x_2^4}{d^2} + \frac{x_2^2}{d} = \frac{x_2^2(x_2^2 + d)}{d^2} = \frac{x_2^2 x_3^2}{d^2} \, ;$$

and

$$x^2 - x = \frac{x_2^4}{d^2} - \frac{x_2^2}{d} = \frac{x_2^2(x_2^2 - d)}{d^2} = \frac{x_2^2 x_1^2}{d^2}.^{[18]}$$

PROPOSITION XVIII. PROBLEM. To solve in rational numbers the pair of equations

$$x^2 + 2x = u^2,$$

$$x^2 - 2x = v^2.$$

The method is similar to that in Proposition XVII, the value of x being found to be $2x_2^2/d$. Leonardo adds, "You will understand how the result can be obtained in the same way if three or more times the root is to be added or subtracted."

PROPOSITION XIX. PROBLEM. To solve (in integers) the pair of equations

$$x^2 + y^2 = u^2,$$

$$x^2 + y^2 + z^2 = v^2.$$

Take for x and y any two numbers that are prime to each other and such that the sum of their squares is a square, let us say u^2. Adding all the odd numbers from unity to $u^2 - 2$,[19] the result is $((u^2 - 1)/2)^2$.

Now

$$\left(\frac{u^2 - 1}{2} \right)^2 + u^2 = \left(\frac{u^2 + 1}{2} \right)^2.$$

Thus

$$z^2 = \left(\frac{u^2 - 1}{2} \right)^2,$$

and

$$v^2 = \left(\frac{u^2 + 1}{2} \right)$$

PROPOSITION XX. PROBLEM. To solve in rational numbers the set of equations

$$x + y + z + x^2 = u^2,$$

$$x + y + z + x^2 + y^2 = v^2,$$

$$x + y + z + x^2 + y^2 + z^2 = w^2.$$

By an extension of the method used in Proposition XIX Leonardo obtains the results $x = 3\frac{1}{5}$, $y = 9\frac{3}{5}$, $z = 28\frac{4}{5}$. He even goes farther and obtains the integral solutions $x = 35$, $y = 144$, $z = 360$. He continues, "And not only can three numbers be found in many ways by this method but also four can be found by means of four square numbers, two of which in order, or three, or all four added together make a square number . . . . I found these four numbers, the first of which is 1295, the second 4566⁶/₇, the third 11417¹/₇, and the fourth 79920." In the midst of the explanation of how these values were obtained, the manuscript of the *Liber quadratorum* breaks off abruptly. It is probable, however, that the original work included little more than what the one known manuscript gives. At all events, considering both the originality and power of his methods, and the importance of his results, we are abundantly justified in ranking Leonardo of Pisa as the greatest genius in the field of number theory who appeared between the time of Diophantus and that of Fermat.

NOTES

Editor's note: Notes have been renumbered. In the original, notes appeared as true footnotes and the numbering was restarted at 1 on each page.

1. This is probably a contraction for "Filiorum Bonacci," or possibly for "Filius Bonacci"; that is, "of the

family of Bonacci" or "Bonacci's son." See Boncompagni, *Della Vita e delle Opere di Leonardo Pisano, matematico del secolo decimoterzo,* Rome, 1852, pp. 8–12.

2. *Scritti di Leonardo Pisano,* 2 vols., Rome, 1857–61. Vol. I, p. 1.

3. Boethius, ed. Friedlein, Leipzig, 1867. The arithmetic occupies pages 1–173. This was the arithmetic that was very generally taught throughout Europe before the thirteenth century, and its use continued to be widespread long after better works were in the field.

4. Smith and Karpinski, *The Hindu-Arabic Numerals,* Boston and London, 1911. Chapter VII gives an account of the first European writings on these numerals.

5. Archibald, *Euclid's Book on Divisions of Figures; with a Restoration based on Woepcke's Text and on the Practica Geometriæ of Leonardo Pisano,* Cambridge, England, 1915.

6. In the introduction to *Flos* we are told that two other problems were propounded at the same time. *Scritti,* II, p. 227.

7. *Scritti,* II, p. 253.

8. See, for example, Woepcke, *Recherches sur plusieurs ouvrages de Leonard de Pise, et sur les rapports qui existent entre ces ouvrages et les travaux mathématiques des Arabes,* Rome, 1859.

9. Heath, T. L., *The Thirteen Books of Euclid's Elements,* Cambridge, 1908. Vol. I, pp. 372–374, 383–385, 386–388; Zeuthen, H. G., *Geschichte der Mathematik im Altertum und Mittelalter,* Copenhagen, 1896, pp. 44–53; Karpinski, L. C., *Robert of Chester's Latin translation of the Algebra of Al-Khowarizmi,* New York, 1915, pp. 77–89.

10. Throughout this article, unless otherwise stated, the word "number" is to be understood as meaning "positive integer."

11. The use of quotation marks indicates a literal translation of Leonardo's words; in other cases the exposition follows his thought without adhering closely to his form of expression.

12. Proclus, ed. Friedlein, Leipzig, 1873, p. 428.

13. This is possible by Proposition II or Proposition III.

14. For instance, letting $a = 6$, $b = 4$, $c = 3$, $d = 2$,

$$(36 + 16)(9 + 4) = 676 = (6 \cdot 3 + 4 \cdot 2)^2$$
$$= (6 \cdot 2 + 4 \cdot 3)^2 + (6 \cdot 3 - 4 \cdot 2)^2 = 26^2 = 24^2 + 10^2.$$

15. This is Euclid's general solution of the problem of finding rational right triangles. Heath, *op. cit.,* III, p. 63. (Euclid's Elements, X, Lemma to Theorem 29.)

16. That is, the three squares in arithmetic progression, whose common difference is the congruum 720. They are obtained by Proposition XIII, thus: Taking $x = 5$ and $y = 4$, $y^2 + 2xy - x^2 = 31$, the root of the first square; $x^2 + y^2 = 41$, the root of the second square; and $x^2 + 2xy - y^2 = 49$, the root of the third square.

17. *Fermat, Oeuvres,* Paris, 1891, vol. 1, p. 340; Heath, *Diophantus of Alexandria,* Cambridge, 1910, p. 293.

18. The simplest numerical example would be $x_1^2 = 1$, $x_2^2 = 25$, $x_3^2 = 49$, and this is the illustration given by Leonardo. It leads to $x = {}^{25}/24$, from which we have $x^2 + x = {}^{1225}/576 = ({}^{35}/24)^2$ and $x^2 - x = {}^{25}/576 = ({}^{5}/24)^2$.

19. Here u^2 is odd, because it is the sum of the squares of two numbers x and y which are prime to each other. It is not possible that both x and y are odd, since $(2m + 1)^2 + (2n + 1)^2 = 4m^2 + 4m + 4n^2 + 2$, and this is divisible by 2 but not by 4, and hence cannot be a square. Thus, of the numbers x and y, one must be even and the other odd, hence $x^2 + y^2$ is odd.

Algorist versus the Abacist

Although algorithmic computational procedures for the four basic operations of arith-metic were introduced into Europe in the ninth century, it was a long time before they were fully accepted or adopted. The controversy over which methods of computation were better, those performed with the counting board or those written out using algorithms, raged for centuries. Abacus techniques were deeply engrained in all commercial and financial enterprises. Business was conducted over the "counter," the table upon which abacus movements were made. Many early arithmetic books contained woodcut prints depicting this controversy. The following fanciful illustration showing an algorist in competition with an abacist comes from Gregor Reisch's, *Margarita philosophica* of 1508.

40

Some Uses of Graphing before Descartes

THOMAS M. SMITH

During the first half of the seventeenth century at least four students of mathematics and natural philosophy were making use of a simple graphing technique.

In 1618, Isaac Beeckman, while writing in his journal on the subject of "a stone falling in a vacuum," employed this graphing technique to represent uniform acceleration analyzed in terms of what might be called "geometric infinitesimals." His approach suggests certain features of the infinitesimal calculus that Von Leibniz and Newton were to develop later in that century.

In 1637, René Descartes presented an application of the same graphing technique (without Beeckman's infinitesimal "individua," as he called them) in his newly published *La géométrie*. Analytical geometry, as it came to be called, properly owes its essential character to the simultaneous, independent discoveries of Descartes and Pierre Fermat, but like many another advance in scientific thought, its origins antedate the men who made it important.

In the following year, 1638, Galileo Galilei made use of the same graphing technique that Descartes, Fermat, and Beeckman had employed. He used it to provide a mathematical description of uniform acceleration in the physical example of a body in free fall.

All of these men were engaged, when using the same graphing technique, in describing two variables that are simple functions of each other. At that time the technique was neither new nor fully developed. Indeed, it appears to have been then not quite three hundred years old, according to present historical evidence, and it grew out of an even older tradition of purely rhetorical discussion that omitted geometry when discussing certain simple functional variables.

A common pair of functional variables that were being discussed by 1350 in many universities of Europe were "extension" and "intension." Richard Swineshead of Oxford—a logician known among later scholars simply as "The Calculator"—pointed out explicitly, for example, that the extension of a thing could be altered dimensionally without altering the *intensity* of the thing. One could, in the mind's eye, extend hotness or whiteness, for example, without altering the intensity of either the over-all hotness or the hotness at any particular point, if one wished. Or one could increase the intensity without increasing the extent.

Swineshead used a geometric analogy to clarify his point: geometrically speaking, one could place one rectangle beside another without increasing the over-all length, if one chose. Or he could place one rectangle next to another in such a way as to add to the over-all length but without affecting the width.

Some time between 1345 and 1365, these notions were systematically explored, developed, and sharpened, especially under the mind and the hand

Reprinted from *Mathematics Teacher* 54 (Nov., 1961): 565–67; with permission of the National Council of Teachers of Mathematics.

of one man, a Parisian scholar named Nicole Oresme. The essence of the technique that he systematized, however, was first employed by an Italian logician, Giovanni di Casali, while Oresme was still a student. Casali remarked in passing, when discussing the traditional topic of "the velocity of motion of alteration," that if one took the example of the quality hotness, one could conceive of a uniform hotness throughout "just as a rectangular parallelogram is formed between two equidistant lines, such that any part you wish is equally wide with another . . . . "

Again, Casali said," . . . let there be throughout a uniformly difform hotness, such that it is a triangle . . . . "

We are quite unable to say whether or not Oresme read Casali's treatise. But about 1350 or 1360 Oresme wrote a lengthy work on "the configurations of qualities," and in this work he gave a detailed, systematic, and exhaustive explanation of how to use geometric figures to depict the extension and the intensity of any quality. "Although indivisible points or lines do not actually exist," he said in his introductory remarks, "yet it is necessary to picture them mathematically for the measure of things and for comprehending their proportions. Therefore, every intensity successively acquirable is to be imagined by a straight line perpendicularly erected upon some point of space or of the subject of that intensible thing."

Oresme also pointed out that his figures could be used to depict local motion. The perpendiculars represented speed, the base on which they were erected, time, and the area of the enclosed figure, distance. Uniform acceleration from rest would be portrayed by a right triangle, uniform speed by a rectangle.

These same geometric representations of motion were employed by Galileo and Beeckman nearly three hundred years later.

And in the meantime, what had happened to this simple graphing technique? Preliminary evidence indicates it persisted, often in the form of marginal illustrations, in a large number of documents. Thus, at the present time, thirteen extant

manuscripts are known of Oresme's long work, *De configurationibus qualitatum,* and two copies are known of another treatise in which he described his graphing technique—his "Questions on the books of Euclid's Elements."

These treatises apparently were never printed after movable type and the printing press became available during the fifteenth century. However, another work, a primer, quite brief, was printed more than once. This little handbook was called "On the latitudes of forms." For a while, modern authorities, such as Pierre Duhem, H. Wieleitner, M. Curtze, Lynn Thorndike, Anneliese Meier, and Marshall Clagett, thought that Oresme had composed the primer. More recently, Miss Meier has suggested it was written by one Jacob of Florence (or Naples or St. Martin). Whoever wrote it, there is no question that it derives from Nicole Oresme's full-length treatise, "On the configurations of qualities."

Twenty-two versions of the short work that we know of are extant. Eighteen of these are manuscript copies; four are printed editions. The earliest dated manuscript is inscribed "1395." The last of the editions was printed in 1515.

Five of the manuscripts were written during the fourteenth century. Eight more copies survive from the fifteenth century; two of these are printed versions, the first dated 1482, the second 1486. Two more printed versions are known to be from the sixteenth century, one published in 1505, the other in 1515.

Seven manuscripts remain that cannot be sharply dated with any assurance at present. Tentatively, their provenance would seem to place them in the fourteenth century or the fifteenth century.

The primer was the most popular of all the treatises using or discussing the Casali-Oresme figures. However, scattered instances of the use of these simple graphs are to be found in other documents of the fourteenth, fifteenth, sixteenth, and seventeenth centuries. It is not always possible to say with assurance in every case who wrote a particular treatise that contains or refers to the figures embodying the crude graphing technique here under discussion, nor is it possible to assert that the figures in the margin were always placed there by the man who inscribed the text. Nevertheless, the fact remains that at present some fifty documents, written or printed, are known which reveal that this protographing technique was in use among European scholars of the fourteenth, fifteenth, sixteenth, and seventeenth centuries as an aid to their exploration and their understanding of certain abstract concepts of the uniform and the nonuniform.

REFERENCES

Clagett, M., *The Science of Mechanics in the Middle Ages.* Madison, Wisconsin: University of Wisconsin Press, 1959.

Duhem, P. *Études sur Leonard de Vinci.* 3 vols. Paris: F. De Nobelle, 1955.

Meier, A., *An der Grenze von Scholastik und Naturwissenschaft.* Rome: Edizioni di Storia e Letteratura, 1952.

SUGGESTED READINGS FOR PART IV

Clagett, Marshall. *Archimedes in the Middle Ages.* The Ababo-Latin Tradition. Vol. 1. Madison: The University of Wisconsin Press, 1964.

Gies, Joseph, and Gies, Frances. *Leonard of Pisa and the New Mathematics of the Middle Ages.* New York: Thomas Y. Crowell, 1969.

Grant, Edward, ed. *Nicole Oresme: De proportionibus proportionum and Ad pauca respecientes* (Madison: The University of Wisconsin Press, 1966).

Hill, G. F. *The Development of Arabic Numerals in Europe.* New York: Oxford University Press, 1915.

Murray, Alexander. *Reason and Society in the Middle Ages.* Oxford: Clarendon Press, 1978. Part II: Arithmetic, pp. 141–203.

Nicomachus of Gerasa. *Introduction to Arithmetic.* Translated by M. L. D'Ooge. In *Studies in Greek Arithmetic,* by F. E. Robbins and L. C. Karpinski. Ann Arbor: University of Michigan Press, 1938.

Swetz, Frank J. *Capitalism and Arithmetic: The New Math of the Fifteenth Century.* La Salle, IL: Open Court Publishing Co., 1987.

PART V

NON-WESTERN MATHEMATICS

*M*ATHEMATICS IS NOT A WESTERN INVENTION. The universality of mathematics is based on humans' need to reach out to, touch, and, in a sense, control their environment. To some degree, all societies seek to achieve this control. Modern scholarship has enhanced an appreciation of non-Western accomplishments in the field of mathematics, many of which eventually influenced European mathematical thinking.

During the period from the sixth to the twelfth century when the civilizations of the West were experiencing a time of transition and reorientation, those of the East were engaging in high intellectual and cultural accomplishment. India under the Gupta rulers had experienced a Golden Age of Sanskrit accomplishment during the period from the fourth to the sixth century. The phenomenal rise and spread of Islam beginning in the seventh century eventually saw a consolidation of power in a Baghdad caliphate. Under the patronage of Caliph al-Ma'mun, a center of learning called the House of Wisdom was founded in Baghdad in A.D. 830. The patronage of al-Ma'mun and his successors saw scholars from the Eastern world attracted to Baghdad to pursue scientific and literary studies and theses of classical knowledge collected and translated for distribution throughout the Islamic world. The Celestial Empire of China stood unchallenged as a cultural and scientific mentor to the kingdoms surrounding it. When the Italian merchant Marco Polo visited China in the thirteenth century, he marveled at the level of civilization he witnessed—it far surpassed any he had known in Europe.

Each Eastern society developed and accumulated mathematical knowledge according to its own priorities and needs. Eastern outlooks in mathematics differed from those that existed in the West. For the Chinese hydraulic society, social stability was based on a Confucian humanism that discouraged a spirit of mathematical adventurism. In China, mathematics was merely a tool of the bureaucratic government used in the efficient running of the Empire. Still, the corpus of Chinese empirically-derived mathematical knowledge was impressively broad and high in quality; it exceeded that followed in the West until about the sixteenth century. Hindu mathematics was almost ritual in its conception and poetic in its problem-solving approaches. A strong involvement with astronomy resulted in the formulation of trigonometric relations and theories. The scholars of Islam preserved and synthesized classical and Far Eastern mathematical theories but also added their own insights and interpretations to the works that came under their purview. Eventually, it would be these modified Arab sources that, beginning in the ninth century, filtered into Europe and revitalized the mathematical outlook of the West.

41

The Status of Mathematics in India and Arabia during the "Dark Ages" of Europe*

F. W. KOKOMOOR

$\mathcal{P}$OLITICAL and economic world historians have found it convenient to divide the time from the fall of Rome to the discovery of America into two periods, and to designate the first of these by the term "Dark Ages." One work accounts for this name by "the inrush into Europe of the barbarians and the almost total eclipse of the light of classical culture." The period covers, roughly, the time from A.D. 500 to 1000. Part of these "barbarians" came down from the north and the rest attacked from the south, the latter bound together politically and religiously by the great, although probably totally illiterate, leader Mohammed into a vast dominion that at one time or another covered all of eastern Asia, northern Africa, Spain, France in part, and the European islands of the Mediterranean Sea. It was during this period that Europe was dark, learning at low ebb, and the development of mathematics almost negligible. The world as a whole was not dark, and as applied to general history the expression "Dark Ages" is a gross misnomer. Throughout the entire period there was consider-able intellectual (including mathematical) activity among the Hindus and, beginning about 750, there developed many centers of Muslim civilization which rose to the very peak of mathematical productivity.

A number of facts combine to account for our heretofore slight emphasis upon oriental mathematical science. (1) We have been so enamored by the story of the Golden Age of Greece and the Modern Period that the Orient failed until recently to divert our attention. (2) There is such an overwhelming mass of mediaeval material left us and so much of it is worthless, due largely to the overabundance of study devoted to Scholasticism, that the search for gems among the rubbish has seemed hardly worth the effort. (3) Then, too, the difficulty of accessibility involved is enormous. Before many Greek and Hindu works could be assimilated in the main scientific current in the West, they had to be translated first into Syriac, then Arabic, then Latin and finally into our own language. Thus the completeness and the accuracy of these transmissions can only be determined by painstaking investigations of historians of science. (4) Furthermore, but few persons have ever been qualified to advance our knowledge of Muslim mathematics, due to the rare qualifications required. Not only does one need to know well mathematics and astronomy, but also Sanskrit, Syriac, Arabic, and Persian, and, in addition, one needs to have a thorough training in paleography

Reprinted from *Mathematics Teacher* 29 (May, 1936): 224–31; with permission of the National Council of Teachers of Mathematics.

* In the preparation of this article the author has drawn heavily from Dr. George Sarton, *Introduction to the History of Science*, volume I, a work of great value to the student of the history of mathematics.

and a keen historical sense. This rare combination, together with a lifetime of ceaseless work, is the price that must be paid to increase our understanding of oriental mathematics.

Modern mathematics is easily transmitted; the process is simple. Articles appearing in any scientific journal are announced in other principal ones, and hence, it is easily possible for a worker in any field to be fully informed on what is being done throughout the world regardless of the language used in the original publication. So simple is it that the modern scientist who lacks historical training can hardly comprehend the difficulties involved in the handing down of knowledge from the early ages to the present.

For our knowledge of Greek mathematics we owe an unpayably large debt to two men who devoted years to the production of numerous accurate translations: J. L. Heiberg of Copenhagen, and T. L. Heath, a great English scholar in both mathematics and Greek. Unfortunately we have neither a Heiberg nor a Heath to enlighten us in Hindu and Muslim mathematics. In our knowledge of the former we are perhaps least fortunate of all. Cantor's three chapters are quite satisfying on the works of Brahmagupta and Bhāskara, being based upon Colebrooke's *Algebra with Arithmetic and Mensuration, from the Sanscrit of Brahmegupta and Bhascara*, but later historians place much dependence upon the interpretations of G. R. Kaye who was for many years a resident of India as a high government official, and whose work on Indian mathematics (1915) is shown to be erroneous in many respects by competent scholars of India today, such as Saradakanta Ganguli. One of the ablest scholars in the field of Muslim mathematics was Carl Schoy (1877–1925), who, between the years 1911 and 1925, contributed many valuable papers and books containing critical translations of Muslim mathematics and astronomy.

Recent scholars such as Schoy have shown us that, just as the greatest achievements of antiquity were due to Greek genius, so the greatest achievements of the Middle Ages were due to Hindu and Muslim genius. Furthermore, just as, for many centuries of antiquity, Greek was the dominant progressive language of the learned, so Arabic was the progressive scientific language of mankind during the period of the Middle Ages. We have learned further that the fall of ancient science and the dampening of the scientific spirit in Europe was far less due to the overrunning of southern Europe by the barbarians than it was to the passive indifference of the Romans themselves, and to the theological domination of a little later time. As soon as the Arabs were apprised of the Greek and Hindu sources of mathematical knowledge they were fired with a contagious and effective enthusiasm that led to numerous remarkable investigations in mathematics prosecuted from a number of cultural centers throughout the Muslim world, and that did not abate until the close of the twelfth century when they had made a permanent impression on mathematics as a whole.

What I wish to do is to give as comprehensive a survey of Hindu and Muslim mathematics as space permits, pointing out principal achievements of leaders in the field, and indicating work still to be done to make the history complete.

Our attention is first called to Hindu mathematics in one of the five Hindu scientific works on astronomy called *Siddhāntas*, which were theoretical as opposed to karaṇas which were practical. Its date is very uncertain, but is placed in the first half of the fifth century. The *Sūrya-Siddhānta*, the only one we have in full, is composed of fourteen chapters of epic stanzas (ślokas) which show decided knowledge of Greek astronomy but also much Hindu originality, especially the consistent use throughout of sines (jyā) instead of chords, and the first mention of versed sines (utramadjyā).

Even more important is the *Pauliśa Siddhānta* which we have only indirectly through the commentator Varāhamihira (ca. 505). It contains the foundation of trigonometry and a table of sines and versed sines of angles between 0° and 90° by intervals of 225′ (kramajyā). The sine and the arc of 225′ were taken to be equal, and sines of multiples of 225′ were obtained by a rule equivalent in our symbolism to

$$\sin (n + 1) x = 2 \sin nx - \sin (n - 1)x$$

$$- \frac{\sin nx}{\sin x} ; \sin x = x = 225'.$$

The next important Hindu advance is due to Āryabhaṭa (The Elder) who wrote in 499 a work, *Āryabhaṭiyam,* of four parts, the second of which—the *Gaṇitapāda*—was a mathematical treatise of 32 stanzas in verse, containing essentially the continued fraction process of solution of indeterminate equations of the first degree; an amazingly accurate value of π, namely 3 177/1250; the solution of the quadratic equation implied in the problem of finding n of an arithmetic series when a, d, and s are known; and the summing of an arithmetic progression after the pth term. These are expressed now by means of the formulas:

(1) $\quad n = \dfrac{d - 2a \pm \sqrt{(2a - d) 8ds}}{2d}$

(2) $\quad s = n \left[a + \left(\dfrac{n - 1}{2} + p \right) d \right].$

To him also were due other startling truths less mathematical in character, among which was the theory that the apparent rotation of the heavens is due to the rotation of the earth about its axis.

Varāhamihira, astronomer-poet and contemporary of Āryabhaṭa, contributed the equivalent of trigonometric facts and formulas as follows:

$$\sin 30° = 1/2; \quad \cos 60° = \sqrt{1 - \tfrac{1}{4}};$$

$$\sin \frac{x}{2} = \sqrt{\frac{1 - \cos x}{2}}$$

$$\sin^2 x + \text{versin}^2 x = 4 \sin^2 \frac{x}{2} ;$$

$$\sin^2 x = \frac{\sin^2 2x}{4} + \frac{[1 - \sin(90° - 2x)]^2}{4} .$$

Then we have a leap of about a century when Brahmagupta (ca. 628), one of India's greatest scientists, and the leading scientist of his time of all races, made his study of determinate and indeterminate equations of first and second degree, cyclic quadrilaterals and combinatorial analysis. He solved the quadratic for positive roots completely, and the Pellian equation $nx^2 + 1 = y^2$ in part (it was finished by Bhāskara, 1150). If a, b, c, d, x, and y are the sides and diagonals of a cyclic quadrilateral, s the half-perimeter, and K the area, his results can be expressed by the equations:

$$K = \sqrt{(s - a)(s - b)(s - c)(s - d)};$$

$$x^2 = \frac{(ab + cd)(ac + bd)}{(ad + bc)} ;$$

$$y^2 = \frac{(ad + bc)(ac + bd)}{(ab + cd)} .$$

Another proposition, "Brahmagupta's trapezium," states that if $a^2 + b^2 = c^2$ and $x^2 + y^2 = z^2$, then az, cy, bz, cx form a cyclic quadrilateral whose diagonals are at right angles. His work in combinations and permutations is much like that now offered in a first course, but is not quite complete. Brahmagupta used three values of π: for rough work, 3; for "neat" work, $\sqrt{10}$; and for close accuracy, the finer value given by Āryabhaṭa.

By the close of the period of Brahmagupta Hindu mathematics had developed to such an extent that its influence reached out far toward both the east and the west. Two Chinese authors are of special value as witnesses of the influence of Hindu mathematics in China. (1) The first is Ch'ü-t'an Hsi-ta, a Hindu-Chinese astrologer of the first half of the eighth century, whose work gives a detailed account of a number of ancient systems of chronology, the most important of which, from our point of view, is the Hindu system, and in the explanation of which is implied the Hindu decimal notation and rules. Thus we think this must have been the very latest date of the introduction of the Hindu numerals and hence other Hindu mathematics into China. Much more probably they were introduced earlier, about the second half of the sixth century, for the catalog of the Sui dynasty (589–618) lists many books devoted to Hindu mathematics and astronomy. Then, too, it was about this time that Buddhism entered China from India. (2) The second author, I-hsing (683–727), or Chang Sui (the first being his religious name) was an astronomer of note who undertook, by order of the emperor, an investigation of chronological and arithmetical systems of India, but he failed to finish the work on account of a premature death.

The westward reach of Hindu mathematics is equally certain. The Syrian philosopher and scientist, Severus Sebōkht (fl. 660), who studied also Greek philosophy and astronomy, was the first to mention the Hindu numerals outside of India, and expressed his full appreciation of Hindu learning in these words: "I will omit all discussion of the science of the Hindus, a people not the same as the Syrians; their subtle discoveries that are more ingenious than those of the Greeks and the Babylonians; their valuable methods of calculation; and their computing that surpasses description. I wish only to say that this computation is done by means of nine signs. If those who believe, because they speak Greek, that they have reached the limits of science should know these things they would be convinced that there are also others who know something." (Quoted from Smith: *History of Mathematics*, v. I, p. 167.)

But few Hindu mathematicians after the day of Brahmagupta stand out prominently. One of uncertain date but probably of the ninth century is Mahāvīra, author of the *Ganita-Sāra-Sangraha*, which deals with arithmetic, including geometric progressions, the relation between the sides of a rational sided right triangle ($2mn$, $m^2 + n^2$, $m^2 - n^2$), and the solution of several types of equations involving the unknown and its square root. Two hundred years later, about 1030, another *Ganita-Sāra* (compendium of calculation) was produced by Śrīdhara, but was quite elementary. However, he wrote a work (now lost) on quadratic equations in which, according to the eminent Bhāskara of the twelfth century, was found our present formula for the quadratic solution.

Now let us set our clock back at the second half of the eighth century. Here we find that practically all the work done in mathematics was done by Arabs. This is the beginning of a long period extending to the close of the eleventh century, during the whole of which there was an overwhelming superiority of Muslim culture. Stimulated from the east by the Hindus and from the west by the eastward transmission of Greek mathematics, the Arabs began a remarkable and altogether too little emphasized flourish of activity. At first there were mainly students and translators of the five Hindu siddhāntas and other works. Chief of these were al-Fazārī, the elder, and his son. Then with the ninth century began a series of very important steps forward, especially in the field of trigonometry and the construction of astronomical tables; but there was also an imposing group of geometers, arithmeticians, algebraists, and translators of Greek works.

The cause of science was greatly enhanced by the caliph al-Ma'mūn (813–833) who, although religiously exceedingly intolerant, was one of the world's greatest patrons of science. He collected all the Greek manuscripts he could, even sending a special mission into Armenia for that purpose, then ordered the translation of these into Arabic. He built two observatories, had made (probably by al-Khwārizmī) a large map of the world which was a much improved revision of Ptolemy's, organized at Bagdad a scientific academy, and stocked a library which was the finest since the Alexandrian (third century B.C.). He then invited many of the world's greatest scientists to his court. Among them was al-Khwārizmī (d. ca. 850) who wrote very important works on arithmetic and algebra and widely used astronomic and trigonometric tables of sines and tangents. He revised Ptolemy's geography, syncretized Hindu and Greek knowledge and is recognized by authorities as having influenced mathematical thought more than any other mediaeval writer.

A very notable astronomer of that period was al-Farghānī (fl. 860) who wrote the first comprehensive treatise in Arabic on astronomy, which was in wide use until the close of the fifteenth century, and was translated into Latin and Hebrew thus influencing European astronomy greatly before Regiomontanus.

Among the notes and extracts left us from Habash al-Ḥāsib (ca. 770 – ca. 870) we find record of the determination of time by the altitude of the sun and complete trigonometric tables of tangents, cotangents and cosecants which are at present preserved in the Staatsbibliothek in Berlin. In connection with the problem of finding the sun's altitude, the cotangent function arose. If x = the sun's altitude, h = the height of a stick, and l = the length of its shadow,

$$ h = l \left(\frac{\cos x}{\sin x} \right), $$

and al-Ḥāsib constructed a table of values of h for $x = 1, 2, 3, \ldots$ degrees, from which either x or h could be read if the other were known. According to Schoy, the Berlin manuscript also contains the equivalent of

$$ \sin x = \frac{\tan y \cos z}{\sin z}, $$

where y is the declination and z the obliquity of the ecliptic.

One of the world's best general scientists and the greatest philosopher of the Arabs was al-Kindī (fl. 813–842), prolific author of between 250 and 275 works on astronomy, geography, mathematics, and physics. He understood thoroughly the Greek mathematical works and influenced widely the early European scientists among whom were Girolamo Cardano, of cubic equation fame, and Roger Bacon.

The Banū Mūsā (Sons of Moses, also known as the Three Brothers) were wealthy scientists and patrons of science of this period. Through their efforts many Greek manuscripts were collected, studied, translated, and thus preserved for us. They were among the earliest to use the gardener's method of the construction of the ellipse (use of a string and two pins) and discovered a trammel based upon the conchoid for trisecting angles. Various problems of mechanics and geometry interested them most.

Perhaps chief among the many translators is to be mentioned al-Ḥajjāj (ca. 825) who first (under Hārūn al-Rashīd) translated Euclid's *Elements* into Arabic and improved it later (under al-Ma'mūn) in a second translation. He also was an early translator of Ptolemy's *Syntaxis,* which he called *al-mijisṭi* (the greatest) from which term the name *Almagest* was derived. He was preceded, however, in this work by the Jewish Arab al-Ṭabari of the same period. By the middle of the ninth century, then, these men and others of their day had made accessible to the Arabs the most important works of the Greeks and the earlier Hindus, had extended the sum total of astronomy and trigonometry, and had given tremendous impetus to independent investigation, which bore its fruit in the century to follow.

There was, however, much translating still to be done. Al-Māhānī (fl. 860) wrote commentaries on at least the first, fifth, and tenth books of the *Elements,* and on Archimedes' *Sphere and Cylinder,* and studied also considerably the *Spherics* of Menelaos which led him to an equation of the form $x^3 + a^2 b = a x^2$, with which he wrestled long enough to cause his successors in the field to refer to it as "Al-Māhānī's equation." He never solved it. Al-Ḥīmṣī (d. ca. 883) translated the first four books of the *Conics* of Apollonius. Al-Nairīzī (d. ca. 922) wrote (both commentaries lost) most authoritatively on the *Quadripartitum* and *Almagest* of Ptolemy as well as on the *Elements.* Thābit ibn Qurra (ca. 826–901) founded within his own family a school of translators and enlisted outside scholars to aid him in producing translations of nearly all the Greek mathematical classics even including the commentary of Eutocius. Without naming others, suffice it to say that by the beginning of the tenth century there was probably not a single important work of the Golden Age of Greece that had not been translated and mastered by the Arabs.

As to the results of independent investigators of this period, the following must be mentioned. The best and most complete study among the Arabs on the spherical astrolabe was produced by Al-Nairīzī. The school of Thābit ibn Qurra wrote about 50 works of independent research, and 150 (about) translations. Many of these works are still extant. The most valuable ones are those on theory of numbers and the study of parabolas and paraboloids. A sample of the former on amicable numbers shows the fine reasoning employed: $2^n pq$ and $2^n r$ are amicable numbers if p, q, and r are prime to each other and if $p = (3)(2^n) - 1$, $q = (3)(2^{n-1}) - 1$, and $r = (9)(2^{2n-1}) - 1$. The latter contains ingenious developments of Archimedes' Method.

In many respects the outstanding scholar of the century was al-Battānī (d. 929), Muslim's [Islam's] greatest astronomer, in whose principal work on astronomy are found numerous important facts. He gave the inclination of the ecliptic correct to 6"; calculated the precession at 54.5" a year; did not believe in the trepidation of the equinoxes, which Copernicus still believed many years later. In trigonometry, to which his fifth chapter is devoted, he gave the equivalent of our formulas:

$$\sin x = \frac{1}{\csc x} \; ; \qquad \cos x = \frac{1}{\sec x} \; ;$$

$$\tan x = \frac{\sin x}{\cos x} \; ; \qquad \cot x = \frac{\cos x}{\sin x} \; ;$$

$$\csc x = \sqrt{1 + \cot^2 x}; \qquad \sec x = \sqrt{1 + \tan^2 x};$$

$$\sin x = \frac{\tan x}{\sec x} \; ;$$

$\cos a = \cos b \cos c + \sin b \sin c \cos A$; and the sine law (doubtful). This work was translated into Latin by Plato of Tivoli in the twelfth century and into Spanish in the thirteenth, and exerted a tremendous influence for 500 years.

Considering now the first half of the tenth century, we note again that almost all the original work was done by Arabs in Arabic, but with the marked difference that there is a decided decline in activity. The development of mathematics may be compared with rainfall, which, after uncertain periods of drouth or average intensity, may burst forth in torrents. Some sort of unknown law of rhythm seems to hold rather than a law of uniform advance. The outstanding men of the period are Muslim. Ibrāhīm ibn Sinān (d. 946) wrote numerous commentaries on astronomy and geometry, but has received high recognition only since 1918 when Schoy's translation of his *Quadrature of the Parabola* revealed the fact that his method was superior to and simpler than that of Archimedes. Abū Kāmil (ca. 925) was an able algebraist, improved essentially the algebra of al-Khwārizmī by some generalizations, development of algebraic multiplication and division, and operations with radicals. He also gave an algebraic treatment of regular inscribed polygons.

Toward the close of the century a decided renewal of creative work is to be observed. Far more names of famous mathematicians could be mentioned than space permits. But Abū-l-Fath (fl. 982), al-Khāzin (fl. 950), al-Kūhī (fl. 988), al-Sijzī (fl. 1000), al-Ṣūfi (fl. 975), al-Khujandī (fl. 990) and Abū-l-Wafā' (fl. 990) cannot be omitted.

In Florence there is an untranslated commentary on the first five books of Apollonius' *Conics*, by Abū-l-Fath, who also wrote a translation of the first seven books of the *Conics*. His work on books V–VII is considered highly important because the original Greek is no longer extant. Al-Khāzin solved the cubic equation of al-Māhānī (mentioned before in this paper). Al-Kūhī is chiefly known for his work on the solution of higher degree equations by means of the intersections of two conics. His problem on trisecting an angle I quote here (from French) from Woepcke's *L'Algèbre d'Omar Alkhayyami* (1851), page 118:

Let the given angle be *CBE*. Take on *EB* produced points *A* and *D*, and on the other side a point *C* so that (1) *AD = DC*, (2) *AB:BC: :BC:BD*. Draw *BP* parallel to *DC*. Then the angle *CBP* = ¹/₃ angle *CBE*.

The geometry of the figure shows clearly that the angle is trisected, but the construction involves the solution of a cubic equation.

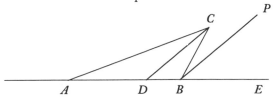

Of the works of al-Sijzī, fourteen are now preserved in Cairo, Leyden, Paris, and the British Museum. All deal with conic sections, trisection problems, and the resulting cubic equations. These works rank him as an outstanding Arabic geometer.

The greatest mathematician of the century, however, was Abū-l-Wafa', who wrote 15 or 20 works (mostly extant) on geometry, geometrical solutions of special fourth degree equations, and trigonometry, to which he added a number of formulas and the line values of the functions.

The eleventh century continues in the first half to show remarkable activity, with an imposing array of first order mathematicians, the principal ones being Al-Bīrunī (973–1048), Ibn Yūnus (d. 109), al-Karkhī (d. 1025), Ibn Sīna (980–1037), al-Ḥusain (?), and al-Nasawī (ca. 1025). Of paramount importance is the work of Al-Bīrunī. Two of his writings are of great mathematical significance. We have known the first, *A Summary of Mathematics,* for some time, but the other, *Al-Qānūn al-Masūdī,* has recently been given us in German, and proves its author to be of considerably more importance in the history of trigonometry than we have suspected.

Noteworthy among the contributions of Ibn Yūnus is the introduction of the prosthaphaeretical

(sum and difference) formulas of trigonometry which were so useful before the time of logarithms. Al-Karkhī, whose work, *al-Fakhrī,* Woepcke has given us in French, was an algebraist of the first rank. He gave a splendid treatment of the solution of Diophantine equations, equations of quadratic form, operations with radicals, and summation of integro-geometric series. In connection with series he gives such results (not symbolically, of course) as

$$\sum_1^n i^2 = \left(\sum_1^n i\right)\left(\frac{2n+1}{3}\right)$$

and

$$\sum_1^n i^3 = \left(\sum_1^n i\right)^2.$$

Ibn Sīna was principally a philosopher and hence emphasized that phase of mathematics; al-Ḥusain wrote one of the few Arabic treatises on the construction of right triangles with rational sides; and al-Nasawī, an able arithmetician, explained extraction of square and cube roots by a method very similar to our own, and furthermore anticipated decimal fractions in the manner indicated in the equations

$$\sqrt{17} = \frac{1}{100}\sqrt{170000} = \frac{412}{100},$$

but he changed them to sexagesimals for the final form of his answer.

There is a distinct decline toward the close of the eleventh century with a notable decrease in the number of mathematicians of the first rank. Of these I mention but one, Omar Khayyam, who, although living in a period of decline, was one of the greatest of the Middle Ages. His chief distinction results from his admirable work on *Algebra,* in which he classified equations by the number and degree of terms, treated 13 types of cubic equations, and referred to the general expansion of the binomial with positive integral coefficients which he treated in another work now unknown. His other mathematical writings dealt with the assumptions of Euclid and a very accurate reform of the calendar.

Thus closes the marvelous periods of scientific activity of the Hindus and Muslims in the field of mathematics. The Hindu period began shortly after the time of Proclus (d. 485), last straggler of the great Greek period. After the Hindu peak comes the tremendous Muslim flourish. Suter in his work, *Die Mathematiker und Astronomen der Araber und ihre Werke*, lists the works of 528 Arabic scholars who were active in mathematics from 750 to 1600. Certainly 400 of these came within the period of the "Dark Ages." We have learned much about them, but there is still much to be done. Tropfke tells us that "ueberreiche Schaetze" still lie untranslated in the large libraries of Europe.

During this same period in Europe, but few names deserve even feeble mention as writers on mathematics. There was no creative work. Boethius (d. 524) wrote texts on the quadrivium (arithmetic, music, geometry and astronomy), which, though comparatively poor, were so widely used in schools that they had tremendous influence. Anthemios (d. 534) is interesting for his history of conic sections and his use of the focus and directrix in the construction of the parabola. Eutocius (b. 480?) is important for his commentaries on the works of Archimedes and Apollonius. Bede (673–735) deserves mention because our information on finger reckoning is almost entirely dependent on his work. Alcuin or Albinus (735–804) wrote a work on puzzle problems which furnished material for textbook writers for ten centuries. Gerbert (Pope Sylvester II) (999) was great and influential as compared with other European writers of his day and doubtless did much to popularize the Hindu-Arabic numerals. But these men were all small as compared with the Muslim giants.

Let me repeat what I said at the beginning: Europe was dark, but India and the Muslim world were not. To quote Dr. George Sarton, "the 'Dark Ages' were never so dark as our ignorance of them."

Where Geometry and Algebra Intersect

ROSHDI RASHED

How would you describe the beginnings of Arab mathematics?

Arab research into mathematics seems to have begun early in the ninth century when, in Baghdad, the movement to translate the great Greek authors into Arabic was at its height. For instance, Al-Hajjaj ibn Matar made two translations of Euclid's *Elements* and Ptolemy's *Almagest*, and Hilal ibn Hilal al-Himsi translated the first four volumes of Apollonius of Perga's *Conics*. Several works by Archimedes, Pappus and Diophantus, among others, were also translated in the same century.

This major undertaking was noteworthy for two very important features, in that the translations were made by leading mathematicians and were inspired by the most advanced research of the time. For example, volumes V to VII of Apollonius' *Conics* were translated by the great mathematician Thabit ibn Qurrah, who died in the year 901. Moreover, all the indications are that Qusta ibn Luqa was prompted to translate the *Arithmetica* of Diophantus around 870 on account of studies already being carried out on indeterminate analysis.[1]

Many other examples could be cited to demonstrate the close links between such translations and the innovative research already being done at this high point in the dissemination of Hellenistic mathematics in Arabic.

Al-Khwarizmi is the best known of the Arab mathematicians. What was his contribution to mathematical thinking?

It was at this time, in the ninth century, at the Baghdad Academy, or "House of Wisdom" as it was known, that al-Khwarizmi wrote a work that represented a new departure in terms of both content and style. This was the *Kitab al-jabr wa'l muqabalah* ("The Book of Integration and Equation"),[2] from which algebra emerged for the first time as a separate and independent mathematical discipline. This was a crucial development, and was recognized as such by al-Khwarizmi's contemporaries, as much for its new mathematical style as for the nature of its subject matter and above all for the rich promise it held out for further advances.

In terms of style, it was both algorithmic, in that the author set out a series of computational procedures, and demonstrative. A new mathematics had to be devised that was general enough to be capable of handling different types of formulation while existing independently of those formulations. In al-Khwarizmi's work, an algebraic expression could refer to a number or to an irrational quantity or to a geometrical magnitude. This new mathematics and its combination of the demonstrative and applied approaches was a striking innovation for the thinkers of the time.

The novelty of the conception and style of al-Khwarizmi's algebra, which did not hark back to any previously known tradition, cannot be over-

Edited from an interview in *UNESCO Courier* (Nov., 1989): 37–41.

emphasized. The new algebra already afforded a glimpse of the enormous potential for applying one mathematical discipline to another that lay ahead from the ninth century onwards. In other words, while algebra, by virtue of its widespread scope and the new concepts it introduced, made such applications possible, their number and variety were thereafter constantly to change the face of mathematics.

Al-Khwarizmi's successors increasingly applied arithmetic to algebra and vice versa, arithmetic and algebra to trigonometry, algebra to the Euclidian theory of numbers, and algebra to geometry and vice versa. All these applications paved the way for new disciplines or at least for new chapters in the history of mathematics.

Could you give us a significant example of the encounter between arithmetic and algebra?

One example that springs to mind is the contribution which Arab mathematics made to classical number theory.

By the end of the ninth century, the most important Greek texts on arithmetic, such as Euclid's own volumes on the subject, the *Introductio arithmeticae* of Nicomachus of Gerasa, and the *Arithmetica* of Diophantus of Alexandria, had all been translated. Further chapters in the theory of numbers were opened in the wake of these translations and could in a sense be said to have been a reply to them. For example, two significant steps were taken in respect of the theory of amicable numbers. The first of these led, in the context of Euclidian arithmetic, to a new set of findings, while, as a result of the application of algebra to the theory of numbers, the second step culminated some centuries later in the creation of a field of number theory that owed nothing to the Greeks. We can look into these two aspects more closely.

Although Euclid put forward a theory of perfect numbers at the end of Volume IX of his *Elements,* neither he nor Nicomachus of Gerasa had developed the theory of amicable numbers.[3] Thabit ibn Qurrah, who translated the work of Nicomachus and revised a translation of the *Ele-*

ments, decided to work on this theory. He came up with and demonstrated, in pure Euclidian style, a remarkable formula for amicable numbers which now bears his name.

If we disregard the mystique which had surrounded amicable numbers and look only at the mathematics, it has to be acknowledged that, up to the end of the seventeenth century at least, the history of amicable numbers merely consisted of a passing reference to Ibn Qurrah's formula and its transmission by later mathematicians. These included such Arabic-speaking mathematicians as al-Antaki, al-Bagdadi, ibn al-Banna, al-Umawi and al-Kashi, whose differing origins in both time and place clearly show the widespread dissemination of Ibn Qurrah's formula, which crops up again in the work of Pierre de Fermat in 1636 and in that of René Descartes in 1638.

The second step is noteworthy for the fact that the celebrated physicist and mathematician Kamal al-Din al-Farisi, who died in 1320, wrote a paper in which he deliberately set out to demonstrate Ibn Qurrah's formula, but by a different path. Al-Farisi based his new demonstration on a systematic knowledge of the divisors of a whole number and of the operations that can be applied to them. However, this demonstration involved a reorganization which gave rise not only to a change in the perspective of Euclidian arithmetic, but to the promotion of new topics in number theory. It accordingly became possible to speak of a non-Hellenistic area in number theory.

In order to make it possible to engage in this new study of divisors, al-Farisi had to establish explicitly certain facts that had only been latent in Euclid's *Elements.* He also had to make use of the advances in algebra since al-Karaji's time in the tenth and eleventh centuries, and especially of combinatory methods. Hence, al-Farisi's approach was by no means confined to demonstrating Ibn Qurrah's formula, but enabled him to embark on a new study involving the first two arithmetical functions: the sum of the divisors of a whole number and the number of those divisors.

This style, which applied algebra and combinatory analysis to Euclidian arithmetic, was still prevalent in Europe in the seventeenth century, at least until 1640. The analysis of al-Farisi's conclusions and of the methods he used thus goes to show that, by as early as the thirteenth century, it had been possible to produce a set of propositions, findings and techniques that had hitherto been ascribed to the mathematicians of the seventeenth century.

What about the new relationship that was established between algebra and geometry?

We have already seen that the mathematical landscape was no longer the same from the ninth century onwards; it was transformed and its frontiers were rolled back. Greek arithmetic and geometry became increasingly widespread. In addition, non-Hellenistic areas were developed within the corpus of Hellenistic mathematics itself. The relationship between the old disciplines was no longer the same, and a host of other groupings were formed. This changing pattern is crucial to an understanding of the history of mathematics generally, for the new relationship between algebra and geometry gave rise to techniques of enormous potential.

The mathematicians of the tenth century embarked on a two-way exercise in conversion which had never previously been envisaged—the translation of geometrical problems into the language of algebra, and vice versa. They translated into algebraic terms solid problems that could not be constructed with ruler and compass, such as the trisection of the angle, the two means, and the regular heptagon. Moreover, algebraists and also geometers such as Abu al-Jud b. al-Leith, when faced with the difficulty of using radicals to solve a cubic equation, were able to turn to the language of geometry and apply the intersecting curves technique to the study of this type of equation.

The first attempts to provide a basis for these conversions were made by al-Khayyam (c. 1048–1131). In his bid to go beyond the specific cases represented by a particular form of cubic equation, al-Khayyam developed a theory of algebraic equations of a degree less than or equal to three which at the same time provided a new model for the formulation of equations. He then studied cubic equations using conic sections in order to reach positive real solutions for them. In order to construct his theory, al-Khayyam had to visualize the new relationship between algebra and geometry in clearer terms before being able to formulate it. From then onwards, the theory of equations appeared, albeit still tentatively, to bridge the gap between algebra and geometry.

In his celebrated treatise *Algebra*, al-Khayyam produced two remarkable findings which historians have wrongly attributed to Descartes. These are the generalized solution of all third-degree equations by means of two intersecting curves, and the possibility of performing a geometric calculation by defining a unit length, which was a fundamental concept.

Some fifty years after al-Khayyam, his successor Sharaf al-Din al-Tusi made a further step forward. In an attempt to demonstrate the existence of the point of intersection of two curves, he arrived at the problems of finding and separating the roots of the equation and of dealing with the conditions under which they existed. In order to find a solution he defined the notion of maximum values for an algebraic expression and tried to find concepts and methods for determining such "maxima."

This not only led al-Tusi to the development of concepts and methods, such as derivatives, that were only to be named as such at a later date, but also compelled him to change his approach. He discovered the need to use local procedures, whereas his predecessors had only considered the overall properties of the objects being studied. All these findings and the theory embracing them are obviously important and have often been attributed to mathematicians who only came several centuries later.

These are the main features of the dialectic between algebra and geometry. To complete the picture, however, two impediments which slowed down the progress of the new mathematics should be mentioned. These were a reluctance to use negative numbers as such, at a time when they had not yet been defined, and the shortcomings of symbolic notation. Both issues were to preoccupy later mathematicians.

In political historiography distinctions are made between Antiquity, the Middle Ages, the Renaissance, and modern times. Do you think that his breakdown is relevant in an account of the history of mathematics, and especially of the Arab contribution?

It is true that "medieval" mathematics has been contrasted with "modern" mathematics. The first historical corpus, covering Latin, Byzantine, and Arab mathematics, and Indian and Chinese mathematics as well, could be distinguished from another body of work which came into being with the Renaissance. I don't feel that this dichotomy is relevant in either historical or epistemological terms. Arab mathematics clearly represents a continuation and outgrowth of Hellenistic mathematics, which was its breeding-ground. This is also true of the mathematics which grew up in the Latin world from the twelfth century onwards. Finally, the work accomplished in both Arabic and Latin (or Italian) between the ninth century and the early seventeenth century cannot be split into separate periods.

On the contrary, all the indications are that the type of mathematics involved was the same. This is borne out by the fact that we can now compare the work on algebra and numerical computation produced by al-Samaw'al in the twelfth century with that of Simon Stevin in the sixteenth century; al-Farisi's findings in number theory with those of Descartes; the methods used by al-Tusi for the numerical resolution of equations with that of François Viète in the sixteenth century or al-Tusi's search for *maxima* with that of Fermat; al-Khazin's work on integral Diophantine analysis in the tenth century with that of Bachet de Meziriac in the seventeenth century, and so on. If we were to disregard the work of al-Khwarizmi, Abu Kamil,

al-Karaji, and others, how could we understand the work of Leonardo of Pisa and the other Italian mathematicians in the twelfth and thirteenth centuries, or the mathematics of the seventeenth century?

True, the later seventeenth century was marked by the emergence of new methods and new areas of mathematical interest in Europe. However, that break did not necessarily occur all of a sudden, nor did it take place at the same time in every discipline. Moreover, the dividing lines seldom coincide with the works of different authors, but often cut right across them. In number theory, for example, the innovation did not lie in Descartes' and Fermat's use of algebraic methods, as has been claimed, since they simply rediscovered al-Farisi's findings. The break can actually be identified within Fermat's own work around 1640, with his invention of the "infinite descent" method and his study of certain quadratic forms.

It was really from the mid-seventeenth century onwards that the tangled threads come together and the main breaks in continuity are identifiable. The contribution made by the Arab mathematicians thus fitted into a coherent pattern which grew up between the ninth century and the first half of the seventeenth century.

NOTES

1. The last five books of Diophantus' *Aritmetica* are principally devoted to the solution of indeterminate equations, i.e. those with more than one variable and a large number of solutions. *Editor*

2. The word *algebra* is derived from the title of this work which was translated into Latin many times in the Middle Ages and had a strong influence on medieval Western science. The word *algorithm*, which designates any method of computation (such as the decimal system) involving a series of steps, is derived from the Latinized form of al-Khwarizmi's name. *Editor*

3. A number is *perfect* if it is equal to the sum of its own divisors (for example $6 = 3 + 2 + 1$; $28 = 14 + 7 + 4 + 2 + 1$). Two integers form a pair of amicable numbers if the sum of the proper divisors of one is equal to the other. This is true of the numbers 220 and 284, which were for a long time the only known pair of amicable numbers.

43

Thâbit ibn Qurra and the Pythagorean Theorem

ROBERT SHLOMING

*D*URING the Dark Ages of Western Europe, Islam was very much mathematically alive. The Arabs were able to solve the most difficult problems of Archimedes and Apollonius at a time when Latin mathematical knowledge was at a level below that of the ancient Egyptians. Indeed, the Arabs were translators and transmitters of mathematical knowledge rather than innovators. Yet the student of mathematics owes them a large debt for preserving a priceless mathematical legacy. For without the contributions of the Arab translators, much of Greek and Hindu mathematics would now be lost. By uniting Greek and Hindu mathematical ideas, the Arabs further developed arithmetic, algebra, geometry, and trigonometry.

Between A.D. 500 and A.D. 600 many of the important Greek mathematical works were translated into Syriac. After A.D. 700, these works were translated into Arabic so that by around A.D. 850 the Arabs had in their possession valuable translations of Aristotle, Ptolemy, Euclid, Archimedes, Menelaus, Apollonius—to name but a few. During the eighth and ninth centuries, the Arabs engaged in considerable trade with India and in so doing obtained much knowledge of Hindu mathematics. But the apogee of Arabic science and mathematics came during the tenth and eleventh

centuries.[1] Particularly in mathematics was this an active and fruitful period. This period witnessed the writing of scholarly translations and critical commentaries on Greek and Hindu sources. In addition, some original works were produced by Arab mathematicians during the ninth through the eleventh centuries. It must be realized, however, that these works reached the Latin West through a meandering linguistic stream, because no less than four languages might be involved in the translation of a single treatise.[2]

Among the Arab scholars of mathematics who devoted their energies to translating and refining Greek manuscripts stands Thâbit ibn Qurra. The works of Thâbit ibn Qurra, "one of the greatest Islamic scholars that ever lived,"[3] are either neglected or not appreciated. It is the purpose of this article to shed some light on Thâbit and his contributions to mathematics. Although Thâbit was a prolific writer in many areas of mathematics (writing on amicable numbers, spherical trigonometry, mechanics, geometry, cubic equations, geometrical constructions, and the gnomon), we will, in this article, examine only his work dealing with the Pythagorean theorem.

Abū' l'Hasan Thâbit ibn Qurra Marwân al' Harrani was born at Harrân, Mesopotamia, in the year 826 and spent most of his life at Bagdad where he died in 901.[4] Nothing is known of his childhood at Harrân. With regard to his adult life we know that Thâbit was a money changer in

Reprinted from *Mathematics Teacher* 63 (Oct., 1970): 519–28; with permission of the National Council of Teachers of Mathematics.

Harrân and was later excommunicated from this town for his unorthodox religious beliefs.[5] On a chance trip to Bagdad, Thâbit met the famed Muhammid ibn Musa ibn Sakir.[6] It was this accidental meeting with Musa ibn Sakir that changed Thâbit's life and brought him into contact with a circle of scholars and translators at Bagdad. At Bagdad, the center of mathematical activity during Thâbit's life, many Greek and Hindu mathematical manuscripts were translated. Indeed, the caliphs were generous patrons of learning. With the enlightenment of the Abbāsids, the Eastern Caliphate whose seat was at Bagdad, a new chapter in the history of mathematics ensued.[7] Carmody writes that Thâbit "was discovered" by Musa ibn Sakir who took him into the caliphate of Mu'tadid.[8] At the court of Mu'tadid, Thâbit distinguished himself as an expert in astronomy and mathematics.

Thâbit is often regarded as an astonishing polymath who excelled not only in mathematics but also in medicine, astronomy, linguistics, philosophy, physics, music, geography, botany, natural history, agriculture, and meteorology.[9] Historical evidence is conclusive in verifying that Thâbit was a competent mathematician. As a matter of fact, his work on amicable numbers is completely original.[10] He wrote a scholarly commentary on the *Liber Assumptia* of Archimedes.[11] His dissection proofs of the Pythagorean theorem are elegant. His work on the generalization of the Pythagorean theorem is indeed sophisticated with respect to the ninth century. Moreover, his proof of the law of the lever, contained in the *Liber karastonis*, is extant today in one Arabic and fifteen Latin copies. The *Liber karastonis* of Thâbit ibn Qurra is perhaps one of the most important scientific works of the Middle Ages.[12] This work combines the Aristotelian and Archimedean traditions of dynamics and mechanics.[13] Recently, the *Liber karastonis* was translated by Professors Clagett and Moody.[14] Returning to Thâbit's contributions in mathematics, mention must be made of his work on the gnomon which was one of the earliest

known works on this subject.[15] With regard to spherical trigonometry, Thâbit's work on the transversal theorem is of vast importance in the history of mathematics.[16] Furthermore, Thâbit's contributions with respect to the paraboloid provide a link between the infinitesimal geometry of Archimedes and the works of Cavalieri, Kepler, and Wallis. In fact, Suter writes that Thâbit's work on the paraboloid is among the "choicest fruits" of Arabic mathematics.[17] Thâbit has made remarkable quadratures of parabolas as well as paraboloids. Smith credits Thâbit with finding the volume of a paraboloid.[18] To appreciate the mathematical contributions of Thâbit ibn Qurra, let us remind ourselves that we are speaking only of the ninth century. His aptness in mathematics is summed up by Clagett in the following words: ". . . ability of Thâbit is clearly attested in his numerous works which have survived."[19]

Many historians of mathematics consider Thâbit one of the most gifted translators from Greek and Syriac into Arabic. His love for translating was so great that he founded a school of translation at Bagdad. The prominent members of Thâbit's translation school were Qusta ibn Luqa, Ishaq ibn Hunain, and al-Kindi. In addition, many of Thâbit's own family were active in translating Greek manuscripts into Arabic. Thâbit's sons and grandsons were first-rate astronomers and mathematicians who belonged to his translation school. For instance, Abū-Su'id Sinain ibn Thâbit ibn Qurra, the son of Thâbit ibn Qurra, followed his father's footsteps as a geometer, astronomer, and physician. Abū wrote an improved version of Thâbit's translation of the *Elements*.[20] Both of Thâbit's grandsons, Thâbit ibn Simain ibn Thâbit ibn Qurra and Ibrahim ibn Sinain ibn Thâbit ibn Qurra, distinguished themselves as scholars in mathematics. In fact, Ibrahim's writings on the quadrature of the parabola were the "simplest ever made before the invention of the integral calculus."[21]

Of the many works attributed to Thâbit ibn Qurra, few have as yet received proper attention.

One reason for this neglect is our relative ignorance of his life and the time in which he lived. On the other hand, we do know that Thâbit translated mathematical works of Menelaus, Archimedes, Euclid, Ptolemy, Galen, Theodosious, Appolonius, Hippocrates (both of Cos and of Chios), Eutocious, Nichomachus, and others.[22] Thâbit also improved some of the translations of Ishaq ibn Hunain.[23] His revised translation of the *Elements*, from a translation by Ishaq ibn Hunain, is considered to be among the best in the world.[24] All in all, Thâbit wrote over 150 works in Arabic on logic, mathematics, astronomy, and medicine. In addition, he wrote fifteen books in Syriac.[25] A large number of Thâbit's manuscripts contain more than one tract; the total number of copies that appear in Latin translations runs as high as 245.[26]

Some of the translations that Thâbit wrote form an integral part of the "Arabic Corpus."[27] There are many fine mathematical works that appear in the Arabic Corpus. The works listed below were translated by Thâbit ibn Qurra and are contained in the Arabic Corpus.[28]

1. *Almagest*—Ptolemy (ca. A.D. 150)[29]
2. *Arithmetica*—Nichomachus (ca. A.D. 100)
3. *Assumpta*—Archimedes (287–212 B.C.)
4. *Catoptrica*—Euclid (ca. 300 B.C.)
5. *Conica*—Appolonius (ca. 225 B.C.)
6. *Data*—Euclid
7. *De leve*—Euclid
8. *De mensura circulii*—Archimedes[30]
9. *De orto*—Autolycus (ca. 330 B.C.)
10. *De sphaera et cylindro*—Menelaus (A.D. 100)
11. *De sphaera mota*—Autolycus
12. *Elements*—Euclid
13. *Hypotheses*—Ptolemy
14. *Liber datorum*—Thâbit ibn Qurra
15. *Optica*—Euclid

Let us now examine the mathematical contributions made by Thâbit ibn Qurra through his dissection proof of the Pythagorean theorem and his generalization of the Pythagorean theorem.

Tradition is unanimous in ascribing to Pythagoras the independent discovery of the theorem for a right triangle, that the square of the length of one side added to the square of the length of the other side equals the square of the length of the hypotenuse. This theorem was known to the Babylonians during the time of Hammurabi, more than a thousand years before the birth of Pythagoras. In any event, the first general proof of this theorem may have been given by Pythagoras. The proof of the Pythagorean theorem is contained in the *Elements* of Euclid (Book I, Proposition 47). Today there are over 300 demonstrations of the Pythagorean theorem.[31]

Thâbit ibn Qurra did considerable work on this theorem and presented proofs that are completely original. His work on this theorem is found in an extant Istanbul manuscript.[32] In this manuscript, one finds a collection of scientific and scholarly articles. Among them are several by Thâbit ibn Qurra. One of them deals with the proof of the Pythagorean theorem by a method of dissection. It is of historical interest to remark that the manuscript containing Thâbit's proof is still in good condition. It is now located in the library of Aya Sofya Museum in Turkey, registered under the number 4832. Another copy of this manuscript is located at Cairo and has been investigated by H. Suter.[33]

Thâbit proves the Pythagorean theorem by a method called reduction and composition or, as it is sometimes referred to, the reduction of triangles and rearrangement by juxtaposition. Dissection proofs of the Pythagorean theorem are quite numerous in the history of mathematics. The main idea in a dissection proof is the concept of congruence. Two areas are said to be congruent by addition if they both can be dissected into corresponding pairs of congruent pieces. They are said to be congruent by subtraction if corresponding parts can be summed together so that the resulting figures, formed by this addition, are congruent to each other. From Figure 1, H. Perigal in 1873 proved the Pythagorean theorem by the method of dissection. But Professor Eves points out that this was only a "rediscovery," for the dissection proof was already known to Thâbit ibn Qurra.[34]

Thâbit considers triangle *ABC* with right angle at *C*, and he wishes to prove that $a^2 + b^2 = c^2$. The proof now follows:

1. Construct on side *AC* the square *A'C* whose area is b^2.

2. Extend *BC* to *F* so that *B'F* = *BC* and construct on side *B'F* the square *B'D* whose area is a^2.

FIGURE I

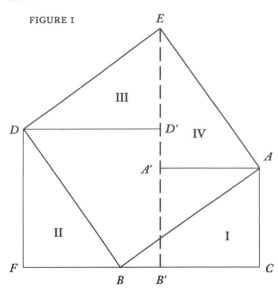

3. *A'D'* is extended to *E* so that *A'E* = *B'D'* (= *BC*). Draw *DE* and *EA*.

4. It now follows that in triangles *ABC, BDF, DED'*, and *AEA'*,

   a) the sides *AC, FB, D'E*, and *A'A* are equal to each other.
   AC = *FB* = *D'E* = *A'A*.
   b) the sides *BC, DF, DD'*, and *EA'* are equal to each other.
   BC = *DF* = *DD'* = *EA'*.

5. Angle *ACB* = angle *DFB* = angle *DD'E* = angle *EA'A* = 90°.

6. By S.A.S., $\triangle ABC \cong \triangle BDF \cong \triangle DD'E \cong \triangle EA'A$.

7. As a consequence of the congruence, *AB* = *BD* = *DE* = *EA*.

8. Moreover, angle *EAA'* = angle *BAC* which implies that angle *EAB* = angle *A'AC* = 90°.

9. Furthermore, angle *BDE* can be shown to be a right angle and thus *ABDE* is a square; that is, the square on the hypotenuse *BA*.

10. The area of square *AD* = c^2.

11. Clearly we can obtain the following with regard to area.
$$\triangle I + \triangle II + ABDD'A' = A'C\,(= b^2) + B'D\,(= a^2);$$
$$\triangle III + \triangle IV + ABDD'A' = AD\,(= c^2).$$

12. Obviously, $\triangle I + \triangle II + ABDD'A' = III + IV + ABDD'A'$ since
$$\triangle I = \triangle II = \triangle III = \triangle IV \text{ (from step 6).}$$

13. Thus, $a^2 + b^2 = c^2$.

That Thâbit had given such a proof is reinforced by yet another written source, al-Nairizi (fl. ca. A.D. 900) in his commentary on Euclid's *Elements*.[35] The Arabic text and Turkish translation of Thâbit's proof have been briefly sketched in *Belletin*, a quarterly journal of the Turkish historical society.[36]

A second proof of the Pythagorean theorem that is attributed to Thâbit[37] is quite subtle. Let us now consider figure 2.

1. $\triangle ABC$ is a right triangle with right angle at *C*.
2. On side *CB* construct square *CD* whose area is a^2.

FIGURE 2

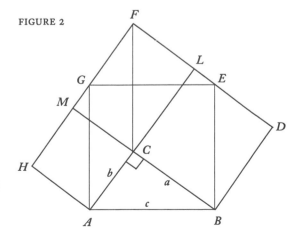

3. On side AC construct square HC whose area is b^2.

4. On hypotenuse AB construct square BG whose area is c^2.

5. It can be shown that $\triangle FLC = \triangle FMC = \triangle ABC = \triangle BED = \triangle AGH = \triangle FGE$.

6. $ABDFH - (\triangle ABC + \triangle FCM + \triangle FLC) =$ square $CD +$ square $CH = a^2 + b^2$.

7. $ABDFH - (\triangle BED + \triangle FGE + \triangle GHA) =$ square $BG = c^2$.

8. From steps 6 and 7 we have square $CD +$ square $CH =$ square BG or $a^2 + b^2 = c^2$.

It is important to note that, in 1910, A. R. Colburn proved the Pythagorean theorem by using Thâbit's construction.[38] The similarity of Colburn's proof to Thâbit's is indeed amazing. One difference, however, is that Colburn proves the Pythagorean theorem algebraically by extending line segments HA and DB to meet at point P. Let us now consider figure 3, which Colburn used to demonstrate the Pythagorean theorem.

FIGURE 3

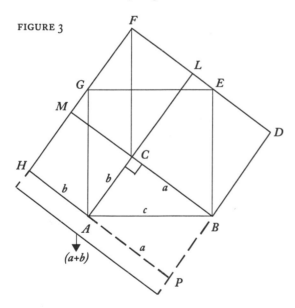

Colburn's proof immediately follows from figure 3.

$(a + b)^2 = PF$, and $PF = AE + 4\triangle ABC$.

$$a^2 + 2ab + b^2 = PF = c^2 + 4\left(\frac{ab}{2}\right) = c^2 + 2ab.$$
$$\therefore a^2 + b^2 = c^2.$$

It is interesting to note that Professor B. F. Yanney in 1903 offered a proof of the Pythagorean theorem that is not unlike Thâbit's dissection proof. Indeed, the diagram that Yanney used is identical with Thâbit's, but Yanney presents three proofs of the Pythagorean theorem based on the same diagram.[39] In Yanney's proof we will deal with parallelograms rather than triangles. Let us now consider figure 4.

FIGURE 4

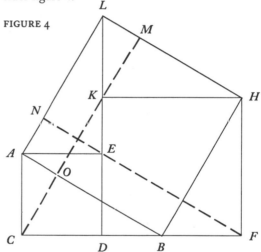

In figure 4, triangle ABC is a right triangle with right angle at C. Yanney proves the Pythagorean theorem in the same manner as Thâbit ibn Qurra but adds the following:

1. Construct $MC \| LA$ and construct $FN \| BA$.

2. It can be shown that $LMOA$ and $LKCA$ are parallelograms and that $LMOA = LKCA = b^2 (ACDE)$.

3. It can be shown that $HMOB$, $HKCB$ are parallelograms and that $HMOB = HKCB = HKDF = a^2$.

4. By the addition axiom we have $LMOA + HMOB = a^2 + b^2$. But $LMOA + HMOB$ is precisely square AH whose area is c^2.

5. Therefore, $a^2 + b^2 = c^2$.

Yet another way of looking at this proof is to make use of the following:

$$ANQB = AEFB = AEDC = b^2;$$
$$LNQH = LEFH = HKDF = a^2.$$

By the addition axiom we have $ANQB + LNQH = a^2 + b^2$ and, of course, $ANQB + LNQH =$ square AH whose area is c^2. Again we conclude that

$$a^2 + b^2 = c^2.$$

According to A. Sayili, the more important contributions of Thâbit ibn Qurra deal with the generalization of the Pythagorean theorem.[40] In connection with the generalization of the Pythagorean theorem, our interest centers around a letter written by Thâbit to a friend who was somewhat disappointed with the so-called Socratic proof of the Pythagorean theorem. Basically, the disappointment stems from the fact that the proof deals with a special case only, namely, that of an isosceles triangle. Thâbit's friend expressed interest in a general proof of the Pythagorean theorem. Sayili claims that Thâbit's communication to his friend constitutes an "important contribution" to the history of mathematics.[41] It is this communication that we will be concerned with. After relating to his friend the two proofs of the Pythagorean theorem (given earlier in this article), Thâbit commends his friend for desiring such mathematical knowledge and further adds that generalizations achieved by the former proofs may not be considered sufficient. Although the request by Thâbit's friend was fairly straightforward, Thâbit offered an elaborate and highly comprehensive answer. For example, one would wish, he says, not to impose the restrictions of working solely with squares and right triangles.[42] In the first case, one could demonstrate, as Euclid had done, that the sum of any similar figures located on the sides of any right triangle is equal to a similar figure placed on the hypotenuse.[43] In the latter case, on the other hand, Thâbit wishes to generalize the theorem to any triangle whatever. This proposal that Thâbit wishes to demonstrate is no small undertaking. The following is the generalization of the Pythagorean theorem as given by Thâbit ibn Qurra. If from vertex A of a triangle ABC two lines AB' and AC' are drawn forming with the base BC the angles $AC'B$ and $AB'C$, respectively, both equal to angle BAC, then the sum of the squares of the sides AB and AC is equal to the rectangle $BB' + CC'$ times BC. If angle BAC is obtuse, then B' and C' will be interchanged. In light of the aforementioned, let us consider figure 5.

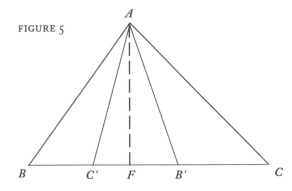

FIGURE 5

We now wish to prove, in figure 5, the following:

$$(AB)^2 + (AC)^2 = BC\,(BB' + CC').$$

Instead of giving a proof of this theorem, Thâbit only states the theorem.[44] The proof that now follows, of course, is not due to Thâbit ibn Qurra and, moreover, uses modern notation.

1. From the law of cosines, for which the Pythagorean theorem is a special case, one can write

$$b^2 + c^2 = a^2 + 2bc \cos A.$$

2. Upon substituting for a, b, and c, respectively, in step 1, the result is:

$$(AB)^2 + (AC)^2 = (BC)^2 + 2(AB)(AC) \cos A.$$

3. Since angle $BAC =$ angle $AC'B =$ angle $AB'C$, then $\cos A = \cos AB'C = \cos AC'B$. Thus,

$$2 \cos A = \cos AC'B + \cos AB'C.$$

4. Substituting step 3 in step 2, the result is:

$$(AB)^2 + (AC)^2 = (BC)^2 + (BA)(AC)$$
$$\cdot (\cos AC'B' + \cos AB'C')$$

5. Altitude AF is drawn on side BC so that (i) $\cos AC'B' = FC'/AC'$ and (ii) $\cos AB'C = FB'/AB'$.

6. Substituting 5 in 4, the result is as follows:

$$(AB)^2 + (AC)^2 = (BC)^2 + (BA)(AC)$$
$$\cdot \left(\frac{FC'}{AC'} + \frac{FB'}{AB'} \right).$$

7. Since triangle $AC'B'$ is isosceles it then follows that side AC' is equal to side AB'. In light of this, the following is then true:

$$(AB)^2 + (AC)^2 = (BC)^2 + (BA)(AC)$$
$$\cdot \left(\frac{FC' + FB'}{AB'} \right).$$

8. Furthermore, triangle ABB' is similar to triangle ABC (a.a. = a.a.) by an identity angle at B and because it is given that angle A is equal to angle $AB'C'$. Consequently,

$$\frac{BA}{AB'} = \frac{BC}{AC} \text{ or } AB' = \frac{(BA)(AC)}{BC}.$$

9. Substituting 8 in 7, the result is:

$$(AB)^2 + (AC)^2 = (BC)^2 + (BA)(AC)$$
$$\cdot \frac{(FC' + FB')(BC)}{(BA)(AC)}.$$

10. Upon simplifying step 9, we have

$$(AB)^2 + (AC)^2$$
$$= (BC)^2 + (BC)(FC' + FB')$$
$$= (BC)[BC + (FC' + FB')]$$
$$= BC(BC + B'C') = BC(BB' + CC').$$

It is interesting to note that Thâbit's generalization of the Pythagorean theorem immediately follows from the following proportions.

$$BC : AC : AB = AC : CC' : AC' = AB : AB' : BB'.$$

The next step would then be the following:

$$(AB)^2 = (BC)(BB'), \text{ and } (AC)^2 = (BC)(CC').$$

By adding, we arrive at the generalization of the Pythagorean theorem:

$$(AB)^2 + (AC)^2 = (BC)(BB' + CC').$$

One should be cognizant of the historical fact that this theorem appeared in the 1790 edition of Jacques Ozanam's *Recréations Mathématiques et Physiques,* volume 1, pp. 289–90.[45] Recent evidence shows that the theorem stated by Thâbit was discovered independently about the year 1665 by John Wallis. In fact, Wallis's *Treatise of Angular Sections* is based on two sources, namely, Ptolemy's lemma and the generalization of Thâbit ibn Qurra.[46]

Thâbit goes on to say that "should one wish to see yet a still more general theorem, one in which neither the type of the triangle nor the shape of the figure drawn on its sides is restricted, then one may say that the sum of any similar figures similarly placed on the two sides of any triangle is equal to the figure whose ratio to the similar figure similarly drawn on the third side is the same as the ratio of $BB' + CC'$ to BC."[47] But this conclusion was known to the Greek mathematician Pappus.[48]

It is clear from a study of Thâbit's theorem that he has heavily relied on the *Elements.* In particular, Sayili states that the above theorem is "reducible or derivable" from the Euclidean propositions 12, II and 13, II. It is highly probable that it was on these propositions that Thâbit based his generalization. The very theorem being dealt with in this article was long nonextant but was rediscovered only as recently as 1953. Thâbit's generalization contains references to various special cases where angle A is a right angle (angle B' and angle C' would have to coincide) or when angle B or angle C is a right angle. In addition, Thâbit also considers the fascinating case in which angle A is less than 60°, with C' or B' situated outside the segment BC, and the case in which angle A is larger than a right angle and $BB' + CC'$ is less than BC.

In concluding his letter, Thâbit remarks that knowledge is perfect when it combines the most general with the particular. For, he goes on to say, in purely general knowledge there is already the knowledge of the particular cases. Moreover, he also states that in the course of instruction one has to follow a procedure in which there is a gradual increase in generalization and comprehension.

In concluding this article, it is important to note that the *Fihrist* mentions that Thâbit ibn Qurra translated the *Elements* on the basis of the version by Ishaq ibn Hunain.[50] There seems to be little doubt that Ishaq, the son of the famous Arabic physician Hunain ibn Ishaq (807–873), knew Greek as well as his father did.[51] Heath claims that there must have been an agreement between Ishaq and Thâbit with respect to the translation of Euclid's *Elements*. Moreover, Thâbit undoubtedly consulted Greek sources for the purposes of his manuscript revision. There is direct evidence that Gerard of Cremona's translation of the *Elements* is a word-for-word translation of the work of Thâbit ibn Qurra. Gerard's translation of the *Elements* contains several notes directly quoted from Thâbit concerning alternative proofs that Thâbit found in other Greek manuscripts. Heath claims that this is further testimony to Thâbit's critical treatment of the text.[52] Clagett points out that Gerard of Cremona based his translation on the version of Ishaq ibn Hunain that was later revised and improved by Thâbit ibn Qurra. For Clagett has compared the Gerard of Cremona text with Thâbit's version and concluded that there is a great similarity between the two works.[53] Gerard's translation is quite different from that used by Adelard and is, in fact, much closer to the "best Greek tradition."[54] In accordance with the Arabic style of writing during the ninth century, Thâbit made his additions in the third person so that there would be a contrast with other editions.[55] The original version of Ishaq has not survived. We only have Thâbit's revision of Ishaq's translation. Two manuscripts of the Ishaq-Thâbit improved version exist today.[56] As strange as it may seem, the famous Arabic commentator and mathematician al-Qifti credits Thâbit with an independent translation of the *Elements* and not with a mere improvement of Ishaq's version.[57]

The preface of the Ishaq-Thâbit version is of great historical interest in mathematics because it uncovers the methods that the Arabic translators and mathematicians used. To be sure, the Ishaq-Thâbit version is not free from errors, obscurities, and omissions. Nevertheless, Nasir ed-Din Ahmed al-Tusi (1201–1274) became familiar with the Ishaq-Thâbit translation, and he, himself, translated two editions of the *Elements* on the basis of this work. In 1594, one of al-Tusi's editions was published in Rome, and remarkably enough some copies of this work are still found today with the following Latin title: *Euclidis elementorum geomerricorum libri tredecim ex traditione doctissimi nasiridinni nunc primum arabice impressi.*[58]

In conclusion, the introduction of Islamic science and mathematics into the Latin world altered the mathematical culture of medieval Latin Christendom. For Western Europe assimilated the main features of the Greco-Arabic mathematical heritage that ultimately resulted in the awakening of critical mathematical thought throughout Europe. The epoch of translations that took place in Islam, mainly in the ninth and tenth centuries, absorbed the classical mathematics of Greece. The numerous translations that were transmitted by the Arabs, Thâbit ibn Qurra in particular, during the course of several centuries reached Europe through Spain, Syria, and Sicily and laid the foundation of mathematical knowledge that was to dominate European thought. Thus, the introduction of Arabic mathematics and science into Western Europe divides the history of mathematics in the Middle Ages into two distinct periods. In the early period, man had to rely on the fragmentary writings of Capella, Bede, and Isidore. In the later period, the Latin West had Arabic commentaries on the original works of the Greek mathematicians.[59] To be sure, Thâbit ibn Qurra is a strong link in the evolutionary mathematical chain between Greek and Latin mathematics.

NOTES

1. Martin Levey and Marvin Petruck, *Kūshyār ibn Labbān: Principles of Hindu Reckoning,* pp. 3–4.

2. Robert E. Bruce, "Sicily and the March of Ancient Mathematics and Science to the Modern World," *Scripta Mathematica* 5 (1938): 117–21.

3. Max Meyerhof, "The 'Book of Treasure' an Early Arabic Treatise on Medicine," *Isis* 14 (1931): 55.

N.B. The importance of Thâbit's mathematical work was fully appreciated by Heinrich Suter, for he wrote: *"Er war ein bedeutender arzt, aber sein hauptgebiet waren die philosophischen und mathematischen wissenschaften, in denen er einen hohen rang in der arabischen littera-turgeschichte einnimmt."* The interested reader should consult the following source: Heinrich Suter, "Mathe-matiker und Astronomen der Araber," *S.B.P.M.S.* 48 (1900): 34. Other sections of interest in Suter are pp. 65–86, 186–227.

4. George Sarton, *Introduction to the History of Science,* vol. 1, p. 600.

5. Lynn Thorndike, *A History of Magic and Experimental Science During the First Thirteen Centuries of our Era,* pp. 660–61.

6. Heinrich Suter, "Mathematiker und Astronomen der Araber," *S.B.P.M.S.,* p. 34.

7. De Lacy O'Leary, *How Greek Science Passed to the Arabs,* pp. 146–49.

8. F. Carmody, *The Astronomical Works of Thâbit ibn Qurra,* pp. 15–17.

9. E. Wiedemann, "Über Thâbit b. Qurra: Sein Leben und Sein Werken," *S.B.P.M.S.* 52 (1922): 210–17.

10. Suter, "Mathematiker und Astronomen der Araber," pp. 34–35.

11. B. L. Van der Waerden, *Science Awakening,* pp. 225–27.

12. George Sarton, "Review on the Work of the Quarastan by Thâbit ibn Qurra," *Isis* 5 (1922): 494.

13. Ernest A. Moody and Marshall Clagett, *The Medieval Science of Weights* (Scientia de Ponderibus, pp. 3–6, 145–49. Cf. H. L. Crosby, Jr., *Thomas of Bradwardine: His Tractatus de Proportionibus, Its Significance for the Development of Mathematical Physics.*

14. Ibid.

15. Suter, "Mathematiker und Astronomen der Araber," p. 34.

16. Axel Björnbo, "Thâbit's Werk über der Transversalensatz (Liber de Figura Sectore)—mit bemerkungen von H. Suter," in *Abhandlungen zur Geschichte der Naturwissenschaften und der Medizin* (Erlangen, 1924).

17. Suter, "Mathematiker und Astronomen der Araber," p. 35 Cf. Idem, "Über die Ausmessung der Parabel von Thâbit," *S.B.P.M.S.* 52 (1921): 141–88.

18. David Eugene Smith, *History of Mathematics,* vol. 2, p. 685.

19. Moody and Clagett, p. 80.

20. O'Leary, pp. 174–75.

21. Heinrich Suter, *"Abhandlung über die Ausmessung der Parabel von Ibrahim ibn Sinain ibn Thâbit ibn Qurra, aus dem Arabischen übersezt und commentient,"* Der *Viertelijahrschrift der Naturforschen den Gesellschaft* 63 (1918): 214–28.

22. Sarton, *Introduction,* p. 601.

23. Son of Hunain ibn Ishaq, famous physician and translator.

24. Howard Eves, *An Introduction to the History of Mathematics,* p. 193.

25. F. Carmody, "Notes on the Astronomical Works of Thâbit ibn Qurra," *Isis* 46 (1955): 235–43.

26. Carmody, "Notes," p. 236.

27. Carmody, *Astronomical Works,* p. 23.

28. Ibid.

29. The *Almagest* (*Mathematical Syntaxis or Megali Syntaxis tes Astronomias*) was later translated from the Arabic of Thâbit into the Latin of Gerard of Cremona, the greatest translator of the Middle Ages, at Toledo in 1175. It is significant to note that Gerard of Cremona is the father of Arabicism in Europe, for "more of Arabic science came into Western Europe at the hands of Gerard of Cremona than in any other way." Cf. Haskins, *The Renaissance of the Twelfth Century,* pp. 286–88.

30. *De mensura cirulii* was translated from the Arabic of Thâbit into the Latin of Gerard of Cremona and perhaps also by Plato of Tivoli. Clagett writes that Archimedes's method of quadrature became available through Thâbit's translations. This was certainly a major step over the vague method of quadrature discussed in Boethius's *Commentary on the Categories of Aristotle.* Cf. Marshall Clagett, *Archimedes in the Middle Ages: The Latin-Arabo Tradition,* pp. 15–58.

31. Elisha S. Loomis, *The Pythagorean Theorem* (pp. 120–21 in particular).

32. A. Sayili, "Thâbit ibn Qurra's Generalization of the Pythagorean Theorem," *Isis* 51 (1960): 35–37.

33. Suter, p. 37.

34. Eves, p. 73.

35. Sir Thomas L. Heath, *The Thirteen Books of Euclid's Elements* translated from the text of Heiberg, I, p. 364.

36. A. Sayili, "Sâbit ibn Kurra' nin Pitagor Teoremini Tamimi," *Belletin* 22 (1958): 527–49.

37. Sayili, "Thâbit ibn Qurra's Generalization," p. 36.

38. A. R. Colburn, *Scientific American Supplement* 70 (1910): 382.

39. B. F. Yanney and J. Calderheath, "New and Old Proofs of the Pythagorean Theorem," *American Mathemtical Monthly* 3 (1903): 65, 110, 169, 299.

40. Sayili, "Thâbit ibn Qurra's Generalization," p. 35.

41. Ibid.

42. Ibid.

43. Heath, II, pp. 268–69 (Book VI: p: 31).

44. C. J. Scriba, "John Wallis' Treatise of Angular Sections and Thâbit ibn Qurra's Generalization of the Pythagorean theorem," *Isis* 57 (1966): 56–67. Cf. Carl B. Boyer, *A History of Mathematics*, p. 259.

45. Carl B. Boyer, "Clairaut le cadet and a Theorem of Thâbit ibn Qurra," *Isis* 55 (1964): 68–76.

46. Scriba, p. 67.

47. Sayili, "Thâbit ibn Qurra's Generalization," p. 38.

48. Eves, p. 179.

49. Sayili, "Thâbit ibn Qurra's Generalization," p. 38.

50. Heath, p. 37.

51. Max Meyerhof, "New Light on Hunain ibn Ishaq," *Isis* 8 (1926): 685–725.

52. Heath, II, p. 93.

53. Marshall Clagett, "The Medieval Latin Translations from the Arabic of the *Elements* of Euclid, with Special Emphasis on the Versions of Adelard of Bath," *Isis* 44 (1956): 16–42.

54. Ibid.

55. Ibid. N. B. *Refert Theibit qui transtulit hunc librum de greco in Arabicam linguam se invenisse quod additur ante figuram tricesiman primam huius partis in quibusdam scriptis graecis cuiusdam joanitii babiloniensis, quod tamen non est de libro.*

56. These MSS are described by Nicoll and Pusey. The MSS are presently located in the Bodelian Library ‡279, ‡280.

57. Heath, p. 94.

58. George Sarton, Ancient and Medieval Science During the Renaissance, pp. 160–161.

59. Carl B. Boyer *A History of Mathematics* (op. cit., n. 44), p. 269.

44

Al-Khwarizmi

A. B. ARNDT

I ENJOY introducing my mathematics classes to the history of mathematics and science. It is a subject they don't seem to encounter anywhere else, not even in their history courses, and it is one of the few topics that span that wide and culturally divisive gap between science and art in our society. In these brief historical excursions, I point out that our word *algorithm* is derived from the name of the ninth-century Arab mathematician Al-Khwarizmi. A teacher is always looking for some fact on the fringes of a student's knowledge to use as a peg on which to hang new information.

When I taught at an industrial training school in Saudi Arabia, I naturally tried the same trick, but it became immediately apparent that my students there were far more familiar with the great Al-Khwarizmi than I. "Oh yes, Abu-Abdullah Muhammed ibn-Musa," they responded, smiling.

"Uhh, right," I said, checking my notes quickly. I resolved on the spot to find out more about the man. My students had found a peg on the fringes of my information, and I was eager to load it with as many facts as possible.

This task wasn't as easy as I had expected. Many books about mathematicians or the history of mathematics, especially the older ones, simply fail to mention Al-Khwarizmi or any of his Arab colleagues at all. For example, H. W. Turnbull's *The Great Mathematicians* (1929), introduced as a "biographical history of mathematics," makes a

quantum leap from Diophantus, who worked in Alexandria in the fourth century A.D., to the Renaissance Europeans Napier and Kepler. More recent historians, however, have begun to abandon the erroneous old notion that the Arabs had merely preserved Greek science for a few centuries before handing it on to Europe and to recognize, instead, original Arab contributions to science, particularly those of Al-Khwarizmi.

Abu Abdullah Muhammed ibn-Musa Al-Khwarizmi (AH-boo ab-DULL-ah moo-HAM-mud IB-ben mū-sah al-Kwār-IZ-mee), that is, Muhammed, the father of Abdullah and the son of Moses, worked in the ninth century under the patronage of the Caliph Al-Ma'mun. He was one of many scholars gathered together by the caliph in his capital city of Baghdad, situated beside the Tigris river in present-day Iraq. His surname indi-

Reprinted from *Mathematics Teacher* 76 (Dec., 1983): 668–70; with permission of the National Council of Teachers of Mathematics.

cates that either he or his family originally came from Khwarizm, a region located east of the Caspian Sea and now incorporated into the U.S.S.R. (see map).

We know little about Al-Khwarizmi's personal life. Some sources give his birth date as A.D. 780; it is thought that he died sometime between the years A.D. 830 and A.D. 850. Some difficulty in searching for the facts about Al-Khwarizmi is caused by another man, also named Muhammed, who worked in Baghdad under a later Abbasid caliph. Since this man was also an astronomer and mathematician and his father, too, was called Musa, this Muhammed ibn-Musa has been frequently confused, even by major reference works today, with our scientist; but he is Abu-Jaffer (the father of Jaffer), not Abu-Abdullah, and he is *not* Al-Khwarizmi!

However little we know of Al-Khwarizmi himself, his scientific work was widely known and well documented. Al-Khwarizmi was the author of many books—in addition to his best-known works, he wrote about the astrolabe and the sun-dial and elaborated on the geometry of Ptolemy. He probably also took part in a project, sponsored by Al-Ma'mun, to measure the length of one degree of the earth's circumference.

Early in his career, Al-Khwarizmi prepared a practical abridgment of the Hindu astronomical tables known in Arabic as the *Sindhind*. This work made him famous almost instantly throughout the Arab world. Caliph Al-Ma'mun then asked him for a popular work on the science of equations, and he produced *Al-Kitab Al-jabr wa'l muqabalah*.

The words *jabr* (JAH-ber) and *muqabalah* (moo-KAH-ba-lah), which do not appear earlier in mathematics, were used by Al-Khwarizmi to designate the two basic operations in solving equations. The title of his work has been variously translated along the lines of "The Book of Completion and Cancellation" and "The Book of Restoration and Balancing."

Jabr is employed in the step in which an equation such as $x - 4 = 10$ becomes $x = 14$, where the left-hand side of the first equation, in which x is diminished by 4, is "restored" or "completed" back to x. In Spain and Portugal today, a somewhat obsolete name for a "restorer" of bones—a bone-setter—is *algebrista*.

The operation Al-Khwarizmi called *muqabalah* leads us from $x^2 + x = x^2 + 4$ to $x = 4$ by "canceling" or "balancing" the two sides of the equation. In time, the word *muqabalah* was dropped, and the name of this branch of mathematics became, in a great many languages, simply *algebra*.

Al-Khwarizmi's popularizing treatise on algebra was extremely successful, and its fame and influence lasted for centuries. Numerous copies of the Arabic work were made, many in the West, and during the early Middle Ages, several translations were made into Latin as well. Later on, as the use of Latin waned, the work appeared in different European languages; the first treatise on algebra written in German, dated 1461, was a translation of a portion of Al-Khwarizmi's book. Even in this century—because so many of the book's examples are drawn from the field of Islamic law—students of law in Cairo and Mecca have been advised to study *Al-Kitab Al-jabr wa'l muqabalah*.

It is in looking at mathematical texts over the years, rather than in histories, that we can see Al-Khwarizmi's profound effect on mathematics. We find his influence on all who followed him. Over the centuries, writers on elementary algebra have borrowed his format, his terminology, and his classification of the types of linear and quadratic equations; often they have even used the very same numerical examples he did in the original *Algebra*.

Al-Khwarizmi's *Algebra*, including its pious introduction, was rendered into modern English in 1831 by Frederic Rosen. Somehow it came as a shock to me, when I looked at this translation of *Al-Kitab Al-jabr wa'l muqabalah*, that this granddaddy algebra textbook, written over eleven hundred years ago, was among the most lucid and useful I'd ever seen. It is straight-forward and practical, full of examples, and, as Al-Khwarizmi himself stated, deals with "what is easiest and most useful in arithmetic." It is a textbook that I would have no difficulty teaching from today.

Al-Khwarizmi wrote with a genuine sympathy for the readers' difficulties. After explaining several ways to find the circumference of a circle, for example, he cautions:

> This is an approximation, not the exact truth itself: nobody can ascertain the exact truth of this and find the real circumference, except the Omniscient: for the line is not straight so that its exact length might be found. . . . The best method here given is that you multiply the diameter by three and one-seventh: for it is the easiest and quickest. God knows best! (Rosen 1831, p. 200.)

All numbers in the *Algebra*, as in the quote above are expressed by words rather than numerals—*three and one-seventh* instead of 3 1/7. The numerals we use today had at that time been developed in India only a couple of centuries earlier and were just beginning to make their way westward.

Al-Khwarizmi's next great work was a treatise on arithmetic, in which the Hindu-Arabic numerals were presented and the place-value system was explained. This textbook was the earliest written on the decimal system. It represents a milestone in the development of mathematics and science. Al-Khwarizmi demonstrated, in the *Arithmetic*, the basic operations of addition, subtraction, multiplication, and division and showed how to work with fractions and how to extract square roots. All these operations were greatly simplified by the new system.

This work, too, was enormously popular; it was avidly studied in the West and was instrumental in effecting Europe's conversion from the cumbersome Roman numerals to the present-day system. Unfortunately, no copies of Al-Khwarizmi's original Arabic version of the *Arithmetic* are extant, but we have several early Latin translations. Along with many other European scholars, the Englishman Adelard of Bath traveled to Spain to study Arabian mathematics. He produced, in 1126, what may be the earliest Latin translation of this work.

Through his *Arithmetic*, Al-Khwarizmi added another word to the vocabulary of the English language. One well-read Latin translation begins with the words *Dixit algorizmi*, or "Algorithm says," and follows with instructions for making various computations. Thus *algorithm*, a Latinized version of the mathematician's name, has come to its present meaning of a general computational procedure.

Issac Newton once said that he was able to see so far because he stood on the shoulders of giants, and surely Al-Khwarizmi stands firm in that tall pyramid of the world's great scientists. Perhaps the best assessment of the man can be found in his own quiet words from the introduction to his *Algebra*:

> The learned in times which have passed away, and among nations which have ceased to exist, were constantly employed in writing books on the several departments of science and on the various branches of knowledge. . . .

> Some applied themselves to obtain information which was not known before them, and left it to posterity; others commented upon the difficulties in the works left by their predecessors, and defined the best method [of study], or rendered the access [to science] easier or placed it more within reach; others again discovered mistakes in preceding works, and arranged that which was confused, or adjusted what was irregular, and corrected the faults of their fellow-laborers, without arrogance towards them, or taking pride in what they did themselves. (Rosen 1831, pp. 2–3)

All the achievements and all the services that he mentions were accomplishments of Al-Khwarizmi himself, although he modestly attributes them to others. These works of a learned man, from a nation that has ceased to exist and in a time that passed away more than a millennium ago, are still benefiting us today.

BIBLIOGRAPHY

Boyer, Carl B. *A History of Mathematics.* New York: John Wiley & Sons, 1968.

Karpinski, Louis Charles, ed. and trans. *Robert of Chester's Latin Translation of the Algebra of Al-Khowarizmi.* New York: Macmillan Co., 1915.

Newton, Issac. *Correspondence.* Vol. I, 1661–75. Cambridge: Published for the Royal Society at the University Press, 1959.

Rosen, Frederic, ed. and trans. *The Algebra of Mohammed Ben Musa.* London: Oriental Translation Fund, 1831.

Scott, J. F. *A History of Mathematics.* London: Taylor & Francis, 1958.

Struik, Dirk J. *A Concise History of Mathematics.* New York: Dover Publications, 1967.

Turnbull, H. W. *The Great Mathematicians.* London: Methuen & Co., 1929.

Vogel, Kurt, ed. and trans. *Mohammed Ibn Musa Alchwarizmi's Algorismus; Das fruehste Lehrbuch zum Rechnen mit indischen Ziffern.* Aalen, W. Germany: Otto Zeller Verlagsbuchhandlung, 1963.

45

The Contributions of Karaji—
Successor to al-Khwarizmi

HORMOZ PAZWASH AND GUS MAVRIGIAN

*I*T IS ENLIGHTENING to notice an increasing number of historical notes on the development of mathematics and to realize their pedagogical value in the classroom (Arndt 1983; Campbell 1977; Flusser 1981; Swetz 1984). Truly, mathematics without historical connotations loses its greatness and realism.

Mention has been made in the *Mathematics Teacher* (Arndt 1983) of the efforts of al-Khwarizmi (al-kwa-riz-me) and his contributions in the subject of algebra. Such significant words as *algebra* and *algorithm* were products of the work of al-Khwarizmi over a millennium ago. Herein, identification is made of another scientist, Karaji (ka-ra-yee), who followed al-Khwarizmi, and of his contributions and efforts in arithmetic, algebra, geometry, and the applied sciences.

Karaji was a Persian scholar who lived and produced the major portion of his work at the end of the tenth and the beginning of the eleventh centuries. Variants of his name include Karagi, al-Karaji, and, as he is known in the mathematical world, al-Karkhi (al-kar-key), although his full name was Mohammed ibn al-Hassan al-Hasib al-Karaji (Boyer 1968; Rashed 1972). He was born in the city of Karaj, near the present-day city of Baghdad (just as al-Khwarizmi was born in the city of Khwarizm, now the modern city of Khiva,

south of the Aral Sea in the Soviet Union). Without adding to the confusion often associated with the identification of Islamic writers, we shall use the name of Karaji in this article.

Karaji offered a new viewpoint with his mathematical symbolism—the arithmetization of algebra. Extending the work of al-Khwarizmi, Karaji revealed his efforts in his treatise on algebra, *Al-Fakhri* (al-fak-ree). This algebra of Karaji is well summarized in the reports by Rashed (1972). In this treatise, Karaji studied exponents and gave the familiar rule: $x^m \cdot x^n = x^{m+n}$; he also discussed solutions to algebraic equations of higher degrees by first examining solutions to "canons" (canonical forms), such as the following:

$$ax = b$$
$$ax^2 = b$$
$$ax^2 = bx$$
$$ax^2 + c = bx$$
$$ax^2 + bx = c$$
$$bx + c = ax^2$$

The major contributions of Karaji center on two treatises: *Kafi fil Hisab* (ka-fee-fil-he-sab), a treatise on arithmetic (ca. A.D. 1010), which translates as "Book of Satisfactions," and the already mentioned *Al-Fakhri,* and advanced treatise on algebra (ca. A.D. 1021), a work dedicated to his patron and protector, the vizir (minister) Fakhr-al-Mulk (fa-ker-al-mulk) (Eves 1964). In the treatise on arithmetic, beautifully translated in German (Hochheim 1878), the calculation skills of Karaji are displayed.

Reprinted from *Mathematics Teacher* 79 (Oct., 1986): 538–41; with permission of the National Council of Teachers of Mathematics.

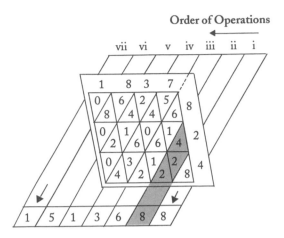

FIGURE 1 *Illustration of Arabic "sieve" multiplication: 1837 × 824 = 1 513 688*

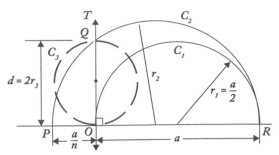

FIGURE 2 *From circle C_1, construction of circle C_3, whose area is 1/nth the area of C_1*

The lattice or sieve multiplication illustrated in Figure 1 shows how Karaji computed the product 1837 × 824 = 1 513 688. Notice that (1) the multiplicand is arranged as the horizontal heading in the tableau; (2) the multiplier is arranged vertically in the extreme right column; (3) the products of respective single digits are then decomposed in terms of tens and units within subdivided rectangles arranged horizontally; and (4) the product of the multiplication is then performed by the additions in the diagonal channels of operation, starting with channel i, through channel vii, allowing for "carry-overs." Such mechanical multiplicative tableaus later surfaced (Smith 1958) as "grating" methods found in *The Treviso Arithmetic* (1478, in Italy), as Napier's "rods" (1617, in Scotland), and as the Sokuchi method (nineteenth century, in Japan).

Perhaps the greatest contributions of Karaji were his arithmetization of algebra, the algebra conceived by al-Khwarizmi, and the geometrical representation of algebraic operations. As a dedicated disciple of Diophantus, Karaji was famous for his delineations of geometrical ideas. One such problem requires the construction of a circle with area equal to $(1/n)$th the area of a given circle, where n is a natural or counting number. Referring to Figure 2, let C_1 denote the given circle of diameter a units, that is, let $OR = a$. Construct perpen-

dicular $\overrightarrow{OT}$ to $\overline{OR}$ at O and extend $\overline{RO}$ through O so that $OP = a/n$. Now with $\overline{PR}$ as the diameter, construct circle C_2. The intersection of circle C_2 with $\overrightarrow{OT}$ yields point Q such that

$$OQ = d = \frac{a}{\sqrt{n}}.$$

A proof of this result is as follows (see Fig. 3):

In right triangle POQ,

$$PQ^2 = OP^2 + OQ^2 = \left(\frac{a}{n}\right)^2 + d^2.$$

In right triangle QOR,

$$QR^2 = OR^2 + OQ^2 = a^2 + d^2.$$

By adding, we get

$$(PQ^2 + QR^2) = \left[\left(\frac{a}{n}\right)^2 + a^2 + 2d^2\right].$$

In right triangle PQR, note that

$$(PQ^2 + QR^2) = PR^2 = \left(a + \frac{a}{n}\right)^2.$$

FIGURE 3

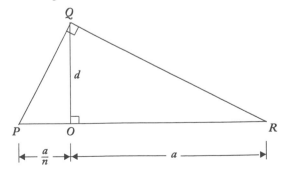

$$\therefore \left(\frac{a}{n}\right)^2 + a^2 + 2d^2 = \left(a + \frac{a}{n}\right)^2$$
$$= \left(\frac{a}{n}\right)^2 + a^2 + \frac{2a^2}{n},$$

or

$$2d^2 = \frac{2a^2}{n},$$

yielding

$$OQ = d = \frac{a}{\sqrt{n}}.$$

Thus, the ratio of the area of circle C_3 with diameter d to the area of circle C_1 is $1 : n$ as required.

Although Karaji proved magnificent in the preservation of mathematics through his efforts in translating the first five books of Diophantus's *Arithmetica*, he also added new problems to those taken from such works. The latter portion of Karaji's life was devoted to works in engineering, in areas of mechanics, hydrology, and surveying, and in the science of tunneling or construction of walls and underground conduits (referred to as *ganats* (ghanats)). Much of his engineering knowledge and productivity is revealed in his third major work, or treatise, entitled *Khafiya* (ka-fee-ga).

As an illustration of his skills in surveying, Karaji (Jam 1966) analyzed the problem of determining the height of a mountain from several recorded lines of sight. In Figure 4, height AB is desired, with point A defining the top of the mountain and point B, the foot. Two points, C and D, at the same elevation as B and planar with $\overline{AB}$ are selected to sight the mountain top. The squarelike

FIGURE 4 *Geometry for calculating mountain height AB*

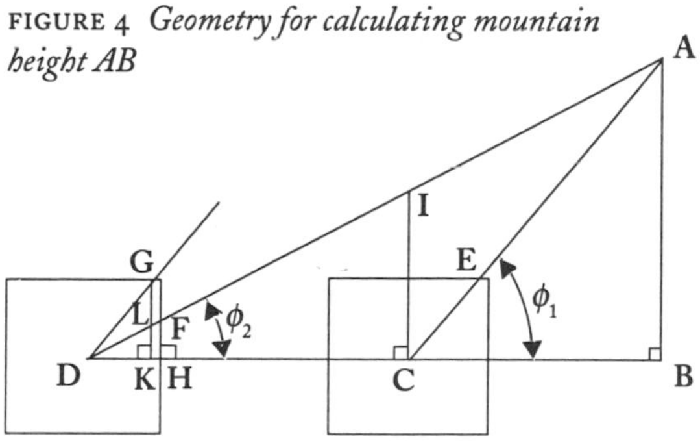

figures, having centers at points C and D, represent instruments (invented by Karaji) for the task. These equal squares, several feet on an edge, were circumscribed by circles with protractor-type graduations, together with a sighting tube. From point C, the line of sight $\overline{AC}$ is taken; then point E is marked on the square form of the instrument. Similarly, from reference point D, the line of sight $\overline{AD}$ gives point F. In the square frame centered at point D, $\overline{DG}$ is drawn parallel (hence equal) to $\overline{CE}$. Now distance CI is determined from similar right triangles DFH and DIC. Distance AC is determined from similar triangles DGL and ACI, and distance BC is determined from similar triangles DGK and CAB. Finally, by the Pythagorean principle, height AB is calculated from right triangle ABC. In a modern-day setting, which would employ elements of plane trigonometry, note that

$$\sin(\phi_1 - \phi_2) = \sin\phi_1 \cdot \cos\phi_2 - \cos\phi_1 \cdot \sin\phi_2$$
$$= \frac{AB}{CA} \cdot \frac{DC + CB}{DA} - \frac{CB}{CA} \cdot \frac{AB}{DA},$$

giving

$$AB = DC \cdot \frac{\sin\phi_1 \cdot \sin\phi_2}{\sin(\phi_1 - \phi_2)}$$

where

$$(\phi_1 - \phi_2) \neq 0.$$

The efforts and transcriptions of Karaji led to the preservation of famous mathematical works. Further, his use of symbolism and his arithmetization of the algebra of al-Khwarizmi led to generalizations in mathematical methodology. These are evident in his detailed accounting of the algebra of polynomials. In the centuries since Karaji lived and worked, it is pleasing to identify him as a great custodian and promoter of mathematics—a true successor of al-Khwarizmi.

REFERENCES

Arndt, A. B. "Al-Khwarizmi." *Mathematics Teacher* 76 (December 1983): 668–70.

Boyer, Carl B. *A History of Mathematics.* New York: John Wiley & Sons, 1968.

Campbell, William L. "An Application from the History of Mathematics." *Mathematics Teacher* 70 (September 1977): 538–40.

Eves, Howard. *An Introduction to the History of Mathematics.* Rev. ed. New York: Holt, Rinehart & Winston, 1964.

Flusser, Peter. "An Ancient Problem." *Mathematics Teacher* 74 (May 1981): 389–90.

Hochheim, Adolf. *Kafi fil Hisab (Genugendes uber Arithmetik) des Abu Bekr Muhammed Ben Alhusein Al-Karkhi.* Berlin: Verlag von L. Nebert, 1878; Ann Arbor, Mich.: University Microfilms International, 1980 (facsimile).

Jam, H. Khadiv. *Extraction of Hidden Water* (Persian translation of Karaji's treatise; ca. A.D. 1000). Tehran, Iran: Bonyad Farhand Iran Publications, 1966.

Rashed, Roshdi. "Al-Karaji (or Al-Karkhi), Abu Bakr Ibn Muhammed Ibn Al Husayn (or Al-Hasan)." In *Dictionary of Scientific Biography*, vol. 7. New York: Charles Scribner's Sons, 1972.

Smith, David E. *History of Mathematics.* 2 vols. New York: Dover Publications, 1958.

Swetz, Frank J. "Seeking Relevance? Try the History of Mathematics." *Mathematics Teacher* 77 (January 1984): 54–62, 47.

46

Omar Khayyam, Mathematician

D. J. STRUIK

Introduction

OMAR'S QUATRAINS, the *Rubáiyát*, are familiar to many of us through the exquisite translation by Edward FitzGerald, often adorned with delicate illustrations. Not all who like the *Rubáiyát*, either because of the verses' melodious English, their philosophy of life often akin to that of Ecclesiastes (or, if one prefers, of Epicurus), or their images of life under Sassanian or Seljuk kings in ancient Persia—not all these admirers are aware that their author, Abu-l-Fath Omar ibn Ibrahim Khayyam, of Nishapur in present North Iran, was also a distinguished philosopher, astronomer, and mathematician. As a philosopher, he was a follower of Aristotle, whom he interpreted with a keen rationalism; as an astronomer, he composed a calendar more accurate than that proposed centuries later by Pope Gregory XIII and now adopted by most people. Omar's mathematics has been brought closer to the English-reading world by the Kasir translation of his *Algebra*, published in 1931, although a French translation by Woepcke has existed since 1851 [1].* Recently a Russian translation has appeared, accompanied by translations of some other of Omar's scientific and philosophical writings, so that now we can obtain a somewhat more definite understanding of Omar's position in the history of thought.

Reprinted from *Mathematics Teacher* 51 (Apr., 1958): 280–84; with permission of the National Council of Teachers of Mathematics.

* Numbers within brackets refer to the Notes and References at the end of the article.

Omar's life is usually placed between 1040 and 1123, but these dates are not certain—let us say that he flourished in the period which witnessed the First Crusade. When FitzGerald, in 1859, published the first edition of the quatrains (an edition which once sold for a penny a copy and now has brought as much as $8000 a copy), he added a preface with a biography of Omar which contains most of what is known about the Persian poet's life. This biography has often been reprinted, and we need not repeat the details. It suffices to recall that Omar, after study at Nishapur, lived a thinker's quiet life near the court of the Seljuk sultans, first at Baghdad, then at Merv, not far from Nishapur and now inside the Soviet Union. He is buried at Nishapur, where the Iranian government, in 1934, erected a marble monument over his tomb.

By birth, Omar was a Tadjik and thus hailed from a region which had figured in antiquity as Bactria, and belonged to a people from which had sprung men of great eminence, such as the Persian poet Firdausi (ca. 940–1020) and the equally great scientist Ibn Sina (980–1037), also known as Avicenna. The influence of both of these men is visible in Omar's work. Another distinguished writer, a fellow student with Omar at Nishapur, was Nizam-al-Mulk, who became vizir at the Seljuk court and whose book on statesmanship, well known in Iran, was in 1949 republished in a Russian translation. Since the Tadjiks now have their own republic, one of the constituent republics of the USSR, academic interest in Omar, already existing in Iran, has increased in the Soviet Union. Popular interest, of course, has long existed, for

Omar's quatrains, like Firdausi's stanzas, have passed from mouth to mouth throughout the ages among the inhabitants of the regions near the borders of Iran, the USSR, and Afghanistan.

The Solution of Cubic Equations

Our understanding of Omar Khayyam as a thinker has lately been enhanced by the publication, in a Russian translation, of two mathematical, one physical, and five philosophical papers [2], followed by a book sketching Omar's life and works on the basis of a thorough knowledge of all sources presently available [3]. For this work we are indebted to Professors S. B. Moročnik, B. A. Rozenfel'd, and A. P. Juškevič.

Here we shall be primarily concerned with mathematical contributions of Omar. First of them is the *Algebra*, or, in full title, *On Demonstrations of Problems of Algebra and Almucabala*, in a translation from the same Leiden manuscript in the Arabic language which was used by Woepcke. (The Kasir translation is from a manuscript that was owned by Professor D. E. Smith of Teachers College, Columbia University.) Its interest for us, who already have access to the English translation, lies mainly in the extensive commentary, which supplements that given by Kasir; in it, 1079 is proposed as the year in which the book was finished. Those who have not read Kasir's translation may like to know that Omar, in his *Algebra*, extended the methods explained in Euclid's *Elements* (300 B.C.) for the solution of second-degree equations to those of the third degree, a considerable achievement.

Euclid's method was geometrical; for him a quadratic equation represented a problem on areas, so that the equation which we write $x^2 - ax + b = 0$ was interpreted as the problem of finding a square (side x) which added to a given area (b) would yield a rectangle of given side a and required side x. Our equation was thus interpreted in the form $x^2 + b = ax$, with a and b positive, and all three terms represented areas. Euclid therefore had to consider three types of quadratic equations: (1) x^2

$+ b = ax$, (2) $x^2 + ax = b$, (3) $x^2 = ax + b$, with only positive roots recognized, while our type, $x^2 + ax + b = 0$, does not occur. This ancient theory remained in force, with certain qualifications, until the days of Descartes (1637).

Khayyam's contribution was to extend this theory to cubic equations, in which he therefore had to think in terms of cubes and parallelopipeds, and in which again only positive roots had a meaning. Omar had thus to consider three binomial equations: $x^3 = a$, $x^3 = ax$, $x^3 = ax^2$; nine trinomial equations: $x^3 + bx^2 = a$, $x^3 + a = bx$, etc.; and seven tetranomial equations: $x^3 + cx^2 + bx = a$, $x^3 + cx^2 = bx + a$, etc.

Where Euclid could accomplish the actual construction of his unknown segment (our x) with compass and straight-edge [5], Omar had to use more advanced methods, and so, following certain special Greek examples (found, for instance, in the works of Archimedes), he obtained his solutions by the intersection of conics. His treatment of cubic equations in the particular sense in which he approached his problem was exhaustive, and he added to it the solution of certain other equations which led to quadratic and cubic equations, such as $1/x^2 + 2/x = 5/4$, $x + 2 + 10/x = 20/x^2$, and similar ones.

We used, above, the words, "with certain qualifications." Parallel with the geometrical theory of equations, which was based on the homogeneity of the terms and not easily generalized to equations of degree greater than three (the notion of a space of four dimensions only became familiar in the nineteenth century), there existed a purely arithmetical-algebraic theory. This theory already flourished in the Mesopotamia of the third millennium B.C., and considered equations as relations among numbers, asking for numbers satisfying the relations with as much precision as was possible or desirable. There was no natural limit at degree three, and in ancient Mesopotamia we even find transcendental equations such as $a^x = b$ (appearing quite naturally in compound interest problems). However, this theory lacked the finesse of Euclid's structure, which was designed, under the influence of Plato and Eudoxus, to deal with both commensurable and incommensurable quantities without concession to rigor.

The arithmetic-algebraic considerations continued to be cultivated through the ages. We meet them in the Hellenistic world despite all exorcisms of the Platonists, and they even reached lofty heights in the work of Diophantus (A.D. 250). This theory also met with success in India and China, and appeared again in Islamic Mesopotamia and Iran. Omar was conscious of this trend, and he refers to the difficulties of finding an algebraic solution of the cubic equation [6]. It was indeed not until the sixteenth century that this solution was found in its generality, and then this happened in Renaissance Italy, where the "Great Art" was first presented to a brave new world by Cardan in the year 1545.

The Binomial Theorem

Omar seems actually to have studied numerical solutions, if only of equations of the form $x^n = a$ (n a positive integer). He refers in his *Algebra* to a book he had written which dealt with this problem. The manuscript of this book has not (yet?) been found. It seems from the context that Omar must have known the formula for the expansion of the binomial $(a+b)^n$, where n is a positive integer. If so, then he has the prior claim to a result which up to recent years was usually credited to the Lutheran minister Michael Stifel (1544), but which already occurs in the works of Al-Kashi, a Persian living in the entourage of the Khan Ulugh Beg at Samarkand during the first decades of the fifteenth century.

The Parallel Axiom

The second of Omar's mathematical works now published in Russian, with an ample commentary, is entitled *Commentaries on the Difficulties in the Postulates of Euclid's Elements* [4]. The original Arabic text, from a Leiden manuscript, had already

been published at Tehernal in 1936 [5]. The content of this book was partly known to the English reader through an informative paper by D. E. Smith in the *Scripta Mathematica* of 1935; Professor Smith had found some quotations from Omar's text in an ancient manuscript he had bought [6]. Now that we have a chance to inspect the whole book, we see that this *Scripta* article gives a good account of the first section of Omar's book, the section in which he deals with the parallel axiom.

The parallel axiom, which states that through a point not on a line one and only one parallel can be drawn to this line, appears (in a slightly different form) in Euclid's *Elements* as the Fifth Postulate. All through the centuries courageous souls have tried to prove this postulate, that is to derive it from other more obvious postulates, or else to replace it by some principles which are less complicated. Omar, trying to connect the axiom with the Fourth Postulate, which asserts that all right angles are equal, tried to bridge the gap between the two postulates with the aid of five principles followed by eight propositions. The principles, or stepping stones in his reasoning, are taken, directly or indirectly, from Aristotle, who lived from two to three generations before Euclid. The first principle, for instance, is that quantities can be divided without end, that is, there are no indivisibles. The second principle states that a straight line can be indefinitely prolonged. The next two principles deal with intersecting lines, and the last one is the axiom of Archimedes.

The propositions contain some interesting material. In the first one, Omar proves that if two equal line segments AC and BD are drawn perpendicular to a segment AB, then angle ACD = angle BDC (Fig. 1). In the second one, he proves that if, in this same figure, E is the midpoint of AB, and EG is perpendicular to AB (Fig. 2), then $GD = CG$ and EG is perpendicular to DC. Then, in the third proposition, he comes to the conclusion that angles ACD and BDC are right angles, which indeed establishes the parallel "axiom," from which the other propositions, such as that $CD = AB$, follow.

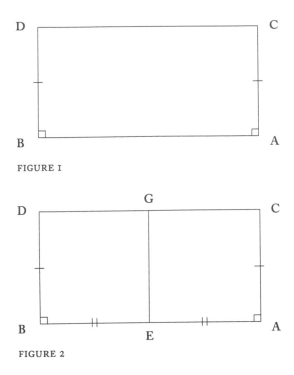

FIGURE I

FIGURE 2

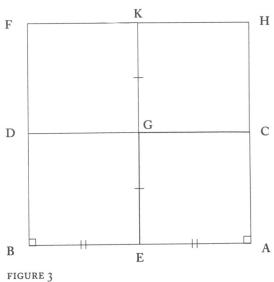

FIGURE 3

In order to prove the third proposition, Omar finds he must consider three cases according as angles *ACD* and *BDC* are each (a) less than a right angle, (b) larger than a right angle, and (c) equal to a right angle. The assumptions (a) and (b) he rejects by constructing *CDFH* on *ABDC* (see Fig. 3) by prolonging *EG* its own length to *K* and drawing *HKF* perpendicular to *GK* to intersect *AC* and *BD* prolonged in *H* and *F*. Then, by folding the figure first about *CD*, then about *AB*, he shows that in both cases (a) and (b) he comes into conflict with his principles (that is, in hidden form, with the Fifth Postulate), so that only case (c) remains.

The interesting aspect of this argument is that we find here, apparently for the first time in history, the three situations later known as the hypothesis of the acute angle (case [a]), that of the obtuse angle (case [b]), and that of the right angle (case [c]). These three situations are now known to lead, respectively, to the non-Euclidean geometry of Bolyai-Lobačevskiǐ, to that of Riemann, and to Euclidean geometry. The three cases were quoted by the Persian mathematician and astronomer Nasir

ed-din (1201–74), in whose work Professor Smith discovered them, and Nasir's studies on Euclid became known in 1594 through an Arabic, and in 1651 through a Latin, version. The Latin version was due to the well-known Oxford mathematician John Wallis, whose work was widely read. The terminology, "hypothesis of the acute, obtuse, and right angles," is due to Girolamo Saccheri (pronounced Sak´ keri), a Jesuit mathematician, some of whose ideas, published in a book of 1736 entitled *Euclid Freed of All Blemish*, anticipate later developments in non-Euclidean geometry. From here the terminology passed into our present books on this subject.

The Real Number Concept

Omar's book on Euclid contains still another contribution. After having discussed the Fifth Postulate, he passes to the theory of proportions. This theory appears twice in the *Elements*, first in Book V in geometrical form, developed on line segments independently of commensurability or incommensurability, and second in Book VII in arithmetical form, derived for "numbers" ("arithmoi"), which are always commensurable since they are defined as "composed of units." Two ratios, which we shall denote by *a/b* and *c/d*, are defined as equal if, whenever $ma<nb$ then also $mc<nd$, whenever $ma=nb$ then also $mc=nd$, and whenever $ma>nb$ then also $mc>nd$. Here m and n are any numbers (arithmoi). Omar is displeased with this definition, since it is impossible to try all numbers m and n; it also leads to difficulties when ratios are multiplied. Instead of Euclid's definition, Omar prefers to apply his first principle and to define equal ratios by what we might describe as something like a limit process—ratios are equal when they can be expressed by the ratio of integer numbers with as great a degree of accuracy as we like [7]. This amounts to the statement that a ratio can be expressed by numbers with any desired degree of accuracy.

We see that Omar is here on the road to that extension of the number concept which leads to the notion of the real numbers. For Euclid, with his "arithmoi" and their ratios, only rational numbers had arithmetical meaning. We see the Renaissance mathematicians further wrestling with this subject, especially Stevin (1585), though with him the number concept is still restricted to radicals. The full correspondence of arithmetical and geometrical continuum is reached by Descartes, although the exact definition of the real number had to await the work of Dedekind and Cantor in the nineteenth century.

Summary

Omar thus emerges as a mathematician of considerable penetration, who contributed to the theory of equations, to the understanding of the parallel axiom, to the development of the concept of real number, and probably to the generalization of the binomial theorem. Together with this emergence of Omar as a scientific figure has come a better understanding of his position as a philosopher and a literary figure [8]. The vague medieval figure, outlined by Woepcke in 1851 and by FitzGerald in 1859, has taken on flesh and blood. We now see before us a Persian philosopher of the school of Avicenna, representing the rationalist and anti-theological wing of the Aristotelians (this wing, to which also Averroes belongs, has been called "the Aristotelian left"), a professor of astronomy and mathematics who softens the hard hours of study and struggle by writing mildly ironic-blasphemous verse in the popular style and with a delicate sense of language. And although we know that his own verses in the *Rubáiyát* were quite different from those of his English interpreter both in spirit and in meter, let us sum up by paraphrasing a FitzGerald quatrain:

Ah, but my Meditations, people say,
Are idle and will perish with my Clay.
Yet, if their Fragile Vessel be well steer'd,
No wind can blow Oblivion in their way.

NOTES AND REFERENCES

1. D. S. Kasir, *The Algebra of Omar Khayyam*, Teachers College, Columbia University, New York, 1931, v + 126 pp. *L'Algèbre d'Omar Alkhayyami*, publiée et accompagnée d'extraits de manuscrits inédits par F. Woepcke. Paris, Duprat, 1851. There also exists another English translation, by H. J. J. Winter and W. Arafat, *Journal Royal Asiatic Society of Bengali Science* 16 (1950), pp. 27–77.

2. B. A. Rozenfel'd and A.P. Juškevič, "The Mathematical Tracts of Omar Khayyam, with Commentaries" (in Russian), *Istoriko: Matematičeskie Issledovanija* 6 (1953), pp. 1–172.

3. S. B. Moročnik and B. A. Rozenfel'd, *Omar Khayyam, Poet, Thinker, Teacher* (in Russian), Samarkand, Tadjjik-gosuzdat, 1957, 208 pp.

4. See R. C. Archibald, "Notes on Omar Khayyam (1050–1122)," *Pi Mu Epsilon Journal* 1 (1953), pp. 351–358.

5. *Discussion of Difficulties of Euclid by Omar Khayyam*, edited with an introduction by Dr. T. Erani, Teheran, 1936.

6. D. E. Smith, "Omar Khayyam and Saccheri," *Scripta Mathematica* 3 (1935), pp. 5–10.

7. On this see also E. B. Plooy, "Euclid's Conception of Ratio and His Definitions of Proportional Magnitudes as Criticized by Arabian Commentators." Thesis, Leiden, 1950, vi + 71 pp.

8. On Omar Khayyam as a poet, and an evaluation of the FitzGerald version, see A. J. Arberry, *Omar Khayyam*, Murray, London, 1952, 159 pp.

47

Omar Khayyam's
Solution of Cubic Equations

HOWARD EVES

OMAR KHAYYAM was the first to solve geometrically, so far as positive roots are concerned, every type of cubic equation. Let us illustrate his procedure for the special type $x^3 + b^2x + a^3 = cx^2$, where a, b, c, x are thought of as lengths of line segments. Omar stated this type of cubic rhetorically as "a cube, some sides, and some numbers are equal to some squares." Stated geometrically, the problem of solving the cubic equation is this: Given line segments a, b, c, construct a line segment x such that the above relation among a, b, c, x will hold. The object is to construct x using only straightedge and compasses as far as is possible. A solution using only straightedge and compasses is in general impossible, and at some point of the construction we must be permitted to draw a certain uniquely defined conic section.

A basic construction used several times in the solution of the cubic is that of finding the fourth proportional to three given line segments. This is an old problem whose solution was known to the ancient Greeks. Suppose u, v, w are three given line segments and we desire a line segment x such that $u{:}v = w{:}x$. Figure 1 will recall how, with straightedge and compasses, one may construct the desired segment x.

We now follow Omar's geometrical solution of the cubic equation $x^3 + b^2x + a^3 = cx^2$. First of all,

Reprinted from *Mathematics Teacher* 51 (Apr., 1958): 285–86; with permission of the National Council of Teachers of Mathematics.

FIGURE I

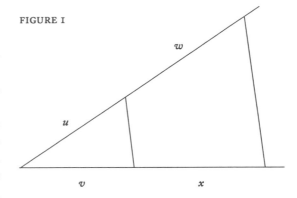

by the basic construction, find line segment z such that $b{:}a = a{:}z$. Then, again by the basic construction, find line segment m such that $b{:}z = a{:}m$. We easily find that $m = a^3/b^2$. Now, in Figure 2, construct $AB = m = a^3/b^2$ and $BC = c$. Draw a semicircle on AC as diameter and let the perpendicular to AC at B cut it in D. On BD mark off $BE = b$ and through E draw EF parallel to AC. By the basic

FIGURE 2

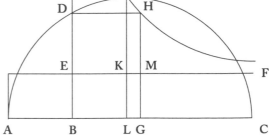

construction, find G on BC such that $ED{:}BE = AB{:}BG$ and complete the rectangle $DBGH$. Through H draw the rectangular hyperbola having EF and ED for asymptotes (that is, the hyperbola through H whose equation with respect to EF and ED as x and y axes is of the form $xy = $ a constant). Let the hyperbola cut the semicircle in J, and let the parallel to DE through J cut EF in K and BC in L. Let GH cut EF in M. Now:

1. Since J and H are on the hyperbola, $(EK)(KJ) = (EM)(MH)$.

2. Since $ED{:}BE = AB{:}BG$, we have $(BG)(ED) = (BE)(AB)$.

3. Therefore, from (1) and (2), $(EK)(KJ)$ $= (EM)(MH) = (BG)(ED) = (BE)(AB)$.

4. Now $(BL)(LJ) = (EK)(BE + KJ)$ $= (EK)(BE) + (EK)(KJ) = (EK)(BE)$ $+ (AB)(BE)$ (by[3]) $= (BE)(EK + AB)$ $= (BE)(AL)$, whence $(BL)^2(LJ)^2$ $= (BE)^2(AL)^2$.

5. But, from elementary geometry, $(LJ)^2 = (AL)(LC)$.

6. Therefore, from (4) and (5), $(BE)^2(AL)$ $= (BL)^2(LC)$, or $(BE)^2(BL + AB)$ $= (BL)^2(BC - BL)$.

7. Setting $BE = b$, $AB = a^3/b^2$, $BC = c$ in (6), we obtain $b^2(BL + a^3/b^2) = (BL)^2(c - BL)$.

8. Expanding the last equation in (7), and arranging terms, we find $(BL)^3 + b^2(BL) + a^3 = c(BL)^2$, and it follows that $BL = x$, a root of the given cubic equation.

It must be admitted that Omar's method is ingenious, and it certainly can interest some high school mathematics students. A point-by-point construction of the hyperbola (utilizing the basic construction) is easy, for if N is any point on EF and if the perpendicular to EF at N cuts the hyperbola in P, then $(EM)(MH) = (EN)(NP)$, whence $EN{:}EM = MH{:}NP$, and NP is the fourth proportional to the three given segments EN, EM, MH. In this way a number of points on the hyperbola can be plotted and then the hyperbola sketched in by drawing a smooth curve through the plotted points. The student might be given the numerical cubic $x^3 + 2x + 8 = 5x^2$. Here $a = 2$, $b = \sqrt{2}$, $c = 5$. The three roots of this cubic are 2, 4, −1. The student should be able to find the two positive roots by Omar's method; perhaps he can extend the method slightly to find the negative root. Here is the start of an excellent so-called "junior research" project. The student may find Omar's geometrical approach to other types of cubics in J. L. Coolidge, *The Mathematics of Great Amateurs*, Oxford (1950), Chapter II (Omar Khayyam), pp. 19–29. The algebraic solution of cubic equations can be found in any text on the theory of equations.

48

Lilavati, Gracious Lady of Arithmetic

FRANCIS ZIMMERMANN

$\mathcal{D}$URING the eighth century A.D., Arab scholars working in India on Sanskrit mathematical texts made two major discoveries which they developed and later transmitted to the Western world: the place-value notation of figures using the decimal system coupled with the "zero" concept, and a trigonometry which incorporated the use of sines.

It was not merely by chance that these major advances in the fields of writing, calculation, and triangulation were made by Indian mathematicians; they all touched on matters of traditional interest in India, whose scholars had always shown a particular taste and talent for grammatical forms.

Mathematics, like all the other scientific disciplines in ancient India, was subject to the constraints and stylistic forms of the Sanskrit language as well as to the demands of verse since most scientific texts were written in verse form.

The great mathematical treatises, written in Sanskrit, usually by a Brahman (a member of the highest, priestly caste), consisted of a basic, often cryptic, text made up of *sutras* or aphorisms, or else of verses that were learned by heart. A stream of prose commentaries explain the full meaning of these ancient texts, confirming that they were aphoristic in nature and deliberately conceived as summaries of a master's teachings, expressed in such a way as to stick in the memory of his pupils.

Edited from *UNESCO Courier* 51 (Nov., 1989): 18–21.

A simple problem taken from the twelfth-century Indian mathematician Bhaskara's Lilavati *('Arithmetic')*

Example of the reduction of fractions to a common denominator:

One-fifth of a swarm of bees flew towards a lotus flower, one-third towards a banana tree. (A number equal to) three times the difference between the two (preceding figures), O my beauty with the eyes of a gazelle, flew towards a Codaga tree (whose bitter bark provides a substitute for quinine). Finally, one other bee, undecided, flew hither and thither equally attracted by the delicious perfume of the jasmine and the pandanus, Tell me, O charming one, how many bees were there?

Let x = the number of bees

$$x = \frac{x}{5} + \frac{x}{3} + 3x\left(\frac{1}{3} - \frac{1}{5}\right) + 1$$

Reducing the fractions to a common denominator, we get

$$x = \frac{3x}{15} + \frac{5x}{15} + 3x\left(\frac{5}{15} - \frac{3}{15}\right) + 1$$

$$x = 15$$

Evidence of the early use of numbers, in the sense of graphic symbols, is to be found in inscriptions on stone or copper that have been studied by archaeologists—such as the numbers 4 and 6, for example, found in the Asoka inscriptions which date back to the third century B.C. They are very rarely to be found, however, in proper mathematical texts. Arabic numerals, so called because they were made known to the rest of the world by Arab authors, are, in fact, of Indian origin. In general, however, they were rarely used in Sanskrit texts, in which numbers are written out in full or symbolized by alphabetical codes. To be more precise, we must distinguish clearly between the basic texts, generally written in verse, and the prose commentaries which alone provide us with information about the way in which numbers were written down during the actual process of calculation.

Numbers were set out vertically over several lines, or at least this is what emerges from a commentary written by Bhaskara the Elder in A.D. 629 on the *Aryabhatiya*. Unfortunately, the life-span of Indian manuscripts was short, averaging only about three centuries. Written on paper or on palm leaves, they were a prey to mildew and insects. The manuscripts of Bhaskara's commentary that have come down to us are modern copies and cannot be taken as evidence of the manner of writing of more ancient times. The twelfth-century Bakhsali manuscript, which is probably the oldest document in this sense, shows calculations in Arabic numerals spread over several lines and contained in cartouches, or boxes, within the body of a mathematical text written in Sanskrit.

The absence of graphic symbols and numbers in the aphorisms and verses of the classic mathematical texts does not mean that symbolism is totally excluded, but that the symbolism employed is grammatical or rhetorical in nature. Thanks to the virtually limitless synonymical possibilities of the Sanskrit language, numbers are represented by literary phrases and metaphors. Thus, *nayana* (eye) or *bahu* (arm) are the names of the number 2. *Agni* (fire) means 3 (by allusion to the three Vedic forms

of ritual fire) and *adri* (mountain) means 7 (a reference to the seven mountains of India of Hindu religious geography). The Sanskrit words for "sky" or "space" stand for zero. The order of the figures making up a number is the reverse of modern numerical systems—for example, the number 23 would be written as *agninayana*.

This symbolism is useful for writing in verse form series of figures which today would be laid out in tabular form. In India, as elsewhere, the astronomical data in almanacs have for centuries been presented in columns of numbers. But this mode was an Arab invention and in the ancient Sanskrit texts numbers were presented in the form of a line or verse of poetry.

Another form of numerical symbolism frequently used in astronomical and mathematical texts is based on the Sanskrit alphabet. Several such systems exist. The *katapayadi* system, which is widely used in southern India, makes it possible for very large numbers and trigonometric tables to be expressed in the form of mnemonic words, aphorisms, or verses. The system is flexible enough to enable such numbers to be expressed in phrases which also have another meaning. For example, the priestly injunction *acaryavag abhedya*, which means literally "The word of the Master must not be betrayed," is coded writing of the number 1434160, a chronogram for the 1,434,160th day of the era of Kali, the day on which the philosopher Sankaracarya introduced certain reforms.

Did this poetic form of expression have an influence on mathematical reasoning? Was there some specific characteristic, something special in the way of thinking or the social status of Indian mathematicians which led them to shape their teachings in a literary mould?

There has never been a caste or even a real school of mathematicians in India. Mathematicians, if we class as such those who wrote or used Sanskrit texts dealing with geometry, arithmetic, or algebra, worked in fairly close collaboration with experts in Vedic and Brahmanic ritual. Brahmans or members of a high caste and steeped in

Sanskrit culture, they were classed among scientists as *jyotirvid*, or "experts on stars." Mathematical texts were usually inserted in treatises on astronomy, and trigonometry only really came into its own in the study of the angular distances between stars.

Like all the Brahman sciences (the *sastra*), mathematics had been developed primarily for religious purposes, as an aid to the proper performance of ritual. We know nothing of the life of the great Indian mathematicians, but we can picture fairly accurately the ritualistic and scholastic setting within which they worked, so heavily coloured by it is the style of Sanskrit texts. After the pupils had memorized a text by repeating it word for word, time and again until it was "clasped to their bosoms" (until they had learned it by heart), the teacher would provide them, orally, with the illustrations, demonstrations and calculations that the text concealed. This was the key that opened up the paths of knowledge and an instrument of spiritual fulfillment.

The *Lilavati*, a text by the twelfth-century mathematician Bhaskara the Learned, was traditionally used for this purpose for arithmetic and this explains why it terminates with a verse with a double meaning in which Bhaskara compares his *Lilavati*, his "gracious one" (it also means arithmetic), to a woman endowed with all the graces of the *jati* (a word which means both of noble lineage and, in a technical sense, the reduction of fractions to a common denominator). "Joy and happiness in this world shall continually increase for those who hold her *kanthasakta*, close in their arms or clasped to their bosoms (learned by heart by repetition)."

From Ritual Geometry to Bhaskara's Treatise

The oldest of these texts that have come down to us are the *sulbasutras*, "maxims concerning measuring cords," which are believed to have been written between the fifth and the first century B.C. They are treatises which lay down the rules for building altars used for Vedic ritual sacrifices and

made of bricks laid in accordance with symbolic forms. The geometric constructions which are taught in these treatises are based on knowledge of several special cases of right-angled triangles (for example, with sides measuring 3–4–5, or 5–12–13, 7–24–25, etc.) and on the general rule according to which "the diagonal of a rectangle produces (by the construction on it of a square) the equivalent of the product of both the rectangle's length and breadth; and the diagonal of a square produces (by the construction on it of a square) twice its own area." Here, however, the rule is not expressed as a theorem, but as a maxim, as a formula to ensure the proper functioning of ritual and as a building guideline. The word *sutra* itself, which at first meant "aphoristic in style," came, in the later treatises, to mean a "rule" in the technical sense of a rule for building.

There are no theorems in Indian mathematics, only rules based on reasoning which starts from an intuitive point of departure. The rules, aphorisms, and the mnemonic verses of the basic texts are not the outcome of a demonstration but rather guidelines for a geometric construction to be carried out by the reader or the commentator. Even in algebra, the typical line of reasoning links areas to the products of factors and implies the construction of a geometric figure.

It has often been said that the Indians were algebraists rather than geometers yet, in fact, throughout all the commentaries on the teachings of Aryabhata (sixth century), Brahmagupta (seventh century), and Bhaskara (twelfth century), geometry was the source of the practical applications of the rules of arithmetic and algebra. A geometric space and a numerical ensemble were taken together as two facets of the same reality. The algebraic solution was grafted on to the geometric construction. To demonstrate was to display the solution, to render it intuitively manifest. As one commentator said: "A demonstration by quantities should be made for the benefit of those who do not understand the demonstration by areas." Thus, in Indian mathematics, to reason is to explain an intuition.

49

The Ganita-Sāra-Sangraha of Mahāvīrācārya

S. BALAKRISHNA AIYAR

*I*N THE HISTORY of human civilization, two countries are prominent in learning and culture—Greece in the West and India in the East. It is to the latter that our attention is now to be directed. History records the great heights achieved by the Hindus in the realm of mathematics. Whatever might be the reason for the stagnation in the stretch of time between the end of the twelfth and the end of the nineteenth centuries, there is no gainsaying the fact that for better than a thousand years prior to this period of stagnation, India produced mathematicians of high rank and solid achievement in many of the branches of that science.

There is, however, considerable speculation and uncertainty regarding the exact dates and periods of early Hindu mathematical achievement. According to G. R. Kaye,[1] three periods can be broadly indicated within which the mathematical literature activity can be spread:

(1) The *Vedic* and *S'ulvasūtras period*, ending about A.D. 200

(2) The *astronomical period*, A.D. 400 to 600

(3) The *Hindu mathematical period proper*, A.D. 600 to 1200

The chief centers of development in Northern India were Ujjain and Pātalīputra (Patna). Down south, the tract of country in South India compris-

ing Mysore and the neighboring districts was inhabited by people whose mother tongue was Kannada. About the middle of the third period referred to above, the Rāshha Kūta line of kings were ruling over the area. To the court of one of them, King Amōghavardana Vripatunga (A.D. 814–877), was attached a man of great learning, the astronomer and mathematician, Mahāvīrācārya. Chronologically, therefore, he belongs to the time between two of the other great Hindu mathematicians, Brahmagupta (ca. 628) and Bhāskarā-Chārya (ca. 1150), which on all accounts is regarded as a fairly accurate conclusion.

Mahāvira was a Jain by religion. The study of mathematics was very popular among the Jain scholars. Indeed it was accorded the status of one of the four *anuyogās*, which were the auxiliary sciences, the study of which helped the aspirant to the attainment of soul-liberation. Mahāvira's claim to fame rests on the firm foundation of his great work, the *Ganita-Sāra-Sangraha*. Its name indicates its technical nature, which is the usual characteristic of books which summarize knowledge of a subject up to the time of its production. The work seems to have been popular in his part of the country and much appreciated by the students and lovers of mathematics.

It is to be noted, however, that, in spite of the technical nature of its content and development, and recognition of its practical helpfulness, it is valued more for its historic significance than as a mathematical treatise. A comprehensive and schol-

Reprinted from *Mathematics Teacher* 47 (Dec., 1954): 528–33; with permission of the National Council of Teachers of Mathematics.

arly edition of the treatise was brought out by the late professor M. Rangācārya, Professor of Sanskrit of the Presidency College, Madras, with an accurate translation and elaborate notes.[2] It is included in the bibliography of books on the history of mathematics in *The Mathematics Teacher,* with the comment that it is "the greatest contribution made in a century to the history of Hindu mathematics. A translation with notes of an extensive work of the ninth century."[3]

By and large, the *Ganita-Sāra-Sangraha* is regarded as the most scholarly treatise in Hindu mathematics. A casual comparison with the works of Brahmagupta and Bhāskara brings out a general similarity of topics with appreciable differences. These differences pertain to methods, rules of work and much of the illustrative material, while similarity is discernible in the general spirit that inspires and pervades their major topics and many problem situations.

The language is elliptical and riddle-like. As F. Cajori says, "The Indians were in the habit of putting into verse all mathematical results they obtained, and of clothing them in obscure and mystic language which, though well adapted to aid the memory of him who has already understood the subject, was often unintelligible to the uninitiated."[4]

The treatment is clothed in verses employing many of the popular metrical modes which, by their variety, relieve monotony and appeal to our aesthetic sense.

Coming to the subject matter itself—Mahāvīrācārya being a staunch Jain by religion, the opening invocation is addressed to Lord Jina. He is extolled as the shining lamp of the knowledge of numbers (*Sankyā-gnāna-pradīpa*), by which is illuminated (*prakāsitam*) the whole universe. In fact, stanzas in praise of Jinadeva occur at the beginning of almost all the sections of the treatise.

After the invocation follows an appreciation of the science of calculation (*ganita-sāstra-prasamsa*). There is no aspect of life and no phenomenon in this world and in all the worlds that is free from the control and application of the principles of

mathematics. After a comprehensive enumeration of these aspects, the author concludes, "What is the good of saying much in vain? Whatever there is in all the three worlds which are possessed of moving and non-moving things—all that indeed cannot exist as apart from measurement."[5]

With an exuberance of poetic imagination characteristic of many Hindu authors, the divisions of the book find enumeration thus: *Ganita-Sāra-Sangraha* is a vast ocean. Terminology (*samguya*) provides its waters, which stand impounded by the eight arithmetical operations (*parikarma*). Fractions and operations with them (*kalāsavarna*) are its innumerable rolling fish, and crocodiles abound in the shape of miscellaneous examples (*prakīrnaka*). Rule of three (*thri-rāsika*) lashes up its waters into waves, and mixed problems (*misraka*) are the gems which by their navigated splendor impart luster to the deep. Its extensive bed itself is formed of area problems (*kshetra-vis-thīrma*) with cubic measurements (*Khāta*) providing mounds upon mounds of sand. Shadow reckoning and connected references to astronomical matters (*chāya*) are the advancing tides. Into this vast ocean of the Sangraha[6] should the arithmeticians dive to gather in abundance the pure gems of their desire.

Indeed, so riotous has been the author's imagination in some places that it oversteps the limits of discretion and introduces situations that call forth our amusement. Particularly in the applications of the principles and rules, there is a strange admixture of the significant and the fanciful. Here is a sample: "A well completely filled with water is ten *dandas* in depth; a lotus sprouting up therein grows from the bottom at the rate of 2 ½ *angulas* in a day and a half; the water flows out through a pump at 2 ½ angulas in 1 ½ days; 1 ⅕ angulas of water are lost in a day by evaporation owing to the rays of the sun; a tortoise below pulls down 5 ¼ angulas of the stalk of the lotus plant in 3 ½ days. By what time will the lotus be on the same level with the water in the well?"[7] (N.B.: 6 angulas = 1 foot, 96 feet = 1 danda.)

In the caustic words of the Arab mathematician, Alberuni, there is to be found in Hindu mathematics and astronomy a mixture of "pearl shells and sour dates, or of pearls and cow dung, or of costly crystals and common pebbles."

Then Mahāvira steps aside to indicate who should be regarded as properly equipped for studying the science of mathematics. It is not to be easily undertaken by all and sundry. "By means of the eight qualities, viz., quick method of working, forethought as to whether a desirable result will be produced, or an undesirable result arrived at, freedom from dullness, correct comprehension, power of retention, and the devising of new means of working, along with getting at those numbers which made (unknown) quantities known—(by means of these qualities) an arithmetician is to be known as such."[8]

Then follow the enumeration and definition of units and the relations among them, tables of weights and measures as we call them. The terminology (*paribhāsha*) includes space (*kshetra*), time (*kāla*), and other entities like quantities of grain, gold, silver, etc.

Mahāvira enumerates 1 to 9 and zero *by name*. In the ascending scale of integers, twenty-four rank-names are given, beginning with one (*éka*) and mounting up to 10^{23} (*mahākshôbhe*, a number formed by 1 followed by 23 zeros), each of which is thus ten times the preceding one.

Nominal numerals are also used. They are the names of objects and object-groups, suggesting the particular manynesses of the corresponding numbers in view. For example, *moon* stands for one, *oceans* for four, *Rudras* for eleven and so on. Zero (*sūnya*, void) is indicated by names which are synonymous with the sky (*akāse, gagara*). The use of such notation made it possible to represent a number in several ways. This greatly facilitated the framing of verses containing arithmetical rules and scientific constants, which could thus be more easily remembered.

In addition to the indicating of the digits by pure number names and nominal numerals exclusively, there is the additional device of combining the two modes in the same sentence, which is adopted obviously for the purpose of "securing metrical convenience and avoidance of cumbrous ways of stating numbers."

The decimal system is followed. The digits are set down in two ways, from left to right or right to left beginning with units, the order being determined by the student from the context largely.

The eight operations dealt with as fundamental are multiplication (*gunakāra*), division (*bhāgahāra*), squaring (*krīti*), square root (*vargamūla*), cubing (*ghana*), cube root (*ghana-mūta*), series (*chiththi* or *samkalita*) and subtraction of parts of series (*sesha* or *vyuthkalita*).

The rules of work (*karana sūtras*) regarding these are given, but the several steps are not detailed. In this sense, the name of the treastise is significant; it is a *sangraha*, a synopsis of the essentials of mathematics. Alternative methods and different applications are also provided. "Although the great Hindu mathematicians doubtless reasoned out most or all of their discoveries, yet they were not in the habit of preserving the proofs, so that the naked theorems and processes of operation are all that have come down to our time."[9]

The operations of addition and subtraction receive no separate attention for the apparent reason that they are obvious and simple.

Three alternative rules of multiplication are stated, two of which are obviously to facilitate working through choice of suitable factors. These rules are: "(1) After placing (the multiplicand and the multiplier one below the other) in the manner of the hinges of a door, the multiplicand should be multiplied by the multiplier, in accordance with (either of) the two methods of normal (or) reverse working, by adopting the process of (i) dividing the multiplicand and multiplying the multiplier by a factor of the multiplicand, (ii) of dividing the multiplier and multiplying the multiplicand by a factor of the multiplier, or (iii) of using them (in the multiplication) as they are (in themselves)."[10]

The product-patterns are often given fanciful names. Thus: "In this (problem), 12345679 multiplied by 9 is to be written down; this (product) has been declared by the holy preceptor Mahāvira to constitute the necklace of Narapala."[11]

The bare rules of division are given, thus: "Put down the dividend and divide it, in accordance with the process of removing common factors, by the divisor, which is placed below that (dividend), and then give out the resulting (quotient)" and, "The dividend should be divided in the reverse way (i.e., from left to right) by the divisor placed below, after performing in relation to (both of) them the operation of removing the common factors, if that be possible."[12]

The division problem situations refer mainly to distribution of coins, gems, fruits, etc., among men and temples.

The next four sections on squaring, square root, cubing, and cube root are more or less alike in treatment. They begin with bare statements of rules of work, with five or six stanzas embodying exercises for application. Thus the rule for squaring numbers is: "The multiplication of two equal quantities: or the multiplication of the two quantities obtained (from the given quantity) by the subtraction (therefrom), and the addition (thereunto), of any chosen quantity, together with the addition of the square of that chosen quantity (to that product): or the sum of a series in arithmetical progression, of which 1 is the first term, 2 is the common difference, and the number of terms wherein is that (of which the square is) required: gives rise to the (required) square."[13]

Many of these rules, if set in modern algebraic notation, bear resemblance to formulas with which we are familiar. In modern notation the rules for squaring a number are:

(i) $a \times a = a^2$;

(ii) $(a+x)(a-x) + x^2 = a^2$

and

(iii) $1+3+5+7+$ to a terms $= a^2$.

The next two sections on series give working rules for finding the sum, number of terms, common difference and first term of an arithmetical progression and the first term and common ratio of a geometrical progression. The exercises that follow are of a mechanical nature.

With regard to operations with zero, we gather that (1) a number remains unchanged when it is combined with or diminished by zero; (2) a number multiplied by zero is zero; (3) every number divided by zero is left unaltered. This last assertion is apparently based on the notion that "division by zero" means that there is no number to divide by, and the original number (dividend) is therefore left alone.

The rules for the product of negative numbers are also stated.

Fractions receive a relatively more elaborate treatment than integers. They are of six types: simple (bhāga), fraction of a fraction (prabhāga), complex fraction (bhāgabhāgya), associated fractions (bhāgyabandha), dissociated fractions[14] (bhāgāpavāha), and combinations of the above varieties (bhāgamātra).

The eight fundamental operations are equally applicable to integers and fractions. The statements of them are often longwinded and complicated. There is practically no attempt to obtain generalized procedures. Each type of fraction is treated on its own. The exercises involve mostly abstract numbers and include no recreational variety as in the case of integers.

The section on miscellaneous problems envisages ten important varieties. They relate to fanciful types and a good number of them are very complicated, so much so, that all the ingenuity of the reader is required to unravel and set down the data figures and the relations among them in a manageable analytical form to aid manipulation and solution, with the help of problem-solving techniques employing equations. Many problems involve squares and square roots of collections (numbers) of things, wholes or parts thereof and intermediate remainders, and solutions can without exaggeration be regarded as computational and manipulational gymnastics.

The rule of three[15] includes direct and inverse varieties of ratio relations and compound proportions involving single, double, treble and quadruple combinations. The problems given are often fanciful as is illustrated by the following: "A powerful unvanquished excellent black snake, which is 32 hastas in length, enters into a hole (at the rate of) 7 1/2 angulas in 5/15 of a day; and in the course of 1/4 of a day its tail grows by 2 3/4 of an angula. Oh ornament of arithmeticians, tell me by what time this same (serpent) enters fully into the hole."[16]

Very complicated applications of the rule of three, involving investments problems, barter and exchange of commodities are given. For example: "Twenty horses, (each) of 16 years (of age), are worth 100,000 gold coins. Oh leading arithmetician, say how much 70 horses, (each) of 10 years (of age), will be (worth) at this (rate)."[17]

Proportionate division of proceeds and profits are then dealt with and involve ratios with integral and fractional terms. They lend themselves to be treated with the help of simple, simultaneous, quadratic and radical equations. We also have a large collection of problems on calculations relating to transactions with gold.

Then there are problems to test logical argument and validity or otherwise of conclusions. Others refer to progressions, time and distance, relative velocity, time and work, and so on.

The rules on areas in this chapter relate to the mensuration of triangles (equilateral, isosceles and scalene), quadrilaterals (squares, rectangles, trapeziums, etc.), circles and their parts, ellipses, conchoidal forms (*kambuka vritta*), areas of concave and convex spherical surfaces, annular areas, segments of circles, and other complex miscellaneous configurations like surfaces of grains (*yava*), drum (*muraja*), trumpet (*panava*), etc. Patterns built up of triangles, squares, and circles are offered for area finding.

In this connection it will be of interest to note the rule for finding the area and circumference of a circle. "The (measure of the) diameter multiplied by three is the measure of the circumference; and the number representing the square of half the diameter, if multiplied by three, gives the (resulting) area in this case of a complete circle."[18]

After dealing with relatively simple cases, the author passes on to the consideration of what he styles as devilishly difficult problems. They require the construction and use of involved formulas. As an example: "In the case of an equilateral quadrilateral figure, the (numerical) measure of the diagonal is equal to (that of) the area. What may be the measure of (its) base?"[19]

Excavations (volume) come next. The sides of pits and cavities illustrate a variety of geometrical shapes. Rules are stated for finding the cubic contents of solids bounded by spherical surfaces (*golakasa-kshetram*), cylinders, triangular pyramids, piles of bricks, walls of different shapes, broken walls, water flowing through pipes, etc. Examples are appended.

We have already seen that Mahāvira was a notable astronomer, like the other great mathematicians of India. The section on shadows furnishes some mathematics for many astronomical problems. The rules relate, among other things, to finding directions relative to the due East, shadow reckoning with the gnomon (sundial) consisting of a circle with a style planted at the center, lengths of shadows near sunrise and sunset and at intermediate times, equatorial shadows (when sun passes through the First Points of Aries and Libra), shadows at places where there are no shadows. Many problems remind us of the common types or heights and distances found in our textbooks. For example: "The shadow of a pillar, 12 hastas (in height), is 24 hastas in measure. At that time, Oh arithmetician, of what measure will the human shadow be?"[20]

This brings us to the end of a brief survey of an outstanding work on Hindu mathematics.

Professor D. E. Smith, American mathematician and historian of mathematics, in his valuable and scholarly introduction to Prof. Rangācārya's edition of the Ganita-Sāra-Sangraha says, ". . . oriental mathematics possesses a richness of imagination, an interest in problem solving and poetry, all of which are lacking in the Treatises of the west, although abounding in the works of China and Japan."[21] The Arabs were particularly attracted by the poetical, the rhetorical and the picturesque in the Hindu treatment of the subject, more than by the abstract approach.

NOTES

1. G. R. Kaye, *Indian Mathematics* (Calcutta: Thacker, Spink and Co., 1915), p.3.

2. M. Rangācārya. *The Ganita-Sāra-Sangraha of Mahāvīrācārya* (Madras: Government Press, 1908).

3. "A Brief List of Mathematical Books Suitable for Libraries in High Schools and Normal Schools," *The Mathematics Teacher*, Vol. XVIII, p. 481.

4. Florian Cajori, *A History of Mathematics* (New York: The Macmillan Company, 1919), p. 83.

5. M. Rangācārya, *The Ganita-Sāra-Sangraha of Mahāvīrācārya* (Madras: Government Press, 1912), p.3.

6. Sangraha means a synopsis of the essentials of mathematics.

7. M. Rangācārya, *op. cit.*, p. 89.

8. *Ibid.*, p. 8.

9. Cajori, *op. cit.*, p. 83.

10. M. Rangācārya, *op. cit.*, p. 9.

11. *Ibid.*, p. 10.

12. *Ibid.*, p. 12.

13. *Ibid.*, p. 13.

14. *Editor's Note:* There are two kinds of associated fractions according to Mahāvīrācārya. The first is a fraction associated with integer, i.e., a mixed number. The second consists of fractions associated with fractions, e.g., $1/2$ associated with $1/2$ means $1/2 + 1/2$ of $1/2$.

There are also two kinds of dissociated fractions. The first is the difference between an integer and a fraction. The second consists of fractions dissociated from fractions, e.g., $2/5$ dissociated from $1/6$ means $1/6$ of $2/5$ is to be subtracted from $2/5$.

15. *Editor's Note:* The rule of three is simply a mechanical device for solving proportions in which one of the four quantities is unknown. Thus a proportion which in modern notation would be $a : b : : c : x$, would be solved by writing $a—b—c$ and the rule is to multiply the second and third numbers and divide by the first. In more complicated cases such as $a : b : c : : x : d : e$ the numbers were written

and the lines indicated that the product of a, d and c were to be divided by the product of b and e. See Vera Sanford, *A Short History of Mathematics* (Boston: Houghton Mifflin Company, 1930), p. 162–163.

16. M. Rangācārya, *op. cit.*, p. 90.

17. *Ibid.*, p. 91.

18. *Ibid.*, p. 189.

19. *Ibid.*, p. 221.

20. *Ibid.*, p. 278.

21. *Ibid.*, p. 24.

π in the Sky

Prediction of Celestial Phenomena Was a Mainspring of Early Chinese Mathematics

JEAN-CLAUDE MARTZLOFF

"CHINESE mathematics," defined by the Chinese themselves in ancient times as the "art of calculation" (*suan chu*), covers a vast range of practices and currents of thought that grew up in China between the first millennium B.C. and the fall of the Manchu dynasty in 1911. After that date Chinese mathematics became westernized and the traditional practices were virtually impenetrable to those without specialized training.

Divination, Astronomy, and Mathematics

Although writing already played an important role in China at the time of the composition of the earliest canonical writings (the "classics" which were always so essential in the training of intellectual élites), mathematics was not then considered to be a body of knowledge that it was necessary to record in specialized texts.

It seems, however, that mathematics was instrumental in the emergence of what the sinologist L. Vandermeersch has judiciously called "divinatory rationality." Beginning with oracles in the form of tortoise shells, animal bones, or the leaves of the Chinese yarrow, the prognostics made using this form of divination were based on the interpretation of all manner of natural signs, especially meteorological and astronomical phenomena such as rainbows, halos, winds, meteors, positions and conjunctions of stars, eclipses, and sunspots. This magical worldview did not however exclude recourse to purely rational investigations. The diviners attempted with some success to fit all their observations into numerical and arithmetical schemes devised for the recording of memorable past phenomena and the prediction of certain recurrent events. Some predictions of periodically recurring celestial phenomena began to be confirmed by observation, and led to the birth of the calendar and mathematical astronomy.

Each successive dynasty in China made its mark by introducing a new system for computing the calendar so that the historical events recorded by annalists could be reassessed and future events predicted. The ruling class therefore needed qualified personnel who specialized in calendrical and astronomical calculations. A body of imperial chronologers thus gradually came into being, serving both as historian-annalists and astronomer-calendrists.

The continuing requirements of Chinese dynasties down the ages explain why the search for appropriate methods of predicting the most visible celestial phenomena (conjunctions of stars, occultations, eclipses) was foremost in the minds of mathematicians.

However, as the astronomer-calendrists of imperial China usually had a lowly social status, and

Edited from *UNESCO Courier* (Nov., 1989): 22–28.

as their knowledge was handed down from father to son, their activities were often denigrated and looked upon simply as a static way of upholding tradition.

The overwhelmingly dominant factor in the history of Chinese mathematical astronomy is the extraordinary durability of conflicts between rival schools. From the beginning of our era to the sixteenth century, the Chinese calendar underwent no less than fifty reforms. However, these conflicts turned out to be more constructive than destructive in that they were almost always settled by the concordance between observed reality and the predictions.

Unfortunately, very few works devoted exclusively to mathematical astronomy have survived. Monographs written by non-specialists, abridged versions of which were included in the annals of successive dynasties, are practically all that remain.

The Social Background of Chinese Mathematics

During the Han dynasty (206 B.C. to A.D. 220), another branch of mathematics appeared, this time recorded in specialized manuals. Collections of problems and their solutions were organized in chapters according to their practical applications. The descriptions in these texts are often so detailed and realistic that it is almost possible to reconstruct from them entire chapters of Chinese social and economic life at a given period. Whether they are about tax collection, corvée labour, weights and measures, currency, the construction of dykes or canals, the management of manpower, land or river transport, policing or military logistics, no practical detail was neglected. Generations of "civil servant-mathematicians" were trained for the imperial bureaucracy by the study of such manuals.

The Tang dynasty (A.D. 618 to 907) set up an examination system whose syllabus included mathematics as well as letters. Although least importance was accorded to mathematics, study of the subject at the *Guozijian*, the "College of the Sons of the State," lasted seven years and was based on a classic textbook, "The Ten Mathematical Manuals" (*Suanjing shi shu*). In 1084, under the Southern Sung dynasty, this text was printed. Yet after 1230 mathematics was banished from the examination syllabus for good, in favour of literary subjects.

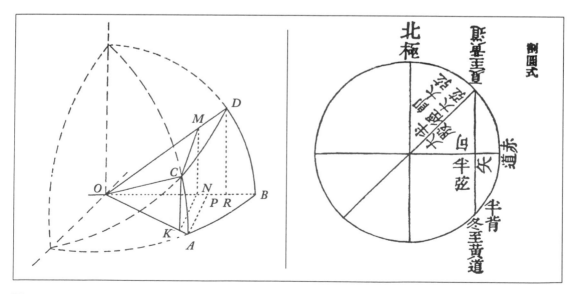

Diagrams illustrating problems in sperical trigonometry examined by the astronomer-mathematicians Kuo Shou Ching (1276, left) and Hsing Yūn Lu (1600, right).

Paradoxically, the high-water marks of mathematics are not to be found in the short periods when the teaching of the subject was institutionalized. On the contrary, the most remarkable advances took place when the empire was divided, in time of war and bureaucratic breakdown. During the Warring States period (453 to 222 B.C.), the Moists (followers of Mo-tzu who challenged the dominant Confucian ideology) developed the rudiments of theoretical geometry, but their method had little influence on the main current of mathematical thought. The greatest mathematician of ancient China, the details of whose life are unfortunately unknown, was Liu Hui, who elaborated several mathematical theorems at the time when China was split into three kingdoms (A.D. 220 to 265). During the Mongol conquest in the thirteenth century, a wealth of original achievements took root, flowered, and were quickly consigned to oblivion.

The last and greatest of the Sung mathematicians was Zhu Shijie (fl. around 1300), a wandering scholar who travelled far and wide teaching the results of his research. Also during these troubled times, the hermit Li Zhi (1192–1279) organized a private group devoted to the study of the mysteries of the world and of numbers, which led to the development of algebra in China. And in the nineteenth century, shortly after the Opium War, when the Chinese empire was overtaken by major disasters, Li Shanlan (1811–1882) failed his literature examinations and dedicated himself to the joys of mathematics. He discovered an astonishing series of perfectly correct formulae which were to baffle even a twentieth-century mathematician of the calibre of Paul Turan of Hungary.

The reasons behind these sudden and short-lived flashes of brilliance are poorly understood. Perhaps political anarchy freed the intellectual élite from the obligation of preparing the sterile mandarinate examinations. Confucianism, the dominant ideology, relegated mathematics to a minor role. Those who were attracted by the subject could therefore give free rein to their intellectual curiosity without great risk. Even in stable times, however, many Confucian scholars used their leisure hours to study mathematics. From the eighteenth century onwards, more and more philologists and specialists in textual criticism dreamed of turning mathematics into a branch of history. Some of them tried their best to use mathematical astronomy to authenticate the classics. They wanted to define by calculation the historical reality of natural events recorded in ancient documents (especially eclipses of the Moon or the Sun). Others enthusiastically studied "concrete sciences" such as economics, hydraulics, civil engineering or architecture. Out of these excursions along lonely and difficult paths there sometimes grew a taste for pure mathematics, independent of any particular application.

Contact with other cultures is also of importance in the history of Chinese mathematics: with India during the first millennium A.D. as Buddhism was being preached abroad; with the Arab and Persian peoples at the time of the Mongol expansion; and with European missionaries from the late sixteenth century onwards. Although many similarities can be seen between mathematics in China and in other cultures before 1600, it is unclear whether this is a sign of actual influence or simply of parallel development.

For example, the Chinese symbol for zero, which was written as a small circle as it is today, appeared for the first time around 1200 in astronomical tables and could have been of Indian origin. A number of ancient and medieval mathematical games, whether from the Hellenistic or the Arab world, India, Europe or China, have a surprising family likeness. Very similar mathematical techniques existed at the same period in China and in Greece. To cite but two examples, Euclid and Liu Hui both discovered methods of determining the volume of a pyramid; Archimedes and Liu Hui found the volume of a solid by the intersection of two orthogonal cylinders.

Even if these parallels do indicate the transmission of influences, Chinese mathematics had an internal coherence and developed independently.

Geometry without Parallel

In traditional Chinese mathematics there is no geometrical reasoning based on axioms, postulates, definitions, and theorems. There are no absolute truths in the manner of Euclid, but only relative and provisional truths. In geometry, there are no angles nor even any parallel lines, but only lengths, areas, and volumes. Neither is there any algebra in the Arab style, nor the search for the roots of equations using radicals or intersecting curves. And there is no "rhetorical" algebra enshrined in prose or verse.

Yet Chinese mathematics is based not on purely empirical procedures but on heuristic principles. The emphasis is on the answer itself, rather than on explaining the intermediate stages in getting there, which are taken for granted. For example, one of the cardinal rules of Chinese geometry stipulates that the surface or the volume of a figure remain the same after it has been broken up and re-assembled, even though the number of the fragments is potentially infinite. This type of principle does not at all exclude the recourse to an axiomatic system, but it is true that Chinese geometric figures were not usually abstract ideals. They were, rather, like tangible pieces of a puzzle that can be distinguished by their colour and manipulated at will. Chinese geometry basically relies on a shrewd and meticulous examination of a problem to reveal certain results. This method is not only important in the calculation of areas and volumes, but also in bringing out certain properties of the right-angled triangle, in calculating the total of a series, in resolving equations or systems of equations, and in the visual proof of mathematical identities.

What is more, Chinese geometry sees nothing wrong (as Euclid would have done) in resorting to calculations or any other method that may be useful to solve a given problem. This attitude shows Taoist influence, and Chinese mathematicians of the third to the fifth centuries A.D. declared their boundless admiration for Zhuangzi, the father of philosophical Taoism. This eccentric sage had rejected the use of language as a notably effective way of reaching the truth, on the grounds that the fallacious arguments of the sophists showed its limits.

He had concluded that all attempts to reach the truth founded on discursive argument, that is to say on a method capable of engendering manifestly false conclusions, would be doomed to failure. Hence the tendency of Chinese mathematicians, influenced by Taoism, to place only limited faith in the power of language. On the other hand they tended to make use of all the means at their disposal, without ever ignoring the evidence of their senses. This is why they were so fond of calculations and all kinds of manipulations, so long as they could avoid putting them into words. Only as a last resort did they have recourse to discursive arguments.

How could mathematicians apparently so obsessed by the concrete have produced sophisticated results? In fact the practical nature of Chinese mathematics does not imply any lack of abstraction in their thinking. On the contrary, some of the results obtained from manipulating their puzzles presuppose great ingenuity and a remarkable capacity for abstraction.

Moreover, Chinese mathematicians often deliberately distorted reality, as they could not easily teach their subject using as examples the increasingly complex problems of daily life. This is why many Chinese mathematical problems conceal purely fictitious situations behind a veneer of practical reality: values which are impossibly large or small, or that make no sense (fractions of men); data arranged in an arbitrary manner, as when areas are added up with volumes and prices; inversion of known and unknown values as when calculating the dimensions of objects from their volume, or capital from the interest gained, or quantities of goods from the share received by each person. Obviously such methods opened up very much more interesting situations from the mathematical point of view.

Chinese algebra developed on this "fictional territory." In the oldest manuals are many calculatory formulae for resolving limited classes of problems. In extreme instances, each problem constitutes a specific case. Later, generalized procedures appear by which an increasingly wide range of problems could be solved, so fictitious situations were no longer required.

All this could not have taken place without Chinese computational methods, which were based on mechanical aids. The abacus is perhaps the best known of these, but it appeared at a late date (around the fifteenth century). Chinese mathematicians relied above all on the manipulation of counting-rods (*chousuan*), which were arranged to represent the various coefficients of numerical equations. Mathematical problems transferred to arrangements of counting-rods were, so to speak, divorced from their practical context and taken into the realm of abstraction.

This method of calculation by tabulation is generally known as *fangcheng* (*fang* meaning "square" or "rectangle" and *cheng* meaning "divide up") and involves arranging the counting-rods to form a square or a rectangle (a matrix). Two types of rod were used, red and black or positive and negative, representing the complementary forces which controlled the Chinese universe, *yin* and *yang*.

Here we have an "instrumental" algebra which dispenses with all discursive methods of argument. This is its strength, but also its weakness, as during the manipulations of the counting-rods, calculations vanish without trace as soon as they are completed. This "art of the rods" is like the art of a

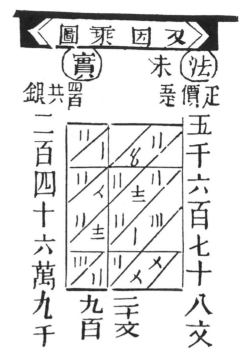

The gelosia or grating method of multiplication as shown in a Chinese book of 1593.

musician who plays without looking at the score, and it is no coincidence that some Chinese mathematicians explicitly compare mathematics with music.

Calculations and manipulations are the basic ingredients of a form of mathematics which has never felt itself to be bound by any dogma, and which at various phases of its history has syncretized many elements from other cultures.

HISTORICAL EXHIBIT 5.1

The Chinese Derivation of π

The earliest known value for π in China is 3. This value was used as far back as the twelfth century B.C. In approximately A.D. 260, the scholar Liu Hui undertook the task of revising the mathematical classic *Jiuzhang Suanshu* [*The Nine Chapters on the Mathematical Art*], a manual of applied mathematics. He was dissatisfied with the approximate value of π, i.e. 3, used in the first chapter of the text which is entitled "Field Measurement." Liu undertook a polygonal approximation to determine the area of a circle with a unit radius [1 Chinese foot = 10 inches]. In an accompanying commentary, he described his theory:

> Starting with an inscribed regular hexagon in a circle, one can double the number of sides to obtain an inscribed 12-sided regular polygon. If the process is repeated until practically one cannot go further, then the area of this "ultimate" polygon is virtually equal to that of the circle.

Liu's technique can be summarized as follows: if 1_n = the length of the side of the regular n-sided inscribed polygon and r the radius of the circle, then the area of the resulting 2n-sided polygon is given by

$$A_{2n} = \frac{n1_n r}{2}$$

Liu used this to obtain

$$A_{192} = 314\frac{64}{625} = 314.1024 \text{ sq. in.}$$

i.e. $\pi r^2 = 314.1024$ or

$$\pi = 3.141024$$

Two hundred years later, the mathematician Zu Chongzhi also used the "cutting the circle" technique devised by Liu Hiu obtaining a value for π:

$$3.1415926 < \pi < 3.1415927$$

Such accuracy was not to be obtained in the West for at least another thousand years.

51

The Evolution of Mathematics
in Ancient China

FRANK SWETZ

$\mathcal{A}$ POPULAR survey book on the development of mathematics has its text prefaced by the following remarks:

> Only a few ancient civilizations, Egypt, Babylonia, India and China, possessed what may be called the rudiments of mathematics. The history of mathematics and indeed the history of western civilization begins with what occurred in the first of these civilizations. The role of India will emerge later, whereas that of China may be ignored because it was not extensive and moreover has no influence on the subsequent development of mathematics.[1]

Even most contemporary works on the history of mathematics reinforce this impression, either by neglecting or depreciating Chinese contributions to the development of mathematics.[2] Whether by ignorance or design, such omissions limit the perspective one might obtain concerning both the evolution of mathematical ideas and the place of mathematics in early societies. In remedying this situation, Western historians of mathematics may well take heed of Whittier's admonition:

> We lack but open eye and ear
> To find the Orient's marvels here.[3]

Language barriers may limit this quest for information; however, a search of English language sources will reveal that there are many "marvels" in Chinese mathematics to be considered.

Reprinted from *Mathematics Magazine* 52 (Jan., 1979): 10–19; with permission of the Mathematical Association of America.

Legend and Fact

The origins of mathematical activity in early China are clouded by mysticism and legend. Mythological Emperor Yü is credited with receiving a divine gift from a Lo river tortoise. The gift in the form of a diagram called the *Lo shu* is believed to contain the principles of Chinese mathematics, and pictures of Yü's reception of the *Lo shu* have adorned Chinese mathematics books for centuries. This fantasy in itself provides some valuable impressions about early Chinese science and mathematics. Yü was the patron of hydraulic engineers; his mission was to control the flood-prone waters of China and provide a safe setting in which a water-dependent civilization could flourish. The users of science and mathematics in China were initially involved with hydraulic engineering projects, the construction of dikes, canals, etc., and with the mundane tasks of logistically supporting such projects. A close inspection of the contents of the *Lo shu* reveals a number configuration (Figure 1) which would be known later in the West as a magic square. For Chinese soothsayers and geomancers from the Warring States period of Chinese history (403–221 B.C.) onward, this square, comprised of numbers, possessed real magical qualities because in it they saw a plan of universal harmony based on a cosmology predicated on the dualistic theory of the *Yin* and the *Yang*.[4]

319

FIGURE I

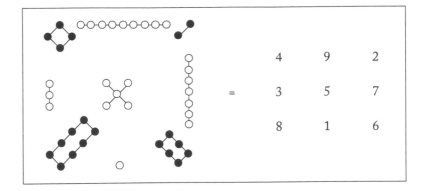

4	9	2
3	5	7
8	1	6

= (to the left of the grid)

FIGURE 2

7	4	1
8	5	2
9	6	3

```
      1
   4     2
7     5     3
   8     6
      9
```

```
      9
   4     2
3     5     7
   8     6
      1
```

4	9	2
3	5	7
8	1	6

Construct a natural square.   Distort it into a diamond.   Exchange corner elements.   Compress back into a square

FIGURE 3

1	10	19	28	37	46	55	64	73
2	11							
3	12							
4	13							
5	14							
6	15							
7	16							
8	17							
9	18							

(a)

55	28	1
64	37	10
73	46	19

(b)

28	73	10
19	37	55
64	1	46

(c)

31	76	13	36	81	18	29	74	11
22	40	58	27	45	63	20	38	56
67	(4)	49	72	(9)	54	65	(2)	47
30	75	12	32	77	14	34	79	16
21	39	57	23	41	59	25	43	61
66	(3)	48	68	(5)	50	70	(7)	52
35	80	17	28	73	10	33	78	15
26	44	62	19	37	55	24	42	60
71	(8)	53	64	(1)	46	69	(6)	51

(d)

Start with a natural square (a) then fold each row into a square (b) of order 3 (example using row 1) and apply the *Lo shu* technique (c). The nine resulting magic squares of order 3 (d) are then positionally ordered according to the correspondence of the central element in their bottom rows with the numbers of the *Lo shu*, i.e., 4,9,2; 3,5,7; 8,1,6.

When stripped of ritualistic significance, the principles used in constructing this first known magic square are quite simple and can best be described by use of diagrams as shown in Figure 2. The construction and manipulation of magic squares became an art in China even before the concept was know in the West.[5] Variations of the *Lo shu* technique were used in constructing magic squares of higher order with perhaps the most impressive square being that of order nine; see Figure 3.

While the *Lo shu* provides some intriguing insights into early mathematical thinking, its significance in terms of potential scientific or technological achievement is negligible. Historically, the first true evidence of mathematical activity can be found in numeration symbols on oracle bones dated from the Shang dynasty (fourteenth century B.C.). Their numerical inscriptions contain both tally and code symbols are clearly decimal in their conception, and employ a positional value system. The Shang numerals for the numbers one through nine were:

$$ -\ =\ \equiv\ \overline{\underline{\equiv}}\ \text{X}\ \wedge\ +\ )(\ \text{5} $$

By the time of the Han Dynasty (second century B.C.–fourth century A.D.), the system had evolved into a codified notation that lent itself to computational algorithms carried out with a counting board and set of rods. The numerals and their computing-rod configurations are shown below. Thus in this system 4716 would be represented as

$$ \text{||||}\ \perp\ \text{|}\ \perp $$.[6] (Occasionally the symbol $\times$ was used as an alternative to $\equiv$.)

Counting boards were divided into columns designating positional groupings by 10. The resulting facility with which the ancient computers could carry out algorithms attests to their full understanding of decimal numeration and computation. As an example, consider the counting board method of multiplying 2 three-digit numbers, as illustrated in Figure 4. The continual indexing of partial products to the right as one multiplies by smaller powers of ten testifies to a thorough understanding of decimal notation. In light of such evidence, it would seem that the Chinese were the first society to understand and efficiently utilize a decimal numeration system.[7] If one views a popular schematic of the evolution of our modern system of numeration (Figure 5) and places the Chinese system in the appropriate chronological position, an interesting hypothesis arises, namely that the numeration system commonly used in the modern world had its origins 34 centuries ago in Shang China!

The Systematization of Early Chinese Mathematics

The oldest extant Chinese text containing formal mathematical theories is the *Arithmetic Classic of the Gnomon and the Circular Paths of Heaven* [*Chou pei suan ching*]. Its contents date before the third century B.C. and reveal that mathematicians of the time could perform basic operations with fractions according to modern principles employing the concept of common denominator. They were knowledgeable in the principles of an empirical geometry and made use of the "Pythagorean theorem." A diagram (see Figure 6) in the *Chou pei* presents the oldest known demonstration of the validity of

1	2	3	4	5	6	7	8	9	
—	=	≡	≣	≣	⊥	⊥	⊥	⊥	for coefficients of 10^{2n-2} n = 1, 2 . . .
		‖	‖‖	‖‖‖	‖‖‖‖	⊤	⊤	⊤	⊤	for coefficients of 10^{2n-1} n = 1, 2 . . . .

Counting board

		2	4	6	(multiplier)
					(product)
		3	5	7	(multiplicand)

		2	4	6
7	1	4		
3	5	7		

			4	6
8	5	6	8	
	3	5	7	

				6
8	7	8	2	2	(answer)
	3	5	7	

Accompanying
rod computations

$2 \times 3 =$ 6
$2 \times 5 =$ 10
        70

$2 \times 7 =$ 14
      714
$4 \times 3 =$ 12
      834

$4 \times 5 =$ 20
      854
$4 \times 7 =$ 28
     8568
$6 \times 3 =$ 18
     8748

$6 \times 5 =$ 30
     8778
$6 \times 7 =$ 42
    87822

FIGURE 4

this theorem. This diagram, called the *hsuan-thu* in Chinese, illustrates the arithmetic-geometric methodology that predominates in early Chinese mathematical thinking and shows how arithmetic and geometry could be merged to develop algebraic processes and procedures. If the oblique square of the *hsuan-thu* is dissected and the pieces rearranged so that two of the four congruent right triangles are joined with the remaining two to form two rectangles, then the resulting figure comprised of two rectangles and one small square have the same area as their parent square. Further, since the new configuration can also be viewed as being comprised of two squares whose sides are the legs of the right triangles, this figure demonstrates that the sum of the squares of the legs of a right triangle is equal to the square of the hypotenuse.[8] The process involved in this intuitive, geometric approach to obtain algebraic results was called *chi-chü* or "the piling up of squares."[9]

The next historical text known to us is also a Han work of about the third century B.C. It is the *Nine Chapters on the Mathematical Art* [*Chiu chang suan shu*], and its influence on oriental mathematics may be likened to that of Euclid's *Elements* on western mathematical thought. The *Chiu chang's* chapters bear such titles as surveying of land, consultations on engineering works, and impartial taxation, and confirm the impression that the Chinese mathematics of this period centered on the engineering and bureaucratic needs of the state. Two hundred and forty-six problem situations are considered, revealing in their contents the fact that the Chinese had accumulated a variety of formulas for determining the areas and volumes of basic geometric shapes. Linear equations in one unknown

FIGURE 5

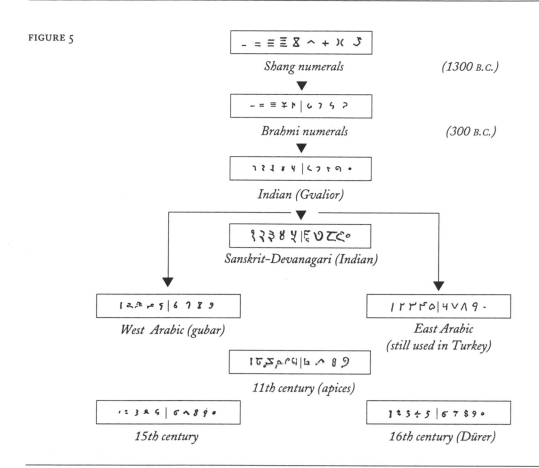

Shang numerals   (1300 B.C.)

Brahmi numerals   (300 B.C.)

Indian (Gvalior)

Sanskrit-Devanagari (Indian)

West Arabic (gubar)   East Arabic
                                       (still used in Turkey)

11th century (apices)

15th century   16th century (Dürer)

FIGURE 6

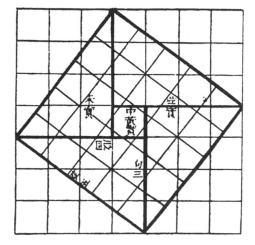

were solved by a rule of false position. Systems of equations in two or three unknowns were solved simultaneously by computing board techniques that are strikingly similar to modern matrix methods. While algebraists of the ancient world such as Diophantus or Brahmagupta used various criteria to distinguish between the variables in a linear equation,[10] the Chinese relied on the organizational proficiency of their counting board to assist them in this chore. Using a counting board to work a system of equations allowed the Chinese to easily distinguish between different variables.

Consider the following problem from the *Chiu chang* and the counting board approach to its solution.

Of three classes of cereal plants, 3 bundles of the first; 2 of the second and 1 of the third will produce 39 *tou* of corn after threshing; 2 bundles of the first;

3 of the second and 1 of the third will produce 34 *tou;* while 1 of the first, 2 of the second and 3 of the third will produce 26 *tou.* Find the measure of corn contained in one bundle of each class.[11]

(1 *tou* = 10.3 liters)

This problem would be set up on the counting board as:

1	2	3	1st class grain
2	3	2	2nd class grain
3	1	1	3rd class grain
26	34	39	Number of *tou*

Using familiar notation this matrix of numbers is equivalent to the set of equations

$$3x + 2y + z = 39$$
$$2x + 3y + z = 34$$
$$x + 2y + 3z = 26$$

which are reduced in their tabular form by appropriate multiplications and subtraction to

$$3x + 2y + z = 30 \qquad 36x = 333$$
$$36y = 153 \quad \text{and} \quad 36y = 153$$
$$36z = 99 \qquad 36z = 99.$$

Thus $x = 333/36$, $y = 153/36$ and $z = 99/36$.

A companion problem from the *Chiu chang* involves payment for livestock and results in the system of simultaneous equations:

$$-2x + 5y - 13z = 1000$$
$$3x - 9y + 3z = 0$$
$$-5x + 6y + 8z = -600$$

Rules provided for the solution treat the addition and subtraction of negative numbers in a modern fashion; however, procedures for the multiplication and division of negative numbers are not found in a Chinese work until the Sung dynasty (+ 1299). Negative numbers were represented in the computing scheme by the use of black rods, while red computing rods represented positive numbers. Zero

was indicated by a blank space on the counting board. This evidence qualifies the Chinese as being the first society known to use negative numbers in mathematical calculations.

The *Chou pei* contains an accurate process of extracting square roots of numbers. The ancient Chinese did not consider root extraction a separate process of mathematics but rather merely a form of division.[12] Let us examine the algorithm for division and its square root variant. The division algorithm is illustrated in Figure 7 for the problem 166536 ÷ 648. The Chinese technique of root extraction depends on the algebraic proposition

$$(a + b + c)^2 = a^2 + 2ab + b^2 + 2(a + b)c + c^2$$
$$= a^2 + (2a + b)b + (2[a + b] + c)c$$

which is geometrically substantiated by the diagram given in Figure 8. This proposition is incorporated directly into a form of division where $\sqrt{N} = a + b + c$. The counting board process for extracting the square root of 55225 is briefly outlined in Figure 9. Root extraction was not limited to three digit results, for the Chinese were able to continue the process to several decimal places as needed. Decimal fractions were known and used in China as far back as the fifth century B.C. Where a root was to be extracted to several decimal places, the computers achieved greater accuracy by use of formulae $\sqrt[n]{m} = \sqrt[n]{m10^{kn}}/10^k$.[13] Cube root extraction was conceived on a similar geometric-algebraic basis and performed with equal facility.

Historians of mathematics often devote special consideration to the results obtained by ancient societies in determining a numerical value for π as they believe that the degree of accuracy achieved supplies a comparative measure for gauging the level of mathematical skill present in the society. On the basis of such comparisons, the ancient Chinese were far superior to their contemporaries in computational mathematical ability. Aided by a number system that included the decimalization of fractions and the possession of an accurate root extraction process the Chinese had obtained by

FIGURE 7 $\qquad$ $166536 \div 648$

Counting board layout	Accompanying rod computations	Explanations

2 (quotient)	166500	200 is chosen
	$- 120000 = (200 \times 600)$	as the first
166536 (dividend)	46500	partial
	$- 8000 = (200 \times 40)$	quotient
648 (divisor)	38500	
	$- 1600 = (200 \times 8)$	
	36900	
25	36930	50 is chosen
	$- 30000 = (50 \times 600)$	as the second
36936	6930	partial
	$- 2000 = (50 \times 40)$	quotient
648	4930	
	$400 = (50 \times 8)$	
	4530	
257	4536	7 is chosen
	$- 4200 = (7 \times 600)$	as the third
4536	336	partial
	$- 280 = (7 \times 40)$	quotient
648	56	
	$- 56 = (7 \times 8)$	
	0	process is finished

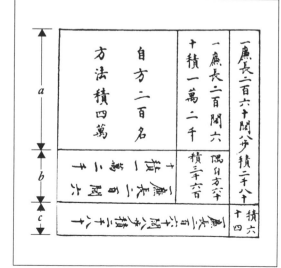

FIGURE 8

A geometric "proof" (Figure 8) of the algebraic proposition (see p. 324) which justifies the calculations (Figure 9) leading to $\sqrt{55225} = 235$. The 1 in the upper box represents an indexing rod that determines the decimal value of the divisors used. At the beginning of the process, it is moved to the left in jumps of two decimal places until it establishes the largest power of ten that can be divided into the designated number. After each successful division, the rod is indexed two positional places to the right.

Algebraic significance	Numerical entries on board
N 1	55225 1
a $N - a^2$ $a \times 10000$ 10000	2 15225 20000 10000
$a + b$ $N - a^2$ $(2a + b)\,b \times 100$ $(2a + b) \times 100$ 100	23 15225 12900 4300 100
$a + b$ $N - [a^2 + (2a + b)b]$ $(2a + b) \times 100$ 100	23 2325 4300 100
$a + b + c$ $N - (a + b)^2 - [2a(a + b) + c]c$ $2(a + b) + c$	235 0 465

FIGURE 9

the first century a value of π of 3.15147. The scholar Liu Hui in a third century commentary on the *Chiu chang* employed a "cutting of the circle method"—determining the area of a circle with known radius by polygonal approximations—to determine π as 3.141024. A successor, Tsu Chung-chih, refined the method in the fifth century to derive the value of π as 355/113 or 3.1415929.[14] This accuracy was not to be arrived at in Europe until the sixteenth century.

Trends in Chinese Algebraic Thought

While the Chinese computational ability was indeed impressive for the times, their greatest accomplishments and contributions to the history of mathematics lay in algebra. During the Han period, the square and cube root extraction processes were being built upon to obtain methods for solving quadratric and other higher order numerical equations. The strategy for extending the square root process to solve quadratic equations was based on the following line of reasoning. If $x^2 = 289$, 10 would be chosen as a first entry approximation to the root, then $289 - (10)^2 = 189$. Let the second entry of the root be represented by y; thus, $x = 10 + y =$ or $(10 + y)^2 = 289$ which if expanded, gives the quadratic equation $y^2 + 20y - 189 = 0$. By proceeding to find the second entry of the square root of 289, 7, we obtain the positive root for the quadratic $y^2 + 20y - 189 = 0$.[15]

By the time of the Sung Dynasty in the thirteenth century, mathematicians were applying their craft to solve such challenging problems as:

> This is a round town of which we do not know the circumference or diameter. There are four gates (in the wall). Three *li* from the northern (gate) is a high tree. When we go outside of the southern gate and turn east, we must walk 9 *li* before we see the tree. Find the circumference and the diameter of the town. (1 *li* = .644 kilometers)

If the diameter of the town is allowed to be represented by x^2, the distance of the tree from the northern gate, a, and the distance walked eastward, b, the following equation results.

$$x^{10} + 5ax^8 + 8a^2x^6 - 4a(b^2 - a^2)x^4 - 16a^2b^2x^2 - 16a^3b^2 = 0$$

For the particular case cited above, the equation becomes

$$x^{10} + 15x^8 + 72x^6 - 864x^4 - 11{,}664x^2 - 34{,}992 = 0$$

Sung algebraists found the diameter of the town to be 9 *li*.[16]

The earliest recorded instance of work with indeterminate equations in China can be found in a problem situation of the *Chiu chang* where a system of four equations in five unknowns results.[17] A particular solution is supplied. A problem in the third century *Mathematical Classic of Sun Tzu* [*Sun Tzu suan ching*] concerns linear congruence and supplies a truer example of indeterminate analysis.

> We have things of which we do not know the number; if we count by threes, the remainder is 2; if we count by fives, the remainder is 3; if we count by sevens, the remainder is 2. How many things are there?[18]

In modern form, the problem would be represented as

$$N \equiv 2 \ (\mathrm{mod}\ 3) \equiv 3 \ (\mathrm{mod}\ 5) \equiv 2 \ (\mathrm{mod}\ 7).$$

Sun's solution is given by the expression

$$70 \times 2 + 21 \times 3 + 15 \times 2 - 105 \times 2 = 23$$

which when analysed gives us the first application of the Chinese Remainder Theorem.

If $m_1 \ldots, m_k$ are relatively prime in pairs, there exist integers x for which simultaneously $x \equiv a_1$ (mod m_1), $\ldots$, $x \equiv a_k$ (mod m_k). All such integers x are congruent modulo $m = m_1 m_2 \ldots m_k$. The existence of the Chinese Remainder Theorem was communicated to the west by Alexander Wylie, an English translator and mathematician in the employ of the nineteenth century Chinese court. Wylie recorded his findings in a series of articles, "Jottings on the Science of the Chinese; Arithmetic" which appeared in the *North China Herald* (Aug.–Nov.) 1852. The validity of the theorem was questioned until it was recognized as a variant of a formula developed by Gauss.[19]

Perhaps the most famous Chinese problem in indeterminate analysis, in the sense of its transmission to other societies, was the problem of the "hundred fowls" (ca. 468).

> A cock is worth 5 *ch'ien*, a hen 3 *ch'ien*, and 3 chicks 1 *ch'ien*. With 100 *ch'ien* we buy 100 fowls How many cocks, hens, and chicks are there?
>
> (*ch'ien*, a small copper coin)

The development of algebra reached its peak during the later part of the Sung an the early part of the following Yuan dynasty (thirteenth and fourteenth centuries). Work with indeterminate equations and higher order numerical equations was perfected. Solutions of systems of equations were found by using methods that approximate an application of determinants, but it wasn't until 1683 that the Japanese Seki Kowa, building upon Chinese theories, developed a true concept of determinants.

Work with higher numerical equations is facilitated by a knowledge of the binomial theorem. The testimony of the *Chiu chang* indicates that its early authors were familiar with the binomial expansion $(a + b)^3$, but Chinese knowledge of this theorem is truly confirmed by a diagram (Figure 10) appearing in the thirteenth-century text *Detailed Analysis of the Mathematical Rules in the Nine Chapters*. [*Hsiang chieh chiu chang suan fa.*]. It seems that "Pascal's Triangle" was known in China long before Pascal was even born.

While mathematical activity continued in the post-Sung period, its contributions were minor as compared with those that had come before. By the time of the Ming emperors in the seventeenth century, Western mathematical influence was finding its way into China and the period of indigenous mathematical accomplishment had come to an end.

Conclusions

Thus, if comparisons must be made among the societies of the pre-Christian world, the quality of China's mathematical accomplishments stands in contention with those of Greece and Babylonia, and during the period designated in the West as pre-Renaissance, the sequence and scope of mathematical concepts and techniques originating in China far exceeds that of any other contemporary society. The impact of this knowledge on the subsequent development of Western mathematical thought is an issue that should not be ignored and

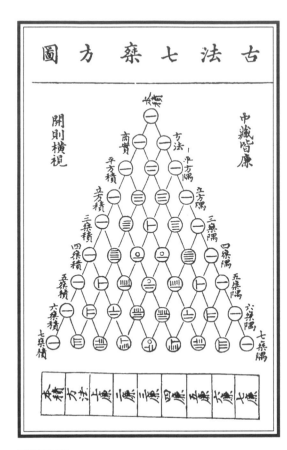

FIGURE 10

can only be resolved by further research. In part, such research will have to explore the strength and vitality of Arabic-Hindu avenues of transmission of Chinese knowledge westward. The fact that Western mathematical traditions are ostensibly based on the logico-deductive foundations of early Greek thought should not detract from considering the merits of the inductively-conceived mathematics of the Chinese. After all, deductive systemization is a luxury afforded only after inductive and empirical experimentation has established a foundation from which theoretical considerations can proceed. Mathematics, in its primary state, is a tool for societal survival; once that survival is assured, the discipline can then become more of an intellectual and aesthetic pursuit. Unfortunately, this second stage of mathematical development never occurred in China. This phenomenon—the

fact that mathematics in China, although developed to a high art, was never elevated further to the status of an abstract deductive science—is yet another fascinating aspect of Chinese mathematics waiting to be explained.

NOTES

1. Morris Kline, *Mathematics: A Cultural Approach* (Reading, Mass.: Addison-Wesley Publishing Co. 1962), p. 12.

2. In his 712 page *A History of Mathematics* (New York: John Wiley & Sons Inc., 1968) Carl Boyer devotes 12 pages to Chinese contributions; the latest revised edition of Howard Eves, *An Introduction to the History of Mathematics,* (New York: Holt, Rinehart and Winston, 1976) contains 6 pages on the history of Chinese mathematics. The contents of these pages are based on information given in an article by D. J. Struik, "On Ancient Chinese Mathematics," *The Mathematics Teacher* (1963), 56: 424–432, and represent little of Eves' own research.

3. John Greenleaf Whittier, "The Chapel of the Hermits."

4. Under this system, the universe is ruled by Heaven through means of a process called the *Tao* ("the Universal way"). Heaven acting through the Tao expresses itself in the interaction of two primal forces, the *Yin* and the *Yang.* The *Yang,* or male force, was a source of heat, light and dynamic vitality and was associated with the sun; in contrast, the *Yin,* or female force, flourished in darkness, cold and quiet inactivity and was associated with the moon. In conjunction, these two forces influenced all things and were present individually or together in all physical objects and situations. In the case of numbers, odd numbers were *Yang* and even, *Yin.* For a harmonious state of being to exist, *Yin-Yang* forces had to be balanced.

5. For a fuller discussion of Chinese magic squares, see Schyler Camman, "Old Chinese Magic Squares", *Sinologica* (1962), 7: 14–53; Frank Swetz, "Mysticism and Magic in the Number Squares of Old China," *The Mathematics Teacher* (January, 1978), 71: 50–56.

6. The evolution of counting rod numerals continued for about 3000 years in China, i.e, 14th century B.C.–13th century A.D. For a discussion of this process, see Joseph Needham, *Science and Civilisation in China* (Cambridge: Cambridge University Press, 1955), vol. 3. pp. 5–17.

7. A strong case for this theory has been made by Wang Ling, "The Chinese Origin of the Decimal Place–Value System in the Notation of Numbers". Communication to the 23rd International Congress of Orientalists, Cambridge, 1954.

8. Although a 3, 4, 5 right triangle is used in the demonstration, the Chinese generalized their conclusion for all right triangles. The 3, 4, 5 triangle was merely a didactical aid.

9. See Frank Swetz, "The 'Piling Up of Squares' in Ancient China," *The Mathematics Teacher* (1977), 70: 72–79. [This article is reproduced as Chapter 53 of the present volume.]

10. Diophantus (A.D. 275) spoke of unknowns of the first number, second number, etc., whereas Brahmagupta (A.D. 628) used different colors in written computations to distinguish between variables.

11. *Chiu chang suan shu*, Fang Cheng (chapter 8), problem 1.

12. For a discussion of the Chinese ability at root extraction, see Wang Ling and Joseph Needham, "Horner's Method in Chinese Mathematics: Its Origins in the Root Extraction Procedures of the Han Dynasty", *T'oung Pao* (1955), 43: 345–88; Lam Lay Yong, "The Geometrical Basis of the Ancient Chinese Square-Root Method", *Isis* (Fall, 1970), pp. 92–101.

13. A lengthy discussion of the use of this formula in Europe is given in D. E. Smith, *History of Mathematics* (New York: Dover Publishing Co., 1958 reprint) vol. II, p. 236.

14. The evolution of π in China is traced out in Lee Kiong-Pong, "Development of π in China", *Bulletin of the Malaysian Mathematical Society* (1975), 6:40–47.

15. An actual computational procedure used in solving quadratics can be found in Ho Peng Yoke, "The Lost Problems of the Chang Ch'iu-chien Sual Ching, a Fifth Century Chinese Mathematical Manual," *Oriens Extremus* (1965), 12.

16. For a detailed discussion of the solution of this problem see Ulrich Libbrecht, *Chinese Mathematics in the Thirteenth Century* (Cambridge, Mass.: The MIT Press, 1973), pp. 134–40.

17. *Chiu chang suan shu*, chapter 8, problem 13:

There is a common well belonging to five families; (if we take) 2 lengths of rope of family X, the remaining part equals 1 length of rope of family Y; the remaining part from 3 ropes of Y equals 1 rope of Z; the remaining part from 4 ropes of Z equals 1 rope of V; the lacking part remaining from 5 ropes of V equals 1 rope of U; the remaining part from 6 ropes of U equals 1 rope of X. In all instances if one gets the missing length of rope, the combined lengths will reach (the water). Find the depth of the well and the length of the ropes.

If we let W equal the depth of the well, the following system of equations result:

$$2X + Y = W$$
$$3Y + Z = W$$
$$4Z + V = W$$
$$5V + U = W$$
$$6U + X = W$$

which are readily reduced to:

$$2X - 2Y - Z = 0$$
$$2X + Y - 4Z - V = 0$$
$$2X + Y - 5V - U = 0$$
$$X + Y - 6U = 0$$

18. *Sun Tzu suan ching*, chapter 3, problem 10.

19. See the discussion of the Chinese Remainder Theorem in Oystein Ore, *Number Theory and its History*, (New York: McGraw-Hill Inc., 1948), pp. 245–49.

52

The Amazing
Chiu Chang Suan Shu

FRANK SWETZ

*I*N THE mid-nineteenth century, China was rudely awakened from her two thousand year old xenophobic slumber by the ironclads and cannons of foreign visitors. Suddenly, the survival of the Middle Kingdom was no longer contingent on the moral ethics of Confucian humanism but rather depended on acquiring technological proficiency. Emissaries of reform pleaded their case before the Dowager Empress. Their memorials[1] urged educational innovations as a basis for industrialization. Principally, they requested that the languages and sciences of the "foreign devils" be studied in order to decipher the secrets of the West's technological prowess. In the study of science, priority was given to mathematics as it was felt that this subject was basic to all other sciences.

Prince Kung, president of the Bureau of Foreign Affairs, in exhorting educational reforms in the 1860's, counseled the throne that:

> Western sciences borrowed their roots from ancient Chinese mathematics. Westerners still regard their mathematics as coming from the Orient. It is only because of the careful, inquiring minds of the Westerners that they are good at developing something new out of the old. . . . China invented the method, Westerners adopted it. . . .[2]

Such attitudes were prevalent among many scholar-officials of Old China. Were Kung's words a mere psychological ploy designed to induce a more ready acceptance of Western learning or an ethnocentric illusion? Could they possibly be true?

During the first decades of the twentieth century, scholarly journals published an abundance of articles on the then newly-emerging field of Sinology. Among the previously unknown aspects of Chinese civilization to be examined were China's mathematical accomplishments. Some writers on this topic conjectured a high state of mathematical development in early China while others subordinated Chinese accomplishments to Western influence.[3] Summarizing the controversy was David Eugene Smith's article entitled, "Unsettled Questions Concerning the Mathematics of China," in which he enumerated the difficulties of undertaking such an investigation.[4] Smith, the only American scholar to research Eastern mathematics in depth, advised that further Sinological studies would be necessary to establish the true historical significance of Chinese mathematics. In the interim since the appearance of Smith's article, war and internal Sino-upheaval have disrupted the advance of this research. During this period, the accepted position of mathematical historians has been either to ignore early Chinese contributions to the field of mathematics or to indicate they are insufficient to warrant serious attention.[5]

The current revival of Sinological studies, highlighted by the monumental work of Joseph Needham, *Science and Civilisation in China*, warrants a return to this moot question.[6] In a brief

Reprinted from *Mathematics Teacher* 65 (May, 1972): 423–30; with permission of the National Council of Teachers of Mathematics.

attempt to better understand the significance of ancient Chinese mathematics, let us examine the contents of the most controversial and influential of all the ancient Chinese mathematical works, the *Chiu Chang Suan Shu* [Nine Chapters on the Mathematical Art]. This work, the main reference of Eastern mathematics for many centuries, was first translated into English by Alexander Wylie, a scholar and competent mathematician in the employ of the Manchu government. Wylie marvelled at his findings.[7]

The Origins and Form of the Chiu Chang Suan Shu

Dating of ancient Chinese literature is extremely difficult. Despotic emperor Shih Huang-ti of the Ch'in dynasty (221–207 B.C.) originated the first burning of books in 213 B.C. Scholars of the following Han period (206 B.C.–A.D. 220) were forced to transcribe China's literary and scientific traditions from memory or remaining scroll fragments. Hyperbole was often used in attributing a work to an ancient sage with the design of increasing its prestige; forgery was a common practice. As a result of such phenomena, the dating of the *Chiu Chang* has at times been confused. However, it is accepted today to be a product of the late Ch'in and early Han dynasties (third century B.C.) and was probably first written by one Chang Tshang, who made use of older works then in existence. The version that survives to the present day is a commentary prepared by a certain Liu Hui in approximately A.D. 250. Thus, the contents of this text is a summary of the mathematical knowledge possessed in China up to the middle of the third century, the nucleus of it having been assembled two hundred years before the Christian era.

The *Chiu Chang* consists of nine distinct sections. Each section deals with a particular topic in mathematics relevant to the Chinese society of the time. Instruction is provided through the presentation of specific mathematical problems and the rules for obtaining their solutions. The Chinese authors did not possess a system of algebraic notation; therefore all exposition is in a literary form. Little indication is given as to how the solution rules were derived, but the general context is algebraic-arithmetic and reveals an empirical methodology.

The Contents of the Chiu Chang Suan Shu

I. *Fang thien* (land surveying).[8] Various formulas are given for elementary engineering work. They include correct computational rules for finding the area of rectangles, triangles, trapezoids, and circles with π being approximated by 3. A trapezoidal approximation for the area of a segment of a circle is given by $\frac{S}{2}(C + S)$, C being the measure of the length of the chord involved and S the measure of the length of the sagitta of the segment (see Fig. 1). A discussion on the adequacy of such an approximation could serve as an interesting present-day classroom exercise.

This section also reveals an arithmetic of fractions employing the principles of lowest common denominators and greatest common factors in accord with modern practices. Such an approach to operations with fractions was not utilized in Europe until the fifteenth and sixteenth centuries.

FIGURE I

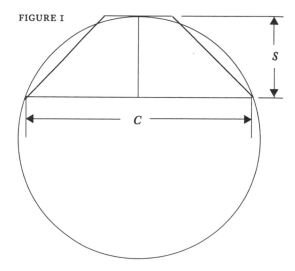

The utilitarian nature of the *Chiu Chang's* presentation fosters an elaborate treatment of common fractions at the expense of decimal fractions which are also used in the work, but to a lesser extent.

Approximation of π by 3 is common in ancient mathematical texts. Wylie, while in the process of translating this work, queried his Chinese colleagues on the adequacy of using this value of π. He was politely informed that the scholars of the ancient period of authorship intended to supply a sufficient approximation rather than a more accurate but unwieldy one. When we remember the nature of this work and the fact that similar Egyptian and Babylonian writings chose "3" over more exact values, credibility can be given to this theory. Needham's research would also seem to bear this out.[9]

II. *Su mi* (millet and rice). This chapter treats questions of simple percentage and proportions.

III. *Tshui fen* (distributions by progressions). Problems involving taxation, partnership, and arithmetical and geometrical progressions are presented. Their solutions are arrived at by methods of proportion and the *rule of three.*

Although the rule of three has generally been considered of Indian origin,[10] the *Chiu Chang's* exposition of this technique predates Sanskrit versions (ca. 628). This rule, held in high regard in the mercantile society of medieval Europe, is simply a proportion involving three knowns from which an unknown quantity is found. A Chinese example involving this principle goes as follows:

> Two and one-half piculs[11] of rice are purchased for ³/₇ of a taiel of silver. How many can be purchased for nine taiels?

Employing modern symbolism and the rule of three the solution can easily be found:

$$\frac{2\,{}^{1}/{}_{2}}{{}^{3}/{}_{7}} = \frac{x}{9}$$

$$\frac{35}{6} = \frac{x}{9}$$

$$x = \frac{(35)(9)}{6}$$

$$x = 52\,{}^{1}/{}_{2}\ \text{piculs}$$

Typical of the problems concerning series found in this section is one involving textile production:

> A girl skillful in weaving doubles the previous day's output of cloth. She produces five feet of cloth in five days. What is the result of the first day, and successive days, respectively?

A modern student with a minimum of algebraic knowledge would interpret the problem symbolically to finding a solution for the expression: $a + 2a + 4a + 8a + 16a = 5$. He would find a to be ⁵/₃₁, the answer supplied by the *Chiu Chang*. The Chinese authors used a rule of false position to arrive at their answer (see below).

IV. *Shao kuang* (diminishing breadths). This section contains twenty-four problems on land mensuration. Some of the examples given illustrate the Chinese process for finding square and cube roots. Wylie noted that the rules prescribed for solution were very similar to those given in contemporary [1876] English school books.

Square and cube roots of numbers were extracted through a computational procedure known as the *Celestial Element Method.* The concept behind this procedure paralleled the method devised by W. G. Horner in the nineteenth century. Twenty centuries preceding Horner, the Chinese in the manipulation of their counting rods were employing Horner's method in solving equations such as $x^3 = 1,860,867$ and $x^2 + 34x = 71,000$. Although later mathematical historians were to affirm Wylie's original observations, it wasn't until 1955, when Wang and Needham's published research revealed and analyzed this method in detail, that the fact was accepted—these two root extracting processes were the same in principle

The *Celestial Element Method*—so named because the Chinese character that represented the word heaven, *t'ien,* also was used to represent the unknown—was derived from the geometrically established relation $(a + b)^2 = a^2 + 2ab + b^2$. Theon of Alexandria (ca. 390) used a similar procedure for arriving at his rules for finding square roots.[12] Although the *Chiu Chang* apparently deals with integral roots in its solutions, there is evidence that scholars of the time could extract roots to several decimal places. Problems 9 and 10 of this section as solved in a first century commentary reveal that the Chinese sometimes extracted square and cube roots employing the computational technique whereby $\sqrt[n]{m}$ could be replaced by $\sqrt[n]{\dfrac{mp}{p}}$. If p is allowed to be a power of ten so that

$$\sqrt[n]{m} = \frac{\sqrt[n]{m10^{kn}}}{10^k},$$

the method assists in the decimalization of roots and permits a greater accuracy in root extraction. Han mathematicians in obtaining solutions for problems such as $\sqrt{0.35}$ would transform them into decimal fraction equivalents such as $\sqrt{\dfrac{350}{1000}}$ hence $\sqrt{\dfrac{350000}{1000}}$ and thus could readily extract the square root of 350,000, using their counting rods and the Celestial Element Method. A generalized version of this technique, transmitted westward by Hindu and Moslem agents, was used in Europe in the late middle ages in the computation of square root tables.

Some controversial problems found in this section concern work with unit fractions. These problems confused early researchers into conjecturing that the ancient Chinese employed a system of unit fractions similar to that of the Egyptians.

> There is a [rectangular] field whose breadth is one pu and a half and a third. If the [area] is one mu, what is the length of the field?
>
> Answer: 130 10/11 pu

The solution rule indicates the Chinese facility for working with fractions:

As there is 3 in the denominator, we have to change 1 to 6, a half to 3 and one-third to 2. Adding these together we get 11, which will be made the *fa* or divisor. Set down the 240 square pu [1 mu] of the field, where every 1 is to be counted as 6, and so this number 6 is to be multiplied. The product is the *shih* or dividend. Thus, carrying out the division we get the number of pu contained in the length.

V. *Shang kung* (consultations of engineering works). Here we are provided with formulas for the computation of volumes encountered in constructing dikes or fortifications; for prisms, cylinders, tetrahedrons, pyramids, circular cones, and various truncated solids. One of the more unusual formulas given is that for the volume of a truncated triangular right prism, $\dfrac{b_1 + b_2 + b_3}{6} DL$, where b_1, b_2, b_3 are measures of the tree breadths of the wedge, D is the measure of the depth, and L the measure of the length (see Fig. 2). Legendre is usually credited with having devised this formula in his *Eléments de Géométrie* published in 1794.[13]

Also noteworthy is the formula $\dfrac{1}{6} abh$ for the volume of a tetrahedron whose two opposite edges, a and b, are at right angles, and where h is the measure of the height of the common perpendicular. It appears highly likely that the formulas such

FIGURE 2

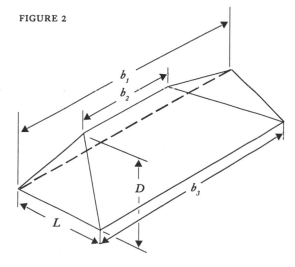

as these were the end product of experimentation employing concrete models of the structures under concern.

VI. *Chun shu* (impartial taxation). Consideration is given to the distribution of taxes and the various difficulties in transporting the taxes, as paid for with grain, to the capital (pursuit problems). Pursuit problems, the nemesis of many a school boy, were rather late in arousing European interest. Smith lists as one of the first Western sources of such problems Alcuin of York's *Propositiones ad Acundas Juvenes* (ca. 775), in which a hound pursuing a hare problem is given.[14] The *Chiu Chang's* version of this problem is:

> A hare runs 50 pu ahead of a dog. The latter pursues the former for 125 pu. When the two are 30 pu apart, in how many pu will the dog overtake the hare?

VII. *Ying pu tsu* (excess and deficiency). This section is devoted to the Chinese algebraic technique commonly called in the West the *rule of false position* and used to solve equations of the form $ax - b = 0$. While to us the solution of such problems would be considered trivial, for centuries before the adoption of a manipulative symbolism, their literal presentation posed a high degree of difficulty. Employing the Chinese method, let us solve $6x - 12 = 0$. We make two guesses, g_1 and g_2, as to the value of x; $g_1 = 3$ (excess) and $g_2 = 1$ (deficiency). They provide two failures, f_1 and f_2, in the attempt to reach a solution.

$$(6)(3) - 12 = 6 = f_1$$

$$(6)(1) - 12 = -6 = f_2$$

Following the scheme

$$x = \frac{f_1 g_2 - f_2 g_1}{f_1 - f_2} = \frac{(6)(1) - (-6)(3)}{6 - (-6)} = \frac{24}{12} = 2$$

From the thirteenth century onward, a variation of this method was widely employed in Europe under the name *Regula Falsae Positionis*. It was formally believed that this rule was transmitted to Europe from India by Arab mathematicians; care-

ful scrutiny of Sanskrit texts reveals the earliest indication of such a method in *Brahmagupta* (ca. 628). The Egyptian *Rhind Papyrus* reveals another method of false position from antiquity; however, this differs from the Chinese version and its European adaptation.

VIII. *Fang chheng* (the way of calculating by tabulation). Eighteen problems are presented dealing with the solution of systems of simultaneous linear equations.

Counting rod techniques require their user to establish a matrix[15] representing the numerical coefficients of the given systems of equations. Elementary column and row operations consistent with modern practices were performed on the matrix to obtain a solution. The examination of a problem from this section would prove enlightening:

> There are three grades of corn. Neither two baskets of the first grade, three baskets of the second, nor four baskets of the third grade taken separately comprise a full required measure. If however, one basket of the second grade were added to the first grade corn, one basket of the third grade corn added to the second, and one basket of the first grade to the third grade baskets, then the grain would comprise one required measure in each case. How many measures of the various grades does each mixed basket contain?

That is,

$$
\begin{array}{rrrl}
2x + & y & & = 1 \\
& 3y & + & z = 1 \\
x + & & 4z & = 1
\end{array}
$$

The matrix formed is:

1		2	1st grade (x)
	3	1	2nd grade (y)
4	1		3rd grade (z)
1	1	1	required measure

A "zero" coefficient was indicated by a blank in the computational matrix. This particular problem requires its solver to subtract a number from nothing! A rule is provided for such a possibility—distinguishing the Chinese as the first known society to have developed an algebra of negative numbers, *Cheng-fu-shu* (positive and negative). Similarly, coefficients in the matrix could also be negative in themselves. They were indicated by black calculating rods in this scheme. Red rods represented positive numbers. Previously the earliest accepted use of negative numbers in a society has been credited to Indian mathematicians (ca. 630).[16] The existence of such problems in the *Chiu Chang* discredits this claim. While certainly advanced in their computational techniques, the Chinese failed to develop this method further into a theory of determinants. An heir to this early work, the Japanese mathematician Seki Kōwa, finally formalized a theory of determinants in 1683, predating a similar European achievement by ten years.

The solution of a system of four equations in five unknowns is considered in the last problem of the section indicating a Chinese investigation of indeterminate equations at an early date. Chinese mathematicians of a later period were to refine indeterminate analysis and christen this area of study *Ta yen chiu shu* or "searching for unity." Typifying the concern of Chinese scholars were problems such as that of the "Hundred Fowls" (ca. 475).

> If a cock is worth 5 coins, a hen 3 coins and three chickens together only 1 coin, how many cocks, hens and chickens, together totaling 100, may be bought for 100 coins?

IX. *Kou ku* (right angles). Twenty-four problems are presented concerning the properties of right triangles. The accurate surveying of the farming lands of China demanded an understanding and application of the Pythagorean theorem at a very early date.

Several of the more interesting problems of this last section will now be given so that the readers may appreciate their novel presentation and perhaps adapt them for their classes. These problems provide an exercise in imagination as well as mathematical reasoning. In instances where rules for solution are provided, the students could supply the missing derivations.

> There is a rope hanging from the top of a pole with 3 feet of it lying on the ground. When it is tightly stretched, so that its end just touches the ground, it reaches a point 8 feet from the base of the tree. How long is the rope?

> There is a cylindrical column just buried beneath the surface of a wall. When the wall is chiseled into a depth of $1/10$ of a foot, the column is seen to have a breadth of 1 foot. What is the diameter of the column? (See Fig. 3.)

Rule for solution: $a + \left(\dfrac{b}{2}\right)^2 \Big/ a = d$

FIGURE 3

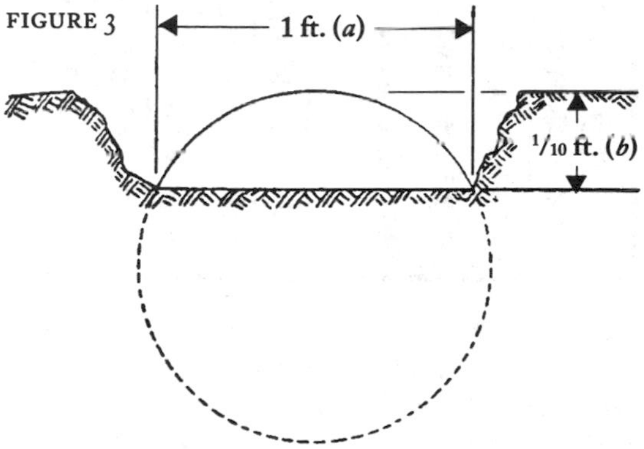

> Under a tree 20 feet high and 3 feet in circumference, there grows an arrowroot vine, which winds seven times around the stem of the tree and just reaches the top. How long is the vine?

> There grows in the middle of a pond 10 feet square a reed, which projects 1 foot above the surface. When it is drawn to the bank, the top of the reed comes even with the surface. What is the depth of the water and the reed's length?

Rule for solution:

$$\text{depth} = \frac{\left(\dfrac{\text{measure of pond's side}}{2}\right)^2 - \left(\text{projection of reed above water}\right)^2}{2\,(\text{projection of reed above water})}$$

> There is a bamboo 10 feet high, the upper end of which being broken touches the ground 3 feet from the stem. What is the height of the break? (see Fig. 4.)

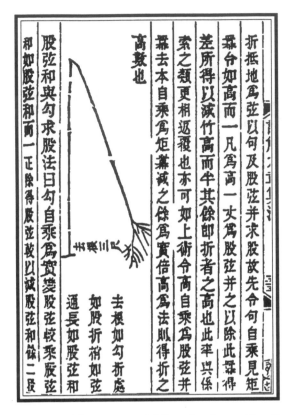

FIGURE 4 *The controversial "bent bamboo problem."*
From Science and Civilisation in China, *by Joseph
Needham; volume III, courtesy of Cambridge
University Press.*

Rule for solution:

$$\text{Height of break} = \frac{\text{height}}{2} - \frac{(\text{distance from stem})^2}{2\,\text{height}}$$

This last problem is an indicator of the migration
of Chinese mathematics westward. Sanskrit works
after the time of Aryabhata (ca. 510) include this
example.[17] Still later in history it is found in European works.

A knowledge of the solution of quadratics is
revealed in the twentieth example of this section.

> A square village of unknown sides is crossed by a
> street that joins the centers of the north and south
> sides. At a distance 20 pu north of the north gate is
> a tree visible from a point reached by proceeding 14
> pu south from the south gate and 1775 pu west.
> What is the length of each side of the village?

This problem reduces to one of finding the
positive root of the equation $x^2 + (20 + 14)x - (2)(20)(1775) = 0$. The side of the village is found
to have a length of 250 pu. Although the exact
method of solution is not given, the reader is told
to use the *ts'ung fa* to solve the problems. The
ts'ung fa was a position on the counting board used
for working the Celestial Element Method, therefore it appears that a variation of this technique
was used for solving quadratic equations.

Conclusions

While the *Chiu Chang Suan Shu* is not a true
deductive thesis on mathematics, it should be remembered that few early works were, with the
exception of Greek contributions. Many years later,
medieval European writers of mathematical tracts
were to employ similar formats. Certainly the early
Han authors demonstrate a high level of calculatory
proficiency. The sequence of mathematical "firsts"
that emanate from the pages of the *Chiu Chang*
give partial credence to Prince Kung's claims that
Chinese scholars at an early stage of development
achieved a relatively high level of sophistication in
utilitarian mathematics. How much of this knowledge was transmitted to the West is still debatable.
As early as the third century B.C., Chinese silk and
fine-cast ironware were prized in the markets of
Imperial Rome. Why then could not some of her
intellectual wares, including mathematical knowledge, also be carried to the West at an early date?

Despite the very early spurt of mathematical
activity in China, as evidenced in the contents of
the *Chiu Chang,* further development of mathematics as anything but a series of empirical techniques was stunted by continuing sociological pressures. A harsh land such as the Chinese possessed
could only accommodate a large population through
the development of a proficient technology or a
rigid social code. The Chinese chose the latter.
Rather than altering nature to suit their society,
they preferred to limit their personal and collective
freedoms in order to achieve reconcilement with

nature. Thus the development of deductive science, including mathematics, was sacrificed for a Confucian humanism that insured social harmony. One cannot help but wonder what the course of history would have been had this policy not been adopted by the Chinese? The amazing *Chiu Chang Suan Shu* provides a partial answer!

NOTES

1. Memorial—a petition presented before the throne of China.

2. John Fairbank and Teng Ssu-Yu, *China's Response to the West: A Documentary Survey 1839–1923* (New York: Atheneum , 1967), p. 74.

3. For example, Albion Fellows, "China's Mathematics—the Oldest," *China Review* (1921), 1:214–15; and Gino Loria and R. B. McClenon, "The Debt of Mathematics to the Chinese People," *Scientific Monthly* (1921), 12:517–21.

4. David E. Smith, "Unsettled Questions Concerning the Mathematics of China," *Scientific Monthly* (1931), 33:244–50.

5. Morris Kline, *Mathematics: A Cultural Approach* (Reading, Mass.: Addison-Wesley, 1962), p. 12.

6. Joseph Needham, *Science and Civilisation in China*, 3 vols. (London: Cambridge University Press, 1959).

7. Alexander Wylie, *Notes on Chinese Literature* (Shanghai: American Presbyterian Press, 1901); *Chinese Researches* (Shanghai, 1897).

8. Needham, *Science and Civilisation in China*.

9. Needham, *Science and Civilisation in China*, vol. 3, p. 99.

10. David E. Smith, *History of Mathematics* (1923; reprint ed., New York: Dover Publications), 2:483.

11. Several different units of Chinese measure are used throughout this article. They are:

picul—a measure of weight carried on a man's back (approximately 136 lbs.)
mu—a measure of area, 240 square paces.

12. For a detailed discussion of the Celestial Element Method and its Greek counterpart, see Lam Lay Yong, "The Geometrical Basis of the Ancient Chinese Square-Root Method," *Isis* (Fall, 1970), pp. 92–101; and Wang Ling and Joseph Needham, "Horner's Method in Chinese Mathematics: Its Origins in the Root Extraction Procedures of the Han Dynasty" (*T'oung Pao*, Leiden, 1955), 43:345–88.

13. J. L. Coolidge, *A History of Geometrical Methods* (Oxford: Clarendon Press, 1940), p. 21.

14. Smith, *History of Mathematics*, vol. 2, p. 546.

15. For more details on Chinese matrix problems see D. J. Struik, "On Ancient Chinese Mathematics," *Mathematics Teacher* 56 (1963) : 424–32; or Carl B. Boyer, *A History of Mathematics* (New York: John Wiley & Sons, 1968).

16. Kline, *Mathematics: A Cultural Approach*, p. 19.

17. Yoshio Mikami, *The Development of Mathematics in China and Japan* (1913; reprint ed., New York: Chelsea Publishing Co.), p. 23. Needham indicates much duplication of *Chiu Chang* material in later Indian texts.

53

The "Piling Up of Squares" in Ancient China

FRANK SWETZ

$\mathcal{T}$HE merits of a mathematics-laboratory approach to problem-solving activities have been widely acclaimed. Indeed, most teachers will readily acknowledge that laboratory learning situations sharpen the appeal to intuition and provide multifaceted learning interactions for the student. It is interesting to note that this modern teaching strategy often employs problem situations and learning devices that are quite historical in their origins. Napier's rods, magic squares, tangrams, the abacus, and the Tower of Hanoi all possess a historical significance that, if known, would enhance their teaching potential. A knowledge of historical origins can aid a teacher in devising a learning task or preparing a teaching strategy, as well as developing student's appreciation for the evolutionary nature of mathematics. The Thirty-first Yearbook of the National Council of Teachers of Mathematics, *Historical Topics for the Mathematics Classroom* (1969), is particularly helpful.

Of the five mathematics-laboratory devices mentioned, three had their historical origins in the Orient, and of these, two—magic squares and tangrams—had their beginnings in China. Yet it is only recently that Western scholars have begun to accord the development of mathematics in China its due attention. Although a variety of mathematical puzzles are ascribed to the Chinese, little

acknowledgment has been made of their more noteworthy accomplishments, which include an early familiarity with the use of the Pythagorean theorem; the most accurate approximation for π obtained in the ancient world, 3.1415929293 (Needham 1959, p. 101); an accurate root-extraction process preceding Horner's method by six hundred years (Needham and Wang 1955); and the use of the Pascal triangle for finding coefficients for a binomial expansion three hundred years before Pascal conceived of it (Needham 1959, p. 47). A testimony to the mathematical ability of the early Chinese scholars is provided by the contents of the *Chiu-chang suan-shu* [Nine chapters on the mathematical art]. The *Chiu-chang* is a manual of utilitarian mathematics compiled during the early Han period (ca. 300 B.C.—A.D. 300) and has been described as the most influential of all ancient Chinese texts (Swetz 1972). Although several parts of this book have been translated into English, a completed translation remains to be undertaken.* In studying the contents and methods of the *Chiuchang*, modern scholars have been intrigued by the depth and resourcefulness of the early Chinese methods of solution. These methods are empirically based and can easily lend themselves to a mathematics laboratory.

It is the object of this paper to examine one algebraic-geometric solution technique employed in the *Chiu-chang*—a solution technique used in ancient Greece and Babylon but developed to a high art in China. The technique called *chi-chü*, or

Reprinted from *Mathematics Teacher* 70 (Jan., 1977): 72–79; with permission of the National Council of Teachers of Mathematics.

the "piling up of squares," employs an intuitive geometric approach to solve algebraic problems. Such techniques were the forerunner of algebraic computation. A knowledge of this method can provide insights into the development of algebra as well as become a source for classroom activities.

The "Piling Up of Squares"

"Kou-ku" [right triangles] is the title of the ninth chapter of the *Chiu-chang*. This chapter presents twenty-four problems whose solutions depend on a thorough understanding of the Pythagorean theorem and the basic mathematical properties of right triangles. Problem narratives proceed from very elementary to rather complex and often picturesque situations. The solutions of these problems are obtained by employing one of two techniques, either *chi-chü* or *ch'ung-cha*. The literal translation of *ch'ung-cha* connotes a use of proportions equivalent to the tangent function (Needham 1959, p. 109).

In the consideration of the actual "Kou-ku" problems of the *Chiu-chang*, the original Chinese format will be followed, which includes, in respective order, a statement of the problem; the answer; a brief description of the solution method; and an explanation, usually supplied by a later commentator on the work. Whereas the first three parts of each question will be directly translated from the old Chinese in deference to the modern reading audience, the explanation will be modified by the use of contemporary algebraic symbolism and additional diagrams. The presence of Chinese characters will designate diagrams preserved from antiquity. In their illustrations the Chinese frequently used 3, 4, 5 right triangles, although the actual dimensions of the problem under consideration are quite different. The conjecture exists that early number-theoretic investigations of the well-known 3, 4, 5 triple resulted in the discovery of the Pythagorean equation (Loomis 1968, pp. 3–4), though this is disputed (Neugebauer 1969, p. 36). The apparent inconsistency of using a 3, 4, 5 right

triangle is justified for its convenience of illustration and didactical appeal. In Chinese the sides of a right triangle are *ku*, the longer leg; *kou*, the remaining leg; and *shian*, the hypotenuse. In my explanations, I shall represent these sides by k_1, k_2, and s, respectively. The problem numbers indicate the order in which the problems appear in "Kou-ku."

Chinese units of linear measure employed in the problems include these:

1 *chang*	=	10 *ch'ih*
1 *pu* (pace)	=	5 *ch'ih*
1 *ch'ih* (foot)	=	10 *ts'un* (inch)

With this perspective established, let us examine some problems involving the "piling up of squares."

(1) *Given: kou* = 3 *ch'ih, ku* = 4 *ch'ih*
What is the length of shian?
 Answer: shian = 5 *ch'ih*
 Method: Add the squares of *kou* and *ku*. The square root of the sum is equal to *shian*.
 Explanation: See Figure 1.

FIGURE I

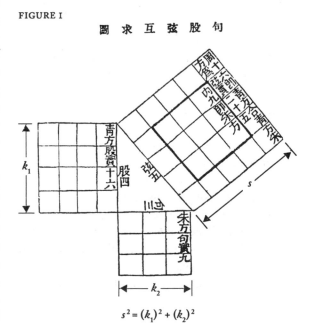

$$s^2 = (k_1)^2 + (k_2)^2$$

(6) *Given:* In the center of a square pond whose side measures 10 *ch'ih* grows a cattail whose top reaches 1 *ch'ih* above the water level. If we pull the reed toward the bank, its top becomes even with the water's surface. What is the depth of the pond and the length of the plant?

Answer: The depth of the water is 12 *ch'ih*, and the length of the plant is 13 *ch'ih*.

Method: (See Fig. 2.) Find the square of half the pond's width: from it subtract the square of 1 *ch'ih*. The depth of the water will be equal to the difference divided by twice the height of the reed above water. To find the length of the plant we add 1 *ch'ih* to the result.

FIGURE 2

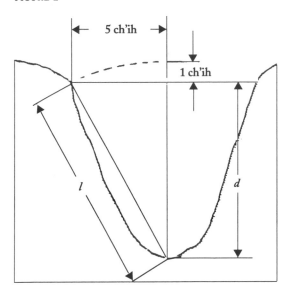

Explanation: It is known that the sum of the squares erected on the legs of the right triangle (see Fig. 3) should equal the area of the square erected on the hypotenuse, that is, $(d + 1)^2 = d^2 + 25$. Manipulating the resulting squares to perform the indicated operations, we obtain the configuration shown in Figure 4. Since the shaded portion of Figure 4 is equal in area to d^2, the remaining two rectangles and square must be equal in area to 25.

$$25 = 2d + 1$$

FIGURE 3

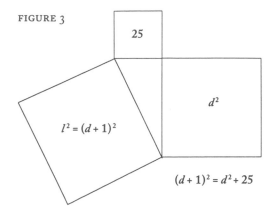

FIGURE 4

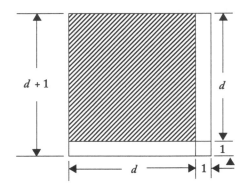

If we subtract the square of 1 *ch'ih* as instructed, then the sum of the areas of the remaining two rectangles are equal to 24, that is, $2d = 24$. The area of one rectangle, d, will then correspond to the depth of the water, 12 *ch'ih*. To obtain the total length of the reed, l, we add the depth of the water, 12 *ch'ih*, to the length of the reed extending above the surface, 1 *ch'ih*, and obtain the desired result, 13 *ch'ih*.

A more poetic rendering of this problem is found in the later writing of the Indian mathematician Bhāskara (1114–ca. 1185):

> In a certain lake, swarming with red geese, the tip of a bud of a lotus was seen a span (9 inches) above the surface of the water. Forced by the wind, it gradually advanced and was submerged at a distance of two cubits (approximately 40 inches). Compute quickly, mathematician, the depth of the pond. [Collidge 1940, p. 17]

The similarity of these two problems, even to the ratio of the distances involved, is striking. Such replication gives rise to a controversy concerning the influences of early Chinese mathematics on later Hindu writings (Needham 1959, pp. 146–50).

(11) *Given:* The height of a door is 6 *ch'ih* 8 *ts'un* larger than the width. The diagonal is 10 *ch'ih*. What are the dimensions of the door?

Answer: width = 2 *ch'ih* 8 *ts'un*

height = 9 *ch'ih* 6 *ts'un*

Method: (See Fig. 5.) From the square of 10 *ch'ih* subtract twice the square of half of 6 *ch'ih* 8 *ts'un*. Take half of this result and obtain its square root. The width of the door is equal to the difference of this root and half of 6 *ch'ih* 8 *ts'un*, and the height of the door is equal to the sum of the root and half of 6 *ch'ih* 8 *ts'un*.

FIGURE 5

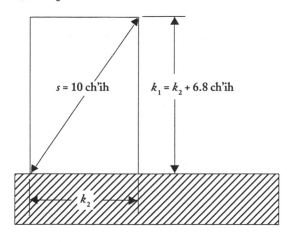

Explanation: Using the original solution diagram as given in Figure 6, we can derive the following equations:

(1) $(k_1 + k_2)^2 = 4(k_1 + k_2) + (k_1 - k_2)^2$;

(2) $s^2 = 2(k_1 \times k_2) + (k_1 - k_2)^2$,

which implies

$$2(k_1 \times k_2) = s^2 - (k_1 - k_2)^2;$$

FIGURE 6

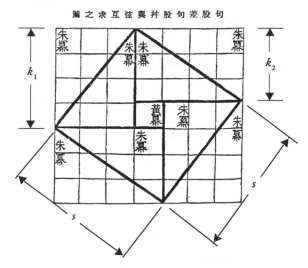

陽 之 求 互 弦 與 并 股 句 差 股 句

(3) $(k_1 + k_2)^2 = s^2 + 2(k_1 \times k_2)$.

Combining (2) and (3), we arrive at

(4) $(k_1 + k_2)^2 = 2s^2 - (k_1 - k_2)^2$.

The last equation supplies us with a readily workable expression involving the unknowns and the given.

$$k_1 + k_2 = \sqrt{2s^2 - (k_1 - k_2)^2}.$$

$$k_1 + k_2 = \sqrt{2(10)^2 - (6.8)^2}$$

$$= \sqrt{153.76}$$

$$\text{width} = \frac{\sqrt{153.76} - 6.8}{2} = 2.8 \ ch'ih$$

$$\text{height} = \frac{\sqrt{153.76} + 6.8}{2} = 9.6 \ ch'ih$$

The diagram of Figure 5 is known as the *Hsuan-thu* and first appeared in the mathematical classic *Chou pei suan ching* [The arithmetical classic of the gnomon and the circular paths of heaven]. It represents one of the earliest known proofs of the Pythagorean theorem. Admired for its simple elegance, it also found its way into the work of Bhāskara.

The Chinese ability to extract the square root of 153.76 attests to their highly developed computational proficiency. In this particular operation, the methods of the Han mathematicians surpassed those of other contemporary societies (Lam 1970).

(14) *Given:* Two men starting from the same point begin walking in different directions. Their rates of travel are in the ratio 7:3. The slower walks toward the east. His faster companion walks to the south 10 *pu* and then turns toward the northeast and proceeds until both men meet. How many *pu* did each man walk? (See Fig. 7.)

Answer: The fast traveler: 24 $\frac{1}{2}$ *pu*

The slow traveler: 10 $\frac{1}{2}$ *pu*

Method: (See Fig. 8.) The circuit traveled forms a right triangle. The three sides of the triangle are in the ratio given by the following magnitudes:

northeast route, $(7^2 + 3^2) \div 2$

southward route, $\dfrac{-(7^2 + 3^2)}{2} + 7^2$

eastern route, 3×7

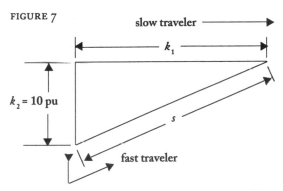

FIGURE 7

The distance traveled to the northeast is then found to be

$$10 \times \frac{(7^2 + 3^2)}{2} \div 7^2 - \frac{(7^2 + 3^2)}{2} \,;$$

and to the east,

$$10 \, (3 \times 7) \div 7^2 - \frac{(7^2 + 3^2)}{2}$$

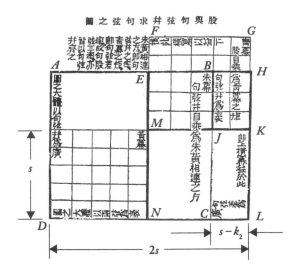

FIGURE 8

Explanation: The area of rectangle $ABCD$ = $(k_2 + s)^2$. The area of rectangle $EFGH$ + area of rectangle $BHKJ$ = $(k_1)^2$; this follows, since the area of the square $FGKM$ = $(s)^2$ and the area of the square $EBJM$ = $(k_2)^2$. The area of rectangle $EFGH$ is seen to equal the area of rectangle $JKLC$; therefore, it can be seen that the area of the square $AHLD$ = $(k_2 + s)^2 + k_1^2$ $(2s)(k_1 + s)$; and therefore one half the area of the square $AHLD$ = $s(k_2 + s)$, which supplies the term of the proportion relative to s. Similarly, (the area of rectangle $EHLN$) = (the area of the rectangle $BHLC$)

$$= (k_2 + s) - (k_1)^2$$

$$= s^2 + (k_2)(s) - (k_1)^2$$

$$= (k_2)^2 + (k_2)(s)$$

$$= k_2(k_2 + s),$$

which supplies a term for the proportion relative to k_2. The term for the proportion relative to k_1 is found to be $k_1(k_2 + s)$; thus

$$\frac{s}{29} = \frac{k_2}{20} = \frac{k_1}{21}.$$

Replacing "k_2" by the distance traveled southward, 10 *pu*, we obtain the distances traveled in the other directions:

$$\frac{s}{29} = \frac{10}{20} = \frac{k_1}{21};$$

northeast distance walked (s)

$$= \frac{29}{2} = 14\,^{1}/_2 \; pu;$$

eastward distance traversed (k_1)

$$= \frac{21}{2} = 10\,^{1}/_2 \; pu;$$

total distance covered by the faster man

$$= 10 + 14\,^{1}/_2 = 24\,^{1}/_2 \; pu;$$

(15) *Given:* (A right triangle with) *kou* = 5, *ku* = 12. What is the largest square that could be inscribed in the triangle?

Answer: A square with side 3 and 9/17 *ch'ih*

Method: The side of the square will be given by the quotient of the product of 5 and 12 and the sum of 5 and 12.

Explanation: The product of *ku* (k_1) and *kou* (k_2) is represented by the rectangle *ACKH* in Figure 9. For illustration purposes the dimensions are changed to 12 and 6, respectively. From the diagram the following relations are obvious:

area of triangle *ABD*

= area of triangle *GKJ*

area of square *BCED*

= area of square *FGJH*

area of triangle *AGF*

= area of triangle *DEK*

Now if the rectangle is dissected and the pieces rearranged as indicated in Figure 10, it becomes obvious that the side of the desired square is given by $\frac{(k_1)(k_2)}{k_1 + k_2}$. Although the Han authors supply a

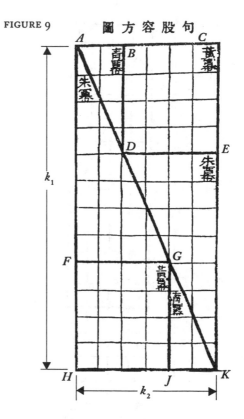

FIGURE 9   句股容方圖

proof for a particular case, with a little effort this method can be altered to prove the general theorem that the measure of the side of the square inscribed in a right triangle is given by the quotient of the product of the measure of the legs of the triangle and the sum of the measure of the legs of the triangle.

Given: (A right triangle with) *kou* 8 units long and *ku* 15 units. What is the largest circle that can be inscribed in this triangle?

Answer: A circle with a diameter of six units.

Method: Find the length of *shian* from *kou* and *ku*. The diameter will be the quotient of twice the product of *kou* and *ku* and the sum of *kou, ku,* and *shian*.

Explanation: From the diagram given in Figure 11, we can obtain the following information: the area of triangle *ACD* equals the sum of the areas of triangles *AFO, AEO, FCO,* and *GCO,* and the area of the rectangle *EOGD*. Since the area of the triangle

FIGURE 10

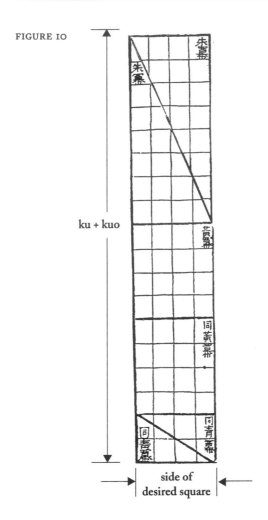

ku + kuo

side of
desired square

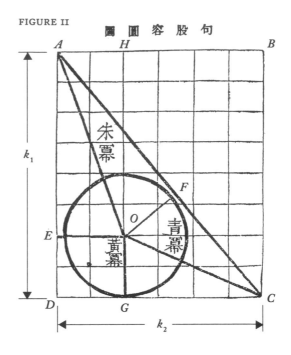

FIGURE 11

圖 圓 容 股 句

ACD is also equal to $\dfrac{(k_1)(k_2)}{k_1 + k_2}$,

$2(k_1)(k_2)$ = 4 × area of triangle ACD

= 4 × (area of triangle AFO + area of triangle AEO) + 4 × (area of triangle FCO + area of triangle CGO) + 4 × (area of triangle $EOGD$).

This in turn equals 4 × (area of rectangle $AHOE$) + 4 × (area of rectangle $OJCG$) + 4 × (area of rectangle $EOGD$) = 2 × (radius) $(k_1 + k_2 + s)$; therefore.

$$2(k_1)(k_2) = 2 \times (\text{radius}) \times (k_1 + k_2 + s),$$

or

$$\text{diameter} = \frac{2k_1 k_2}{k_1 + k_2 + s} \ .$$

Conclusions

The *chi-chü* method of solution is applied to eleven of the twenty-four exercises in "Kou-ku." The traditional Chinese proof of the Pythagorean theorem the *Hsuan-thu* employed in obtaining a solution to problem 11 is admired by Coolidge in his history of geometry for its simple elegance, as is also the solution for the inscribed square in problem 15 (Coolidge 1940, pp. 21–22).

The high degree of facility with which the Han mathematicians could employ such geometrically conceived algebraic solution schemes attests to a keen sense of spatial perception and an appreciation of the additive property of area measure. One cannot help but wonder about the influence of manipulative tangram exercises on the development of this ability. Recent research on tangrams attributes their origin to the Chou dynasty (Li and Morrill 1971), predating the appearance of the *Chiu-chang*, and implies certain psychological and educational designs in their conception. Although the conjecture of tangram exercises being more than mere children's games is intriguing, its resolution awaits the results of more detailed research on the intent and origin of Chinese tangrams (Gardner 1974).

The methods of "piling up squares" are not completely unknown in American classrooms. Commercial models do exist for the demonstration of the Pythagorean theorem given in problem 1, and for years many teachers have employed concrete models to justify intuitively the expansions of such expressions as $(a + b)^2$, $(a - b)^2$, $(a + b) (a - b)$, and their cubic analogues. Multibase arithmetic blocks used by Dienes or Montessori also employ a technique involving manipulation of squares; however, such demonstrations are usually ends in themselves and do not lead into problem-solving situations. Once students have been introduced to this technique and are given a supply of large-grid graph paper and cardboard, they can explore and devise their own algebraic-geometric solutions for selected problems. It is hoped that this exposition will help in broadening the scope of possible mathematics-laboratory activities.

NOTE

*Several problems of the *Chiu-chang* have been translated in Yoshio Mikami's *Development of Mathematics in China and Japan* (New York: Chelsea Publishing Co., 1913) and in Lam Lay Yong's "Yang Hui's Commentary on the *Ying-nu* of the *Chiu chang suan shu*" (*Historia Mathematica*, February 1974, pp. 47–64). A complete translation of the ninth chapter has been done by Frank Swetz and T. I. Kao in "Right Triangle Computational Techniques of Early China as Revealed by the Kou-ku of the Chiu Chang Suan Shu" (unpublished manuscript, Harrisburg, Pa., 1973). Translations of the complete work are available in Russian (E. I. Berezkina, "Drevnekitajsky Traktat Matematika v devjati Knigach," *Istoriko-matematiceski issledovaniya* 10 [1957]: 423–584) and in German (Kurt Vogel, *Chiu Chang Suan Shu: Neun Bucher Arithemetischer Technik* [Brunswick, Germany: Friedrick Vieweg & Sohn, 1968]).

REFERENCES

Coolidge, Julian Lowell. *A History of Geometrical Methods.* New York: Oxford University Press, 1940.

Gardner, Martin. "Mathematical Games: On the Fanciful History and the Creative Challenges of the Puzzle Game of Tangrams." *Scientific American,* August 1974, pp. 98–103B.

Lam, Lay Yong. "The Geometrical Basis of the Ancient Chinese Square-Root Method." *Isis,* Fall 1970, pp. 92–101.

Li, H. Y., and Sibley S. Morrill. *I Ching Games.* San Francisco: Gadlean Press, 1971.

Loomis, Elisha Scott. *The Pythagorean Proposition,* 1927, 1940. Reprint (Classics in Mathematics Education, vol. 1). Washington, D.C.: National Council of Teachers of Mathematics, 1968.

National Council of Teachers of Mathematics. *Historical Topics for the Mathematics Classroom.* Thirty-first Yearbook. Washington, D.C.: The Council, 1969.

Needham, Joseph. *Science and Civilisation in China,* vol. 3. Cambridge: At the University Press, 1959.

Needham, Joseph, and Wang Lin. "Horner's Method in Chinese Mathematics: Its Origins in the Root Extraction Procedures of the Han Dynasty." *T'oung Pao* 43 (1955): 345–88.

Neugebauer, Otto. *The Exact Sciences in Antiquity.* New York: Dover Publications, 1969.

Swetz, Frank. "The Amazing *Chiu Chang Suan Shu.*" *Mathematics Teacher* 65 (May 1972): 423–30. [This article appears as chapter 52 in the present volume.]

54

From Ancient China 'til Today!

PHILLIP S. JONES

Sun-Tsu, a Chinese of the first century A.D., proposed the problem of finding a number which when divided by 3, 5, and 7 would give the remainders 2, 3, 2 respectively. He gave in verse a rule for this problem which is equivalent to saying that $23 + 3 \cdot 5 \cdot 7 n$ is a solution for all integral values of n.

This rule did not apparently become known in Europe until after 1852. But in the meantime Nichomachus, a Greek neo-Pythagorean of about A.D. 100, had given the same problem and the answer 23, and the Chinese Yih-hing of the seventh century had generalized the problem. Similar remainder problems had been worked on by the Hindus Aryabhata (ca. 500), Brahmagupta (ca. 620), Bhaskara (ca. 1130), the Arab Ibn Al-Haitam (ca. 1000), Leonard of Pisa (1201), and many others.[1]

Our chief purpose here, however, is not to trace the detailed history of the Chinese problem of remainders, but to point out, in more or less chronological order, its connections with a number of other elementary topics. Some of these topics are of current interest because of their possible utility as enrichment for superior students or as simple examples of "different" mathematical systems such as are being called for in some proposals for curriculum revision. In particular, we shall point out relationships to linear equations, congruences in algebra (and so-called finite arithmetics or "arithmetics on a dial"), the Euclidean algorism, and continued fractions. Perhaps some of our readers with interests in these matters will send us further details of the history of the remainder problem, of the theories mentioned above, or of some of the other fanciful and amusing related problems.

Sun-Tsu's problem when written in modern symbolism would be to find an integral value for x satisfying the equations

$$\frac{x}{3} = y + \frac{2}{3}, \quad \frac{x}{5} = z + \frac{3}{5}, \quad \frac{x}{7} = w + \frac{2}{7},$$

or

$$x - 3y = 2, \quad x - 5z = 3, \quad x - 7w = 2$$

where y, z, w are also integers. Each of these is a linear equation which has an infinite number of solutions and which graphs as a straight line until we restrict ourselves to integral values. When we do this to the first of the equations, we see that since $x = 3y + 2$, every integral value assigned to y will lead to a corresponding integral value for x and this equation has an infinite number of both positive and negative integral solutions. However, if these solutions were graphed they would lie on the line represented by $x - 3y = 2$ at its so-called lattice points; i.e., those of the points on the line whose co-ordinates are integers.

From graphical considerations one can visualize that there could be linear indeterminate equations with only a finite number of lattice points in the first quadrant and even lines with no lattice

Reprinted from *Mathematics Teacher* 49 (Dec., 1956): 607–10; with permission of the National Council of Teachers of Mathematics.

points at all in any quadrant. The problem of determining the number and nature of the positive integral solutions of $ax + by = c$ was studied by Paoli in 1780, and it was Euler who gave, in 1760, the simplest proof that this equation is always solvable if a and b are relatively prime.[2] On the basis of this it is easily shown that $ax + by = c$ is solvable in integers if and only if the greatest common divisor of a and b also divides c.

To illustrate a method which works even when the problem does not have a unit coefficient to make it simple, as in the case of $x - 3y = 2$, let's take a problem given by the sixteenth-century Hindu writer Paramesvara to illustrate a rather obscurely described process given by both Aryabhata and Brahmagupta.[3] The complete problem is to find an integer which when multiplied by 8 and divided by 29 gives a remainder of 4, and which when multiplied by 17 and divided by 45 gives a remainder of 7.

The first condition is expressed by the equation $8x - 29y = 4$. The solution may be obtained by dividing 29 by 8, 8 by the remainder of this division, etc., until a remainder of 1 is obtained; then, in a sense, reversing the process to find values of x and y such that $8x - 29y = 1$; and finally multiplying these values by 4, thus:

$$
\begin{array}{r}
8 \underline{|\,29\,|3} \\
24 \\
\hline
5 \underline{|\,8\,|1} \\
5 \\
\hline
3 \underline{|\,5\,|1} \\
3 \\
\hline
2 \underline{|\,3\,|1} \\
2 \\
\hline
1
\end{array}
$$

(1)

The last remainder may be rewritten in the form

(2) $1 = 3 - 1 \cdot 2,$

and the preceding steps as:

(3)
$$
\begin{aligned}
2 &= 5 - 1 \cdot 3 \\
3 &= 8 - 1 \cdot 5 \\
5 &= 29 - 3 \cdot 8
\end{aligned}
$$

Substituting from (3) into (2) we get:

(4)
$$
\begin{aligned}
1 &= 3 - 1 \cdot (5 - 1 \cdot 3) \\
&= 2 \cdot 3 - 1 \cdot 5 \\
&= 2(8 - 1 \cdot 5) - 1 \cdot 5 \\
&= 2 \cdot 8 - 3 \cdot 5 \\
&= 2 \cdot 8 - 3(29 - 3 \cdot 8) \\
&= 11 \cdot 8 - 3 \cdot 29.
\end{aligned}
$$

Thus $x = 11$, $y = 3$ satisfies $8x - 29y = 1$, and $x = 44$, $y = 12$ will satisfy $8x - 29y = 4$.

The process displayed in step (1) is also the Euclidean algorism for finding the greatest common divisor of 8 and 29 (see Euclid's *Elements*, Book VII, Proposition 2, for its geometric origin and any theory of equations or higher algebra book for its algebraic proof and use). The fact that this g.c.d. is 1 means both that 8 and 29 are relatively prime and that the substitution process (4) which gave the solution of $8x - 29y = 1$ will be possible.

Of course, 44 and 12 are not the only integral solutions of our equation. It can be shown that if (x_0, y_0) is an integral solution of $ax + by = c$ then $x = x_0 + bt$, $y = y_0 - at$ is a solution for all integral values of t.[4] Thus, for our example $x = 44 - 29t$, $y = 12 - 8t$, and for $t = 1$, we have $x = 15$, $y = 4$. This last is the result originally obtained by Paramesvara by using a "chain" described rather vaguely by both Aryabhata and Brahmagupta. Öre gives an algorism for constructing a chain which is essentially the same as the Hindu chain, but easier to set up and explain.[5] His chain is constructed by writing the quotients from the algorism (1) as in the center, or q_i column here and then constructing the chain in a parallel c_i column by arbitrarily writing 1 and 0 in the c_i column below the level of the last q_i and then filling in the c_is *from the bottom up* by multiplying each q_i times the c_{i+1} on the line below it, then adding the c_{i+2} on the second line below it; i.e., $c_i = q_i c_{i+1} + c_{i+2}$. The two top links in the chain, 11, 3, are a solution of the equation $8x - 29y = 1$. These produce a solution of the original equation when multiplied by 4.

i	q_i	c_i
1	3	11
2	1	3
3	1	2
4	1	1
5	1	
6		0

Öre's chain can be derived as a general rule by generalizing the g.c.d. process and the "reverse" substitutions which we used in our first solution. Paramesvara's interpretation of Aryabhata's process begins with the g.c.d. division process and constructs a chain by the same rule as Öre's except that by properly choosing two numbers different from the 0 and 1 with which we began our second column he ends with a solution of $8x - 29y = 4$ immediately.

We leave it to the reader to complete the solution of the original problem, and instead note that the g.c.d. process could have been replaced by the conversion of the fraction 8/29 into a simple continued fraction thus:

$$(5) \qquad \frac{8}{29} = \cfrac{1}{\cfrac{29}{8}}$$

$$= \cfrac{1}{3 + \cfrac{5}{8}}$$

$$= \cfrac{1}{3 + \cfrac{1}{\cfrac{8}{5}}}$$

$$= \left\{ \cfrac{1}{3 + \cfrac{1}{1 + \cfrac{1}{1 + \cfrac{1}{2}}}} \right.$$

The convergents of the final continued fraction are $1/3$, $1/4$, $2/7$, $3/11$, $8/29$. Note that the denominator and numerator of the next to the last convergent are again our solution, $(11, 3)$ of $8x - 29y = 4$.

This continued-fraction approach will work on all linear indeterminate equations in two unknowns. Such equations are one type of Diophantine equation, indeterminate equations for which all their integral solutons are sought. These are named after the Greek algebraist Diophantus, who lived in the first century after Christ. However, Diophantus himself left no writings on linear indeterminate equations, although he did work with second degree and other "Diophantine equations." But in general he was satisfied with finding a single solution where later mathematicians have sought to find all of the solutions.

Although some idea of continued fractions has been read into this work of Aryabhata's and into the early Greek treatment of quadratic irrationalities, it seems an exaggeration to attribute to them such a general and theoretical knowledge of continued fractions as makes it possible to say that the penultimate convergent of such a fraction will always give a solution of the corresponding linear indeterminate equation. Italian Pietro Antonio Cataldi (1548–1676) [sic] and Swiss Leonard Euler (1737) were chief among the early workers who had a real conception of continued fractions.[6]

Finally, we note that C. F. Gauss in his *Disquisitiones Arithmeticae* of 1801 defined two integers a and b to be *congruent* modulo m when their difference is divisible by m. That is, $a \equiv b \bmod m$ if and only if $(a - b) = my$ or $my + b = a$ for some integer y. Hence the congruence $x \equiv b \bmod m$ can be written $my + b = x$ and the solution of the congruence becomes equivalent to the solution of a linear indeterminate equation. Thus linear equations, their graphs and the related lattices are related to "modular arithmetics" or "arithmetics on dials," and in the related algebra of congruences some linear "equations" (congruences) may have no (integral) solutions or a finite or infinite number of positive (or negative) solutions. Modular arithmetics and congruences furnish rather simple and elementary illustrations of the existence of different but comparable structures in algebra, just as non-Euclidean geometries illustrate the existence of different structures in geometry. Modular arithmetics and congruences also demonstrate the nature and role of definitions and axioms in the foundations of mathematics.

NOTES

1. L. E. Dickson, *History of the Theory of Numbers* (Washington: The Carnegie Institution, 1920), vol. [II, pp. v–vi, 57–64.

2. *Loc. cit.*

3. Walter Eugene Clark, *The Aryabhatiya of Aryabhata* (Chicago: University of Chicago Press, 1930), p. 44 ff. See also Henry Thomas Colebrooke, *Algebra with Arithmetic and Mensuration from the Sanskrit of Brahmegupta and Bhaskara* (London: 1817), p. 325 ff.

4. Oysteen Öre, *Number Theory and Its History* (New York: McGraw-Hill Book Co., 1948), p. 149.

5. Öre, *op. cit.*, pp. 142–149.

6. See H. S. Hall and S. R. Knight, *Higher Algebra* (London: Macmillan, 1932), pp. 111–112, Chaps, XXV, XXVI for a simple explanation of continued fractions and indeterminate equations. For historical comments and references see E. T. Bell, *The Development of Mathematics* (New York: McGraw-Hill Book Co., 1940), pp. 277, 444, and D. E. Smith, *History of Mathematics* (Boston: Ginn and Co., 1923), vol. I, pp. 156, 303. [According to the *Dictionary of Scientific Biography*, ed. Charles C. Gillespie, 1970–1980, Cataldi lived from 1552 to 1626.—F. S.]

SUGGESTED READINGS FOR PART V

Al-Daffa, Ali A. *The Muslim Contributions to Mathematics*. Atlantic Heights, NJ: Humanitites Press, 1977.

Berggren, J. L. *Episodes in the Mathematics of Medieval Islam*. New York: Springer-Verlag, 1986.

Clark, W. E., ed. *The Aryabhatiya of Aryabhata*. Chicago: University of Chicago Press, 1930.

Datta, B. and Singh, A. N. *History of Hindu Mathematics*. Bombay: Asia Publishing House, 1962.

Institute of the History of Natural Sciences, Academy of Sciences. *Ancient China's Technology and Science*. Beijing: Foreign Languages Press, 1983.

Joseph, George Gheverghese. *The Crest of the Peacock*. London: I. B. Tauris, 1991.

Karpinski, L. C., ed. *Robert of Chester's Latin Translation of the Algebra of al-Khowarizmi*. New York: Macmillan, 1915.

Kasir, D. S., ed. *The Algebra of Omar Khayyam*. New York: Columbia Teachers College, 1931.

Li Yan and Du Shiran. *Chinese Mathematics: A Concise History*. Translated by John Crossley and Anthony Lun. Oxford: Clarendon Press, 1987.

Libbrecht, Ulrich. *Chinese Mathematics in the Thirteenth Century*. Cambridge, MA: MIT Press, 1973.

Needham, Joseph. *Science and Civilisation in China*. "Mathematics." Vol. 3, pp. 1–168. Cambridge: Cambridge University Press, 1959.

Swetz, Frank J. "Mysticism and Magic in the Number Squares of Old China." *The Mathematics Teacher* 71 (January, 1978): 50–56.

Swetz, Frank J., with Kao, T. I. *Was Pythagoras Chinese? An Examination of Right Triangle Theory in Ancient China*. University Park, PA: The Pennsylvania State University Press, 1977.

PART VI

THE
REVITALIZATION
OF
EUROPEAN
MATHEMATICS

*F*ROM THE THIRTEENTH CENTURY onwards, increased technological advances and the expansion of European mercantile activities spurred an interest in applications of mathematics and improved methods of computation. These needs were satisfied by the appearance of a new profession; that of reckoning master. Reckoning masters were skilled in the techniques of arithmetic and counter computation; they could figure tariff and exchange rates, compute interest charges, do bookkeeping, gauge barrels and casks, survey property lines, and even cast horoscopes upon demand. This new breed of entrepreneur was versatile and talented. In Italy, where this profession first appeared, reckoning masters were called *maestri d'abacco*. As trade activities moved northwards, the Germanic territories had their *Rechenmeister* and later in France *maîtres d'algorism* appeared. Reckoning masters not only practiced in the lucrative field of mathematics but they also taught it to a growing number of youths attracted to the uses of mathematics and wrote manuals of arithmetic for a waiting reading audience. Knowledge of the Hindu-Arabic numerals, their use in written algorithms, and the problem solving techniques of arithmetic transmitted through a flood of manuals and tracts helped further popularize the use of mathematics. The newly developed art of printing with movable type and the practice of publishing materials in local languages rather than scholastic Latin helped achieve a rapid dissemination of this knowledge and forced a standardization of terms and usage. Gradually, the repeated printed image of certain words encouraged a use of abbreviation and symbolism. For example, the Italian word for "the unknown" or "thing" was *cosa;* on a single typeset page of an arithmetic book *"cosa"* could appear a dozen times during an explanation. Such repetition prompted Luca Pacioli in writing his *Summa de arithmetica, geometrica, proportiono et proportionalita* (1494) to introduce the abbreviation *co* for *"cosa."* Thus the rhetorical algebra that had persisted from classical times now advanced to a syncopated stage from which it would soon move into a symbolic form. Printed or written computations allowed for examination, retrospection, and possible computational or algebraic experimentation and the development of new mathematical theories and techniques. Pacioli, whose *Summa* was believed to be the compendium of fifteenth-century mathematical knowledge, noted in its closing that an algebraic solution for the cubic equation was impossible. But within forty years solution techniques for the cubic equation were developed and perfected by such mathematicians as Scipione del Ferro, Nicolo Tartaglia, Girolamo Cardano, and Rafael Bombelli. In the work done in obtaining the solution process for cubic equations, a foundation was laid for the development of a theory of equations. Algebraic manipulations now encompassed the use of imaginary numbers. Improved astronomical techniques and measurements prompted a maturing of trigonometry and highlighted an urgency for improved computational efficiency and mathematical accuracy. In part to meet these needs, a variety of physical computing devices were invented and computational capacities were strengthened by Simon Stevin's systematization of decimal fractions (1585) and the appearance of John Napier's logarithms (1614).

During the period of European history from the thirteenth to the sixteenth century, the use of mathematics increasingly found its way into the activities of daily life and mathematics became essential to people's resurrected quest to understand the world. As Galileo Galilei (1594–1642) noted: "This grand book, the universe . . . cannot be understood unless one first learns to comprehend the language and read the letters in which it is composed. It is written in the language of mathematics." No longer would this comprehension be the sole preserve of the clergy and academics. Mathematical knowledge was broadened, popularized, and raised to a status of high expectations. These expectations would rapidly be expanded and realized in the following centuries.

55

Adam Riese

DOROTHY I. CARPENTER

REFLECT, FOR A MOMENT, on the reeducation and records-changing entailed if we were to replace our illogical weights-and-measures systems by systems employing metric units. Such a Gargantuan task probably dooms us to the status quo. Yet a comparable change was effected in Europe during the period from the twelfth century to the sixteenth century as people gradually rejected Roman numerals in favor of the Hindu-Arabic. Many advocates of the new system have long been forgotten by even their own countrymen, but not Adam Riese. It was this arithmetic teacher and textbook writer who was largely responsible for the German acceptance of computation with numerals in lieu of the counting board. So influential was he that now, more than four centuries later, his name is still being perpetuated in the folk phrase *nach Adam Riese* used in connection with results in arithmetic computation, much as we might say that something is "according to Hoyle."

Riese was born in 1492 in Staffelstein, a town in Upper Franconia. Nothing is known of his early childhood. When he was fourteen his father died; and two years later he and a younger brother, Conrad, who had attended the famous Zwickauer Latin School, left home to make their own way. Although there is no record of formal education, Riese's arithmetics credit certain problems as being received from various individuals as early as 1509. This would indicate that by the time he

reached seventeen he was becoming proficient in the art of reckoning. His aptness in geometry is attested by this anecdote which has survived: A self-confident surveyor-engineer once wagered that he could construct more right angles in a given length of time than could Riese. Before the surveyor had completed the construction of his first perpendicular, Riese had drawn a great quantity of right angles in a semicircle.

FIGURE I
Portrait of Adam Riese from a woodcut on the title page of the 1550 edition of Riese's *Rechnung nach der lenge/auff den Linien vnd Feder.*

For a number of years Riese seems to have wandered about, returning once to Staffelstein for the contested settlement of his brother Conrad's estate. But there is no evidence that he received his just share, and he never went back again. During these years he visited arithmetic schools, examined their books, and was generally dissatisfied with the quality of instruction.

Reprinted from *Mathematics Teacher* 58 (Oct., 1965): 538–43; with permission of the National Council of Teachers of Mathematics.

His own first arithmetic appeared in 1518, a second in 1522. There is a record of his being comptroller of a mine in Epperstein in the spring of 1525. Later that year he married Anna Lewber, of Freiberg, bought a house, and took a freeman's oath. From 1529 to 1532 he was employed as mine comptroller in Marienberg. There is also evidence that he served as a *Rechenmeister*, or "master arithmetician," in Erfurt and Annaberg. In the smaller cities of the sixteenth century a *Rechenmeister* often assisted the town clerk in computational duties, serving as superintendent of measures, examiner of casks, etc. In 1536, at the request of city officials of Annaberg, Riese compiled and published a table of "bread regulations," whereby for a fixed price the weight of bread and rolls must vary with the price of grain. This booklet served as basis of the city's bread laws for over a century.

By 1550, Riese's reputation was so well established that he could risk asking the king for a five-year privilege for his arithmetic book. Along with this recognition, assistance came from Elector Moritz von Sachsen, who advanced the cost of printing.

To appreciate Riese's significance as "the arithmetic teacher of Germany" [2],* let us remind ourselves of several facts. First, printing had been known in Europe for fewer than seventy-five years when Riese's first book was published. Thus relatively few books on any subject were yet available to the common people, and this was especially true of textbooks in arithmetic. (Among the more important predecessors of Riese were Johannes Widman, who had published in 1489, and Jacob Koebel, whose text of 1514 ran through 22 editions.) Secondly, although Hindu-Arabic numerals had been introduced into Germany over a hundred years before, few but the monks and scholars were acquainted with them, and the common people were scarcely aware they existed. The latter used the Roman numerals and the counting board almost exclusively, and regarded the mysterious

new ciphers with considerable apprehension and superstition. Thirdly, some officials of Riese's day commanded that Roman numerals be used because alteration of records was less easily accomplished thereby. In fact, Riese himself had to keep his mining accounts in Roman numerals at the same time that he was advocating the Hindu-Arabic numerals in his books and using them in the bread regulations. Furthermore, it was the custom of the period to guard one's skills rather zealously. Since the thirteenth century, arithmetic masters had tended to settle in the large cities, where they often united in guilds, kept the city accounts, and were frequently well paid for lessons in arithmetic.

Many of the schools of Riese's day were conducted in Latin by the clergy and were only incidentally concerned with the practical art of reckoning. In a Wittenberg Church Order of 1533 one finds the statement: "After they can read, write and sing, one ought also in time teach them ciphers and something from arithmetic." While there were also "German schools" (for reading and writing, but usually not for arithmetic), and writing and arithmetic schools, one can see that, in general, the populace had little opportunity to learn the new symbols and how to compute with them. Typical of the situation of the day is this quotation from a French arithmetic published near the end of the fifteenth century: "... for there are many merchants who cannot read and write but yet must reckon." So as trade expanded, there was increased demand for those trained in the indispensable art of reckoning.

Adam Riese was one of the most successful of those who set up arithmetic schools to meet the above need. Then, as now, a natural consequence was to put courses of instruction into printed form. The oldest arithmetic book printed in Italian (1478) and the first one in German (1482) were both by a *Rechenmeister* of Nuremberg. Thus it is not surprising that Riese's books show direct evidence of teaching experience. In particular, in the 1529 edition, he explains that he has found that the

---

* Numerals in square brackets refer to the bibliography at the end of the article.

young people who begin with instruction "on the lines" (that is, with the equivalent of the use of a counting board) "obtain a better foundation" and "ought to execute their computation with the ciphers with less labor" [4]. His texts were designed to be used by the teachers in the arithmetic schools rather than by the pupils, but they were directed to the level of the people's education.

In his first arithmetic book, published in 1518, Riese himself used only "line reckoning." But his second book, *Rechnung auff der Linien vnd Federn,* published in 1522, began with a brief explanation of the familiar line reckoning and then led its readers into detailed and systematic use of the new "ciphers." He contended that they "will not be wearisome to learn, but ought to be comprehended with joy and gladness" [4]. This text was the most used of Riese's four arithmetics, and was published in eight different German towns. His 1550 edition is described as the "best exponent of the practical arithmetic of the middle of the century in Germany" [8]. In all, over forty editions of his textbooks appeared in the sixteenth century, and several more were published in the seventeenth century and were in use until the middle of that century. Their quality and popularity are attested to by a contemporary, Michael Stifel, who called them "elegant." He and others liked them so well that they "borrowed" some of Riese's illustrative examples—without due credit!

In 1524 Riese began an algebra text, *Die Coss,* completing it only through methods of solving first-degree equations. This book was never printed, and the manuscript was not found until 1855. Hence it exerted no influence. But it does serve now as representative of the state of algebra at the beginning of the sixteenth century.

Since the various editions differed primarily in the extent of their practical problems, one table of contents will suffice to show the organization and general scope of Riese's books: numeration, addition, subtraction, duplation, mediation, multiplication, division, progressions, rule of three, money exchange, profit, silver and gold computa-

tion, partnership, reduction (of fractions). Other editions included problems of inheritance, guardianship, coinage, alloys, proportion, and magic squares.

Compared with other texts, Riese's arithmetics were "head and shoulders above those of his predecessors and contemporaries in arrangement and preparation of subject matter, in the form of his presentation, and in the logical and superior execution of his pedagogical principles" [4]. These principles included: (1) proceeding from the concrete to the abstract (reckoning with counters to computation with ciphers), (2) passing from the simple to the complex, without and with short cuts, (3) passing from the particular to the general, and (4) offering numerous exercises to perfect a given technique.

While giving due credit to the above now commonly accepted teaching procedures, we must note that all of Riese's solutions omit development and proof, and instead are completely dogmatic. After the statement of each problem, he regularly continues, "Do it thus," or "Set it so," as in the following coin-exchange problem, literally translated from the 1559 edition:

> Seven florin from Padua make 5 at Venice, and 10 at Venice make 6 at Nuremberg, and 100 from Nuremberg make 73 at Köln. How many make 1000 fl. from Padua to Köln. It makes 312 and 6/7. Set it so:

7 Padua	5 Venice
10 Venice	6 Nurenberg 1000 Pad.
100 Nuremberg	73 Köln

> Multiply the first ones with one another, the same also the middle ones. Stand it so:

$$7000 \ldots 2190 \text{ fl.} \ldots 1000$$

In modern notation we might write: Let

x = value of one florin from Padua,

y = value of one florin from Venice,

z = value of one florin from Nuremberg,

w = value of one florin from Köln.

Then we have the system of equations

$$7x = 5y,$$
$$10y = 6z,$$
$$100z = 73w,$$

from which we obtain

$$7000x = 2190w,$$

and

$$x = 312\frac{6}{7}w.$$

Likewise, from the 1559 edition comes this problem, solved by the method of double false position:

A son asks his father how old he is. The father answers him saying: When you were yet as much older, half as old, and $^1/_4$ as old, and a year older, so were you already 100 years old [That is: When you shall be yet as much older, half your age still older, one fourth your age still older, and a year still older, you will be 100 years old.] The question is how old is the son?

Make thus. Take for yourself two numbers which encompass in themselves a half and a fourth [that is, are divisible by 2 and by 4], as 40 and 48. Examine those in the problem according as the 40 also. Say 40 but 40, half 20, the fourth are 10 and one year more make in sum 111 years. From that take the 100 years, leaving 11 years more. Likewise examine also the 48. Stand thus:

40	more	11
		22
48	more	33

Compute, so comes 36 years. So old is the son.

Using the 48 as one possible age, we obtain 48 + 48 + 24 + 12 + 1 = 133, which exceeds the 100 by 33. Previously, Riese had explained the remainder of such a solution as follows: "Take 11 from 33, leaving 22 the divisor. Thereafter cross-multiply, take one from the other, and divide." That is,

$$48 \times 11 = 528,$$
$$40 \times 33 = 1320,$$
$$1320 - 528 = 792,$$
$$792 \div 22 = 36.$$

For the most part, Riese's calculations and techniques are similar to ours. Among the exceptions may be noted that he taught subtraction by the equal additions method, that the divisor was written below the dividend and the remainders above it, and that the names million, billion, etc., were not used. Instead, he instructed that the number 86,789,325,178 be read: "86 thousand thousand times thousand, 700 thousand times thousand, 89 thousand times thousand, 300 thousand, 25 thousand, 178" [1].

The accompanying Figure 2 shows a page, from Riese's 1559 edition, devoted to division by what has come to be known as the "scratch method."** The division, at the top of the page, of 95,472 by 12, is accomplished by moving the divisor from left to right as the multiplication and subtraction are performed mentally and the remainders written above. The separate steps would appear as follows:

```
1
2 1
9 5 4 7 2 (7
1 2

1 2
2 1 6
9 5 4 7 2 (7 9
1 2 2
1

1 2 1
2 1 6
9 5 4 7 2 (7 9 5
1 2 2 2
1 1
```

** Sometimes called the "galley method" because of the pattern's resemblance, when viewed from the left, to a ship with a mast.

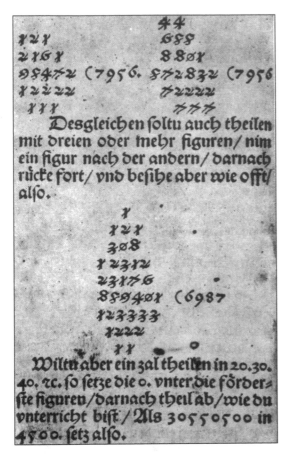

FIGURE 2
Scratch division problem from the 1559 edition of Riese's *Rechnung auff der Linien vnd Federn auff allerley Handtierung.* Courtesy of the University of Michigan Library, Rare Book Room.

In the example shown in Figure 2, Riese illustrates division by a three-digit number: 859,401 divided by 123.

In the case of rather involved computations, Treutlein conjectured that Riese made use of his knowledge of algebra and then explained the solution in words, as in this example:

21 persons, men and women, have spent 81 pf. for drinking. A man ought to give 5 pf. and woman 3 pf. Now I ask how much have been each in particular. Set it so:

man 5

21 persons   81 pf.

woman 3

Take 3 pf. from 5 pf., leaving 2 the divisor. Now multiply 3 with 21, come 63, which take from 81 pf., leaving 18, which dividing with 2 some 9 men, which take from 21 persons, leaving 12. So much are the women [1].

We see that, if x denotes the number of men and $21 - x$ the number of women, the above explanation fits exactly the solution of the linear equation $5x + 3(21 - x) = 81$.

As Berlet assesses Rieses's greatness, "it consists not alone of his dexterity in arithmetic, but especially in that he recognized the adaptability of Hindu-Arabic numerals to calculation and demonstrated it to the German people by spoken and written word." Truly do his books "guard a part of their cultural heritage" [3].

BIBLIOGRAPHY

1. Berlet, Bruno. *Adam Riese, sein Leben, seine Rechenbücher und seine Art zu rechnen.* Leipzig: E. V. Mayer, 1892.

2. Cantor, Moritz. *Vorlesungen über Geschichte der Mathematik* (2nd ed.). Leipzig: B.G. Teubner, 1913. Vol. II, pp. 385–89.

3. Deubner, Fritz. "Adam Riese der Rechenmeister des deutschen Volkes," Sachsische Heimblatter (Dresden), V (1959), 602-8.

4. Falckenberg, Hans. "Adam Riese, ein deutsche Rechenmeister," *Deutsche Mathematik* (Leipzig), III (February 1938), 1–8.

5. Menninger, Karl. *Zahlwort und Ziffer.* Breslau: Vandenhoeck & Ruprecht, 1934. Pp. 252-54, 265, 340–45.

6. ———. *Zahlwort und Ziffer.* Göttingen: Vandenhoeck & Ruprecht, 1957. Vol. I, pp. 154–55.

7. Riese, Adam. *Rechnung auff der Linien vnd Federn auff allerley Handtierung.* Leipzig: Hans Rhambaw, 1559.

8. Smith, David Eugene. *Rara Arithmetica.* Boston: Ginn & Co., 1908. Pp. 138–40, 250.

9. ———. *History of Mathematics.* Boston: Ginn & Co., 1923. Vol. I, pp. 337–38.

10. Unger, Friedrich. *Die Methodik der praktischen Arithmetik in historischer Entwicklung vom ausgange des Mittelalters bis auf die Gegenwart 1888.* Leipzig: B. G. Teubner, 1888. Pp. 48–53.

56

Tangible Arithmetic IV: Finger Reckoning and Other Devices

PHILLIP S. JONES

$\mathcal{T}$HE FINGER SYMBOLS SHOWN in our Figure 1 were once used to communicate at international fairs and as an aid in remembering numbers when using an abacus. This picture is from Philippi Calandri's *Arithmetice* published in Florence in 1518. The first edition of this book is an incunabulum which means that it was printed during the "cradle" days of printing, which are taken to be its first fifty years, 1450–1499. The first edition was printed in 1491, but finger symbols existed in Greece as early as the fifth century B.C. and it has been stated that some also existed in China as early as the time of Confucius (551–478 B.C.)[1] Smith notes that finger symbols even appear in art and literature. For example, he cites a passage from

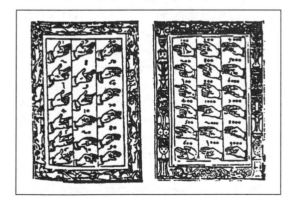

FIGURE I

Reprinted from *Mathematics Teacher* 48 (Mar., 1955): 153–59; with permission of the National Council of Teachers of Mathematics.

Juvenal who said, "Happy is he who . . . numbers his years upon his right hand," referring to the fact that the numbers less than one hundred were on the left hand and greater than one hundred on the right.[2]

There were even devices for using fingers to operate with numbers as well as to record them. For example, a device for multiplying integers between 5 and 9 was to hold up fingers on one hand to show by how much the multiplier exceeded 5 and on the other hand to show similarly the amount by which the multiplicand exceeded 5. The product of the original numbers was the number whose tens digit was given by the sum of the fingers held up and whose units digit was the product of the fingers turned down on the two hands. For example 7×9 would lead to 2 and 4 fingers up and 3 and 1 fingers down. This would give the product 63. For this process one needed to know the multiplication tables only up to 5×5. The explanation of this device is enrichment material for algebra as well as for arithmetic since it hinges on the identity

$$ab = 10\,[(a-5) + (b-5)] \\ + [(10-a)(10-b)].$$

Other devices for recording numbers which should be included in an exhibit, display, report, or assembly program on tangible arithmetic are the *quipu*[3] used by Peruvian natives before the time of the Spanish conquerors, and *tally sticks*[4] used in

FIGURE 2

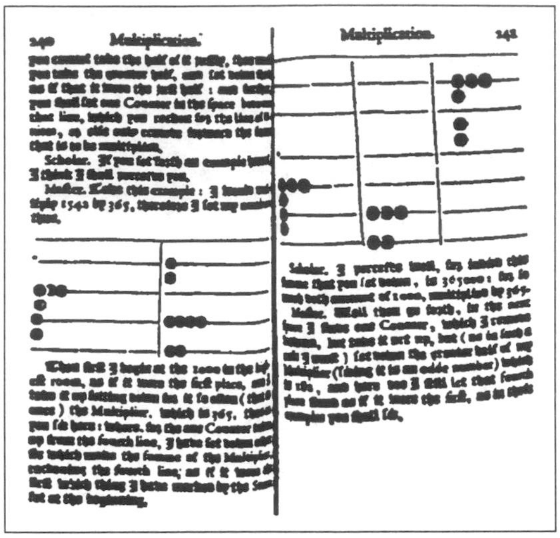

FIGURE 3

medieval England. Encyclopedias will give some help on these items as well as the references cited in the notes.

Among computing devices the *abacus*[5] is so well known that we will only mention it here with a few references, and the *counting board* has been so recently discussed in this journal[6] that we will only add Figures 2 and 3 to the earlier data. Figure 2 is the title page of a German book on "Rechnen-pfenigen" by J. Köbel published in 1514. It shows a merchant at his counting board with the lines drawn or cut into it. Note the title which reads "A new kind of little reckoning book (telling of reckoning) on lines with reckoning pennies." These

"pennies" were often little stones from whose Latin name, "calculi," we get our words "calculate" and "calculus" just as our word "counter" for a table or bench in a store comes from the medieval merchant's counting board.

Figure 3 shows an explanation of the multiplication of 1542×365 on a counting board from the most famous early English arithmetic. This is Robert Recorde's *The Grounde of Arts* first published about 1542. Our picture is from the 1654 edition and shows one of the many interesting features of Recorde's works; namely, they were written in the form of a dialogue between the "Master" and his "Student." The arithmetics we have pictured display three significant common and related characteristics: their period, the late fifteenth and early sixteenth century; their use of the common or "vulgar" languages rather than Latin, the scholarly international language; and their concern with the computations of commerce.

Figures 4 and 5 relate to two other well-known and recently discussed computing devices, *Gunter's scale* and the *slide rule*.[7] Figure 4 shows the scale as it appeared in the sixth (1698) edition of William Leybourn's *The Line of Proportion or Numbers, Commonly Called Gunters Line, Made Easie. . . .* Although Edmund Gunter was an Englishman, he

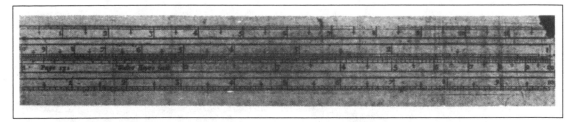

FIGURE 4

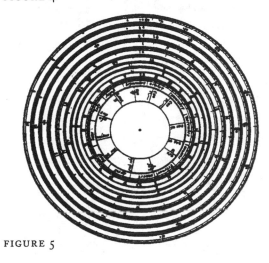

FIGURE 5

published his first discussion of his scale in French in Paris in 1624. The first discussion of the straight slide rule, which would seem the natural next step after Gunter's scale, was published by William Oughtred in 1633.

However, the first slide rule invented was a circular rule described by Richard Delamain in 1630. Oughtred, Delamain's teacher at one time, claimed to have invented a circular rule earlier, but he did not publish a description of it until 1632. Our Figure 5 is a picture of this rule from the 1660 edition of his book, *The Circles of Proportion*.

The *nomogram*[8] is a recently developed computing chart which uses logarithmic scales and could well be included in a tangible arithmetic project. Frenchman Maurice d'Ocagne began its modern phase in 1891.

Computing machines are so in the public eye and press today that it is not necessary to do more than mention them in the connection. Clippings, pictures, demonstrations of desk calculators and an outline of their historical development[9] would be part of a "tangible arithmetic" project.

Calandri's Arithmetic:

The book from which our Figure 1 came is such a fascinating little work that we include a few more pictures from it. Figure 6 is its title page. This illustrates a false notion of that day, that Pythagoras introduced the Hindu-Arabic numerals and their arithmetic into Europe. The square form of the multiplication table was also often

FIGURE 6

FIGURE 7

FIGURE 8

called the table of Pythagoras. Actually, the Greek "arithmetica" which owes much of its origins to the early Pythagoreans was the beginnings of what today we call number theory. "Arithmetica" neither used Hindu numerals nor concerned itself with "logistica," the Greek counterpart of modern elementary arithmetic.

Figures 7 and 8 show some of the illustrated pages from Calandri. These are of interest not only because Calandri's was the first printed Italian arithmetic with illustrations accompanying the problems, but also because the problems themselves are so similar to current ones that one can figure them out from the pictures, the numbers, and a few hints. Try them on some of your students. Figure 7 shows a serpent who each day climbs 1/7 "bracchia" (a unit of measure) out of a well only to slip back 1/9 bracchia. If the well is 50

bracchia deep how long does it take him to climb out?

The other picture in Figure 7 asks how long does it take to fill a trough or tub if water enters at a rate such that it would be full in 4 days if it weren't leaking at a rate which would empty it in 11 days?

Figure 8 asks at what height above the ground the 50 bracchia tree should be cut in order that its tip should touch the bank of the river at a distance of 30 bracchia as shown. Clearly this involves the Pythagorean theorem, but one might challenge students to set up algebraic expressions which would explain the process by which Calandri arrived at the correct answer, 16.

Also to be found in this book are "work problems," a lion, leopard, and wolf who eat at different rates, and ants who travel toward each other at

362

different rates. Problems on the altitude of an equilateral triangle and circumference and area of a circle are also included in this book which was the first to give our modern form of long division and to omit the old "galley" type of division.[10]

NOTES

1. D. E. Smith, *History of Mathematics*, Vol. II (Boston: Ginn and Co.), p. 197. Chapter III, pages 156-207, titled "mechanical aids to computation" discusses many tangible computing devices.

2. Smith, *loc. cit.*

3. Smith, *op. cit.*, p. 195.

4. Smith, *op. cit.*, p. 192. Vera Sanford, "Art of Reckoning: I. Tallies," *The Mathematics Teacher*, Vol. XLIII (Oct. 1950), p. 292.

5. Smith, *op. cit.*, pp. 156-192. Vera Sanford and Yen Yi-Yun, "The Chinese Abacus," *The Mathematics Teacher*, Vol. XLIII (Dec. 1950), p. 402 ff. T. Kojima, *The Japanese Abacus, Its Use and Theory* (Rutland, Vt.: 1954). Jerry Adler, "So You Think You Can Count!" *Mathematics Magazine*, Vol. 28 (Nov.–Dec. 1954), pp. 83–86.

6. Smith, *op. cit.*, p. 181. Vera Sanford, "Counters: Computing If You Can Count to Five," *The Mathematics Teacher*, Vol. XLIII (Nov. 1950), p. 368.

7. Phillip Jones, "The Oldest American Slide Rule," *The Mathematics Teacher*, Vol. XLVI (Nov. 1953), p. 501. Florian Cajori, "On the History of Gunter's Scale and the Slide Rule During the Seventeenth Century," *University of California Publications in Mathematics*, Vol. 1 (Feb. 17, 1920), pp. 187–209. Smith, *op. cit.*

8. Florian Cajori, *A History of Mathematics* (New York: Macmillan, 1919), p. 481. Douglas P. Adams, "The Preparation and Use of Nomographic Charts in High School Mathematics," *Eighteenth Yearbook*, National Council of Teachers of Mathematics, pp. 164-181.

9. Smith, *op. cit.*, p. 202. Phillip S. Jones, "The Binary System," *The Mathematics Teacher*, Vol. XLVI (Dec. 1953), p. 575. Philip and Emily Morrison, "The Strange Life of Babbage," *Scientific American*, Vol. 186 (April 1952), pp. 66–73.

10. D. E. Smith, *Rara Arithmetica* (Boston: Ginn and Co., 1908), p. 47 ff.

57

The Cardano-Tartaglia Dispute

RICHARD W. FELDMANN

Introduction

As the Middle Ages came to a close, a rebirth of scientific inquiry occurred. Most large cities had universities, but these consisted primarily of lecturers, as there were few books beyond those of the classical authors. Hence the reputation of a university depended upon its ability to provide the best lecturers. But who was the best lecturer? One way to settle this was in a public debate with the winner gaining prestige and academic positions while the loser was ignored.

Since a scholar had to outperform all challengers to maintain his position, he needed every trick he knew. If someone knew something that no one else knew, his fame was insured. One assurance of success would be a knowledge of the general solution to the previously unsolved cubic equation. It was the search for this solution that led to the dispute between Gerolamo Cardano and Niccolo Tartaglia. But it is impossible to discuss the debate without first giving brief biographies of the antagonists to show why they acted as they did.

The Life of Tartaglia

Niccolo Tartaglia—his actual family name was Fontana—was born in Brescia, Italy, about the turn of the sixteenth century.[1] When the French sacked Brescia in 1512, the youthful Fontana suf-

Reprinted from *Mathematics Teacher* 54 (Mar., 1961): 160-63; with permission of the National Council of Teachers of Mathematics.

fered severe saber cuts about the face and mouth which caused a speech impediment. Because of this he was nicknamed Tartaglia, "the stutterer." He later wore a long beard to hide the scars, but he could never overcome the stuttering.

His education was very meager. In fact he tells, in one of his books, that his mother had accumulated a small amount of money so that he might be tutored by a writing master. The money ran out, so he says, when the instruction reached the letter K and his education stopped before he could write his own initials. But being an enterprising youth, he stole the master's lecture notes and completed the course on his own. In times of extreme poverty he even used tombstones in place of writing slates.

His mathematical knowledge was completely self-taught, but his attempts to teach himself Latin resulted in failure and he was forced to write his treatises in Italian instead of the accepted Latin. He taught primarily in Venice, and, once established, lived quite comfortably on his wages and the wagers he won in public disputes and challenges.

The Life of Cardano

The other participant, Gerolamo Cardano,[2] was born in Milan in 1501. His father, a lawyer and lecturer in geometry, married Cardano's mother, who came from a "socially unacceptable" family, a few years after Cardano's birth. This led to later complications.

He started his studies at the University of Pavia, but transferred to the University of Padua when war broke out. At college he supplemented his meager finances by constant gambling, at which he became very proficient. He wrote a treatise, *Liber de ludo aleae,*[3] which not only introduced the idea of probability as we use it today, but also included ways to cheat in these games.

At twenty-five, he graduated as a doctor of his chosen profession, medicine. Due to his illegitimate birth, he was not allowed to practice in Milan. Later he was recognized as a doctor and proved his ability by becoming the second most renowned medical expert in Europe at that time. Most of his 412 written works[4] were in the field of medicine, popular science, and astrology. One of these was an autobiography in which he describes himself as follows: "living from day to day, outspoken, despising religion, mindful of injuries caused by others, envious, melancholy, a spy, a betrayer of trusts, a deviner, a caster of charms, subject to frequent failures, hateful, given to shameful lust, solitary, unpleasant, strict, foretelling the future willingly, jealous, obscene, lascivious, abusive in speech, inconsistent, two-faced, dishonest, a slanderer, entirely unparalleled in vices, and by nature incompatible even with those with whom he converses daily."[5]

His sons were "chips off the old block," as one robbed his father and the other married and later murdered an immoral girl.

In 1570, Cardano was arrested and jailed on a charge of heresy. The charge could be substantiated by a horoscope of Christ, and by a book which Cardano had published praising Nero, the Roman emperor well known for his persecution of the early Christians. After he was convicted, he was forced to admit and renounce his heresies. As a punishment he was denied the right to lecture publicly and was ordered to refrain from writing and publishing books. Heartbroken, he accompanied one of his students to Rome, where he received an invitation to become a consultant to the College of Physicians. A pension was received from the Pope and soon, possibly insane,[6] he died.

A student of Cardano's who received fame as a mathematician was Ludovico Ferrari. He entered Cardano's house as a servant, soon was elevated to secretary, and became a public lecturer before he was twenty years old. His major contribution to mathematics was a general solution to the biquadratic equation $x^4 + ax^2 + b = cx$. Unfortunately his career was cut short by premature death when in his early forties.

The Dispute[7]

About 1510, Scipione del Ferro found a general solution to $x^3 + ax = b$, but he died before he could publish his discovery. His student, Antonio Maria Fiore, knew the solution and attempted to gain a reputation by exploiting his master's discovery. He challenged Tartaglia with thirty questions, all of which reduced to the solution of $x^3 + ax = b$. Tartaglia had the general solution to $x^3 + ax^2 = b$, so he responded with thirty questions of a more general theoretical nature, although some resolved to this equation. Besides the prestige to be gained, the winner and his friends were to receive thirty banquets from the loser. Just before the time limit elapsed, Tartaglia found general solutions to both $x^3 + ax = b$ and $x^3 = ax + b$.[8] With these, Tartaglia solved all of Fiore's problems, but Fiore was unable to solve any of the questions proposed by Tartaglia and so was vanquished. The banquets were not collected.

At this time Cardano was writing the *Practica arithmeticae generalis,* which would encompass arithmetic, geometry, and algebra. Inasmuch as Fra Luca Pacioli had earlier stated that there could not be a general solution to the cubic, Cardano had ignored this topic. Upon hearing that Tartaglia had a solution for $x^3 + ax = b$, he tried to find one. Failing, he asked Tartaglia for the solution so that he might publish it in a special section of the *Practica arithmeticae generalis* under Tartaglia's name. Tartaglia refused, stating he would publish the solution himself at a later date. This prompted Cardano to label Tartaglia as greedy and unwilling to help mankind.

Because of these insults, a correspondence developed between the two mathematicians which resulted in Tartaglia visiting Cardano in Milan. During this visit Tartaglia relented and offered Cardano a cryptic poem containing a solution to $x^3 + ax = b$, provided Cardano swore an oath that he would never reveal the solution. Cardano accepted the terms, but was unable to decipher the code, so he asked for and received the necessary clue from Tartaglia. Using the solution to $x^3 + ax = b$, Cardano and Ferrari found solutions for $x^3 + ax^2 = b$, $x^3 = ax^2 + b$, and $x^3 + b = ax^2$ by employing substitutions which reduced them to the known case.

In 1545, Cardano published the *Ars magna*, which contained Tartaglia's solution of the cubic with a statement that del Ferro and Tartaglia had each found solutions by independent research. Cardano also included some of his own discoveries, including the idea that every cubic should have three roots. Cardano also published Ferrari's solution to the biquadratic equation here, with due credit to Ferrari.

When he had seen the *Ars magna*, Tartaglia publicly denounced Cardano for breaking an oath sworn on the Gospels, and he ridiculed Cardano's mathematical ability.

Cardano disdained to refute the slur, but Ferrari attacked Tartaglia, charging that Tartaglia had built up his reputation by defaming others, had stolen one proof in his new book without giving credit, and in addition had at least one thousand errors in the text. Ferrari ended his published response by challenging Tartaglia to a public debate on mathematics and all related subjects.

Tartaglia answered with further insults and refused the debate on the grounds that Cardano knew the men who would be the judges. Perhaps he really feared a public debate because he stammered.

After a further exchange of insults, each proposed thirty-one questions which were exchanged, answered, and returned. However, no decision was reached, because each tore the other's answers to shreds.

Then for no given reason, Tartaglia accepted a debate to be held in Milan, Cardano's stronghold.[9]

On August 10th, 1548, Tartaglia and Ferrari met in combat, Cardano having left town.

Very little is known about the actual debate, but it appears to have degenerated into an invective match, with Tartaglia doing most of the shouting. Tartaglia left after the first day, claiming to have won, although it seems Ferrari won by default.

An indication of Ferrari's triumph is that Tartaglia lost his teaching post in Brescia, and Ferrari was invited to lecture in Venice, Tartaglia's stronghold.

Tartaglia died in 1557 without publishing his solution to the cubic, and when an attempt was made to publish his unpublished papers, none could be found which even mentioned the solutions to the cubic.

With Cardano's death in 1576, one of the most interesting and colorful episodes in the history of mathematics ended.

NOTES

1. The actual year is in doubt; Oystein Ore lists it as 1499 in *Cardano, the Gambling Scholar*, and D. E. Smith gives 1506 in his *History of Mathematics*.

2. Hieronymus Cardanus in Latin; also translated into English as Jerome Cardan.

3. *A Book on Games of Chance.*

4. At the time of his death, 131 of his works had been published, 111 existed in manuscript, and he claimed to have burned 170 which he found unsatisfactory.

5. A quote taken from the introduction by Gabriel Naude in Cardano's autobiography, *Liber de propria vita*, Amsterdam, 1654.

6. Charles W. Burr, *Jerome Cardan as Seen by an Alienist*, in University of Pennsylvania: University Lectures Delivered by Members of the Faculty in the Free Public Lecture Course, 1916–1917, vol. 4.

7. The two main sources for the dispute are Oystein Ore, *Cardano, the Gambling Scholar*, Princeton:

(Princeton University Press, 1953) and M. A. Nordgaard, "Sidelights on the Cardan-Tartaglia Dispute," *National Mathematics Magazine,* XII (1937), 327–346.

8. Due to the limited use of symbols and a lack of understanding of negative quantities, these equations had to be treated separately. For example, Cardano wrote $x^3 + 6x = 20$ as cub⁵ p: 6 reb⁵ aeqlis 20. Other methods of writing equations during this period can be seen in D. E. Smith's *History of Mathematics,* II, pp. 427–431.

9. Perhaps Brescia, Tartaglia's city, demanded it because of civic pride.

HISTORICAL EXHIBIT 6.1

Cardano's Technique for the Solution of a Reduced Cubic Equation

A solution is sought for an equation of the form

$$x^3 + ax = b, \, a > 0, \, b > 0.$$

It is known that

$$(p - q)^3 + 3 \, pq(p - q) = p^3 - q^3 \, ;$$

therefore, if we let

$$x = (p - q) \text{ then } a = 3 \, pq \text{ and } b = p^3 - q^3.$$

It follows then that:

$$p = \frac{a}{3q} \text{ and } b = \left(\frac{a}{3q} \right)^3 - q^3 \text{ or}$$

$$27b \ q^3 + a^3 - 27 \left(q^3 \right)^2 \text{ which can be rewritten as}$$

$$27 \left(q^3 \right)^2 + 27b \ q^3 - a^3 = 0.$$

This last equation is a biquadratic for which use of the existing quadratic solution scheme could supply a solution for q^3, that is,

$$q^3 = \frac{-b \pm \sqrt{b^2 + \dfrac{4a^3}{27}}}{2} \qquad \text{and}$$

q is found to be

$$q = \sqrt[3]{-\frac{b}{2} + \sqrt{\frac{b^2}{4} + \frac{a^3}{27}}} \, .$$

In a similar manner, the value for p is also obtained:

$$p = \sqrt[3]{\frac{b}{2} + \sqrt{\frac{b^2}{4} + \frac{a^3}{27}}} \, ,$$

and finally

$$x = p - q = \sqrt[3]{\frac{b}{2} + \sqrt{\frac{b^2}{4} + \frac{a^3}{27}}} - \sqrt[3]{\frac{-b}{2} + \sqrt{\frac{b^2}{4} + \frac{a^3}{27}}} \, .$$

58

Complex Numbers: An Example of Recurring Themes in the Development of Mathematics —I

PHILLIP S. JONES

*T*HE STORY OF THE DEVELOPMENT of the concept and uses of complex numbers provides a fine example of how the history of mathematics may shed light on the meaning of terminology, the relative roles of "practical" needs and intellectual curiosity in the motivation of mathematicians, the utility of pure mathematics, and the development not only of mathematics itself but also of the concepts of rigor and proof. The story also involves intrigue and illustrates the international nature of mathematical scholarship. Hence, properly told, this story does much more than merely provide interest and motivation; in any event, it surely does no less.

Some historians[1] trace, in a negative sort of way, the beginnings of the concept of an imaginary number to Heron's *Sterometria* (ca. A.D. 75). In this Heron states a problem about a pyramid which properly computed would have led to an imaginary number for the length of a line. Heron (or a later copyist) reversed the order of subtraction, and thereby missed both the imaginary number and the correct answer. To the writer this hardly seems to represent a phase in the development of complex numbers, but it does illustrate that in Heron's day, as in our own textbooks, not all apparently "real" problems were actually based upon real-life situations or measurements. Further this illustrates a situation which was later emphasized by René Descartes, that imaginary numbers or roots may be the algebraic counterpart of non-existent intersections in geometry, or of impossible physical conditions. In these situations the imaginary numbers obtained may however tell a real story since their mere occurrence may be significant in itself.

Still negative in its nature, but of real significance in the story of the growth of the idea because of its explicit recognition of the problem, is the statement of the Hindu Mahavir (ca. 850), "as, in the nature of things, a negative is not a square, it has no square root."[2]

The first printed book to contain algebra, Luca Pacioli's *Summa de Arithmetica Geometria* (Venice: 1494), in discussing quadratic equations, explained in words that a solution was possible only if the constant term was less than or equal to the square of one-half the coefficient of the first degree term (the coefficient of the second degree term being unity).

The real beginning of complex numbers is to be found in the work of the Italian, Jerome Cardan. His *Ars magna* was first printed in Nuremberg in 1545. Our Figure 1 showing the title page of its 1570 edition printed in Basel suggests many other things, however. Not only does it suggest the

Reprinted from *Mathematics Teacher* 61 (Feb., 1954): 106–14; with permission of the National Council of Teachers of Mathematics.

HIERONYMI

CARDANI MEDIO

LANENSIS, CIVISQV'E BONO,
NIENSIS, PHILOSOPHI, MEDICI ET
Mathematici clarissimi,

OPVS NOVVM DE

PROPORTIONIBVS NVMERORVM, MO
TVVM, PONDERVM, SONORVM, ALIARVMQV'E RERVM
mensurandarum, non solùm Geometrico more stabilitum, sed etiam
uarijs experimentis & obseruationibus rerum in natura, solerti
demonstratione illustratum, ad multiplices usus ac
commodatum, & in V libros digestum.

PRAETEREA
ARTIS MAGNÆ, SIVE DE REGVLIS
ALGEBRAICIS, LIBER VNVS, ABSTRVSISSIMVS
& inexhaustus planè totius Arithmeticæ thesaurus, ab
authore recens multis in locis recogni-
tus & auctus.

ITEM.
DE ALIZA REGVLA LIBER, HOC EST, ALGEBRAICAE
logisticæ suæ, numero s recondita numerandi subtilitate, secundum Geometricas quantitates inquirendis, necessaria Coronis,
nunc demum in lucem edita.

Opus Physicis & Mathematicis imprimis
utile & necessarium.

Cum Cæf. Maieft. Gratia & Priuilegio.

BASILEÆ.

FIGURE I

internationalism of printing revealed by the sources of these two editions, but also it points out the cosmopolitan nature of Cardan who is here titled *Mediolanensis* (of Milan) as well as *civisque Bononiensis* (citizen of Bologna). This recalls the heyday of the city-states of Italy, and the fact that Cardan was a contemporary of Machiavelli (1469–1527). This latter fact suggests that the famous "steal" of the solution of the general cubic by Cardan from Nicholas of Brescia (Tartaglia), using as a ruse the story of a noble sponsor, a possible patron, who was interested in Tartaglia's discovery, was quite in keeping with the political morals portrayed in Machiavelli's *The Prince,* as well as displaying something of the relationships between the persons of wealth, the nobility, and the scholars of the day.

Of further interest on the title page is the characterization of Cardan as *philosophi, medici, et mathematici clarissimi.* For not only was Cardan himself a remarkable person, famous in the history of medicine, cryptography, and gambling,[3] but also this suggests the parallel growth of philosophy and science with mathematics, and the continuing relationships which exist today among these disciplines.

Lastly, the words *Ars magnae, sive de Regulis Algebraicis* followed by a reference to *Arithmeticae* in contrast to which algebra was the "great art," with the added comment "*fecundum geometricas quantitates inquirentis*" (fruitful in inquiries concerning geometric quantities) shows that the interrelatedness of these three parts of mathematics was recognized then as it should be emphasized in our teaching now.

Figure 2 shows page 131 of the book whose title page we have just discussed. The first two lines state Cardan's famous problem "divide 10 into two parts such that the product of one times the remainder is (30 or) 40." Cardan immediately says *manifestum est—impossibilis.* He does not explain why this is manifestly impossible, but our students can be led to do good "functional" thinking by suggesting that they consider the products of pairs of numbers whose sum is 10, or on a higher level, that they graph $y = x(10 - x)$ and observe or even prove that its maximum is $y = 25$ for $x = 5$.

However, the fascinating thing is to note in the next step one way in which a mathematician, motivated by intellectual curiosity in this case, may move on to discover something new. Cardan says in line 3 (Fig. 2) *sic tamen operabimur,* "nevertheless we will operate," i.e., he says let's follow the procedures we used in other cases and see what happens. By completing the square he obtains $5 + \sqrt{-15}$ and $5 - \sqrt{-15}$. This is stated in line 8 using the abbreviations which characterized the "syncopated" algebra that was intermediary between the earliest period (sometimes called "rhetorical"), when all algebra was written out in words in full, and the modern "symbolic" period.

DE ARITHMETICA LIB. X.   111

exemplum, si quis dicat, diuide 10 in duas partes, ex quarum vnius in reliquam ductu, prodicat 30, aut 40, manifestum est quod casus seu quæstio est impossibilis, sic tamen operabimur, diuidemus 10 per æqualia, & fiet eius medietas 5, duc in se fit 25, auferes ex 25, ipsum producendum, ut pote 40, ut docui te, in capitulo operationum, in quarto libro, fiet residuum m:15, cuius R addita & detracta à 5, ostendit partes, quæ inuicem ductæ producunt 40, erunt igitur hæ, 5 p:R m:15, & 5 m:R m:15.

DEMONSTRATIO.

Vt igitur regulæ uerus pateat intellectus, sit a b linea, quæ dicatur 10, diuidenda in duas partes, quarum rectangulum debeat esse 40, est autem 40 quadruplum ad 10, quare nos uolumus quadruplum totius a b, igitur fiat a d, quadratum a c, dimidii a b, & ex a d auferatur quadruplum a b, absque numero, R igitur residui, si aliquid maneret, addita & detracta ex a c, ostenderet partes, at quia tale residuum est minus, ideo imaginaberis R m:15, id est differentiæ a d, & quadrupli a b, quam adde & minue ex a c, & habebis quæsitum, scilicet 5 p:R v: 25 m:40, & 5 m:R v: 25 m:40, seu, 5 p:m m:15, & 5 m:R m:15, duc 5 p:R m:15 in 5 m:R m:15, dimissis incruciationibus, fit 25 m:m:15 quod est p:15, igitur hoc productum est 40, natura tamen a b, non est eadem cum natura 40, nec a b, quia superficies est remota à natura numeri, & lineæ, proximius tamen huic quantitati, quæ uere est sophistica, quoniam per eam, non in puro m:nec in alijs: operationes exercere licet, nec uenari quid sit est, ut addas quadratum medietatis numeri numero producendo, & à R aggregati minuas ac addas dimidium diuidendi. Exemplum, in hoc casu, diuide 10 in duas partes, producentes 40, adde 25 quadratum dimidij 10, ad 40, fit 65, ab huius R minue 5, et adde etiam 5, habebis partes secundum similitudinem, R 65 p:5 & R 65 m:5. At hi numeri differunt in o, non iuncti faciunt 10, sed R 260, & hucusque progreditur Arithmetica subtilitas, cuius hoc extremum ut dixi, adeo est subtile, ut sit inutile.

QVÆSTIO IIII.

Fac ex 8 duas partes, quarum quadrata iuncta sint 50, hæc soluitur per primam, non per secundam regulam, est enim de puro m: ideo duc 3 dimidium 6 in se fit 9, minue ex dimidio 50, quod est 25, fit residuum R 2

5 p:R m:15
5 m:R m:15
25 m:m:15 q̄d. est 40

From Arabic times the proof of algebraic procedures had been given using geometric diagrams such as one sees in the *"Demonstratio"* in our Figure 2. In this case, however, Cardan had to resort in line 9 of this *"Demonstratio"* to *imaginaberis* ("You will imagine"). Although he gives below the diagram a computational check of his results by showing that $5 + \sqrt{-15}$ times $5 - \sqrt{-15}$ is $25 - (-15)$, which is 40, in the discussion he notes that these quantities are *uere sophistica* ("truly sophisticated"), and in the last line of the *"Demonstratio"* he states that continuing to work with these numbers would involve one in an *Arithmetica subtilitas*— "as subtle as it would be useless."

Thus we see a man proceeding to obtain by analogy and formalism results which he finds interesting but which he does not fully comprehend nor accept himself. In more recent times Oliver Heaviside had a similar experience when he developed and used his operational calculus in spite of the criticism of some of his mathematical contemporaries.

The first statement of the rules for operating with square roots of negatives is shown in our Figure 3 taken from the 1579 edition of Rafael Bombelli's *L'Algebra parte maggiore del arithmetica* published in Bologna (1st edition, 1572). Cardan's formulas for the roots of cubic equations involve complex numbers, oddly enough, in the so-called

PRIMO.   169

Ho trouato un'altra sorte di R.c. legate molto differenti dall'altre, laqual nasce dal Capitolo di cubo eguale à tanti, e numero, quando il cubato del terzo delli tanti è maggiore del quadrato della meta del numero come in esso Capitolo si dimostrarà, laqual sorte di R. q. hà nel suo Algorismo diuersa operatione dall'altre, e diuerso nome; perche quando il cubato del terzo delli tanti è maggiore del quadrato della meta del numero; lo eccesso loro non si può chiamare ne più, ne meno, però lo chiamarò più di meno, quando egli si douerà aggiongere, e quando si douerà cauare, lo chiamerò men di meno, e questa operatione è necessarissima più che l'altre R.c.L. per rispetto delli Capitoli di potenze di potenze, accompagnati cō li cubi, ò tanti, ò con tutti due insieme, che molto più sono li casi dell'agguagliare doue ne nasce questa sorte di R. che quelli doue nasce l'altra, la quale parerà à molti più tosto sofistica, che reale, e tale opinione hò tenuto anch'io, sin che hò trouato la sua dimostratione in linee (come si dimostrarà nella dimostratione del detto Capitolo in superficie piana) e prima trattarò del Moltiplicare, ponendo la regula del più & meno.

Più uia più di meno, fa più di meno.
Meno uia più di meno, fa meno di meno.
Più uia meno di meno, fa meno di meno.
Meno uia meno di meno, fa più di meno.
Più di meno uia più di meno, fa meno.
Più di meno uia men di meno, fa più.
Meno di meno uia più di meno, fa più.
Meno di meno uia men di meno fa meno.

irreducible case in which all three roots are real. He did not solve this case.

If $y^3 + py + q = 0$ Cardan's formula in modern form and symbols is $y = A + B$, where

$$A = \sqrt[3]{-\frac{q}{2} + \sqrt{\left(\frac{q}{2}\right)^2 + \left(\frac{p}{3}\right)^3}},$$

$$B = \sqrt[3]{-\frac{q}{2} - \sqrt{\left(\frac{q}{2}\right)^2 + \left(\frac{p}{3}\right)^3}}.$$

The other roots are $\omega A + \omega^2 B$ and $\omega^2 A + \omega B$ where $\omega = (-1 + \sqrt{3i})/2$. The "irreducible case" is that in which

$$\Delta = \left(\frac{q}{2}\right)^2 + \left(\frac{p}{3}\right)^3 < 0.$$

Bombelli showed how to combine the two complex numbers given by Cardan's formula in this latter case to obtain finally the correct real root. To do this he had to develop rules for operating with square roots of negatives. One can read his statement of these rules at the bottom of his page 169 by use of the following vocabulary:

> piu: a positive (quantity)
> uia: times
> di meno: a square root of a negative number
> meno: a negative (quantity)

Some perception of the difficulties involved in the "rhetorical" stage of algebraic symbolism may be obtained by looking at the end of the second and third lines of the page. Here Bombelli writes that he is treating "cubes equal to quantities and numbers" (i.e., $x^3 = px + q$). Note that he uses a different word for each power of the unknown, and that he deliberately writes his equations so that there are no negative signs. He goes on to say, beginning in line 7, that he will treat the case in which *il cubato del terzo del li tanti è maggiore del quadrato della metà del numero* ("the cube of one third of the [coefficient of] x is greater that the square of one half the constant"). Since Bombelli's

DISCOURS
DE LA METHODE
Pour bien conduire sa raison,& chercher
la verité dans les sciences.
PLUS
LA DIOPTRIQVE.
LES METEORES.
ET
LA GEOMETRIE.
Qui sont des essais de cete METHODE.

A LEYDE
De l'Imprimerie de IAN MAIRE.
cIɔ Iɔ c xxxvII.
Auec Priuilege.

FIGURE 4

original equation is set up such that the p of our modern symbolic form of Cardan's formulas would be negative, this condition amounts to requiring that $\Delta < 0$, the irreducible case noted above.

Figure 4 is the title page of René Descartes's most famous work. It emphasizes the connections between mathematics, philosophy, and logic, and the interesting fact that his *La Geometrie* was merely the third illustrative appendix to his *Discourse on the Method of Reasoning and Seeking Truth in Science*. (Note the old use of cIɔ and Iɔ for M and D in writing Roman numerals.)

Figure 5 shows page 380 of this book and points out further that *La Geometrie* was far from being anything like our modern analytic geometry texts and that, in fact, it contained much that was purely algebraic. There are three further things to be remarked about this page. The middle para-

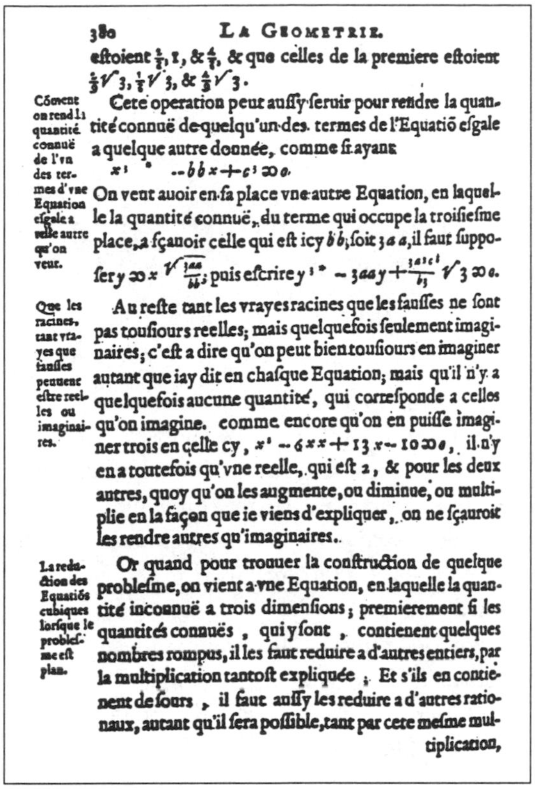

(La Geometrie, p. 380)

> 380   LA GEOMETRIE.
>
> estoient $\frac{1}{2}$, 1, & $\frac{4}{5}$, & que celles de la premiere estoient $\frac{1}{3}\sqrt{3}$, $\frac{1}{2}\sqrt{3}$, & $\frac{2}{5}\sqrt{3}$.
>
> *Cōment on rend la quantité connuë de l'vn des termes d'vne Equation esgale à l'autre qu'on veut.*   Cete operation peut aussi seruir pour rendre la quantité connuë de quelqu'vn des termes de l'Equatiō esgale a quelque autre donnée, comme si ayant
>
> $x^3 \ast -bbx + c^3 \infty 0.$
>
> On veut auoir en sa place vne autre Equation, en laquelle la quantité connuë, du terme qui occupe la troisiesme place, a sçauoir celle qui est icy bb, soit $3aa$, il faut supposer $y \infty x\sqrt{\frac{144}{bb}}$; puis escrire $y^3 \ast - 3aay + \frac{3a^3c^3}{b^3}\sqrt{3} \infty 0$.
>
> *Que les racines, tant les vrayes que fausses peuuent estre reelles ou imaginaires.*   Au reste tant les vrayes racines que les fausses ne sont pas tousiours reelles, mais quelquefois seulement imaginaires; c'est a dire qu'on peut bien tousiours en imaginer autant que iay dit en chasque Equation; mais qu'il n'y a quelquefois aucune quantité, qui corresponde a celles qu'on imagine. comme encore qu'on en puisse imaginer trois en celle cy, $x^3 - 6xx + 13x - 10 \infty 0$, il n'y en a toutefois qu'vne reelle, qui est 2, & pour les deux autres, quoy qu'on les augmente, ou diminue, ou multiplie en la façon que ie viens d'expliquer, on ne sçauroit les rendre autres qu'imaginaires.
>
> *La reduction des Equatiōs cubiques lorsque le problesme est plan.*   Or quand pour trouuer la construction de quelque problesme, on vient a vne Equation, en laquelle la quantité inconnuë a trois dimensions; premierement si les quantités connuës, qui y sont, contienent quelques nombres rompus, il les faut reduire a d'autres entiers, par la multiplication tantost expliquée; Et s'ils en contienent de sours, il faut aussy les reduire a d'autres rationaux, autant qu'il sera possible, tant par cete mesme multiplication,

FIGURE 5

graph and its marginal note contain the first use of "imaginary" as the name of these new numbers. They had earlier been described variously. (We have seen them called "sophisticated" and "subtle.") Unfortunately the term "imaginary" stuck and has come down to us as an historical hangover—and as a headache to teachers, since students naturally think of "imaginary" as a descriptive adjective rather than as merely an arbitrary name for a mathematical object which is as real in its existence as a mathematical point or line. Note further that Descartes's *fausses racines* ("false roots") denoted negative, not imaginary, numbers. He apparently was not entirely willing to accept negatives as being as completely valid as positive numbers and did most of his geometrical drawing in what we would call the first quadrant. This illustrates how each extension of our number system from integers to fractions, to irrationals, to complex, to

quaternions, to transcendentals has been first viewed with scepticism as an unreal or inapplicable abstraction. Acceptance and, eventually, fairly general understanding of such extensions have grown with familiarity, improved logical rigor and organization, graphical representation, and applications in both the mathematical and physical worlds. But these concepts, in their earliest days, bothered such geniuses as Descartes and Gauss.

All of these aspects appear in the growth of the acceptance of complex numbers. Their use in extending mathematics appears implicitly in the middle paragraph of our Figure 5 where Descartes is discussing the number of roots which an equation may have. Although, even before Descartes, Albert Girard and others had suggested that an *nth* degree equation may have *n* roots, it required a further expanded theory of complex numbers and their graphical representation before C. F. Gauss could prove the fundamental theorem of algebra, that every rational integral polynomial with real or complex coefficients has at least one root. From this the *n* root theorem follows directly.

Figure 6 serves to illustrate several phases in the development of ideas, as well as to point up again the continuity and internationalism so common in the history of mathematics. This picture is from the 1693 Latin edition of John Wallis's *Algebra*, which first appeared in English in 1685. The lower diagrams speak for continuity and internationalism because they appeared first in the work of Descartes. They show a geometrical construction for the roots of a quadratic equation of the type $x^2 \mp bx - c = 0$. (However, Wallis did not use x as Descartes did, but wrote $aa \mp ba - æ = 0$, using vowels for the unknown as Viete had done earlier. Wallis regarded the arbitrary constants as always representing positive numbers.)

If $CP = {}^1\!/_2 b$, and $PB = \sqrt{c}$ is perpendicular to CP, then the circle on CP determines the secant αB on which AB and $B\alpha$ are equal in length to the roots of the equation. (Note the connection between the geometric theorem that the product of the segments of a secant equals the square of the

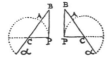

FIGURE 6

tangent from an external point, and the algebraic theorem that the product of the roots of this quadratic is c.)

The left-hand diagram at the top of the page gives Wallis' construction for the roots of $x^2 \pm bx + c = 0$ if $(b/2) \geq \sqrt{c}$. Take $AC = \frac{1}{2}b = C\alpha$ and $PC = \sqrt{c}$ perpendicular to $A\alpha$ at C. The radius $PB = \frac{1}{2}b$ with center at P determines the points B. The roots are then the numerical values of the two lengths AB.

If $(b/2) < \sqrt{c}$, the above construction fails, in which case Wallis used PC and PB to construct a right triangle with the right angle at B, as shown in the upper right hand diagram, rather than at C. He then says that the points B which are *above the line* may be regarded as representing the solution in

this case just as the B's *on the line* determined the solution when the roots were real. This, we see, was a step toward the modern graphical representation of complex numbers. Incidentally, recall that Descartes, whom Wallis had studied and admired, had interpreted imaginary roots for algebraic equations as indicating that the related geometric construction was impossible. Wallis also suggested representing $\sqrt{-1}$ by applying the Euclidean construction for a mean proportional to directed line segments representing $+1$ and -1, and he had argued that, just as people had become reconciled to representing both positive and negative numbers on a line and associating them with distances forward and backward, so one could imagine negative areas as land covered by rising water,[4] and use such a visualization to encourage the acceptance of negative squares.

Not long after Wallis, Lambert used complex variables in the development of map projections, and D'Alembert used them in the study of hydrodynamics. The stories of these and other applications, the story of the geometric representation of complex numbers, and the story of a modern rigorous formulation of an arithmetic for these numbers are interesting tales to which we will return later; but we hope that the story so far offers a way to an understanding of the numbers, their name, how and why they developed and the relative roles in their growth of application and abstraction, representation and generalization, practical necessity, and intellectual curiosity.

NOTES

1. D. E. Smith, *History of Mathematics* (Boston: Ginn and Co., 1925), p. 261. Smith also gives several references to other discussions of Heron's problem.

2. M. Rangacarya, *The Ganita-Sara-Sangraha of Mahaviracarya with English Translation and Notes* (Madras: 1912), p. 7 of the English translation. The sentence before the one quoted above was, "The square of a positive as well as of a negative (quantity) is positive; and the square roots of those (square quantities) are positive and negative in order."

3. There have been several biographies of Cardan. The most recent is by Oystein Öre, *The Gambling Scholar* (Princeton University Press, 1953). Cardan's own autobiography has been translated as *The Book of My Life* (New York: 1930, translated by Jean Stover; and London: J. M. Dent and Son, 1931), while his medical achievements are emphasized in Henry Morley's *Jerome Cardan: The Life of Girolamo Cardano of Milan, Physician* (London: Chapman and Hall, 1854). In addition to W.G. Walters, *Jerome Cardan a Biographical Study* (London: 1898), there have been many journal articles about him and his work as well as short biographies in books and in *Portraits of Eminent Mathematicians, Portfolio II* published by *Scripta Mathematica*.

4. A reprint of these discussions may be found in *A Source Book in Mathematics* edited by D. E. Smith (New York: McGraw-Hill and Co., 1929), pp. 46 ff.

Multiplication Algorithms of the Fifteenth and Sixteenth Centuries

The introduction of algorithmic computing schemes using "Hindu-Arabic" numerals in fifteenth century Europe was a complex and often confusing process. Algorithms for the multiplication of two multi-digit numbers had to build upon the basic multiplication facts and insure the proper preservation of place-value in the computation. Luca Pacioli in his *Summa de Arithmetica Geometria Proportioni et Proportionalita* (1492) described eight different algorithms for multiplication. Four of these methods are illustrated below for finding the product of 1234 and 56789.

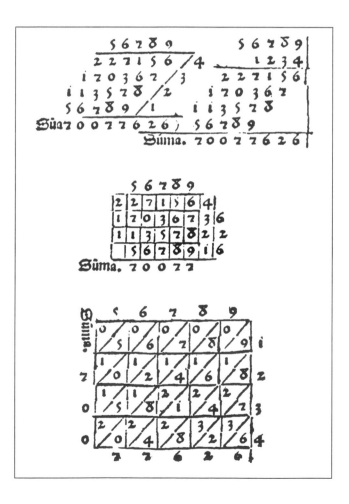

59

Robert Recorde's
Whetstone of witte, 1557

VERA SANFORD

$\mathcal{T}$HE YEAR 1957 is the four hundredth anniversary of the publication of Robert Recorde's *Whetstone of witte,* the first algebra printed in English, the book in which the equality sign (=) was used for the first time. It is appropriate to mark this anniversary by quoting Recorde's explanation of his invention and by outlining its later history.

Contrary to the natural assumption that the value of this symbol would have been recognized at once, it develops that fully sixty years elapsed before its second appearance, although in less than a century after that time, it had general acceptance. This situation suggersts that we examine the state of mathematics in England in the 1550s, that we review what we know about the author, and study the contents of the book itself.

I. The Equality Sign

At the beginning of his work with equations, Recorde says

> And to auoide the tediouse repetition of these woordes: is equalle to: I will sette as I doe often in woorke vse, a pair of paralleles, or Gemowe (twin) lines of one lengthe, thus: ＝＝ bicause noe .2. thynges can be moare equalle.

The equality sign is Recorde's single contribution to the symbolism of algebra. Although the signs + and − were already in use, having letters represent

unknown quantities was in the future. Algebra looked very different from the way it looks today.

In writing equations, Recorde's contemporaries used the word *equal* in its different forms as "aequales," "faciunt," "gleich." Sometimes they abbreviated these words. There was considerable experimentation with symbols, governed in part by the type the printers had on hand. In the years following the publication of the *Whetstone of witte,* different writers used a number of symbols to represent equality and the symbol (=) stood for a variety of meanings.[1] In view of this competition and in consideration of the confusion of meanings assigned to the twin parallels, it is all the more remarkable that Recorde's invention ultimately won universal adoption. The process was slow.

The second appearance of the equality sign was in an appendix to a translation (1618) of Napier's first work on logarithms. It is supposed that this appendix was the work of William Oughtred, whose *Clavis mathematicae* (1631), a closely-packed work on arithmetic and algebra, was influential in winning for Recorde's symbol a general acceptance in England in the seventeenth century. By the eighteenth century, it had come into use on the continent.

II. Robert Recorde and His Times

Robert Recorde (ca. 1510–1558) was born shortly after Henry VIII came to the throne of England. He died just before the accession of Elizabeth I.

Reprinted from *Mathematics Teacher* 50 (Apr., 1957): 258–66; with permission of the National Council of Teachers of Mathematics.

His contemporaries on the continent included Cardan and Tartaglia who were doing pioneer work in mathematics in Milan and Venice. There were the German algebraists Rudolff, Stifel, and Scheubel. Copernicus, from Poland, published his great work in 1543. In 1538, Mercator, in Flanders, produced a map of the world on a projection used earlier by a Spaniard, but now called by his name. Methods of navigation were being improved. Almanacs of a sort were being published. The variation and the dip of the magnetic needle were being studied. The problem of the determination of longitude baffled mathematicians and other scientists. Sebastian Cabot maintained that he had solved it, but he died in 1557 without divulging his secret, and it was supposed that his boasts were the imaginings of an octagenarian.

England lagged behind in this activity. In fact the one important publication in mathematics prior to Recorde was Cuthbert Tunstall's *De arte Svppvtandi* (1522), a scholarly arithmetic in Latin based largely on continental models. So far as navigation and exploration were concerned, it is true that in 1497 the king had financed the voyage of the Genoese mariners John and Sebastian Cabot to the New World. John Cabot died the next year. Royal support stopped at this point, and Sebastian Cabot entered the employ of the King of Spain.

It was in the first half of the sixteenth century, however, that the center of world trade began to shift. Prior to this time, goods from China, India, and the East Indies had been brought by caravan to the ports on the eastern shores of the Mediterranean where they were loaded on ships from Genoa and Venice and other Italian cities for distribution to different parts of Europe. The Mediterranean was still the great highway it had been in the days of Greece and Rome.[2] But in the period under consideration, sea routes were being developed. England no longer lay at the periphery of the world but was shortly to assume an important position as a maritime power.

So far as mathematics was concerned, Englishmen were not isolated from scholars on the continent. Tartaglia had dedicated a book to Henry VIII in 1546. John Dee (1527–1608) studied on the continent and knew some of the outstanding mathematicians personally. Cardan visited London on his way to Scotland in 1552. Robert Recorde himself seems to have read the works of a number of these men, and we know that he made considerable use of Scheubel's algebra which was published in Paris in 1551.

But England had no mathematician to match the best of those in Europe, and so far as applied mathematics was concerned, the situation was deplorable. In England, navigation had not become a science. Map making was primitive. Surveying was casual. Little mathematics was taught. Arithmetic was neglected. Roger Ascham, tutor to the Princess Elizabeth, said that unless it was taught in moderation, mathematics overcharged the memory. It was a time when mathematics was needed to develop new techniques in surveying, navigation, and gunnery, but the workers knew no mathematics and the mathematicians, such as they were, lacked practical experience. The universities were of no help in the matter. The gap was bridged by a group, many of them amateurs, whom Dr. E. G. R. Taylor calls "Mathematical Practitioners."[3] Among them was Robert Recorde.

Recorde, a native of Pembrokeshire in Wales, studied at Oxford, received the degree of Doctor of Medicine from Cambridge, and taught mathematics privately at both places, making the subject "clear to all capacities to an extent wholly unprecedented." He wrote a medical treatise of considerable importance, for it had at least ten editions from 1547 to 1665. He is reported to have practiced medicine in London, and it is claimed that he was physician to Edward VI and to Queen Mary. About 1551 he was appointed Surveyor of the Mines and Monies of Ireland and was made Comptroller of the Mint at Bristol. The *Whetstone of witte* ends abruptly with the entrance of a messenger sent to fetch the author to answer charges presumably of mismanagement of one of these political jobs. Within a year he died in the King's Bench Prison.

III. *Recorde's Mathematical Books*

Recorde was an opportunist. In his mathematical books, his aim was to provide the reader with needed information in palatable form. Recorde's four published books in mathematics were:

> The *Grovnd of Artes*, an arithmetic, 1542 or perhaps 1540
> The *Pathewaie to Knowledge: First Principles of Geometry*, 1551
> The *Castle of Knowledge*, an astronomy, 1551
> The *Whetstone of witte*, an algebra, 1557

In the prefaces of these volumes he refers to others he intends to write, in one case speaking of "other sundrye woorkes partly ended and partly to bee ended."

He took a realistic view of his qualifications as a writer. In the Preface of the *Grovnd of Artes*, he says

> . . . I know that no man can satisfie euery man, and therefore like as many do esteeme greatly other bookes, so I doubt not but some will like this my booke aboue any other English Arithmeticke hitherto written, & namely such as shal lacke instructers, for whose sake I haue plainly set forth the exa[m]ples, as no book (that I haue seene) hath hitherto.

His books were in the form of a dialogue between a Master and a Scholar, with the Scholar asking the leading questions in just the proper places. The books were well adapted to "such as shal lacke instructers."

The *Pathewaie to Knowledge* was dedicated to Edward VI with this explanation or apologia,

> Excuse me, Gentle Reader, if ought bee amisse straunge pathes are not trode[n] al truly at the first: the way muste needes be comberous, wher none

hathe gone before. . . . For neithe is my witte so finely filed, neither my learnyng so largely lettered neither yet my laisure so quiet and vncombered, that I maie performe iustely so learned a labour. . . . Yet may I thinke thus: This candle did I light: this light haue I kindeled: . . . I drew the platte [plan] rudelie wheron they maye builde, whom God hath indued with learnyng. . . . And this Gentle Reader I hartelie protest, where erroure hath happened I wishe it redreste.

As for the *Whetstone of witte*, Recorde is modest, but he felt that an algebra should be written and that it behooved him to do it. He says

> For better is it that a simple Coke doe prepare thy brekefast then that thou shouldest goe a hungered to bedde.

The *Grovnd of Artes* was a practical arithmetic. It found great use, as is witnessed by the fact that it had many editions over a period of a hundred and fifty years. It was quite in keeping with Recorde's purpose that the editors of later editions incorporated new things in them.

The *Pathewaie to Knowledge*, dedicated to Edward VI, was an introduction to geometry with definitions and conclusions in the first part, and proofs of the conclusions in the second. Third and fourth parts giving applications of these principles were promised but were never printed.

Recorde's astronomy, the *Castle of Knowledge*, dedicated to the Princess Mary, was both practical and theoretical. Dr. Taylor states that, "Like most physicians of his day, Recorde believed that the aspects of the heavens determined the correct times for taking medicines and other remedies, and he emphasized this equally with the needs of navigation for the study of astronomy." It has been claimed that Recorde's astronomy set forth the Copernican hypothesis, but Dr. Taylor says that Recorde "expressed himself cautiously . . . but held no brief for an unmoveable earth circled by the stars and planets." Recorde might well be excused for being cautious for it must be remembered that the date of Copernicus's great work was 1543, and that this had a preface explaining that the hypothesis that the earth moved round the sun was merely a convenience in computation. It was after Recorde's time that Kepler detected that this was an interpolation.

IV. Recorde's Algebra

The full title of Recorde's algebra is

The whetstone
of witte
whiche is the seconde parte of
Arithmetike: containyng the extrac
tion of Rootes; the *Cossike* practise,
with the rule of Equation: and
the woorkes of *Surde*
Nombers

Augustus de Morgan shows that this title is a play on words. The word *cos,* meaning a thing, was derived from the Latin *causa* by way of the Italian *cosa.* German and English writers used it to represent an unknown quantity. Consequently "Cossike practise" meant algebra. In Latin, the word *cos* means a grindstone. "Cossike practise" might be put into Latin as *cos ingenii* which in turn could be translated into English as the *Whetstone of witte.*

The preface is dated November 11, 1557. It should be remembered that at that time, the new year in England began March 25. There was time to print the volume before the end of 1557.

The book is dedicated to the "Venturers into Muscovia." In his introduction, Recorde says that he will

> . . . shortly set forthe soche a booke of Nauigation as I dare saie shall partly satisfie and contente, not onely your expectation but also the desire of a greate nomber beside wherin I will not forgette specially to touche bothe the olde attempte for the Northlie Nauigators, and the later good aduenture.[4]

In writing this algebra, Recorde referred to the work of Johannes Scheubel (1494–1570), profes-

sor of mathematics in the University of Tübingen, whose *Algebrae compendiosa facilisque descriptio* was published in Paris in 1551. Recorde objects to Scheubel's classification of equations, substituting one which he considered much simpler. Having thus improved upon Scheubel, Recorde proceeded to use direct translations of Scheubel's verbal problems. In fact, not only was Recorde acquainted with Scheubel's work but he plagiarized a considerable part of it. This is true not only in the section on algebraic equations, but also in the work on surds.[5]

V. The Extraction of Rootes

The first part of the *Whetstone of witte* considers numbers: whole numbers and broken numbers, "abstracte" and "contracte," evenly even and evenly odd. These last were of the type 2^n and $2(2^n + 1)$. The ratios of numbers are classified with special names for each type. The subject of diametral numbers is treated in detail. A diametral number is the product of two integers which have the property that the sum of their squares is itself a perfect square. In other words, when a, b, and c are integers such that $a^2 + b^2 = c^2$, then ab is a diametral number. Recorde shows that diametral numbers must end in 0, 2, or 8. They must be divisible by 12 and they cannot themselves be square numbers.

Recorde devoted over fifty pages to the extraction of roots. Here he gives particular attention to what he calls "lower ma[t]ters in warre." For example,

> A citie should bee scaled, beyng double diched. And the inner diche .32. foote broade. And the walle .21. foote high. The captain commaundeth ladders to be made of that iuste lengthe, that maie reche from the utter brow of the inner diche, to the toppe of the wal.

> A bollette of yron of .7. inches diameter, doeth waie .27. pounds weighte: what shall be the diameter to that bollette that shall wai .125. pounde to the weighte?

VI. The Cossike Practice

The second part of the *Whetstone of witte* deals with algebraic numbers and with equations.

Following Scheubel's example, Recorde uses the symbol Q to denote a number, $\mathcal{C}$ for the unknown quantity or root, $\mathcal{Z}$ for its square, and $\mathcal{C}$ for its cube. The fourth order of these numbers was indicated by repeating the symbol for a square, i.e., a square of a square. Recorde's predecessors had carried this system to great lengths, but Recorde outdid them by taking it to the eightieth order. It should be noted that in any equation every term, even the constant one, had its symbol. Like Scheubel, Recorde also made use of abbreviations for these quantities—N, Ra., Pri., Se., Ter., and so on (the number, a root, the first product, the second product, etc). As an aid to the reader, Recorde lists the symbols in order, numbering them 1, 2, 3, ... and says "By this table, maie you easily knowe the signe that shall serve for your newe somme in multiplication."

The Scholar had difficulty in grasping the fact that .12.$\mathcal{Z}$ multiplied by .6.$\mathcal{C}$ makes .72.$\mathcal{C}$. He says

> This passeth my cunnynge, for the findyng of the newe signe although the multiplication of the nombers be as easie as can be.

The Master answers

> If you did well remeber what you haue learned before; the mater would not seme so harde.

In dealing with "Cossike numbers," Robert Recorde introduced two signs, familiar to German writers, and familiar also to readers of the *Grovnd of Artes*. He says

> ... touchyng these twoo signes + and − which bee the figures of more and lesse, you must giue regard whether thei bee like or unlike, in those numbers that must be added: For if thei be like in nombers of one denomination, then muste thei so remain as thei be. But if thei be vnlike, euermore abate the smaller number of theim that followe those unlike signes out of the greater, and sette doune the reste with the signe of the greater number.

The Scholar is then confronted with eight examples in which cossike numbers are to be added. These are the different arrangements of two basic problems.[6]

$$10x \pm 12 \qquad 10x \pm 8$$
$$4x \pm 8 \qquad\quad 4x \pm 12$$

He says

> Here haue I varied one example diuersely, to the intente you maie marke the vse of your rules in theim.

He explains the addition of $4x - 8$ to $10x + 12$ as follows:

> this sum is not fully $4x$ but wanteth of it 8 and therefore if you put downe $4x$ fully, you must abate 8 out of the 12 in the larger summe.

Recorde had a liking for putting things that were to be memorized into rhyme. The results were interesting. Here is his verse for the rules of signs in multiplication and division.

> who that will multiplie
> or yet divide trulie:
> shall like stille to haue more
> and mislike lesse in store.

The Scholar summarized it in these words:

> So meane you that like signes multiplied together, doe make more, or + and vnlike sines multiplied together doe yelde lesse or − ?

In Recorde's work with polynomials, no powers of the cossike number are omitted. For example, $8x^3 + 0x^2 + 0x + 64$ is to be divided by $2x + 4$.

Having practiced the fundamental operations with polynomials, the Master shows how to add fractions whose numerators and denominators are algebraic quantities. By his scheme, horizontal lines are drawn above and below the fractions that are to be added. The common denominator is written below the bottom line and the new numerators are written above the top line.

VII. "The Rule of Equation Commonly Called Algebers Rule"

> This Rule is called the Rule of *Algeber,* after the name of the inuentoure, as some men thinke . . . but of his vse it is rightly called the rule of *equation:* bicause that by the *equation* of nombers, it doeth dissolue doubtful questions: and vnfolde intricate ridles.

Recorde's statement about the use of equations in solving problems is involved but worth considering.

> When any question is propounded . . . you shall imagin a name for the nomber, that is to bee soughte, as you remember that you learned in the rule of false position. And with that nomber shall you procede, accordyng to the question, vntil you find a Cossike nomber equalle to that nomber that the question expresseth, whiche you shal reduce euer more to the leaste nomber.

The Scholar compares this with the rule of false, the method of solving problems by guessing the answer and then adjusting the guess until it fits the problem. The Master goes on

> . . . it mai bee thoughte to bee a rule of wonderful inuention that teacheth a manne at the firste worde to name a true nomber before he knoweth resolutely [surely] what he hath named. But bicause that name is common to many nombers [although not in one question] and therefore the name is obscure till the worke doe detect it, I thinke this rule might well bee called the rule of darke position, or of strange position, but not of false position.

> And for the more easie and apte worke in this arte wee doe commonly name that dark position .1. ℞ and with it doe we worke as the question intendeth till we come to the equation.

At this point Recorde introduced the sign of equality, following it at once, as is shown in the accompanying facsimile, by a series of equations which he then considers in detail. In the case of $14x + 15 = 71$, he says

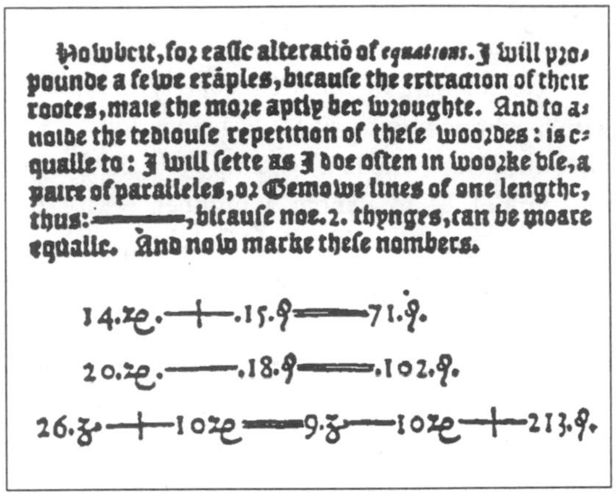
Howbeit, for easie alteratiō of *equations.* I will propounde a fewe exāples, bicause the extraction of their rootes, maie the more aptly bee wroughte. And to auoide the tediouse repetition of these woordes: is equalle to: I will sette as I doe often in woorke vse, a paire of paralleles, or Gemowe lines of one lengthe, thus: ═══, bicause noe. 2. thynges, can be moare equalle. And now marke these nombers.

FIGURE I

> . . . you mai see one denomination on both sides of the *equation* which neuer ought to stand. Wherfore abating the lesser, that is 15, out of both nombers, there will remain
>
> $$14x = 56$$
>
> this is by reduction x = 4 according to the third common sentence in the *Pathewaie*—If you abate euen portions from thynges that bee equalle, the partes that remain shall be equall also.

Other "common sentences" are used in solving other equations.

When confronted with the equation $26x^2 + 10x = 9x^2 - 10x + 213$, the Scholar wonders whether to add $10x$ or to "abate them." The Master says

> In soche a case, you maie dooe either of bothe at your libertie and all will be to one ende. . . . And euermore when occasion serueth to translate nombers compounde, − on the one side is equalle to + on the other side.

Robert Recorde classifies equations into two groups:

1. Where one number is equal to another number,
2. Where one number is equal to two other numbers.

Here he has departed from Scheubel whose detailed classification can be represented as follows:

First type $bx = N$

Second type   (1) $ax^2 + bx = N$
                 (2) $x + N = x^2$ or
                      $bx + N = x^2$
                 (3) $bx + N = ax^2$

Third type $x^2 + N = bx$

Neither writer classified equations according to their degree. Recorde includes the equation $6x^3 = 24x$ under his first heading, one number equal to another number, but he shows that the answer is not 4 but the square root of 4. He did not recognize roots that were zero or negative.

No reasons or explanations are given. Quadratic equations are solved by completing the square.

When the Master has completed the solution of a number of equations, the Scholar says "I doe couette some apte questions, appertainyng to these equations." The Master supplies them. They are not easy.

In one of these problems two men have silk to sell, one man has 40 ells, the other 90. The first man's silk is of the poorer quality, so he has to sell a third of an ell more for an angel than does the second man. If their total receipts are 42 angels, how many ells of silk did each sell for an angel? The Scholar makes a false start. He lets his unknown quantity represent the first man's receipts. The Master does not approve and suggests that he divide each man's ells by the number he sold for an angel and the quotient will be the man's receipts. Working this way, the Scholar finally comes to the conclusion that one man sold three ells for an angel and the other three and one third, so their receipts were 12 angels and 30 angels. Not content with this, the Scholar reverses the problem. Instead of using x and $x + 1/3$, he uses $x - 1/3$ and x. This time he gets one root correctly, but he also finds the other one, namely 2/21, and says

> But how I maie frame that roote to agree to this question, I doe not see.

The Master is no help to him, except to say that

> . . . the forme of the question maie easily instruct you whiche of these .2. rootes you shall take for your purpose.

As an example of a question where both roots fit the problem, the Master gives the following:

> A gentilman, willyng to proue the cunnyng of a braggyng *Arithmetician*, saied thus: I haue in bothe my handes .8. crounes: But and if I accoumpte the somme of eche hande by it self seuerally and put therto the squares and the cubes of bothe, it will make in nomber 194. Now tell me (quod he) what is in eche hande: and I will giue you all for your laboure.

As would be expected, the answers 3 and 5 are the numbers in the man's hands.

In another case, the problem states that if 8 is added to a number and if 16 is taken from the square of the number, the product of the two results is 2560. This yields the equation $x^3 + 8x^2 - 16x = 2688$. The Scholar is baffled. It is above his cunning, for here two numbers are equal to two others, or one number is equal to three numbers. The Master does not solve the problem. One suspects that he concocted the question by starting with the result. He simply says that the required number is 12 and has the Scholar show that this answer fits the equation. At this point the Master says

> But to put you out of doubte, this equation is but a trifle to others that bee untouched.

There is no lack of variety in the problems. A man travels $1\frac{1}{2}$ miles the first day and increases his day's journey by $\frac{1}{6}$ mile each day. How long does it take him to go 2955 miles? In another case, the day's mileage increases in a geometric progression.

A herald is offered a bribe if he will tell the number of his king's army. The problem goes on this way—

The Heraulte lothe to lease those giftes, and as lothe to bee vntrue to his Prince, diuiseth his answere, wiche was true, but yet not so plain, that the aduersarie could thereby vnderstand that whiche he desired. And that aunswere was this. Looke how many Dukes there are, and for eche of them, there are twise so many Erles. And vnder euery Erle, there are fower tymes so many souldiars, as there be Dukes in the field. And when the muster of the soldiers was taken, the .200. parte of them was .9. tymes so many as the nomber of the Dukes.

That is the true declaratio of eche number, quod the Heraulte: and I haue discharged my othe. Now guesse you how many of eche sorte there was.

A man dies leaving 72 crowns to his four children in this way: the second and third together were to have seven times as much as the first. The third and fourth were to have five times as much as the second. The first and fourth were to have twice as much as the third. It might be well to note that the children received the following sums: $4\frac{1}{2}$, $11\frac{1}{4}$, $20\frac{1}{4}$, 36.

A captain marshals his army in a square formation. When the square was of one size, he had 284 men too many, so he tried to arrange them in a square one man more on a side than before. This time he lacked 25 men. How many men did he have?

VIII. *The woorkes of Surde Nombers*

The third part of the *Whetstone of witte* is devoted to irrational numbers, which Recorde calls surds. He says

> Nombers *radicalle*, which commonly bee called nombers *irrationalle:* bicause many of them are soche, as can not bee expressed, by common nombers *abstracte*, nother by any certain ratiionalle nomber. Other men call them more aptly *surde* nombers.... A *surde* number is nothyng els, but soche a number set for a roote, as can not be expressed by any number absolute, as $\sqrt{10}$ or $\sqrt{18}$ or any number, that is not a square.

Surds are commensurable if they can be expressed as multiples of the same root; otherwise they are incommensurable.

Binomials made up of a rational number and an irrational one, are classified in two groups: "Nombers that be compounded with + be called Bimedialles and with − Residualles."

Accordingly, to divide by a binomial, dividend and divisor must be multiplied by the residualle of the divisor if the divisor is a bimedialle, or by the bimedialle of the divisor if the divisor be a residualle.

Recorde's square root sign was much like ours, but his signs for the cube root and the fourth root were perplexing. These are the symbols used by Scheubel, which he in turn had taken from the work of Rudolff (1525). When confronted with the symbol .∧∧∧. for cube root and .∧∧. for the fourth root, the Scholar speaks his mind.

> It were against reason, to take treason for these signes, which be set voluntarily to signifie any thyng; although some tymes there bee a certain apte conformitie in soche thynges. And in these figures, the number of their minomes [i.e. upstrokes] seameth disagreable to their order.

The Master replies:

> In that there is some reason to bee thewed [instructed]: for as .√. declareth the multiplication of a nomber, ones by itself; so .∧∧∧. representeth that multiplication *Cubike*, in whiche the roote is represented thrise. And .∧∧. standeth for .√.√. that is .2. figures of Squarc multiplicatiun: and is not expressed with .4. minomes. For so should it seme to expresse moare then .2. *Square* multiplications. But voluntarie [i.e., arbitrary] signes, it is inough to knowe that thie doe signifie. And if any Manne can diuise other, moare easier or apter in use, that mai well be received.

IX. *Appraisal of the* Whetstone of witte

The chances are that the *Whetstone of witte* had few if any readers on the continent. Anyone interested in algebra would have found more satisfaction in Scheubel's concisely worded, elegantly printed treatise in Latin than in Recorde's popularization of it in English. So far as the equality sign was concerned, it is more than likely that

Recorde had no particular pride in his invention. It was a thing he had found convenient. On the other hand, his simplification of Scheubel's classification of equations, in which he took apparent satisfaction, is of little interest today.

The *Grovnd of Artes* was closely connected with practical affairs. The *Whetstone of witte* seems to contain no practical applicatons, the reason being that algebraic problems which we would solve by equations of the first degree were treated by the rule of false in Recorde's arithmetic. The *Whetstone of witte* certainly had a smaller public. Had Recorde been able to complete the books that were "partely to bee ended," the algebra might have had a wider circulation. As it was, the last of Recorde's books was the only one to have but a single edition.

Frequently the introduction of a book is best read after reading the book itself. So here, it seems appropriate to quote as a conclusion the verses on Recorde's title page:

> Though many stones doe beare great price,
> The whetstone is for exercise
> As neadfulle, and in woorke as straunge:
> Dulle thinges and harde it will so chaunge,
> And make them sharpe, to right good vse:
> All artesmen know, thei can not chuse,
> But vse his helpe: yet as men see,
> Noe sharpnesse semeth in it to bee.
> The *grounde of artes* did brede this stone:
> His Vse is greate, and moare then one.
> Here if you list your wittes to whette,
> Moche sharpnesse therby shal you gette.
> Dull wittes hereby doe greatly mende,
> Sharp wittes are fined to their full ende
> Now proue, and praise, as you doe find,
> And to yourself be not vnkinde.

NOTES

1. Florian Cajori, *History of Mathematical Notation* (Chicago, 1928), Vol I.

2. See Chapter 6 of G. M. Trevelyan, *History of England* (New York: Doubleday Anchor Books, 1953), Vol. II.

3. E. G. R. Taylor, *Mathematical Practitioners of Tudor and Stuart England* (Cambridge, 1954).

4. The reference here is to expeditions sent out by the "mystery and Company of Merchant Venturers" with the co-operation of Sebastian Cabot, newly returned from Spain. The purpose was to find a route to India and to wider markets for English products. The 'olde attempte' was an expedition to find the North West Passage. The 'later good aduenture' was sent out in 1553 to find a North East passage to India. It reached Russia instead. By penetrating the White Sea to the place where the port of Archangel now stands, the leader went overland to Moscow, where he met the tsar and obtained trading concessions. He thus circumvented the Hanseatic League which held a monopoly of trade through the Baltic, and opened the way for later commercial agreements between Russia and England. The backers of this expedition received a charter for trade with Russia and became the Moscovy Company. Robert Recorde was one of their advisers, presumably in questions of navigation.

5. See Mary S. Day, *Scheubel as an Algebraist* (Contributions to Education No. 219 [Teachers College, Columbia University, 1926]).

6. For convenience, modern symbols are used in these illustrations and in the balance of this paper.

The Evolution of Algebraic Symbolism

Since the time of the ancient Egyptians and Babylonians, mathematical problems, even situational problems, were completely written out in words as were their solution procedures. This phase of algebraic thinking is usually called the "rhetorical stage." Due to printing and the repeated use of certain terms and words, mathematicians began to use abbreviations to form mathematical relationships. At first, each mathematician or local group of mathematicians had their own system of symbolization but gradually the symbols as well as the procedures became standardized. Below are various examples of how different mathematicians employed symbols to express the modern equation $4x^2 + 3x = 10$.

Nicolas Chuquet	(1484)	$4^2\,p3^1$ égault 10^0
Vander Hoecke	(1514)	$4\ Se + 3$ Pri dit is ghelijc 10
F. Ghaligai	(1521)	$4\ \square\ e\ 3c° - 10$ numeri
Rudolff	(1525)	Sit $4\,\mathcal{X} + 3\,\mathcal{C}$ aequatus 10
Jean Buteo	(1559)	$4\ \lozenge\ p\,3\ p\ [\,10$
R. Bombelli	(1572)	$4\ \overset{\smile}{p}\ 3$ equals á 10
Simon Stevin	(1585)	$4\,②+ 3\ ①$ egales 10
Ramus and Schoner	(1586)	$4\ q \longmapsto 3\ \mathcal{R}$ aequatus sit 10
François Viète	(c1590)	$4Q + 3N$ aequatur 10
Thomas Harriot	(1631)	$4aa + 3a === 10$
René Descartes	(1637)	$4ZZ + 3Z\ \infty\ 10$
John Wallis	(1693)	$4XX + 3X = 10$

60

The Teaching of Arithmetic in England
from 1550 until 1800 as Influenced by Social Change

JAMES KING BIDWELL

*I*N 1542 THERE APPEARED IN LONDON a book by Robert Recorde entitled "The Grounde of artes: Teaching the worke and practice of Arithmetike, both in whole numbers and Fractions, after a more easer and exacter form than any like hath hitherto been sette forthe." This publication was the first significant arithmetic book printed in English. Its printing marked the beginning of the dissemination in England of arithmetic techniques that we today would consider typical in our elementary schools. In the middle of the sixteenth century, however, these techniques were not widely known and were not taught in the vernacular English in the formal grammar schools.

The Sixteenth Century

The mathematics taught at that time was considered a part of the formal education of the classical "seven liberal arts" and was taught in grammar schools from Latin texts. Besides this, the so-called arithmetic was theoretical (what we would now call number theory) and was not computational. It amounted to classical mathematics essentially unchanged since the time of Greece. The teaching concerning numbers in English was quite limited. John Brinsley could say in the sixteenth century:

> In a word, to tell what any of these numbers stand for, or how to set down any of them; will performe

Reprinted from *Mathematics Teacher* 62 (Oct., 1969): 484–90; with permission of the National Council of Teachers of Mathematics.

> fully so much as is needfull for your ordinaire Grammar scholler. If you do require more for any, you must seeke Record's Arithmetique, or other like Author's and set them to the Cyphering school.[1]

And even this small amount of number work was done in the late afternoon of Saturdays or on half-holidays for a total of one hour a week.

> That arithmetic was not taught in the Latin Schools in order to make proficient reckoners is shown by the lack of practice problems in their textbooks; and, likewise, the lack of vital commercial problems of that day show that it was not taught in order to prepare for a business life.[2]

On the continent the development of Reckoning Schools coincided with the commercial development and the growth of the guilds. The demands for merchants trained in bookkeeping and reckoning, reading and writing became so great that the merchants themselves could not instruct their apprentices. So Reckoning Masters and eventually Reckoning Schools developed to fill this need. There was no corresponding development in England prior to the sixteenth century. No manuals for arithmetic are extant in English. It is clear, however, that Italian methods of accounts were known by the English merchant class. Charlton tells us:

> In 1476, for example, James Harrison was apprenticed to Christopher Ambrose, a Florentine by birth, who traded from Southampton and took English apprentices into his household where he undertook to introduce them to the mysteries of his trade.[3]

In the sixteenth century the growth of commerce, discovery of the larger geographic world, and increased "industrial" type production led to more and more technical training of the merchant class. Improved techniques in navigation, surveying, horology, cartography, gunnery, and fortification were needed, and all of these required good mathematics knowledge. Edmund Worsop, a land surveyor, wrote a book in 1582 with the fantastic title, worded as follows:

> A Discoverie of sundrie errors and faults daily committed by Landmeaters, ignorant of Arithmeticke and Geometrie, to the damage, and prejudice of many of her Maiestris subjects, with manifest proofe that none ought to be admitted to that function, but the learned practitioners of those Sciences.[4]

Worsop demanded better training of surveyors and suggested that they be licensed to insure more effective land measuring practices. W. H. G. Armytage notes that "mining, navigation, river improvement and the building of Elizabethan country houses stimulated mathematics."[5] Others have mentioned the stimulus of the growth of coal mining. The managing of estates by the gentry also required much practical knowledge in order for the landowner to compete for his revenues and preserve his estate. None of these needs could be met in the existing formal schools or universities.

The increased power and restrictiveness of the guild supervision in the sixteenth century cut off the former educative function of the guilds. They instead turned to endow grammar schools, which did not provide at all for the apprenticeship program. So the work of providing arithmetic education fell to individual tutors, or as they were called in England, "mathematical practitioners." These tutors instructed and wrote manuals in English based on previous manuals or books in Latin or another foreign language, from which they had learned themselves. These men were not, of course, university trained. Among these men was Robert Recorde. E. G. R. Taylor tells us:

They were alamanack-makers, astrologers, retired seamen, surveyors, gunners, gaugers—in fact they were themselves mathematical practitioners who simply handed on their art. But, as might be expected, they all worked in close association with the instrument makers, and as the handling of instruments was the very badge of this new profession, it was quite usual for a teacher to design a novel one of his own. . . .[6]

To supply the needs of these technical occupations, more and more printed books dealing with arithmetic, accounts, and general mathematics were printed. Karpinski mentions that in the sixteenth century 45 English arithmetic editions were printed. He goes on to note that "popular interest in arithmetic and general instruction in the subject increased so rapidly after the sixteenth century that hundreds of books appeared (in Europe) to supply the new demand."[7] These books greatly varied in their effectiveness, but they all aimed at giving practical rules for solving problems that the man with commercial or technical interests might have to solve. They were, in fact, compendiums of different problem types. Since these books were generally adapted from earlier books, the standard evil of material, which was in these arithmetics simply because it had long been in arithmetics, was perpetrated. None of these books were, of course, designed for school use; rather, they were designed to be "do-it-yourself" books.

Since these newly printed books were for private use, the authors had to sell their content to the general public. The prefaces of some of these early arithmetics make interesting reading and point out the intended practicality of the arithmetic content:

> Howe profitable and necessary this feat of Algorism is to all maner of persons, which have reckenyings or accountes, other to make, or else to receive, needyth no declaration. Neither is this arte only necessary to those, but also in maner to all manner of sciences and artificies.[8]

> Arithmetique needes not the Logicians arguments, nor the Rhetoricians Eloquence to prove or parswade the vsefulnesse thereof to the world, every mans particular occasion, to vse it, is sufficient to satisfie any man in that point. . . .[9]

For if numbring be so common (as you grant it to be) that no man can do anything alone, and much lesse talk or bargain with others, but he shall still have to do with number; this proveth not number to be contemptible and vile, but rather right excellent and of high reputation, sith it is the ground of all mens affairs, . . . [10]

These books were intended to be complete in themselves, requiring no teacher; the authors tried various methods of exposition, each designed to make the work clear for the reader. They failed miserably if judged by today's teaching methods. The dialogue method was especially common and rules stated in verse were used frequently. Consider as one example this versified rule for the addition of fractions:

Addition of fractions and likewise subtraction
Requireth that first they all have like basses
Which by reduction is brought to perfection
And being once done as ought in like cases,
Then adde or subtract their tops and no more
Subscribing the basse made common before.[11]

The Seventeenth Century

In the seventeenth century, a movement developed in favor of further formal education along mathematical lines. However, more was said than done. Dr. John Pell wrote about reform of education in 1639; he envisioned a public mathematical library where anyone could study on his own. But suggestion of any change in grammar school teaching was lacking. Other plans to include arithmetic and mathematics as part of liberal education were put forth but came to nothing. But Taylor points out:

Even though all these plans failed of realization, they were symptomatic of a changing climate of thought and opinion, and the mathematical practitioners, even to the vulgar mind, were becoming distinguishable from the conjurors and the quack astrologers as useful members of society. Only a few university die-hards continued to maintain that mathematics was no study for a Christian man.[12]

But change came slowly even in urban centers, and the outlying areas were even further behind the needs of the time. Also, the clergy fought the advancement of mathematics and science in general; many of them still treated arithmetic, mathematics, chemistry, and the like as works of the devil.

In 1655 a book was published in London with the title: "An idea of Arithmetick at first Designed for the use of the Free-Schoole at Thurlow in Suffolk By R. B. Schoolmaster there." In 1668 John Newton published *The Scale of Interest: Or the Use of Decimal Fractions*. DeMorgan notes that this book "was expressly intended for a schoolbook, though it is a strange one for the time."[13] These editions mark the beginning of school texts in English. What schools were then beginning to teach the practical art of arithmetic?

The lack of mathematically trained personnel greatly affected navigation. The techniques of recent theoretical advances were beyond most shipmasters of the day. The pressures were such that King Charles ordered the development of a mathematical school at the Christ's Hospital School in 1672. This was to be a school for boys passing from ordinary grammar school at the age of $14\frac{1}{2}$. The most immediate difficulty of the school was to find a master. The requirements were severe; the master was to have knowledge of Latin and mathematics and have experience at sea. The need for such schools for navigation as well as commercial interests led to more such schools. Even though these were schools for older boys who had completed a classical study, they led to further elementary study of arithmetic. In Scotland the demands also were met:

A feature of the last two decades of the seventeenth century was the establishment in Glasgow and Edinburgh of commercial schools and academies which were modelled on the 'Reckoning Schools' of Italy, Germany, France and the Netherlands. The first of these was started in Edinburgh in 1680. . . . In 1695 'a teacher of the art of navigation, bookkeeping, arithmetic and writing' was appointed by the Town Council of Glasgow.[14]

These private schools, offering a curriculum geared to the commercial needs of the times, were established quite naturally in industrial towns. The Manchester School was established in 1666, the Dartmouth School in 1679, and the Rochester School in 1701. Some of these schools lasted many years (some into the nineteenth century). Some existed only a few years and then died from lack of funds.

The Eighteenth Century

The development of specialized schools like Christ's Hospital Mathematical School led to the establishment of schools with a broader curriculum, still oriented to vocational training, but offering foreign languages and some classical studies. These private schools were in direct competition with the grammar schools, and the best of the new schools sent a certain percentage of graduates to the universities. This kind of private school was commonly called an "Academy." Nicholas Hans suggests that these schools were fashioned after the so-called courtly Academies that flourished in Germany, France, and England in the seventeenth century. They were designed to "prepare the noble youth for his profession as a courtier and soldier and introduced military subjects, mathematics, physical training and accomplishments." [15]

The curriculum of these private academies had four groupings of subjects: literary, mathematics-science, vocational-technical, and accomplishments–physical training.

> From textbooks on various subjects published by most of them, and from the description of some of the Academies, it is evident that their methods were not "bookish" but practical and whenever possible approximated to an actual situation of business or technical vocation. It was true not only of vocational and technical training; the same methods were applied in teaching mathematics and languages. [16]

Such academies were founded all during the eighteenth century, and although some were short-lived, approximately 200 such schools were operating in England throughout the latter part of the century.

Most of these schools drew their students from the lower middle class and craftsmen. Thus finally the needs of that group for educational opportunity were being met. In fact, the mathematical part of these schools was so vital a part of the needs of the times

> That when the Schism Act was drafted in 1714 to curb the dissenting academies, special exemptions were made in cases where 'any part of mathematics relating to navigation, or any mechanical art' was taught. [17]

Thus the need for individual tutors of mathematics and small schools for mathematics developed into a need for vocational-technical schools where classical studies formed only a small part of the curriculum. All of this occurred chiefly because the formal schools of the times were still too classically and religiously centered to meet the demands of social and industrial change.

The education in the grammar schools of the eighteenth century was the same as education in the grammar schools in the sixteenth century. These schools were bound by their foundation statutes and hence were unable to modify their classical structure. Some did add modern subjects such as elementary mathematics, French, and German. Some schools merely restricted their program to elementary education. Due to the competition from other private schools, their enrollments dropped and many went out of existence.

Even in the face of this demand for nonclassical education the classical teachers defended their cause, but from a new point of view:

> Joseph Cornish in *An attempt to Display the Importance of Classical learning* (1783), argued that most technical terms used in physics, botany, chemistry, astronomy, architecture, mechanics, mathematics, rhetoric and grammar 'and almost every art and science are of a Greek original and numerous others of a Roman'. That he should be justifying Greek on these grounds is significant. [18]

Those grammar schools that wanted to augment their curriculums created disputes that were taken to the law courts. Consider the Leeds Grammar School, which wished to include practical subjects like mathematics and foreign languages. The Governor of the school took the case to court in 1795. After ten years of indecision, in 1805 the court decided that:

> The intention of the founder was to establish a grammar school and . . . a grammar school was defined as an institution 'for teaching grammatically the learned languages.' The court could not sanction 'the conversion of that Institution by filling a school intended for that mode of Education with Schoolars learning the German and French languages, Mathematics, and anything save Greek and Latin.' [19]

In fact, it was only in 1840 that the Parliament passed the Grammar Schools Act, which allowed for modernization of the curriculum.

In the so-called public schools it was not any better. Classical studies dominated the education curriculum. This is illustrated by Curtis, discussing the writings of Dr. Thomas James, Headmaster of Rugby from 1778 to 1794:

> On holidays the boys attended school from ten to eleven in the morning and two to three in the afternoon. On half-holidays they attended from two to three o'clock. In these periods, the lower forms learnt writing and arithmetic, and the Vth form geography and algebra. It is probable that the latter subject was taught by Dr. James himself. At any rate he afterwards confessed that this weekly mathematical period 'wearied his body to excess and made it hot, or at any rate, perspire too much.' [20]

Curtis mentions that the same kind of arithmetic teaching was common in the grammar schools of the 1590s. Since the needs of the public centered on commercial subjects, the importance of these schools faded for all but the nobility and some of the gentry who desired to maintain the status quo.

Conclusion

We see therefore that the arithmetic taught in the eighteenth century was almost exclusively a vocational subject. Hence all the arithmetic the students ever learned was a set of rules that produced the answers to their problems. Texts had repeated editions with little changes in content and usually considerable additions of specialized commercial problem types.

No accomplished mathematicians wrote the arithmetics of the eighteenth century. DeMorgan, writing reasonably close to this time, greatly condemned the editions of Cocker's Arithmetic. Cocker's book had an enormous popularity, although it was one of the most offensive texts from a mathematical point of view. It was first published in 1677 and was still in print in the early nineteenth century. Hence DeMorgan was well acquainted with it in 1847, and after six pages of disparagement he notes: "I am of the opinion that a very great deterioration in elementary works on arithmetic is to be traced from the time at which the book called after Cocker began to prevail." [21]

Such books as Cocker's were common enough to cause widespread complaints about the mathematics they contained. DeMorgan also comments on the large number of criticizers who can offer nothing better. Cajori, writing in 1896, also comments on the same problem:

> To summarize, the causes which checked the growth of demonstrative arithmetic are as follows:
>
> (1) Arithmetic was not studied for its own sake, nor valued for the mental discipline which it affords, and was, consequently learned only by the commercial classes, because of the material gain derived from a knowledge of arithmetical rules.
>
> (2) The best minds failed to influence and guide the average minds in arithmetical authorship. [22]

By the beginning of the nineteenth century, the dichotomy of teaching practices was at its worst. The grammar schools were slowly breaking from the classical narrowness on the one hand and the technical schools were teaching a rigid rule-dominated commercial arithmetic which had little mathematical content. Surely this was a low point in arithmetic teaching in England. In the nineteenth century, men like DeMorgan advocated and effectively initiated reforms that began the way towards the modern arithmetic teaching of today.

As the years from 1550 to 1800 are reviewed, even in this narrow field, the dynamics of interplay can clearly be seen between the demands of the rising middle class, the clergy, and the government's national needs. The emergence of arithmetic teaching can be seen as a slow and painful development. That is to say, the development of arithmetic practices followed the same patterns that all educational practices follow. When the change involves the integration of many different social points of view, it traditionally follows that the change is slow. Such is the way of all education.

NOTES

1. John Brinsley, *The Grammar Schoole*, ed. E. T. Campagnere (Liverpool: University of Liverpool Press, 1917), p. 26.

2. L. L. Jackson, *Educational Significance of Sixteenth Century Arithmetic* ("Contributions of Education," No. 8 [New York: Teacher's College, Columbia University, 1906]), p. 178.

3. Kenneth Charlton, *Education in Renaissance England* (London: Routledge and Kegan Paul, 1965), p. 258.

4. Edmund Worsop, *Discoverie of Sundrie Errours Committed of Landmeaters . . .* (London: 1582, University Microfilms, S.T.C. Case 404).

5. W. H. G. Armytage, *Four Hundred Years of English Education* (Cambridge: University Press, 1964), p. 9.

6. E. G. R. Taylor, *Mathematical Practitioners of Tudor and Stuart England* (Cambridge: University Press, 1954), pp. 9–10.

7. Louis Karpinski, *The History of Arithmetic* (Chicago: Rand McNally, 1925), p. 73.

8. Dorothy Yeldham, *The Teaching of Arithmetic Through Four Hundred Years* (London: Harrap, 1936), p. 11.

9. Edmund Wingate, *Arithmetique made easie* (London: 1630, Rare Book Room), preface.

10. Robert Recorde, *The Ground of Artes* (1646), preface.

11. T. Hylles, "The arte of vulgar arithmetique," 1600 as quoted in Augustus DeMorgan, *Arithmetical Books* (London: Tahlor and Walton, 1847).

12. E. G. R. Taylor, *op. cit.*, p. 82.

13. Augustus DeMorgan, *op. cit.*, p.46.

14. Duncan Wilson, *The History of Mathematical Teaching in Scotland* (London: University of London Press, 1935), pp. 26–27.

15. Nicholas Hans, *New Trends in Education in the Eighteenth Century* (London: Routledge and Kegan Paul, 1951), p. 64.

16. *Ibid.*, p. 68.

17. W. H. G. Armytage, *op. cit.*, p. 31.

18. *Ibid.*, p. 67.

19. S. J. Curtis, *History of Education in Great Britain* (London: University Tutorial Press, 1948), p. 60.

20. *Ibid.*, p. 62.

21. Augustus DeMorgan, *op. cit.*, p. 62.

22. Florian Cajori, *A History of Elementary Mathematics* (New York: Macmillan, 1896), p. 211.

61

Tangible Arithmetic I:
Napier's and Genaille's Rods

PHILLIP S. JONES

*H*AROLD LARSON COMMENTED that Napier's "bones" or "rods" are well known with several accessible articles on them when he published the drawing of Genaille's rods which is our Figure 4.[1] This is true, and this is our reason for showing without discussion (Fig. 1) the title page of the posthumous (1617) book in which Napier explained them, and a page from it (Fig. 2) which shows the four faces of a few of these bones as Napier designed them.

The word *Rabdologia* is probably compounded from Greek words meaning *a collection of rods*. William Leybourn, who published a translation, *The Art of Numbering by Speaking-Rods; Vulgarly Termed Napier's Bones*, in 1667, thought the latter part of *Rabdologia* was from *logos, speech*, rather than *logia, a collection*. Translations also appeared in Verona and Berlin in 1623 and a second Latin edition in 1626. Similar devices appeared in China later in the seventeenth century and again in the nineteenth.

The diagonal line separating the units and tens digits on the bones and the method of using the bones are counterparts of the popular *gelosia* or *jealousy* method of multiplication. This came into Europe from Arabic writers shortly after the introduction of the Hindu-Arabic numerals, and can be traced back to the Hindu Bhaskara (1150) or

earlier. Wöpcke notes that the notion of a separate multiplication table for each of the nine digits can also be found in the fifteenth century writings of the Arab Alkalsadi, who "taught multiplication by separating the columns of the table of Pythagoras."[2] ("The table of Pythagoras" referred to the 9×9 or 10×10 square multiplication table which appeared in many books on the Hindu-Arabic arithmetic. Neither Pythagoras nor later Greeks used the Hindu-Arabic number system, and Pythagoras' greatest interest was in the "arithmetica" which today we call number theory. Nevertheless the term "table of Pythagoras" illustrates the importance of the Greek contributions to many areas of early mathematics.)

Although the bones and their use are well known today as enrichment-teaching aids especially related to multiplication, it is not so generally known that Napier designed special rods for square roots and cube roots. A later variation of these is shown at the top of our Figure 3. It is also interesting to note the connection of Napier's work with the spread of the idea of decimal fractions. Although Simon Stevin published his basic work in 1585, his notation (a small zero in a circle to mark the units place, a small one in a circle to mark the tenths place, etc.) was awkward. The first publication of a decimal point as we know it occurred in a 1616 translation into English of Napier's work on logarithms. On pages 21–22 of the *Rabdologia*, Napier discusses Stevin's work and uses a comma (as still used on the Continent) for our decimal

Reprinted from *Mathematics Teacher* 47 (Nov., 1954): 482–87; with permission of the National Council of Teachers of Mathematics.

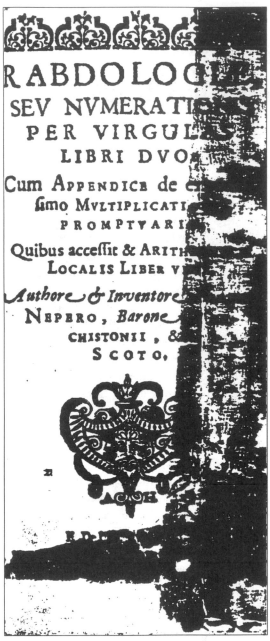

FIGURE 1

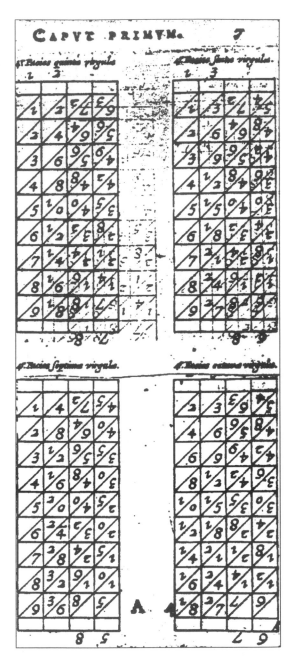

FIGURE 2

point. No doubt the popularity of Napier's works assisted in the spread of these improved notations as well as in the use of decimal fractions.

The *Rabdologia* is a small book (3 in. × 5.5 in.) of 154 pages. "Liber primus" tells how to con-struct the rods and to do multiplication, division, square, and cube root, and the rule of three with them. "Liber secundus" tells how to use them in computing the solutions to a number of different geometric problems relating to regular

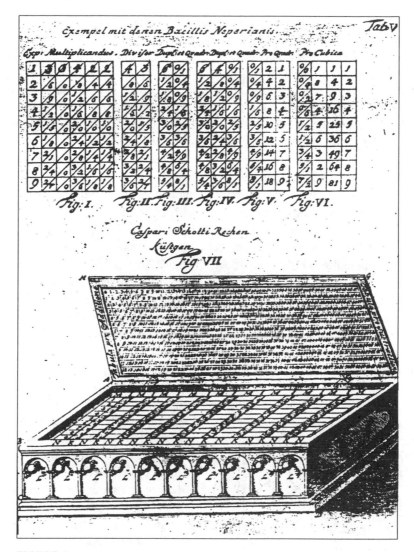

FIGURE 3

polygons and polyhedra and to weights and measures.

In an appendix beginning on page 91, Napier describes more complicated apparatus and procedures for further speeding up computations, especially trigonometric and astronomical computations.

Many variations on Napier's fundamental design grew up in the years after 1617. For example, about 1668 Gaspar Schott mounted cylinders, each of which carried Napier-type tables for 0–9, in a box as shown in our Figure 3. This figure is taken from Jacob Leupold's *Theatrum Arithmetico Geometricum* published in Leipzig in 1727.

A set of bones dating to about 1680 has the tables on flat sticks which had been described by Leybourn in 1667[3] and which were also advocated by Schott.[4]

Figure 4 is Professor Larsen's drawing of the "Réglettes Multiplicatrices" invented by Henri Genaille and "perfected" by Edouard Lucas in 1885.[5] Figure 5 is a photograph of a companion

395

set of "Réglettes Multisectrices" in the University of Michigan library. To multiply 471,963 by 6 one uses the data between the horizontal lines bounding 6 on the left-hand or index scale in Figure 4. Going to the extreme right, one begins with the number on the "3" rod below the upper of horiziontal lines bounding 6. This is 8.

One then proceeds to the number on the next rod which is pointed out by the left-hand vertex of the angle which includes 8. This next number is 7. From there, by going from right to left and always reading the number on the next rod at the vertex of the angle including the last number, we read: 7, 1, 3, 8, 2. Thus, finally, 471,963 × 6 = 2,831,778. As with Napier's bones, to multiply by 526 one would have to write down separately and then add the partial products of 471,963 by 6, 2, and 5 which may be read from the rods. On these rods the numbers in the narrow vertical columns opposite 6 are the units digit of the product by 6 of the number of the rod and this units digit plus 1, 2, 3,

4, 5. These latter are the numbers which might have been "carried" from an earlier multiplication on a previous rod. The vertex of the angle on the previous rod points out which of these is to be used in a particular case. (Professor Larsen writes that he believes one line is wrong in his drawing. Use this as a test of your understanding of the construction of the rods.)

The rods of Figure 5 are used similarly except that they are read from left to right and give the quotient. Thus 1,234,567,890 ÷ 6 = 0,205,761,315 with a remainder of 0 (the remainder is read from the right-hand index rod).

As you can see, these Genaille-Lucas rods actually eliminated one step in the use of Napier's rods, the adding in one's head of the amounts "carried" to the next rod at each step of the reading. The earlier variations we mentioned merely changed the shape, or size and arrangement of Napier's rods.

The next steps were to devise ways of multiplying by more than one digit at a time and then to simplify or eliminate the need to add separately the partial products. Several solutions to the former problem were proposed, some before Genaille's rods were invented. The latter problem was solved in part by M. Rous in 1869 by combining an abacus with a set of Napier's bones. Probably the last step in this process was L. Bollée's "Arithmografo" a device related to the rods but linking multiplication and addition semi-mechanically in such a way as to be considered, historically, a link between Napier's rods and true arithmetic machines. These latter have an interesting story of their own going back to Pascal and Leibnitz. Perhaps some of our readers have data on the latter or on other early machines, or on some of the variations and improvements on Napier's and Genaille's rods which we did not mention. In fact an exposition of Napier's procedures for division, square and cube root might be fun though perhaps they would only rarely be useable as teaching aids or enrichment and have no modern practical value.

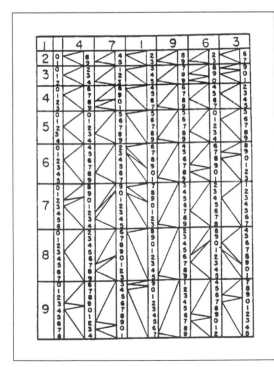

FIGURE 4

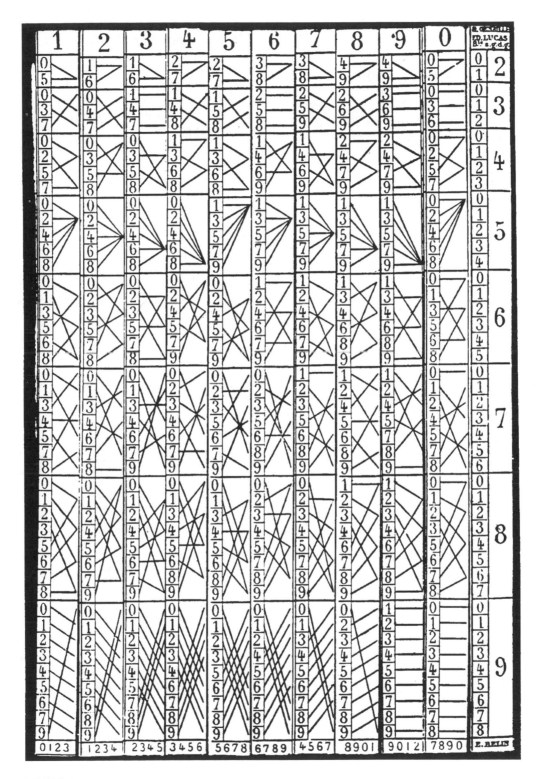

FIGURE 5

NOTES

1. H. D. Larsen, "Genaille's Rods," *American Mathematical Monthly*, Vol. 60 (Feb. 1953), pp. 140–141. He cited "The Pentagon," VIII (Spring 1949), pp. 98–100. We might add D. E. Smith, *History of Mathematics* (Boston: Ginn and Co., 1925), Vol II, pp. 202–203; D. E. Smith, *A Source Book in Mathematics* (New York: McGraw-Hill Book Co., Inc., 1929), pp. 182–185; Vera Sanford, *A Short History of Mathematics* (Boston: Houghton Mifflin Co., 1930), pp. 339–340.

2. F. Wöpckc in *Atti Accademia Pontificiana Nuovi Lincei*, 12 (1858–59), p. 245. This is cited in *Enciclopedia delle Matematiche Elementari e Complementi* (Milan, 1950), Vol. I, Parte 1, pp.415–416, which in turn makes much use of R. Mehmke, M. d'Ocagne, "Calculs Numériques," *Encyclopédie des Sciences Mathematiques* (Paris, 1908), Tome I, Vol. 4, Fascicule 2, pp. 230–234. The last two articles contain many references and were the source of much of the data compiled here.

3. E.M. Horsburgh, *Modern Instruments and Methods of Calculation. A Handbook of the Napier Tercentenary Exhibition* (London and Edinburgh, 1914), pp. 18–19.

4. Gaspar Schott, *Cursus Mathematicus* (Francofurti ad Moenum, 1674), p. 50.

5. H. D. Larsen, *loc. cit.*

62

Life and Times of Johann Kepler

BERNARD H. TUCK

*T*HE YEAR 1571, when Johann Kepler was born, had little to mark it as exceptional in a time when events moved but little faster than in ages past. The shadows of the Dark and Middle Ages still spread the heavy, restraining cloak of authority over the lives of people. The advent of the printing press had made the expression of personal opinions more than personal, and sober thought must now be given to the expression of an idea. But it had also promulgated knowledge, and new discoveries were imminent. An interesting age into which to be born—and safe enough, if one used one's eyes much, and one's tongue but little.

The small town of Weil, Württemberg, in which Kepler was born, was enmeshed in the ideological difficulties of the Reformation, as was most of the north of Europe. Luther had died twenty-five years before, and Calvin but a scant seven years before.

In England, Elizabeth was not yet "mistress of the seas," and Shakespeare was an urchin of seven. In Italy a boy named Galileo was Shakespeare's match in age—and more than his match in curiosity.

Servetus had been burned at the stake some nineteen years before (we will have no shedding of blood, they said), and Giordano Bruno had found it expedient to leave Italy. Not for another twenty-nine years would Bruno suffer the same fate suffered by Servetus—in a different country—persecuted by advocates of a rival religious persuasion, but persecuted by authorities who knew how to build a fire just as hot as that of their ideological opponents.

Copernicus, Kepler's hero, was twenty-eight years in his grave. John Napier, gadfly of Popes, was twenty-one. His logarithms would appear too late to assist at the birth of Kepler's first two laws, but would do yeoman service in lightening the tasks of Kepler's later years.

This, then, in some small part, was the time of Kepler. The spirit of the Renaissance was struggling valiantly, but cautiously, with the institutions of privilege, tradition, and authority. And although the privileged and the supporters of institutions quarreled bitterly (and often violently) among themselves, in the face of innovations the political axiom, "Rumps together, Horns outward," was the rule of the times.

The proponents of the doctrine of *Reason* were themselves captives of habits and customs and beliefs of the centuries past (as, no doubt, are we). Their dedication to the seeking of a small grain of truth was accompanied by acceptance of age-old superstitions and judgments; their hopes were those of the eclectic, to modify, not revolutionize. The shoulders upon which they stood to see better the truth were the shoulders of the ancients—authority here, too, to be respected, but authority to be accepted with judicious restraint.

These people were, then, an intellectually active, vigorous, partly civilized people, ruled by custom, yet adventurous of spirit. Believing in law and discipline, they yet revolted occasionally against the rigidity of the law.

Reprinted from *Mathematics Teacher* 60 (Jan., 1967): 58–65; with permission of the National Council of Teachers of Mathematics.

Kepler's Early Life

The young Johann seems to have spent a joyless youth; yet when one notes the gentleness, courage, and freedom of manner and speech that were his as a young man and as an adult, one wonders if bare circumstance can merit such conclusions.

One of his forefathers had been knighted by the Emperor Sigismund in 1430 while on a campaign in Italy. Perhaps Kepler's father, Henry, a soldier of fortune, sought a like route to privilege and wealth, as did many in those days. He met with no success, but did succeed in drawing Kepler's mother from the young boy's side to keep him company in the Netherlands, leaving the youngster in charge of his grandfather, burgomaster of the Free City of Weil of the Holy German Empire.

Kepler's mother, Catherine Guldenmann, daughter of the burgomaster, seems to have been undisciplined and uneducated as a girl—and later, a woman whose unbridled tongue won her much enmity and trouble. But, at the same time, one has indications of courage that might perhaps make one wish to reserve judgment about her character.

An infection with smallpox when Kepler was but four years of age left his eyesight much impaired, and added to his general weakness. He attended grammar school at Weil, and later (age six) moved with his parents to the small town of Leonberg. Here he attended the local Latin school and graduated at the age of thirteen.

Kepler spent the next two years at the seminary at Adelberg, where the rigorous discipline and renewed attacks of minor ailments left him even more unwell. However, his stay at the higher seminary at Maullbronn (about three years) brought him better health, mastery of Latin, and admittance to Tübingen University. The year was 1589; Kepler was eighteen.

Tübingen and Graz

Theology, philosophy, mathematics, and astronomy: the greatest of these was the first. Kepler prepared himself for the Protestant church. But the mind of a Kepler seems never to be the mind of the orthodox.

Mastlin, his revered professor of astronomy, taught Ptolemy's theory of the solar system but privately believed that the earth went around the sun. So did Kepler. Most unorthodox. Should a young man who seems unable to accept intellectually all the tenets of a religion be permitted the pulpit? Those who felt the heavy responsibility religion demanded in those times believed not. Much respected for his brilliance and for his character (flawed only by his liberal views), he was urgently recommended for a professorship of morals and mathematics at the Protestant Gymnasium (High School) in Graz, the capital of Styria in Austria.

Kepler was now twenty-three. His teaching duties in mathematics were trivial. He lectured on rhetoric and Virgil. He wrote the yearly almanac (*Calendar and Prognostications*) for five years, and gained an increase in salary thereby.

Astrology—"the foolish daughter of the respectable, reasonable mother of astronomy," he once called it, and again indicated that it was mostly foolishness that nonetheless permitted the astronomer to earn a livelihood—was a much respected art, and Kepler's good sense brought forth predictions that often came true. He gained much in respect, authority, and prestige. But, while playing with astrology, he was working, and working hard, on cosmology.

During the later part of 1595 his observations and work were being brought together, and he was eager to publish the results. He wrote Mastlin:

> I strive to publish them in God's honor who wishes to be recognized from the book of nature. But the more others continue in these endeavors, the more shall I rejoice; I am not envious. . . .

This eagerness for truth rather than place, for honesty rather than honors, was not at all the spirit of the time—nor is it entirely the spirit of today.

To the Baron von Herberstein
and
The Estates of Styria
Graz, May 15, 1596

What I promised seven months ago . . . .

His first book, *Mysterium cosmographicum*, was published.

The book was brilliant. The structures it brought forth had no lack of beauty. Mathematics and imagination and aesthetics were brought together in a manner that Aristotle himself would have approved. Perhaps this was its fatal flaw. It was, almost, more Aristotelian than Aristotle himself. It portrayed the power of an active mind, but not enough of nature.

The loveliest gem in Kepler's *Mysterium* was the hypothesis of the five regular polyhedra—or Platonic solids, as we often call them today. The distances between the planets were such that if the sphere containing the earth's orbit as a great circle should be inscribed within a dodecahedron, then the orbit of Mars would be a great circle on the sphere circumscribing the dodecahedron. And if the sphere of Mars' orbit should be circumscribed by a tetrahedron, the Jupiter's course would lie on the sphere circumscribing the tetrahedron. Should the sphere of Jupiter's orbit be circumscribed by a cube, the orbit of Saturn would lie on the sphere that encloses the cube. Uranus, Neptune, and Pluto were, of course, not yet known to exist, so Kepler then looked inward from the earth.

The orbit of the earth lies on a sphere that encloses an icosahedron, which itself envelops the sphere containing the orbit of Venus. And the sphere upon which lies the path of Venus circumscribes an octahedron which encloses the sphere upon which lies the road of Mercury as he goes around the sun.

There are but five regular convex polyhedra. (Kepler himself was to discover two of the four other regular polyhedra which exist.) So there could be but six planets. The orbits of the planets, being planar sections of spheres, must be circles, just as Aristotle said.

So *Mysterium cosmographicum* was published, the small town of Graz prepared itself for the winter, and life proceeded, for a little while, in its usual manner.

Kepler had reached the ripe old age of twenty-five. And bachelorhood, it seems, was a joyless estate.

Goodbye to Graz

The publishing of his first book may well mark the beginning of maturity in Kepler's life. Not, in truth, the maturity of solemn, heavy, cautious nature, but a maturity that finally accepts reverses with understanding and proceeds to make the best of a most imperfect world.

He fell in love with a twice-married noblewoman, Barbara Muller von Mühleck, some three years his junior. To prove his own noble descent he was forced to journey home and spend some seven months in order to procure the necessary papers. They were married February 9, 1597.

On August 4, 1597, Galileo wrote the "highly learned gentleman," thanking him for the copy of *Mysterium cosmographicum* that had been sent to him.

> I would certainly dare to approach the public with my ways of thinking if there were more people of your mind. As this is not the case, I shall refrain from doing so. . . .
>
> Yours in sincere friendship,
> GALILACUS GALILAEUS
> *Mathematician at the Academy of Padua*

To Galileo, October 13, 1597, Kepler sent a letter urging his fellow Copernican to proceed openly with his beliefs.

> Be of good cheer, Galileo, and appear in public. If I am not mistaken, there are only a few among the distinguished mathematicians of Europe who would disassociate themselves from us. . . .

But Galileo chose a later day for his day of reckoning.

The next several years were years of trouble and worry for Kepler. His first son, Heinrich, died, and his wife was inconsolable. The young Archduke Ferdinand instituted such steps to force all Austria back to the Catholic church that Kepler himself had to flee to Hungary for a month. Not one to shift his beliefs for personal advantage, Kepler tried to prepare for the inevitable by writing his friend and teacher, Mastlin, entreating him for assistance in obtaining a position at his beloved Tübingen. Mastlin answered some of these pleas in an offhand manner, and some letters he answered not at all.

Another friend, however, did assist him in his difficulties. Johann Herwart von Hohnburg, a Bavarian diplomat, who had also been a student of Mastlin, suggested a possible collaboration with Tycho Brahe, and may even have suggested the same to Tycho. (Tycho had received and praised Kepler's work, while disagreeing with Kepler's conclusions.) At any rate, Tycho's invitation to visit him in Prague was accepted by Kepler. He stayed with the eminent astronomer from February to June, 1600, succeeding with great difficulty in gaining access to some mathematical data on the planets. It must have been a frustrated Kepler who returned to face the increasing difficulties at Graz.

The intolerance of the times made the end almost inevitable. On August 1, 1600, Kepler and many others were forced to leave Styria forever. His possessions had to be sold hastily, and heavy taxes took most of what he had left. His financial position was most precarious. His wife and her family were in little better position, since their fortune was in real estate, which brought little under conditions of forced sale. Kepler wrote:

> All this is rather hard. But I should not have believed that in the communion of brethren it is so sweet to suffer loss or insult for our faith and Christ's honor, and to abandon home, fields, and country . . . .

Later he was to write:

> I will not take part in the fury of the theologians. I will not stand as a judge over my brethren. . . .

And later still, when refusing to abandon his Protestant beliefs for the privilege of continuing his work in peace and with "honor" :

> I cling to the Catholic Church. Even if she rages and beats my heart I remain united with a heart full of love, so far as human weakness allows . . . .

An appointment at Tübingen was not forthcoming, so Kepler returned to Prague. With some reluctance, it seems, he became associated with Tycho Brahe. But even then he wrote affectionately to Mastlin asking for assistance for an appointment to the university.

The Tables of Tycho and the Years at Prague

The year with Tycho must have been one of continual frustration for Kepler. Tycho was most parsimonious with his astronomical observations. And Kepler was at his wit's end to accomplish all he wished to accomplish in the face of such obstruction. To Mastlin he wrote:

> His observations . . . are accessible to me, but first I had to promise to keep them secret. I have complied with this as far as it befits a philosopher . . . .

Tycho Brahe died in October 1601. Emperor Rudolph II appointed Kepler to the position of "Mathematician of His Holy Christian Majesty." Kepler's major work had begun.

The prelude to the Thirty Years War was being enacted. The troubled times would add to the burdens imposed by Kepler's personal difficulties, but his work would progress and would occupy him for the rest of his life.

Kepler spoke at Tycho's funeral, and later often expressed admiration for the work he had accomplished. He managed to gain custody of Tycho's *Rudolphine Tables*, and spent much of his time editing them and using them to work out the orbit of Mars. In any plane, at least two coordinates are needed to locate oneself: a street and an address, the intersection of two streets, an angle from a place along with a distance away, or (as with sailors) bearings from two objects. Kepler had only one in space, the sun. But Brahe's tables had such voluminous data on Mars that he hoped to use this planet for his second signpost. The difficulties were tremendous and took years to overcome.

At the beginning of the seventeenth century the earth-centered Ptolemaic theory was the theory of the day. But so inaccurate were the predictions that the theory permitted that many astronomers changed it about in minor ways to permit better, though still grossly inaccurate, predictions to be made. Thus Tycho accepted the idea that the sun circled the earth, but believed that the other planets circled the sun. This idea was an improvement on most, but was still inadequate to explain the positions of the planets as measured.

Kepler's almost mystic faith in the Copernican heresy (not yet so-called) urged him toward a hypothesis that would accept the main features of the Copernican scheme, but would improve upon it in such a way that the stars themselves would affirm the plan.

Copernicus believed that the stars were so far away that the distance from the earth to the sun was insignificant in comparison. Brahe did not, and since he could measure no parallax for any star, he disregarded Copernicus' theory, though it led to predictions which were more accurate than his own.

Kepler's first great stride was to abandon the accepted idea that the velocity of the planets is constant. He intuitively arrived at the idea that sources of velocity of planets were rays emanating from the sun. Therefore, as a planet moves farther from the sun, fewer rays reach it, and it slows down proportionally to the square of the distance. (Or, as would usually be stated, its velocity is inversely proportional to the square of its distance from the sun.) Almost eight years were spent trying various ovals and other geometrical figures which might match the observed positions of Mars. At last Kepler tried one which worked, the ellipse.

Kepler published his findings, along with many other observations, in *The New Astronomy*. The place, Prague; the year, 1609; the reaction, far from earth-shaking, the publication causing scarcely a tremor through the intellectual life of the day.

When Copernicus published his *Books on the Revolutions* in 1543, only a few days before his death, his sun-centered ideas were not taken too seriously by the intelligentsia of the day. Its tables and drawings were copied assiduously, his methods of measuring distances evoked the greatest admiration, his standards revolutionized ideas as to what astronomers could do; but since his theory did not explain the erratic movement of, say, Mars, much better than other theories, nor much more accurately, his sun-centered hypothesis was accepted by only a few during the following decades.

Kepler's book caused much less stir among the learned. His position as the court mathematician seemed to earn him more respect than did his work. Yet, as he later wrote as part of as poetic an outburst as may be found in the history of science *(Harmony of the Worlds)*, this was in no way a matter of concern:

> I am writing a book for my contemporaries or—it does not matter—for posterity. It may be that my book will wait for its readers for a hundred years. Has not God himself waited for 6000 years for an observer?

Kepler's first two laws of planetary motion are

1. The planets move about the sun in elliptical orbits with the sun at one focus.

2. The radius vector joining a planet to the sun sweeps over equal areas in equal intervals of time.

These laws seem to have been little noticed by the great Galileo. Full of the importance of his discovery of the "Medician Stars" (the four largest satellites of Jupiter), he wrote the following to Kepler in 1610:

> What do you say to the main philosophers of our school, who, with the stubbornness of vipers, never

wanted to see the planets, the moon, or the telescope, although I offered them a thousand times to show them the planets and the moon.... With logical reasons, as if they were magic formulas, he wanted to tear the planets from the heavens and dispute them away. . . .

Kepler's work at the time encompassed much more than astronomy and astrology. His work on optics, ephemerides, gravity, and tides, and his mathematics took much of his time. And his private life was none too smooth.

Kepler was forty-one. The year was 1612. One of his sons died and his wife passed away. Prague itself was a battleground, and subject to the roughness of the soldiers in the town. Kepler's salary was always in arrears, and his patron, the emperor, pushed aside by his brother, died later in the year. Matthias, Rudolph's successor, confirmed him in his office as "Imperial Mathematician," and granted him permission to seek employment outside the court.

Kepler left Prague for Linz.

The Years at Linz

Leaving his remaining two children with relatives, Kepler took a teaching position in a small college in Linz. This small Austrian city, on the banks of the Danube, remained his home for over fourteen years. Here he remarried (and fathered seven more children, five of whom survived), was denied communion because of his liberal views, and labored industriously at his work.

His work on cubical contents of wine casks, initiated by an unusually heavy vintage year, contained some of the early ideas of the integral calculus. He continued his work on Tycho's tables and published, in 1619, *Harmony of the Worlds.* This publication contained, along with other observations and conjectures, his third, and most beloved, law of planetary motion:

3. The square of the time of one complete revolution of a planet about its orbit is proportional to the cube of the orbit's semi-major axis.

The stage, of course, had been set for Newton in the earlier book containing the first two laws of planetary motion (as well as by Galileo's two laws of motion), but as an embellishment, Kepler's third law was a pretty thing.

During all this time, and more, Kepler was much troubled by charges of witchcraft leveled against his mother. During January 1616 he wrote to the authorities of Leonberg, hoping to settle the affair once and for all. But the proceedings dragged on. Despite all his efforts, and despite the ridiculous nature of the charges, his mother was imprisoned in August 1620, and remained imprisoned for fourteen months.

Kepler, admitting to the Duke of Württemberg that his mother had for years been quite difficult, talkative, malicious, and a maker of trouble, pressed with every means at his command for her release and, later, her acquittal. Since one of her associates had left a thumb in the rack, there can be no doubt that he was very much concerned about her fate.

Kepler spent almost a year in Guglingen trying to save his mother from torture and execution. Her famous son could do nothing to save her from being threatened in the torture chamber, but here she resolutely refused to confess and was released. She died six months later, leaving the learned mathematician with more rumors to combat than those which would have him half popish, half Lutheran, half Calvinist, and a complete eclectic in religious beliefs. Yet the fact that Kepler continued to take delight in his work can scarcely be refuted. His letters well reflect his pride and happiness in his many occupations.

Kepler's *De Cometis* of 1619 was dedicated to James I of England. Although invited to come to England, he did not accept. Perhaps he refused, as he had refused a post in Bologna, because of a fear that freedom of expression would be denied him. Yet the city of Linz itself was becoming more and more untenable.

Protestant rule of Linz itself was overturned in 1620, and a growing oppression of Protestants began. Kepler, however, was permitted to go on with his work. He did much work on Napier's logarithms, and continued his work in astronomy.

However, the climate of intolerance at Linz eventually made work impossible for Kepler. In 1626 his private library was sealed up and it became evident that it would be impossible to print the *Rudolphine Tables* in Linz, as had been commanded by the Emperor Ferdinand II. He therefore saw his wife and children safely away and left for Ulm, lived through a siege for fourteen days, had type set for the tables at his own expense, and had them published. He wrote the following to a friend:

> I long for a place where I can teach them. If possible, in Germany; if not there, in Italy, France, Belgium, or England; but only if an adequate salary is available for the stranger....

The next several years were spent mostly upon work connected with the *Rudolphine Tables*, and with his personal affairs—his children, his wife, his never-collected salaries. In 1628 the emperor found him worthy of favor, and Count Albert of Friedland and Sagan granted him a yearly allowance and promised him a printing press and a quiet place in Sagan. Kepler's life seemed almost secure.

The good fortune, however, was short-lived. The 11,817 gulden owed Kepler by the court were not forthcoming. The work on a supplement to the tables (movements of the planets for 1631) was completed, and other work of like nature was brought to a conclusion.

Kepler left Sagan in the fall for Leipzig, and thence to Regensberg, hoping he might receive his salary from the German Reichstag. The bitterness of the weather during the journey, which he made on horseback, brought him to Regensberg with little life left to live.

Kepler died here, after a short illness, on November 15, 1630. "Your, mine, our Sun," wrote a friend. "The Sun of all astronomers has set."

BIBLIOGRAPHY

1. Baumgardt, Carola. *Johannes Kepler: Life and Letters*. New York: Philosophical Library, 1952.

2. Eves, Howard. *An Introduction to the History of Mathematics* (rev. ed.). New York: Holt, Rinehart & Winston, 1964.

3. Gade, John Allyne. *The Life and Times of Tycho Brahe*. Princeton, N.J.: Princeton University Press, 1947.

4. Gebler, Karl von. *Galileo Galilei*. London: C. Kegan Paul & Co., 1897.

5. Hogben, L. T. *Science for the Citizen*. New York: W. W. Norton & Co., 1938.

6. "Kepler, Johannes," *Encyclopaedia Britannica* (1964).

7. Kuhn, Thomas S. *The Copernican Revolution*. New York: Modern Library, 1957.

8. Physical Science Study Committee. *Physics*. Boston: D. C. Heath & Co.

Simon Stevin and the Decimal Fractions

D. J. STRUIK

I

SIMON STEVIN, the Flemish-Dutch engineer and mathematician, is best known in the history of mathematics as the inventor of the decimal fractions. This is substantially true, even if we acknowledge the fact that decimal fractions were used before Stevin. It was primarily through Stevin's little book, really no more than a pamphlet, entitled *De Thiende (The Tenth)*, published at Leiden in 1585, that decimal fractions became a regular part of the curriculum in arithmetics.

This pamphlet was republished, in Dutch, French, English, and Latin, several times during Stevin's lifetime and shortly thereafter. Its text has now been made available again in a photostatic reproduction, together with a contemporary English translation by Richard Norton, as part of the second volume of the new edition of Stevin's principal works, prepared under the auspices of the Physics Section of the Royal Netherlands Academy of Sciences.[1] This edition contains an introduction devoted to the history of decimal fractions which gives us an opportunity to evaluate Stevin's achievement in this area.

Simon Stevin was a native of Bruges; it is fairly certain that 1548 was the year of his birth. He started as a bookkeeper, but left Flanders in the

period of unrest and persecution which opened the war with Spain. In 1581 we find him at Leiden, and from that time until his death he lived in Holland. Here, like thousands of other immigrants from the Southern Netherlands, he had to find a living under new circumstances. With his particular talents as engineer, accountant, and mathematician, his welcome in the new and striving republic was assured. For many years he was an adviser to Prince Maurice of Orange, one of the most accomplished military commanders of his age, whom he served not only as military engineer, but also as a tutor in different fields of mathematics. Much of Stevin's later work bears the imprint of these tutoring sessions, which covered a wide field of practical and applied geometry, trigonometry, perspective, and bookkeeping. After his death at La Haye in 1620, his admirer Albert Girard, also an immigrant mathematician and engineer, edited his collected mathematical works in French.[2] This edition was published as a big folio in 1634, and has been the main source of our knowedge of Stevin as a mathematician until 1958.[3] The new edition of *The Principal Works* is planned in five volumes, of which Volume I, dealing with Mechanics, and Volume II, dealing with Mathematics, have already appeared. When the five volumes are published they will offer an excellent insight into the status of the exact sciences in Europe in the period just before Fermat and Descartes opened the new era of calculus and coordinate geometry.

Reprinted from *Mathematics Teacher* 52 (Oct., 1959): 474–78; with permission of the National Council of Teachers of Mathematics.

A decimal fraction has a positive integral power of 10 in its denominator, and can be written in many ways, e.g.: 71/100, 0.71, 0,71, or .71, to use notations in use at the present time. It is not easy to say where decimal fractions were first used in any systematic way, but the priority seems to go to the Chinese.[4] With them we find the decimal place-value notation as early as the fourteenth century B.C. and the use of decimal fractions in metrology as early as the third century A.D. Liu Hui, who lived in this period, expresssed a length of 1.355 feet as 1 chhih, 3 tshun, 5 fen, 5 li. The same Liu Hui also performed root extractions in decimals; in this process, which we find in the Middle Ages in use among Arabic, Jewish, and Latin authors, one writes, for instance, $\sqrt{17} = \sqrt{170{,}000}/100 = 412/100$. When we come to the Sung Dynasty, we find computation with decimal fractions already well developed; Yang Hui, in 1261, multiplies 24.68 by 36.56 and finds 902.3008. From the Chinese and the Indians the decimal notation came to the Islamic writers; our example of $\sqrt{17}$ is found in the writings of Al-Nasawi (Persian, ca. 1030), who translated 412/100 again into sexagesimal fractions as 4°7′12″, meaning 4 + 7/60 + 12/3600. The familiarity of Yang Hui with decimal fractions was shared, one and a half centuries later, by the Persian astronomer Jamshid Al-Kashi,[5] who multiplied 25.07 by 14.3 to obtain 358.501. Whether the first notion of decimal fractions in Europe came through contact with the Orient or whether it arose spontaneously is difficult to say, but there is little doubt that wherever computations in the decimal system were used on a wide scale, decimal fractions were sooner or later bound to appear through the logic of the computations themselves. As it was, it is rather astonishing that with the decimal system in use since the Stone Age in so many parts of the world, the regular use of this system for fractions appeared at so late a period.

Readers may ask: if decimal fractions appeared outside of China at such a relatively late date, how did people get along before their introduction?

The answer is that they used either fractions like 3/7, 7/13, etc., with any kind of denominator, or the sexagesimal fractions based on the number 60. An example of a quite complicated reckoning of the first kind is Archimedes's approximation of π as a ratio between 3 10/71 and 3 1/7; examples of sexagesimal reckoning can be found on cuneiform clay tablets dating back to Iraq (Mesopotamia) of the fourth millennium B.C. Ptolemy, in his astronomical handbook known as the *Almagest* (ca. A.D. 150), also used sexagesimal fractions. And so do we when we express the magnitude of an angle as 29°51′32″; here minutes and seconds are expressed in the sexagesimal system, while the numbers 29, 51, 32 are written in the decimal system. Variations of these two methods of the fractional calculus are known to have existed, such as the ancient Egyptian method of expressing fractions as the sum of unit fractions (fractions such as 1/3, 1/7, with numerator 1). Strictly speaking, there was, and still is, another way of coping with the problem, popular wherever fractional calculus is considered difficult, and that is the avoidance of all fractions by the choice of an appropriate scale of measurement, e.g., 60, so that 1/4 is expressed by 15, 2/5 by 24, etc. We do the same when we say 475 m instead of .475 km, or 3 quarts instead of 3/4 gallon. We shall see how all these concepts played a role in Stevin's work.

2

During the fifteenth and sixteenth centuries, European computers began to use the Hindu-Arabic system of decimal notation—that is, our present positional system—with ever-increasing efficiency, stimulated by the spread of a mercantile civilization. Gradually the influence of the system began also to be felt in computation with fractions.

We can clearly see this process going on in the trigonometric tables of the Nuremberg astronomer and mathematician Regiomontanus, who died in 1476, and whose books and tables were standard even in the days of Stevin. In those days, and also much later until the eighteenth century, the sines, as well as the tangents and the other trigonometric entities, were conceived as lines and not as ratios, so that they were expressed in terms of a circle radius R of given length. In the sine table of Regiomontanus we find $R = 60,000$, so that the sine of 30° is 30,000; later he used $R = 6,000,000$. Both data show the influence of the sexagesimal system. (Ptolemy's tables are based on $R = 60$.) The different sines are thus expressed as integers. But Regiomontanus also had a tangent table in which the tangent of 45° is 100,000, so that here $R = 10^5$, and he had another table with $R = 10^7$. The decimal system thus became established as the base for the computation of trigonometric tables. The great tables of Rhaeticus (1551), which contain to seven decimal places the values of all six trigonometric functions for angles ascending by 10" intervals, are based on $R = 10^7$.[6]

These tables contain no decimal fractions in the strict sense of the word. Actual fractions based on powers of 10 as denominator occur in some of the many books on arithmetic which appeared during the sixteenth century. For instance, we find, in a book of 1530 written by the widely read German teacher of arithmetic, Christopher Rudolff, a table for compound interest, in which the values of $375 \times (1 + 5/100)^n$ for $n = 1, 2, \cdots,$ 10 are written in a form which differs from our present notation only by the use of a vertical dash instead of a point as decimal separatrix, e.g., 413|4375 for $n = 2$.[7] There are several similar cases, but none of these authors used decimal fractions consistently, and where they used them they varied their notation.

Stevin was the first in the Occident to divest the decimal fraction of its casual character. Appealing to the learned as well as the practical man, to the teacher as well as the merchant and the wine gauger, he advertised the advantages of his notation as "teaching how to perform with an ease, unheard of, all computations necessary between men by integers without fractions." In doing this he used a notation which reminds us strongly of

the sexagesimal one; where we write 47.58, he wrote 47⓪, 5①, 8②, where the unit ⓪ is called "commencement," the tenth ① is called "prime," the hundredth ② is called "second," etc. When we write 27.847 + 37.675 + 875.782 = 941.304, Stevin wrote

$$
\begin{array}{ccccccc}
 & & ⓪ & ① & ② & ③ \\
 & 2 & 7 & 8 & 4 & 7 \\
 & 3 & 7 & 6 & 7 & 5 \\
8 & 7 & 5 & 7 & 8 & 2 \\
\hline
9 & 4 & 1 & 3 & 0 & 4
\end{array}
$$

In a similar way he deals with subtraction, multiplication, and division—all performed "without fractions." He ends his pamphlet with a plea to introduce the decimal system also in the measurement of lengths, areas, volumes, etc.

Stevin had the correct idea, but his notation seems clumsy to us and less elegant than that which Rudolff used half a century earlier. The circle notation was taken from the Italian mathematician Bombelli, who had used a similar notation in his *Algebra* of 1572 for the powers of the variable (anticipating our exponents in x^1, x^2, x^3, etc.). He might have used the sexagesimal notation in the form 47°5′8″, for our 47.58, as some of his followers did; he himself exchanged his notation occasionally and wrote 732② for our 7.32. His notation may have had some advantages for inexperienced pupils, since the circle notation allows intermediate steps: 7⓪5①8② plus 4⓪7①5② is equal to 11⓪12①13②, which reduces to 11⓪13①3②, and this again to 12⓪3①3②. Stevin could also do away with zeros: 2③7⑤ means .00207. But the notation remained unwieldy, and Stevin's work would not have had its lasting influence if it had not been for Napier and his logarithms.

3

Logarithms, invented by John Napier, the Scottish nobleman, were first presented in his Latin *Descriptio* of 1614. These first logarithms are not the logarithms that we use, but are certain numbers defined with the aid of sines, which in Napier's exposition were based on $R = 10^7$. This edition of 1614 had no decimal fractions. These appear in a 1616 English translation of the *Descriptio*, with a point as decimal separatrix. This notation was adopted by Napier in his Latin *Rabdologia* of 1617, the book in which he showed how to perform computations with his "rods," the so-called "rods of Napier." Here Napier quotes Stevin's *Arithmetica Decimalis* and proposes the notation of 1993,273 (with point or comma) for 1993 273/1000, although he also uses 821, 2′5″ for 821 25/100. Then, in the posthumous *Constructio* of 1619, the notation has become consistent: "whatever is written after the period is a fraction." Thus 25.803 means 25 803/1000.

The great tables of logarithms, based on 10, which now appear, take the decimal notation of fractions for granted, with dot or period. With tables of such logarithms, in which the decimal part of such numbers as 43, 430, 4300 is the same, the decimal notation of fractions is only natural. Henry Briggs, in his table of 1624, and Adrian Vlacq, in his tables of 1627, use this notation consistently, and from then on the decimal fractions with comma or dot as separatrix were generally accepted, at any rate in computations with logarithms.

Stevin's, Napier's and Briggs's contributions were combined in two Dutch books by the surveyor Ezechiel De Decker, entitled *The New Arithmetic, First Part* and *Second Part*[8] (1626, 1627). Here we find together Stevin's *Thiende*, Vlacq's translation of the *Rabdologia*, and the Briggsian logarithms of all integers from 1 to 100,000. These two books by De Decker are a kind of glorification of the triumph of the decimal system. They stress three essential aspects of this victory: the Hindu-Arabic notation with the modern digits, the decimal fractions, and the logarithms to base 10. One change was still due, although it was already implied in the whole framework of the system, namely the rewriting of the trigonometric tables to a unit

$R = 1$, a thing that even Stevin did not do. The systematic introduction of this unit $R = 1$ had to wait until Leonard Euler's *Introductio in Analysin Infinitorum* of 1748, and from that time on trigonometric entities were no longer considered as line segments but as dimensionless ratios.

4

The triumph of the decimal fractions through the work of Stevin and Napier did not mean that notation and use became immediately standardized. There were loyal Stevin followers who preferred his notation, or a slight modification of it. As late as 1739 we find the Abbé Deidier teaching that decimal fractions should be written as $89 \cdot 5^I 2^{II} 7^{III} 6^{IV}$ or 895276^{IV}; he used, however, the ordinary point notation for logarithms. Many such inconsistencies remained, again until Euler, in his *Introductio* of 1748, standardized our present notations.

The use of the decimal notation for weights and measures, also proposed by Stevin, had to wait in the West until the French revolution introduced meters, ares, and liters, and also standardized the monetary system.[9] We know that this was only partially accepted in England and in the United States, though the United States introduced a decimal monetary system as a result of the efforts of Robert Morris, Thomas Jefferson, and Alexander Hamilton.[10] As to angular measurement, the struggle is still on, and when we use our ordinary notation of degrees, minutes, and seconds, we pay our respects to a system which can boast of an age of five thousand years, certainly one of the longest ages in our whole scientific heritage.

NOTES

1. *The Principal Works of Simon Stevin.* Amsterdam: Swets and Zeitlinger. Vol. I, *Mechanics,* edited by E. J. Dyksterhuis, v + 617 pp. (1955); Vol. II, *Mathematics,* edited by D. J. Struik, in two parts, v + 976 pp. (1958).

2. Albert Girard (1595–1632) is best known as the author of the *Invention nouvelle en algebre* (1629, repub-

lished 1884), in which we find the theorem that an algebraic equation of degree *n* has *n* roots (Girard's formulation is different).

3. There have been two facsimile editions of *The Tenth* before 1958, one by H. Bosmans of the Dutch edition of 1585 (La Haye: Anvers, 1924), and one by G. Sarton of the French edition of 1585 (*Isis* 33 (1935)).

4. J. Needham, *Science and Civilisation in China*, vol. III. Cambridge: Cambridge University Press, 1959. The first section of this standard work deals with mathematics.

5. *D.G. Al-Kaši Ključ Arifmetiki, Traktat ob Okružnosti*, translated and edited by B. A. Rozenfel'd (Moscow, 1956), especially p. 62; Y. Mikami, *The Development of Mathematics in China and Japan* (Leipzig, 1913), especially p. 26. This is the way in which Al-Kashi multiplies 25.07 by 14.3 to get 358.501:

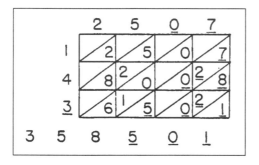

(The integers are in black and the fractional parts in red. The digits are Arabic ones, which differ from our present digits, the latter being in use only since the time of the European Renaissance.)

6. These tables were expanded by Valentin Otho into the *Opus Palatinum* of 1596, a classic among the tables which has many values in ten decimals and the sines up to 15 decimals.

7. *Exempel Buechlin Rechnung belangend darbey* (Augsbury, 1530). The place where decimal fractions are introduced has been more than once reproduced, e.g., in D. E. Smith, *History of Mathematics*, vol. II, p. 241.

8. *Eerste Deel van de Nieuwe Telkonst* (Gouda, 1626); *Tweede Deel van de Nieuwe Telkonst* (Gouda, 1627). The second part was always known, but the first part disappeared until it was rediscovered by M. van Haaften in 1920; see *Nieuw Archief voor Wiskunde*, 15 (1928), pp. 49–54; 31 (1942), 59–64. This discovery has shown that it was not Vlacq in 1628, but De Decker in 1627, who first published a complete table of logarithms.

9. The Chinese, as mentioned in the text, had a decimal system at least as early as the third century A.D.

10. The dollar with its decimal division was introduced by the Coinage Act of 1792, sponsored by Hamilton. This act was preceded by a resolution of Congress of 1785, the result of a report to the President of Congress by Robert Morris (1782), endorsed by Jefferson. See A. Nussbaum, *The History of the Dollar* (New York, 1957, viii + 308 pp.), Chapter II; C. D. Hellman, "Jefferson's efforts toward the decimalization of the U.S. weights and measures," *Isis*, 16 (1931), pp. 266–314.

La Disme of Simon Stevin—
The First Book on Decimals

VERA SANFORD

$\mathcal{T}$HE HISTORY of the decimal fraction has two distinct sections: the evolution of the idea of decimals, and the evolution of a convenient symbolism. The idea might, indeed, be traced to the very beginning of the place-value of the Arabic notation. The symbolism has not been standardized today, as witness the English 3·14, the French 3,14 and our 3.14. Yet, strangely enough, the theory of operations with decimals appeared in complete form in the first work devoted wholly to the subject, *La Disme* (1585) by Simon Stevin of Bruges.

It is the purpose of this sketch to show, by excerpts from *La Disme,* what Stevin did with decimals and what uses he anticipated men would make of them. Since his introduction, definitions, and symbols give a clue to the way in which he came to invent these numbers, a reconstruction of his line of thought is pertinent to the case, even though this may degenerate into merely what one thinks he would have thought had he been Stevin. And to give *La Disme* its proper setting, there must be added a note on computations with fractions in the sixteenth century.

The devices that might have developed into the decimal fraction fall into two groups: those which depend specifically upon the decimal arrangement of the Arabic notation, and those which involve symbols that are happy accidents or actual

forerunners of the decimal point, according to one's interpretation of their use. In the first group was a scheme for finding the nth root of a number. This may be expressed in modern symbols as

$$\sqrt[n]{a} = \frac{\sqrt[n]{a \cdot 10^{kn}}}{10^k}.$$

Its value lies in the fact that it enables one to find a root to any required degree of accuracy and that fractions enter only in the last step of the work. In a similar way, interest was computed on a principal 100,000 times too large, and trigonometric functions were found from a circle of radius 10,000,000. Symbols typical of the second group are the following: in dividing by multiples of 10, 100, 1,000, Pellos (1492) used a period to mark off one, two, or three places in the dividend as if to indicate that digits to the right belonged inevitably to the remainder; and Rudolff (1530) printed a compound interest table much as we would, save that a bar (/) acts as a decimal point. Several other writers used symbols of this sort, but no one before Stevin explained what they actually signified.[1]

Any one of these devices might have been the beginning of the decimal fraction, but a careful study of *La Disme* indicates that Stevin approached the subject from a different standpoint—that of adapting to everyday use and to the current number system a method devised by scholars of an earlier time. Stevin was well qualified to appreciate the learning of the past and the difficulties of his

Reprinted from *Mathematics Teacher* 14 (Oct., 1921): 321–33; with permission of the National Council of Teacher of Mathematics.

411

contemporaries. He lived in the Netherlands in the days of the great struggle against Spain. His ability in mathematics and physics, and his skill in applying his knowledge to military and civil matters, won for him the post of adviser to Maurice of Nassau, the son of William the Silent. Thus Stevin was in close touch with the affairs of a country which necessity had driven to seize every aid that science could give, and he was collaborator with a prince who realized this necessity most keenly.

In all such work, numerical computations must have played a large part, and although methods such as are noted above were in use for special cases, there was no satisfactory general scheme for dealing with fractions.

Common fractions were difficult to use and unsatisfactory in expressing results. The extremes to which they were carried in an effort for accuracy is shown in the case where Stevin gives 233 $^{2356847}/_{23476796}$ lb. for the present value of an annuity of 54 lb. yearly for 6 years, interest at 12%. Stevin was not alone in using large denominators. He makes occasional attempts to reach significant figures in his results, as in computing his interest tables where he counts $^{100}/_{101}$ as 1 "for it is greater than one-half," but in *La Disme* he considers that the 0.00067 in 301.17167 is of no account, and he writes his result as 301.171.

Stevin's decimals seem to have evolved from the sexagesimal fractions of the scientists. Sexagesimals were really a series of denominate numbers in which a unit was 60 minutes (primes), a minute was 60 seconds, a second was 60 thirds, and so on.[2] These numbers were inherited from the Babylonians and the Greeks, and they are still used in our divisions of the hour and the angle. The advantages of sexagesimals in expressing results are obvious. Two such numbers could be readily compared. Many fractions could be expressed exactly in sexagesimals of the first order, and others could be given in combinations of minutes, seconds, and thirds. In one case, for instance, an approximate root of an equation is carried to the sexagesimal of the tenth order. These numbers

had grave disadvantages, however. They were of little use if one wished to extract roots, and the multiplication or division of two sexagesimals was difficult.

It is not surprising that Stevin felt obliged to write a careful introduction to the work that was to revolutionize all computations. One can imagine the scepticism that would greet a pamphlet of 36 octavo pages labeled with that ambitious purpose, and Stevin's attempts to reconcile the size of his book to its sub-title, and his desire to clear himself from implications of undue boasting make his introduction fully as interesting as his theory. As for the book itself—it first appeared in Flemish under the title *La Thiende* (Leyden, 1585). In the same year, a French translation *La Disme* was printed as an appendix to Stevin's *l'Arithmetique*. In the next half century, the work went through at least six editions, one of them a translation into English (1608).[3]

Let us suppose for the moment that we have never heard of decimals and that we are among the people to whom Stevin dedicated *La Disme*. In this case his introduction may sound like a fairy tale that we wish were true, but that we fear is not. We shall be interested, however, to learn what it is that makes the author at once so confident and so modest. Any theory that will take such hold of a man is worth considering, so let us begin *La Disme* which Stevin called "La Practique des Practiques"— we might call it the "super-method."

LA DISME

Teaching how all Computations that are met in Business may be performed by Integers alone without the aid of Fractions.

Written first in Flemish and now done into French
by
Simon Stevin of Bruges
To Astrologers, Surveyors, measurers of Tapestry, Gaugersa[4], Stereometers[5] in general, Mint-Masters, and to all Merchants.
Simon Stevin sends Greeting.

Any one who contrasts the small size of this book with your greatness, my most honorable sirs to whom it is dedicated, may think my idea absurd, especially if he imagines that the size of this volume bears the same ratio to human ignorance that its usefulness has to men of your outstanding ability; but, in so doing, he will have compared the extreme terms of the proportion which may not be done. Let him rather compare the third term with the fourth.

But what is it that is here propounded? Some wonderful invention? Scarcely that, but a thing so simple that it hardly deserves the name Invention; for it is as if some stupid country lout chanced upon great treasure without using any skill in the finding. But, if anyone thinks that, in explaining the usefulness of *La Disme*, I am boasting of my cleverness in devising it, he shows without doubt that he has neither the judgment nor the intelligence to distinguish simple things from difficult, or else that he is jealous of the common good. However this may be, I shall not fail to mention the usefulness of this thing even in the face of this man's empty calumny. But since the mariner who has found by chance an unknown isle, may declare all its riches to the king as for instance, its having beautiful fruits, pleasant plains, precious minerals, etc., without its being reputed to him as conceit; so may I speak freely of the great usefulness of this invention, a usefulness greater than I think any one of you anticipates, without constantly priding myself on my achievements.

Stevin here enumerates cases which show the usefulness of number in the work of the astronomer, the surveyor, the mint-master, and the merchant. He stresses the difficulties of manipulations with sexagesimals and with denominate numbers and speaks of the almost inevitable errors in calculation that vitiate excellent work. Stevin claims that *La Disme* teaches how these computations may be performed by whole numbers and he says that work tedious even to a skilled computer may now be accomplished with the same ease as in reckoning with counters. The force of the comparison is apparent only when one reflects that counter-reckoning in Stevin's time was a device used principally by people little skilled in the finer points of working with Arabic numerals—its greatest value was in addition and subtraction. Then comes Stevin's reason for presenting this to the public:

If by these means, time may be saved which would otherwise be lost, if work may be avoided, as well as disputes, mistakes, lawsuits, and other mischances commonly joined thereto, I willingly submit *La Disme* to your consideration. Someone may raise the point that many inventions which seem good at first sight are of no effect when one wishes to use them, and as often happens, new methods, good in a few minor cases, are worthless in more important ones. No such doubt exists in this instance, for we have shown this method to expert surveyors in Holland and they have abandoned the devices which they had invented to lighten the work of their computations and now use this one to their great satisfaction. The same satisfaction will come to each of you, my most honorable sirs, who will do as they have done.

The compactness of the book implies no lack of formality. The *Argument* outlines the topics to be treated under *Definitions* and *Operations* and promises that "At the end of this discussion there will be added an *Appendix* setting forth the use of *La Disme* in real problems." How modern it all is! Motivation that would whet anyone's interest, a thorough preliminary testing out which we are assured was successful, and, as a climax, real problems!

We come then to the actual theory.

DEFINITION I

La Disme is a kind of arithmetic based on the idea of the progression by tens, making use of the ordinary Arabic numerals, in which any number may be written and by which all computations that are met in business may be performed by integers alone without the aid of fractions.

EXPLANATION

Let the number one thousand one hundred eleven be written in Arabic numerals as 1111, in which form it appears that each 1 is the tenth part of the next higher figure. Similarly, in the number 2378, each unit of the 8 is the tenth part of each unit of the 7, and so for all the others. But since it is convenient that the things which we study have names, and since this type of computation is based solely upon the idea of the progression by tens or disme,[6] as is seen in our later discussion, we may properly speak of this treatise as *La Disme*, and we shall see that by it we may perform all the computations we meet in business without the aid of fractions.

DEFINITION II

A whole number is called the Commencement and has the symbol ⓪.

DEFINITION III

The tenth part of a unit of the commencement is called a *Prime* and has the symbol ①, and the tenth part of a unit of prime is called a *Second* and has the symbol ②. Similarly for each tenth part of the unit of the next higher figure.

EXPLANATION

Thus 3① 7② 3③ 9④ is 3 primes 7 seconds, 3 thirds, 9 fourths, and so we might continue indefinitely. It is evident from the definition that the latter numbers are $^3/_{10}$, $^7/_{100}$, $^3/_{1000}$, $^9/_{10000}$, and that this number is $^{3739}/_{10000}$. Likewise 8⓪ 9① 3② 7③ has the value 8 $^9/_{10}$, $^3/_{100}$, $^7/_{1000}$, or 8 $^{937}/_{1000}$. And so for other numbers. We must also realize that in *La Disme* we use no fractions and that the number under each symbol, except the commencement, never exceeds the 9. For instance, we do not write 7① 12②, but 8① 2② instead, for this has the same value.

Stevin's choice of names and symbols for these numbers is explained in the first pages of *l'Arithmetique*, which was published in the same volume as *La Disme*. Here, he calls the terms of a geometric progression by the ordinals: prime, second, third, . . . with the signs ①, ②, ③ . . . giving as examples ①2 ②4 ③8, ①3 ②9 ③27. In *l'Astronomie*, he uses these symbols in writing both decimals and sexagesimals. In *l'Arithmetique*, he extends this to the case where the ratio is an unknown quantity. Thus, ①, ②, ③ stand for x, x^2, x^3. The commencement, as defined in *l'Arithmetique*, was an integer or an irrational number used in algebraic computation. The symbol ⓪ was to enter only when the commencement was an abstract number. Denominate numbers in Stevin's other works and in the Appendix to *La Disme*, appear as 1 hour 3① 5②, 5 degrees 4① 18②, 2790 verges 5① 9②. The origin of the symbol for the commencement has no connection with the zero exponent. It is probably a direct consequence of the fact that the mediaeval writers left a blank space between the units and the first place of sexagesimals. These writers had de-noted the orders of the fractions by suitable abbreviations or sometimes by numerals, either written above or designated by some distinguishing mark. The symbol ° for a degree first appeared in print in 1586. In *l'Arithmetique*, Stevin notes that Bombelli uses the same symbols with the exception of the ⓪. [These quantities were written by Bombelli (1572) as ↡, ↢, ↣.] One concludes that Stevin borrowed his symbolism directly from Bombelli, but as the latter writer had felt no need for a symbol corresponding to the ⓪, Stevin was forced to invent one for himself, and the zero was the logical outcome, as it fitted with the rules which he laid down for the multiplication and division of the numbers of *La Disme*. Yet he never seems to have connected these symbols with the exponents of the tens in the denominators they replaced!

In 1525, Finaeus had unwittingly provided a precedent for Stevin's use of the same symbols for decimals and sexagesimals, by writing the product of 8 by 42 (minutes?) as 5.36, *i.e.*, 5 $^{36}/_{60}$. Subsequently, Beyer (1616) used both the comma and the ',' notation for decimals. Napier used ',", etc., in his *Rabdologia* (1617) but in his *Constructio* he shifts to the comma which Burgi had used in 1592. Dr. Glaisher[7] gives the opinion that Stevin used his array of symbols to make the numbers of *La Disme* more like those to which his contemporaries were accustomed. But one point should be mentioned here. In *La Geometrie*, expressions such as these occur: 1875③ = $^3/_4 \times {}^9/_4 \times {}^5/_6$ and 7 × $^2/_5$ = 280②. It is possible that this omission of the signs was made by the editor Girard; but by 1625, the date of the publication of his first edition of Stevin's works, Girard would have had ample opportunity to see the advantage of the comma separatrix which Burgi was using. Had he been making changes in Stevin's symbolism, these influences would have led him to retain the ⓪ and drop the ③ or the ②. Similarly, if Stevin was influenced by the bar of Rudolff etc., we would have expected him to do likewise. But Stevin appears to have thought of his 1875 as 1875 thirds rather than as 1.875.

Stevin's explanation of Definition I and his choice of names bear out our hypothesis that his work with decimals resulted from an effort to devise a system of numbers analogous to sexagesimals in which the place-value of the Arabic notation would function. His retention of the symbol of the last number, rather than that of the commencement, seems to indicate that he was not guided by the sporadic inventions of Rudolff and the rest. Yet, at first sight, these appear to be the more probable group of ancestors for the decimal fraction!

Operations with the numbers of *La Disme* are treated in four propositions: Addition, Subtraction, Multiplication, and Division. Each contains a formal statement of what is given and of what is to be proved. Careful reference is made to the corresponding work with integers in *l'Arithmetique*. The rigid uniformity of these propositions permits us to judge their style from a single example, and their content may be summarized without including the rather tedious proofs which are identical in method with the proof in Proposition I.

Proposition I. To add numbers of *La Disme*.

Given three numbers of *La Disme,* 27⓪ 8① 4② 7③, 37⓪ 6① 7② 5③, 875⓪ 7① 8② 2③.

Required: to find their sum.

Construction: Arrange the numbers in order as in the accompanying figure, adding them in the usual manner of adding integers.

			⓪	①	②	③
	2	7	8	4	7	
	3	7	6	7	5	
8	7	5	7	8	2	
9	4	1	3	0	4	

This (by the first problem of *l'Arithmetique*) gives the sum 941304, which, as the symbols above the numbers show, is 941⓪ 3① 0② 4③. And this is the sum required.

Proof: By the third definition of this book, the given number 27⓪ 8① 4② 7③ is **27** $^8/_{10}$, $^4/_{100}$, $^7/_{1000}$ or 27 $^{847}/_{1000}$. Similarly, the 37⓪ 6① 7② 5③ is **37** $^{675}/_{1000}$ and the 875⓪ 7① 8② 2③ is **875** $^{782}/_{1000}$.

These three numbers **27** $^{847}/_{1000}$, **37** $^{675}/_{1000}$, **875** $^{782}/_{1000}$, added according to the tenth problem of *l'Arithmetique* give **941** $^{304}/_{1000}$, but 941⓪ 3① 0② 4③ has this same value, and is, therefore, the true sum which was to be shown.

Conclusion: Having been given numbers of *La Disme* to add, we have found their sum, which was to be done.

Note: If, in the numbers in question, some figure of the natural order be lacking, fill its place with a zero. For example, in the numbers 8⓪ 5① 6② and 5⓪ 7② where the second number lacks a figure of order prime, insert 0①, and take 5⓪ 00① 7② as the given number and add as before.

⓪	①	②
8	5	6
5	0	7
1 3 6 3		

Stevin has three ways of writing these numbers, the symbols may be placed above or below the numbers or after the digit which they qualify. It should be noticed that Stevin shifts decimals to common fractions, never vice versa. And although he uses a zero to fill a gap in his decimals, he feels no obligation to begin with the prime. In Proposition IV, for instance, he writes 3④ 7⑤ 8⑥.

In multiplication, the symbol of the last number of the product is found by adding the symbols of the last terms of multiplicand and multiplier. Stevin gives as proofs the case of 2① multiplied by 3②. The product, according to the rule, should be 6③. The values of the given numbers are $^2/_{10}$ and $^3/_{100}$ and the value of the supposed product is $^6/_{1000}$, but this is identical with the product of 2① by 3②, which was to be shown.

In division, the symbol of the last term of the quotient is found by subtracting the symbol of the last term of the divisor from that of the dividend. If the numbers of the divisor should be of higher order than those of the dividend, zeros must be added to the latter.

It sometimes happens that the quotient cannot be expressed by whole numbers, as in the case of 4③ divided by 3②. Here, it appears that the quotient will be infinitely many threes with always one-third in addition. In such a case, we may approach as near to the real quotient as the problem requires and omit the remainder. It is true indeed that 13⓪ 3① 3 $\frac{1}{3}$②, or 13⓪ 3① 3② 3 $\frac{1}{3}$③ is the exact result, but in *La Disme*, we propose to use whole numbers only, and, moreover, we notice that in business one does not take account of the thousandth part of an ounce or of a grain. Omissions such as these are made by the principal geometers and arithmeticians even in computations of great consequence. Ptolemy and Jehan de Montroyal, for instance, did not make up their tables[8] with the utmost accuracy that could be reached with mixed numbers, for, in view of the purpose of these tables, approximation is more useful than perfection.

Finally, square roots may be extracted by first making the last symbol an even number by the addition of a zero if necessary, then getting the root as with integers. The symbol of the last number of the root will be one-half that of the given number. "Similarly, for all other roots."

The *Appendix* proposes to show how all computations that arise in business may be performed by the numbers of *La Disme*. Its greatest interest to us is the treatment of units of area and volume, and the extension of the numbers of *La Disme* to less obvious applications. We may safely guess that the sections of the least interest to the readers of the sixteenth century are of the greatest interest to us today, and conversely.

ARTICLE ONE
OF THE COMPUTATIONS OF SURVEYING

When the numbers of *La Disme* are used in surveying, the verge[9] is called the commencement, and it is divided into ten equal parts or primes, each prime is divided into seconds, and, if smaller units are required, the seconds into thirds, and so on so far as may be necessary. For the purposes of surveying, the divisions into seconds are sufficiently small, but in matters that require greater accuracy as in the measuring of lead roofs, one may need to use thirds. Many surveyors, however, do not use the verge, but a chain of three, four or five verges, and a cross-

staff[10] with its shaft marked in five or six pieds with their doigts. These men may follow the same practice here, substituting five or six primes with their seconds. They should use these marks without regard to the number of pieds and doigts that the verge contains in that locality, and add, subtract, multiply and divide the resulting numbers as in the preceding examples. . . . To find the number of pieds and doigts in 5 primes 9 seconds (the result in one of the examples cited) look on the other side of the verge to see how many pieds and doigts match with them; but this is a thing which the surveyor must do but once, i. e., at the end of the account which he gives to the proprietaries and often not then, as the majority of them think it useless to mention the smaller units.

This last illustrates a consequence of the tremendous divergence in units of measure resulting from the feudal conditions of Mediaeval Europe, where each petty ruler ordained such measures as he saw fit—standards that might or might not bear recognizable relation to other units of the same name in other parts of the same country.

Stevin's examples of the use of the numbers of *La Disme* in surveying consist of adding and subtracting areas, finding the area of a rectangle, and determining where a line should be drawn to cut a rectangle of given area from a given rectangle. In all of this work, the prime of a unit of square measure is one-tenth the unit itself.

Article Two advises that the aune of the measurer of tapestry be divided into tenths as was done with the verge of the surveyor.

Article Three deals with the measurement of casks. Stevin says that he has made his demonstration brief because he is writing to master gaugers, not to apprentices. As a result it is almost unintelligible. He gives great emphasis to the division of the measuring rod by points spaced according to the square roots of 0.1, 0.2, 0.3, . . . with intermediate points corresponding to the square roots of 0.11, 0.12, . . . This division is made by finding the mean proportional between the unit and its half which gives a point answering to $\sqrt{0.5}$, and the mean proportional between this segment and its fifth part gives $\sqrt{0.1}$. The method of using this is not explained, but it is evident that the diameters measured by this rod would have the same ratio as the areas of their respective sections when measured by the decimally divided rod.

Article Four discusses the numbers of *La Disme* in volume measurement. The illustrative problem is to find the volume of a rectangular column of dimensions 3① 2②, 2① 4②, 2⓪ 3① 5②. The volume is 1① 8② 0③ 4④ 8⑤ 0⑥.

> **Note:** Someone who is ignorant of the fundamentals of stereometry, for it is such a man that we are addressing now, may wonder why the volume of the above column is but 1①, etc., for it contains more than 180 cubes of sides 1①. He should realize that the cubic verge is not 10 but 1000 cubes of side 1①. Similarly, the prime of the volume unit is 100 cubes of side 1①. If, however, the question had been how many cubes of side 1① are in the above column, the result would have been altered to conform to this requirement, bearing in mind that each prime of volume units is 100 cubic primes, and that each second is 10 cubic primes.

Thus, Stevin makes his units proceed strictly in accordance with the decimal scheme. In this way he forestalls the objection to the metric system that its linear, square, and volume units follow the progressions by tens, hundreds, and thousands respectively, but he sacrifices the simple connection between the submultiples of the units of the three systems. The prime and second of his units of area are squares whose sides are $\sqrt{0.1}$ and $\sqrt{0.01}$; the prime and second of his units of volume are cubes whose edges are $\sqrt[3]{0.1}$ and $\sqrt[3]{0.01}$. This is startling, for we think at once of the difficulty of teaching it to an immature person, but it would be more convenient than the metric idea if we, like Stevin, felt a need of labeling each of our subdivisions and wrote 2 cubic decimeters 55 cubic centimeters instead of 2055 cm³.

Article Five explains that the division of the angle into minutes and seconds was devised so that astronomers might work with integers. The sexagesimal scheme was used "because 60 is com-mensurate with many whole numbers." Paying all due reverence to the past, Stevin maintains that the decimal progression is even more convenient. He says that he contemplates publishing astronomical tables using the decimal division of the degree, and he ends with a characteristic peroration on the beauties of the Flemish language in which these tables are to be written. As a matter of fact, the tables which Stevin published (1608) were on the old basis. Father Bosmans (*La Thiende*, Louvain, 1920) explains this on two grounds—Stevin did not rejoice in computations as did van Ceulen (which is fortunate for the world, for the impatience which may have cost us a set of tables, gave us *La Disme*); and secondly, large errors would arise in shifting instrument readings to the decimal system for computation, and in then changing the results back again.

Article Six, "On the Computations of Mint-Masters, Merchants, and in general of all States," begins with the thesis that all measures may be divided decimally, and that the largest unit of each denomination should be the commencement. In the case of money, the limit for the divisions would be the first sub-unit that is less than the smallest coin. Instead of half pound, ounce, and half ounce weights, the smaller weights should be the 5, 3, 2, 1 of each order, and the names prime, second, third, etc., should be retained because of the aid which they offer in computation. Stevin illustrates the advantage of decimal division of money by a problem of exchange—1 marc of gold is worth 36 lb. 5① 3②, how much is 8 marcs 3① 5② 4③ worth? The value of a simple method of treating such a problem is obvious when we consider that the value of the marc varied in the different cities, and that the number of the smaller units in a marc varied also.

> We might give examples of all the common rules of arithmetic that pertain to business as the rules of partnership, interest, exchange, etc., to show how they may be carried out by integers alone and also how they may be performed by the easy manipulation of counters, but as these may be deduced from

the preceding, we shall not elaborate them here. We might show by comparison with vexing problems with fractions the great difference in ease between working with ordinary numbers and with the numbers of *La Disme*, but we omit this in the interest of brevity.

Finally, we must speak of one difference between the sixth article and the five preceding articles, namely that any individual may make the divisions set forth in the five articles but this is not the case in the last where the results must be accepted by everyone as good and lawful. In view of the great usefulness of the decimal division, it would be a praiseworthy thing if the people would urge having this put into effect so that, in addition to the common divisions of measures, weights, and money that now exist, the state would declare the decimal division of the large units legitimate, to the end that he who wished might use them. It would further this cause also if all money that is newly coined should be based on this system of primes, seconds, thirds, etc. But if this is not put into operation so soon as we might wish, we have the consolation that it will be of use to posterity, for it is certain that if men of the future are like men of the past, they will not be neglectful of so great an advantage. Secondly, it is not the most discouraging thing to know that men may free themselves from such great labor at any hour they wish. Lastly, though the sixth article may not go into effect for some time, individuals may always use the five preceding articles as it is clear that some are already in operation.

The end of the Appendix.

Stevin's scheme for the decimal division of measures is unlike that of the metric system in three respects:

1. He did not reach the idea of a universal, invarient standard.

2. All of his units proceed according to the progression by tens with the disadvantage noted above.

3. He avoids the multiple units by taking the large unit as the commencement, but, in Article Four, he provides for a change in the unit according to the exigencies of the problem.

These are minor matters, however, in comparison with the idea itself!

The suggestion that individuals may use the decimal divisions of measures brings to mind our use of the decimally divided mile on the speedometer and the milepost, the decimally divided foot of the surveyor's tape, and the grinding of piston heads to the thousandth of an inch. From a scientist of Stevin's type, we would expect a rigorous demonstration of his device; from a person of his prominence in affairs, we would anticipate an application to real problems. But who would have imagined that the first man to write on decimals would have had the vision to see decimal weights and measures, and that, pending the adoption by the state of such a system, he should today recommend the very compromises used by individuals today!

NOTES

Editor's note: Original notes have been renumbered consecutively.

1. A discussion of these devices is given in *The Invention of the Decimal Fraction*, by D. E. Smith, Teachers College Bulletin, First Series No. 5.

2. *Para minuta prima, para minuta secunda*, etc.

3. The translation quoted in this article was made from Girard's second edition of *Les Oeuvres Mathematiques de Simon Stevin* (Leyden, 1634).

4. Gaugers: men whose business was the measuring of wine-casks, an important thing in connection with the excise duties, and necessitated by the great divergence in the size and shape of the barrels of the time. Gauguing lingered in the American arithmetics well into the last century.

5. Stereometers: men who measured volumes of all sorts. Stevin later is at great pains to show that, although all gauging is stereometry, all sterometry is not gauging.

6. Disme: "tithe," later the word was contracted into dime, so our coin has this connection with the first decimals.

7. Logarithms and Computation, Napier Tercentenary Volume p. 77.

8. The tables of Ptolemy to which Stevin had access were probably those in the translation of the Almagest made by Peurbach and Regiomontanus (Stevin's Montroyal) in the fifteenth century. In these tables trigonometric functions were computed from a circle of radius 10,000,000.

9. The verge was a unit of both linear and square measure, and the word was also used for the surveyor's rod.

10. The cross-staff is a piece of wood mounted at its midpoint perpendicular to a shaft and free to move along this shaft to other positions parallel to the first one. To use this instrument, the observer adjusts it so that the lines of sight from the end of the staff to the tip of the cross-piece coincide with the endpoints of the lines to be measured. Distances are computed from one measurement and similar triangles. When neither end of the required line is accessible, the distance is computed from two observations at known distances from each other. The pied and doigt were units approximately equal to our foot and inch.

65

Viète's Use of Decimal Fractions

CARL B. BOYER

THERE IS NO SUCH PERSON as the inventor of decimal fractions. It is well known that the earliest appearances of decimal fractions were incidental and inadvertent and that their use developed slowly.[1] The positional system had entered mathematics in conjunction with the sexagesimal system of numeration—in Babylonia of some four thousand years ago—and sexagesimal fractions had preempted the field of accurate computation until some four hundred years ago. Even when the principle of local value came to be associated with a decimal system of integers, more than a thousand years ago, the decimal fraction did not form part of the new system (generally referred to as Hindu-Arabic). The fractional domain continued to be made up of two principal parts. The layman still generally made use of the common or vulgar fractions (occasionally also of Egyptian unit fractions); the applied mathematician adhered to the established use of sexagesimals so single-mindedly that these latter came to be known as *fractiones astronomiae or fractiones physicae.* Occasional use of the equivalent of decimal fractions can nevertheless be found during the early centuries of our era. In the third century in China[2] the use of decimal metrological units can be regarded as a forerunner of decimal fractions, and square roots were found by a process tantamount to the rule

$$\sqrt{a} = \frac{\sqrt{10^{2n}a}}{10^n}.$$

Rules such as this were known to the Hindus and were used in medieval Europe, keeping alive the spore from which the systematic use of decimal fractions developed during the sixteenth century.

Trigonometry has been a fertile source for innovation in computation, and here, too, one finds germs of the decimal idea prior to the definitive development. In order to avoid difficulties inherent in the "astronomer's fractions," it became customary in early modern trigonometry to replace Ptolemy's radius of 60 by a still larger integer, such as 10^8. In this case the trigonometric lines could be expressed to a high degree of approximation without resorting to fractions of any kind, and their modern equivalents are easily read off today by a simple shift of the decimal point eight places to the left. That is, sin 45° would in this case appear as 70,710,678 instead of 0.70710678. Nevertheless, the concept of decimal fraction was in these cases not specifically invoked before the sixteenth century. Occasionally a closer approach to the decimalization of fractions is found in problems involving division by multiples of ten. In 1492 Pellos, for example, carried out the division of 5836943 by 30 through a process akin to our shift of a decimal point.[3] He first separated the final digit from the others by a dot, divided the left-hand number by three to obtain 194546, and then expressed the remainder as $^{23}/_{30}$. The form of the answer betrays clearly that he was not thinking of a decimal fractional algorithm.

It is common knowledge that the first book written with the purpose of popularizing the decimal fraction was published by Stevin in 1585 as *La*

Reprinted from *Mathematics Teacher* 55 (Feb., 1962): 123–27; with permission of the National Council of Teachers of Mathematics.

Disme or *De Thiende*. Despite the clumsy notation Stevin employed, he has been referred to repeatedly as the inventor of the decimal fraction, and his little work has been so thoroughly described, and so frequently reproduced,[4] that there is no need to duplicate this effort. It will be well to remark, however, that Stevin's point of view is not ours. Instead of believing that he was introducing a new type of fractional calculation, Stevin claims in *La Disme* that he has done away with the need for fractions of any sort. The first part of Stevin's work opens with, "The First Definition: Disme is a kind of arithmetic, invented by the tenth progression, consisting in characters of ciphers, whereby a certain number is described and by which also all accounts which happen in human affairs are dispatched by whole numbers without fractions or broken numbers."[5] That is, Stevin regards his tenths, hundredths, etc., as units of a new order of magnitude, similar to the subdivisions in the metric system. His successors, however, soon shifted from this view to the modern concept of decimal fraction.

Stevin is by no means the only claimant for the invention. David Eugene Smith wrote, "If one man were to be named as the best entitled to be called the inventor of decimal fractions, Rudolff might properly be the man."[6] Christoff Rudolff operated with decimal fractions fifty-five years before Stevin, using a vertical bar as a separatrix, but he made no special plea for the decimal fraction. Pages of his *Exempel Buechlin* (1530) and *Kunstliche Rechnung* (1626) have been reproduced by Smith and others. Smith went further and reproduced facsimile pages from other original sources in order to illustrate "some of the steps in the invention of the decimal fractions that have not been recognized as fully as they deserve." Among these is, of course, a page from Adam Riese, *Rechnung auff der Linien und Federn* (1522), for the name Adam Riese has been a household term in Germany indicating correctness of computation. There is also a page from Pellos' arithmetic of 1492. Overlooked entirely, in Smith's account, is a celebrated

mathematician whose contribution to the rise of decimal fractions has too often been neglected by historians, and it is the purpose of this note, through a brief excerpt from Viète's work and the facsimile reproduction of a page, to call attention to the fact that, Stevin excepted, no one made a more significant contribution to the rise of decimal fractions than did Viète, the leading mathematician of the sixteenth century.

In 1579 Viète published a work entitled *Canon mathematicus* in which, following the practice of his day, he adopted a radius so large that for ordinary trigonometric purposes fractions are avoided. With a radius of 100,000,000 parts, he gives the sine of 45°, for example, as 70,710,678. However, in a supplementary work, *Universalium inspectionum ad canonem mathematicum, liber singularis*, bound with the *Canon* and of the same date, one finds not only the systematic use of decimal fractions, but also a strong plea for their adoption throughout mathematics. Sarton years ago remarked that "Viète renounced sexagesimal fractions in favor of decimal ones," but Sarton's statement was submerged in a glowing eulogy of Stevin's *La Disme*.[7] Incidental allusions to Viète's contribution have appeared since but they have not attracted attention.[8] In the thought that seeing might lead to believing, we therefore reproduce[9] page 51 of Viète's *Universalium inspectionum*. Here more than two dozen decimal fractions, in an eminently clear and appropriate notation, appear on a single page. Among these is the approximate value

$$314{,}159{,}\underline{265.36}$$

for the circumference of a circle whose diameter is 200,000. Slight modifications of this form appear elsewhere in the book, and on page 69 it is published simply as

$$314{,}159{,}265{,}36.$$

Sometimes a vertical bar replaces the comma as a decimal separatrix, as on page 65 when the apothem of a regular polygon of 96 sides in a circle of radius of 100,000 is given as

99,946 | 458,75.

Forms of the decimal fractions in this work, both in the supplement and in the *Canon* itself, may differ slightly, but their use is far from incidental. Viète, on page 17 of *Universalium inspectionum*, wrote: "Finally, sexagesimals and sixties are to be used sparingly or never in mathematics, and thousandths and thousands, hundredths and hundreds, tenths and tens, and similar progressions, ascending and descending, are to be used frequently or exclusively."

A biographer of Stevin wrote some years ago that "if one understands by inventor the one who crystallizes, in an attractive and harmonious form, the timid and scattered attempts of his predecessors, if one understands by inventor the one who proposes the systematic use rather than one who had employed them incidentally, Stevin is incontestably the inventor of decimal fractions."[10] And another expositor has categorically attributed to Stevin the first systematic use of decimal fractions to the exclusion of ordinary fractions.[11] It is clear that the name of Viète could very appropriately be substituted for that of Stevin in either case without falsifying the proposition, and Viète's work antedated that of Stevin by half a dozen years.

Viète was a deep and original research mathematician who did not institute a campaign to popularize decimal fractions, nor did his works enjoy a circulation commensurate with his reputation. Thus it was that the world of science adopted decimal fractions chiefly after Stevin, in 1585, had explained their use in language readily comprehended by the multitude. By this time decimal fractions were in the air, and their use was not in all cases traceable to Stevin. Kepler, for example, believed that the use of decimal fractions was to be attributed to Jobst Bürgi's trigonometry. The spread of decimalization in fractions was spurred not only through trigonometry, but also by the seventeenth-century invention of logarithms, in which Bürgi shared. Notations similar to Viète's were among the varied forms which were preferred to the clumsy indices of Stevin, but even to this day uniformity has not been achieved. The use of a comma as a decimal separatrix can be traced back to G. A. Magini's *De planis triangulis* of 1592, and a year later the decimal dot appeared in Clavius' *Astrolabium*.[12] Each of these notations is with us today, the former chiefly on the European continent, and the latter among Anglo-Saxons. No single notation can be said to have won the field, and no one mathematician can claim the decimal fraction as his own. But any list of names of men who shared in the invention must provide an honorable place for François Viète. Decimal fractions were by no means the chief contribution of Viète to mathematics; their use was but a small part of the work of one whom the late E. T. Bell appropriately described[13] as "the first mathematician of his age to think occasionally as mathematicians habitually think today."

NOTES

1. An excellent account by D. J. Struik of the early history of decimal fractions, together with further references, is to be found in *The Principal Works of Simon Stevin*, edited by D. J. Struik, II (Amsterdam: 1958), 373–85.

2. Joseph Needham, *Science and Civilisation in China*, III (Cambridge: 1959), 46, 82, Cf Wang Ling, "The development of decimal fractions in China," *Acts of the International Congress of the History of Science*, VIII (1956), Vol. I, 13–17.

3. David Eugene Smith, "The invention of the decimal fraction," *Teachers College Bulletin*, No. 5 (1910), 11–21.

4. See especially George Sarton's two articles, "Simon Stevin of Bruges," *Isis*, XXI (1934), 241–303, and "The first explanation of decimal fractions and measures," *Isis*, XXIII (1935), 153–244. For further references see the work of Struik cited in Note 1, or his article "Simon Stevin and the decimal fractions," *The Mathematics Teacher*, LII (1959), 474–78.

5. *The Principal Works of Simon Stevin*, II 403.

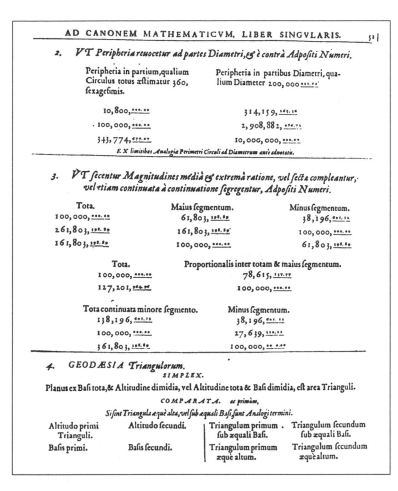

Reproduction of page 51 of Viète's Universalium inspectionum, *through the courtesy of the Burndy Library.*

6. Smith, *op. cit.* This view is echoed in Carl N. Shuster, *A Study of the Problems in Teaching the Slide Rule* (New York: 1940), p. 9.

7. *Isis*, XXIII (1935), 173–74.

8. Casual notice of Viète's use of decimals is found in Florian Cajori, *History of Mathematical Notations* (Chicago: 1928–29), II, 316, and in *Encyclopédie des Sciences Mathématiques*, I (1), p. 53, note 180, and in Struik's article cited in Note 1. In the *Encyclopaedia Brittanica* article in "Mathematics, History of," Otto Neugebauer has Viète share with Stevin credit for the development of a consistent decimal place-value notation. Forceful statements of Viète's role are found in K. Hunrath, "Zur Geschichte der Decimalbrüche," *Zeitschrift für Mathematik und Physik*, XXXVIII (1893), Hist. lit. Abt., pp. 25–27, and Pierre Dedron and Jean Itard, *Mathématiques et mathématiciens* (Paris: *c.* 1959), p. 290.

However, I have nowhere seen a reproduction of a page from Viète's work showing his use of decimal fractions.

9. We have used, through the kindness of Colonel Bern Dibner, the copy of Viète's *Canon* at the Burndy Library, and the accompanying reproduction has been provided through the courtesy of the Burndy Library.

10. Robert Depau, *Simon Stevin* (Bruxelles: 1942), p. 62.

11. Henri Bosmans, *La Thiende de Simon Stevin* (Andvers and La Haye: Édition de la Société des Bibliophiles Anversois, No. 38, 1924), 1–5.

12. Jekuthiel Ginsburg, "On the early history of the decimal point," *Scripta Mathematica*, I (1932), 84–85, 168–69.

13. E. T. Bell, *The Development of Mathematics* (New York: 1940), p. 99.

66

John Napier
and His Logarithms

C. B. READ

*I*T IS DOUBTFUL that any teacher would today try to introduce the subject of logarithms other than by means of some use of exponents. It is, then, indeed hard to realize that Napier constructed his logarithms before the concept of exponents as we know them was developed.

John Napier was born in 1550, probably when his father was about sixteen years of age. He is known to have matriculated as a student at St. Andrews University, but records fail to show that he graduated. His first published work was *A Plaine Discovery of the Whole Revelation of St. John*. However, it appears that from an early age he was interested in mathematics as well as theology. He apparently did some early work in algebra and in arithmetic. Toward the end of the sixteenth century, Napier became perturbed at the fact that scientific progress was hindered by the great labor involved in numerical calculation. He deliberately attacked the problem of developing some means of lessening the work involved. He devised several mechanical aids, among these being sets of small rods often called "Napier's bones." In his approach it seems evident that Napier was (at least at first) interested in trigonometric computation, for he deals almost exclusively with sines.

The great contribution of Napier was the invention of what we now know as logarithms, al-

though at first these were called by Napier "artificial numbers" or simply "artificials." The number corresponding to the logarithm was called a "natural number."

The exact date of the discovery, if indeed one can say there was a definite date of discovery or invention, is uncertain. Kepler tells us that in 1594 Tycho Brahe had information from a Scotch friend that there was a possibility of publication of the material. Actually, the computation of the table, or "canon" as the author called it, probably took several years. There are two books which are explanatory of the system, generally abbreviated as the *Constructio* and the *Descriptio*. Of the two, the *Mirifici Logarithmorum Canonis Constructio* is the most important, in fact the most important of all Napier's works.

The *Contructio* presents clearly his original conception of logarithms. The "canon," with instructions for its use, was published in 1614. The method of construction, although probably written several years earlier, was not published until 1619. It was published in Edinburgh, some two years after Napier's death (which may have been hastened by the strain involved in the development and computation of the "canon"). In the body of this book, the term "artificial number" is frequently but not exclusively used instead of "logarithm." The term "logarithm" is used in the title page, the headings, and in the appendix, which discusses the advantages of logarithms to the base ten.

Reprinted from *Mathematics Teacher* 53 (May, 1960): 381–85; with permission of the National Council of Teachers of Mathematics.

This original work is relatively rare, few writers having had the privilege of examining a copy. In addition, the fact that the work is in Latin makes it less easy for a present-day student or teacher to follow, even if a copy were available. An English translation was published by W. R. Macdonald in 1889.

The book consists of sixty numbered paragraphs, occupying, in the English translation, less than fifty pages. (This does not include the appendix, which is entitled "On the construction of another and better kind of Logarithms, namely one which the Logarithms of unity is 0," nor a supplement dealing with methods of solving spherical triangles with the use of logarithms.) The entire book needs to be studied to understand Napier's approach; a few necessarily brief comments may be helpful.

The first numbered paragraph defines a logarithmic table as " . . . a small table by the use of which we can obtain a knowledge of all geometrical dimensions and motions in space, by a very easy calculation."

As the treatment proceeds, the author finds it necessary to define arithmetical and geometrical progressions. Since accuracy is needed, he suggests taking large numbers for a basis, "but large numbers are most easily made from small by adding ciphers." He then says "these large numbers may again be made still larger by placing a period after the number and adding ciphers." It is found necessary to explain that 10000000.04 is the same as $10000000 \ ^4/_{100}$. It is perhaps not generally known that Napier was the first to use our present notation for decimal fractions. It is not impossible that this was a by-product of his working out the invention of logarithms, although in another work Napier makes reference to the contributions of Simon Stevin. However, in spite of the simplicity of the notation proposed by Napier, it did not come into general use until long after his death.

Napier next presents a discussion of the accuracy of working with his artificial numbers—a discussion which has much resemblance to our present-day work on the accuracy of computation with approximate numbers.

To follow the text, it is necessary to understand that Napier's canon was not a table of logarithms of numbers as we know it (that is, of equally spaced numbers) but of "sines of arcs" for every minute from 0 to 90 degrees. It may be easier to understand Napier's approach if it is known that at this period the "sine of an arc" (which we call the sine of an angle) was the length of a line equal to the half-chord of a double central angle in a circle whose radius was unity. Hence Napier's explanations frequently represent both sines and logarithms by lines. Instead of calling the radius unity, he adds seven ciphers. From this radius he suggests subtracting its 10000000th part, obtaining 9999999; then from this he subtracts the 10000000th part, obtaining 9999998.0000001; again subtracting the 10000000th part he obtains 9999997.0000003. Continuing this process he creates, as he calls them, "a hundred proportionals." The "hundredth proportional" is 9999900.0004950. The resulting table is called the "First Table."

For his "Second Table" Napier points out the difficulty of forming fifty proportional numbers between the first and last numbers of his First Table (10000000.0000000 and 9999900.0004950). However, as he puts it, "A near and at the same time an easy proportion is 100000 to 99999," which will yield "sufficient exactness." So for the Second Table he starts by "adding six ciphers to radius and continually subtracting from each number its own 100000th part," obtaining fifty other proportional numbers, the last being 9995001.222927 (an unfortunate error in computation—the correct value being 9995001.224804).

The Third Table "consists of 69 columns, and in each column are placed 21 numbers, proceeding in the proportion which is easiest, and as near as possible to that subsisting between the first and last numbers of the Second Table." For ease in calculation, the proportion is considered to be 10000 to 9995.

Essentially these three tables constitute a table of sines, or natural numbers, "progressing in geometrical proportion" with no two numbers differing by more than unity (after rounding). Napier now proceeds to show how to place, beside the sines, or "natural numbers" (which decrease geometrically), their logarithms, or "artificial numbers" (which increase arithmetically). To do this he makes use of the concept of a "geometrically moving point" which, when approaching a fixed point, has its "velocities proportionate to its distances from the fixed point." With this concept we have essentially Napier's definition of a logarithm: "that number which has increased arithmetically with the same velocity throughout as that with which the radius began to decrease geometrically."

Having shown that "nothing is the logarithm of the radius," he discusses "the limits of the logarithm," which covers roughly what we would call the maximum error in a computed logarithm. He then develops certain laws relating to proportions, expressed in terms of logarithms. In modern notation, for example, if $a : b = b : c$ and we know the logarithms of any two of the three quantities, we may obtain the logarithm of the third; similarly for $a : b = c : d$. The explanation proceeds to show how to find logarithms of sines outside the limits of the table. There are numerous examples.

The above explanation seems quite involved. It may be of some help to note that in the First Table, the logarithm of 10000000 is zero; the logarithm of the next number 9999998.0000001 is 1; that of 9999997.0000003 is 3, and of 9999900.0004950 is 100. Rounding for ease in computation, gives 100 as the logarithm of 9999900. Then, using a new ratio for the Second Table, more values can be inserted.

Napier then used existing tables of sines (disregarding decimal points which would appear in modern tables). He chose the number in his tables nearest the desired sine, and by use of "limits of its logarithm" (essentially interpolation) found the desired logarithm. Paragraph 43 demonstrates the method of finding the logarithm when the given

sine is 9995000.000000. The nearest sine in the Second Table is 9995001.222927, from which he obtains, in modern notation: log 9995000.000000 = 5001.2485387.

Paragraph 58 shows that if the logarithm of "all arcs not less than 45 degrees are given, the logarithms of all less arcs are very easily obtained," thus paralleling our present-day tables which tabulate values to 45 degrees, then use the complementary angle. Paragraph 60 gives details for forming a logarithmic table—employing 45 pages, each capable of holding sixty lines of figures, and with seven columns of figures on each page. Once the pages are prepared, reference is made to paragraphs 49 and 50, which explain in detail how the calculations are performed.

Although it is very unlikely that anyone at the present time would wish to make up a "canon" of logarithms using Napier's method, there are several points of definite historic interest. First, with Napier's logarithms, as the natural numbers *decrease*, their logarithms *increase*. The logarithm of 10,000,000 is zero, the logarithm of 9,995,000 is 5,001.2, the logarithm of 9,900,000 is 100,503.3, the logarithm of 4,998,609 is 6,934,250, etc. Second, in the sense that we know it, Napier's logarithms actually had no base. Third, it is interesting to note that although decimal fractions were used in the construction of the canon, the logarithms given are large whole numbers—for example, paragraph 57 computes the logarithm of an arc of 34° 40' as 5,642,242. (For brevity Napier speaks of the logarithm of an arc rather than, as in modern usage, of the logarithm of the sine of an angle.) Finally, we note with interest that Napier's construction or demonstration was based on the concept of moving points, using what might be termed arithmetical and geometrical motion. Arithmetical motion implies constant velocity; Napier's geometrical motion involved variable velocity, continually decreasing.

By no means is it to be implied that credit should be withheld from Napier. On the other hand, when we read that Napier invented logarithms, it should be recognized that his canon, or table, differed in several important essentials from what we now recognize as a logarithmic table, and that his method of construction differs markedly from the explanation now given (as, for example, in a course in calculus) relative to the computation of logarithms. Finally, what we now often call natural or Naperian logarithms are not identical with those first developed by John Napier of Merchiston.

NOTE

* See, however, D. E. Smith, *History of Mathematics*, II (Boston: Ginn and Co., 1925), pp. 238 and 244.

Naperian Logarithms and Natural Logarithms

HOWARD EVES

ONE FREQUENTLY READS, in calculus textbooks, that logarithms having the base e are called *natural*, or *Naperian (or Napierian), logarithms,* leaving the impression with the reader that the logarithms devised by John Napier are the natural logarithms, or logarithms having the base e, and that Napier should be credited with the development of this system of logarithms. Actually, Naperian logarithms and natural logarithms are quite different, and it is the purpose of this note to emphasize the difference by obtaining an expression for the logarithm function of Napier in terms of the familiar logarithm function of today. We shall denote the Naperian logarithm of y by Nap log y, the familiar logarithm of y to base b by $\log_b y$, and the natural logarithm of y (the special case where $b = e$) by ln y.

As we know, the power of logarithms as a computing device lies in the fact that by them multiplication and division are reduced to the simpler operations of addition and subtraction. A forerunner of this idea is apparent in the formula

$$\sin A \sin B = \tfrac{1}{2}[\cos (A - B) - \cos (A + B)],$$

newly discovered in Napier's time, and it could be that Napier's line of thought started with this formula. In any event, Napier at first restricted his logarithms to those of the sines of angles.

Napier labored at least twenty years upon his theory, and, whatever the genesis of his idea, his final definition of a logarithm is as follows: Con-

Reprinted from *Mathematics Teacher* 53 (May, 1960): 381–85; with permission of the National Council of Teachers of Mathematics.

sider a line segment AB and an infinite ray DE, as shown in the accompanying figure. Let points C and F start moving simultaneously from A and D, respectively, along these lines, with the same initial rate. Suppose C moves with a velocity always

numerically equal to the distance CB, and that F moves with a uniform velocity. Then (according to Napier) DF is the logarithm of CB. That is, setting $DF = x$ and $CB = y$, $x = $ Nap log y.

To avoid the nuisance of fractions, Napier took the length of AB as 10^7, for the best tables of sines available to him extended to seven places. Then we have $AC = 10^7 - y$, whence velocity of $C = -dy/dt = y$. That is, $dy/y = -dt$, or integrating, ln $y = -t + C$. Evaluating the constant of integration by substituting $t = 0$, we find that $C = $ ln 10^7, whence ln $y = -t + $ ln 10^7.

Now velocity of $F = dx/dt = 10^7$, so that $x = 10^7 t$. Therefore

$$\begin{aligned} \text{Nap log } y &= x = 10^7 t \\ &= 10^7 (\ln 10^7 - \ln y) \\ &= 10^7 \ln (10^7/y) \\ &= 10^7 \log_{1/e} (y/10^7), \end{aligned}$$

and we have an expression for the Naperian logarithm in terms of the familiar log function (with base $1/e$).

It follows that the frequently made statement that Naperian logarithms are natural logarithms is without basis.

Projective Geometry

MORRIS KLINE

*I*N THE HOUSE OF MATHEMATICS there are many mansions and of these the most elegant is projective geometry. The beauty of its concepts, the logical perfection of its structure and its fundamental role in geometry recommend the subject to every student of mathematics.

Projective geometry had its origins in the work of the Renaissance artists. Medieval painters had been content to express themselves in symbolic terms. They portrayed people and objects in a highly stylized manner, usually on a gold background, as if to emphasize that the subject of the painting, generally religious, had no connection with the real world. An excellent example, regarded by critics as the flower of medieval painting, is Simone Martini's "The Annunciation." With the Renaissance came not only a desire to paint realistically but also a revival of the Greek doctrine that the essence of nature is mathematical law. Renaissance painters struggled for over a hundred years to find a mathematical scheme which would enable them to depict the three-dimensional real world on a two-dimensional canvas. Since many of the Renaissance painters were architects and engineers as well as artists, they eventually succeeded in their objective. To see how well they succeeded one need only compare Leonardo da Vinci's "Last Supper" with Martini's "Annunciation."

The key to three-dimensional representation was found in what is known as the principle of projection and section. The Renaissance painter imagined that a ray of light proceeded from each point in the scene he was painting to one eye. This collection of converging lines he called a projection. He then imagined that his canvas was a glass screen interposed between the scene and the eye. The collection of points where the lines of the projection intersected the glass screen was a "section." To achieve realism the painter had to reproduce on canvas the section that appeared on the glass screen.

Two woodcuts by the German painter Albrecht Dürer illustrate this principle of projection and section. In "The Designer of the Sitting Man" the artist is about to mark on a glass screen a point where one of the light rays from the scene to the artist's eye intersects the screen. The second woodcut, "The Designer of the Lute," shows the section marked out on the glass screen.

WOODCUT I

From *Scientific American* 192 (January, 1955): 80–86. Reprinted with permission. Copyright © 1955 by Scientific American, Inc. Illustrations supplied through the courtesy of Morris Kline.

WOODCUT 2

Of course the section depends not only upon where the artist stands but also where the glass screen is placed between the eye and the scene. But this just means that there can be many different portrayals of the same scene. What matters is that, when he has chosen his scene, his position, and the position of the glass screen, the painter's task is to put on canvas precisely what the section contains. Since the artist's canvas is not transparent and since the scenes he paints sometimes exist only in his imagination, the Renaissance artists had to derive theorems which would specify exactly how a scene would appear on the imaginary glass screen (the location, sizes, and shapes of objects) so that it could be put on canvas.

The theorems they deduced raised questions which proved to be momentous for mathematics. Professional mathematicians took over the investigation of these questions and developed a geometry of great generality and power. Let us trace its development.

Suppose that a square is viewed from a point somewhat to the side [Figure 1]. On a glass screen interposed between the eye and the square, a section of its projection is not a square but some other quadrilateral. Thus square floor tiles, for instance,

are not drawn square in a painting. A change in the position of the screen changes the shape of the section, but so long as the position of the viewer is kept fixed, the impression created by the section on the eye is the same. Likewise various sections of the projection of a circle viewed from a fixed position differ considerably—they may be more or less flattened ellipses—but the impression created by all these sections on the eye will still be that created by the original circle at that fixed position.

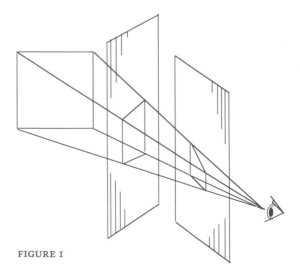

FIGURE I

"The Annunciation" by Simone Martini is an outstanding example of the flat, stylized painting of the medieval artists. The figures were symbolic and framed in a gold background.

"The Last Supper" by Leonardo da Vinci utilized projective geometry to create the illusion of three dimensions. Lines have been drawn on this reproduction to a point at infinity.

Drawing by da Vinci, made as a study for his painting "The Adoration of the Magi," shows how he painstakingly projected the geometry of the entire scene before he actually painted it.

To the intellectually curious mathematicians this phenomenon raised a question: Should not the various sections presenting the same impression to the eye have some geometrical properties in common? For that matter, should not sections of an object viewed from different positions also have some properties in common, since they all derive from the same object? In other words, the mathematicians were stimulated to seek geometrical properties common to all sections of the same projection and to sections of two different projections of a given scene. This problem is essentially the one that has been the chief concern of projective geometers in their development of the subject.

It is evident that, just as the shape of a square or a circle varies in different sections of the same projection or in different projections of the figure, so also will the length of a lone segment, the size of an angle, or the size of an area. More than that, lines which are parallel in a physical scene are not parallel in a painting of it but meet in one point; see, for example, the lines of the ceiling beams in da Vinci's "Last Supper." In other words, the study of properties common to the various sections of projections of an object does not seem to lie within the province of ordinary Euclidean geometry.

Yet some rather simple properties that do carry over from section to section can at once be discerned. For example, a straight line will remain a line (that is, it will not become a curve) in all sections of all projections of it; a triangle will remain a triangle; a quadrilateral will remain a quadrilateral. This is not only intuitively evident but easily proved by Euclidean geometry. However, the discovery of these few fixed properties hardly elates the finder or adds appreciably to the structure and power of mathematics. Much deeper insight was required to obtain significant properties common to different sections.

The first man to supply such insight was Gérard Desargues, the self-educated architect and engineer who worked during the first half of the seventeenth century. Desargues's motivation was to help the artists; his interest in art even extended to

writing a book on how to teach children to sing well. He sought to combine the many theorems on perspective in a compact form, and he invented a special terminology which he thought would be more comprehensible than the usual language of mathematics.

His chief result, still known as Desargues's theorem and still fundamental in the subject of projective geometry, states a significant property common to two sections of the same projection of a triangle. Desargues considered the situation represented here by two different sections of the projection of a triangle from the point O [Figure 2]. The relationship of the two triangles is described by saying that they are perspective from the point O. Desargues then asserted that each pair of corresponding sides of these two triangles will meet in a point, and, most important, these three points will lie on one straight line. With reference to the figure, the assertion is that AB and $A'B'$ meet in the point R; AC and $A'C'$ meet in S; BC and $B'C'$ meet in T; and that R, S, and T lie on one straight line. While in the case stated here the two triangle sections are in different planes, Desargues's assertion holds even if triangles ABC and $A'B'C'$ are in the same plane, e.g., the plane of this paper, though the proof of the theorem is different in the latter case.

The reader may be troubled about the assertion in Desargues's theorem that each pair of corresponding sides of the two triangles must meet in a point. He may ask: What about a case in which the sides happen to be parallel? Desargues disposed of such cases by invoking the mathematical convention that any set of parallel lines is to be regarded as having a point in common, which the student is often advised to think of as being at infinity—a bit of advice which essentially amounts to answering a question by not answering it. However, whether or not one can visualize this point at infinity is immaterial. It is logically possible to agree that parallel lines are to be regarded as having a point in common, which point is to be distinct from the usual, finitely located points of

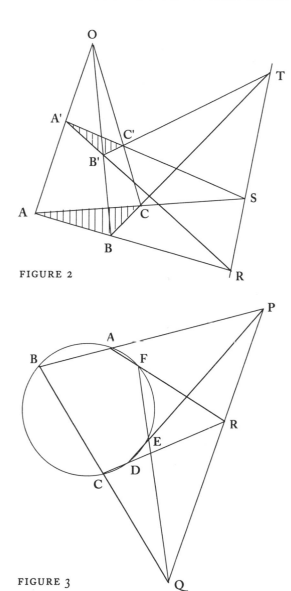

FIGURE 2

FIGURE 3

These conventions or agreements not only are logically justifiable but also are recommended by the argument that projective geometry is concerned with problems which arise from the phenomenon of vision, and we never actually see parallel lines, as the familiar example of the apparently converging railroad tracks reminds us. Indeed, the property of parallelism plays no role in projective geometry.

At the age of 16 the precocious French mathematician and philosopher Blaise Pascal, a contemporary of Desargues, formulated another major theorem in projective geometry. Pascal asserted that if the opposite sides of any hexagon inscribed in a circle are prolonged, the three points at which the extended pairs of lines meet will lie on a straight line [Figure 3].

As stated, Pascal's theorem seems to have no bearing on the subject of projection and section. However, let us visualize a projection of the figure involved in Pascal's theorem and then visualize a section of this projection [Figure 4]. The projec-

FIGURE 4

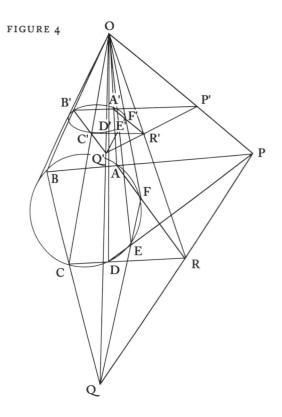

the lines considered in Euclidean geometry. In addition, it is agreed in projective geometry that all the intersection points of the different sets of parallel lines in a given plane lie on one line, sometimes called the line at infinity. Hence even if each of the three pairs of corresponding sides of the triangles involved in Desargues's theorem should consist of parallel lines, it would follow from our agreements that the three points of intersection lie on one line, the line at infinity.

433

tion of the circle is a cone, and in general a section of this cone will not be a circle but an ellipse, a hyperbola, or a parabola—that is, one of the curves usually called a conic section. In any conic section the hexagon in the original circle will give rise to a corresponding hexagon. Now Pascal's theorem asserts that the pairs of opposite sides of the new hexagon will meet on one straight line which corresponds to the line derived from the original figure. Thus the theorem states a property of a circle which continues to hold in any section of any projection of that circle. It is indeed a theorem of projective geometry.

It would be pleasant to relate that the theorems of Desargues and Pascal were immediately appreciated by their fellow mathematicians and that the potentialities in their methods and ideas were eagerly seized upon and further developed. Actually this pleasure is denied us. Perhaps Desargues's novel terminology baffled mathematicians of his day, just as many people today are baffled and repelled by the language of mathematics. At any rate, all of Desargues's colleagues except René Descartes exhibited the usual reaction to radical ideas: they called Desargues crazy and dismissed projective geometry. Desargues himself became discouraged and returned to the practice of architecture and engineering. Every printed copy of Desargues's book, originally published in 1639, was lost. Pascal's work on conics and his other work on projective geometry, published in 1640, also were forgotten. Fortunately a pupil of Desargues, Philippe de la Hire, made a manuscript copy of Desargues's book. In the nineteenth century this copy was picked up by accident in a bookshop by the geometer Michel Chasles, and thereby the world learned the full extent of Desargues's major work. In the meantime most of Desargues's and Pascal's discoveries had had to be remade independently by nineteenth-century geometers.

Projective geometry was revived through a series of accidents and events almost as striking as those that had originally given rise to the subject.

Gaspard Monge, the inventor of descriptive geometry, which uses projection and section, gathered about him at the Ecole Polytechnique a host of bright pupils, among them Sadi Carnot and Jean Poncelet. These men were greatly impressed by Monge's geometry. Pure geometry had been eclipsed for almost 200 years by the algebraic or analytic geometry of Descartes. They set out to show that purely geometric methods could accomplish more than Descartes's.

It was Poncelet who revived projective geometry. As an officer in Napoleon's army during the invasion of Russia, he was captured and spent the year 1813–14 in a Russian prison. There Poncelet reconstructed, without the aid of any books, all that he had learned from Monge and Carnot, and he then proceeded to create new results in projective geometry. He was perhaps the first mathematician to appreciate fully that this subject was indeed a totally new branch of mathematics. After he had re-opened the subject, a whole group of French and, later, German mathematicians went on to develop it intensively.

One of the foundations on which they built was a concept whose importance had not previously been appreciated. Consider a section of the projection of a line divided by four points [Figure 5]. Obviously the segments of the line in the section are not equal in length to those of the original line. One might venture that perhaps the ratio of two segments, say $A'C'/B'C'$, would equal the corresponding ratio AC/BC. This conjecture is incorrect. But the surprising fact is that the ratio of the ratios, namely $(A'C'/C'B') / (A'D'/D'B')$, will equal $(AC/CB) / (AD/DB)$. Thus this ratio of ratios, or cross ratio as it is called, is a projective invariant. It is necessary to note only that the lengths involved must be directed lengths; that is, if the direction from A to D is positive, then the length AD is positive but the length DB must be taken as negative.

The fact that any line intersecting the four lines OA, OB, OC, and OD contains segments possessing the same cross ratio as the original segments suggests that we assign to the four projection lines meeting in the point O a particular cross ratio, namely the cross ratio of the segments on any section. Moreover, the cross ratio of the four lines is a projective invariant, that is, if a projection of these four lines is formed and a section made of this projection, the section will contain four concurrent lines whose cross ratio is the same as that of the original four [Figure 6]. Here in the section $O'A'B'C'D'$, formed in the projection of the figure $OABCD$ from the point O'', the four lines $O'A'$, $O'B'$, $O'C'$, and $O'D'$ have the same cross ratio as OA, OB, OC, and OD.

The projective invariance of cross ratio was put to extensive use by the nineteenth-century geometers. We noted earlier in connection with Pascal's theorem that under projection and section a circle may become an ellipse, a hyperbola, or a parabola, that is, any one of the conic sections. The geometers sought some common property which would account for the fact that a conic section always gave rise to a conic section, and they found the answer in terms of cross ratio. Given the points O, A, B, C, D, and a sixth point P on a conic section containing the others [Figure 7], then a remarkable theorem of projective geometry states that the lines PA, PB, PC, and PD have the same cross ratio as OA, OB, OC, and OD. Conversely, if P is any point such that PA, PB, PC, and PD have the same cross ratio as OA, OB, OC, and OD, then P must lie on the conic through O, A, B, C, and D. The essential point of this theorem and its converse is that a conic section is determined by the property of cross ratio. This new characterization of a conic was most welcome, not only because it utilized a projective property but also because it opened up a whole new line of investigation on the theory of conics.

The satisfying accomplishments of projective geometry were capped by the discovery of one of the most beautiful principles of all mathematics—the principle of duality. It is true in projective geometry, as in Euclidean geometry, that any two points determine one line, or as we prefer to put it,

FIGURE 5

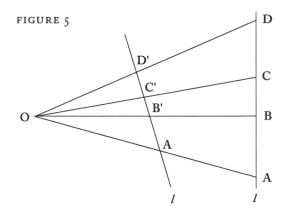

FIGURE 6

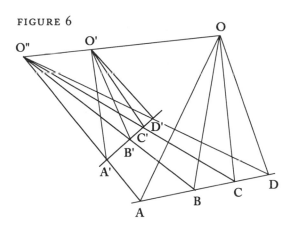

FIGURE 7

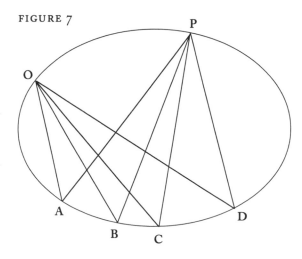

any two points lie on one line. But it is also true in projective geometry that any two lines determine, or lie on, one point. (The reader who has refused to accept the convention that parallel lines in Euclid's sense are also to be regarded as having a point in common will have to forego the next few paragraphs and pay for his stubbornness.) It will be noted that the second statement can be obtained from the first merely by interchanging the words point and line. We say in projective geometry that we have dualized the original statement. Thus we can speak not only of a set of points on a line but also of a set of lines on a point [Figure 8]. Likewise the dual of the figure consisting of four points no three of which lie on the same line is a figure of four lines no three of which lie on the same point [Figure 9].

Let us attempt this rephrasing for a slightly more complicated figure. A triangle consists of three points not all on the same line and the lines joining these points. The dual statement would

FIGURE 8

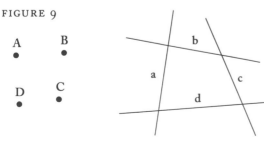

FIGURE 9

read: three lines not all on the same point and the points joining them (that is, the points in which the lines intersect). The figure we get by rephrasing the definition of a triangle is again a triangle, and so the triangle is called self-dual.

Now let us rephrase Desargues's theorem in dual terms, using the fact that the dual of a triangle is a triangle and assuming in this case that the two triangles and the point O lie in one plane. The theorem says:

"If we have two triangles such that lines joining corresponding vertices pass through one point O, then the pairs of corresponding sides of the two triangles join in three points lying on one straight line."

Its dual reads:

"If we have two triangles such that points which are the joins of corresponding sides lie on one line O, then the pairs of corresponding vertices of the two triangles are joined by three lines lying on one point."

We see that the dual statement is really the converse of Desargues's theorem, that is, it is the result of interchanging his hypothesis with his conclusion. Hence by interchanging point and line we have discovered the statement of a new theorem. It would be too much to ask that the proof of the new theorem should be obtainable from the proof of the old one by interchanging point and line. But if it is too much to ask, the gods have been generous beyond our merits, for the new proof can be obtained in precisely this way.

Projective geometry also deals with curves. How should one dualize a statement involving curves? The clue lies in the fact that a curve is after all but a collection of points; we may think of a figure dual to a given curve as a collection of lines. And indeed a collection of lines which satisfies the condition dual to that satisfied by a conic section turns out to be the set of tangents to that curve [Figure 10]. If the conic section is a circle, the dual figure is the collection of tangents to the circle [Figure 11]. This collection of tangents suggests the circle as well as does the usual collection of

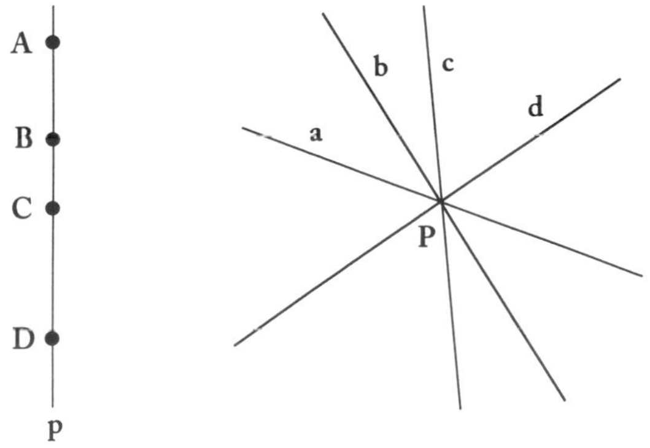

FIGURE 10

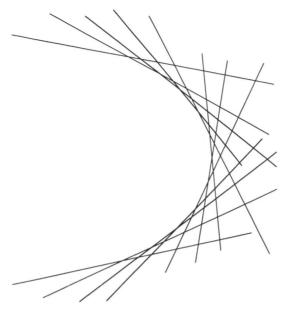

FIGURE 11

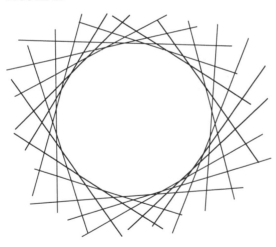

points, and we shall call the collection of tangents the line circle.

Let us now dualize Pascal's theorem on the hexagon in a circle. His theorem goes:

"If we take six points, *A, B, C, D, E,* and *F,* on the point circle, then the lines which join *A* and *B* and *D* and *E* join in a point *P;* the lines which join *B* and *C* and *E* and *F* join in a point *Q;* the lines

which join *C* and *D* and *F* and *A* join in a point *R.* The three points *P, Q,* and *R* lie on one line *l.*"

Its dual reads:

"If we take six lines, *a, b, c, d, e,* and *f,* on the line circle, then the points which join *a* and *b* and *d* and *e* are joined by the line *p;* the points which join *b* and *c* and *e* and *f* are joined by the line *q;* the points which join *c* and *d* and *f* and *a* are joined by the line *r.* The three lines *p, q,* and *r* lie on one point *L.*"

The geometric meaning of the dual statement amounts to this: Since the line circle is the collection of tangents to the point circle, the six lines on the line circle are any six tangents to the point circle, and these six tangents form a hexagon circumscribed about the point circle. Hence the dual statement tells us that if we circumscribe a hexagon about a point circle, the lines joining opposite vertices of the hexagon, lines *p, q,* and *r* in the dual statement, meet in one point [Figure 12]. This dual statement is indeed a theorem of projective geometry. It is called Brianchon's theorem, after Monge's student Charles Brianchon, who discovered it by applying the principle of duality to Pascal's theorem pretty much as we have done.

It is possible to show by a single proof that every rephrasing of a theorem of projective geometry in accordance with the principle of duality must lead to a new theorem. This principle is a remarkable possession of projective geometry. It reveals the symmetry in the roles that point and

FIGURE 12

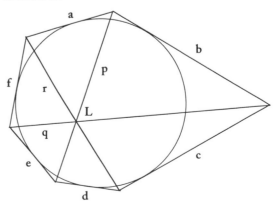

line play in the structure of that geometry. The principle of duality also gives us insight into the process of creating mathematics. Whereas the discovery of this principle, as well as of theorems such as Desargues's and Pascal's, calls for imagination and genius, the discovery of new theorems by means of the principle is an almost mechanical procedure.

As one might suspect, projective geometry turns out to be more fundamental than Euclidean geometry. The clue to the relationship between the two geometries may be obtained by again considering projection and section. Consider the projection of a rectangle and a section in a plane parallel to the rectangle [Figure 13]. The section is a rectangle similar to the original one. If now the point O moves off indefinitely far to the left, the lines of the projection come closer and closer to parallelism with each other. When these lines become parallel and the center of the projection is the "point at infinity," the rectangles become not merely similar but congruent [Figure 14]. In other words, from the standpoint of projective geometry the relationships of congruence and similarity, which are so intensively studied in Euclidean geometry, can be studied through projection and section for special projections.

FIGURE 13

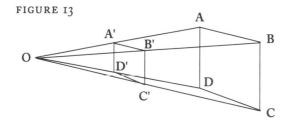

FIGURE 14

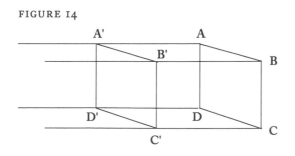

If projective geometry is indeed logically fundamental to Euclidean geometry, then all the concepts of the latter geometry should be defined in terms of projective concepts. However, in projective geometry as described so far there is a logical blemish: our definition of cross ratio, and hence concepts based on cross ratio, rely on the notion of length, which should play no role in projective geometry proper because length is not an invariant under arbitrary projection and section. The nineteenth-century geometer Felix Klein removed this blemish. He showed how to define length as well as the size of angles entirely in terms of projective concepts. Hence it became possible to affirm that projective geometry was indeed logically prior to Euclidean geometry and that the latter could be built up as a special case. Both Klein and Arthur Cayley even showed that the basic non-Euclidean geometries could be derived as special cases of projective geometry. No wonder that Cayley exclaimed: "Projective geometry is all geometry!"

It remained only to deduce the theorems of Euclidean and non-Euclidean geometry from axioms of projective geometry, and this geometers succeeded in doing in the late nineteenth and early twentieth centuries. What Euclid did to organize the work of three hundred years preceding his time, the projective geometers did recently for the investigations which Desargues and Pascal initiated.

Research in projective geometry is now less active. Geometers are seeking to find simpler axioms and more elegant proofs. Some research is concerned with projective geometry in *n*-dimensional space. A vast new allied field is projective differential geometry, concerned with local or infinitesimal properties of curves and surfaces.

Projective geometry has had an important bearing on current mathematical research in several other fields. Projection and section amount to what is called in mathematics a transformation, and it seeks invariants under this transformation. Mathematicians asked: Are there other transformations more general than projection and section whose invariants might be studied? In recent times one new geometry has been developed by pursuing this line of thought, namely, topology. It would take us too far afield to consider topological transformations. It must suffice here to state that topology considers transformations more general than projection and section and that it is now clear that topology is logically prior to projective geometry. Cayley was too hasty in affirming that projective geometry is all geometry.

The work of the projective geometers has had an important influence on modern physical science. They prepared the way for the workers in the theory of relativity, who sought laws of the universe that were invariant under transformation from the coordinate system of one observer to that of another. It was the projective geometers and other mathematicians who invented the calculus of tensors, which proved to be the most convenient means for expressing invariant scientific laws.

It is of course true that the algebra of differential equations and some other branches of mathematics have contributed more to the advancement of science than has projective geometry. But no branch of mathematics competes with projective geometry in originality of ideas, coordination of intuition in discovery and rigor in proof, purity of thought, logical finish, elegance of proofs, and comprehensiveness of concepts. The science born of art proved to be an art.

69

Pisa, Galileo, Rome

EDMOND R. KIELY

*I*N PROFESSOR W. R. CARNAHAN's article, "Mathematics an Essential of Culture," which appeared in the November, 1950 issue of this magazine, there occur certain oft-repeated statements which, in the light of recent research, appear in need of revision. These statements concern Galileo's experiments with falling bodies from atop the Tower of Pisa and Galileo's clash with ecclesiastical authorities. As a matter of fact the reiterated statements concerning Galileo and Aristotle, and Galileo and the Catholic Church tend to obscure Galileo's real merits as a scientist. Even in Professor Carnahan's article we do not have as detailed a discussion of Galileo's contributions to the science of mechanics, on which his right to a place in the history of science properly rests, as we should have expected.

Concerning "the famous gravity demonstration at the Tower of Pisa" the writer would like to recommend an admirable little book by Professor Cooper.[1] There one learns that the story of these experiments was first written by Viviani, Galileo's earliest biographer, about twelve years after Galileo's death and more than sixty years after the assumed date (ca. 1590) of the experiments. Since Galileo meticulously committed to writing and published[2] the results of his physical experiments and deductions it is strange that he should have omitted any mention of these "famous" experiments. It is true that in Galileo's writings we find

the material on which, in all probability, Viviani built his story. In criticizing Aristotle's ideas on motion Galileo asks the following question in *De Motu:*

> . . . if two stones were flung at the same moment from a high tower, one stone twice the size of the other, who would believe that when the smaller was half-way down the larger had already reached the ground?[3]

Again in Galileo's last and most important work, *Discorsi e Dimostrazioni Mathematiche intorno a Due Nuove Scienze,* we are told in the words of Salviati, one of the characters carrying on the dialogue:

> Aristotle says "An iron ball of one hundred pounds, falling from a height of one hundred cubits, reaches the ground before a one-pound ball has fallen a single cubit." I say that they arrive at the same time. You find on making the experiment, that the larger precedes the smaller by two finger-breadths; that is, when the larger one has struck the ground, the other is short of it by two fingers. Now you would not conceal behind these two fingers the ninety-nine cubits of Aristotle."[4]

Nowhere does Galileo state that he himself carried out such experiments. But we know definitely that others did.[5]

However, more important than the actual performance of such experiments is the defamation which the story throws on Aristotle as a physicist. Nowhere in Aristotle's works can be found the statement attributed to him by Galileo through the mouth of Salviati and which has been continuously associated with the various ramifications of Viviani's story. Arguments against Aristotle are usually based on quotations from his *Physica* and *De Caelo* where he proposes proofs for the non-

Reprinted from *Mathematics Teacher* 45 (Feb., 1952): 173–82; with permission of the National Council of Teachers of Mathematics.

existence of a void or vacuum. But as Hardcastle pointed out:

> What he [Aristotle] taught was that the terminal velocity of a heavy body was greater than the terminal velocity of a light body. . . . Aristotle throughout treats only of the motion of projectiles, and of that only in a resisting medium, and then only of that part of the vertical motion when the projectile has attained that constant speed known to ballisticians as "terminal velocity" which can be as readily observed in rising smoke as in falling rain.[6]

One from many possible quotations from Aristotle should suffice to substantiate this interpretation:

> We see the same weight or body moving faster than another for two reasons, either because there is a difference in what it moves through, as between water, air and earth, or because, other things being equal, the moving body differs from the other owing to excess of weight or of lightness. Now the medium causes a difference because it impedes the moving thing, most of all if it is moving in the opposite direction, but in a secondary degree even if it is at rest, and especially a medium that is not easily divided, i.e., a medium that is somewhat dense.[7]

It is true there are passages in Aristotle's physics which seem to offer indirect evidence in proof of Galileo's viewpoint. But it must be remembered that, first and foremost, Aristotle was a realist, and not a dreamer, as he has been sometimes portrayed. From ordinary experience it should have been obvious to him that similar bodies of different weight would fall through a given distance in practically the same time. It is difficult to imagine that the man whose physical theories have recently been so highly estimated by Professor Boyer[8] could have decided otherwise and with such a discrepancy from reality. It may be pointed out here that Aristotle's meaning in many passages has been obscured. Through errors in the copying of early manuscripts and the difficulty of determining the actual meaning of many of the technical terms employed a variety of translations is possible, as may be seen by even a comparison of the translations given here with those in The Loeb Classical Library editions.

In any examination of Aristotle's physical theories it is important to bear in mind, as Professor Boyer points out, that to the Greeks "*physis* meaning the essence or nature of things, was more concerned with the explanation of essential properties and relations than with quantitative description." Professor Boyer's estimate of Aristotle's scientific achievements is of importance in the light of later developments in the physical sciences:

> The physical science of Aristotle is a coherent and systematic treatment which, while less accurate than that of Archimedes, is of far wider scope. In method there is a striking similarity between these two great scientists of antiquity. Both men began with careful observation unaided by the use of instruments, framed inductive generalizations and with these as premises built a deductive science. But while Archimedes, a mathematician, limited himself to a few observations comparable in exactitude to the axioms of Euclid's *Elements* of geometry, the philosopher Aristotle surveyed the whole of nature so that he might disclose the rational order in cause and purpose. For this reason Aristotle placed the inner consistency of his system above accuracy in detail, and in his eager search for certainty he failed to exercise the needed suspension of judgment.[9]

Criticisms of Aristotle's statements on falling bodies and other sections of his physics did not begin with Galileo, they date back at least as far as the period of Philoponus in the sixth century. The modern concept of the content of physics as a science which seeks proximate causes in contrast to Aristotle's concept which embraced ultimate as well as immediate causes came prominently to the fore during the thirteenth century. Two important innovations of this century were the increased emphasis on the mathematical or quantitative aspects of the subject matter of physics and the systematic development of the experimental method for the verification of scientific hypotheses. The earliest important depositions on these fundamental changes are to be found in the writings of Bishop Grosseteste of Lincoln,[10] for many years prominently connected with Oxford University. But it was in connection with the quantitative study of motion that the most fundamental rupture occurred between the advocates of the new physics and the adherents of Aristotelian physics. Through the critical attacks on Aristotelian ideas on motion, carried on particularly at the Universities of Oxford and Paris by such men as Occam, Buridan, Duns Scotus, Suiseth, Bradwardine, Roger Bacon, Albert of Saxony, and Bishop Nicholaus of Oresme, there arose the mathematical theory of accelerated motion which reached its final form with the development of the calculus by Newton and Leibniz, and also the theory of impetus which Descartes was later to develop into the theory of inertia.[11] In the meantime, during the fifteenth and sixteenth centuries, the study of natural phenomena by the so-called adherents of Aristotelian-Scholasticism had sunk to a low ebb and as a result the whole science of Aristotle was repudiated. As Whittaker, the well-known British mathematical-physicist, has expressed it:

> From the fourteenth century onwards, Scholasticism was decadent and by the end of the sixteenth it had become thoroughly debased. The love of nature that had been so vital in Aristotle had almost perished; the practice of observation and experiment, on which he and St. Thomas had so strongly insisted, was neglected save by a few solitary workers; and the degenerate Schoolmen occupied themselves with futile subtleties that bore no relation to life and reality.[12]

Or again quoting Professor Boyer:

> By the time of Galileo and Newton the science of Aristotle had been thoroughly discredited. It became the fashion to ridicule Aristotelian science for its errors, more putative than real.[13]

The fashion still seems to persist, but in view of the high esteem in which the work of Aristotle is held by many modern historians of the physical sciences and mathematics, future critics of Aristotelian physics should be deterred from endeavoring to bury "The Philosopher" beneath hundred-pound balls dropped from atop the Tower of Pisa. All the more so when it is kept in mind that the greatest derider of Aristotle in the seventeenth

century was himself a rather haphazard physicist even by medieval standards. It is rather difficult to overlook Galileo's disregard of experimental verification as exemplified in the following:

> Galileo worked out his physics by thought, by correct reasoning and mathematics, not by induction from experiments. During his days at Pisa, before he went to Padua, he wrote: "But, as ever, we employ reason more than examples (for we seek the causes of effects, and these are not revealed by experiment)." Galileo liked to use what he called "Thought Experiments," imagining the consequences rather than observing them directly. Indeed, when he described the motion of a ball dropped from the mast of a moving ship, in his *Dialogue on the Two Great Systems of the World*, he then had the Aristotelian, Simplicio, ask whether he had made an experiment, to which Galileo replied: "No, and I do not need it, as without any experience I can affirm that it is so, because it cannot be otherwise."[14]

Before considering Galileo's astronomical work it is necessary to clarify the meaning of scientific truth as developed by Grosseteste and the Medieval Schoolmen who broke away from Aristotelian concepts of science in the thirteenth century. To them physics or more generally the physical sciences, as such, were no longer concerned with ultimate reality. All questions on primal substances, essences, and ultimate causes were relegated to the science of metaphysics. The objective of the physical sciences was henceforth to answer questions on the immediate efficient causes of phenomena. Briefly, the usual order of the process by which such answers were to be determined consists, in the first place, in intelligent observation. The second step is the setting up of hypotheses as a result of inductive thought on these observations. As part of such hypotheses or auxiliary to them, there may be set up a mathematical or mechanical model to simplify and clarify the hypotheses. The third step consists in the execution of experiments (controlled if possible) which test the hypotheses directly or deductions from them. The more experimental evidence acquired in favor of the hypotheses the stronger the hypotheses become. One piece of confirmed experimental evidence which contradicts a hypothesis is sufficient to cause either the complete rejection or the modification of the hypothesis. Finally, a hypothesis which has been proved to be in agreement with experimental evidence or, as more generally happens, a combination of such hypotheses, helps to formulate by logical deduction into an intelligible system or set of laws a series of physical facts which in themselves are seemingly isolated. These laws should be suggestive of further experimentation and expansion in accordance with the deductions which can logically be derived from them; such deductions are usually expressible in mathematical form. Our scientific truths are contained in such laws which, because of the method of derivation, can never be final or absolute but must always remain subject to revision. It is also possible to set up hypotheses on purely intuitional ideas, but the evaluation of all hypotheses depends primarily on the amount of agreement between them and the phenomena which they are supposed to represent as disclosed by experimentation.

The history of scientific progress or man's endeavor to improve his knowledge of nature is marked by the tombstones of abandoned hypotheses. What was thought to be true at one period is later shown, usually as a result of better experimentation, to be either totally at variance with the natural phenomena—in which case the former knowledge must be considered as false—or only partially in agreement with phenomena—in which case there is proportionally partial truth. A scientific theory is true only in so far as it remains uncontradicted by experimentation. Two or more hypotheses on the same phenomena could, in this sense, be considered as simultaneously true. Since experimentation depends on measurement and measurement of its nature is subject to error, scientific knowledge, as now conceived, cannot aspire to produce the complete truth about any natural phenomena.[15] If two or more hypotheses appear to agree equally well with observable facts or, as the Greeks said, "save the phenomena," a scientist is likely to choose the simpler hypothesis, not because it is more significant of reality but because it more easily leads to new experiment. A simpler hypothesis is not necessarily in better agreement with the actual facts than a more complex one; neither is the simpler mathematical relationship of a physical law necessarily closer to actuality than a more complex one, even though this idea, Pythagorean in origin, has often supplied the necessary impetus to formulate new hypotheses.[16]

In 1609 when Galileo's astronomical work really began he had a choice of three hypothetical systems recently proposed as replacements for the Ptolemaic system. Up to the fifteenth century Europeans, in general, found that the geocentric system, as explained in such popular treatises as the *Sphere* of Sacrobosco, was sufficiently accurate for the needs of the times. But by the sixteenth century matters had considerably changed. The tremendous impetus given to navigation in the previous century had made acute the need for more accurate astronomical data. Developments in surveying and navigation instruments had reciprocally aided in the production of better instruments for astronomical observation. The work of the Portuguese school of navigation, established in 1416, and the astronomical studies of the mid-Europeans under the influence of Cardinal Nicholaus de Cusa, Purbach, and Regiomontanus had prepared the way for new developments. Another impetus was given by the need felt by the Catholic Church authorities for a reform of the calendar. It was becoming more and more apparent that the complicated Ptolemaic system was unsatisfactory.

Such were the conditions under which Copernicus undertook to promulgate once more the heliocentric hypothesis of Aristarchus. He was actuated primarily by the desire to provide a simpler mathematical model than that of the eighty-odd spheres necessary for explanation and calculation in the Ptolemaic system. For Copernicus, the heliocentric system represented nothing more than a simplified mathematical hypothesis of thirty-four spherical orbits which would also serve to save the phenomena. For him it was not the result of induction from observational data. He spent many years of his life pondering over this mathematical hypothesis but did little practical observation. Whereas in terms of Aristotelian physics there had been a plausible explanation of the Ptolemaic system, Copernicus was unable to offer any acceptable scientific explanation of his system either in terms of the anti-Aristotelian physics of motion or in terms of the older concepts of motion. For this reason and also on religious grounds Tycho Brahe, the greatest astronomical observer of the ensuing period, rejected the Copernican hypothesis, but since he also saw defects in the Ptolemaic system, he established a compromise system of his own. Kepler, the great computer and enthusiastic follower of Pythagorean number mysticism, who fell heir to the observational data of Brahe was an ardent supporter of the heliocentric system, but found it necessary to change Copernicus' circular orbits into elliptical ones so that the mathematical model would be in better agreement with the calculations he had laboriously worked out from the data of Brahe. All three systems still needed explanation and confirmation in terms of anti-Aristotelian physics. Let us now examine what Galileo actually did with these hypotheses.

Galileo claimed and most of his biographers to date have also claimed that, as a result of his astronomical observations and experiments on motion, he had proved the Aristarchian-Copernican hypothesis to be scientifically true. If this were so he should have been able to produce experimental data either in proof of the two main physical facts implied by the hypothesis, namely, that the earth revolved daily on its own axis and that it revolved annually round the sun, or at least in proof of deductions from these facts. Such proofs, as given by Galileo, are to be found in his *Dialogo sopra i due Massimi Sistemi del Mondo, Tolemaico e Copernicano*.[17] In 1597 Galileo stated in a letter to Kepler that he was a firm believer in the Copernican system; therefore, *The Two Systems* may justly

be considered the result of thirty-three years of research with the objective of establishing the scientific truth of the hypothesis. These proofs as given by Galileo break down into three groups. The first group is in terms of the physics of motion. They consist, in the main, of arguments against Aristotelian ideas on motion which had been bandied across Europe for centuries since the rise of anti-Aristotelian physics. Despite his work on dynamics Galileo was unable to furnish a convincing argument in favor of either motion of the earth because he had failed to bring to completion the theory of impetus. With regard to Galileo's proofs from motion Butterfield points out:

> In his mechanics he was a little less original than most people imagine, since, apart from the older teachers of the impetus theory, he had had more immediate precursors, who had begun to develop the more modern views concerning the flight of projectiles, the law of inertia and the behavior of falling bodies. He was not original when he showed that clouds and air and everything on the earth—including falling bodies—moved around with the rotating earth, as part of the same mechanical system. . . . His system of mechanics did not quite come out clear and clean, did not even quite explicitly reach the modern law of inertia, since even here he had not quite disentangled himself from obsessions concerning circular motion.[18]

Galileo's second group of proofs are in terms of his telescopic discoveries. From observations of sun-spots he concluded that the sun was revolving on its axis. Parenthetically it may be mentioned that the revolution of the sun on its own axis was not part of the Copernican hypothesis. Galileo discovered that the moons of Jupiter revolved round that planet and explained the phases of Venus in terms of its revolution round the sun. Granted that his explanations of these phenomena were correct, they are nothing but proofs from analogy so far as the motions of the earth are concerned and could not be accepted as proofs from experiment in any system of physical sciences. Galileo knew only too well that these first two groups of proofs would not and could not be acceptable as scientific proofs for the Copernican hypothesis, and therefore, as he says himself, he reserved to the end of his book his clinching argument for the rotation of the earth. This, his main proof, turns out to be nothing more than the erroneous doctrine that the main factor in the production of the tides was the diurnal rotation of the earth.

> I have been induced upon no slight reasons to omit these two conclusions (having made withal the necessary presupposals) that in case the terrestrial Globe be immoveable, the flux and reflux of the Sea cannot be natural; and that, in case these motions be conferred upon the said Globe, which have been long since assigned to it, it is necessary that the Sea be subject to ebbing and flowing, according to all that which we observe to happen in the same.[19]

It should be borne in mind that Galileo arrived at this conclusion despite the fact that Kepler and others had, years before, suggested lunar influences as the main cause of the tides. As an argument against one of these lunar advocates Galileo writes:

> To that Prelate I would say that the Moon moveth every day along the whole Mediterrane, and yet its waters do not rise thereupon, save only in the very extream bounds of it Eastward and here to us at Venice.[20]

Is it then any wonder that the following estimate of Galileo's astronomical work is to be found in the *Encyclopaedia Britannica:*

> The direct services of permanent value which Galileo rendered to astronomy are virtually summed up in his telescopic discoveries. To the theoretical perfection of the science he contributed little or nothing.[21]

Butterfield's recent conclusion on the value of Galileo's proofs is unequivocal:

> At the end of everything Galileo failed to clinch his argument—he did not exactly prove the rotation of the earth—and in the resulting situation a reader could adopt his whole way of looking at things or could reject it *in toto*—it was a question of entering into the whole realm of thought into which he had transposed the question.[22]

As an astronomer Galileo ignored the work of Kepler whose first and second mathematical hypotheses of planetary motion were made known in 1609, to be followed by a third ten years later. He also ignored the life-time accumulation of observational data of Brahe, and as Dreyer points out:

> In the whole book [*The Two Systems*] there is no allusion whatever to the Tychonic system although it is scarcely too much to say that about the year 1600 nobody, whose opinion was worth caring about, preferred the Ptolemaic to the Tychonic system.[23]

As a matter of historic fact the heliocentric hypothesis was not even completely plausible until Newton had explained it in terms of his synthesis of mathematics, mechanics, gravitation through a hypothetical medium, and Kepler's elliptical orbits. Newton's explanation of the solar system appeared in 1687 in his famous *Principia.* The first experimental proof that the earth was revolving on its own axis occurred, more or less accidentally, when the French astronomer Jean Richer found that his pendulum clock, regulated for Paris, lost about two and a half minutes per day when set up at Cayenne, South America, in 1671. The change in the time of the pendulum oscillation could be explained in terms of a revolving, oblate, spheroidal earth flattened at the poles, an idea which both Newton and Huygens had derived by mathematical deduction. The conclusion from Richer's experience was contradicted in the ensuing years by the Cassini surveys, but was later reaffirmed by the more elaborate surveys carried on by the French Academy of Science in Peru, Ecuador, and Lapland during the years 1735 to 1743. Experimental proof of the earth's rotation round the sun was a more difficult problem. The first tangible evidence for it came when the astronomer Bradley had observed and calculated the phenomena of aberration of light in 1729.

In view of all this evidence it is not surprising that theologians of the seventeenth century should have been reluctant to give unequivocal consent to Galileo's arguments in *The Two Systems* as proof of the scientific truth of the Copernican hypothesis.

This is not the place to reiterate at length the facts leading to Galileo's condemnation by an ecclesiastical commission. If any reader wishes a clear, concise account of the whole affair he will find it in an article by Father Conway.[24] If he wishes for more details he has but to delve into the works of Von Gebler[25] and Favaro.[26] Here it will suffice to state that Galileo was twice tried by Church tribunals, the first time at his own request. As a result of the 1616 trial he was admonished privately by Cardinal Bellarmine, at the request of Pope Paul V, to abandon his opinions on the heliocentric system. To this admonition Galileo aquiesced. In 1633 he was publicly condemned after the publication of *The Two Systems*. The primary reason for the first trial was Galileo's provocative attitude on the question of interpretation of the Scriptures. Professor Carnahan's statement that "Theologians discovered that their doctrines and Galileo's science were not in agreement" is a strange twist to put on the controversy, but again, oft-repeated like the Tower of Pisa story. Far nearer the truth is the statement by Strong in his study of the development of science in the sixteenth and seventeenth centuries:

> The furore raised against Galileo in connection with the Copernican hypothesis cannot be taken to be the attitude of the Church against the mathematical-physical sciences in general. The pursuit of physical science involved Galileo in controversies with men opposed to his physical views, but this was not a clash over the religious consequences involved in astronomical questions.[27]

The results of both trials show that the examiners were satisfied that Galileo's proofs were insufficient to raise the Copernican hypothesis to the level of a scientific truth. With these decisions the calmer judgments of modern scientists must corroborate.

If Galileo had presented his scientific theories without being so arrogant as regards the positiveness of his proofs, and had avoided being drawn into the trap of Scriptural controversy, set up by some of his more wily opponents after the publica-

tion in 1613 of his *Istoria e Dimostrazioni intorno alle Macchie Solari*[28] in which he first publicly stated his decided Copernican views, the trial of 1616 would probably never have occurred. And if there had been no directive from the Congregation of the Index, as a result of the 1616 trial, that Copernicus' book was to be banned until its hypothetical nature was made clearer—which correction was carried out within a short while—the inquiry of 1633 before the Inquisition would have been unnecessary. At the second trial Galileo was accused of breaking his promise of 1616, since any unintelligent person could see that *The Two Systems* in no way favored the hypothetical nature of the Copernican system, despite the ruse which Galileo used in trying to cover up his real opinions on the subject. That there are reasonable grounds for this assumption may be seen from the attitude of Cardinal Bellarmine who was regarded as anti-Copernican. In a letter written to the Carmelite, Father Foscarini, an advocate of Copernicanism, shortly before the 1616 trial he states:

> I say that when there shall be a real demonstration that the sun stands in the center of the universe and the earth revolves around it, it will then be necessary to proceed with great consideration in explaining those passages of Scripture which seem to be contrary to it, and rather to say we do not understand them, than to say that a thing which is demonstrated is false. But I will not believe that there is such proof until it is shown to me; nor is it the same thing to show that the phenomena are saved by assuming that the sun is in the center and the earth revolves round it, and to show that in reality the sun is in the center and the earth revolves.[29]

These are the words of an outstanding theologian of the period but surely, no one can read into them that he is trying to direct the course of scientific investigation so as to make it conform to a religious belief; rather is it the statement of a careful scientist. Similar opinions can be found in the letters of many prominent Catholic ecclesiastics of the period.

Besides, since the publication of Copernicus' *De Revolutionibus Orbium Coelestium* in 1543 there had been no opposition to the heliocentric hypothesis on the part of the Catholic Church; whatever opposition existed came from individuals of various religious persuasions. In fact many of the higher Catholic clergy were known to favor it openly. Was it not through the instrumentality of the Catholic clergy that the Aristarchian heliocentric system had been revived, first by Cardinal Nicholaus de Cusa in his *De Docta Ignorantia* (ca. 1450) and again, nearly a century later, by the canon Copernicus? When Galileo visited Rome in 1611 to demonstrate his telescopes he was enthusiastically received by the Church authorities and especially by the future Pope Urban VIII. It is also important to notice that, despite the 1616 trial, Galileo's ecclesiastical supporters did not abandon him; he even became the recipient of a life pension from his very personal friend Urban VIII in 1624, a reward for his scientific endeavors, setting the precedent, in modern times, for scientific stipends! There was no action on the part of the Catholic Church during this period to show that it was in opposition to scientific progress. Any unbiased history of the period shows the opposite to be true. Nevertheless Galileo knew that, unless he could produce the experimental evidence, anything beyond a hypothetical statement of the Copernican system would run counter to the then accepted interpretation of the Scriptures. He also knew that there was nothing in Catholic theology or tradition against the Church's sanctioning a change in interpretation if necessity arose. What really led to Galileo's troubles was the method he adopted to bring about this change. Like many another highly gifted man who lacked tact he found himself before long antagonizing people who were formerly his best friends. If he had approached the matter cautiously and with less demand that his reasoning be accepted as demonstrative, which it was not, Cardinal Bellarmine and others would undoubtedly have used their influence to allow Galileo to pursue his astronomical work unmolested by any clerical interference, and the Church authorities would have been his best supporters if he had ever

reached the stage of producing what the Cardinal called "a real demonstration," that is an actual proof by experimentation of some deduction from the hypothesis.

It is often assumed that the philosophical system of Aristotle and St. Thomas Aquinas was irreconcilable with the researches undertaken by Galileo. But just as Aristotle was admittedly more realistic in certain respects than Galileo, so also St. Thomas might have given him timely—and modern—advice on the necessity of distinguishing between a scientific hypothesis and a scientific demonstration, as the following excerpt shows:

> The assumptions made by the astronomers are not necessarily true. Although these hypotheses seem to be in agreement with the observed phenomena we must not claim that they are true. Perhaps one could explain the observed motion of the celestial bodies in a different way which has not been discovered up to this time.[30]

A recent statement of Whittaker on the influence of the work of Brahe and Kepler is also apposite here:

> At this point it may be observed that, while the Scholastic cosmology was thereby completely disproved and overthrown, there was nothing in the new methods and discoveries that was inherently irreconcilable with the Scholastic metaphysics; the whole of Tycho's and Kepler's work might conceivably have been absorbed into the philosophy of the Schoolmen by a peaceful and conservative revolution. If this had happened, we in the twentieth century should have been spared the necessity of readjusting our position by a movement back towards Aristotelianism.[31]

But this did not happen because of the lack of true Aristotelianism at the time. Statements found in many of our popular histories of science to the effect that the experimental work of Galileo was completely at variance with the methods of the Aristotelian philosophers of the Italian universities of the seventeenth century are a calumny on Aristotle. A professor of natural philosophy who would refuse to look through a telescope, as some of Galileo's scientific opponents did, should not for that reason be classified as a follower of Aristotle. Not only was Aristotle one of the greatest exemplars of the use of deductive logic but, and this is too often forgotten, he was also well aware of the value and pitfalls of induction, as a reading of the first section of his *Metaphysics* will show. His insistence on the necessity of observation and experiment may be seen from what he has to say of those who

> . . . had certain predetermined views and were resolved to bring everything into line with them. . . . As though some principles did not require to be judged from their results, and particularly from their final issue! And that issue, which in the case of productive knowledge (i.e. in the case of art) is the product, in the knowledge of nature is the unimpeachable evidence of the senses as to each fact.[32]

Why then do we continue to associate the name of Aristotle with a system of natural philosophy which had become atrophied by lack of experimentation? The conflict between Galileo and his opponents in the realm of natural philosophy was caused by an absence of true Aristotelian principles and this was in no way confined to the members of one religious group. Even Galileo admitted that Aristotle would undoubtedly change his ideas about certain physical phenomena if he were then living and could have participated in the experimental work being carried on.

Milder judgments on the decisions of the Inquisition in 1633 than that implied by Professor Carnahan have long since been passed by men of high standing in the scientific world who cannot be accused of bias toward the Catholic Church. Thomas Huxley, in a letter to Professor Mivart in 1885 writes:

> I have looked into the matter [the Galileo controversy] when I was in Italy, and I have arrived at the conclusion that the Pope and the College of Cardinals had rather the best of it.[33]

The following from the latest edition of Dampier shows the change in attitude from the unwarranted statements of condemnation of the Catholic Church to be found in so many of the nineteenth century histories of science, and unfortunately still appearing from time to time.

> In spite of Whewell's clear and fair account of the incident some more recent writers have made too much of the persecution of Galileo for his Copernican views. As Whitehead [A. N.] says "In a generation which saw the Thirty years War and remembered Alva in the Netherlands, the worst that happened to men of science was that Galileo suffered an honorable detention and a mild reproof, before dying peacefully in his bed."[34]

Butterfield concluding his essay on the science of this period says:

> Aristotelian physics were clearly breaking down, and the Ptolemaic system was split from top to bottom. But not until the time of Newton did the satisfactory alternative system appear. . . . The long existence of this dubious, intermediate situation brings the importance of Sir Isaac Newton into still stronger relief. We can better understand also, if we cannot condone, the treatment which Galileo had to suffer from the Church for a presumption which in his dialogues on *The Two Principal World-Systems* he had certainly displayed in more ways than one.[35]

The man who, in Professor Carnahan's words, had to live "the last ten years of his life under the sternest requirement to say nothing, write nothing, think nothing of which entrenched authority does not approve" produced during that time the treatise on which his proper claim to fame rests, that is, the *Two New Sciences*, a compendium of his life's work on mechanics and allied topics. He also had time to work in conjunction with Torricelli and Viviani, and to write numerous letters on various scientific topics.

NOTES

1. Lane Cooper, *Aristotle, Galileo and The Tower of Pisa* (Cornell University Press, 1935).

2. Among the few exceptions is *De Motu*, written about 1590 but not published until 1883.

3. Galileo Galilei, *Le Opere di Galileo Galilei* (ed. A. Favaro), (Florence: 1890–1907). Reprint 1929; Vol. 1, p. 263. Trans. by Cooper, *op. cit.*, p. 83.

4. *Ibid.*, Vol. 8, p. 109. Trans. from *Dialogues Concerning Two New Sciences* by H. Crew and A. De Salvio (New York: The Macmillan Co., 1914), pp. 64–65. Hereafter referred to as *Two New Sciences*.

5. See Cooper, *op. cit.*, for details and also interesting examples of variations of the story. See also, H. Butterfield, *The Origins of Modern Science* (New York: The Maximillan Co., 1951), p. 59 ff.; H. T. Pledge, *Science Since 1500* (London: Ministry of Education, 3rd reprint, 1947), p. 61.

6. J. H. Hardcastle, Correspondence Section, *Nature* (June 25, 1914), p. 429. See also (January 22, 1914), pp. 584–85.

7. Aristotle *Physica*, 4, c. 8, 215ᵃ 25–31. Trans. by Hardie and Gaye, *The Works of Aristotle* (ed. W. D. Ross), Oxford University Press, 1930.

8. Carl B. Boyer, "Aristotle's Physics," *The Scientific American* (May, 1950), pp. 48–51. See also by the same author "Quantitative Science without Measurement: The Physics of Aristotle and Archimedes," *The Scientific Monthly* (May, 1945), pp. 358–64.

9. *Ibid.*, p. 50.

10. G. Sarton, *Introduction to the History of Science* (Baltimore: The Williams and Wilkins Co.), Vol. 2, p. 583 ff.

11. P. Duhem, *Études sur Léonardo de Vinci*, Vol. 3, "Les Précurseurs de Galiléi" (Paris, 1913). A. Koyré, *Études Galiléennes* (Paris, 1939). Carl B. Boyer, *The Concepts of the Calculus* (New York: Hafner Co., 1949), p. 71 ff. H. Butterfield, op. cit., Ch. 1.

12. E. Whittaker, "Aristotle, Newton, Einstein," *Science* (Sept. 17, 1943), p. 250.

13. Carl B. Boyer, *op. cit.* (8), p. 50.

14. I. B. Cohen, "Galileo," *Scientific American* (August, 1949), p. 45. For further elaboration on Galileo's "Thought Experiments" see, E. A. Burtt, *The Metaphysical Foundations of Modern Physical Science*, London, 1949, p. 65 ff. and H. Butterfield, *op. cit.*, p. 61. It is strange to find A. Koyré in agreement with Galileo's attitude in the case mentioned above. Mathematical deduction of itself is not sufficient to establish the truth of natural phenomena. See his article "Galileo and the Scientific Revolution of the Seventeenth Century," *The Philosophical Review*, July, 1943, p. 347.

15. " . . . no contingent, hypothetical laws, however wide, can offer an ultimate explanation of concrete facts. Laws are but the expression of the *modus agendi*, the manner of acting, of causes or combinations of causes. And we shall not have fully explained any concrete fact in the universe until we know why the agencies of nature act according to those widest laws—why, for example, matter gravitates, or why life comes only from life, or why natural causes act uniformly." P. Coffey, *The Science of Logic* (New York: Longmans, Green and Co., 1918), Vol. 2, p. 240.

16. For further analysis of scientific truth, see E. F. Caldin, *The Power and Limits of Science* (London: Chapman and Hall, 1949).

17. G. Galilei, *op. cit.*, Vol. 7, English translation by Thomas Salusbury, *Mathematical Collections and Translations*, Vol. 1 (London: 1661). Hereafter referred to as *The Two Systems*.

18. H. Butterfield, *op. cit.*, p. 54.

19. G. Galileo, *op. cit.* (17), Salusbury translation, p. 380.

20. *Ibid.*, p. 383.

21. Agnes M. Clerke, "Galileo Galilei," *Encyclopaedia Britannica* (14th ed., 1946), vol. 9, 980.

22. H. Butterfield, *op. cit.*, p. 54.

23. J. L. E. Dreyer, *History of the Planetary Systems from Thales to Kepler* (Cambridge University Press, 1906), p. 416.

24. Pierre Conway, O.P. "Aristotle, Copernicus, Galileo," *The New Scholasticism*, Vol. 23, Nos. 1 and 2, 1949.

25. Carl Von Gebler, *Galileo Galilei und die römische Curie* (Stuttgart: 1876). English translation by Mrs. G. Sturge, London, 1879. This was the first historical research to make use of all the Vatican records relative to Galileo's trials.

26. See note 3. This work, in twenty volumes, contains all Galileo's scientific treatises and available letters together with numerous other letters and documents relative to Galileo's work.

27. E. W. Strong, *Procedures and Metaphysics* (California University Press, 1936), p. 137.

28. G. Galilei, *op. cit.*, Vol. 4.

29. D. Berti, *Copernico e le vicende del sistema Copernicano in Italia* (Rome, 1876), p. 123.

30. St. Thomas Aquinas, *Commentary on Aristotle's De Caelo II*, lectio 17, no. 2 (Rome, 1886).

31. E. Whittaker, *op. cit.*, p. 251.

32. Aristotle, *op. cit.*, De Caelo III, 306ᵃ.

33. Thomas H. Huxley, *Life and Letters* (ed. L. Huxley), London, The Macmillan Co., 1900, Vol. 2, p. 122. It is interesting to compare this with Huxley's statement on the same subject in his essay "Descartes' Discourse on Method" written in 1870.

"At that time, physical science suddenly strode into the arena of public and familiar thought and openly challenged not only Philosophy and the Church, but that common ignorance which often passes by the name of Common Sense. The assertion of the motion of the earth was a defiance of all three, and Physical Science threw down her glove by the hand of Galileo. . . . Charity children would be ashamed not to know that the earth moves; while the Schoolmen are forgotten; and the Cardinals—well, the Cardinals are at the Oecumenical Council, still at their old business of trying to stop the movement of the World." *Collected Essays*, New York, D. Appleton and Co., 1894, Vol. 1, pp. 179, 180.

34. Dampier, W. C., *A History of Science*, Cambridge University Press, 1949, p. 113.

35. Butterfield, H., *op. cit.*, p. 55.

70

Irrationals or Incommensurables IV
The Transitional Period

PHILLIP S. JONES

*T*HE EARLIER NOTES in this series told the story of incommensurables in Greek mathematics from Hippasus (500 B.C.) to Euclid (300 B.C.).[1] we saw how Eudoxus' theory of proportion for "magnitudes" suppressed the logical scandal caused by the discovery that there are geometric quantities or magnitudes whose ratio can not be represented by the ratio of two "numbers" where "number" at this time meant "integer." Eudoxus' theory made it possible to establish many synthetic geometry theorems without actually defining an irrational number. The need for a clear concept of *irrational numbers* as opposed to a theory of proportion for *incommensurable magnitudes* began to be felt with the development of analytic geometry and became essential as the theory of limits and continuity was developed with the nineteenth-century drive to give a rigorous foundation for the calculus.

This note is to summarize the status of incommensurables and irrationals in the transitional period between Euclid and Descartes (1637) or perhaps Cauchy (1825). This is a long period, and even though it is true that in it real developments with respect to irrationals were few and took place slowly, it is still a bit of an oversimplification to condense the entire period into such a short note as this. However, the chief growth in this period was in the development of computa-

tional devices for finding numerical approximations for roots of numbers and in operations with and manipulations of expressions involving radicals. Neither of these developments either represented or stimulated a recognition that here was an entirely new class of numbers which once defined and studied would be a mathematical system in its own right and not merely a device for treating ratios of "magnitudes." In fact the very words used implied a denial to these quantities of status as numbers.

Although we can see that the Eudoxian-Euclidean theory of proportion applies to all types of what we today call irrationals, Euclid used this theory only in situations where geometric magnitudes were involved, and in Book X treated only types which were readily obtainable by geometric construction and which in modern symbolism would be represented by $\sqrt{\sqrt{a} \pm \sqrt{b}}$. In fact the Greek words, which became, via Latin translations, our "rational" and "irrational," meant *expressible* and *inexpressible* in the original. Further, in Euclid, *irrational* was used only for quantities which were incommensurable "in square." Thus, although the diagonal of a square would be "incommensurable" with its side, it was not "irrational" in a Euclidean sense since $(\sqrt{2})^2$ is commensurable with $(1)^2$. Cassiodorius (ca. 585) is said to be the first to use "irrational" for the numbers to which it is now applied.[2]

Surd is a second word which is used confusingly even today and hence is probably best not used at all. This word came from the use of the

Reprinted from *Mathematics Teacher* 49 (Oct., 1956): 469–71; with permission of the National Council of Teachers of Mathematics.

Latin *surdus,* meaning silent or mute, to translate an Arabic word meaning "deaf" which the Arabs had in turn used in their translation of Euclid. The nonmathematical word *absurd* means "out of harmony" with reason; *illogical* has a similar origin in the Latin for "inaudible." Although *absurd* is not a mathematical word and *surd* did not come from it directly, it is interesting to note that it was first used in this sense in English by Robert Recorde in his *Whetstone of Witte* (1557) where he said, "8 − 12 is an absurde number. For it betokeneth less than nought by 4."[3]

In all usages $\sqrt{2}$ would be termed a *surd,* but there have been writers who would have excluded $\sqrt{6}$ because it can be written as $\sqrt{2} \cdot \sqrt{3}$, while many, such as Chrystal, would exclude $\sqrt{2} + \sqrt{3}$, reserving surd for the "incommensurable root of a commensurable number."[4]

It should not be thought that there was no progress at all from the Euclidean concept of incommensurable magnitudes toward a modern concept of irrational numbers in the years from 300 B.C. to A.D. 1800. Diophantos (ca. 250?) developed some algebraic symbolism and thought of numbers as distinct from geometric magnitudes in his work on theory of equations and number theory. He did not, however, accept either negatives or irrationals as numbers or as solutions to his problems. Hindu and Arabic writers recognized that quadratic equations have two roots which might be negative or surd. This led them to state rules for operating with surds.

For example, the Hindu Bháscara (ca. 1150) gave rules for the four fundamental operations with surds[5] which may provide some interesting enrichment material for a class. He said, "Term the sum of two irrationals the great surd; and twice their product the less one. The sum and the difference of them reckoned like integers are so [of the original surd roots]." He clarifies this with the example which we write with modern symbols as follows:

To find $\sqrt{2} + \sqrt{8}$, take the square root of the sum of the great surd, 2 + 8 = 10, and the less surd,

$2\sqrt{2 \cdot 8} = 8$. This finally gives $\sqrt{2} + \sqrt{8} = \sqrt{(2 + 8)} + 2\sqrt{2 \cdot 8} = \sqrt{10 + 8} = \sqrt{18}$. In symbols this says $\sqrt{a} + \sqrt{b} = \sqrt{a + b + 2\sqrt{a \cdot b}}$ which today is easily verified by any student of second-year algebra and which will always suffice to write the sum of two square roots as a single radical, although the latter may also itself contain a second radical.

Bháscara's second rule for the sum and difference of two radicals was "the root of the quotient of the greater irrational number divided by the less, being increased by one and diminished by one; the sum and remainder being squared and multiplied by the smaller irrational quantity, are respectively the sum and difference of the two surd roots." Applying this to the first example, he wrote the equivalent of

$$\sqrt{2} + \sqrt{8} = \sqrt{\left[\sqrt{\frac{8}{2}} + 1\right]^2 \cdot 2} = \sqrt{18}.$$

We leave a literal formulation and verification of this to the reader.

Bháscara wrote similarly involved rules for the multiplication and division and involution of irrational expressions. However, although they were being operated with and were recognized as the solutions of equations, such expressions can hardly be said to have had the status of numbers.

Another development which we can now view as to some extent tending toward an idea of irrational number was the derivation of several processes for finding successive rational approximations to irrationals. Of this nature were the side and diagonal numbers and the related continued fraction expansion of the $\sqrt{2}$ which we cited in an earlier article.[6] Of course, approximations to irrationals date back to Babylonian times, and mathematicians, such as Hero, in the later periods of Greek mathematics, developed repeated processes which served practical needs. These processes shifted thinking from purely geometric considerations toward the arithmetical aspects of irrationals. However, since many persons regarded them as merely practical calculations which determined rational numbers which approximately satisfied the conditions represented by a symbol, these techniques, from another viewpoint, tended to turn attention away from the more basic and theoretical problem of defining a new kind of number.

Slight steps which from a modern viewpoint can be seen to be progress in the right direction are found in the work of Nicholas Chuquet in the thirteenth century who replaced the terms *square* and *cube* roots by *second* and *third* roots, and in the fourteenth century Nicholas Oresme simplified calculations with radical expressions by introducing fractional exponents. In the sixteenth century Michael Stifel made more substantial progress toward a general theory of irrational numbers by distributing numbers into different classes, by replacing lines by numbers in some of his discussions, and by formulating an analogy between "true numbers" and the symbols for irrationals. However, he still denied to the latter status as true numbers and remained attached to the ideas of Euclid.

In the same century Simon Stevin sought to show that roots and fractions have similar properties, and soon thereafter the development of the theory of equations and algebraic symbolism at the hands of Cardan, Viete, and Descartes accelerated the separation of number ideas from geometric interpretations as lines, areas, and volumes. This together with the calculation of logarithmic and trigonometric tables prepared the way for the treatment which the development of rigorous notions of limits and continuity required and which, as we will see in the final article of this series, grew out of the efforts of Cantor, Dedekind, and others in the nineteenth century.

NOTES

1. P. S. Jones, "Irrationals or Incommensurables," *The Mathematics Teacher*, XLIX (February, 1956), pp. 123–27 (March, 1956), pp. 187–91 (April, 1956), pp. 282–85.

2. Florian Cajori, *A History of Mathematics* (New York: Macmillan, 1918), p. 68.

3. *The Oxford English Dictionary*, I (Oxford: Clarendon Press, 1955), p. 43.

4. D. E. Smith, *History of Mathematics II* (Boston: Ginn & Co., 1925), p. 253.

5. H. T. Colebrook, *Algebra, with Arithmetic and Mensuration from the Sanskrit of Brahmegupta and Bháscara* (London: 1817), p. 145 ff.

6. P. S. Jones, "The $\sqrt{2}$ in Babylonia and America," *The Mathematics Teacher*, XLII (October, 1949), pp. 307–8.

SUGGESTED READINGS FOR PART VI

Cardano, Girolamo. *The Great Art, or the Rules of Algebra.* Translated by Richard Witmer. Cambridge, MA: MIT Press, 1968.

Flegg, G., Hay, C., and Moss, B. *Nicolas Chuquet, Renaissance Mathematician.* Boston: Reidel, 1985.

Galilei, Galileo. *Discourses on the Two Chief Systems.* Ed. Stillman Drake. Berkeley, CA: University of California Press, 1953.

Gies, Joseph, and Gies, Francis. *Leonardo of Pisa and the New Mathematics of the Middle Ages.* New York: Thomas Y. Crowell Co., 1969.

Hay, Cynthia, ed. *Mathematics from Manuscript to Print 1300–1600.* Oxford: Clarendon Press, 1988.

Jackson, Lambert L. *The Educational Significance of Sixteenth Century Arithmetic.* New York: AMS Press, 1972, reprint of 1906 edition.

Kearney, Hugh. *Science and Change 1500–1700.* New York: World University Library, 1971.

Kline, Morris. *Mathematics: A Cultural Approach.* Reading, MA: Addison-Wesley, 1963.

Kuhn, Thomas. *The Copernican Revolution.* New York: Modern Library, 1957.

Rose, Paul. *The Italian Renaissance of Mathematics.* Geneva: Librairie Droz, 1975.

Sarton, George. *Six Wings: Men of Science in the Renaissance.* Bloomington: Indiana University Press, 1957.

Swetz, Frank J. *Capitalism and Arithmetic: The New Math of the 15th Century.* La Salle, IL: Open Court, 1987.

PART VII

MATHEMATICAL RESPONSES TO A MECHANISTIC WORLD OUTLOOK

$\mathcal{T}$HE ECONOMIC ADVENTURISM of the Early Renaissance spread northwards from Italy so that by the seventeenth century institutions of trade and commerce flourished throughout continental Europe and the British Isles. Accompanying and assisting in the increase in regional and international trade was an emphasis on manufacturing and the processing and refinement of basic commodities. The rise of a European textile industry, mining activities, irrigation, and land reclamation required the harnessing of wind and water power, the improvement of existing machines, and the invention of new ones. Increased maritime activities prompted innovations in navigation: the development of precise maritime clocks and more accurate charts. Adoption of the cannon as a tool of warfare resulted in the study of projectile trajectories as well as alterations in the architectural theories of fortification design. It was an age of change: cultural, political, economic, and scientific.

Perhaps, this period of change was best exemplified by the machine itself. Machines were admired for their efficiency, their purposeful organization, and their predictability. They were concrete manifestations of the skill and intelligence of their makers. Machines operate on certain principles or laws—there is an order in their behavior. This fact was both fascinating and comforting to a people who had often suffered from the uncertain whims of nature. Many natural scientists began to view both nature and the universe as a great machine. A mechanistic world outlook is evident in some of the existing sketches of Leonardo da Vinci, the theories of Galileo Galilei, and the writings of René Descartes. Scientists sought to gain an understanding of the new machines and mechanistic theories, and an understanding of the processes of change and motion. Classical theories of dynamics were inadequate for the new needs. The geometry of Euclid was static in its conception, bound to the plane, therefore new geometries allowing for motion in space had to be devised. Fermat and Descartes laid the foundations for a coordinate geometry that allowed algebraic interpretations of geometric situations. Pascal and Desargues improved the visualization of space with their contributions to projective geometry. Considerations by Fermat, Pascal, and Huygens of the phenomenon of change and attempts to determine its predictability led to a new mathematics of probability. Greek number mysticism was replaced by a more mathematically based theory of numbers. It was a time of great mathematical activity. Intellectual nationalism found an outlet in scientific discussion groups and the establishment of national academies. The English Royal Society was founded in 1660 and the French Academy five years later in 1665. An established printing industry helped in the dissemination of the new information and theories. The appearance of scientific periodicals such as the Royal Society's *Philosophical Transactions* helped spread the new knowledge. It was an age of scientific experimentation, skepticism, and rationalism as advocated by works such as William Gilbert's *De Magnete* (1600), Galileo's *Dialogue* (1632), and Descartes's *Discourse on Method* (1637).

Mathematicians strove to understand the dynamics of motion and change. Galileo's experimental work on trajectories proved that motion could be resolved into two components, one vertical, the other horizontal. Thus, motion along a curve could also be separated into components. To do this, one would have to isolate motion at a point and consider instantaneous change. Techniques had to be devised to allow for infinitesimal processes. This study or analysis of infinitesimal change became the basis of the calculus and had long been a focus of attention of such mathematicians as: Galileo, Descartes, Torricelli, Cavalieri, Barrow, and Wallis. But it was in the specific genius of two men, Isaac Newton and Gottfried Wilhelm Leibniz, that the calculus found consolidation and articulation as a mathematical theory. With the development of the differential and integral calculus a new era of mathematical understanding and great potential began.

71

Analytic Geometry: The Discovery of Fermat and Descartes

CARL B. BOYER

*U*NTIL a century ago the discovery of analytic geometry was ascribed categorically to the father of modern philosophy—René Descartes. *Proles sine matre creata*, Chasles called it, but this famous characterization is now quoted only to be refuted. The offspring is now recognized not only as having arrived as twins—for Pierre de Fermat had developed substantially the same method at about the same time—but also as having descended from a long line of legitimate antecedents. Recently attempts have been made to shift responsibility for the birth from the shoulders of Descartes and Fermat to those of one or more of the ancestors—Harriot or Cataldi or Ghetaldi or Viète or Benedetti or Oresme or Apollonius or the two thousand year old grandfather Menaechmus. In the case of any one of these individuals arguments may be advanced in his support. These can be based upon a definition of analytic geometry which is so framed as to include the work of the candidate in question and to exclude that of his rivals with respect to priority. The plausibility of the arguments is easily enhanced through an ingenious interpretation of the man's statements in terms of later symbolisms and concepts, thus implying a specious modernity of viewpoint. Few branches of mathematics lean as heavily upon a felicitous choice of notations as do those two powerful tools of science which we call

analytic geometry and the calculus. It is in fact precisely the algorithmic nature of these subjects which represents the secret of their effectiveness. In view of this, great caution must be exercised in superimposing modern notations upon ancient treatises. Conventional symbolisms have been devised ad hoc to express contemporary concepts which have developed slowly and laboriously, and through repeated use such notations have come to be associated implicitly with the ideas which they are now intended to represent. Projection of such notations into antiquity leads easily to a corresponding anachronism with respect to concepts, as the history of analytic geometry clearly demonstrates.

The query, "What constitutes analytic geometry?" has been variously answered. It is frequently characterized roughly as the combination of algebra and geometry. The utter inadequacy of this statement becomes apparent when one recalls that the ancient Pythagorean solution of quadratic equations through the application of areas would fall within this description and yet bears no significant resemblance to our analytic methods. Nor is one justified in holding that analytic geometry consists simply of the introduction of analytic methods of reasoning into the subject matter of geometry. This would result in applying the name to work which is far removed from Cartesian geometry. Plato is commonly regarded as having suggested the use of the analytic type of argument—of proceeding from the conclusion to the premise—although in a broad sense it can be said to have been

Reprinted from *Mathematics Teacher* 37 (Mar., 1944): 99–105; with permission of the National Council of Teachers of Mathematics.

employed much earlier. Plato, however, had no vision of the part this was to play later in Cartesian geometry.

The use of coordinate systems often has been regarded as the feature distinguishing analytic methods from other approaches in geometry. That this is a necessary aspect of the subject can scarcely be denied, but in no sense is it to be regarded as a sufficient criterion. The ancient Egyptians apparently made use of schemes of rectangular coordinates in their cadastral surveys; and Hipparchus applied such methods more systematically, both to the geography of the Mediterranean region and in the construction of star charts. In the field of geometry coordinate methods were certainly used by Apollonius in his study of the properties of conic sections, if indeed these may not have been employed in this connection a century earlier by Menaechmus. Yet the point of view and the method of attack of Apollonius and Menaechmus were essentially different from those indicated by Fermat and Descartes. A closer examination of their work will reveal that the coordinate frame was in every case simply an auxiliary construction superimposed a posteriori on a given curve in order to *study* its properties. It was not used as a means a priori of *defining* the locus or of determining the generation of the curve. That is, Menaechmus, in attempting to solve the problem of the duplication of the cube by means of the continued proportion of Hippocrates (expressed in modern symbolism

as $\dfrac{a}{x} = \dfrac{x}{y} = \dfrac{y}{b}$), had no conception of the

relationships $ay = x^2$, $y^2 = xb$, and $ab = xy$ as in themselves *defining* certain plane loci, but sought rather to *discover* curves otherwise defined, the points of which should possess these properties. This search led him, as is well known, to the sections of a cone. The very fact that conics were known in Greek geometry as "solid loci" indicates that they were defined stereometrically rather than analytically. More than 600 years later Pappus spoke likewise of "linear loci"—i.e., curves other

than straight lines, circles, and conics—as "being generated from more irregular surfaces and intricate movements." They were not given by equations. Once a curve had been thus defined, it was of course not difficult to discover innumerable relationships in terms of lines associated with the curve. Such properties when symbolically expressed may in a sense be regarded as equations corresponding to the curves, and for this reason the origin of analytic geometry recently has been claimed for Greek antiquity. But according to this point of view the Babylonians possessed analytic geometry by 2000 B.C.! They were familiar with the relationship $4r^2 = c^2 + 4(r - s)^2$, where s is the sagitta of any chord of length c in a circle of radius r. This relationship *may* be regarded—although there is no evidence that the Babylonians did so—as the equation defining the circle in terms of the rectangular coordinates c and s. Thus interpreted, the Babylonians would be regarded as having used analytic geometry. Similarly the Euclidean propositions on the straight line and the circle can be looked upon as equations of these curves, and the *Elements* would then be an example of analytic geometry. One will protest that Euclid did not make use of the modern algebraic symbolism which is used in equations. This is true, but then neither did Menaechmus before him nor Apollonius somewhat later. Moreover, they are alike in that invariably *the equations are derived from the curve*, and not conversely. It is true that Apollonius and other Greek geometers used the stereometric origin of the conics only so far as was necessary to deduce a single fundamental plane property (equation) for each curve, and that this latter result was then made basic in establishing further properties, such as those of the asymptotes. Nevertheless, it is significant that they felt it necessary to give the curves a geometrical definition distinct from that of points the coordinates of which satisfy a given equation. In fact, by "finding" a conic for which a given relationship held (i.e., satisfying a given equation) they understood not plotting it by points but by determining it as a section of a cone. While one may safely assert that in a broad sense Menaechmus and Apollonius made use of analytic methods, it is nevertheless not correct to say that they understood the fundamental principle of analytic geometry—that, in general, an equation in two variables defines a plane curve in a given coordinate system such that all pairs of values satisfying the equation are coordinates of points on a curve; and, conversely, all points on this curve have coordinates which satisfy the given equation. In part the failure of Menaechmus and Apollonius to appreciate this principle may have resulted from the weakness of contemporary algebra. It may also have been the consequence of a peculiar deficiency in Greek thought—the lack in mathematics of the notion of a variable. The paradoxes of Zeno had left a profound impression which caused the generic ideas of change and variability to be related to metaphysics. Geometric magnitudes were static and continuous; algebraic quantities were discrete constants. The symbols in the *Arithmetic* of Diophantus represent unknown numbers rather than variables in the sense of elementary algebra. Modern higher analysis may indeed be founded logically upon the Weierstrassian "static theory of the variable" which represents in a sense of partial return to the ancient Greek view; but the considerations leading to the original development of analytic geometry (as also of the calculus) were definitely phoronomic. During the later Scholastic period, and particularly during the fourteenth century, there arose at Paris and Oxford lively discussions on questions concerning rates of change, both uniform and nonuniform. The ideas of inertia and of acceleration, ubiquitously but erroneously ascribed today to Galileo, were developed at this time. In connection with this work Nicole Oresme made one of the earliest attempts to represent graphically the manner in which one quantity varied with another. For example, the velocity of a freely falling body, which was taken to be proportional to the time of fall, he plotted as a function of the time. Marking off points on a horizontal straight line as time units, the velocity corresponding to any time

was represented as an ordinate at this point, and the resulting graph was an oblique straight line. Oresme attempted to picture also certain functions which were "difformly difform" (that is, the rate of change was not constant), the graphs in these cases being necessarily curvilinear.

In certain quarters Oresme has been vigorously acclaimed as the inventor of analytic geometry. His work undoubtedly represents a very definite step in this direction, for he may be looked upon as the father of systematic graphical representation, but his view is seen to fall short of the conceptions of Descartes and Fermat. In a sense one may say that Oresme showed that a simple proportion can be represented by a straight line and that certain other types of variation are associated with characteristic diagrams. This may be regarded as an advance beyond the coordinate considerations of Apollonius in that the curve is here determined by the coordinate system and by the law of variation; the coordinate frame is not introduced as a device for studying a curve already given. Nevertheless, Oresme was prevented from taking full advantage of his novel idea by deficiencies in geometrical knowledge and algebraic technique. Unfortunately the mathematical interest of the age took the form of abortive speculations (resumed half a millennium later by Cantor) on infinity and the continuum, and the Greek classics were neglected. Consequently there is in the work of Oresme no systematic association of algebra and geometry in which an *equation* in two variables determines a specific curve and conversely.

The shortcomings in algebra which are apparent in the work of Menaechmus, Apollonius, and Oresme were not removed until the sixteenth century. The arithmetic of Diophantus (and, a fortiori, that of the Babylonians) had remained largely unknown to the Latin medieval world, but by the Arabs it had been translated, amplified, and transmitted to Europe. Although Leonardo of Pisa appreciated the significance of this work, it remained latent in Europe for almost three centuries before algebra was systematically developed by

EXAMPLES OF EARLY GRAPHS

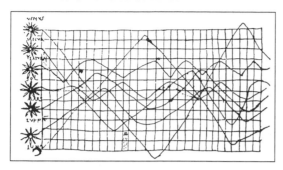

Illustration from a tenth-century manuscript depicting the positions of the "seven heavenly wanders," i.e., five planets, the sun, and the moon, in the night sky during the period of one month. The dependence on a coordinate system is obvious.

Sketches on a fourteenth-century manuscript on kinetics. The sketches relating time and distance were composed by the theologian–natural philosopher, Nicole Oresme, or one of his students.

Chuquet, Pacioli, Cardan, Stifel, and many others. This development happily coincided with a strong revival of interest in ancient geometry, including the *Conics* of Apollonius. Not unnaturally the gap between geometric and algebraic methods became less pronounced than it had been in the works of the classical Greek period. Tartaglia, Cardan, and Bombelli solved equations geometrically and, inversely, algebraic methods were widely advocated by Benedetti, Clavius, Viète, Stevin, Girard, Ghetaldi, Cataldi, Harriot, and Oughtred as a means of simplifying geometrical problems. Viète realized as well the advantage to be gained by Plato's analytic approach. Moreover, he contributed to algebra a significant point of view: the subject was to be looked upon not simply as numerical logistic but as the logic of magnitudes in general—*logistica speciosa*. These magnitudes he represented by letters, vowels for the unknown and consonants for those assumed known. It is to be remarked clearly, however, that these symbols denoted determinate quantities rather than variables. The geometric problems were reducible to equations in a single unknown, and the value of this unknown was determined by solving this equation. In fact, the algebra of the time was predominantly concerned with the solution of such equations. Diophantus had considered equations in several unknowns from the point of view of the theory of numbers; but these did not attract the algebraist of the sixteenth century, for they could not be "solved" in the ordinary sense. When a geometric problem led to an equation in two unknowns, it was abandoned as unsolvable. This is where the genius of Fermat and Descartes appeared. These men saw that lack of a unique solution did not in this case render the problem devoid of interest. They interpreted the one unknown as a variable horizontal line segment (abscissa) with a fixed initial end-point (origin), and at the other extremities of this they erected perpendicular line segments (ordinates) of lengths corresponding to the values determined for the second unknown in terms of the first. The extremities of these ordinates, infinite in number, formed a curve which, for a given equation and coordinate system, was uniquely determined. As Fermat stated it (*Oeuvres*, I, 91; III, 85): "whenever in a final equation two unknown quantities are found, we have a locus, the extremity of one of these [unknown magnitudes] describing a line, straight or curved."

That the crux of the matter lies in the interpretation of problems leading to a final equation in more than one unknown is quite clear also from the work of Descartes: Book I deals with determinate problems; i.e., those which lead to "as many such equations as there are supposed to be unknown lines." (*Geometry*, p. 6.) Book II, on the other hand, deals with plane problems on loci in which "this [final] condition can be expressed by a single equation in two unknown quantities." (*Geometry*, p. 34.) In suggesting the extension of this discovery to three dimensions he adds: "If two conditions for the determination of the point are lacking [i.e., if the final equation has three unknowns], the locus of the point is a surface." (*Geometry*, p. 80.) Fermat stated the principle still more sharply as follows: "There are certain problems which involve only one unknown, and which can be called determinate, to distinguish them from the problems of loci. There are certain others which involve two unknowns and which can never be reduced to a single one: these are the problems of loci. In the first problems we seek a unique point, in the latter a curve. But if the proposed problem involves three unknowns, one has to find, to satisfy the question, not only a point or a curve, but an entire surface." (*Oeuvres*, III, 1161.)

The object of this paper has not been to decide whether or not Descartes and Fermat invented analytic geometry. An answer to this problem depends on one's definition and sense of proportion; but *de gustibus non est disputandem*. The aim has been the simpler one of indicating just what it was that these men did discover in this connection. Neither one of them invented the use of coordinates or of the analytic method. Neither was first in applying algebra to geometry or in graphically

representing variables. Moreover, it had long been known that, for a given curve, certain distance relationships are determined which may be interpreted as equations of the curve with respect to coordinate systems. However, there appears to have been no conception before their time of the converse—the fact that in general an arbitrary *given equation* involving *two unknown quantities* can be regarded as determining *per se*, with respect to a coordinate system, a plane curve. This latter recognition, together with its fabrication into a formalized algorithmic procedure, constituted the decisive contribution of Fermat and Descartes. This is the sense in which these men may be regarded as the founders of analytic geometry, much as Newton and Leibniz are generally regarded as the inventors of the calculus through precise formulation of the mutually inverse nature of area and tangent problems in terms of a definitely regularized operational procedure. The publication of Descartes's *Géométrie* in 1637 preceded by 42 years that of Fermat's *Isagoge*, but both men were in independent possession of their methods well before this time—about 1619 for Descartes and 1629 for Fermat. As in the calculus the root idea was due to the two greatest figures of the second half of the seventeenth century, so likewise in analytic geometry the fundamental principle was recognized by the two greatest mathematicians of the first half of this same century.

REFERENCES

1. Apollonius of Perga, *Treatise on Conic Sections* (ed. by T. L. Heath, Cambridge, 1896). [Heath points out that the method of Apollonius does not differ essentially from that of modern analytic geometry except that geometrical operations take the place of algebraical calculations.]

2. Bosmans, Henri, "La première édition de la 'Clavis mathematica' d'Oughtred, son influence sur la 'Géométrie' de Descartes," *Annales de la Société Scientifique de Bruxelles*, XXXV (1910–1911), 24–78. [Oughtred as a link from Viète to Descartes.]

3. Chasles, Michel, *Aperçu historique sur l'oringine et le développement des methodes en géométrie* (Paris, 1875). [Strong claim for Descartes as sole inventor (pp. 94–95).]

4. Coolidge, J. L., "The origin of analytic geometry," *Osiris*, I (1936), 231–250. Appears also as a section in his *History of Geometrical Methods* (Oxford, 1940). [Defends thesis that analytic geometry was an invention of the Greeks, perhaps of Menaechmus.]

5. Descartes, René, *The Geometry of René Descartes* (transl. by D. E. Smith and M. L. Latham, Chicago and London, 1925). [Excellent English edition.]

6. Duhem, Pierre, *Études sur Léonard de Vinci* (3 vols., Paris, 1906–1913). [Fullest available account of the medieval precursors of Galileo and Descartes.]

7. Duhem, Pierre, "Oresme," *Catholic Encyclopedia*, XI (1911), 296–297. [Asserts that Oresme "forestalls Descartes in the invention of analytic geometry."]

8. Fermat, Pierre de, *Oeuvres* (ed. by Paul Tannery and Charles Henry, 4 vols. and supp., Paris, 1891–1922). [See vol. I for Latin of "Introduction to plane and solid loci," vol. III for French translation.]

9. Funkhouser, H. G., "Historical development of the graphical representation of statistical data," *Osiris*, III (1937), 269–404. [Includes reference to Oresme's work.]

10. Gelcich, E., "Eine Studie ueber die Entdeckung der analytischen Geometrie mit Berücksichtigung eines Werkes des Marino Ghetaldi Patrizier Ragusaer aus dem Jahre 1630," *Abhandlungen zur Geschichte der Mathematik*, IV (1882), 191–231. [Excellent critical account of relations between algebra and geometry at that time. Says Ghetaldi lacked principle of coordinates.]

11. Günther, Sigismund, "Le origini ed i gradi di sviluppo del principio delle coordinate," *Bullettino di Bibliografia e di Storia delle Scienze Matematiche e Fisiche*, X (1877), 363–406. [Shows that use of coordinates goes back to Greek times, and that work of Oresme and Kepler resembles analytic geometry in some respects.]

12. Karpinski, L. C., "Is there progress in mathematical discovery and did the Greeks have analytic geometry?", *Isis*, XXVII (1937), 46–52. [Rejects Coolidge's thesis that Greeks had analytic geometry. Emphasizes idea of progress and development of algebraic notations.]

13. Libri, Guillaume, *Histoire des sciences mathématiques in Italie, depuis la renaissance des lettres jusqu'a la fin du dix-septième siècle* (vols. III and IV, Paris, 1840–1841). [Emphasizes Benedetti (III, 124) and Cataldi (IV, 95).]

14. Loria, Gino, "Da Descartes e Fermat a Monge e Lagrange. Contributo alla storia della geometria analitica," *Reale Accademia dei Linceri*. Atti. *Memorie della classe di scienze fisiche, matematiche e naturali* (5), XIV (1923), 777–845. [Excellent general summary with emphasis on development after Descartes.]

15. Loria, Gino, "Descartes géomètre," *Etudes sur Descartes* (Paris, 1937), pp. 199–220. [Summary and analysis of the Géométrie.]

16. Loria, Gino, *Il passato e il presente delle principali teorie geometriche; storia e bibliografia* (4th ed., Padova, 1931). [Excellent and ample account of the development of geometry in modern times.]

17. Loria, Gino, "Pour une histoire de la géométrie analytique," *Verhandlungen des dritten internation-alen Mathematiker-Kongresses in Heidelberg vom 8. bis 13. August 1904* (Leipzig, 1905). [Critical analysis of contributions of Fermat and Descartes and of subsequent development.]

18. Loria, Gino, "Qu'est-ce que la géométric analytique?", *L'Enseignement Mathématique*, XIII (1923), 142–147. [Emphasizes Euler's *Introductio in analysin infinitorum* for use of formulas in solving geometrical problems.]

19. Milhaud, Gaston, *Descartes savant* (Paris, 1921). [Excellent analysis of the development of Descartes' scientific and mathematical thought.]

20. Morley, F. V., "Thomas Hariot," *Scientific Monthly*, XIV (1922), 60–66. [Claims analytic geometry for Harriot. This claim is no longer substantiated.]

20a. Müller, Felix, "Zur Literatur der analytischen Geometrie und Infinitesimalrechnung von Euler," *Jahresbericht der Mathematiker-Vereinigung*, XIII (1904), 247–253. [Emphasizes the use of analytic methods between Fermat and Euler.]

21. Ritter, F., "Première série de notes sur la logistique specieuse par François Viète," *Bullettino di Biblio-grafia e di Storia delle Scienze Matematiche e Fisiche*, I (1868), 245–276. [On Viète's combination of algebra and geometry.]

22. Saltykow, N., "'La géométrie' de Descartes. 300•anniversaire de géométrie analytique," *Bulletin des Sciences Mathématiques* (2), LXII (1938), 83–96, 110–123. [Emphasizes originality of Descartes in combining many elements previously known.]

23. Strong, E. W., *Procedures and Metaphysics. A Study in the Philosophy of Mathematical-Physical Science*

in the Sixteenth and Seventeenth Centuries (Berkeley, Calif., 1936). [A critical discussion of tendencies, some of which concern analytic geometry. Work of Cardan, Tartaglia, Benedetti, Stevin, Viète discussed.]

24. Tropfke, Johannes, *Geschichte der Elementarmathematik* (vol. VI, Berlin and Leipzig, 1924). [One of the fullest available histories of analytic geometry is included in pp. 92–169.]

25. Wieleitner, Heinrich, "Der 'Tractatus de latitudinibus formarum' des Oresme," *Bibliotheca Mathematica* (3), XIII (1912–1913), 115–145. [Oresme's graphical representation.]

26. Wieleitner, Heinrich, "Marino Ghetaldi und die Anfänge der Koordinatengeometrie," *Bibliotheca Mathematica* (3), XIII (1912–1913), 242–247. [Sees combination of algebra and geometry, but not analytic geometry, in Ghetaldi.]

27. Wieleitner, Heinrich, "Ueber den Funktionsbegriff und die graphische Darstellung bei Oresme," *Bibliotheca Mathematica* (3), XIV (1914), 193–243. [Wieleitner rejects Oresme as the founder of analytic geometry but sees a likelihood that he influenced Descartes. (See p. 241.)]

28. Zeuthen, H. G., *Geschichte der Mathematik im XVI und XVII. Jahrhundert* (German ed. by Raphael Meyer, Leipzig, 1903). [A brief summary of the geometry of Fermat and Descartes is given on pp. 192–233.]

See also the standard histories of mathematics by Archibald, Ball, Bell, Cajori, Cantor, Hankel, Heath, Kaestner, Marie, Montucla, Smith, Wieleitner, Zeuthen; also notes in *Encyclopédie des sciences mathématiques.*

A further contribution by Gino Loria, "Perfectionnements, évolution, métamorphoses du concepte de coordonnées. Contribution à l'histoire de la géométrie analytique," was scheduled for publication in volume VIII of *Osiris*, but the appearance of this volume has been held up by the German invasion of Belgium.

HISTORICAL EXHIBIT 7.1

Mathematical Considerations on the Trajectory of a Cannon Ball

Early cannons were quite primitive and their range limited but as the technology of warfare improved by the fifteenth century, they were hurling their projectiles beyond the view of their gunners. The questions of the range of a cannon shot became important. Early mathematical models for the trajectory of a cannon ball were mostly speculative. The first such model to be used was a right triangle with the path of the shot tracing out the hypotenuse until a maximum height is reached at which time the shot would fall vertically downwards completing a leg of the right triangle. In this instance, the length of

the horizontal leg supplied the range of the cannon shot. By the early sixteenth century, this model was modified to include a circular arc connecting the ascending and descending paths of a cannon ball.

Niccola Tartaglia investigated the paths of cannon shot and published his finding in *Nova Scientia* [The New Science], 1537. While Tartaglia related the properties of a shot to the angle of barrel elevation, he still employed a line-arc-line model for trajectory. Galileo Galilei took up the challenge of determining the geometrical path of a cannon ball trajectory. Through experimentation, he determined that if a body is projected horizontally, its descent in successive time intervals t_1, t_2, $t_3 \ldots t_k$ would be in the ratios given by the number sequence 1, 4, 9, 16, $\ldots d_k$. Galileo concluded that the path would be a parabola (i.e., $d = kt^2$). Evangelista Torricelli (1608–1647), a pupil of Galileo, refined these theories further to place the science of artillery on a firm mathematical basis.

465

72

The Young Pascal

HAROLD MAILE BACON

$\mathcal{V}$ERSATILITY is not inevitably the companion of genius. It is not altogether common to find a man who is at the same time a clever experimental physicist, a creative mathematician, an inventor with an eye to money-making, a gifted writer whose artistry places him among the foremost French stylists, and a religious philosopher of singular originality and ardor. Blaise Pascal was such a man. He could write an important treatise on the vacuum as well as produce those incomparable examples of controversial literature, the famous *Provincial Letters*. He invented the first adding machine of practical consequence and tried (in vain) to realize a profit from its sale. In the fragments of his projected apology for the Christian Religion, left unfinished by his death at the early age of thirty-nine, are to be found many evidences of his mathematical genius as well as of a remarkable piety and zeal. A clear and complete picture of his early life and education would not only be of rare interest, but it could not fail to contain many suggestions of value to the modern teacher. It is indeed a pity that such incomplete information is available. The story is soon told, but it is well worth the telling, and perhaps it holds some inspiration or lesson for our own times.

Blaise Pascal belonged to a family of provincial officials. Although it could boast a title of nobility granted by Louis XI in recognition of the faithful services of one Etienne Pascal, an important fiscal

officer in the King's entourage, it was of parliamentary rather than noble condition, and the Province of Auvergne provided places in its local magistracy and revenue offices for many of the younger members of the family. The Pascals were well-to-do, substantial people of considerable local importance, and had for generations audited the accounts of the province or presided over its courts.

Another Etienne Pascal was President of the *Cour des Aides*, a court having jurisdiction in all matters pertaining to indirect taxation, when his son, Blaise, was born at Clermont-Ferrand on June 27th, 1623. This strong-minded and talented magistrate had married in 1618 Antoinette Begon, the pious daughter of a family of the merchant class. A son born in 1619 lived only long enough to be baptized, but 1620 saw the arrival of a daughter, Gilberte, who was to become the biographer of her famous younger brother and sister, Jacqueline, born in 1625.

This was a period of scientific discovery and advance. The first half of the seventeenth century was to see the death of Francis Bacon, the work of Galileo, Kepler, Descartes, Roberval, and Fermat, the birth of Newton, Huygens, and Leibniz. We shall find Etienne Pascal and his son taking their parts in the work of the age. But in spite of the spirit of true scientific thought, there was still a wide-spread faith in astrology, "spells," witchcraft, signs, and omens. Even Kepler had to make astrological predictions to help to earn his living, and it was only by exerting considerable influence that he was able to save his mother from conviction of witchcraft. A curious story is told of the infant

Reprinted from *Mathematics Teacher* 30 (Apr., 1937): 180–85; with permission of the National Council of Teachers of Mathematics.

Blaise Pascal by his niece, Marguerite Perier,[1] which illustrates how even a man of his father's sagacity was forced to take account of the prevailing superstitions.

According to this story, when Blaise was between one and two years of age, he fell into a strange sort of languor attended by two surprising symptoms. The sight of water threw him into convulsions, and, although he enjoyed the caresses of his mother or of his father separately, their approach together sent him into transports of childish rage. This illness lasted for over a year during which time he grew so much worse that his parents began to despair of his life. Everybody said that this malady was the result of a "spell" cast over the child by a woman who had received charity from his mother. Although his father was skeptical, he seems to have sent for the women and to have wrung from her a confession of the sorcery. She was able to explain how the spell could be removed, but it would require the sacrifice of the life of another to whom the enchantment could be transferred. Fortunately an animal would do. The anxious father offered his horse. But, the witch remarked, the thing could be accomplished at much less sacrifice: a cat would suffice. So a cat was produced. When thrown out a window only six feet from the ground, it instantly died. Next morning a poultice was made from three leaves each of three different herbs gathered before sunrise by a girl less than seven years of age. Upon application of this poultice to his body the child fell into a coma from which he awoke—just at midnight—completely cured.

Unlikely as this incident appears, it may well symbolize the spirit of the times, while the illness itself was the forerunner of that ill health which persisted throughout so much of Pascal's life.

When Blaise was but five years old his mother died. The father, a man of about forty, thenceforth lavished his care and affection upon his three children. Two years later, in 1631, he sold his magistracy at Clermont, and moved with his family to Paris where he planned to devote his time to the education of his children, especially his son. He would permit no one to help or to interfere with his original plan of teaching the boy. Although the Jesuit schools were then flourishing and providing what was regarded as a well balanced and methodical course of instruction, Etienne Pascal would have none of it. He preferred to keep Blaise at home and to teach him in his own way. He failed to count upon his own insufficiency in many branches of knowledge. He knew the Law as a former official, could use Latin about as well as most educated men of the day, was acquainted with mathematics and physics, and was a fervent believer in experimental science. He made up for the gaps in his knowledge and for his lack of experience by a method of teaching which was of his own personal invention. His basic principle was that the child's lesson should always be entirely within his grasp, easily and completely understood. A glimpse of the whole subject under consideration was to suggest simple and easily discerned general laws and to arouse curiosity. From these general principles were to be deduced explanations of particular facts observed or questioned by the pupil. Etienne Pascal would make no attempt to force or strain an inquiring mind.

That the youthful Blaise had an inquiring mind—not easily restrained by his father's notion that the child should be "held well beyond his work"—is evident from the biography written by his sister Gilberte.[2] He wanted to know the reasons for everything, and if good ones were not given he would seek them for himself and not give up until he had a satisfactory explanation. For instance, we are told that when he was eleven years old he noticed that a china dish struck by a knife produced a loud sound which ceased when the hand touched the dish. He was not content until, after several experiments, he had discovered the reason. He then wrote, no doubt with parental encouragement, a little essay on sound.

There was an incessant inquiry into the origin and nature of things in general, into their properties and their uses. For example, what is grammar?

How does it happen that all languages are communicable from one country to another? What of the extraordinary effects of nature such as gunpowder exploding in a cannon? It was expected that these inquiries would last until Blaise was twelve years old when he was to be put to the study of Latin and Greek. He was also to acquaint himself with Spanish and Italian, and always by methods and according to rules imposed by his father. When he should have mastered languages, he was to take up mathematics. According to this plan he would have been about fifteen or sixteen years old when introduced to geometry, and should by that time have reached a maturity and orderliness of mind properly fitting him to appreciate this absorbing subject. The whole scheme was intended to accustom the child to seek out knowledge for himself while guided in broad paths of learning, to render account of his work to himself, and never to take preconceived notions and hypotheses for truths.

While this method of education suited the boy's nature admirably, while it developed his good qualities and nourished his originality, it overlooked many important matters. History was neglected. Every idea was a personal discovery of the pupil and appeared to him as his own exclusive property. Such a lack of historical perspective encourages a man to exaggerate the consequences of his own ideas and to disdain the work of others to whom he feels he owes nothing. Who can say that Pascal did not at times exhibit evidences of a lofty egotism? "He who reads the *Thoughts* of Pascal should give a thought to the president of the *Cour des Aides* of Clermont-Ferrand who taught his son so well while teaching him so badly."[3]

That the method had surprising effects upon the lad is clear if we can believe an often repeated story[4] told by his sister Gilberte. One day his father happened to enter the play room unobserved. What was his surprise to find the boy of twelve engaged in drawing figures upon the stone floor with a piece of charcoal. Although he had not been allowed to study Euclid, when questioned

the boy explained that he was trying to prove that the sum of the angles of a triangle is two right angles. His father then drew from him step by step the method by which he had used his "bars" (lines) and "rounds" (circles) to arrive at this thirty-second proposition of Euclid. Overcome with emotion, Etienne Pascal withdrew, and soon Blaise was given mathematical books to study. He had, no doubt, overheard in the conversation of his father's friends some references to geometry, and, not content to regard lessons in geometry as simply a promised reward for proficiency in Latin and Greek, had set out to investigate the subject for himself. This story may be largely the product of an adoring sister's imagination, and yet there must be at least some element of truth in it. She was older than her brother and was taking the place of a mother and a housewife. A young woman matured beyond her years by this experience, while perhaps possessing strong family partiality, must have seen clearly the remarkable ability of the boy. Whether in later years she dramatized this ability in some such imaginary incident, or whether the story is strictly accurate in all details, unlikely as this may seem, cannot be definitely said. However, there can be no doubt that young Blaise showed such proficiency in mathematics that his father, perhaps surprised and gratified at this result of his educational method, felt constrained to revise his plan to withhold mathematical studies until languages had been mastered.

Etienne Pascal had found his level in Parisian society. He was admitted to a circle of scientific men such as Roberval, Carcavi, Le Pailleur, Desargues, and gifted amateurs of high spirit. The central figure of the little group was Father Mersenne, a clever and agreeable Minimist friar who was in close touch with Galileo, Descartes, Torricelli, and other scientific men of the day. The evident genius of young Blaise prompted his father to introduce him to this circle which was the immediate ancestor of the French Academy of Sciences. Here, although he was but in his early teens, he took part in work and discussions. And here, too, he found a school of manners. The right to speak depended upon the amount of information. Older men who knew little of a subject had to be content to listen to younger men who knew much. Such experiences led him to attempt to judge everyone by purely intellectual standards. It did not take him many years to discover that the world does not employ such a standard of judgment, for he observes in the *Thoughts*, "Rank is a great advantage, for it gives to a man of eighteen years of age a degree of acceptance and respect which another man can scarcely obtain by merit at fifty. Here is a gain, then, of thirty years without difficulty."[5]

It must not be supposed that this gifted youth led a life of solemn study with no amusement. While his father was a former magistrate and a *savant*, he was not solely a meditative philosopher. He was an intimate friend of the versatile and volatile Le Pailleur who ran about France and England, changed his religion from Protestant to Catholic and back again, taught himself mathematics, sang songs—he is said to have sung eighty-eight songs during one gay evening—danced, and perpetrated practical jokes upon unsuspecting friends. The mere fact that a man of this character was an intimate friend of the family is enough to dispel the picture of an austere cultivation of the intellectual at the expense of all other powers. A rare cordiality and familiarity characterized the family life. President Pascal was a most affectionate father. Gilberte was a sagacious and practical young housekeeper who, when away from her household duties, took a quiet place. Jacqueline, though much younger, was much more noticed.

The Pascals had visited Auvergne only once, in 1636, since their establishment in Paris. Most of their interests were to be found in the capital where their fortune had been invested in municipal bonds. But unfortunately Richelieu's policy of opposition to the House of Hapsburg had involved a considerable drain on French finances, and in 1638 in order to meet certain exigencies,

the interest rate on the Parisian bonds was arbitrarily cut. This of course embarrassed and angered Etienne Pascal who associated himself with a group of protesting investors. Unhappily there was some show of violence, and two of the leaders were arrested and summarily removed to the Bastille. Pascal was able to escape and to take refuge in Auvergne. He risked return only when Jacqueline was dangerously ill with the smallpox. In the meantime the pathetic situation of the three children had struck the fancy of the fashionable world. In February, 1639, Jacqueline was encouraged to take part in the play, *L'Amour Tyrannique*, which was performed before the Cardinal. She so charmed Richelieu that his attention could be favorably attracted to the gifted son and exiled father, and Jacqueline was instructed to tell her father to return.

The evident ability and well-known integrity of President Pascal so appealed to the Cardinal that he resolved to send the former magistrate to Rouen to act as Intendant. Like most people, the French did not like to pay taxes. In 1639 the national treasury was seriously depleted, and in spite of the most determined efforts of the Minister, the provincial parlements failed to squeeze any more money out of the people. The local officials sympathized with their neighbors and did nothing. The Cardinal, resolving to put an end to such nonsense, created the office of Intendant whose occupant was to act as a sort of vice-regent in the matter of tax collections. His power was as great as his popularity was minute. Normandy was one of the most unruly of the provinces and needed a firm hand to bring it up to the mark, and the capable and courageous Etienne Pascal was the man for the job. Accordingly he arranged to remove his family to Rouen in the autumn of 1639.

Meanwhile Blaise had been thinking about mathematics and joining in the discussions of Mersenne's circle of friends. He had undoubtedly studied carefully many of the mathematical books in his father's library, and had also made himself familiar with the new geometrical methods of

Desargues. His father's educational method had encouraged original work on his part, and at the age of sixteen he was to be found engaged in the production of a new theory of the conic sections, a forerunner of which appeared in 1640 as *Essay pour les coniques*. In this little handbill was announced a lemma which has become famous as "Pascal's Theorem" and amounts to the statement that the intersections of the three pairs of opposite sides of a hexagon inscribed in a conic are collinear. The youthful investigator acknowledged his indebtedness to the methods of Desargues who had endeavored to reduce the properties of the conic sections to a small number of propositions. The improvement over the old method of considering each conic as a separate curve was in treating these curves simply as various perspectives of a circle. That Pascal had carried this new method much farther than had its inventor was at once acknowledged by Desargues who gave the "mystic hexagram" the name *La Pascale*. Great praise immediately fell to the lot of the sixteen year old mathematician. Everyone was enthusiastic—everyone but Descartes who grumbled to Mersenne, "It is just as I thought! I had not read half the little essay on Conics by M. Pascal's son before I saw that he had taken most of his ideas from M. des Argues, and this was confirmed soon afterwards by his own confession."[6] But there can be no doubt of Pascal's originality. The *Essay pour les coniques* was the announcement of a treatise which he was preparing. It concludes with the modest statement

> We have several other problems and theorems and several consequences deducible from the preceding, but the mistrust which I have of my slight experience and capacity does not permit me to advance more until my present effort has passed the examination of able men who may oblige me by looking at it. Afterwards, if they think it has sufficient merit to be continued, we shall endeavor to push our studies as far as God will give the power to conduct them.[7]

Mersenne reported that in this treatise Pascal had deduced the properties of the conic sections from his "mystic hexagram" in four hundred corollaries.

While the work was never published, Leibniz knew of it and gave a summary of it in his letter of August 30th, 1676 to Etienne Perier, Pascal's nephew.

This discovery of Pascal indicates clearly his view of mathematics. It rests upon the idea that the properties of a complex figure may be considered as a modification of those of a more simple figure. The effort of his father to have him always take a comprehensive and unifying view of everything helped to save him from the all too common error of the student who so concentrates upon some small details that he misses the simplicity and unity of his subject.

The removal to Rouen when Blaise was seventeen soon put a stop to these extended researches in pure theory. There his father plunged into the business of straightening out the chaotic finances of Normandy, and his son was expected to help him as much as possible. But his native genius, combined with the encouragement which his education had given to original work of all sorts, soon (1643) led to the idea of an "arithmetical machine." Before this adding machine could be finally perfected, more than fifty models had to be constructed, and it was not until some years later that a patent was secured which forbade the copying or construction of the machine by anyone else.

The achievements of Pascal in the fields of mathematics, physics, literature, and philosophy, and the story of his later life are too well known to be repeated here. But it is well to pause to consider the remarkable effect which the manner of his education must have had upon him. It fostered original investigation—even of simple things well known to the world of his day—above all else. We see its flowering, if not in the apocryphal "discovery" of geometry, certainly in the theorem of the hexagon, at an early age. Undoubtedly his genius would have asserted itself under any circumstances, but it was surely awakened and nourished by that schooling so well adapted to his needs. We should be fortunate indeed if today we could make available such an opportunity for every talented young man and woman.

NOTES

1. Marguerite Perier, "Mémoire de la vie de M. Pascal, écrit par mademoiselle Perier, sa nièce" in *Oeuvres de Pascal* ed. by Brunschvicg and Boutroux, Paris, 1908, 14 vols., vol. I, pp. 125–36.

2. Gilberte Perier, "La vie de Monsieur Paschal, escrite par Madame Perier, sa soeur, femme de Monsieur Perier, conseiller de la Cour des Aides de Clermont" in *Oeuvres de Pascal,* vol. I, pp. 50–114. This biography, written soon after his death, is the chief source of information concerning his early life.

3. Fortunat Strowski, *Pascal et son Temps, deuxième partie, l'histoire de Pascal,* Paris, 1907, p. 9.

4. It appears on pp. 53–56 in vol. I of *Oeuvres de Pascal.*

5. Blaise Pascal, *Oeuvres de Pascal,* vol. XIII, pp. 240–41.

6. René Descartes, Letter of April 1, 1640, to Mersenne in *Oeuvres de Descartes* ed. Charles Adam and Paul Tannery, Paris, 1899, 12 vols., vol. III, p. 47.

7. Blaise Pascal, *Oeuvres de Pascal,* vol. I, pp. 259–60.

73

Early Teachers of Mathematics

EDITED BY WILLIAM L. SCHAAF

ODAY, if one wishes to learn mathematics, there are ample opportunities in the schools, colleges, and universities throughout the land. One rarely seeks such instruction from private tutors, except under unusual circumstances. But in the seventeenth and eighteenth centuries it was otherwise. There were, for example, many outstanding mathematical lecturers in English universities, notably at Cambridge, as well as in universities scattered over the Continent. One recalls men like Isaac Barrow, a forceful and popular lecturer at Cambridge who voluntarily relinquished the Lucasian chair of mathematics to his pupil Isaac Newton, whose superior abilities he was quick to recognize and frank enough to acknowledge. Another well-known teacher at Cambridge was the blind lecturer in optics, Nicholas Saunderson. And there were many others, such as John Wallis of Oxford, Cavalieri of Bologna, and Gauss of Göttingen. Not infrequently mathematics was taught by well-known scholars under the patronage of royalty. Euler, the Swiss mathematician, was invited to Russia by the Empress, and later brought to the court of Frederick the Great. Descartes, the French soldier-philosopher-mathematician, was in later life called to the court of Sweden by Queen Christine, a headstrong young woman who insisted that he tutor her daily in philosophy at five o'clock in the morning.

More interesting, perhaps, was the rather common practice of teaching elementary mathematics

privately. These teachers were, for the most part, mathematicians of lesser note; in some cases, self-styled mathematicians. Many of them were authors of textbooks. The following is a reproduction of a typical eighteenth-century advertisement of a private teacher:

MATHEMATICKS, with their Use and Application to Natural Philosophy, Taught by the Author, at the *Hand and Pen* in *Little-Moorfields* : Where are taught Writing, Arithmetick, and Merchants Accompts ; Also Youth Boarded, by *Ralph Snow*.

Here is another example, equally naive:

At the *Hand* and *Pen* in *Barbican*, are Taught, *viz*. Writing in all Hands, Merchant's Accounts, Bookkeeping, Algebra, Geometry, Measuring, Surveying, Gauging, Mechanicks, Fortification, Gunery, Navigation, Dialling, and other Parts of Mathematicks; also the Use of the Globes and Maps, after a Natural, Easy, and Concise Method, without Burthen to the Memory.

One of the most colorful of these early self-styled teachers was a William Leybourn, who wrote more than twenty textbooks. One of these, published in 1694, called *Pleasure with Profit*, I have already alluded to in an earlier issue. Among the "arts and sciences mathematical professed and taught" by the author were "arithmetick, geometrie,

Reprinted from *Mathematics Teacher* 48 (May, 1955): 348–51; with permission of the National Council of Teachers of Mathematics.

astronomie, and, upon these foundations, the following superstructures: the use of geometrical instruments, trigonometria, navigation, and horologiographia, or Dialling." Note the absence of any allusion to algebra. His advertisement was frank enough:

> The Place of the Author's Residence is about Ten miles from *London* Westward, at a Place called Southal, in the Road between *Action* and *Uxbridge,* and Three miles from *Brainford:* Where he intends to *Read* the *Mathematicks,* and Instruct young Gentlemen, and others: And to Board upon reasonable Terms, all such as shall be pleased to make a more close Application to those Studies: Where such Boarders, and others, (during their time of Residence with him) shall have the Use of all *Books, Maps, Globes* and other *Mathematical Instruments,* as are necessary for their Instruction, till they provide themselves of such as they shall have occasion for afterwards.

A further sidelight on this vogue of private teaching is revealed by the observation of William Webster (1740), himself a teacher, and author of *Arithmetick in Epitome.* Said he:

> When a Man has tried all Shifts, and still failed, if he can but scratch out anything like a fair Character, tho' never so stiff and unnatural, and has got but *Arithmetick* enough in his Head to compute the number of Minutes in a Year, or the Inches in a Mile, he makes his last Recourse to a Garret, and, with the Painter's Help, sets up for a Teacher of *Writing* and *Arithmetick;* where, by the Bait of low Prices, he perhaps gathers a Number of Scholars.

It must not be inferred, however, that all such private teaching was of an elementary sort or of an inferior grade. An outstanding exception was the skillful and scholarly instruction of William Oughtred, a great seventeenth-century teacher of mathematics. Oughtred's chief concern was the ministry. But his love for mathematics (for which he had considerable aptitude) and his resourcefulness in making mathematical instruments consumed much of his time and energy. He gave private lessons, without remuneration, to pupils genuinely interested in mathematics. Among his noted pupils were John Wallis, whose book, *Arithmetic of Infinites,* was one of the forerunners of the calculus; Sir Christopher Wren, the famous architect who designed St. Paul's Cathedral in London; and Seth Ward, the celebrated astronomer. Oughtred was a staunch advocate of the use of symbols in mathematics. He used more than 150 symbols, most of which he invented. Only three of these have survived to modern times, namely, × for multiplication, :: for proportion, and ~ for difference; of these, the last two have almost completely fallen into disuse. Oughtred is also to be remembered for his invention of a rectilinear and a circular slide rule. His most famous book, the *Clavis Mathematicae,* or *Key to the Mathematicks,* was the outgrowth of a manuscript prepared for the instruction of one of his pupils. This popular book, which first appeared in 1631, passed through many editions. It was probably the most influential mathematical book published in England for nearly fifty years from the time that Napier published his treatise on logarithms.

While on the subject of early teachers of mathematics, a word about early arithmetic books may also be of interest. Perhaps the reader shares our penchant for delving into books of bygone days. If so, he will understand the pleasures of the antiquarian. Not the least rewarding is an examination of arithmetic textbooks of several hundred years ago. The contrast with modern schoolbooks is almost unbelievable. One wonders how anyone could actually learn arithmetic with some of these old books. Among some of the features of these early arithmetics which will impress the modern reader are the naive definitions, the quaint language, the strange and obsolete weights and measures once in common use, the utterly unrealistic and whimsical problems, the frequent use of verse, and the vanity of many of the authors, to say nothing of their craving for publicity.

The very titles themselves have preserved the flavor of the times in which they were written. We reproduce herewith a few of them for your perusal:

The Well Spryng of Science, which teacheth the perfecte woorke and practise of Arithmeticke—beautified with moste necessary rules and questions.

Humfrey Baker; London, 1568

The arte of vulgar arithmeticke, both in integers and fractions, devided into two bookes—whereunto is added a third book entituled *Musa Meratorum,* comprehending all rules used in the most necessarie and profitable trade of merchandise—Newly collected, digested, and in some part devised, by a wel willer to the Mathematicals.

Thomas Hylles; London, 1600

The Hand-Maid to Arithmetick refined: Shewing the variety and facility of working all Rules in whole Numbers and Fractions, after most pleasant and profitable waies. Abounding with Tables above 150. for Monies, Measures and Weights, tale and number of things here and in forraigne parts; verie usefull for all Gentlemen, Captaines, Gunners, Shopkeepers, Artificers, and Negotiators of all sorts: Rules for Commutation and Exchanges for Merchants and their Factors. . . .

Nicholas Hunt; London, 1633

A

Platform			Purchasers
Guide	}	for {	Builders
Mate			Measurers

William Leybourn; London, 1668

Cocker's Arithmetick: Being A plain and familiar Method, suitable to the meanest Capacity for the full understanding of that Incomparable Art, as it is now taught by the ablest School-Masters in City and Country. Composed by Edward Cocker, late Practitioner in the Arts of Writing, Arithmetick, and Engraving. Being that so long since promised to the World.

Edward Cocker; London, 1700

Arithmeticke Made Easie for the Use and Benefit of Trades-Men.

Joh. Ayres; London, 1714

An Essay to facilitate Vulgar Fractions; after a New Method, and to make Arithmetical Operations Very Concise.

Wm. Bridges; London, 1718

The Schoolmasters Assistant: being a Compendium of Arithmetic, both Practical and Theoretical—The Whole being delivered in the most familiar Way of *Question* and *Answer,* is recommended by several eminent *Mathematicians, Accompants,* and *Schoolmasters,* as necessary to be used in *Schools* by all Teachers, who would have their *Scholars* thoroughly understand, and make a quick Progress in ARITHMETIC. To which is prefixt, an ESSAY on the *Education* of YOUTH; humbly offer'd to the Consideration of PARENTS.

Thomas Dilworth; London, 1762

An Introduction to so much of the Arts and Sciences, More immediately concerned in an Excellent Education for Trade In its lower Scenes and more genteel Professions.

J. Randall; London, 1765

An Essay on Arithmetic—Briefly, Shewing, First, the Usefulness; Secondly, It's extensiveness; Thirdly, The Methods of it.

William Wallis; London, 1790 (?)

The Young Arithmetician's Assistant; Comprising a Select Compilation of useful Examples in the Minor Rules of Arithmetic, Simplified and Adapted to the capacities of Young Ladies and Gentlemen composing the Junior Classes in Academical Establishments.

H. Marlen & W. Smeeth; Canterbury, 1823

A New System of Mental Arithmetic, by the Acquirement of which all numerical questions may be promptly answered without recourse to pen or pencil, &.

Daniel Harrison; London, 1837

HISTORICAL EXHIBIT 7.2
Roberval's Quadrature of the Cycloid

As a circle rolls along a horizontal plane, a fixed point on its circumference traces out a curve known as a cycloid. This name was given to the curve by Galileo who was fascinated by its properties.

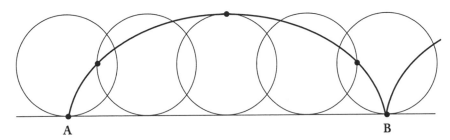

The first mathematician known to study the properties of this mechanical curve was the Frenchman Charles Bouvelles (ca. 1470–1553). Galileo admired the graceful arches of the cycloid and sought to incorporate them into bridge construction. He attempted to approximate the area contained within one arch by balancing a cycloidal template against circular templates of the generating circle. Through this experimentation he concluded that the cycloidal area was about three times that of the generating circle. Further investigation of this question was needed. In 1630, Père Marin Mersenne (1588–1648) suggested to his community of mathematical correspondences which included such notables as Descartes and Fermat that finding the quadrature of the cycloid would provide a good test for the newly-devised techniques of infinitesimals. This challenge was undertaken by Gilles Personne de Roberval (1602–1675), a profesor of mathematics at the Royal College. Roberval's method is outlined below:

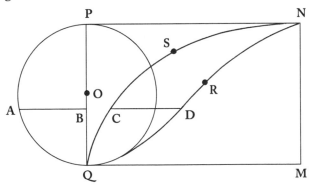

Let *QMNS* be half an arch of the cycloid generated by circle *O*. The area of *QMNP* equals twice the area of circle *O*. Construct infinitesimals (line segments) parallel to *QM* with lengths determined by the horizontal distance between diameter *PQ* and the circumfer-

(continued)

ence of a semi-circle, that is, *AB*. For each line in the semi-circle, a corres-ponding line of the same length is constructed from the cycloid as shown, that is, *CD*. The sequence of such infinitesimals with the cycloid, by their end points determine a curve *NRS* called the companion to the cycloid.

The area contained between the cycloid and its companion is composed of infinitesimals corresponding to those of the semi-circle *PAQ*, therefore this area is equal to that of the semi-circle or one-half the area of circle *O*. Now, by Cavalieri's Theorem [If two areas are everywhere of the same width, then the areas are equal], *NRQ* bisects *QMNP*; therefore, the area under *NRQ* equals the area of circle *O*. Adding together the area of these two regions, *QMNR* and *QRNS*, we find that the area under half the arch of the cycloid is equal to $^3/_2$ area of circle *O*. Thus it can be concluded that the area contained in an arch of a cycloid is three times the area of its generating circle.

74

Controversies on Mathematics between Wallis, Hobbes, and Barrow

FLORIAN CAJORI

*I*N THE WRITINGS of John Wallis, Thomas Hobbes, and Isaac Barrow, persons proficient in the dramatic art will find rich material for a mathematical tragedy or comedy. Hobbes, the aged philosopher superficially versed in all fields of human knowledge, presumes to show mere professors of mathematics that a philosopher from his wider outlook easily disposes of problems which had baffled mathematicians for ages—such problems as squaring the circle, trisection of an angle, and the duplication of the cube. Wallis, young and ambitious, a highly trained mathematician and gifted also in the power of satire and ridicule, demolishes Hobbes's geometric structure, leaving no stone unturned.[1] Barrow, the theologian and mathematician, assuming the position of a superior judge, takes under advisement the various mathematical views advanced by Hobbes and Wallis, and finds something to praise and much to criticize in the philosophy of both. Hobbes the septuagenarian emerges from the conflict battered and torn. Wallis shows little outward evidence of the struggle, but inwardly repents for the mass of ridicule which he had heaped upon the aged philosopher; Wallis refuses to have his controversial articles against Hobbes included in his collected works. Such in crude outline is the material available for the dramatist. We note a few of the topics discussed by these distinguished seventeenth-century thinkers.

Reprinted from *Mathematics Teacher* 22 (Mar., 1929): 146–51; with permission of the National Council of Teachers of Mathematics.

Foundations of Mathematics

John Wallis of Oxford founded mathematics on arithmetic; Barrow founded it on geometry. The very title of Wallis's most original book, the *Arithmetica infinitorum,* indicates his arithmetical procedure, even though the main subject of investigation in the book is the quadrature (i.e., determination of areas) of geometric figures. The work in which Wallis gives more explicit expression to his views on the philosophy of mathematics is entitled *Mathesis universalis: sive arithmeticum opus integrum.* With Wallis arithmetic was an abstract science to which geometry was subordinate. To Hobbes arithmetical and algebraic reasoning made little appeal, and he found fault with Wallis for excessive use of algebraic symbols. He insisted that Wallis "mistook the study of *symbols* for the study of *geometry,*"[2] and referred to the "scab of symbols" in Wallis's geometry of the conic sections. On the relative place of arithmetic and geometry, Barrow also disagreed with Wallis, but on other grounds. The viewpoints of both may be presented most compactly by reproducing Barrow's quotations from Wallis's *Mathesis universalis,* and also Barrow's remarks on these quotations. The following extract from Wallis shows that he considered arithmetic the more general of the two sciences: "simply because a line of two feet added to a line of two feet makes a line of four feet, it does not follow that two and two make four; on the contrary, the former follows from the latter."[3] To this Barrow answers:

"Whence comes it to pass that a line of two feet added to a line of two palms cannot make a line of four feet, four palms, or four of any denomination, if it be abstractedly . . . true that 2 + 2 makes 4? You will say, that these numbers are not applied to the same matter or measure; and so say I too; from whence I infer that 2 + 2 makes 4, not from the abstract reason of the numbers, but from the condition of the matter to which they are applied."[4] Barrow starts from the consideration that only concrete numbers can be added, and those must be of the same denomination. From these concrete relations Barrow infers that abstractly 2 and 2 make 4. Wallis, on the other hand, begins with the concept of the numbers 1, 2, 3, . . . disassociated from any particular objects or units of measure; with him 2 + 2 = 4 is an abstract relation which can be applied to special objects or units of the same kind. Wallis's point of view is brought out somewhat more sharply to the foreground by the following quotation from his book:[5] "The assertion concerning the equality of the number 5 with the numbers 2 and 3 taken together, is a general assertion, which is applicable not only to geometry but to all other things; for also 2 angels and 3 angels are 5 angels." Barrow on the other hand maintains that "no particular number taken separately and absolutely can signify anything certain. . . . I note that numbers of themselves can neither be added to nor subtracted from one another, so as in one case to compose a sum or in the other case leave a difference."[6] "Mathematical number has no real existence proper to itself." After showing how the square root of 3 years can be found geometrically with the aid of a circle 4 yards in diameter, Barrow says that numbers like $\sqrt{3}$ "cannot even in thought itself be abstracted from all magnitude." While he would not go so far as to expunge arithmetic out of the list of mathematical sciences, he affirmed "the whole of mathematics to be in some sort contained and circumscribed within the bounds of geometry." The verdict of time has gone against the views of Barrow. Arithmetical analysis has been found a much keener and more powerful procedure than geometry. For example, the impossibility of the quadrature of the circle, the trisection of an angle, and the duplication of a cube, by the use only of a pair of compasses and an unmarked ruler, has been established conclusively, not by reasoning involving geometric figures, but by reasoning involving arithmetical steps and the study of the properties of algebraic equations. Wallis anticipated some nineteenth-century ideas on arithmetization. The most rigorous modern expositions of the theory of limits and of the differential and integral calculus rest on the modern number-system with its careful definition of irrational number.

Geometrical Proportion

Hobbes declares that Euclid's definition of proportion is impossible of application, because it requires an infinite number of tests.[7] Four magnitudes a, b, c, and d are in proportion, according to Euclid,[8] if simultaneously one or the other of the relationships, $ma \gtreqless nb$, $mc \gtreqless nd$, holds for any positive integral values of m and n. That is, if for a pair of values of m and n, we have $ma > nb$, then we must have also $mc > nd$, otherwise a, b, c, and d could not be proportional. But we are not certain that a, b, c, and d are proportional, unless we know that these inequalities or equalities hold simultaneously for *all* possible pairs of value of m and n. Hobbes insists that multiplication by all numbers m and n is impossible, for there are infinitely many of them. Indeed, since the span of our lives is *finite*, it is not possible to carry out separately each of these trials. Does not Hobbes's criticism of Euclid appear altogether valid? The nature of this criticism is, by the way, the same as that levelled by certain modern mathematical philosophers against other workers in the field of transfinite numbers. Nevertheless, in spite of Hobbes's criticism, Euclid's definition has been widely admired. It has been declared to be equivalent to the "Dedekind cut," famous in the theory of irrational number.[9] But to return to the seventeenth century, Euclid's definition was staunchly defended by Barrow. He says truly:[10] "This simultaneous Defect, Excess or Equality may in some cases happen to quantities not proportional, but it happens to proportionals alone universally." Barrow proceeds to explain, by example, how a single test may suffice, for the same reason that proving that one ray of light travelling parallel to the principal axis of a parabolic mirror is reflected through the focus, proves that all such rays are reflected through the focus. The reduction of an infinite number of proofs to a single proof results from physical and geometric properties applicable to all rays. Barrow refers to Euclid's proof that two triangles having the same altitude have their areas proportional to their bases, even when the bases are incommensurable.[11] Barrow is too verbose for quotation, but he shows that the infinity of trials are in this instance reduced to one trial because of the theorem previously established in Euclid that, in two triangles, having the same height, the area of the first triangle is equal to or greater than the area of the second, according as the base of the first is equal to or greater than the base of the second.

Hobbes's Claim that Algebra and Geometry May Yield Contradictory Results

If a is a number, the interpretations of a, a^2, a^3, as "line," "square," "cube," respectively, caused Hobbes much trouble. So did $\sqrt{a}$ as the side of a square and $\sqrt[3]{a}$ as the side of a cube. He seriously objected to $\sqrt{8} = (\sqrt{2})^3$, for "here the root of 8 is put for the cube of the root of 2. Can a line be equal to a cube?"[12] Considering $9\sqrt{2} = \sqrt{162}$ he says, "How does 9 roots of 2 make the root of 162?", for 9 multiplied into $\sqrt{2}$ gives a rectangle, but $\sqrt{162}$ is a line. "I see the calculation in numbers is right, though false in lines. The reason whereof can be no other than some difference between multiplying numbers into lines or planes, and multiplying lines into the same lines or planes." He was very realistic (should say materialistic?) in assuming a

point and a line as having some thickness; "every one line has some latitude." As Wallis expressed it:[13] "Since it is, with Hobs, not conceivable for anything *to be*, which is not *Body*, it must be as impossible, for a Point to be if it be not *Great*." Hobbes's assumption established a cleavage between arithmetical and geometrical processes. He had given a construction effecting, as he thought, the duplication of the cube.[14] A French critic pointed out Hobbes's error by assuming suitable arithmetical values and proceeding arithmetically. There arose a discrepancy equal to the difference between $\sqrt{1682}$ and $\sqrt{1681}$. Hobbes claims that his "demonstration is not confuted; for the point Y will have latitude enough to take in that little difference which is between the root of 1681 and the root of 1682." His geometry, he tells us, is not refuted by any but mathematicians, whose judgment in this case is not to be credited. He felt that he had "wrested out of the hands of [his] antagonists this weapon of algebra, so as they can never make use of it again." Barrow did not go to such extremes: "Be it far from me to take away or seclude a science so excellent and profitable as that of numbers from the Mathematics. I will rather restore it into its lawful place, as being removed out of its proper seat, and ingraff and unite it again into its native geometry, the stock from whence it has been plucked (p. 29)." Hobbes thought he saw an impassable chasm between algebra and geometry. He had lost sight of the advantages resulting from the abstraction made by the Greeks which considers points and lines as having no thickness. In 1904 Horace Lamb, addressing Section A of the British Association, said: "If any scientific invention can claim pre-eminence over all others, I should be inclined myself to erect a monument to the unknown inventor of the mathematical point, as the supreme type of the process of abstraction which has been a necessary condition of scientific work from the very beginning."

The reader will find much in Wallis, Hobbes, and Barrow, touching the philosophy of mathematics, which we omit for lack of space, such as the definition of parallel lines, the recognition of irrationals as numbers, the definition of an angle, the angle of contact, the possibility of applying the method of superposition to solid bodies, change of magnitude with change of place, the existence of indivisible lines, whether magnitude be peculiar to bodies or belong to spirits also, the cycloid, and infinity.

NOTES

1. A good account of the controversial writings between Wallis and Hobbes is found in the articles "Hobbes" and "Wallis" in the eleventh edition of the Encyclopaedia Britannica.

2. Thomas Hobbes, "Six Lessons to the Professors of the Mathematics," in *The English Works of Thomas Hobbes,* edited by W. Molesworth, Vol. VII, London, 1845, p. 187.

3. John Wallis, *Mathesis universalis,* Oxford, 1657, p. 69.

4. Isaac Barrow, Mathematical Lectures, transl. from the Latin by John Kirkby, London, 1734, p. 37.

5. J. Wallis, *Mathesis universalis,* 1657, pp. 73, 226.

6. I. Barrow, *op. cit.,* pp. 29, 35, 41, 45.

7. Th. Hobbes, *op. cit.,* Vol. VII, p. 243.

8. Euclid's *Elements,* Book V, Definition 5.

9. J. W. Young, *Fundamental Concepts of Algebra and Geometry,* New York, 1911, pp. 103–106.

10. Barrow, *op. cit.,* p. 393.

11. Euclid's *Elements,* Book VI, Prop. I.

12. Th. Hobbes, *op. cit.,* Vol. VII, p. 66.

13. Hobbius, *Heauton-timorumenos, or A Consideration of Mr. Hobbes his Dialogues,* by John Wallis, Oxford, 1662, p. 31.

14. Th. Hobbes, *op. cit.,* Vol. VII, p. 60.

75

John Wallis and Complex Numbers

D. A. KEARNS

Fiction Is a Form in Search of an Interpretation[1]

In the long history of the development of the number concept, the square roots of negative numbers had occurred in the solution of quadratic and, later, of cubic equations. But they puzzled mathematicians and were considered useless and merely symbols devoid of any significance. The present-day sophisticated technique of neglecting meaning and developing an arithmetic of symbols is largely a device to systematize what has already been given meaning and proven useful. In the seventeenth century, however, forms such as $\sqrt{-1}$ were ignored because they stood for no concept.

Perhaps the first to attempt the task of providing specific realizations of the complex numbers was John Wallis (1612–1703), a contemporary of Newton and one of the geniuses of the early renaissance of mathematics. He pointed out that negative numbers "as to bare Algebraick Notation . . . import a Quantity less than nothing,"[2] but that when interpreted as directed distances along a line, they become useful and can be given meaning. He argued, similarly, that complex numbers, in themselves meaningless in terms of quantity, can also be given significance and use in a geometric context.

Reprinted from *Mathematics Teacher* 51 (May, 1958): 373–74; with permission of the National Council of Teachers of Mathematics.

Wallis had an ingenious algebraic interpretation of a pure imaginary number as the mean proportional between a positive and a negative number, but it is only by considering his geometric realizations of this idea that it becomes convincing. The following is a brief discussion of these "geometric effections."

To illustrate the above interpretation, Wallis considers a circle with diameter AC. (See Fig. 1.) Segments on the diametral line measured in the direction AC are positive, and those measured in the opposite direction are negative. If we erect, at a point B between A and C, a perpendicular which cuts the circle in point P, then $(BP)^2 = (AB)(BC)$,

FIGURE I

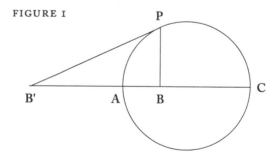

and BP is the mean proportional between AB and BC. Letting $AB = b$, $BC = c$, we have $BP = \sqrt{+bc}$.

On the other hand, let the tangent to the circle at P cut CA produced in B'. Then, as before, $(B'P)^2 = (AB')(B'C)$, and the tangent $B'P$ is the mean proportional between AB' and $B'C$. Letting $B'C = c$ and $AB' = -b$, a negative number, we see that $B'P$ corresponds to $\sqrt{-bc}$.

Wallis's originality is exhibited in another and more exciting illustration. Suppose we consider the problem of constructing a triangle when two sides and an angle, not included between the two sides, are given—the so-called ambiguous case. Let AP, BP, and $\angle PA\,\alpha$ be the given elements.

FIGURE 2

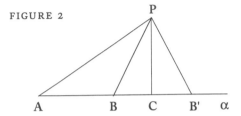

Then the altitude PC (see Fig. 2) is determined, and the construction of a triangle upon the base $A\alpha$ is possible if $PC \leq PB$ but impossible if $PC > PB$. If PC is actually less than PB, two triangles, $\triangle APB$ and $\triangle APB'$, are found which have the given elements as components. Furthermore, the following three relations hold:

$$AB = \sqrt{(AP)^2 - (PC)^2} \qquad (1)$$
$$-\sqrt{(PB)^2 - (PC)^2},$$
$$AB' = \sqrt{(AP)^2 - (PC)^2} \qquad (2)$$
$$+\sqrt{(PB)^2 - (PC)^2},$$
$$AB + AB' = 2(AC). \qquad (3)$$

Now, for the case in which $PC > PB$, $\sqrt{(PB)^2 - (PC)^2}$ is pure imaginary. But, Wallis points out, two points B and B' are still determined by the given sides and angle, *these points no longer lying on $A\alpha$, but lying elsewhere in the plane.*

On PC as a diameter, construct a circle tangent to $A\alpha$, and then with PB (which is given) as a radius, construct points B and B' on this circle (Fig. 3). Triangles PBC and $PB'C$ now have their right angles at B and B' rather than at C. Therefore BC and $B'C$ can be thought of as tangents to a circle with center at P and radius PB, and thus as representing the *pure imaginary numbers* which are given algebraically by the second radical in the

right members of equations (1) and (2). Triangles APB and APB' do not have $\triangle PAC$ as an element, but the line PC determined by $\triangle PAC$ is essential to their construction. AB and AB' can be used to represent the complex numbers given by the right members of equations (1) and (2) and are, in a sense, the vector difference and vector sum of AC and CB and of AC and CB' respectively. Moreover, if we add (1) and (2) in a formal manner, we obtain (3); this can be interpreted as the addition of the segments $A\beta$ and $A\beta'$ obtained by projecting AB and AB' on $A\alpha$. In the case where $PC < PB$, AB and AB' are their own projections on $A\alpha$ and this same interpretation is therefore shown to be valid.

FIGURE 3

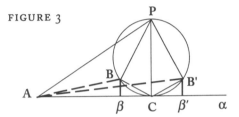

Thus Wallis generalizes the meaning of equations (1), (2), and (3) so that operations with them hold in both cases. In so doing, he makes it possible to realize the existence of complex numbers in a geometric way.

Then too, these constructions give the solutions of the quadratic equation $x^2 - bx + a = 0$, since if we let (in either Fig. 2 or Fig. 3) $x = AB$ (or AB'), $AC = PB = PB' = b/2$, and $PC = \sqrt{a}$ and substitute these in equations (1) and (2), we obtain

$$x = b/2 \pm \sqrt{(b/2)^2 - a},$$

the quadratic formula applied to the equation being considered.

NOTES

1. T. Dantzig, *Number, the Language of Science* (New York: The Macmillan company, 1939), p. 205.

2. J. Wallis, *Algebra* as given in D. E. Smith, *Source Book in Mathematics* (New York: McGraw-Hill Book Company, Inc., 1929), p. 46.

76

Isaac Newton:
Man, Myth, and Mathematics

V. FREDERICK RICKEY

THREE hundred years ago, in 1687, the most famous scientific work of all time, the *Philosophiae Naturalis Principia Mathematica* of Isaac Newton, was published. Fifty years earlier, in 1637, a work which had considerable influence on Newton, the *Discours de la Méthode*, with its famous appendix, *La Géométrie*, was published by René Descartes. It is fitting that we celebrate these anniversaries by sketching the lives and outlining the works of Newton and Descartes.

In the past several decades, historians of science have arranged the chaotic bulk of Newton manuscripts into a coherent whole and presented it to us in numerous high quality books and papers. Foremost among these historians is Derek T. Whiteside, of Cambridge, whose eight magnificent volumes overflowing with erudite commentary have brought Newton to life again.

> By unanimous agreement, the *Mathematical Papers* [of Isaac Newton] is the premier edition of scientific papers. It establishes a new criterion of excellence. Every further edition of scientific papers must now measure itself by its standard. [26, p. 87]

Other purposes of this article are to dispel some myths about Newton—for much of what we previously "knew" about him is myth—and to encourage the reader to look inside these volumes

Reprinted from *College Mathematics Journal* 60 (Nov., 1987): 362–89; with permission of the Mathematical Association of America.

and to read Newton's own words, for that is the only way to appreciate the majesty of his intellect.

Newton's Education and Public Life

Isaac Newton was born prematurely on Christmas Day 1642 (O.S.), the "same" year Galileo (1564–1642) died, in the family manor house at Woolsthorpe, some 90 km NNW of Cambridge. His illiterate father—a "wild, extravagant, and weak man"—had died the previous October. His barely literate mother, Hanna, married the Reverend Barnabas Smith three years later, leaving Newton to be raised by his aged grandmother Ayscough.

Newton attended local schools and then, at age 12, traveled 11 km north to the town of Grantham, where he lived with the local apothecary and his books while attending grammar school. The town library had two or three hundred books, some 85 of which are still chained to the walls. Of course he studied Latin, also some Greek and Hebrew. Four years later, in 1658, he returned home to help his now twice-widowed mother manage the farm. Recognizing that Newton was an absent-minded farmer, his uncle William Ayscough (M.A. Cambridge, 1637) and former Grantham schoolmaster, Henry Stokes, persuaded his mother to send him back to Grantham to prepare for Cambridge. Judging by a mathematical copybook in use at Grantham in the 1650s, Stokes was a most unusual schoolmaster. The copybook contained arithmetic through the extraction of cube roots, surveying, elementary mensuration, plane trigonometry, and elaborate geometric constructions, including the Archimedean bounds for π. This went far beyond anything taught in the universities of the period; consequently, contrary to tradition, Newton had a superior knowledge of mathematics before he went to Cambridge [33, pp. 110–111; 34, p. 101; updating 20, I, p. 3].

In 1661, eighteen-year-old Newton matriculated at Trinity College, the foremost college at Cambridge, as a subsizar (someone who earned his way by performing simple domestic services). This position reflected his wealthy mother's reluctance to send him to the university. At that time, Cambridge was little more than a degree mill. Lectures were seldom given. Fellows tutored primarily to augment their income. Although Newton did not finish any of the books from the established curriculum, which consisted mostly of Aristotelian philosophy, he did learn the patterns of rigorous thought from Aristotle's sophisticated philosophical system. A chance encounter with astrology in 1663 led him to the more enlightened "brisk part of the University" that was interested in the work of Descartes [28, p. 90]. The laxity of the university allowed him to spend the last year and a half of his undergraduate studies in the pursuit of mathematics. In 1665, Newton received his B.A. "largely because the university no longer believed in its own curriculum with enough conviction to enforce it." [28, p. 141].

In the summer of 1665, virtually everyone left the university because of the bubonic plague. The next March the university invited its students and Fellows to return for there had been no deaths in six weeks, but by June it was clear that the plague had not left, so the students who had returned left again. The university was able to resume again in the spring of 1667. Newton had left by August 1665 for Woolsthorpe. He returned on 20 March 1666, probably left again in June, but not until he had written his famous May 1666 tract on the calculus. He did not return to Cambridge until late April 1667, having revised the May tract into the October 1666 tract while back on the farm. "For whatever it is worth, the papers do not indicate that anything special happened at Woolsthorpe." [27, p. 116]. Much has been written about these plague years as the *anni mirabiles* of Newton, but the record clearly shows that he wrote the bulk of his mathematical manuscripts on the calculus while he was at Cambridge.

Myth: At the Woolsthorpe farm, during the plague years, Newton invented the calculus so that he could apply it to celestial mechanics.

The primary source for the myth [27, p. 110] of Newton's miracle years is this 1718 (unsent?) letter from Newton to Pierre DesMaizeaux:

> In the beginning of the year 1665 I found the Method of approximating series & the Rule for reducing any dignity [= power] of any Binomial into such a series. The same year in May I found the method of Tangents . . . , & in November had the direct method of fluxions & the next year in January had the Theory of Colours & in May following I had entrance into y^c inverse method of fluxions. And the same year I began to think of gravity extending to y^c orb of the Moon & (having found out how to estimate the force with wch globe revolving within a sphere presses the surface of the sphere) from Keplers rule . . . I deduced that the forces wch keep the Planets in their Orbs must [be] reciprocally as the squares of their distances from the centers about wch they revolve: & thereby compared the force requisite to keep the Moon in her Orb with the force of gravity at the surface of the earth, & found them answer pretty nearly. All this was in the two plague years of 1665 & 1666. For in those days I was in the prime of my age for invention & minded Mathematicks & Philosophy [= Science] more then at any time since. [27, p. 109]

LUCASIAN PROFESSOR

At Trinity College, Newton became a Minor Fellow in 1667 and a Major Fellow in 1668. On 29 October 1669, at the age of 26, Newton became the second Lucasian Professor of Mathematics at Cambridge, succeeding Isaac Barrow (1630–1677). This post gave him security, intellectual independence, and a good salary. According to the Lucasian statutes, Newton was to lecture once a week during each of the three terms and to deposit ten of the lectures in the library. Even though this position had been designed by its founder Henry Lucas as a teaching post, not a research position [20, V, xiv], Barrow had already turned the position into a sinecure and Newton did not work much harder at the teaching aspects of the post. He deposited 3–10 lectures per year for the first seventeen years as Lucasian Professor, and none thereafter.

As a teacher, Newton left no mark whatsoever. Years later, when he was duly famous, one would expect that many people would have claimed to have attended his lectures, yet we know of only three. Perhaps the situation is best summed up by Newton's amanuensis (a human wordprocessor), the unrelated Humphrey Newton:

> He seldom left his chamber except at term time, when he read in the schools as being Lucasianus Professor, where so few went to hear him, and fewer that understood him, that ofttimes he did in a manner, for want of hearers, read to the walls. [6, X, 44]

London and Beyond

In 1696, Newton accepted the post of Warden of the Mint (moving to London in March or April of 1696) and four years later became Master. In 1701, Newton resigned the Lucasian professorship. In 1703, he was elected President of the Royal Society, which he ruled with an iron hand until his death. In 1705, Newton was knighted by Queen Anne—not for his scientific advances, but for the service he had rendered the Crown by running (unsuccessfully) for Parliament in 1705 [28, p. 625]. For the rest of his life, Newton looked after the Mint and the Royal Society, twice revised his *Principia* (1713 and 1726), engaged in the infamous priority dispute with Leibniz, and toiled on secret research in religion and church history. His creative scientific life essentially ended when he left Cambridge.

Newton died 20 March 1727, at the age of 84, having been ill with gout and inflamed lungs

for some time. He was buried in Westminister Abbey.

Newton's Nachlass

At the time of his death Newton was wealthy. Income from the Lucasian Chair and farm rents brought £250 per year, sufficient for a handsome living for a bachelor don. When he became Master of the Mint, his salary jumped to £600 and he also received the perquisite of a commission on the amount of coinage. This amounted to some £1500 pounds per year, thus bringing his income to over £2000 pounds per year, a very substantial figure at that time. On his death his estate was valued at £30,000.

Newton left his library of some two thousand volumes to his nieces and nephews. The books were quickly sold to the Warden of Fleet Prison for £300 for his son Charles Huggins who was a cleric near Oxford. On Huggins's death in 1750 they were sold to his successor, James Musgrave, for £400. They remained in the Musgrave family until 1920, when some of them were sold at auction as part of a "Library of miscellaneous literature," fetching only £170. Although the family

FIGURE I

Isaac Newton

didn't know what they had sold, the book dealers knew what they had bought. Newton's annotated copy of Barrow's *Euclid,* which sold for five shillings, was soon in a bookseller's catalogue for £500. In 1927, the remaining 858 volumes were offered for £30,000 but remained unsold until 1943 when they were purchased for £5,500 and donated to the Wren Library at Trinity College. Of the thousand or so that were dispersed in 1920, some still show up unrecognized in bookshops. As recently as 1975, one was purchased in a Cambridge bookshop for £4. The books are easily identified by Newton's peculiar method of dog-earing by folding a page down to point to the precise word that interested him.

From Newton's library, 1736 books have now been located. Since his was a working library, a subject classification of the nonduplicates provides some information about Newton's interests. (For additional details, see [10, p. 59], from which the table below is condensed.) Newton also had access to the library of Barrow until Barrow's death in 1677, and to the Cambridge libraries until he moved to London in 1696.

Whiteside has tracked down every available scrap of material on Newton's mathematics and published it in *The Mathematical Papers of Isaac Newton* [20]. To really appreciate Newton's mathematical genius, one must grapple with his mathematics as he wrote it. The best place to gain an overview for this project is in Whiteside's wonderful introductions to these volumes and to the various papers in them. They have been used extensively in preparing this paper.

This biographical sketch has been intentionally kept short. For further details about Newton and his work, see the article by I. B. Cohen in the *Dictionary of Scientific Biography* (DSB) [6, X, pp. 42–103]. This is the single most authoritative reference work about the lives and contributions of deceased scientists. To avoid frequent references to it, we give dates after the first occurrence of an individual's name if the DSB contains an article about him. Two excellent biographies of Newton are Westfall's full scientific biography, *Never at Rest* [28], and Manuel's psychobiography *A Portrait of Isaac Newton* [15], some conclusions of which must be taken with care. For mathematical details, consult the many papers of Whiteside, only a few of which are cited here.

Newton's Mathematical Readings

The year 1664 was a crucial period in Newton's development as a mathematician and scientist, for it was then that he began to extend his readings beyond the traditional Aristotelian texts of the moribund curriculum to the new Cartesian ideas. (For details of Newton's nonmathematical readings, see McGuire [16].) According to Abraham DeMoivre (1667–1754), the expatriate French intimate of Newton during Newton's last years, the immediate impulse for Newton taking up mathematics was:

> In 63 [Newton] being at Sturbridge [international trade] fair bought a book of Astrology, out of a curiosity to see what there was in it. Read in it till he came to a figure of the heavens which he could not understand for want of being acquainted with Trigonometry.
>
> Bought a book of Trigonometry, but was not able to understand the Demonstrations.
>
> Got Euclid to fit himself for understanding the ground of Trigonometry.
>
> Read only the titles of the propositions, which

Subject	Mathematics	Physics and Astronomy	Alchemy	Theology
Number of titles (%)	126 (7.2)	85 (4.9)	169 (9.6)	477 (27.2)

Subject	History	Other Science	Other
Number of titles (%)	143 (8.2)	158 (9.0)	594 (33.9)

he found so easy to understand that he wondered how any body would amuse themselves to write any demonstrations of them. Began to change his mind when he read that Parallelograms upon the same base & between the same parallels are equal, & that other proposition that in a right angled Triangle the square of the Hypothenuse is equal to the squares of the two other sides.

Began again to read Euclid with more attention than he had done before & went through it.

Read Oughtreds [Clavis] which he understood tho not entirely, he having some difficulties about what the Author called Scala secundi & tertii gradus, relating to the solution of quadratick [&] Cubick Equations. Took Descartes's Geometry in hand, tho he had been told it would be very difficult, read some ten pages in it, then stopt, began again, went a little farther than the first time, stopt again, went back again to the beginning, read on till by degrees he made himself master of the whole, to that degree that he understood Descartes's Geometry better than he had done Euclid.

Read Euclid again & then Descartes's Geometry for a second time. Read next Dr Wallis's Arithmetica Infinitorum, & on the occasion of a certain interpolation for the quadrature of the circle, found that admirable Theorem for raising a Binomial to a power given. But before that time, a little after reading Descartes Geometry, wrote many things concerning the vertices Axes [&] diameters of curves, which afterwards gave rise to that excellent tract de Curvis secundi generis.

In 65 & 66 began to find the method of Fluxions, and writt several curious problems relating to that method bearing that date which were seen by me above 25 years ago. [20, I, pp. 5–6]

These words of DeMoivre, which agree with the report of Conduitt [20, I, pp. 15–19], certainly have an air of authenticity to them, and we know, based on extant manuscripts, that they are substantially correct (modulo Stokes's copybook). In the years 1664–1665, Newton made detailed notes on the following contemporary high level books, which influenced him at the very beginning of his mathematical studies.

- Barrow's *Euclidis Elementorum* (1655)
- Oughtred's *Clavis Mathematicae* (1631) in the 1652 edition

- *Geometria, à Renato des Cartes,* 1659–1661 edition of Schooten
- Schooten's *Exercitationum Mathematicarum Libri Quinque* (1657)
- Viète's *Opera Mathematica,* 1646 edition of Schooten
- Wallis's *Arithmetica Infinitorum* (1655)
- Wallis's *Tractatus Duo* (1659).

Let us look carefully at each of them to see what Newton learned.

EUCLID (FL. CA. 295 B.C.)

As DeMoivre indicated, Newton read Euclid as a student, although he did not develop any deep knowledge of the work then. Recall the story [28, p. 102] that Barrow examined Newton on Euclid and found him wanting. Newton was mainly influenced by books II (geometrical algebra), V (proportion), VII (number theory), and X (irrationals). The primary thing that he learned from Euclid was the traditional forms of mathematical proof [20, I, p. 12].

WILLIAM OUGHTRED (1575–1660)

At age fifteen, Oughtred went to Cambridge where he studied mathematics diligently on his own, for there was then hardly anyone there to teach him. He graduated B.A. in 1596 and M.A. in 1600. In 1603, he became a (pitiful) preacher and soon settled in as rector at Albury where he remained until his death.

It was as a teacher that he was renowned. He taught privately and for free. People came from the continent to talk to him, so wide was his reputation in mathematics. To instruct a young Earl, Oughtred wrote a little book of 88 pages that contained the essentials of arithmetic and algebra. *Clavis Mathematicae* (*Key to Mathematics*) published in 1631, was "a guide for mountain-climbers, and woe unto him who lacked nerve." [2, p. 29]. The style was obscure, the rules so involved they were difficult to comprehend. Oughtred carried symbolism to excess, a habit acquired by his most famous pupil, John Wallis. Nonetheless, *Clavis* established Oughtred as a capable mathematician and exerted a considerable effect in England, for it was a widely studied book in higher mathematics [32, p. 73].

Oughtred's Clavis, in the 1652 edition, was one of the first mathematical books that Newton read. From it he learned a very important lesson: Oughtred taught that *algebra was a tool for discovery* that did not need to be backed up by geometry [13, p. 408]. Newton held Oughtred in high regard, describing him as "a Man whose judgment (if any man's) may be safely relied upon." [19, III, p. 364]

RENÉ DESCARTES (1596–1650)

René du Perron Descartes was born 31 March 1596 in La Haye (now La Haye-Descartes), France, a small town 250 km SSW of Paris. At the age of eight, he enrolled in the new Jesuit *collège* at La Flèche. There Descartes received a modern education in mathematics and physics—including the recent telescopic discoveries of Galileo—as well as more traditional schooling in the humanities, philosophy, and the classics. It was there, because of his then delicate health, that he developed the habit of lying abed in the morning in contemplation. Descartes retained an admiration for his teachers at La Flèche but later claimed that he found little of substance in the course of instruction and that only mathematics had given him any certain knowledge.

Descartes graduated in law from the University of Poitier in 1616, at age 20, but never practiced law as his father wished. By this time, his health improved and he enjoyed moderately good health for the rest of his life. Because he decided that he could not believe in what he had learned at school, he began a ten year period of wandering about Europe, spending part of the time as a gentleman soldier. It was during this period that Descartes had his first ideas about the "marvelous science" that was to become analytic geometry.

Although we have little detail about this period of his life, we do know that he hoped to learn

Pólya was very much influenced by Descartes [22, I, p. 56].

DESCARTES'S RULES:

The first was never to accept anything as true that I did not know evidently to be such; that is to say, carefully to avoid haste and bias, and to include nothing more in my judgments than that which presented itself to my mind so clearly and so distinctly that I had no occasion to place it in doubt.

The second was to divide each of the difficulties that I examined into as many parts as possible, and according as such division would be required for the better solution of the problems.

The third was to direct my thinking in an orderly way, by beginning with the objects that were simplest and easiest to understand, in order to climb little by little, gradually, to the knowledge of the most complex; and even for this purpose assuming an order among those objects which do not naturally precede each other.

And the last was at all times to make enumerations so complete, and reviews so general, that I would be sure of omitting nothing. [4, p. 16]

FIGURE 4

René Descartes

from "the book of the world." Descartes reached two conclusions. First, if he was to discover true knowledge he must carry out the whole program himself, just as a perfect work of art is the work of one master. Second, he must begin by methodically doubting everything taught in philosophy and looking for self-evident, certain principles from which to reconstruct all science.

In November 1628, Descartes had a public encounter with Chandoux, who felt that science was founded only on probability. By using his method to distinguish between true scientific knowledge and mere probability, Descartes easily demolished Chandoux. Among those present was the influential Cardinal de Bérulle, who charged Descartes to devote his life to working out the application of "his manner of philosophizing . . . to medicine and mechanics." To execute this design, Descartes moved to the Netherlands in 1628, where he lived for the next twenty years.

In Holland, Descartes worked at his system and, by 1634, had completed a scientific work

entitled *Le Monde.* He immediately suppressed the book when he heard about the recent condemnation of Galileo by the inquisition. He learned this from Marin Mersenne (1588–1648), a fellow student at LaFlèche and later the hub of the scientific correspondence network in Europe. This reveals Descartes's spirit of caution and conciliation toward authority (he was a lifelong devout Catholic). Later he took care to present his less orthodox views more obliquely.

Three hundred and fifty years ago, in 1637, the *Discours de la Méthode* [Figure 2], with appendices *La Dioptrique, Les Meteores,* and *La Géométrie,* appeared anonymously in Leyden, although it was soon widely known that Descartes was the author. The opening *Discours* is notable for its autobiographical tone, compressed presentation, and elegant French style. It was written in French since he intended—as did Galileo—to aim over the heads of the academic community to reach the educated people. Today, it is this opening *Discours,* with its problem-solving techniques [Figure 3], that is read.

In 1644, Descartes published *Principia Philosophiae,* a work in which he presented his views on cosmology. He expounded a mechanical philosophy in which a body could influence only those other bodies that it touched. Thus, for example, Descartes imagined space filled with "vortices" that moved the planets. This world view quickly became dominant in Europe. After the publication of Newton's *Philosophiae Naturalis Principia Mathematica,* the two scientific outlooks competed until well into the eighteenth century. Significantly—and this is reflected in the titles—Newton made mathematics indispensable for understanding the universe.

Queen Christiana of Sweden, ambitious patron of the arts and collector of learned men for her court, had seen the works of Descartes and pleaded with him to join her and teach her philosophy. She sent a man-of-war to fetch him but he was loath to go, in his words, to the "land of bears between rock and ice." But go he did. Being more of an athlete than a scholar, the 23-year-old

Queen wanted her lessons at five in the morning in a cold library with windows thrown wide open. This harsh land, where "men's thoughts freeze during the winter months," was too much for Descartes. A few months later he caught pneumonia and died on 11 February 1650.

CONTENTS OF THE *GEOMETRY*

The *Geometry* of Descartes is available to us in two English editions, the well known Smith-Latham translation [3] and the only complete English translation of the whole *Discours de la Méthode* by Olscamp [4]. The latter should be consulted since the appendix on *Optics* contains much interesting material on the conics.

In the first book of the *Geometry*, Descartes gave new geometric solutions of quadratic equations. For example [Figure 5], to solve the equation $z^2 = az - b^2$ (where a and b are both positive), Descartes drew the base line LM of length b and a perpendicular line LN of length $a/2$. Then he drew the circle with center N and radius NL. This circle cut the line perpendicular to LM at M in two points. The line segments MR and MQ are the solutions of the equation, as the reader can easily check. Descartes was aware that if the circle misses (only touches) the perpendicular to LM at M, then there is no (only one) solution to the equation.

Observe [Figure 5] that we have adopted Descartes's notation. In fact, his *Geometry* is the oldest mathematics text that we can read without having great difficulties with the algebraic notation. Descartes introduced the use of x, y, z for variables and a, b, c for constants, and he also introduced the exponential notation (except that he sometimes writes "aa" for our "a^2"). The only significant difference is that Descartes uses the symbol ∞ for equality.

Another problem Descartes dealt with in the first book was the problem of Pappus (fl. A.D. 300–350), which he mistakenly believed was still open. The problem asks for the locus of points such that the product of the distances (measured at fixed angles) to half of a fixed set of lines is equal to the

FIGURE 5

From p. 303 of Descartes's Géométrie

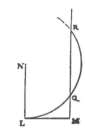

product of the distances to the other half (times a constant if the number of lines is odd). If there are three or four lines, Descartes showed that the locus is a conic. As an example with five lines, Descartes considered one horizontal line and four equally spaced vertical lines (Figure 6).

He set the product of the distances to the first, third, and fourth vertical lines equal to the product

FIGURE 6

Cartesian Parabola

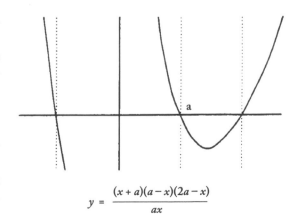

$$y = \frac{(x+a)(a-x)(2a-x)}{ax}$$

of the constant distance *a* between the lines, the distances to the second vertical line and the horizontal line, and obtained the equation $axy = (x + a)(a - x)(2a - x)$. Newton later called this curve the Cartesian Parabola. Since there were very few curves in Descartes's day, each received its own fancy name. This curve was only the second cubic (that is, a polynomial in two variables of degree three) ever discussed. The first was the Cissoid of Diocles (fl. ca. 190 B.C.). Descartes used his new curve extensively in his third book to solve equations of the fifth and sixth degrees as intersections of it and a circle.

GEOMETRICAL VS. MECHANICAL CURVES

The second book of Descartes's *Geometry* begins with a discussion of those curves which Descartes believed should be admitted into geometry. He does not consider the equation to be a sufficient representation of a curve, for equations are clearly algebraic objects. This forced him to always define curves by giving some geometric criterion. Later he derived the equation.

Descartes made a strict distinction between the curves that he called "geometrical" and those which he called "mechanical," but his explanation was none too clear. It has turned out that Descartes's geometrical (mechanical) curves are just the graphs of our algebraic (transcendental) functions. See Bos [1] for a full discussion. Descartes said that a curve is geometrical if it "can be conceived of as described by a continuous motion" [3, p. 43]. This excludes the spiral and the quadratrix because "they must be conceived of as described by two separate movements whose relation does not admit of exact determination" [3, p. 44]. Descartes allowed the use of a loop of thread to trace out a geometrical curve, as long as the shape of the string remained polygonal [3, p. 91]. Thus, the ellipse is a geometrical curve since it can be traced out using the familiar gardener's construction using string and pegs. In *La Dioptrique*, Descartes showed how to construct the hyperbola using straightedge and string [4, p. 135]. However, the curve generated

by the moving end of a piece of thread as it unwinds from a spool is a mechanical curve, for the thread was curved while wound around the spool and straight after it unwinds.

> On the other hand, geometry should not include lines that are like strings, in that they are sometimes straight and sometimes curved, since the ratios between straight and curved lines are not known, and I believe cannot be discovered by human minds, and therefore no conclusion based upon such ratios can be accepted as rigorous and exact. [3, p. 91]

That straight and curved lines cannot be compared is an old dictum of Aristotle. Descartes's adoption of it was important for it set up the question of rectification of curves—that is, the problem of finding arc length of curves.

Let us now consider Descartes's argument for the Cartesian Parabola being a geometrical curve. He gave the following definition of a geometrical curve, then found its equation. Since its equation is the same as that of the Cartesian Parabola, the Cartesian Parabola is a geometric curve.

> I shall consider next the curve *CEG* [Figure 7], which I imagine to be described by the intersection of the parabola *CKN* (which is made to move so that its axis *KL* always lies along the straight line *AB*) with the ruler *GL* (which rotates about the point *G* in such a way that it constantly lies in the plane of the parabola and passes through the point *L*). [3, p. 84]

FIGURE 7

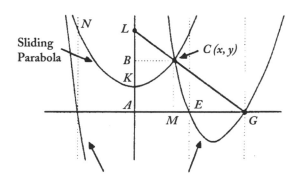

Cartesian Parabola

If we let AB be the y-axis and AG be the x-axis (Descartes used the opposite convention), then the Cartesian Parabola is the locus of all points $C(x,y)$ of intersection of the parabola that slides up and down the y-axis and the ruler that pivots at the fixed point $G(2a, 0)$ and passes through the point L moving along the y-axis with the parabola. The parabola has equation $x^2 = az$, where $a = KL$ and $z = BK$ (the focus of the parabola is one-fourth of the way from K to L). Descartes found the equation of the curve using classical geometry: Since the triangles GMC and CBL are similar, $GM/MC = CB/BL$, that is, $(2a - x)/y = x/BL$. Thus, we have

$$BK = a - BL = a - \frac{xy}{2a - x} .$$

But the equation of the parabola CKN can be written $BK = x^2/a$. Equating these expressions for BK, and simplifying, we obtain,

$$x^3 - 2ax^2 - a^2x + 2a^3 = axy,$$

which is the equation of the Cartesian parabola. (Note that the name comes from the fact that a parabola is sliding up and down the line.)

DESCARTES'S SUBNORMAL METHOD

In our calculus classes, one important problem is to find an equation of the tangent line to a curve at a given point on the curve. Problems were not phrased this way in the seventeenth century, because equations of lines was not a well developed topic. They asked (equivalently) for the subnormal for a given point on the curve, that is, the length of the segment on the x-axis between the abscissa of a point on the curve and the x-intercept of the normal line at that point. The subtangent was defined analogously.

Descartes presented a method for finding the subnormal [Figure 8]. If we can find a circle, with center P on the x-axis, that cuts a curve in precisely one point C, then the radius at that point is normal to the curve. But if the center of the circle through the point C be moved "ever so little" along the x-axis, the circle will cut the curve at two points.

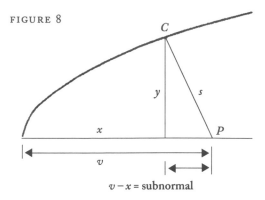

$v - x = $ subnormal

This idea provided a means of finding the subnormal for any point (x_0, y_0) on the curve. Starting with the equation of the curve and the equation of a variable circle with center $P = (v, 0)$, find the equation giving their intersection. Then choose P so that the intersection equation has a double root.

Let us consider the case of the parabola $y^2 = kx$. The circle having center $P = (v, 0)$ and radius s that passes through the point (x_0, y_0) has equation $(v - x_0)^2 + y_0^2 = s^2$. Since (x_0, y_0) is on the parabola, $y_0^2 = kx_0$, and we obtain

$$x_0^2 + (k - 2v)x_0 + (v^2 - s^2) = 0.$$

This equation will have a double root if and only if the discriminant is zero; in which case, $x_0 = -(k - 2v)/2$, or

$$v - x_0 = k/2.$$

This looks mysterious today, but any mathematically literate contemporary of Descartes would know that the parabola has constant subnormal. Perhaps we should check this result using the new calculus: If $y^2 = kx$, then $2yy' = k$. So $y' = k/(2y)$. Thus, the normal line at (x_0, y_0) has slope $-y_0/(k/2)$. To plot the normal line, we go down y_0 from the point (x_0, y_0) to land on the x-axis, and then go right the constant distance $k/2$. Thus, the subnormal for any point on this parabola does indeed have constant length, $k/2$.

Descartes was justly proud of this work, for he wrote:

I have given a general method of drawing a straight line making right angles with a curve at an arbitrarily chosen point upon it. And I dare say that this is not only the most useful and most general problem in geometry that I know, but even that I have ever desired to know. [3, p. 95]

There is one final quotation from Descartes that is important here, for it deceived Newton—in a positive way:

When the relation between all points of a curve and all points of a straight line is known [that is, when we have the equation of the curve] . . . it is easy to find . . . its diameters, axes, center and other lines [e.g., tangent and normal lines] or points which have especial significance for this curve . . . By this method alone it is then possible to find out all that can be determined about the magnitude of their areas, and there is no need for further explanation from me. [3, p. 92]

Newton believed Descartes's claim, that from the equation of a curve one can tell everything about it. This encouraged Newton to develop the variety of ad hoc techniques which he learned from the works of Descartes and Wallis into algorithms for solving problems about all curves. This was just one of the motivations that Newton had for inventing the calculus.

For further information about Descartes, see the DSB article by Crombie, Mahoney, and Brown [6, IVC, pp. 51–65]. The book by Scott [23] contains a detailed discussion of his mathematical work. Bos [1] gives an interesting study of Descartes's concept of curve. Of course, one should read the *Geometry* itself [3], [4].

FRANS VAN SCHOOTEN (1615–1660)

Schooten enrolled at the University of Leiden at age 16, where he was carefully trained by his father in the Dutch school of algebra. He met Descartes when the latter was in Leiden to supervise the printing of the *Discours de la Méthode* (1637). Schooten recognized the value of the work but had difficulty mastering its contents. So he went to Paris for further study, where he was cordially welcomed by Mersenne.

While in Paris, Schooten read the manuscripts of Pierre de Fermat (1601–1665) and Françoise Viète (1540–1603), and under commission of the famous Leiden publishing house of Elsevier, gathered all the printed works of Viète. This included Viète's most famous work, *In Artem Analyticam Isagoge* (*Introduction to the Analytic Art*) of 1591, which dealt mainly with the theory of equations. Because of this work, Viète is known as the father of algebra. Conscious of the great importance of the scattered works of Viète on algebra, geometry, and analysis, which had been published separately from 1579 to 1615, Schooten republished them with commentary as *Francisci Vietae Opera Mathematica* (1646). The work quickly became an indispensable collection of mathematical source materials, and Newton carefully studied a copy from the Cambridge libraries [20, I, p. 21]

Schooten returned to Leiden in 1643 and began working on a Latin translation of Descartes's *Géométrie*, which he published in 1649. Descartes had been dissatisfied with the form and argument of his *Géométrie* from the very day of its publication, and therefore encouraged the writing of commentaries clarifying its obscurities and developing its approach. Because of its valuable commentary and excellent figures, Schooten's edition was enthusiastically received. This success led him to prepare a much enlarged second edition that appeared in two volumes (1659–1661). It contained about 800 pages of commentary and new work, in addition to the 100 page translation of Descartes's *Géométrie*, and included [20, I, pp. 19–20]:

- Schooten's extremely valuable commentaries. Many of these details were derived directly from Descartes's own criticisms made in correspondence with Schooten.

- Florian Debeaune's (1601–1652) *Notae Breves*, a work which Descartes welcomed as a perceptive exposition of the more elementary aspects of his work. Debeaune posed the first inverse tangent problem.

- Jan Hudde's (1628–1704) studies on equations and extreme values. His rule for locating double roots of equations was useful in applying Descartes's tangent method. It was an important precursor of the derivative.

- Jan de Witt's (1629–1695) excellent tract on conic sections.

- An example of Fermat's extreme value and tangent method.

- Christiaan Huygens's (1629–1695) first publication, an improved method for finding the tangent to the conchoid.

- Hendrik van Heuraet's (1633–ca. 1660) rectification method, of which we shall say more below.

All of this shows the great effort that Schooten devoted to the training of his students and to the dissemination of their findings. Much of their work is available only in correspondence, careful studies of which are currently being made. It was from these editions of Schooten that mathematicians learned of the work of Descartes. It was the second Latin edition that Newton borrowed and annotated in the summer and autumn of 1664 (the copy he bought the following winter may have been the 1649 edition). It had an immense impact on his mathematical development; for after mastering it, he was current with research in the new analysis.

JOHN WALLIS (1616–1703)

Before attending Emanuel College, Cambridge, the only mathematics Wallis knew was what he learned from his brother who was preparing for a trade. At Cambridge, mathematics "were scarce looked upon." He took his M.A. in 1640 and was ordained. In 1649, he was appointed Savilian professor of Geometry at Oxford, an appointment that must have surprised those who thought the only mathematics he had done was to decode a few messages for the Parliamentarians.

This is not quite true, but, wrote Wallis "I had not then [in 1648] seen Descartes' Geometry." [20, III, p. xv]. In 1647 or 1648, he chanced upon Oughtred's *Clavis*, mastering it in a few weeks, and then rediscovered Cardano's formula for the cubic. In 1648, at the request of Cambridge professor of mathematics John Smith, he reworked Descartes's treatment of the fourth degree equation by factoring it into two quadratics. As soon as he was appointed Savilian professor at Oxford, he took up the study of mathematics, with rare energy and perseverance, and soon became one of the best mathematicians in Europe. He held the post for 50 years.

Wallis's *Operum Mathematicorum Pars Altera* (Oxford, 1656) was a fat and rather motley two-part collection of his early mathematical lectures, commentaries, and researches [20, I, p. 23]. It contained his *De Sectionibus Conicis* (dated 1655), a treatise of 110 pages that was the first elementary text on the conics treated from the Cartesian viewpoint. In an appendix, Wallis tried to extend the approach to higher plane curves, especially the cubical parabola $a^2y = x^3$, where the constant a^2 was used to preserve dimensionality. He successfully found the subtangent, but had trouble with the graph because he did not feel comfortable with negative numbers. He also introduced the semicubical parabola $ay^2 = x^3$, a curve that played a very important role in the development of the calculus [30, p. 295–298]. Quite suddenly the mathematical world had been presented with a powerful analytic geometry, only to find that there were few curves on which to practice it. The new perspective of Wallis—which took some time to be adopted by the mathematical community—was that any algebraic equation in two variables defines a curve [13, p. 238].

Together with his conic sections, Wallis published the work on which his fame rests, *Arithmetica Infinitorum* (dated 1656; printed 1655). This volume developed from his study of the *Opera Geometrica* (1644) of Torricelli (1608–1647). Wallis tried to apply these methods to the quadrature of the circle, but not even the study of the voluminous *Opus Geometricum* (1647) of Gregorius Saint Vincent (1584–1667), helped. Out of the project of squaring the circle, he did get his famous infinite product for $\pi/4$.

The *Arithmetica Infinitorum* exerted a singularly important influence on Newton when he studied it in the winter of 1664–1665. From it, Newton learned of the problem of quadratures, or, as we now say, finding areas under curves. Newton probably also read Wallis's *Tractatus Duo* (1659) that presented his research on the cycloid, cissoid, and other geometrical figures.

RECTIFICATION OF CURVES

By 1638, Descartes suspected that the logarithmic spiral might be rectifiable; that is, the length of an arc of the curve could be computed. Even if correct, this would not cause him any difficulties because the spiral is a mechanical curve, and Descartes only accepted Aristotle's dictum that straight lines and curved lines could not be compared for geometrical curves. In 1657, Huygens found the length of an arc of a parabola; but he used a mechanical curve in his solution, and thus Descartes's version of Aristotle's dictum was still intact. Also Huygen's method did not generalize.

The first geometrical curve to be rectified in a geometric way was Wallis's semi-cubical parabola $ay^2 = x^3$. As often happens, several people solved the problem simultaneously: William Neil in 1657, Hendrick van Heureat in 1659 [14], and Pierre de Fermat in 1660. Of course, a priority dispute erupted. Heureat's solution was the most influential because it was published in Schooten's second Latin edition of Descartes's *Geometry*. The proof used the new classical geometry of the seventeenth century and was fairly intricate (for details, see [8] or [13]). The method of proof was to replace the problem of rectification of the semi-cubical parabola by a simpler problem, the quadrature of an ordinary parabola.

This transformation of the problem to a simpler one shows up even when we do the problem today with the calculus, but it is so slick that it is easy to miss what happens. Starting with $y^2 = x^3$ (it is no accident that we still do this first today), we obtain $(y')^2 = 9x/4$. Thus, the arc length from, say, $(0, 0)$ to $(4, 8)$, is

$$L = \int_0^4 \sqrt{1 + (9x/4)}\, dx.$$

The substitution $u = 1 + (9x/4)$ transforms this into

$$L = (4/9)\int_1^{10} \sqrt{u}\, du.$$

The first of these integrals represents an arc length, whereas the second stands for the area under a parabola. Today, we just look at these as two simple integration problems, but in the old days B(efore) C(alculus), these were viewed as two separate kinds of problems.

Heuraet's method was entirely general. When Newton saw the proof, he realized the value of transforming one type of problem into another. This is one of the roots of the Fundamental Theorem of Calculus. It is the biggest swap of all—we trade integration for anti-differentiation. This is precisely what Newton did soon after he read Heureat's proof. (For a full history of the rectification problem, see Hofmann [11, Ch. 8].)

CONCLUDING REMARKS ABOUT NEWTON'S READINGS

In order to do creative work, a mathematician "needs an adequate notation, a competent knowledge of mathematical structure and the nature of axiomatic proof, an excellent grasp of the hard core of existing mathematics and some sense of promising line for future advance." [20, I, p. 11]. The works that Newton chose to read in 1664 and 1665 magnificently met these needs. He took his arithmetic symbolism from Oughtred, his geometrical form from Descartes. Of course, he grafted on new modifications of his own while creating the calculus. He learned elementary scholastic logic in grammar school and traditional forms of mathematical proof from Euclid. He learned the new analytic geometry of the seventeenth century from Schooten and de Witt, topics in algebra and the theory of equations from Viète, Oughtred, Schooten, and Wallis. Most importantly, he learned of the twin problems of infinitesimal analysis: From Descartes, the method of tangents; from Wallis, quadrature. There were plenty of open problems for Newton to attack. Without doubt, the two strongest influences on Newton were Descartes and Wallis. [20, I, pp. 11–13]

It is of as much interest to note *what Newton did not read.* We miss the names of Napier, Briggs, Harriot, Desargues, Pascal, Fermat, Stevin, Kepler, Cavalieri, and Torricelli. Among the Greeks there is only Euclid, not Apollonius nor Archimedes. In fact, Newton seemed to dislike the method of exhaustion. There is great significance in this lack of knowledge of ancient mathematics and of the new classical (as opposed to analytic) geometry of the seventeenth century. He was not hampered by its knowledge. Had Newton gained a deep knowledge of classical geometry and the new classical geometry of his century, I conjecture it would have hindered his invention of the calculus (and similarly for Leibniz who was also ignorant of classical geometry).

As Westfall points out [28, p. 100] about Newton's readings: "In roughly a year, without benefit of instruction, he mastered the entire achievement of seventeenth-century analysis and began to break new ground." In fact, by mid-1665, Newton's urge to learn from others seems to have abated [20, I, p. 15].

Newton's Works

Newton was an extraordinary scientist because he made so many fundamental contributions to different fields:

- Mathematics, both pure and applied
- Optics and the theory of light and color
- Design of scientific instruments
- Synthesis and codification of dynamics
- Invention of the concept and law of universal gravity.

In addition, we now know, and are willing to admit, that he spent immense amounts of time working on:

- Alchemy
- Chronology, church history, and interpretation of the Scriptures.

The range and depth of Newton's intellectual pursuits never ceases to amaze us.

As a first step in understanding Newton's contributions, consider the chart below that indicates when Newton was involved in various research areas. One might think that Newton thought about everything all of the time, but the manuscript record shows that he worked on only a few areas at any one time, and these were not necessarily—in his mind at least—disjoint.

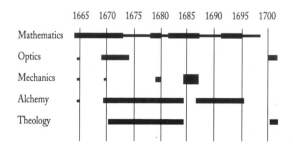

FIGURE 9

Newton's areas of activity

We begin with a synopsis of Newton's mathematics as presented in Whiteside's edition of Newton's *Papers* [20]. This will be followed by a thumbnail sketch of each of these areas of Newton's intellectual efforts. Since it is impossible to discuss all of his contributions here, only a few examples of Newton's mathematical work will be discussed in detail. These were chosen with the teacher in mind, to provide examples that can be used in the classroom.

Volume I. (1664–1666). The volume begins with Newton's annotations on the works of Oughtred, Descartes, Schooten, Viète, and Wallis. The bulk consists of research on analytic geometry and the calculus. Newton turns Descartes's subnormal technique into the notion of curvature, and Hudde's rule for double roots into fluxions (differentiation). We see the calculus become an algorithm in mid-1665. This early work on the calculus was summarized in the October 1666 tract on fluxions. In a schematic diagram, Whiteside [20, I, p. 154] shows how all of these ideas came together to give birth to the calculus. The volume ends with miscellaneous work on trigonometry, the theory of equations, and geometrical optics.

Volume II. (1667–1670). Work on classification of cubics begins here and was published as an appendix to his *Optics* (1704). In this volume, we see Newton struggling with the graphs. The most important work on the calculus is the hastily composed 1669 tract *De Analysi* that summarizes all of his work thus far. He gave a copy of this to Barrow in 1669 to assert his priority over Nicolaus Mercator (1619–1687) whose *Logarithmotechnia* (1668), with its infinite series for the logarithm, had just appeared. Half the volume consists of his annotations on the *Algebra* of Kinckhuysen. One piece of Newton's advice here is too good not to pass on to our students:

> After the novice has exercised himself some little while in algebraic computation . . . I judge it not unfitting that he test his intellectual powers in reducing easier problems to an equation, even though perhaps he may not yet have attained their resolution. Indeed, when he is moderately well versed in this subject . . . then will he with greater profit and enjoyment contemplate the nature and properties of equations and learn their algebraic, geometrical and arithmetical resolutions. [pp. 423–425]

Volume III. (1670–1673). Although Barrow encouraged Newton to revise *De Analysi* for publication, the booksellers were uninterested. But he did combine the two earlier works on the calculus and

many new results in a 1671 tract, with an important foundational change: he postulated a fluent variable of time for his fluxions; that is, all his derivatives are time derivatives. Also, here is an investigation of Huygen's pendulum clock and more research on geometric optics.

Volume IV. (1674–1684). Research in theology and alchemy kept him busy (Figure 9), though his work on mathematics never entirely stopped. This volume contains some of Newton's research on algebra, number theory, trigonometry, and analytical geometry. In the middle of this period, he became fascinated with the classical geometry of the Greeks. Only at the end of this period did Newton show great interest in fluxions and infinite series.

Volume V. (1683–1684). The bulk of this volume consists of Newton's ninety-seven self-styled "lectures," deposited as his Lucasian lectures on algebra for the period 1673–1683. The *Arithmetica Universalis* given here is an incomplete revision of the algebra lectures. Its published version was his most read work, not the papers on calculus.

Volume VI. (1684–1691). Halley's visit in August 1684 turned Newton's interest to the geometry and dynamics of motion, the subject of this entire volume. The work dates from the period 1684–1686, and is arguably as creative as the miracle years of 1664–1666.

Volume VII. (1691–1695). In the early winter of 1691–1692, Newton wrote *De Quadratura Curvarum*, on the quadrature of curves. He also dealt with classical geometry (1693), higher plane curves, and finite-difference approximations (1695). As always, Whiteside has "taken care to preserve all the significant idiosyncrasies, contractions, superscripts and archaic spellings" of the "ink-blobbed, much-cancelled and often rudely scrawled manuscripts." [p. ix]

Volume VIII. (1697–1722). Most mathematicians will find this the most interesting volume after the first, for it contains Newton's solution (simply stated without proof) of the brachistochrone problem as well as documents related to the priority dispute. (To see that this dispute involved much more than mathematics, read Hall's *Philosophers at War* [9].)

We calculus teachers should refrain from telling our students that Newton invented the calculus because he was motivated by physical considerations. Although applications are an excellent reason for studying the calculus, in Newton's case the record is clear: first mathematics, then applications.

THE BINOMIAL THEOREM

On the frontispiece of the first volume of Newton's *Papers* we see the manuscript where he took up the age old problem of squaring the circle, or (to make the activity sound more respectable) the quadrature of the circle. He became interested in this problem after reading Wallis's *Arithmetica Infinitorum*. Newton learned there how to evaluate the integrals (here expressed in Leibniz's notation) $\int_0^x (1-x^2)^{n/2}dx$, where n is an even integer. Newton tabulated the values of these integrals in his attempt to find the area of a circle ($n = 1$). To see how he did this consider the case when $n = 6$:

$$\int_0^x (1-x^2)^{6/2}dx = 1(x) + 3(-x^3/3) + 3(x^5/5) + 1(-x^7/7).$$

The factors in parentheses are recorded in the rightmost column of the table below. The coefficients, 1, 3, 3, 1, are recorded in the column labeled $n = 6$. In general, to evaluate $\int_0^x (1-x^2)^{n/2}dx$, sum the products of the values in the nth-column by the corresponding terms in the rightmost column.

$n=0$	$n=2$	$n=4$	$n=6$	$n=8$	...	times
1	1	1	1	1	...	x
	1	2	3	4	...	$-x^3/3$
		1	3	6	...	$x^5/5$
			1	4	...	$-x^7/7$
				1	...	$x^9/9$
					⋮	⋮

Wallis had also tabulated these integrals, but since he used 1 rather than x as an upper limit, he did not see the pattern. But Newton recognized it as "Oughtreds Analyticall table," from his readings of Oughtred's *Clavis* [20, I, p. 452]. We, of course, now call this Pascal's triangle. Newton knew that each number in the table is the sum of the number to its left and the one above that, so he decided to extend the pattern backwards for all even values of n. Thus he obtained:

... $n=-2$	$n=0$	$n=2$	$n=4$	$n=6$	$n=8$	...	times
1	1	1	1	1	1	...	x
−1	0	1	2	3	4	...	$-x^3/3$
1	0	0	1	3	6	...	$x^5/5$
−1	0	0	0	1	4	...	$-x^7/7$
1	0	0	0	0	1	...	$x^9/9$
⋮	⋮	⋮	⋮	⋮	⋮	⋮	⋮

To extend this table to odd values of n, Newton used a complicated proportionality argument (see [31] for details). Later, in a letter to Leibniz [19, I, pp. 130–131], Newton provided an easier explanation for the extension. When n is even, say, $n = 2m$, the kth entry in the nth column is given by the binomial coefficient ( ) $= m!/k!(m-k)!$. Newton ignored the restriction that n must be even and used the formula for binomial coefficients when n was odd. For example, the fourth entry in the $n = 1$ column is given by

$$\binom{1/2}{4-1} = \frac{(1/2)(1/2-1)(1/2-2)}{(1)(2)(3)}$$

Thus, he obtained the results shown at the top of page 499.

Now, from the $n = 1$ column, Newton was able to draw the conclusion that he sought:

$$\int_0^x (1-x^2)^{1/2}dx = x + (1/2)(-x^3/3) + (-1/8)(x^5/5) + (1/16)(-x^7/7) + \cdots .$$

For $x = 1$, this gives an infinite series for the area of (a quadrant of) a circle. From this, Newton jumped to the conclusion that a similar "interpolation" could be done on curves (we would say, on func-

...	$n=-3$	$n=-2$	$n=-1$	$n=0$	$n=1$	$n=2$	$n=3$	$n=4$	$n=5$	$n=6$	$n=7$	$n=8$	...	times
	1	1	1	1	1	1	1	1	1	1	1	1	...	x
	$-\frac{3}{2}$	-1	$-\frac{1}{2}$	0	$\frac{1}{2}$	1	$\frac{3}{2}$	2	$\frac{5}{2}$	3	$\frac{7}{2}$	4	...	$-\frac{x^3}{3}$
	$\frac{15}{8}$	1	$\frac{3}{8}$	0	$\frac{1}{8}$	0	$\frac{3}{8}$	1	$\frac{15}{8}$	3	$\frac{35}{8}$	6	...	$\frac{x^5}{5}$
	$-\frac{35}{16}$	-1	$\frac{5}{16}$	0	$\frac{1}{16}$	0	$\frac{1}{16}$	0	$\frac{5}{16}$	1	$\frac{35}{16}$	4	...	$-\frac{x^7}{7}$
	$\frac{315}{128}$	1	$\frac{35}{128}$	0	$\frac{-5}{128}$	0	$\frac{3}{128}$	0	$\frac{-5}{128}$	0	$\frac{35}{128}$	1	...	$\frac{x^9}{9}$
	$\vdots$	$\vdots$	$\vdots$	$\vdots$	$\vdots$	$\vdots$	$\vdots$	$\vdots$	$\vdots$	$\vdots$	$\vdots$	$\vdots$		$\vdots$

tions) as well as on their quadratures (integrals), and then guessed the Binomial Theorem for fractional exponents. He checked this result several ways. First, he formally used the square root algorithm to obtain the series

$$(1 - x^2)^{1/2} = 1 - (1/2)x^2 - (1/8)x^4 - (1/16)x^6 - \cdots.$$

Then he checked that it agreed with the Binomial Theorem. Next, he squared both sides of the above equation to see that an equality resulted. As a further check, he used formal long division to obtain an infinite series for $(1 + x)^{-1}$. Note the wonderful research techniques he is using. Nonetheless,

> The paradox remains that such Wallisian interpolation procedures, however plausible, are in no way a proof, and that a central tenet of Newton's mathematical method lacked any sort of rigorous justification ... Of course, the binomial theorem worked marvelously, and that was enough for the 17th century mathematician. [31, p. 180]

Newton became tremendously excited with his new tool, the Binomial Theorem, which became a mainstay of his newly developing calculus. He also did such bizarre computations as approximating $\log(1.2)$ to 57 decimal places.

The Binomial Theorem was Newton's first mathematical publication. It appeared in Wallis's *Treatise of Algebra* (Figure 10) in a summary of Newton's two famous letters to Leibniz in 1676 [24, pp. 330–331]. These letters are readily available, with ample commentary, in Newton's *Correspondence* [19, II, pp. 20–47 and 110–161].

FIGURE 10   *First publication of the Binomial Theorem, 1685*

$$\overline{P + PQ}\Big|_{n}^{m} = P_{n}^{m} + \frac{m}{n}AQ + \frac{m-n}{2n}BQ + \frac{m-2n}{3n}CQ + \frac{m-3n}{4n}DQ + \&c.$$

Where $P + PQ$ is the Quantity, whose Root is to be extracted, or any Power formed from it, or the Root of any such Power extracted. P is the first Term of such Quantity; Q, the rest (of such proposed Quantity) divided by that first Term, And $\frac{m}{n}$ the Exponent of such Root or Dimension sought. That is, in the present case, (for a Quadratick Root,) $\frac{1}{2}$.

OPTICS

Newton's earliest work on optics was done at Cambridge and the experiments continued at Woolsthorpe during the plague, but was not put in near final form until he was preparing his Lucasian lectures for 1670–1672. It had long been known (see, for example, Descartes [4, p. 335]) that when light passed through a prism it was dispersed into a colorful spectrum. Newton was able to give a quantitative analysis of this behavior and to devise a new theory of light. In February 1671/72 (the slash date was used because England had not yet adopted the Gregorian calendar), this resulted in Newton's first publication in optics, the lengthy title of which also provides an abstract:

> A Letter of Mr. Isaac Newton, Mathematick Professor in the University of Cambridge; containing his New Theory about Light and Colors: Where Light is declared to be not Similar or Homogeneal, but consisting of difform rays, some of which are more refrangible than others: And Colors are affirm'd to be not Qualifications of Light, deriv'd from Refractions of natural Bodies, (as 'tis generally believed;) but Original and Connate properties, which in divers rays are divers: Where several Observations and Experiments are alleged to prove the said Theory. [18, p. 47].

This work engendered a controversy with Robert Hooke (1635–1703), who claimed to have published the ideas earlier. As a consequence, Newton became extremely reluctant to publish. In fact, the *Optics* was not published until 1704, the year after Hooke's death.

In developing his theory of light, Newton realized that lenses caused chromatic aberration. This set him thinking about telescope design, and he concluded that the problem could be avoided by using mirrors instead of lenses. Consequently he designed a reflecting telescope, built one himself, and then described it in the March 25, 1672 issue of the *Philosophical Transactions*. These first papers of Newton have been photoreproduced by I. B. Cohen [18], along with a valuable introduction by Thomas Kuhn. Rather than describe Newton's theory of light (which has been done by Alan

Shapiro in the first volume of *The Optical Papers of Isaac Newton* [21]), we shall briefly discuss telescope design. This provides an interesting classroom example of the reflective properties of the conics.

The first reflective telescope was designed by James Gregory (1638–1675) and published in his *Optica Promota* of 1663, a work which Newton did not read until after he had invented his own telescope. Gregory's telescope consists of a concave primary mirror (on the right in Figure 11a) that is parabolic in shape, and a concave secondary mirror that is elliptical (strictly speaking, the surfaces generated by rotating these conics about the axis of the telescope). The incoming rays of starlight bounce off the parabolic mirror and are reflected through its focus. Beyond that focus is an elliptical mirror that shares a focus with the parabola and has its other focus behind a small hole in the primary mirror. Thus, after the reflected rays of starlight pass through the common focus of the parabola and ellipse, they are reflected off the elliptical secondary mirror and converge at the second focus of the ellipse. Gregory tried to have a telescope built to his design, but the opticians were unable to polish the mirrors properly.

In 1668, Newton placed a flat secondary mirror between the primary parabolic mirror and its focus [Figure 11b]. The eyepiece was located at the side of the telescope. Incoming rays of starlight reflect off the parabolic mirror and head for its focus F. Before they get there, they are reflected off the flat mirror. Then they converge toward F', the point symmetric to F with respect to the plane of the flat mirror. This invention remained unknown until Newton made another one (casting and polishing the mirrors himself) and presented it to the Royal Society of London on 11 January 1672. This so impressed the members that they elected him a Fellow of the Royal Society at that very same meeting.

Later in 1672, another telescope design [Figure 11c] was published by Guillaume Cassegrain (fl. ca. 1672) in France and abstracted in the *Trans-*

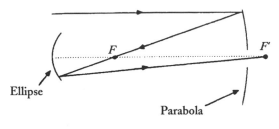

Ellipse

Parabola

FIGURE 11a *Gregorian Telescope, 1663*

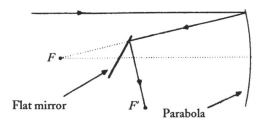

Flat mirror

Parabola

FIGURE 11b *Newtonian Telescope, 1668*

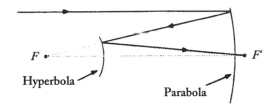

Hyperbola

Parabola

FIGURE 11c *Cassegrain Telescope, 1672*

actions of the Royal Society. The concave primary mirror is again a parabola with a hole in the center, and the secondary is a convex hyperbolic mirror which shares a focus with the parabola and has its other focus behind the hole in the parabola. Rays of starlight reflected from the primary parabolic mirror head toward the focus of the parabola. Before reaching that focus, they are reflected by the hyperbolic mirror toward the other focus of the hyperbola.

Cassegrain claimed that his design was superior to Newton's. In what was to become his typical style, Newton marshalled his evidence and attacked furiously. He claimed that Cassegrain's idea was not only a minor modification of Gregory's, but also optically inferior. Cassegrain retreated into anonymity. But Newton was wrong

about the superiority of the Gregorian telescope. In 1779, Jesse Ramsden (1735–1800) showed that the combination of a concave and a convex mirror partially corrects the spherical aberrations, whereas in the Gregorian telescope, the aberrations of the two concave mirrors are additive. Today the Cassegrain model is used in most large reflectors.

A mathematical result of Newton's work on optics grew out of the problem of grinding a hyperbolic mirror (although he did not use one, the possibility of a hyperbolic lens was noted by Descartes [4, p. 139]). Newton, and independently Christopher Wren (1632–1723), discovered that the hyperboloid of one sheet was a ruled surface. Newton used this result to show how to make a hyperboloid of one sheet on a lathe by holding the chisel obliquely to the axis of the lathe.

RELIGION.

Despite inheriting his stepfather's theological library and buying several theological books when he came to Cambridge, Newton's serious study of theology began only in the early 1670s. No doubt this came about because the position of Fellow at Trinity required that one had to be ordained in the Anglican Church within seven years of receiving the M.A. In Newton's case, this was by 1675. Not being one to do anything halfway, Newton became engaged in an extensive reading program that took him through all the early Church Fathers. As a result, ordination became impossible for he had become a heretic.

Newton became an Arian or Unitarian—he denied the Trinity—of deep conviction and remained so for the rest of his life. His argument was: "Though Christ was the only begotten son of God, and hence never merely a man, he was not equal to God, not even after God exalted him to sit at his right side as a reward for his obedience unto death." [29, p. 130]. Newton arrived at this position through a careful analysis of Scripture. He believed that a deceitful Roman Church had manipulated the Emperor Theodosius to introduce the false doctrine of the Trinity into the Scriptures in the fourth century. The *Book of Revelation* was crucial to Newton's interpretation. He believed that the Roman Church was the "Great Apostasy" and never ceased to hate and fear it [28, p. 321].

By 1675, Newton was making plans to leave Cambridge for he knew that as a Fellow at the College of the Holy and Undivided Trinity he could not reveal his Arian views. To do so would be socially unacceptable, and he never in his life did so, except obliquely to a few people of similar persuasion. That he read the situation correctly is indicated by the dismissal of William Whiston (1667–1752), Newton's successor in the Lucasian Chair, for the uncompromising expression of Unitarian views. But just at this time, the Crown granted a special dispensation that the occupant of the Lucasian Professorship was not required to be ordained. Thus, Newton could stay at Cambridge. Newton's theological studies continued until work on the *Principia* interrupted [Figure 9]. In London, he was able to take up his theological studies again, and they continued for the rest of his life. (For further details, see [28, pp. 309–334] or [29].)

ALCHEMY

Newton's interest in alchemy has long been embarrassing to some scholars, while others delight in this trace of hermeticism and dub him a mystic. But there is now no doubt that he was a serious practitioner [Figure 9]. From 1669 (when he bought his first chemicals) until 1684 (when work on the *Principia* interrupted), Newton spent long hours in the "elaboratory." Newton again practiced alchemy from 1686 until 1696, but after he moved to London he never took it up again seriously [27, p. 121]. Newton did plan on adding alchemical references to the second edition of the *Principia* although he never did so. (For details on his alchemical work, see [5].) One benefit of this work was that he was able to cast the speculum for his first telescope.

In 1693, Newton suffered a nervous breakdown of uncertain duration and severity. There is no doubt that he frequently tasted his chemicals,

but whether it was caused by mercury poisoning is debatable [7, pp. 88–90].

When Newton wrote to Oldenburg in 1673 that he intended "to be no further solicitous about matters of Philosophy" [19, I, p. 294], and to Hooke in 1679 that "I had for some years past been endeavouring to bend my self from Philosophy to other studies in so much yt I have grutched the time spent in yt study" [19, II, p. 300], we must take him at his word. During most of the decade of the 1670s, Newton preferred theology and alchemy to physics and mathematics.

THE PRINCIPIA

In his old age, Newton liked to reminisce and he himself started the story of the falling apple. We have four independent accounts of the tale [7, p. 29–31]. Here is Conduitt's [28, p. 154]:

> In the year 1666 he retired again from Cambridge . . . to his mother in Lincolnshire & whilst he was musing in a garden it came into his thought that the power of gravity (wch brought an apple from the tree to the ground) was not limited to a certain distance from the earth but that this power must extend much farther than was usually thought. Why not as high as the moon.

So let us grant that a falling apple started Newton thinking about gravity during the plague years; even if he made up the story it is harmless. But his retelling of this event, in his 1718 letter to DesMaizeaux (which we quoted earlier), is not harmless. Newton attempted to push back the date of his discovery of the law of universal gravitation to the plague years. His papers tell quite a different story.

In late 1664, Newton learned Kepler's third law: the square of the time that it takes a planet to make one elliptical revolution around the sun is proportional to the cube of the mean distance from the sun, that is, $T^2 \sim R^3$. The following January, Newton discovered the Central Force Law (see the letter to DesMaizeaux), which Huygens independently discovered and first published without proof in his *Horologium Oscillatorium* (1673) (see [12]). The Central Force Law states that the centrifugal (center fleeing) force acting on a body traveling about a central point is proportional to the square of the speed and inversely proportional to the radius of the orbit: $F = S^2/R$. Strictly speaking, this "force" is an acceleration, but we shall follow Newton's usage.

Newton was able to discover that the gravitational force between a planet and the sun must be inversely proportional to the square of the distance between them. If a planet travels with uniform speed around a circular (not elliptical) orbit of radius R in time T, then its speed S is $2\pi R/T$. Thus,

$$F = \frac{S^2}{R} = \frac{(2\pi R/T)^2}{R} = 4\pi^2 \left(\frac{R^3}{T^2}\right)\left(\frac{1}{R^2}\right).$$

By Kepler's third law, R^3/T^2 is constant, and hence, $F \sim 1/R^2$. Newton left off at this point, devoting most of the next decade to alchemy and theology, though he was never completely divorced from mathematics (Figure 9).

Hooke, Halley, and Wren were able to make this same deduction by 1679, but the problem of explaining the elliptical orbits remained. On 24 November 1679, Hooke wrote to Newton suggesting a private "philosophical," that is scientific, correspondence on topics of mutual concern. In this letter, Hooke mentioned *his* hypothesis of "compounding the celestiall motions of the planetts [out] of a direct motion by the tangent & an attractive motion towards the centrall body." [19, II, p, 297]. This does not seem to be much of a hint for proving that if the inverse square law holds, then the planets must move in elliptical orbits. But it started Newton thinking about the question again. Hooke gave further encouragement on January 17, when he wrote "I doubt not but that by your excellent method you will easily find out what that Curve must be." [19, II, p. 313]. Newton did succeed in finding the answer, but he kept it to himself.

It was also in 1679 that Newton learned of Kepler's second law: the radius from the sun to a planet sweeps out equal areas in equal times. It seems strange that Newton would have learned of the third law as a student in 1664, but not about the second until years later. The explanation is that the third law was generally accepted in the scientific community because it could be empirically verified, whereas the second was much more of a conjecture.

In August 1684, Edmond Halley (1656–1743) visited the 41-year-old Lucasian Professor at Cambridge. He asked the question that had been consuming him and his friends Hooke and Wren at the Royal Society in London: What path would the planets describe if they were attracted to the sun with a force varying inversely as the square of the distance between them? Newton replied at once that the orbits would be ellipses. Since this was the expected answer, Halley asked Newton how he knew. Newton astonished him by answering that he had calculated it. Halley asked to see Newton's computation, but as Newton seemingly saved every scrap of paper he ever wrote on, he (not surprisingly) could not find it. Perhaps he did not want to find it; his desire to be left alone to pursue his own interests, his fear of controversy, and his reluctance to publish would all make Newton want to carefully check his proof over again before he showed it to anyone. In November of 1684, Newton did send the computation to Halley in London, who was so excited that he prompted Newton to expand his work. (Weinstock [25] has challenged the common view that this proof actually appears in the *Principia*.)

Newton put aside his alchemical and theological studies to work on what was to become the most significant scientific treatise ever written: *Philosophiae Naturalis Principia Mathematica*. It took Newton eighteen months of intense intellectual effort to compose his masterpiece, but the time was not just spent in writing up results that he had completed long ago. Many critical ideas in the *Principia* were not developed until the treatise

FIGURE 12

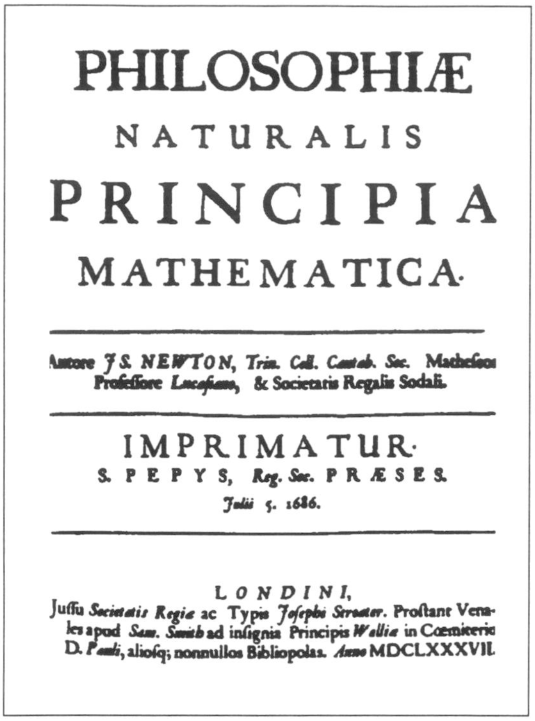

PHILOSOPHIÆ
NATURALIS
PRINCIPIA
MATHEMATICA.

Autore JS. NEWTON, Trin. Coll. Cantab. Soc. Mathefeos
Profeffore Lucafiano, & Societatis Regalis Sodali.

IMPRIMATUR·
S. PEPYS, Reg. Soc. PRÆSES.
Julii 5. 1686.

LONDINI,
Juffu Societatis Regiæ ac Typis Jofephi Streater. Proftant Vena-
les apud Sam. Smith ad infignia Principis Walliæ in Cœmiterio
D. Pauli, aliofq; nonnullos Bibliopolas. Anno MDCLXXXVII.

was being written. In particular, Newton created the concept of universal gravity during this period.

Myth: *Though Newton used the notation of the calculus in arriving at his results, he was careful in the* Principia *to recast all the work in the form of classical Greek geometry understandable by other mathematicians and astronomers.*

Newton started this myth himself in the midst of the priority dispute with Leibniz. If he could argue that he had used his calculus in composing the *Principia*, then he could claim that he did not steal the calculus from Leibniz who published his first paper on the (differential) calculus in 1684.

> The method of fluxions [Newton's calculus] is intrinsically algebraic rather than geometrical, and there is not the slightest reason—in the historical evidence or in logic—to suppose that the argument of the *Principia* was ever cast in an algebraic rather than the geometric mode in which it was published. [9, p. 28]

The geometrical format of the *Principia* is explained by the fact that around 1678, Newton became fascinated with classical geometry [Figure 9]. The *Principia* appears to be densely packed classical geometry, but that is only a façade. One need only read a bit to realize that it is packed with the informal geometrical *ideas* of the new analysis, the calculus. However, the formal machinery of the algebraic algorithms of the calculus is not to be found there. In order to make this point clear, it would help to look at the proof of a proposition from the *Principia*. Book I, Section II, Proposition I, Theorem I says:

> The areas which revolving bodies describe by radii drawn to an immovable centre of force do lie in the same immovable planes, and are proportional to the times in which they are described. [17, p. 40]

That is, if the gravitational force (whatever it might be) always acts toward a fixed point S, then Kepler's equal area law holds.

Newton's proof begins with classical geometry [Figure 13]. Suppose we consider equal time intervals, and that the body moves from A to B in one of those intervals. In the next interval it would move, on the same straight line, from B to c if no external force acts on it. The triangles SAB and SBc that are swept out in these equal time intervals have equal areas since the bases AB and Bc are equal, and the triangles have the same altitude. However, if at B "a centripetal [center seeking] force acts at once with a great impulse," then the body moves to some other point C (in the same plane) in the next time interval. The Parallelogram Law of Forces determines the location of the point C; the lines Cc and SB are parallel. Triangles SBc and SBC also have the same area, since they have a common base SB and their altitudes are equal, namely, the distance between the parallel lines SB and Cc. By transitivity, the triangles SAB and SBC have equal areas. Similarly for other triangles in the diagram. So far the proof was easy geometry. Next, Newton used an idea from his calculus: "Now let the number of those triangles be augmented,

504

FIGURE 13   *Newton's* Principia, *p. 37*

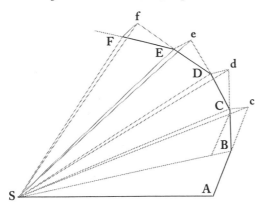

and their breadth diminished *in infinitum* . . . their ultimate perimeter *ADF* will be a curved line: and therefore the centripetal force, by which the body is continually drawn back from the tangent of this curve, will act continually" and the areas traced out in equal times will be equal [17, pp. 40–41].

So that was really quite easy. The geometric ideas of the calculus are used constantly in the *Principia*, but the algebraic notations are not.

CONCLUSION

On 5 February 1675/76, Newton wrote to Hooke [19, I, p. 416]:

> What Des-Cartes did was a good step. You have added much . . . If I have seen further it is by standing on ye shoulders of Giants.

While it is important to realize that Newton recognized the contributions of his predecessors, we must by now feel that Newton was the greatest giant of all. Just as Westfall, after twenty years of effort preparing *Never at Rest,* was more in awe of Newton when he finished than when he began, we too may realize that the closer we get to Newton, even when standing on the shoulders of Whiteside, the bigger the giant becomes.

Yes, Newton was a genius. That is undeniable. But he was not a Greek god. For all his faults, he displayed characteristics that we should tell our students about, for they are the keys to his, and their, success:

- He built on the best work of the past.

- He had brilliant insights.

- He worked "by thinking continually."

* He had stubborn perseverance.

- He steadily expanded his inquiries.

- He made mistakes—and learned from them.

Newton's success was a synergistic combination of innate genius and immense effort. This is the lesson of history.

ACKNOWLEDGMENTS

Portions of this survey paper were presented as lectures at an Ohio Section MAA Short Course on "The History of the Calculus" (Ashland College, July 1986), the University of Wisconsin/ Green Bay, Denison University, and Allegheny College. The many questions of those audiences and the detailed comments of friends and referees were most helpful.

Figures 2, 5, 12, and 13 were obtained courtesy of The Department of Rare Books and Special Collections, The University of Michigan Library; Figure 10 was obtained courtesy of the Center for Archival Collections, Bowling Green State University. [Figures 1 and 4 are reproductions of engravings from *A Portfolio of Eminent Mathematicians,* edited by David Eugene Smith (Chicago: Open Court Publishing Company, 1896).—*Frank Swetz*]

REFERENCES

1. H. J. M. Bos. "On the Representation of Curves in Descartes' *Géométrie.*" *Archive for History of Exact Sciences* 24 (1981): 295–338.

2. Florian Cajori. *William Oughtred, a Great Seventeenth-Century Teacher of Mathematics.* Chicago: Open Court: 1916.

3. René Descartes. *The Geometry of René Descartes.* Translated by D. E. Smith and Marcia L. Latham. Chicago: Open Court, 1925. Reprinted by Dover, 1954.

4. ———. *Discourse on Method, Optics, Geometry, and Meteorology.* Translated by Paul J. Olscamp. Bobbs–Merrill, Indianapolis: The Library of Liberal Arts, 1965.

5. Betty Jo Teeter Dobbs. *The Foundations of Newton's Alchemy or "The Hunting of the Green Lyon."* Cambridge University Press, 1975.

6. Charles Coulston Gillispie (editor). *Dictionary of Scientific Biography.* 16 vols. New York: Scribners, 1970–1980. If there are dates following an individual's name, then there is a signed article in this work concerning him.

7. Derek Gjersten. *The Newton Handbook.* London: Routledge and Kegan Paul, 1986. See abstract in the Media Highlights Column of this CMJ issue.

8. A. W. Grootendorst and J. A. van Maanen. "Van Heureat's Letter (1659) on the Rectification of Curves. Text, Translation (English, Dutch), Commentary." *Nieuw Archief voor wiskunde* 3:30 (1982): 95–113.

9. A. Rupert Hall. *Philosophers at War. The Quarrel Between Newton and Leibniz.* Cambridge University Press, 1980.

10. John Harrison. *The Library of Isaac Newton.* Cambridge University Press, 1978.

11. Joseph E. Hofmann. *Leibniz in Paris, 1672–1676. His Growth to Mathematical Maturity.* Cambridge University Press, 1974.

12. Christiaan Huygens. *The Pendulum Clock or Geometrical Demonstrations Concerning the Motion of Pendula as Applied to Clocks.* Translated by Richard J. Blackwell. Iowa State University Press, 1986.

13. Timothy W. Lenoir. *The Social and Intellectual Roots of Discovery in Seventeenth Century Mathematics.* Ph.D. dissertation. Indiana University, 573 pp., 1974. University Microfilms order number 75–1718.

14. Jan A. van Maanen. "Hendrick van Heureat (1634–1660?): His Life and Mathematical Work." *Centaurus* 27 (1984): 218–79.

15. Frank E. Manuel. *A Portrait of Isaac Newton.* Harvard University Press, 1968; and Washington, D.C.: New Republic Books (paperback), 1979.

16. J. E. McGuire and Martin Tamny. *Certain Philosophical Questions: Newton's Trinity Notebook.* Cambridge University Press, 1983.

17. Isaac Newton. *Sir Isaac Newton's Mathematical Principles of Natural Philosophy and his System of the World.* Revision of Andrew Motte's 1729 translation by Florian Cajori. University of California Press, 1934.

18. ———. *Isaac Newton's Papers and Letters on Natural Philosophy and Related Documents.* Edited by I. B. Cohen. Harvard University Press, 1958.

19. ———. *The Correspondence of Isaac Newton.* Edited by H. W. Turnbull et al. 7 vols. Cambridge University Press, 1959–1977.

20. ———. *The Mathematical Papers of Isaac Newton.* Edited by D. T. Whiteside. 8 vols. Cambridge University Press, 1967–1981. References are by volume and page number.

21. ———. *The Optical Papers of Isaac Newton.* Edited by Alan E. Shapiro. Cambridge University Press, 1984.

22. George Pólya. *Mathematical Discovery.* 2 vols. New York: John Wiley, 1962.

23. J. F. Scott. *The Scientific Work of René Descartes (1596–1650).* London, 1952.

24. John Wallis. *A Treatise on Algebra Both Historical and Practical . . . ,* London, 1685.

25. Robert Weinstock. "Dismantling a Centuries-Old Myth: Newton's *Principia* and Inverse-Square Orbits." *American Journal of Physics,* 50 (1982): 610–17). See abstract in the Media Highlights column of this CMJ issue.

26. Richard S. Westfall. "Award of the 1977 Sarton Medal to D. T. Whiteside." *Isis* 69 (1978): 86–87.

27. ———. "Newton's Marvelous Years of Discovery and Their Aftermath: Myth versus Manuscript." *Isis* 71 (1980): 109–121.

28. ———. *Never at Rest, A Biography of Isaac Newton.* Cambridge University Press, 1980.

29. Richard S. Westfall. "Newton's Theological Manuscripts." *Contemporary Newtonian Research.* Edited

by Z. Bechler. Dordrecht: D. Reidel, 1982, pp. 129–43.

30. Derek Thomas Whiteside. "Patterns of Mathematical Thought in the later Seventeenth Century." *Archive for History of Exact Sciences* 1 (1961): 179–388.

31. ———. "Newton's Discovery of the General Binomial Theorem." *Mathematical Gazette* 45 (1961): 175–80.

32. ———. "Sources and Strengths of Newton's Early Mathematical Thought." *The Annus Mirabilis of Sir Isaac Newton, 1666–1966.* Edited by Robert Palter. M.I.T. Press, 1970, pp. 69–85.

33. ———. "Newton the Mathematician." *Contemporary Newtonian Research.* Edited by Z. Bechler. Dordrecht: D. Reidel, 1982, pp. 109–27.

34. ———. "Newtonian Motion." *Isis* 73 (1982): 100–107. Essay review of [28].

Newton's Method of Fluxions

In perfecting his techniques of "differentiation," Isaac Newton borrowed heavily on the work of his predecessors, particularly John Wallis and Isaac Barrow. Combining their results with the insights of his own genius, Newton produced a technique of differentiation openly based on change. In his *Method of Fluxions* written in 1671, he considers a curve as the path of a moving point, thus the coordinates of the point are in a constant state of change. Newton termed a changing quantity a *fluent* and its rate of change the *fluxion* of the fluent. If a fluent is given by y, then its fluxion would be represented by $\dot{y}$ (in modern terms dy/dt). The fluxion of $\dot{y}$ would be represented by $\ddot{y}$ etc. Using the concept of fluent, Newton then devised the notion of the moment of a fluent, that is, the infinitely small amount of change experienced by a fluent in an infinitely small time interval O. He represented the moment of a fluent by $\dot{x}O$. In calculations, $\dot{x}O$ raised to powers 2 or higher would be zero. Consider how Newton's method of fluxions could be used to obtain the derivative of y with respect to x for $y = 3x^2 + 2x$.

Under the process of change y becomes $(y + \dot{y}O)$ and x becomes $(x + \dot{x}O)$ thus

$$(y + \dot{y}O) = 3(x + \dot{x}O)^2 + 2(x + \dot{x}O) \text{ expanding}$$

$$y + \dot{y}O = 3x^2 + 6x\,\dot{x}O + (\dot{x}O)^2 + 2x + 2\dot{x}O$$

now $(\dot{x}O)^2 = 0$ and $3x^2 + 2x = y$ so

$$\dot{y}O = 6x\,\dot{x}O + 2\dot{x}O \text{ or}$$

$$\dot{y}O = (6x + 2)\,\dot{x}O \text{ and}$$

$$\frac{\dot{y}O}{\dot{x}O} = 6x + 2 \text{ which in modern form}$$

is recognized as $\dfrac{dy}{dx} = 6x + 2$.

77

The Newton-Leibniz Controversy Concerning the Discovery of the Calculus

DOROTHY V. SCHRADER

The State of Analysis in the Seventeenth Century

The seventeenth century was one of activity and advancement in the world of mathematics. Analytic methods had become familiar tools to most of the mathematicians of the period; geometry was being employed to verify and demonstrate analytic conclusions; and special attention was focused on problems dealing with the infinite.

> The time was indeed ripe, in the second half of the seventeenth century, for someone to organize the views, methods, and discoveries involved in the infinitesimal analysis into a new subject characterized by a distinctive method of procedure.[1]

Unfortunately, not one but two men did just that. We say unfortunately, because the methods of the calculus developed by Sir Isaac Newton in England and Gottfried Wilhelm Leibniz on the continent were essentially the same, yet the dispute over the rights of the two discoverers developed into a controversy which has not yet been settled. Both these mathematicians and their followers stooped to tactics which were most unworthy of men of intelligence and honor; as a result, the development of mathematics in England was brought almost to a standstill for a full century.

Reprinted from *Mathematics Teacher* 55 (May, 1962): 385–96; with permission of the National Council of Teachers of Mathematics.

The judgment of history seems to be that credit belongs to both individuals equally.

> [I]t might be far better to speak of the evolution of the calculus. Nevertheless, inasmuch as Newton and Leibniz, apparently independently, invented algorithmic procedures which were universally applicable and which were essentially the same as those employed at the present time in the calculus, . . . there will be no inconsistency involved in thinking of these two men as the inventors of the subject.[2]

It must be remembered, however, that these two inventors are not responsible for the definitions and ideas which are considered basic to the calculus today. Only in the present generation, after more then two centuries of development, has there been laid the foundation of mathematical rigor on which the calculus now rests.

Newton's Method of Fluxions

Newton called his discovery the Method of Fluxions and described it in terms of geometry. The direct method of fluxions can be summarized in the solution of the mechanical problem: "The length of the Space described being continually given, to find the Velocity of the Motion at any time proposed,"[3] or "The relation of the flowing Quantities being given, to determine the relation of their Fluxions."[4] Elsewhere, Newton refers to these flowing quantities as Fluents. The inverse method of fluxions can be summarized in the inverse of the problem given above: "The Velocity of the Motion being continually given, to find the Length of the Space described at any Time proposed,"[5] or "An Equation being proposed including the Fluxions of Quantities, to find the Relation of those Quantities to one another."[6] These direct and inverse methods of fluxions are, of course, the familiar differential and integral calculus. Newton further described his fluents as quantities which are to be considered as gradually and indefinitely increasing; these he represented by the last letters of the alphabet: x, y, z. The velocities by which every fluent is increased by its generating motion, he called fluxions and designated by "pointed" or "prickt" letters, corresponding to the fluents involved: $\dot{x}$, $\dot{y}$, $\dot{z}$. The moment of a fluent, its velocity multiplied by an infinitely small quantity, o, he represented in the fluxional notation as $\dot{x}o$.[7]

Leibniz's Differential and Integral Calculus

Instead of the flowing quantities and velocities of Newton, Leibniz worked with infinitely small differences and sums. He used the now familiar $\frac{dy}{dx}$ instead of Newton's dotted letters for the derivative symbol; he used $\int$ for his integration symbol while Newton used either words or a rectangle enclosing the function. Newton himself asserted that his "prickt" letters were equivalent to Leibniz's $\frac{dy}{dx}$ and that by $\boxed{\frac{aa}{64x}}$ he meant the same thing that Leibniz meant by $\int \frac{aa}{64x}$.[8]

Leibniz was more interested in developing a notation for his new method than was Newton. The Englishman used the dot symbolism in his fluxionary calculus but did not employ it in his treatise on analysis nor in his famous *Principia*. In one report, printed anonymously but commonly believed to be written by Newton, we read, "Mr. Newton doth not place his Method in Forms of Symbols, nor confine himself to any particular

Sort of symbols for Fluents and Fluxions."[9] This independence of symbols, which Newton apparently thought praiseworthy, is today considered one of the major weaknesses of his method. Modern mathematics is almost wholly dependent upon the symbols by which it is expressed, so much so that someone has characterized mathematics as the science in which one operates with and on symbols, neither knowing nor caring what these symbols mean, if indeed they have any meaning. Felix Klein has lauded Leibniz for that very type of symbolism. He notes that $\int y$ and not $\int y \, dy$ was used in Leibniz's manuscripts, and credits him with being the founder of modern, formal mathematics for recognizing that it makes no difference what, if any, meaning is attached to the differentials, but that, if appropriate rules of operation are defined for them and the rules correctly applied, something reasonable and correct will result.[10]

Whether or not Klein errs in attributing too deep a perception to Leibniz is difficult to determine, but it is true that Leibniz was much concerned with finding the best possible notation for his calculus. He experimented with various symbols, explained them to different people, asked advice of a number of mathematicians, and used the dx and dy notation for a long time before he published it. He finally selected the particular form of notation because he saw the great need of being able to identify easily the variable and its differential.[11] Johann Bernoulli, who worked with him on integration, wanted to call the new branch of mathematics "calculus integralis" and use I as an integration symbol; Leibniz preferred the name "calculus summatorius" and the symbol $\int$. Today, we use Bernoulli's name and Leibniz's symbol.

The method of fluxions and the differential and integral calculus differ in more than their notation. While the two methods are essentially the same in that they can be reduced to a common method, they start from different principles. Newton made use of infinitely small quantities to find time derivatives, which he called fluxions; he stated specifically that his mathematical quantities were to be considered as described by continuous motion, not as existing in infinitely small parts. Leibniz, on the other hand, made the infinitely small quantities themselves the basic concepts in his differentials.[12] Newton dealt with a finite quantity which is the ratio of two infinitely small quantities, the ratio of velocities; Leibniz dealt with the finite sum of an infinite number of infinitely small quantities.[13]

The Quarrel

The famed Newton-Leibniz controversy concerning the discovery of the calculus involves more than merely the question of priority of time. Mutual accusations of plagiarism, secrecy which manifested itself in cryptograms, letters published anonymously, treatises withheld from publication, assertions of friends and supporters of the two men, national jealousies, and the efforts of would-be peacemakers, all serve to complicate the situation and make such a tangle of truth and falsehood, information and misinformation, that it can probably never be solved conclusively. We can at best review the major events and draw a few general conclusions.

In June, 1669, Newton, who had become interested in mathematical analysis during his days as an undergraduate at Cambridge, sent to Isaac Barrow, the geometer, a manuscript entitled "Analysis per Equations Numero Terminorum Infinitas," in which the underlying principles of the theory of fluxions were indicated. Barrow was impressed with the work, and, in a letter dated June 20, 1669, mentioned it to a mathematician friend, Collins. On July 31 of the same year, he sent the manuscript to Collins, who copied it and returned the original to Barrow. Collins was in correspondence with many of the leading mathematicians of England and the continent; in letters dated from 1669 to 1672, he communicated Newton's discoveries to Gregory, Bertet, Vernon, Slusius, Borelli, Strode, Townsend, and Oldenburg.[14]

At about the same time, Leibniz was working on problems suggested by the theories of Cavalieri; in 1671, he dedicated to the French Academy a paper, "Theoria Motus Abstracti," in which he showed that he was considering the use of infinitely small quantities in these problems.[15] This was not a well-developed theory of differential calculus, but it does indicate that Leibniz's mind was working along the lines of infinitesimal analysis, as was Newton's.

In 1672, Newton composed a treatise on fluxions, which, however, was not published until John Colson translated it from Latin into English and published it in London in 1736 under the title, *Method of Fluxions*. Why it was not published at the time it was written seems not to be known.

Early in 1673, Leibniz was in London where he visited Oldenburg, a fellow-countryman and the secretary of the Royal Society. As he attended meetings of the society and read mathematical papers before it, it is quite possible that he met Collins, who was a friend of Oldenburg. In 1890, notes of this London visit were discovered in the royal library at Hanover, where Leibniz had been librarian. These notes show extracts from Newton's work on optics, which was not published until 1704, but make no mention of mathematics. Hathaway[16] finds this omission very strange and somehow indicative that Leibniz saw Newton's paper, "De Analysis," during this visit.

By March of 1673, Leibniz was back in Paris, where he began a serious study of geometry with Huygens. In July of the same year, he wrote to Oldenburg, discussing his work on series. In reply, Oldenburg told him of some of Newton's and Gregory's discoveries on series and tangents. Leibniz, working with infinitely small sums and differences, defined the general problem of the area of the curve and from it developed the algorithm of the differential and the integral calculus, a logical outcome of the studies on which he had been engaged for several years. Manuscripts in the library at Hanover show that by the end of 1675 he had a clear idea of the principles of the calculus and had invented the notation.[17]

On June 13, 1671, Newton, answering a request from Leibniz, wrote a brief account of his method of quadrature by means of infinite series and discussed the binomial theorem. Leibniz replied, August 27, 1676, and asked for more details. In September of that year, Leibniz was again in London, where he spent a week with Collins, saw Newton's manuscript of the "De Analysis," and made copious notes from it. One author asserts that "since its pages were open freely to him at that time it is constructive proof that they were as freely open to him for the two months in 1673 that he was in London."[18] This reasoning seems obscure, but it may be that Hathaway's testimony is colored by anti-German feeling,[19] which is perhaps understandable, as he wrote shortly after World War I.

After his return to Paris, Leibniz received another letter from Newton, dated October 24 and sent through Oldenburg. This was fifteen closely written pages, discussing series and mentioning fluxions, but giving no detailed information. Newton himself said later that he had told Leibniz of his method of fluxions but disguised it in an anagram of transposed characters under the sentence, "Data aequatione quotcunque fluentes quantitates involvente, fluxiones invenire, et vice versa," which, transliterated, reads, "Given any equation whatsoever involving flowing quantities, to find the fluxions, and vice versa." This as an anagram would be: 6a 2c d ae 13e 2f 7i 3l 9n 4o 4q 2r 4s 9t 12v x. One wonders how much Leibniz could ever make out of that anagram, and having reconstructed Newton's sentence, could he deduce the method of fluxions from it? "Whoever can form a certain sentence properly out of 6 a's, 2 c's, a d and so on, will see as much as one sentence can show about Newton's mode of proceeding."[20]

The question immediately arises as to why Newton bothered to tell Leibniz anything at all if he was going to conceal it so thoroughly in his anagram. The device is not unprecedented. Galileo used to give his discoveries to his friends in the form of carefully dated cryptograms in order to establish priority. Sometimes academies and learned societies were the trustees for intellectual secrets. Such secrecy was considered necessary to protect the rights of the inventors. There was much jealousy among the learned who, in their desire to conceal their discoveries, were wont to publish theorems without proof or demonstration. Newton deposited his method, by anagram, in the hands of his rival.[21]

There is another question concerned with this famous anagram. If, as has been frequently stated, Newton had given a clear indication of his method of fluxions in his "De Analysis" of 1669, which was, with his knowledge, being circulated and discussed by Collins, why did he consider it necessary seven years later to conceal the method? Could it be that he had merely hinted at it in 1669 and did not have it so well-developed as has been supposed?

After receiving Newton's letter, Leibniz replied on June 21, 1677, through Oldenburg, that he too had a method of drawing tangents, not by fluxions but by differentials; he quite frankly and openly explained the differential calculus and its applications. However, he did not tell Newton that he had seen the 1669 manuscript in London but a few weeks before, that he knew of Newton's work, and that the anagram was useless. Was Leibniz being unfair to Newton or was the 1669 treatise less complete than it is reputed to have been? Newton did not answer and the correspondence ceased, perhaps due to the death of Oldenburg which occurred in August of 1678.

In 1683 Collins died; in 1684, in the *Acta Eruditorum* of Leipzig, Leibniz published his first paper on the calculus, an account similar to that which he had given Newton. If he had obtained his initial ideas from Newton via Collins, that might explain his desire to conceal the theft from Collins by not publishing the ideas as his own until after the death of Collins. Or is there another reason for the long delay in publication? Leibniz made a vague reference to Newton's having a method similar to his own but he made no claim to being the first or sole inventor. The first "Nova Methodus pro Maximis et Minimis" merely developed the rules for differentiation. Later works in the same publication gave an exposition of the principles from a formal viewpoint.[22]

Leibniz and the two Bernoullis were making rapid progress with the new and powerful analytic method when the first edition of Newton's *Principia* was published in 1687. In Book II, Lemma II, Newton explained the fundamental principle of the fluxionary calculus and in a scholium added:

> In letters which went between me and that most excellent geometer, G. W. Leibniz, ten years ago, when I signified that I was in the knowledge of a method of determining maxima and minima, of drawing tangents, and the like, and when I concealed it in transposed letters involving this sentence (Data aequatione, quotcunque, fluentes quantitates involunte, fluxiones invenire, et vice versa; that is, Having given any equation involving ever so many flowing quantities, to find the fluxions, and vice versa) that most distinguished man wrote back that he had also fallen on a method, which hardly differed from mine, except in his forms of words and symbols.[23]

Later, when the controversy was at its height, this scholium was quoted as evidence that Newton recognized Leibniz's rights as a second or simultaneous inventor. Such is not entirely the case; all Newton admitted was that Leibniz did have a method, however he learned it. Newton, after Leibniz's death, asserted that the scholium had been intended as a challenge to Leibniz to prove his priority if he could, not as an admission of his equality. In the second edition of the *Principia*, Newton added a phrase to the scholium, making it a bit more accurate. In that edition, the last words read, " . . . which hardly differed from mine except in the forms of words and symbols, and the concept of the generation of quantities."[24] In the third edition, the scholium was changed entirely and another subject inserted; neither Leibniz's name nor his work was mentioned.

In 1695, Dr. John Wallis chided Newton for being in possession of the method of fluxions for thirty years and never publishing it in its entirety; he had made reference to it in his own complete works, published in 1693, but felt that his treatment of the subject was most inadequate.

In the same year, John Bernoulli challenged Europe with two problems, to be solved in six months. To allow the mathematicians of the world time to work on these problems, Leibniz requested an additional year, which was granted. During this extension, Newton heard of the problems and solved them both in a single evening, sending his solutions to the president of the Royal Society the day after he had received the problems. Acknowledging the receipt of Newton's solutions, Leibniz, in the Leipzig *Acts*, managed to convey the impression that Newton was a pupil of his who, because he had mastered the calculus, was able to solve the problems.[25]

It was four years later, in 1699, that the hidden rivalry flared into open hostility. Fatio de Duillier, a Swiss mathematician from Geneva, who had been living in England for about ten years and was a close friend of Newton, published a memoir in which he claimed for himself independent invention of the calculus (which claim seems to have been completely ignored) and implicitly accused Leibniz of plagiarizing from Newton.

> I am bound to acknowledge that Newton was the first, and by many years the first, inventor of this calculus: from whom, whether Leibniz, the second inventor, borrowed anything, I prefer that the decision should lie, not with me, but with others who have had sight of the papers of Newton, and other additions to this same manuscript.[26]

Leibniz answered the charge, referred to Newton's scholium in the *Principia*, and, ignoring the question of priority, insisted upon his right to credit for the invention of the differential calculus. Newton ignored the entire situation and the Leipzig *Acts* refused to print De Duillier's reply to Leibniz.

Here the matter rested until Newton's *Opticks* came out in 1704. Published with the text on optics was a short treatise explaining the method of fluxions and commenting that, since theorems from the 1669 manuscript were appearing in various guises, it seemed best to the author to make the method public now. The next January, the *Acta Eruditorum* of Leipzig carried an anonymous review, later shown to have been written by Leibniz, stating:

> [t]he elements of this calculus have been given to the public by its inventor, Dr. Gottfried Wilhelm Leibniz in these *Acts*. . . . Instead of the Leibnizian differences, then, Dr. Newton employs and always has employed, fluxions, which are very much the same as the augments of fluents produced in the least intervals of time, and these fluxions he has used elegantly in his *Mathematical Principles of Natural Philosophy* and in other later publications, just as Honoratus Fabri, in his *Synopsis of Geometry* substituted progressive methods for the method of Cavalieri.[27]

This innocent-sounding review provoked a storm of opposition in England. Far from being a compliment to the acuteness of the Englishman, it was, by the comparison with Fabri, a scarcely veiled attack upon Newton's integrity. Fabri had been a notorious plagiarist, discredited and dishonored in the mathematical world for his theft of the ideas of another.[28] At the time, Leibniz denied authorship of the review, although later[29] he tacitly admitted it and gave his approval. While this attack may not have been entirely unprovoked, it most certainly was cowardly and unworthy of a man of Leibniz's standing.

John Keill, a friend of Newton, in a letter to Edmond Halley, on the laws of centripetal force, directly and openly accused Leibniz of plagiarism.

> All these laws follow from that very celebrated arithmetic of fluxions which, without any doubt, Dr. Newton invented first, as can readily be proved by anyone who reads the letters about it published by Wallis; yet the same arithmetic afterwards, under a changed name and method of notation, was published by Dr. Leibniz in *Acta Eruditorum*.[30]

When Leibniz received his copy of *Transactions*, in which the letter was published (Volume XXVI, # 317) he wrote to Sloane, the secretary of the Royal Society, demanding that Keill retract his accusation; he added that he felt sure that Keill had acted from rashness and not from improper motives, and that he would not consider the attack a matter of calumny. When the letter was read to the Royal Society, Newton, who was then the president, expressed displeasure at Keill's action. Keill justified himself by producing the review of Newton's *Opticks* in the Leipzig *Acts*, of which Newton had apparently been unaware until this time. Keill wrote to Leibniz on May 24, 1711, saying that he had not accused Leibniz of knowing the name or notation of Newton's method but that he had merely stated that Leibniz must have seen something from which he had been led to his own method. Instead of being mollified, Leibniz was outraged and declared that his honor was attacked even more openly than before; that Keill was an upstart and an unqualified judge; that he was acting without any authority from Newton; that it was the duty of the Royal Society to silence Keill; that he wanted Newton's own opinion directly expressed; and that in Leipzig *Acts* review, "no injustice had been done to any party, as everyone had received only his due."[31] By this last comment, Leibniz made that opinion his own and at once became the aggressor instead of the injured victim in the controversy.

An Attempt to Settle the Question

A committee of the Royal Society was appointed to investigate the situation, examine the documents which had been placed in the archives, and make a decision. The committee members, appointed on March 6, 20, 27, and April 17, 1712, were Halley, Jones, De Moivre, and Machin, friends of Newton and mathematicians; Brook Taylor, a friend of Keill; Robarts, Hill, Burnet, Ashton, Arbuthnot; and Bonet, the Prussian minister. With the exception of de Moivre and Bonet, they were all Englishmen, and almost all were friends of Newton; it was scarcely an unbiased group. There was no chance for Leibniz to give his side of the story or to produce any papers he might have to substantiate his claims. Burnet wrote to Bernouilli that the committee was busy about proving that Leibniz might have seen Newton's papers.[32] The business was handled quickly, and on April 24 of the same year, the report was read. On January 8, 1713, it was published under the title, *Commercium Epistolicum D. Johannis Collinsii et aliorum de analysi promota*. The report, anonymous but probably written by Newton,[33] contains the findings of the committee and copies of the letters and papers involved in the dispute. The actual report of the committee includes a chronology of Leibniz's contacts with Newtonian influences, an assertion that Newton was the prior inventor, and a statement that Keill had not injured Leibniz by his accusation. The plagiarism issue is not touched, except to clear Newton of any possibility by declaring his priority. The report does tell of one incident in Leibniz's career which, while it is not a direct accusation, implies that he is capable of stealing another's ideas.

> February 1672/3, meeting Dr. Pell at Mr. Boyle's, he pretended to the differential method of Mouton, on being shown that it was Mouton's, insisted that it was his because he hadn't known of Mouton's doing it and had much improved it.[34]

In retrospect, the *Commercium Epistolicum* seems to have been a grossly unfair way of handling the situation. It avoided the main issue and, in effect, told Leibniz that there had been no injustice done to him because Newton had had the method before the time when Leibniz was accused of having stolen it from him; this is a meaningless and utterly illogical conclusion.

Leibniz did not see a copy of the report, but Bernoulli wrote to him about it in a letter dated June 7, 1713, adding that Newton had admitted that he didn't think of his method of fluxions until he read of Leibniz's calculus, and that Newton, when he wrote the *Principia*, had no idea of how to find the fluxions of fluxions; Bernoulli also said that Leibniz might publish his letter if he chose, as

long as he did not disclose the identity of the author. Leibniz printed the letter, with comments, anonymously, without any hint as to where or by whom it was printed, under the title, *Charta Volans*. He circulated the little book among the continental mathematicians and scientists.[35] He seemed convinced that he had been injured and wrote to various mathematicians trying, unsuccessfully, to enlist them on his side; he even attacked, in a letter to the Princess of Wales, Newton's philosophy and religious orthodoxy. He wrote, but never published, *Historia et Origo Calculi Differentialis*[36] sometime between 1714 and his death in 1716, in which, writing in the third person, he gave his version of the discovery and the attempted theft of his method of the differential and integral calculus; this was apparently intended as an answer to the *Commercium Epistolicum*.

A Mr. Chamberlayne tried to make peace between the two warring mathematicians but received the comment from Newton that Leibniz had started the fuss in 1705 with his review of the *Opticks* and a similar statement from Leibniz that it was all Newton's fault due to the *Commercium Epistolicum*. Leibniz tried one last challenge in 1716 when, through Abbe Conti, a Venetian nobleman and priest, he sent a problem to test the ability of the English mathematicians. Newton solved the difficult problem in one evening.

The quarrel gradually subsided after Leibniz's death on November 14, 1716. Bernoulli made advances to Newton, vigorously declaring that he had never said anything against the Englishman and insisting that he had not given Leibniz permission to publish any of his letters. A reconciliation seems to have followed, but there is no record of any further correspondence.

The Judgment of History

It seems that the whole unsavory situation could have been avoided if both the men involved had been frank in their statements and prompt in publishing their findings. How can one account for

Newton's extreme secrecy? Did he hide his methods in order to perfect them before giving them to the public? A laudable but highly imprudent motive. Did he wish to have the methods for his own exclusive use? Inexcusable selfishness. Was he trying to avoid disputes and unpleasantness? Total failure. And how can one account for Leibniz's underhanded methods? Was he striving for the honor and glory of the fatherland? Hardly, since he spent much of his time in Paris and other cities outside of Germany. Was he determined to achieve fame at any cost? But he had won renown for his work. Was he utterly devoid of honor? This is scarcely possible, for he was respected and trusted by eminent friends.

It is obvious from the confused tangle of events, the accusations and counteraccusations, the doubtful statements and bold lies, that any definitive statement or decision is impossible. At best, one can say that Newton was probably the first inventor while Leibniz and Bernoulli were promoters and developers of the calculus. Newton seems to have invented fluxions at least ten years before Leibniz developed the calculus. There is no evidence that Newton borrowed from Leibniz; there is little evidence that Leibniz borrowed from Newton. In the hands of Newton and his followers, fluxions remained a relatively sterile theory, while Leibniz and his followers made of the calculus a powerful means of mathematical progress. To Leibniz and the Bernoullis belongs the credit for most of the vast superstructure which has been erected on the foundation laid by Newton.

> Both men owed a very great deal to their immediate predecessors in the development of the new analysis, and the resulting formulations of Newton and Leibniz were most probably the result of a common anterior, rather than a reciprocal coincident influence.[37]

Claims to the invention of the calculus have been made by or for several mathematicians. Fatio de Duillier's claim was apparently ignored by his contemporaries; indeed, he seemed never to have pressed the issue himself. It has been said that credit for the calculus should go to Barrow or even to Fermat, both of whom did admirable work in laying the foundation for the later discoveries of Newton and Leibniz. But the judgment of history still stands and Sir Isaac Newton and Gottfried Wilhelm Leibniz are generally considered as the two inventors of the differential and integral calculus.[38]

A Century of Isolation in England

Newton's influence on English mathematics was so great that, during the entire eighteenth century, especially at Cambridge, mathematics was confined to the study of optics, gravitation, geometry, and fluxional calculus. The English savants had regarded the struggle with Leibniz and his supporters as an attempt, even a plot, of the Germans to rob Newton of the credit for his invention. In loyalty to its famous son, Cambridge chose Newton's fluxional methods in preference to Leibniz's analytical ones. For some problems, either notation may be used, but for the calculus of variations and for most modern theoretical work, Newtonian notation is impossible. In fact, even Leibnizian notation is proving inadequate for some phases of modern calculus, and that on a fairly elementary level.[39] The relative merit of the two methods was completely obscured by the quarrel over the right to credit for the invention. Personal feelings and national jealousies made the decisions, and as a result, Cambridge withdrew into a sterile isolation.

A common language is essential in the development of a science. By accepting the isolation attendant on its adoption of the Newtonian notation and refusing to make an effort to keep up with the advances made on the continent, Cambridge rendered almost sterile the efforts of the group of truly brilliant followers who had gathered about Newton. The continental mathematicians kept up with whatever English advances there were, translating the Newtonian notation into Leibnizian and thus making the work available to all. However, the English, in the isolation of injured national pride, would not do likewise as the various continental developments were published. Cambridge was out of touch with the continental mathematicians for almost a century, although the journals in which the continental findings were published were circulated widely and gratuitously.[40] As Leibniz's work was interpreted by Bernoulli, Euler, D'Alembert, Lagrange, Laplace, and others, the knowledge of the calculus spread widely among those who would listen and learn. But England was passed by; the history of English calculus led nowhere.[41]

It is not true that the differential notation was entirely unknown in England. John Craig, a friend of Newton, used $dx, dy, dz,$ and $\int$ in articles printed in 1685, 1693, 1701, 1703, 1704, 1708, and 1713. Yet in 1718, he wrote *De Calculo Fluentium* using exclusively Newtonian notations.[42]

De Moivre, in 1702 and 1703, and John Keill, in 1713, used a mixed notation of the form $\int \dot{o}x$, which notation was still being used in some publications as late as 1815.[43] Joseph Fenn, an Irish writer with a fine disregard to national feelings, in a *History of Mathematics*, published in Dublin sometime after 1768, used the Newtonian terms, fluxion and fluent, and the Leibnizian notation. [44]

Maclaurin in Scotland and Clairaut in France seem to be the only non-English mathematicians who used the Newtonian notation. The works of Newton and other Englishmen appearing on the continent were published as they were written, in the fluxional notation, at least until they had been "translated." It is interesting to note that there was one Dutch journal, *Maandelykse Mathematische Liefhebbery* ("Monthly Mathematical Recreations"), published in Amsterdam from 1754 to 1769, which used the Newtonian notation exclusively.[45]

The quality of the work of the English mathematicians declined rapidly after the break with the continent. Isolation accounted for part of that decline, but there was also another reason. When Newton began his work on fluxions, he was aware

that he was dealing with new concepts which would be accepted only if the proofs were unimpeachable. In order to avoid having his ideas rejected because of questionable proofs, he shunned the new (1637) analytic geometry of Descartes and used only the methods of classical geometry.[46] At times he used other methods to discover theorems and derive proofs, but always he confined his final demonstrations to geometry and elementary algebra. Geometric proofs are, in themselves, adequate, but they are often labored and unnecessarily complex. Moreover, separate demonstration is required for each kind of problem; the processes are not general as they are in analysis.[47] However, long after the principles of analytic geometry and analysis were commonly accepted and freely used by the continental mathematicians, the English analysts remained true to the traditions of the master. Thus the situation stood for almost a century.

Some records of the Senate House examinations at Cambridge have come down to us, showing the general tenor of the calculus work being done at the university which was Newton's Alma Mater, a stronghold of Newtonian mathematics and physics. The 1772 examinations included:

> the doctrine of fluxions, and its application to the solution of questions de maximis et minimis, to the finding of areas, . . . as unfolded and exemplified, in the fluxional treatises of Lyons, Saunderson, Simpson, Emerson, Maclaurin, and Newton . . .[48]

In 1785, the examinees were required to

find the fluent of $\dot{x} \sqrt{a^2 - x^2}$

and to find, by the method of fluxions,

> the number from which, if you take its square, there shall remain the greatest difference possible.[49]

The 1786 examination contained problems which were a little more difficult:

> To find the fluxion of $x^2(y^n + z^n)^{1/q}$.
> To find the fluxion of the mth power of the Logarithm of x.

To find the fluent of $\dfrac{ax}{a+x}$.[50]

By 1802, the students were subjected to the following types of problems:

Find the fluents of the quantities

$$\frac{d\dot{x}}{x(a^2 - x^2)} \quad \text{and} \quad \frac{h\dot{y}}{y(a+y)^{3/2}} \; .$$

Given the fluent: $(a + cz^n)^m \times z^{pn+n-1}\,\dot{z}$, find the fluent: $(a + cz^n)^{m+1} \times z^{pn-1}\dot{z}$.

Required also the fluent of

$$\frac{\dot{x}\,\sqrt{a^2 + x^2}}{x^3} \quad \text{and of} \quad \frac{z^\theta \dot{z}}{1 + mz} \; ,$$

θ being a whole positive number.[51]

The Return to Analysis

Towards the end of the eighteenth century, the more thoughtful mathematicians at Cambridge began to suspect the evils that were consequent upon their separation from news of continental mathematics. The logical thing to do was to adopt the Leibnizian notation and methods, but there was a sentimental objection to such an action; would not such a move be an act of disloyalty to the memory of the great Newton? Finally, in 1803, Robert Woodhouse, then a tutor and later a professor at Cambridge, wrote *Principles of Analytic Calculation,* a work which explained the differential notation and advocated its adoption. Woodhouse criticized the continental methods in some points, especially in the use of principles which were neither obvious nor enunciated. By exposing some of the errors of the system, he gave the impression that he was as much against as he was for the analytic system, thus gaining a hearing among those who would have opposed the system on the basis of those very errors. His writings seem to have been ignored by most of the professors but some of the more serious students read and wondered. As soon as they were aware of the great amount of mathematics which had been closed to them, they obtained the continental books and

began to read and study. Unusual answers began to appear on some of the examinations.[52]

"A man like Woodhouse, of scrupulous honor, universally respected, a trained logician and with a caustic wit, was well fitted to introduce the new system."[53] Nevertheless, the movement might have died with him if it had not been for George Peacock, who, with Herschel and Babbage, formed an Analytical Society in 1812. As undergraduates, they habitually breakfasted together on Sunday mornings, and out of these meetings and their common interest in mathematics, the society grew.

George Peacock (1791–1858) received his B.A. from Trinity College in 1813 as second wrangler.[54] He received a fellowship in 1814 and later became a tutor. Well loved by his students, he was a brilliant lecturer and a kindly and practical tutor, indeed a rare combination of qualities. The establishment of the University Observatory was largely due to his efforts.[55]

Sir John Frederick William Herschel (1792–1871), the son of an astronomer, entered St. John's College at Cambridge in 1809 and graduated as senior wrangler in 1813. While still an undergraduate, he wrote a paper on Cotes's theorem; he published several other papers later. He left the University about 1816 and became an astronomer and chemist.[56]

Charles Babbage (b. 1792), who entered Trinity in 1810, had had a good mathematical education before his arrival at Cambridge, having studied the works of Ditton, Maclaurin, and Simpson on fluxions, Agnesi's *Analysis* in fluxional notation, Woodhouse's *Principles of Analytic Calculation,* and Lagrange's *Théorie des Fonctions.* In 1813, Babbage transferred to Peterhouse because he wanted a chance to be first in the examinations and knew that Peacock and Herschel would surpass him if he tried to compete against them. A many-sided personality, he held a professorship, invented a machine for arithmetical processes, and wrote several scientific papers.[57]

These three eager young students, together with Maule, Ryan, Robinson, and D'Arblay, formed the original membership of the Analytical Society, whose aim was, according to Babbage, to advocate "the principles of pure d-ism as opposed to the dot-age of the university."[58] The Society published, in 1819, a translation of Lacroix's *Elementary Differential Calculus*. Peacock, who, as moderator of the Senate House examinations, was in a position to advance the cause, introduced the differential notation in the examination of 1817. In the same year, a colleague, John White, used the fluxional notation. Peacock was criticized but went on his way, feeling that the younger generation of students was ready to accept the change and that the time was right to

> reduce the many-headed monster of prejudice and make the university answer her character as the loving mother of good learning and science.[59]

The differential notation was again used by Peacock in the 1819 examinations, by Whewell in 1820, and by Peacock again in 1821. Whewell published a work on mechanics in 1819, using the differential notation; Peacock's volume on differential and integral calculus was published by the Analytical Society in 1820; Herschel's work on the calculus of finite differences, illustrative of the new method, came out in the same year; Airy, a pupil of Peacock, published *Tracts* in 1826, a work in which the new method was successfully applied to mechanics. By this time, the exclusive use of fluxions had disappeared among all but a few of the older professors.[60]

From Cambridge, the use of analytical methods spread rapidly through the rest of England. By 1830, the fluxional methods and geometric proofs had very largely disappeared, for, while the geometric demonstration is a useful auxiliary to analysis, it is almost useless as a research device.

The results of the Newton-Leibniz controversy in terms of the personal pain and mental disturbance suffered by the two principal protagonists cannot, of course, be adequately judged. The effect of the controversy on the mathematical world seems to be twofold. As far as credit for the discoveries is concerned, today the two men are honored equally as two independent inventors. Concerning the development of analysis, England seems to have been the loser. The world of mathematics progressed. German and French mathematicians established reputations for themselves and their countries, while England remained insular and isolated. What was the cost to English mathematics? Perhaps no one will ever know. British mathematicians and scientists have done much in the last hundred years for the honor of England and the advancement of knowledge. Nevertheless, one cannot but regret the irreparable loss occasioned by that "dark age" in intellectual history.

NOTES

1. Carl B. Boyer, *The Concepts of the Calculus* (Wakefield, Mass.: 1949), p. 187.

2. *Ibid.*, pp. 187–88.

3. Sir Isaac Newton, *The Method of Fluxions* (London: 1736), p. 19.

4. *Ibid.*, p. 21.

5. *Ibid.*, p. 19.

6. *Ibid.*, p. 25.

7. *Ibid.*, p. 20.

8. J. Edleston (ed.), *Correspondence of Sir Isaac Newton and Professor Cotes* (London: 1850), p. 169.

9. "Commercium Epistolicum," *Philosophical Transactions*, No. 342, January, February 1714/15 (London: 1717), p. 204.

10. Felix Klein, *Elementary Mathematics from an Advanced Standpoint*, Part I (New York: 1932), p. 215.

11. Florian Cajori, *A History of Mathematical Notation*, Vol. II (Chicago: Open Court Publishing Co., 1929), p. 181.

12. Sir Isaac Newton, *Mathematical Principles of Natural Philosophy* (Berkeley, Calif.: University of California, 1947), pp. 655–56.

13. John Theodore Merz, *Leibniz* (New York: 1948), p. 60.

14. David Brewster, *The Life of Sir Isaac Newton* (New York: 1831), pp. 175–76.

15. Augustus De Morgan, *Essays on the Life and Work of Newton* (Chicago: Open Court Publishing Co., 1914), pp. 95–96.

16. Arthur S. Hathaway, "The Discovery of the Calculus," *Science,* New Series, Vol. L, No. 1280 (July–December, 1919), pp. 41–43.

17. Merz, *op. cit.,* pp. 54–55.

18. Hathaway, *op. cit.,* p. 42.

19. Hathaway says that Leibniz's methods here described are like the methods of German propaganda in world War I and that Leibniz deserves no honor at all even if he were an independent discoverer because "he does not come into court with clean hands." *Op. cit.,* p. 43.

20. De Morgan, *op. cit.,* p. 25 (note).

21. Merz, *op. cit.,* pp. 58–59.

22. Klein, *loc, cit.*

23. Newton, *Mathematical Principles, loc. cit.*

24. *Ibid.*

25. J. W. N. Sullivan, *Isaac Newton* (New York: 1938), p. 229.

26. Gottfried Wilhelm Leibniz, *The Early Mathematical Manuscripts of Leibniz,* J. M. Child (ed.) (Chicago: Open Court Publishing Co., 1920), p. 8.

27. Sullivan, *op. cit.,* p. 232.

28. *Ibid.*

29. See below.

30. Sullivan, *op. cit.,* p. 234.

31. Brewster, *op. cit.,* p. 189.

32. De Morgan, *op. cit.,* pp. 27–28 (note).

33. Augustus De Morgan, "On the Authorship of the Account of the *Commercium Epistolicum* in the *Philosophical Transactions,*" in *Philosophical Magazine,* 4th series, Vol. 3 (1852), pp. 440–43.

34. "Commercium Epistolicum," *Philosophical Transactions,* p. 183.

35. Brewster, *op. cit.,* pp. 192–93.

36. Leibniz, *op. cit.,* pp. 22–57.

37. Boyer, *op. cit.,* p. 188.

38. For a discussion of the validity of other claims, see Florian Cajori, "Who Was the First Inventor of Calculus?" *American Mathematical Monthly,* XXVI (1919), 15–20; and J. M. Child, "Barrow, Newton, and Leibniz, in Their Relation to the Discovery of the Calculus," *Science Progress,* XXV (1930–31), 295–307.

39. R. Creighton Buck, *Advanced Calculus* (New York: McGraw-Hill Book Co., Inc., 1956), pp. 58–59.

40. W. W. Rouse Ball, *A History of the Study of Mathematics at Cambridge* (Cambridge: 1889), p. 98.

41. *Ibid.*

42. Florian Cajori, "The Spread of Newtonian and Leibnizian Notations of the Calculus," *Bulletin of the American Mathematical Society,* XXVII (June–July, 1921), 453.

43. *Ibid.,* p. 454.

44. *Ibid.*

45. *Ibid.,* p. 455.

46. Ball, *op. cit.,* p. 69.

47. *Ibid.,* p. 98.

48. *Ibid.,* p. 192.

49. *Ibid.,* pp. 195–96.

50. *Ibid.*

51. *Ibid.,* pp. 200–09.

52. *Ibid.,* p. 119.

53. *Ibid.*

54. Wrangler—an honors student, in the first class in the mathematical Tripos. First ranking man is designated as senior wrangler, next as second, third, fourth, etc., wranglers.

55. Ball, *op. cit.,* p. 124.

56. *Ibid.,* pp. 126–27.

57. *Ibid.,* pp. 125–26.

58. *Ibid.,* p. 126.

59. *Ibid.,* p. 121.

60. *Ibid.,* p. 122.

Mengoli's Proof for the Divergence of the Harmonic Series

A nemesis of every first year calculus student is the harmonic series, an innocent looking sum of the form:

$$1 + \frac{1}{2} + \frac{1}{3} + \frac{1}{4} + \frac{1}{5} + \ldots$$

which proceeds forever in the same pattern. Its individual terms become very small approaching zero but what of the sum itself? Can it be found?—that is, is there a finite number that this sum approaches in its total? If such a finite number exists then the series is said to *converge;* if no such number exists, then the series is said to approach infinity or to *diverge.* Intuition would lead many people to believe that this series approaches some definite sum and therefore converges; however, this is not the case. Since at least the fourteenth century, mathematicians have attempted to ascertain the behavior of the harmonic series. An early proof for the divergence of the harmonic series was given by the French cleric and natural philosopher Nicole Oresme (ca. 1323–1382) and later in the eighteenth century one was devised by the Swiss mathematician Jakob Bernoulli (1645–1705). Over the years, these have remained the most popular two proofs used to demonstrate the divergence of the harmonic series; however, other proofs exist. In 1647, the Italian mathematician Pietro Mengoli (1625–1686) devised a noteworthy proof. It goes as follows:

Given a number n, where $n > 1$, then $\dfrac{1}{n-1} + \dfrac{1}{n} + \dfrac{1}{n+1} > \dfrac{3}{n}$

The truth of this statement can be shown by simple algebra. Let us call this mathematical fact proposition *A.* Now consider the sum of the harmonic series as symbolized by S where

$$S = 1 + \frac{1}{2} + \frac{1}{3} + \frac{1}{4} + \frac{1}{5} + \frac{1}{6} + \frac{1}{7} + \frac{1}{8} + \frac{1}{9} + \frac{1}{10} + \ldots$$

Using brackets, the sum, S, can be divided into a series of numerical groupings.

$$S = 1 + \left(\frac{1}{2} + \frac{1}{3} + \frac{1}{4} \right) + \left(\frac{1}{5} + \frac{1}{6} + \frac{1}{7} \right) + \left(\frac{1}{8} + \frac{1}{9} + \frac{1}{10} \right) + \ldots$$

(continued)

Applying proposition A to each of these groupings, we find:

$$\left(\frac{1}{2} + \frac{1}{3} + \frac{1}{4} \right) > \frac{3}{3} = 1$$

$$\left(\frac{1}{5} + \frac{1}{6} + \frac{1}{7} \right) > \frac{3}{6} = \frac{1}{2}$$

$$\left(\frac{1}{8} + \frac{1}{9} + \frac{1}{10} \right) > \frac{3}{9} = \frac{1}{3}$$

therefore:

$$S > 1 + 1 + \frac{1}{2} + \frac{1}{3} + \ldots$$

This process can be repeated over and over again, theoretically forever showing that S is larger than any finite number. Conclusion, the harmonic series diverges.

78

The Bernoulli Family*

HOWARD EVES

$\mathcal{T}$HERE is a general rule to the effect that any given family possesses at most one outstanding mathematician and that, in fact, most families possess none. Thus a search through the ancestors, descendants, and relatives of Isaac Newton fails to turn up any other great mathematician. There are exceptions to this general rule. For example we have, here in the United States, the two Lehmers (father and son) and the two Birkoffs (father and son). One also recalls the two Cassinis (father and son) of the late seventeenth and early eighteenth centuries, and perhaps one can build a case for the two Clairaut children of the eighteenth century. And of course there were Theon and Hypatia (father and daughter), who lived during the closing years of ancient Greek mathematics. But such cases are relatively rare. All the more striking, then, is the Bernoulli family of Switzerland, which in three successive generations produced no less than eight noted mathematicians.

The Bernoulli family record starts with the two brothers Jakob Bernoulli (1654–1705) and Johann Bernoulli[†] (1667–1748). These two men gave up earlier vocational interests and became mathematicians when Leibniz's papers began to appear in the *Acta eruditorum.* They were among the first mathematicians to realize the surprising power of the calculus and to apply the tool to a great diversity of problems. From 1687 until his death, Jakob occupied the mathematical chair at Basel University. Johann, in 1697 , became a professor at Groningen University, and then, on Jakob's death in 1705, succeeded his brother in the chair at Basel University, to remain there for the rest of his active life. The two brothers, often bitter rivals, maintained an almost constant exchange of ideas with Leibniz and with each other.

Among Jakob Bernoulli's contributions to mathematics are the early use of polar coordinates, the derivation in both rectangular and polar coordinates of the formula for the radius of curvature of a plane curve, the study of the catenary curve with extensions to strings of variable density and strings under the action of a central force, the study of a number of other higher plane curves, the discovery of the so-called *isochrone*—or curve along which a body will fall with uniform vertical velocity (it turned out to be a semicubical parabola with a vertical cusptangent), the determination of the form taken by an elastic rod fixed at one end and carrying a weight at the other, the form assumed by a flexible rectangular sheet having two opposite edges held horizontally fixed at the same height and loaded with a heavy liquid, and the shape of a rectangular sail filled with wind. He also proposed and discussed the problem of isoperimetric figures (planar closed paths of given species with fixed perimeter which include a maximum area), and was thus one of the first mathematicians to work in the calculus of variations. He was also one of the

* Largely adapted from the author's *An Introduction to the History of Mathematics* (revised edition), Holt, Rinehart and Winston, Inc., 1964.

[†] Referred to by various authors as Johann, Jean, and John.— F.S.

Reprinted from *Mathematics Teacher* 59 (Mar., 1966): 276–78; with permission of the National Council of Teachers of Mathematics.

JACQ. BERNOULLI

FIGURE I   *Jakob Bernoulli*
(Source: *A Portfolio of Eminent Mathematicians*, edited by
David Eugene Smith (Chicago: Open Court Publishing
Company, 1908)

FIGURE 2   *Johann Bernoulli*
(Source: *A Portfolio of Eminent Mathematicians*, edited by
David Eugene Smith (Chicago: Open Court Publishing
Company, 1908)

early students of mathematical probability; his book in this field, the *Ars conjectandi,* was posthumously published in 1713.

There are several things in mathematics which now bear Jakob Bernoulli's name. Among these are the *Bernoulli distribution* and *Bernoulli theorem* of statistics and probability theory, the *Bernoulli equation* met by every student of a first course in differential equations, the *Bernoulli numbers* and *Bernoulli polynomials* of number-theory interest, and the *lemniscate of Bernoulli* encountered in any first course in the calculus. In Jakob Bernoulli's solution to the problem of the isochrone curve, which was published in the *Acta eruditorum* in 1690, we meet for the first time the word "integral" in a calculus sense. Leibniz had called the integral calculus *calculus summatorius;* in 1696

Leibniz and Johann Bernoulli agreed to call it *calculus integralis.*

Jakob Bernoulli was struck by the way the equiangular spiral reproduces itself under a variety of transformations and asked, in imitation of Archimedes, that such a spiral be engraved on his tombstone, along with the inscription *Eadem mutata resurgo* ("I shall arise the same, though changed").

Johann Bernoulli was an even more prolific contributor to mathematics than was his brother Jakob. Though he was a jealous and cantankerous man, he was one of the most successful teachers of his time. He greatly enriched the calculus and was very influential in making the power of the new subject appreciated in continental Europe. It was his material that the Marquis de l'Hospital (1661–1704), under a curious financial agreement with

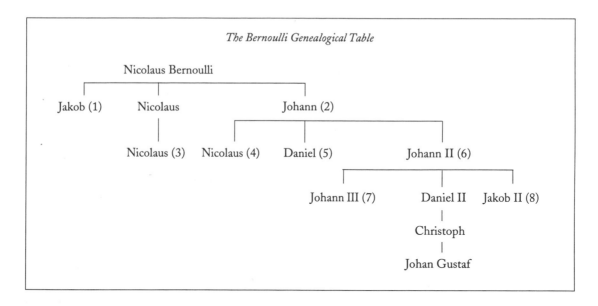

The Bernoulli Genealogical Table

Johann, assembled in 1696 into the first calculus textbook. It was in this way that the familiar method of evaluating the indeterminate form 0/0 became incorrectly known in later calculus texts, as *l'Hospital's rule*.

Johann Bernoulli wrote on a wide variety of topics, including optical phenomena connected with reflection and refraction, the determination of the orthogonal trajectories of families of curves, rectification of curves and quadrature of areas by series, analytical trigonometry, the exponential calculus, and other subjects. One of his more noted pieces of work is his contribution to the problem of the *brachystochrone*—the determination of the curve of quickest descent of a weighted particle moving between two given points in a gravitational field; the curve turned out to be an arc of an appropriate cycloid curve. This problem was also discussed by Jakob Bernoulli. The cycloid curve is also the solution to the problem of the *tautochrone*— the determination of the curve along which a weighted particle will arrive at a given point of the curve in the same time interval no matter from what initial point of the curve it starts. This latter problem, which was more generally discussed by Johann Bernoulli, Euler, and Lagrange, had earlier been solved by Huygens (1673) and Newton

(1687), and applied by Huygens in the construction of pendulum clocks.

Johann Bernoulli had three sons, Nicolaus (1695–1726), Daniel (1700–1782), and Johann (II) (1710–1790), all of whom won renown as eighteenth-century mathematicians and scientists. Nicolaus, who showed great promise in the field of mathematics, was called to the St. Petersburg Academy, where he unfortunately died—by drowning— only eight months later. He wrote on curves, differential equations, and probability. A problem in probability, which he proposed from St. Petersburg, later became known as the *Petersburg paradox*. The problem is: If A receives a penny should heads appear on the first toss of a coin, 2 pennies if heads does not appear until the second toss, 4 pennies if heads does not appear until the third toss, and so on, what is A's expectation? Mathematical theory shows that A's expectation is infinite, which seems a paradoxical result.

The Petersburg paradox problem was investigated by Nicolaus's brother Daniel, who succeeded Nicolaus at St. Petersburg. Daniel returned to Basel seven years later. He was the most famous of Johann's three sons, and devoted most of his energies to probability, astronomy, physics, and hydrodynamics. In probability he devised the concept of *moral expectation*, and in his *Hydrodynamica*, of 1738, appears the principle of hydrodynamics that bears his name in all present-day elementary physics texts. He wrote on tides, established the kinetic theory of gases, studied the vibrating string, and pioneered in partial differential equations. He was awarded the prize of the French Academy no less than ten times.

Johann (II), the youngest of Johann Bernoulli's three sons, studied law but spent his later years as a professor of mathematics at the University of Basel, succeeding his father in that position in 1743. He was particularly interested in the mathematical theory of heat and light. He received the prize of the French Academy on three occasions.

There was another eighteenth-century Nicolaus Bernoulli (1687–1759), a nephew of Jakob and Johann, who achieved some fame in mathematics. This Nicolaus held, for a time, the chair of mathematics at Padua once filled by Galileo. He wrote extensively on geometry, differential equations, infinite series, and probability. Later in life he taught logic and law.

Johann Bernoulli (II) had three sons, Johann (III) (1744–1807), Daniel (II) (1751–1834), and Jakob (II) (1759–1789). Johann (III), like his father, studied law but then turned to mathematics. When barely 19 years old, he was called as a professor of mathematics to the Berlin Academy. He wrote on astronomy, the doctrine of chance, recurring decimals, and indeterminate equations. Jakob (II) also first studied law, but then became a professor of mathematics at the St. Petersburg Academy. His works are related to those of his uncle and teacher, Daniel Bernoulli.

Lesser Bernoulli descendants are Daniel (II), Christoph (1782–1863), a son of Daniel (II), and Johann Gustav (1811–1863), a son of Christoph.

79

The Bernoullis and the Harmonic Series

WILLIAM DUNHAM

$\mathscr{A}$NY INTRODUCTION to the topic of infinite series soon must address that first great counterexample of a divergent series whose general term goes to zero—the harmonic series $\sum_{k=1}^{\infty} 1/k$. Modern texts employ a standard argument, traceable back to the great fourteenth-century Frenchman Nicole Oresme (see [3], p. 92), which establishes divergence by grouping the partial sums:

$$1 + \frac{1}{2} > \frac{1}{2} + \frac{1}{2} = \frac{2}{2}$$

$$1 + \frac{1}{2} + \left(\frac{1}{3} + \frac{1}{4}\right) > \frac{2}{2} + \left(\frac{1}{4} + \frac{1}{4}\right) = \frac{3}{2}$$

$$1 + \frac{1}{2} + \frac{1}{3} + \frac{1}{4} + \left(\frac{1}{5} + \frac{1}{6} + \frac{1}{7} + \frac{1}{8}\right) > \frac{3}{2} + \left(\frac{1}{8} + \frac{1}{8} + \frac{1}{8} + \frac{1}{8}\right) = \frac{4}{2},$$

and in general

$$1 + \frac{1}{2} + \frac{1}{3} + \dots + \frac{1}{2^n} > \frac{n+1}{2},$$

from which it follows that the partial sums grow arbitrarily large as n goes to infinity.

It is possible that seasoned mathematicians tend to forget how surprising this phenomenon appears to the uninitiated student—that, by adding ever more negligible terms, we nonetheless reach a sum greater than any preassigned quantity. Historian of mathematics Morris Kline ([5], p. 443) reminds us that this feature of the harmonic series seemed troubling, if not pathological, when first discovered.

JACOBI BERNOULLI,
Profeſſ. Baſil. & utriuſque Societ. Reg. Scientiar. Gall. & Pruſſ. Sodal.
MATHEMATICI CELEBERRIMI,

ARS CONJECTANDI,
OPUS POSTHUMUM.

Accedit

TRACTATUS
DE SERIEBUS INFINITIS,
Et EPISTOLA Gallicè ſcripta
DE LUDO PILÆ
RETICULARIS.

BASILEÆ,
Impenſis THURNISIORUM, Fratrum.
cIↄ Iↄcc xIII.

Reprinted from *College Mathematics Journal* 60 (Jan., 1987): 18–23; with permission.

Courtesy of the Lilly Library, Indiana University, Bloomington, IN

So unusual a series could not help but attract the interest of the preeminent mathematical family of the seventeenth century, the Bernoullis. Indeed, in his 1689 treatise "Tractatus de Seriebus Infinitis," Jakob Bernoulli provided an entirely different, yet equally ingenious proof of the divergence of the harmonic series. In "Tractatus," which is now most readily found as an appendix to his posthumous 1713 masterpiece *Ars Conjectandi,* Jakob generously attributed the proof to his brother ("Id primus deprehendit Frater"), the reference being to his full-time sibling and part-time rival Johann. While this "Bernoullian" argument is sketched in such mathematics history texts as Kline ([5], p. 444) and Struik ([6], p. 321), it is little enough known to warrant a quick reexamination.

The proof rested, quite unexpectedly, upon the *convergent* series

$$\frac{1}{1} + \frac{1}{6} + \frac{1}{12} + \frac{1}{20} + \ldots = \sum_{k=1}^{\infty} \frac{1}{k(k+1)}.$$

The modern reader can easily establish, via mathematical induction, that

$$\sum_{k=1}^{n} \frac{1}{k(k+1)} = \frac{n}{n+1},$$

and then let n go to infinity to conclude that

$$\sum_{k=1}^{\infty} \frac{1}{k(k+1)} = 1.$$

Jakob Bernoulli, however, approached the problem quite differently. In Section XV of *Tractatus,* he considered the infinite series

$$N = \frac{a}{c} + \frac{a}{2c} + \frac{a}{3c} + \frac{a}{4c} + \ldots,$$

then introduced

$$P = N - \frac{a}{c} = \frac{a}{2c} + \frac{a}{3c} + \frac{a}{4c} + \frac{a}{5c} + \ldots.$$

and subtracted termwise to get

$$\frac{a}{c} = N - P = \left(\frac{a}{c} - \frac{a}{2c}\right) + \left(\frac{a}{2c} - \frac{a}{3c}\right) + \left(\frac{a}{3c} - \frac{a}{4c}\right) + \ldots$$
$$= \frac{a}{2c} + \frac{a}{6c} + \frac{a}{12c} + \frac{a}{20c} + \ldots \qquad (1)$$

Thus, for $a = c$, he concluded that

$$\frac{1}{2} + \frac{1}{6} + \frac{1}{12} + \frac{1}{20} + \ldots = \frac{1}{1} = 1. \qquad (2)$$

Unfortunately, Bernoulli's "proof" required the subtraction of two divergent series, N and P. To his credit, Bernoulli recognized the inherent dangers in his argument, and he advised that this procedure must not be used without caution ("non sine cautela"). To illustrate his point, he applied the previous reasoning to the series

$$S = \frac{2a}{c} + \frac{3a}{2c} + \frac{4a}{3c} + \ldots,$$

and

$$T = S - \frac{2a}{c} + \frac{3a}{2c} + \frac{4a}{3c} + \frac{5a}{4c} + \ldots.$$

Upon subtracting termwise, he got

$$\frac{2a}{c} = S - T = \frac{a}{2c} + \frac{a}{6c} + \frac{a}{12c} + \frac{a}{20c} + \ldots, \qquad (3)$$

which provided a clear contradiction to (1).

Bernoulli analyzed and resolved this contradiction as follows: the derivation of (1) was valid since the "last" term of series N is zero (that is, $\lim_{k\to\infty} a/(kc)=0$), whereas the parallel derivation of (3) was invalid since the "last" term of series S is non-zero (because $\lim_{k\to\infty}(k+1)a/(kc) = a/c \neq 0$). In modern terms, he had correctly recognized that, regardless of the convergence or divergence of the series $\sum_{k=1}^{\infty} x_k$, the new series $\sum_{k=1}^{\infty} (x_k - x_{k+1})$ converges to x_1 *provided* $\lim_{k\to\infty} x_k = 0$. Thus, he not only explained the need for "caution" in his earlier discussion but also exhibited a fairly penetrating insight, by the standards of his day, into the general convergence/divergence issue.

Having thus established (2) to his satisfaction, Jakob addressed the harmonic series itself. Using his brother's analysis of the harmonic series, he proclaimed in Section XVI of *Tractatus:*

> **X V I.** *Summa seriei infinita harmonicè progressionalium,*
> $\frac{1}{1} + \frac{1}{2} + \frac{1}{3} + \frac{1}{4} + \frac{1}{5}$ *&c. est infinita.*

He began the argument that "the sum of the infinite harmonic series

$$\frac{1}{1} + \frac{1}{2} + \frac{1}{3} + \frac{1}{4} + \frac{1}{5} \quad \text{etc.}$$

is infinite" by introducing

$$A = +\frac{1}{2} + \frac{1}{3} + \frac{1}{4} + \frac{1}{5} + \frac{1}{6} + \frac{1}{7} \ldots,$$

which "transformed into fractions whose numerators are 1, 2, 3, 4 etc" becomes

$$\frac{1}{2} + \frac{2}{6} + \frac{3}{12} + \frac{4}{20} + \frac{5}{30} + \frac{6}{42} + \ldots.$$

Using (2), Jakob next evaluated:

$$C = \frac{1}{2} + \frac{1}{6} + \frac{1}{12} + \frac{1}{20} + \ldots = 1$$

$$D = \frac{1}{6} + \frac{1}{12} + \frac{1}{20} + \ldots = C - \frac{1}{2} = 1 - \frac{1}{2} = \frac{1}{2}$$

$$E = \frac{1}{12} + \frac{1}{20} + \ldots = D - \frac{1}{6} = \frac{1}{2} - \frac{1}{6} = \frac{1}{3}$$

$$F = \frac{1}{20} + \ldots = E - \frac{1}{12} = \frac{1}{3} - \frac{1}{12} = \frac{1}{4}$$

$$\vdots \qquad \vdots \qquad \vdots \qquad \vdots$$

By adding this array columnwise, and again implicitly assuming that termwise addition of infinite series is permissible, he arrived at

$$C + D + E + F + \ldots$$

$$= \frac{1}{2} + \left(\frac{1}{6} + \frac{1}{6}\right) + \left(\frac{1}{12} + \frac{1}{12} + \frac{1}{12}\right) + \ldots$$

$$= \frac{1}{2} + \frac{2}{6} + \frac{3}{12} + \frac{4}{20} + \ldots$$

$$= A.$$

On the other hand, upon separately summing the terms forming the extreme left and the extreme right of the arrayed equations above, he got

$$C + D + E + F + \ldots$$

$$= 1 + \frac{1}{2} + \frac{1}{3} + \frac{1}{4} + \ldots = 1 + A.$$

Hence, $A = 1 + A$. In Jakob's words, "The whole" equals "the part"—that is, the harmonic series $1 + A$ equals its part A—which is impossible for a finite quantity. From this, he concluded that $1 + A$ is infinite.

Jakob Bernoulli was certainly convinced of the importance of his brother's deduction and emphasized its salient point when he wrote: "The sum of an infinite series whose final term vanishes perhaps is finite, perhaps infinite."

Obviously, this proof features a naive treatment both of series manipulation and of the nature of "infinity." In addition, it attacks infinite series "holistically" as single entities, without recourse to the modern idea of partial sums. Before getting overly critical of its distinctly seventeenth-century flavor, however, we must acknowledge that Bernoulli devised this proof a century and a half before the appearance of a truly rigorous theory of series. Further, we cannot deny the simplicity and cleverness of his reasoning nor the fact that, if bolstered by the necessary supports of modern analysis, it can serve as a suitable alternative to the standard proof.

Indeed, this argument provides us with an example of the history of mathematics at its best—paying homage to the past yet adding a note of freshness and ingenuity to the modern classroom.

XVI. Summa seriei infinitæ harmonicè progressionalium, $\frac{1}{1} + \frac{1}{2} + \frac{1}{3} + \frac{1}{4} + \frac{1}{5}$ &c. est infinita,

Id primus deprehendit Frater: inventa namque per præced. summa seriei $\frac{1}{2} + \frac{1}{6} + \frac{1}{12} + \frac{1}{20} + \frac{1}{30}$, &c. visurus porro, quid emergeret ex ista serie, $\frac{1}{2} + \frac{2}{6} + \frac{3}{12} + \frac{4}{20} + \frac{5}{30}$, &c. si resolveretur methodo Prop. XIV. collegit propositionis veritatem ex absurditate manifesta, quæ sequeretur, si summa seriei harmonicæ finita statueretur. Animadvertit enim,

Seriem A, $\frac{1}{2} + \frac{1}{3} + \frac{1}{4} + \frac{1}{5} + \frac{1}{6} + \frac{1}{7}$, &c. ∞ (fractionibus singulis in alias, quarum numeratores sunt 1, 2, 3, 4, &c. transmutatis)

seriei B, $\frac{1}{2} + \frac{2}{6} + \frac{3}{12} + \frac{4}{20} + \frac{5}{30} + \frac{6}{42}$, &c. ∞ C+D+E+F, &c.

C. $\frac{1}{2} + \frac{1}{6} + \frac{1}{12} + \frac{1}{20} + \frac{1}{30} + \frac{1}{42}$, &c. ∞ per præc. $\frac{1}{1}$
D. .. $+ \frac{1}{6} + \frac{1}{12} + \frac{1}{20} + \frac{1}{30} + \frac{1}{42}$, &c. ∞ C $- \frac{1}{2}$ ∞ $\frac{1}{2}$
E. ... $+ \frac{1}{12} + \frac{1}{20} + \frac{1}{30} + \frac{1}{42}$, &c. ∞ D $- \frac{1}{6}$ ∞ $\frac{1}{3}$
F. ..... $+ \frac{1}{20} + \frac{1}{30} + \frac{1}{42}$, &c. ∞ E $- \frac{1}{12}$ ∞ $\frac{1}{4}$
&c. ∞ &c.

∞ G; unde sequitur, seriem G ∞ A, totum parti, si summa finita esset.

Ego

Perhaps, in contemplating this work, some of today's students might even come to share a bit of the enthusiasm and wonder that moved Jakob Bernoulli to close his *Tractatus* with the verse [7]

> So the soul of immensity dwells in minutia.
> And in narrowest limits no limits inhere.
> What joy to discern the minute in infinity!
> The vast to perceive in the small, what divinity!

Remark. Jakob Bernoulli, eager to examine other infinite series, soon turned his attention in section XVII of *Tractatus* to

$$1 + \frac{1}{4} + \frac{1}{9} + \frac{1}{16} + \ldots = \sum_{k=1}^{\infty} \frac{1}{k^2}, \qquad (4)$$

the evaluation of which "is more difficult than one would expect" ("difficilior est quam quis expectaverit"), an observation that turned out to be quite an understatement. He correctly established the convergence of (4) by comparing it termwise with the greater, yet convergent series

$$1 + \frac{1}{3} + \frac{1}{6} + \frac{1}{10} + \ldots$$

$$= 2\left(\frac{1}{2} + \frac{1}{6} + \frac{1}{12} + \frac{1}{20} + \ldots\right) = 2(1) = 2.$$

But evaluating the sum in (4) was too much for Jakob, who noted rather plaintively

> If anyone finds and communicates to us that which up to now has eluded our efforts, great will be our gratitude.

The evaluation of (4), of course, resisted the attempts of another generation of mathematicians until 1734, when the incomparable Leonhard Euler devised an enormously clever argument to show that it summed to π²/6. This result, which Jakob Bernoulli unfortunately did not live to see, surely ranks among the most unexpected and peculiar in all of mathematics. For the original proof, see ([4], pp. 83–85). A modern outline of Euler's reasoning can be found in ([2], pp. 486–487).

REFERENCES

1. Jakob Bernoulli. *Ars Conjectandi.* Basel, 1713.

2. Carl B. Boyer. *A History of Mathematics.* Princeton University Press, 1985.

3. C. H. Edwards. *The Historical Development of the Calculus.* Springer-Verlag, New York, 1979.

4. Leonhard Euler. *Opera Omnia* (1), Vol. 14 (C. Boehm and G. Faber, editors), Leipzig, 1925.

5. Morris Kline. *Mathematical Thought from Ancient to Modern Times.* Oxford University Press, New York, 1972.

6. D. J. Struik (editor). *A Source Book in Mathematics (1200–1800).* Harvard University Press, 1969.

7. Translated from the Latin by Helen M. Walker, as noted in David E. Smith's *A Source Book in Mathematics.* New York: Dover, 1959, p. 271.

80

The First Calculus Textbooks

CARL B. BOYER

THE YEAR 1946 marks just a quarter of a millennium since the appearance of the *Analyse des infiniment petits* of L'Hospital.[1] This book, the first published text on the calculus, could boast truly that in its day it filled with distinction that ubiquitous lure of textbook writers—the long-felt need. Moreover, its influence and popularity dominated the whole of the eighteenth century, the period during which the new analysis developed until it completely overshadowed other branches of mathematics.

Historically integration antedates differentiation by some two thousand years. The ancient Greek method of exhaustion and the infinitesimal mensurations of Archimedes represent early examples of limits of integral sums; but it was not until the seventeenth century that Fermat found tangents and critical points through methods equivalent to evaluating limits of difference quotients. It was the discovery of the inverse nature of these two processes, together with the consequent exploitation of the anti-derivative in determining limits of sums, which leads one to call Newton and Leibniz the inventors of the calculus. Differentiation, both direct and inverse, became the basic algorithm in a new and powerful part of mathematics. Integration was nothing but "the memory of differentiation," and it was not until a century and a half later that some attention again was directed to the summation concept in the calculus.

The earliest expositions of the new analysis were but the barest outlines of the rules of differentiation. The calculus first appeared in print in a six-page memoir by Leibniz in the *Acta Eruditorum* of 1684. This contained a definition of the differential and gave brief rules for determining the differentials of sums, products, quotients, power, and roots. It included also a few applications to problems of tangents and critical points. This account was so short and contained so many misprints as to be largely enigmatic to the best mathematicians of the age.[2] The basis of Newton's method of fluxions was first published unobtrusively in the form of lemmas in the *Principia* of 1687, more than a score of years after his discovery of the calculus. Here one finds some properties of limits, as well as cryptically brief directions for finding infinitely small "moments" of products, powers, and roots.[3]

The early years of the calculus resemble the infancy of analytic geometry in that little was done to popularize the subject. It is said that when Huygens about 1690 wished to master the new methods, there were not half a dozen men qualified to expound this analysis. Newton between 1669 and 1676 had composed several treatises on the elements of fluxions, but these did not begin to appear in print until 1704. John Craig in 1685 and 1693 published two works based in part on the method of Leibniz, but these were not intended as introductions to the subject and were moreover difficult to read because of the geometrical form in which they were cast. On the Continent the Bernoulli brothers with some effort achieved a thor-

Reprinted from *Mathematics Teacher* 39 (Apr., 1946): 159–67; with permission of the National Council of Teachers of Mathematics.

ANALYSE

DES

INFINIMENT PETITS,

Pour l'intelligence des lignes courbes.

A PARIS,

DE L'IMPRIMERIE ROYALE.

M. DC. XCVL.

*Title-page of the first edition of
L'Hospital's textbook*

ough understanding of the subject, and it was Jean Bernoulli who then instructed L'Hospital, while the latter in turn passed his knowledge on to Huygens. That there was an atmosphere of enthusiasm surrounding the calculus toward the close of the seventeenth century is to be ascribed in large part to the research articles published in the learned periodicals of their day by Leibniz, L'Hospital, and the Bernoullis. The large mathematical public had, before 1696, no easy introduction to the subject. Jean Bernoulli about 1691–1692 had composed two little textbooks, but their publication was long delayed: that on the integral calculus finally appeared in 1742,[4] and that on the differential calculus almost two hundred years later still.[5] That the text of L'Hospital achieved such prompt success is ascrib-

able largely to the fact that it was the first book to supply what the many had long awaited—an elementary introduction to a new and fascinating branch of mathematics.

Guillaume-François-Antoine de L'Hospital, Marquis de St. Mesme,[6] was born in 1661. His interest in geometry was aroused at an early age, and by the time he was fifteen years old he solved difficult problems on the cycloid which Pascal had proposed. He became a captain of cavalry, but later gave up army life to devote more time to the study of mathematics. When Jean Bernoulli was in Paris in 1692, L'Hospital for four months studied the new infinitesimal geometry under him. From that time on he joined the ranks of the mathematical elite and became the chief exponent of the calculus in France. Recognizing the lack of elementary expository works, he wrote to Bernoulli in 1695 that he was on the point of publishing a work on conic sections and that he proposed to add a little treatise on the differential calculus. L'Hospital indicated that he would render his master the justice he deserved, and he added modestly that the projected work would serve merely as an introduction to a more elaborate treatise, *De scientia infiniti,* which Leibniz planned to write.[7]

The promised book on conics was delayed by the author's illness, and appeared posthumously in 1707; but the *Analyse des infiniment petits* was published promptly in 1696. In the preface L'Hospital points out that he owed much to Leibniz and the Bernoullis, especially "the young professor at Groningen"; and he says that they can take whatever credit they wish, leaving to him what they will.[8] After Jean Bernoulli had received a copy of the book, he wrote and thanked L'Hospital for mentioning him in the volume, and he promised to return the compliment when he in turn should have composed something. He wrote that the work was admirably done, and praised it for the sound arrangement of propositions and for the intelligibility of the exposition. To L'Hospital's suggestion that he should write an integral calculus as a sequel to the *Analyse,* in view of the protracted delay on the part of Leibniz, Bernoulli replied that he was so distraught by domestic embarrassments that he could not carry out the project.[9]

Seldom has a textbook in mathematics been received by savants with such eagerness as was that of L'Hospital. It opened the door to a powerful tool which previously had been accessible only to the very gifted, and thus made it possible for the average reader to solve problems which had defied the best efforts of the geometers of Greece. But the reception was not one of unmixed acclaim. Echoing the contemporary literary controversy on "ancients vs. moderns," some admirers of the classic synthetic geometry cast doubt on the soundness of the new analytic approach. The Académie des Sciences in 1701 was sharply divided over the *Infiniment petits*. The Abbé Gallois led a polemic attack on the methods of the book, ably supported by Rolle.[10] L'Hospital, although he was a member of the Académie, did not reply; but his cause was warmly and effectively sustained by Varignon, Saurin, and others. The controversy reached such a pitch that the Académie finally felt obliged to name a commission to judge the question. Interest in the attack now began to wane, but L'Hospital did not witness the ultimate triumph of the infinitely small, for he died in 1704.

Jean Bernoulli meanwhile had observed with apparent jealousy the growing success of his protégé's book. During the years immediately following the author's death, he wrote letters attacking the work and practically accusing L'Hospital of plagiarism. L'Hospital's general prefatory acknowledgment, Bernoulli argued, did not do him justice inasmuch as he was in effect the true author of almost all the substance of the book. Historians of mathematics have been divided in judgment on this matter, but a thorough analysis of the question made by Eneström reaches the conclusion that the broad claims of Bernoulli with respect to the authorship of the material are not substantiated.[11]

The *Analyse des infiniment petits* represents the first systematic treatment of the calculus, and as such it presents a good picture of the level of the subject at that time. The preface includes a brief historical account in which the author, while admitting that "Newton also had a kind of calculus," prefers and emphasizes the contributions of Leibniz. The treatise itself is divided into ten chapters or sections, of which the first (pp. 1–10) is devoted to the fundamental definitions, assumptions, and rules of procedure. A variable quantity is defined loosely as one which continually increases or diminishes; and the infinitely small portion by which a variable quantity changes is called the "difference" or differential. Two postulates are laid down: that one can take as equal two quantities which differ only by an infinitely small quantity; and that a curved line can be considered as an infinite assemblage of infinitely small straight lines which determine, by the angles which they make with each other, the curvature of the curved line. To L'Hospital these premises appeared to be "so self-evident as not to leave the least scruple about their truth and certainty on the mind of an attentive reader," but he adds that if space permitted he could prove them in the manner of the ancients. The crudeness of these fundamental principles is an indication of the infancy of the subject rather than of carelessness on the part of the author. The language of L'Hospital is eminently restrained in comparison with that of contemporaries who wrote such things as

$$\frac{d^3y}{d^2x} = d^3y \int^2 x \text{ or } \infty \cdot \infty^{m-1} = \infty^m.$$

The basic differential formulas for algebraic functions—sums, products, quotients, powers, and roots—are derived by L'Hospital in the customary Leibnizian manner, infinitesimals of higher order being neglected. Independent and dependent variables were not always clearly distinguished at the time, and here L'Hospital took a step in advance by stating formally the essential implicit function formula in the case of powers of expressions involving one or more variables. A modern reader will be struck by the absence of rules for the differentiation of transcendental functions. These were known at the time, but they had not been popularly formalized. L'Hospital had suggested to Bernoulli that he might wish to append to the *Analyse* his rules for the calculus of logarithms, but this was not done. However, the equivalent of these was implied by the well-known results on the area between the hyperbola and its asymptotes. The differentials of the trigonometric functions were at the time likewise geometrically inferred from the application of the characteristic triangle to the circle. The inverses offered no difficulty in view of the known formula for implicit functions.

The second section of the *Infiniment petits* (pp. 11–40) is devoted to problems of tangency. Inasmuch as the tangent is defined as the prolongation of an infinitely small side of the polygon which makes up the curved line, one sees immediately from the similarity of triangles that the subtangent is given by $y(dx/dy)$. Through this relationship, the author finds the tangents to the curves of his day: the higher hyperbolas and parabolas, the spirals of various kinds, the quadratrix of Dinostratus, the conchoid of Nicomedes, the cissoid of Diocles, the cycloid, the tractrix, and the curve of logarithms. Some of these curves are defined analytically, while others are described kinematically or in terms of geometrical properties. L'Hospital follows the Cartesian classification of curves into geometrical and mechanical, but inasmuch as no special notations are used for transcendental functions, only in the former category is "the nature of a curve" given by an ordinary equation.

The third chapter (pp. 41–54), on maxima and minima, opens with a definition of the basic terms: If the ordinate (appliquée) increases, as the abscissa (coupée) increases to a given value E, and thereafter decreases, then the ordinate at E is called the greatest; and similarly, *mutatis mutandis*, for the least ordinate. The treatment here differs from that in modern textbooks not only in being less rigorous, but also in that cusp maxima and minima

are treated as extensively as extrema having horizontal tangents. L'Hospital points out that the differential of an increasing quantity is positive, that of a decreasing quantity is negative. Moreover, no quantity which varies continuously can go from positive to negative without going through zero or infinity. Hence to find maxima and minima of an algebraic expression, L'Hospital equates to zero both the numerator and the denominator of the differential, discarding results which lead to a contradiction. A few cases of applications in geometry and science are included. He does not explicitly state the various tests distinguishing maxima from minima, but these were well-known to the author and his times.

Points of inflection and of *rebroussement* (cusp points for which the tangents are horizontal) are treated in section four (pp. 55–70), and here the thorny question of second order differences arose. Leibniz had been unable to give a satisfactory definition of differentials of order greater than one, and the first attack on the calculus, launched by Nieuwentijdt in 1695–1696 while L'Hospital was preparing the *Analyse*, was centered on this inadequacy. In his preface (p. ciij) L'Hospital claimed that Leibniz had satisfactorily answered Nieuwentijdt's doubts about the existence of higher differentials, but this is belied by the author's own definition: "The infinitely small portion by which the difference of a variable quantity increases or decreases continually is called the difference of the difference of this quantity, or its second difference." That he obtained correct results in this connection is due less to his definition than to the pragmatic rule which follows it: "One takes as constant whichever difference one wishes, and treats the others as variable quantities." For example, the differential of the function xy being $xdy + ydx$, the second differential is either $dd(xy) = xddy + 2dxdy$ or $dd(xy) = yddx + 2dxdy$. Here one sees the need to distinguish between the differentials of independent and dependent variables. This distinction was not at the time sufficiently emphasized, and even today numerous textbooks founder on the differential through neglect in this connection. L'Hospital then gives the geometrical description of points of inflection and *rebroussement* in terms of convexity and concavity, and he points out that these points are determined by making the second differential either zero or infinitely great. A more rigorous treatment is not, of course, to be expected of that period.

The concept of radius of curvature as such is not presented in the modern manner but is given in the context in which it arose historically. Huygens had been led by his pendulum clock to study the subject of evolutes and involutes (*developpées* and *developpants*), and this material is presented by L'Hospital in section five (pp. 71–103). Toward the close of this chapter, the longest of all, the author points out in a casual sort of way that $dx^2+dy^{2\frac{3}{2}}/-dxddy$ is a general expression for "the radius of an evolute." This is, of course, the equivalent of the modern formula for radius of curvature, where y is a function of x.

Sections six and seven (pp. 104–119, 120–130) are on caustics by reflection and refraction. These are not now traditional topics in elementary calculus, but they were popular new topics in the late seventeenth century. Moreover, L'Hospital and his master, Bernoulli, had shared with Tschirnhaus in their development. Section eight (pp. 131–144) is on envelopes of families of straight lines, and includes the well-known method of Leibniz of differentiating with respect to the parameter.

The portion of the book which in its day roused most discussion was section nine (pp. 145–163). This chapter carries the unenlightening heading, "Solution of some problems depending on the preceding methods," but it involves what would now be called indeterminate forms. The presentation is largely geometrical, but the basic result is the so-called "rule of L'Hospital." To find "the value" of a rational expression in x which for the abscissa in question takes the form 0/0, he determines the ratio of the "differences" of the numerator and denominator for this abscissa. This rule appears not to have been familiar at the time, and

so the author here might well have indicated the source of his material. The author never claimed the rule as his own, but when his friend Saurin implied that it was due to Leibniz, protest was raised by Jean Bernoulli that here, as elsewhere, L'Hospital had not given him due credit. Bernoulli's claim to the rule seems to be substantiated, and his name, rather than L'Hospital's, should be applied to it. But it should be pointed out that the traditional nomenclature is due to the fickleness of fortune rather than to any ill intent on the part of L'Hospital. The *Analyse des infiniment petits* occupies a place in the calculus somewhat analogous to that of Lavoisier's *Traité élémentaire* in chemistry almost a century later. Both books would have been more attractive had they included a substantial historical background, but in neither case should the omission of this be interpreted as an argument *e silencio* that the author claims credit for more than the general arrangement and exposition of the subject matter. One does not demand of writers of textbooks the same scrupulous regard for the citation of sources that one expects in the publication of the results of research work.

The tenth and last section of L'Hospital's treatise (pp. 164–181) is somewhat out of harmony with the remainder. It is a clever bit of salesmanship in which the author contrasts the elegant methods of the calculus with the awkward anticipatory procedures of Descartes and Hudde for determining maxima and minima. On the basis of this comparison L'Hospital emphasizes that the method of Leibniz "gives general solutions, whereas the other furnishes only particular ones, and it extends to transcendental curves and it is not at all necessary to avoid incommensurables, which often is impracticable." This closing sentence of the first textbook on the calculus reminds one of the title of Leibniz's first paper on the new analysis a dozen years before: "A new method for maxima and minima, as well as tangents, which is not impeded by irrational quantities."

The *Analyse* of L'Hospital was composed two hundred and fifty years ago, and hence the mate-

Marquis de L'Hospital

dynamics afforded a golden opportunity for apt and timely illustrations of the new calculus, but we are told in the preface (p. eij) that illness prevented the author from fulfilling his intention to add a section on applications to physics. The language and notation of the *Analyse* differ slightly from that now in use; there is an anachronistic emphasis on the ideas of ratio and proportion; and the material is divided into propositions, corollaries, and scholia. Nevertheless, a modern reader should have no difficulty in following the author's exposition.

The *Analyse des infiniment petits* appeared in a second edition at Paris in 1715. This was in reality a reissue of the original, but it contained numerous typographical errors. The work was reprinted the following year, and again in 1720. By this time the popularity of the calculus was such that a textbook on a lower level was desired. Publishers sought to capitalize on this desire and on the popularity of L'Hospital's work by providing commentaries. In 1721 Crousaz issued at Paris his *Commentaire sur l'analyse des infiniment petits*, consisting of an elementary introduction and a series of supplementary notes. This book was damned by faint praise on the part of Jean Bernoulli and met with severe criticism from Saurin. Four years later there appeared a posthumous and successful supplementary volume by Varignon, *Eclaircissemens sur l'analyse des infiniment petits*. The author appears to have had in mind the preparation of a new edition of L'Hospital's work, together with added notes. The publisher, however, felt that inasmuch as L'Hospital's *Analyse* already "had appeared several times in France and in foreign countries," and since "most men in the profession already had copies," he could save the public some expense by issuing Varignon's *Eclaircissemens* as a separate volume rather than as part of a new edition. This volume is on a much higher level than that of Crousaz, and although it follows L'Hospital section by section, it adds to this fresh problems and methods resulting from later research.

The rift between British and Continental mathematicians during the eighteenth century was not

rial it contains does not in all cases parallel that which appears in modern textbooks. One fails to find, for example, the notions of function and limit. The first of these was in the air in L'Hospital's day, but it was popularized half a century later in Euler's *Introductio ad analysin infinitorum*; the latter was likewise known in a general sense through the calculus of Newton, but it was a hundred years before it became basic in the treatises of Cauchy. One misses also the now traditional work on Taylor's series. This, too, was known to James Gregory and Jean Bernoulli, but it was formalized after L'Hospital's time through Brook Taylor's work on finite and infinitesimal differences. Curve tracing is a further topic which is lacking in the *Analyse*. The seventeenth century here invented the analytical tools, but it was the eighteenth century which first applied them to the systematic study of a wide variety of curves. It is strange to relate also that L'Hospital's text is deficient in applications to science. Seventeenth-century sufficiently wide to prevent Stone in 1730 from publishing an English translation of L'Hospital's *Infiniment petits*, ten years before Buffon reciprocally prepared a French translation of Newton's *Method of Fluxions*. The full title of Stone's book is, *The method of fluxions both direct and inverse. The former being a translation from the celebrated Marquis de L'Hospital's Analyse des infiniments petits: and the latter supply'd by the translator*. The translator opens with praise of the author's work, "the Character whereof is so well establish'd," but adds that out of regard to Sir Isaac Newton he has altered some of the language and notation of the original to conform to the Newtonian. Increments or differentials, for example, are taken as fluxions, inasmuch as they are proportional to the latter; and these are denoted by $\dot{x}, \ddot{x}, \dddot{x}$, etc., instead of by *dx, ddx, dddx*, etc. A few years later Bishop Berkeley took full advantage, in the famous *Analyst* controversy of the confusion then existing in Britain between fluxions and infinitely small quantities.

Stone's translation, in part I, does not depart much from the original, and the diagrams are faithful reproductions of those of L'Hospital. Some new material is added, however, on logarithms, exponentials, and trigonometric functions. The second part or appendix (separately paginated) is on the inverse method of fluxions, and this constitutes a new work. L'Hospital apparently had hoped that his treatise would sometime be completed by a second part on the integral calculus. The pages of the *Analyse* (even in later editions) carry the heading *I. Part.*, but the author never completed another *partie*. Stone was but one of several who undertook to compose a suitable second half. His attempt was so successful that it was translated into French and published in 1735, "servant de suite aux infiniment petits de M. le Marquis de l'Hôpital."

The *Analyse des infiniment petits* continued throughout the eighteenth century to be the standard elementary manual on higher mathematics.[12] During the latter half of the century it found a

worthy rival in the *Instituzioni analitiche* of Maria Agnesi (Milan, 1748), but frequency of publication indicates that the vogue for L'Hospital's work continued. It appeared in new French editions, revised and enlarged, in 1758 (Paris), 1768 (Avignon), and 1781 (Paris), and in Latin translations in 1764 and 1790 (both at Vienne).[13]

Toward the turn of the century, however, Lacroix[14] began the publication of his renowned series of textbooks, and one of these, the *Traité élémentaire de calcul différentiel et de calcul intégral*, (Paris, 1802) took the place during the nineteenth century which L'Hospital had held a hundred years earlier. The concepts of function and limit now took over the field which the infinitesimal "ghosts of departed quantities" had so long dominated. But the work of the first textbook on the calculus had been well done. The enthusiasm for the subject which it kindled and nourished has never abated, and modern textbooks are but new monuments to the influence of the *Analyse des infiniment petits*.

NOTES

1. G.F.A. de L'Hospital, *Analyse des infiniment petits, pour l'intelligence des lignes caurbes*, Paris, 1696. Page references indicated in this paper refer to the original edition.

2. See Gustav Eneström, "Über die erste Aufnahme der Leibnizschen Differentialrechnung," *Bibliotheca Mathematica* (3), IX (1908–1909), 309–320. An English translation of Leibniz's paper of 1684 is found in D. E. Smith, *A Source Book in Mathematics* (New York, 1929), pp. 619–626.

3. Sir Isaac Newton, *Opera quae exstant omnia* (ed. by Samuel Horsley, 5 vols., Londini, 1779–1785), I, 251, II, 277–280.

4. Jean Bernoulli, *Opera omnia* (4 vols., Lausanne and Geneva, 1742), vol. III; also *Die erste Integralrechnung. Eine Auswahl aus Johann Bernoullis mathematischen Vorlesungen über die Methode der Integrale und anderes aufgeschrieben zum Gebrauch des Herrn Marquis de l'Hospital in den Jahren 1691 und 1692*. Translated from the Latin, Leipzig and Berlin, 1914.

5. Jean Bernoulli, *Die Differentialrechnung aus dem Jahre 1691/92*, Ostwald's klassiker, No. 211, Leipzig, 1924.

6. A satisfactory biographical account is found in J. F. Michaud, *Biographie Universelle* (2nd ed., Paris, 1880), XXIII, 448–451. In later years the family name came to be spelled L'Hôpital instead of L'Hospital.

7. See L'Hospital, *op. cit.*, preface.

8. *Ibid.*

9. Eneström, "Sur le part de Jean Bernoulli dans la publication de l'Analyse des infiniment petits," *Bibliotheca Mathematica* (new series), VIII (1894), 65–72. Cf. O. J. Rebel, *Der Briefwechsel zwischen Johann* (I). *Bernoulli und dem Marquis de l'Hospital*, Bottrop i.w., 1934.

10. Académie des Sciences, *Histoire et mémoires*, 1701 (Amsterdam, 1735), Histoire, pp. 114 ff. This controversy was the second of three important attacks on the calculus. The first was that of Nieuwentijdt against the work of Leibniz in 1695–1696. The third was the famous *Analyst* controversy of 1734–1735 concerning the calculus of Newton.

11. Eneström, *op. cit.*

12. The Abbé Sauri in his *Cours complèt de mathématiques* (5 vols., Paris, 1774), III, xij, refers to L'Hospital's *Analyse* as "connue de tout le mond."

13. This paper does not pretend to give a complete listing of editions. I have examined those of 1696, 1715, 1730, 1768, and 1781, and the commentaries by Crousaz and Varignon, all of which are available at the New York Public Library or at Columbia University. Editions other than these have been listed on the basis of references found in standard sources, such as Poggendorff and the catalogues of the British Museum and the Bibliothèque Nationale. It is of interest to note that L'Hospital's *Traité analytique des sections coniques* likewise enjoyed a considerable popularity during the eighteenth century, appearing in editions of 1707, 1720, 1723 (English translation by Stone), 1770 (Venice), and 1776.

14. S. F. Lacroix was perhaps the most prolific textbook writer of modern times, if allowance is made for multiple editions. In 1848 there appeared at Paris the 20th edition of his *Traité élémentaire d' arithmétique* and the 16th edition of his *Élémens de géométrie*. The 20th edition of his *Élémens d' algèbre* was published at Paris in 1858, and the ninth edition of the *Traité élémentaire de calcul* in 1881. And the above figures do not include the large number of editions in other languages.

81

The Origin of L'Hôpital's Rule

D. J. STRUIK

$\mathcal{T}$HE SO-CALLED RULE of L'Hôpital, which states that

$$\lim_{x \to a} \frac{f(x)}{g(x)} = \frac{f'(a)}{g'(a)}$$

when $f(a) = g(a) = 0$, $g'(a) \neq 0$, was published for the first time by the French mathematician G. F. A. de l'Hôpital (or De Lhospital) in his *Analyse des infiniment petits* (Paris, 1696).[1] The Marquis de l'Hôpital was an amateur mathematician who had become deeply interested in the new calculus presented to the learned world by Leibniz in two short papers, one of 1684 and the other of 1686. Not quite convinced that he could master the new and exciting branch of mathematics all by himself, L'Hôpital engaged, during some months of 1691–92, the services of the brilliant young Swiss physician and mathematician, Johann Bernoulli, first at his Paris home and later at his château in the country.[2] When Bernoulli left for his home town Basel, the Marquis kept up correspondence with his tutor, at the same time publishing some original contributions of his own findings. When, in 1696, L'Hôpital's book appeared, he acknowledged his indebtedness to Leibniz and Bernoulli, but only in general terms: "I have made free with their discoveries (*je me suis servi sans façon de leur découvertes*), so that whatever they please to claim as their own I frankly return to them."

The question of the actual dependence of L'Hôpital on Bernoulli remained unanswered, and acquired in the course of the years somewhat the character of a mystery. Bernoulli, after L'Hôpital had sent him a copy of the book, thanked him courteously and praised it. But subsequently, in some private letters written during the lifetime of the Marquis, he claimed that much of the content of the *Analyse des infiniment petitis* was really his own property. In 1704, after L'Hôpital's death, he made a public claim to that section, No. 163, which contains the rule for 0/0.[3] Mathematicians interested in such kind of priority puzzles have been philosophizing about this supposed dependence of L'Hôpital on Johann Bernoulli ever since, weighing Bernoulli's acknowledged reputation for nastiness.[4] A generally acceptable conclusion was not reached until recent times.

Considerable clarification came in 1922, when Johann Bernoulli's manuscript on the differential calculus, dating from 1691–92, was at last published (the corresponding manuscript on the integral calculus was known from Johann Bernoulli's *Opera*, published in 1742 during the lifetime of the author).[5] Comparison of these notes by Bernoulli and the text of L'Hôpital's book revealed that there was a considerable overlapping, so that it seemed that Bernoulli had fathered much of the nobleman's intellectual offspring. But the true situation came to light only in 1955, when Bernoulli's early correspondence was published.[6] It then appeared that in 1694 a deal was actually made

Reprinted from *Mathematics Teacher* 56 (Apr., 1963): 257–60; with permission of the National Council of Teachers of Mathematics.

between the Marquis and his former tutor, by which L'Hôpital offered him a yearly allowance of 300 livres (and more later) provided that Bernoulli agreed to three conditions:

1 To work on all mathematical problems sent to him by the Marquis,
2 To make all the discoveries known to him, and
3 To abstain from passing on to others a copy of the notes sent to L'Hôpital.

This settled the priority question. Here is a translation of the section of the letter which contains the unusual proposition, sent by L'Hôpital in Paris to Johann Bernoulli in Basel, March 17, 1694:

> I shall give you with pleasure a pension of three hundred livres, which will begin on the first of January of the present year, and I shall send two hundred livres for the first half of the year because of the journals that you have sent, and it will be one hundred and fifty livres for the other half of the year, and so in the future. I promise to increase this pension soon, since I know it to be very moderate, and I shall do this as soon as my affairs are a little less confused. . . . I am not so unreasonable as to ask for this all your time, but I shall ask you to give me occasionally some hours of your time to work on what I shall ask you—and also to communicate to me your discoveries, with the request not to mention them to others. I also ask you to send neither to M. Varignon[7] nor to others copies of the notes that you let me have, for it would not please me if they were made public. Send me your answer to all this and believe me, *Monsieur tout à vous.*

le M. de Lhopital

Bernoulli's answer has not been found, but from a letter of July 22, 1694, we know that he had accepted the proposal. It must have been a little windfall for the impecunious young scientist, just married and still looking for a position (which he obtained the following year at the University of Groningen in the Netherlands). How long this interesting relationship lasted we do not know, but Bernoulli's finances improved and those of L'Hôpital did not become any better. By 1695 it may have come to an end.

Several letters from Bernoulli to his patron with answers to questions have now been published, and the one dated July 22, 1694, contains the rule for 0/0. The formulation is very much like the one we find in the *Analyse des infiniment petits*, and is based on a geometrical consideration. In our words, if

$$y = \frac{f(x)}{g(x)}$$

and both curves $y = f(x)$ and $y = g(x)$ pass through the same point P on the x axis, $OP = a$, so that $f(a) = g(a) = 0$, and if we take an ordinate $x = a + h$, then the figure shows immediately that

$$\frac{f(a + h)}{g(a + h)}$$

is almost equal to the quotient of $hf'(a + h)$ and $hg'(a + h)$ when h is small. In the limit we find, now in Bernoulli's words:

"In order to find the value of the ordinate (*appliquée*) of the given curve

$$\left[y = \frac{f(x)}{g(x)} \right]$$

in this case it is necessary to divide the differential (*la différentielle*) of the numerator of the general fraction by the differential of the denominator."

Bernoulli's examples are almost the same that L'Hôpital uses:

1) $y = \dfrac{\sqrt{2a^3x - x^4} - a\sqrt[3]{a^2x}}{a - \sqrt[4]{ax^3}}$ for $x = a$. Then $x = a$

$y = \left(\dfrac{16}{9} \right) a.$

This example is used by both Bernoulli and L'Hôpital.

2) $y = \dfrac{a\sqrt{ax} - xx}{a - \sqrt{ax}}$ for $x = a$. Then $y = 3a$.

This example of Bernoulli is changed by L'Hôpital into

541

$$y = \frac{aa - ax}{a - \sqrt{ax}} \text{ for } x = a. \text{ Then } y = 2a.$$

The situation has thus been clarified. When L'Hôpital's book appeared, Bernoulli was bound by his promise not to reveal which sections of the book belonged to him. He could only express himself privately. Then, after the death of the Marquis, he felt that he need not be so silent any more, and claimed as his own the most striking result of the book—the rule for 0/0. But he could not prove his assertion. At present he stands vindicated.

From this discovery of the origin of L'Hôpital's rule we should not conclude that from now on it should be called after Bernoulli. First of all, there are already plenty of rules and theorems called after Bernoulli (due to at least three members of the Bernoulli family—Jakob, Johann, and Daniel). But there is a more weighty consideration. When we begin to change the names of rules and theorems in accordance with the strict laws of priority, we soon come to the dreary conclusion that our science will lose many of its most familiar expressions. Pythagoras's theorem was known to the Babylonians more than a millennium before the sage of Crotona lived. The Cauchy-Riemann equations were known to D'Alembert and Euler. Taylor's theorem would be Gregory's until another claimant pops up—as a matter of fact, he already exists in the person of the Indian Nilakantha (ca. 1500). The Indians also played with Pell's equation long before John Pell studied it—or better, did not study it, since Pell's connection with the equation is rather remote. Fourier's series were used by Euler and Daniel Bernoulli. Pascal's triangle was known to the Chinese mathematician Yang Hui (thirteenth century), and probably is even older, and his contemporary Chhin Chin-Shao worked with Horner's method in the theory of algebraic equations as with an ancient tool. And so on.

The names attached to mathematical discoveries are often the names of persons who made these results better known or understood through their own outstanding work. Any new historical discovery may disturb the delicate balance of no-menclature again. Let the good Marquis keep his elegant rule; he paid for it and made it public property. After all, he deserves some fame; his book on the new calculus was not only the first to be published, and contained contributions of his own, but it was good enough to hold its prominent position for half a century and longer. Even after other and better textbooks appeared, it continued to be used as a good first introduction into the calculus; we know of an edition as late as 1790. It appeared in an English and a Latin translation, and there also exist commentaries, like that of L'Hôpital's friend Varignon. We should have some respect for it.

NOTES

1. Details are given in C. B. Boyer, "The First Calculus Textbooks," *The Mathematics Teacher*, XXXIX (April, 1946), 159–67.

2. Johann Bernoulli (1667–1748) and his older brother Jakob (1655–1705) were the earliest and most prominent students of Leibniz's mathematical discoveries. Jakob was a professor at the University of Basel until his death, when he was succeeded by his brother, who, from 1695, had been teaching at the University of Groningen. Between ca. 1685 and ca. 1700, Leibniz and the Bernoulli brothers developed most of our elementary differential calculus, advancing also into the integral calculus, differential geometry, and even the calculus of variations.

3. *Acta eruditorum*, 1704; more details in the *Acta eruditorum*, 1721, pp. 223–27.

4. Eneström, G. *Bibliotheca mathematica*, VIII (1894), 65–72.

5. Bernoulli, Jakob, "Die Differential-rechnung (aus dem Jahre 1691/92)," Leipzig, *Ostwald's Klassiker 211* (1922), 56 pp., editor: J. Schafheitlin.

6. *Der Briefwechsel von Johann Bernoulli*, Band I, Basel 1955, 531 pp., editor: O. Spiess. Section B, pp. 121–383, contains the correspondence of L'Hôpital with Bernoulli; it is introduced by a detailed account of the relationship between the two men.

7. Pierre Varignon (1654–1722), French mathematician, was, like his friend L'Hôpital, an early student of the calculus. In 1725, his commentary on L'Hôpital's book was published.

82

Maclaurin and Taylor and Their Series

G. N. WOLLAN

$\mathcal{B}$ROOK TAYLOR (1685–1731) was an eminent British mathematician whose relatively short lifetime roughly coincided with the latter half of the lifetime of Isaac Newton. He was an unusually talented youth with a wide range of interests, including music and art as well as mathematics and philosophy. A derivation of the series which bears his name was included in his principal mathematical work, a book *Methodus Incrementorum directa et inversa* which was published in London in 1715. At that time the mathematical world was deeply embroiled in the Newton-Leibniz priority controversy. Taylor was an ardent supporter of Newton's claim, and the book is filled with biased references to Newton and his work, no mention being made of other contributors. The book was primarily devoted to the development of a branch of mathematics known today as the calculus of finite differences. Much of its content was then new, and the book acquired considerable renown in spite of the fact that the author's obscure style, his complicated notation, and other defects made it very difficult to read. Put into modern notation, Taylor's derivation of his series as given in *Methodus Incrementorum* was approximately as follows.

Suppose $y(x)$ is a function of x, Δx is an arbitrary real number, and

$$\Delta y(x) = y(x + \Delta x) - y(x)$$

is the change in the value of the function corresponding to a change of Δx in the value of the argument x. We also define higher order differences of the function:

$$\begin{aligned}
\Delta^2 y(x) &= \Delta(\Delta y(x)) \\
&= \Delta y(x + \Delta x) - \Delta y(x), \\
\Delta^3 y(x) &= \Delta(\Delta^2 y(x)) \\
&= \Delta^2 y(x + \Delta x) - \Delta^2 y(x), \\
\Delta^n y(x) &= \Delta(\Delta^{n-1} y(x)) \\
&= \Delta^{n-1} y(x + \Delta x) - \Delta^{n-1} y(x).
\end{aligned}$$

It follows of course that if $z(x)$ is any other function of x then

$$\Delta(y(x) + z(x)) = \Delta y(x) + \Delta z(x).$$

Now then, a being an arbitrary number, we have

$$\begin{aligned}
y(a + \Delta x) &= y(a) + \Delta y(a); \\
y(a + 2\Delta x) &= y(a + \Delta x) + \Delta y(a + \Delta x) \\
&= (y(a) + \Delta y(a)) + \Delta(y(a) + \Delta y(a)) \\
&= y(a) + \Delta y(a) = \Delta y(a) + \Delta^2 y(a) \\
&= y(a) + 2\Delta y(a) + \Delta^2 y(a); \\
y(a + 3\Delta x) &= y(a + 2\Delta x) + \Delta y(a + 2\Delta x) \\
&= (y(a) + 2\Delta y(a) + \Delta^2 y(a)) \\
&\quad + \Delta(y(a) + 2\Delta y(a) + \Delta^2 y(a)) \\
&= y(a) + 2\Delta y(a) + \Delta^2 y(a) + \Delta y(a) \\
&\quad + 2\Delta^2 y(a) + \Delta^3 y(a) \\
&= y(a) + 3\Delta y(a) + 3\Delta^3 y(a) + \Delta^3 y(a);
\end{aligned}$$

and, in general,

Reprinted from *Mathematics Teacher* 61 (Mar., 1968): 310–12; with permission of the National Council of Teachers of Mathematics.

$$y(a + n\Delta x) = y(a + (n-1)\Delta x + \Delta y(a + (n-1)\Delta x)$$

$$= y(a) + n\Delta y(a) + \frac{n(n-1)}{2!}\ \Delta^2 y(a)$$

$$+ \frac{n(n-1)(n-2)}{3!}\ \Delta^3 y(a)$$

$$+ \ldots + \Delta^n y(a).$$

Letting $n\,\Delta\,x = v$ then gives

$$y(a + v) = y(a) + \frac{v}{\Delta x}\ y(a)$$

$$+ \frac{\dfrac{v}{\Delta x}\left(\dfrac{v}{\Delta x} - 1\right)}{2!}\ \Delta^2 y(a)$$

$$+ \frac{\dfrac{v}{\Delta x}\left(\dfrac{v}{\Delta x} - 1\right)\left(\dfrac{v}{\Delta x} - 2\right)}{3!}\ \Delta^3 y(a)$$

$$+ \ldots + \Delta^n y(a)$$

$$= y(a) + v\ \frac{y(a)}{\Delta x} + \frac{v(v - \Delta x)}{2!}\ \frac{\Delta^2 y(a)}{(\Delta x)^2}$$

$$+ \frac{v(v - \Delta x)\,(v - 2\Delta x)}{3!}\ \frac{\Delta^3 y(a)}{(\Delta x)^3} + \ldots + \frac{\Delta^n y(a)}{(\Delta x)^n}.$$

At this point Taylor argued that if we consider v fixed and let n become infinite, then Δx approaches 0,

$$\frac{\Delta^K y(a)}{(\Delta x)^K}$$

and

$$\frac{v(v - \Delta x)(v - 2\Delta x) \ldots (v - (K-1)\Delta x)}{K!}$$

become

$$\frac{d^K y(a)}{dx^K}$$

and

$$\frac{v^K}{K!},$$

respectively, for each K, and the number of terms becomes infinite so that we get

$$y(a + v) = y(a) + v\ \frac{dy(a)}{dx} + \frac{v^2}{2!}\ \frac{d^2 y(a)}{dx^2}$$

$$+ \frac{v^3}{3!}\ \frac{d^3 y(a)}{dx^3} + \ldots,$$

which of course, is Taylor's series.

It was this sort of fuzzy argument regarding limits which left not only Taylor but also Newton, Leibniz, and all their contemporaries vulnerable to the attack of Bishop Berkeley and led later mathematicians to develop the proofs that students learn today.

Colin Maclaurin (1698–1746), a brilliant Scottish mathematician who became professor of mathematics at the University of Aberdeen at age nineteen on the basis of a competitive examination, devised a different derivation of the series which appeared in his book *Treatise on Fluxions*, published in Edinburgh in 1742. Maclaurin's argument was intended to meet the objections raised by Berkeley and in modern notation was essentially as follows.

Assume that

$$y(x) = A_0 + A_1 x + A_2 x^2 + A_3 x^3 + \ldots,$$

where the coefficients A_0, A_1, A_2, A_3, ... are fixed numbers whose values are to be determined. Then

$$\frac{dy(x)}{dx} = A_1 + 2A_2 x + 3A_3 x_2 + \ldots,$$

which, setting $x = 0$, gives

$$\frac{dy(0)}{dx} = A_1.$$

Similarly,

$$\frac{d^2 y(x)}{dx^2} = 2A_2 + 3 \cdot 2A_3 + 4 \cdot 3A_4 x^2 + \ldots,$$

so that

$$\frac{d^2 y(0)}{dx^2} = 2A_2,$$

and it is clear that

$$\frac{d^n y(x)}{dx^n} = n!A_n + (n+1)n(n-1)\ldots 2A_{n+1}x$$

$$+ (n+2)(n+1)\ldots 3A_{n+2}x^2 + \ldots,$$

from which

$$\frac{d^n y(0)}{dx^n} = n!A_n.$$

Thus

$$y(x) = y(0) + x\,\frac{dy(0)}{dx} + \frac{x^2}{2!}\,\frac{d^2 y(0)}{dx^2}$$

$$+ \frac{x^3}{3!}\,\frac{d^3 y(0)}{dx^3} + \ldots.$$

It is apparent that in this argument the following questions are ignored: What functions have such series representations? For what values of x does the series represent the function? Does term-by-term differentiation of the series yield the derivative of the given function?

Neither Taylor nor Maclaurin has a valid claim to having discovered the series to which their names are attached since they appeared in a previously published work of John Bernoulli with which both men were probably familiar. However, both men made significant contributions to mathematics and probably deserve the measure of immortality which the association of their names with the series has given them.

REFERENCE

M. Cantor, *Geschichte der Mathematik,* Vol. III.

83

Euler, the Master Calculator

JERRY D. TAYLOR

*O*N THE BICENTENNIAL of Leonard Euler's (pronounced oiler) death, we should review some of his contributions, both to get an idea of the debt we owe to Euler and to get a feeling for the methods of this eighteenth-century mathematician. I shall present some problems and then suggest how Euler might have solved them.

Euler could have worked all the problems in table 1 without pen and paper and without the use of any mathematical tables. It has been said that he could calculate as easily as others breathe. His feats of memory are legendary—he once learned by heart a complete book of Virgil, and in his youth he prepared himself to teach in a medical school in only a few months (Bell 1937, p. 149). He memorized all the logarithmic and trigonometric tables as aids in mathematical calculations (Kline 1972, p. 401).

A good memory was especially helpful to Euler because he was blind for more than one-fourth of his adult life. Being blind did not stop him from becoming the most published mathematician in history. Although estimates vary a little from source to source, his collected works number about a hundred volumes. They include textbooks in algebra, analytic geometry (plane and solid), calculus, and differential equations. He also published about five hundred articles on research in such areas as complex variables, topology, algebra, probability, number theory, mechanics, and optics. Euler's out-

Leonard Euler
(15 April 1707–18 September 1783)

put was amazing, especially when one realizes that during the midyears he wrote while tending to thirteen children and that during his later years he did the mathematics mentally and dictated the results.

In one respect Euler was fortunate to have lived in the eighteenth century. During his lifetime it was fashionable for governments to support royal academies, where some of the best scholars

Reprinted from *Mathematics Teacher* 76 (Sept., 1983): 424–28; with permission of the National Council of Teachers of Mathematics.

TABLE I
Some Easy Problems for Euler

1. Simplify.

$$3 + \cfrac{1}{1 + \cfrac{1}{1 + \cfrac{1}{1 + \cfrac{1}{1 + \cfrac{1}{6}}}}}$$

2. Find the sum.

$$1 - \frac{1}{4} + \frac{1}{16} - \frac{1}{64} + \frac{1}{256} - \cdots$$

3. Approximate to five decimal places
$$(1.01)^{10}$$

4. Find the sum.

$$1 + \frac{1}{2} + \frac{1}{3} + \frac{1}{4} + \frac{1}{5} + \frac{1}{6} + \frac{1}{7} + \frac{1}{8} + \frac{1}{9} + \frac{1}{10}$$

5. A dodecahedron contains how many vertices?

6. Simplify.
$$\ln \ln \ln e^{1/e}$$

7. Simplify.
$$e^{i\pi}$$

8. Find the last two digits of 17^{42}

in Europe were brought together. Euler was a member of the St. Petersburg Academy in Russia from 1727 to 1741, the Berlin Academy from 1741 to 1766, and the St. Petersburg Academy again from 1766 to 1783. For this entire fifty-six-year period he received at least one stipend from an academy. While in Berlin, he drew salaries from both academies.

At the Berlin Academy, Euler was friendly with fellow mathematicians like D'Alambert and Lambert. However, he was not sophisticated enough for the philosopher Voltaire or Frederick the Great, who referred to Euler as the *mathematical cyclops* (Boyer 1949, p. 483). In contrast, Catherine the Great treated Euler like royalty on his return to St. Petersburg, where he remained until his death in 1783 (Bell 1937, pp. 148–49). With this condensed biography of Euler in mind, let us return to the problems listed in the table to see how he might have solved them.

Solutions

1. Simplify.

$$3 + \cfrac{1}{1 + \cfrac{1}{1 + \cfrac{1}{1 + \cfrac{1}{1 + \cfrac{1}{6}}}}}$$

This problem might be found in an algebra text today and called a *complex fraction*. Euler would have recognized this problem as a *continued fraction*. Because he was so familiar with continued fractions, Euler would have instantly recognized that the answer was 119/33. Moreover, he could have started with the answer and found the problem. Note that

$$\frac{119}{33} = 3 + \frac{20}{33}$$
$$= 3 + \cfrac{1}{\cfrac{33}{20}}$$

$$= 3 + \cfrac{1}{1 + \cfrac{13}{20}}$$

$$= 3 + \cfrac{1}{1 + \cfrac{1}{\cfrac{20}{13}}}$$

$$= 3 + \cfrac{1}{1 + \cfrac{1}{1 + \cfrac{7}{13}}}$$

$$= 3 + \cfrac{1}{1 + \cfrac{1}{1 + \cfrac{1}{\cfrac{13}{7}}}}$$

$$= 3 + \cfrac{1}{1 + \cfrac{1}{1 + \cfrac{1}{1 + \cfrac{6}{7}}}}$$

$$= 3 + \cfrac{1}{1 + \cfrac{1}{1 + \cfrac{1}{1 + \cfrac{1}{\cfrac{7}{6}}}}}$$

$$= 3 + \cfrac{1}{1 + \cfrac{1}{1 + \cfrac{1}{1 + \cfrac{1}{1 + \cfrac{1}{6}}}}}$$

We sometimes write the final form of this continued fraction as 3;11116 for brevity. If the fraction were extended to the repeating continued fraction 3;111161111611116 ... , the value would be $\sqrt{13}$ (NCTM 1969, pp. 269–70). The finite continued fraction 3;11116 = 119/33 = 3.6060 . . . is a reasonably good approximation for $\sqrt{13} \doteq 3.6055$....

Euler was very proficient with continued fractions and could easily generate them. Continued fractions were helpful in obtaining approximations for $\sqrt{13}$, e, π, or other irrational numbers. Euler showed that rational numbers can always be represented as finite or terminating continued fractions, and irrational numbers always have infinite continued fractions. Euler was the first to prove, from their expansions, that numbers like e and e^2 are irrational.

2. Find the sum.

$$1 - \frac{1}{4} + \frac{1}{16} - \frac{1}{64} + \frac{1}{256} - \; ...$$

Most students today would think of this problem as a geometric series with ratio − 1/4. Although Euler was familiar with this method, he preferred to divide 1 by (1 + x), using the common division algorithm to obtain the results

$$\frac{1}{1 + x} = 1 - x + x^2 - x^3 + \; ...$$

If x is replaced by 1/4 in this expression, the original problem is solved quickly. This formula appeared to be general and was certainly obtained by legitimate means. Hence, Euler and many of his contemporaries were puzzled by the fact that replacing x by 1 gives

$$\frac{1}{2} = 1 - 1 + 1 - 1 + \; ...$$

Many eighteenth-century mathematicians were convinced that the sum had to be zero, whereas others thought that the answer was one, a solution arrived at by regrouping as follows:

$$1 - (1 - 1) - (1 - 1) - \ldots = 1.$$

We know that none of these answers is correct, because the series does not converge, but presenting the possibilities to a class can be interesting.

3. Approximate to five decimal places.

$$(1.01)^{10}$$

Problems of this nature have been popular for a long time, since they arise in compound interest and related problems. The problem can be done mentally by thinking in terms of $(1 + .01)^{10}$ and using the binomial theorem:

$$(a + b)^n = a^n + na^{n-1}b$$

$$+ \frac{n(n-1)}{2} a^{n-2}b^2$$

$$+ \frac{n(n-1)(n-2)}{2 \times 3} a^{n-3}b^3 + \ldots + b^n.$$

When $a = 1$, $b = 0.01$, and $n = 10$, the expansion becomes

$$(1 + 0.01)^{10} = 1 + (10)(0.01) + (45)(0.01)^2$$
$$+ (120)(0.01)^3 + \ldots$$
$$= 1 + 0.1 + 0.0045 + 0.00012 + \ldots$$
$$\doteq 1.10462.$$

Being a formalist, Euler liked to explore such expressions. He knew the generalized form, of course, but he was concerned about the result

$$(1 - 2)^{-1} = 1 + 2 + 4 + 8 + \ldots$$

For some mathematicians of Euler's time, such expressions simply confirmed their belief that negative numbers often led to absurdities and should be avoided whenever possible. Some even wanted to avoid multiplying two negative quantities, since no physical interpretation of this phenomenon was known. Euler was not troubled by the lack of physical models for negative or even imaginary numbers, and he used both freely. He found expressions like

$$(1 - 2)^{-1} = 1 + 2 + 4 + \ldots$$

to be intriguing, and our students might also. They might be interested to discover that sometimes even well-known formulas have restrictions. In this expression, if 2 is replaced by a number between 1 and $^-1$, then the formula works.

4. Find the sum.

$$1 + \frac{1}{2} + \frac{1}{3} + \frac{1}{4} + \frac{1}{5} + \frac{1}{6} + \frac{1}{7} + \frac{1}{8} + \frac{1}{9} + \frac{1}{10}$$

Euler would have worked this problem just like the rest of us, albeit faster. If you solve the problem by getting a common denominator, you will better understand why Euler tried but failed to find a formula for the sum of the harmonic progression

$$1 + \frac{1}{2} + \frac{1}{3} + \ldots + \frac{1}{n}.$$

No one else has found a formula either. Euler did show that

$$\lim_{n \to +\infty} \left(1 + \frac{1}{2} + \frac{1}{3} + \ldots + \frac{1}{n}\right) - \ln n = \gamma,$$

called *Euler's constant*. Euler found the approximate value of γ to be 0.577218, but it is not yet known whether the actual constant is rational or irrational. This observation is interesting in view of the fact that he did find the exact sum for such related problems as

$$1 + \frac{1}{2^2} + \frac{1}{3^2} + \ldots = \frac{\pi^2}{6}$$

and

$$1 + \frac{1}{2^4} + \frac{1}{3^4} + \ldots = \frac{\pi^4}{90}.$$

5. A dodecahedron contains how many vertices? The problem can be simplified and solved without counting the vertices on a model, if the Euler-Descartes formula

$$V + F = E + 2$$

is used. A dodecahedron has twelve pentagonal faces and, hence, five edges per face. Since each edge is on exactly two faces,

$$E = \frac{5 \times 12}{2} = 30.$$

Thus, $V + 12 = 30 + 2$, or $V = 20$. Euler's proof of the formula $V + F = E + 2$ and his work on the now-famous Königsberg bridge problem mark him as a pioneer in topology.

6. Simplify.

$$\ln \ln \ln e^{1/e}$$

When entered on a calculator, this problem causes the calculator to display an error message. The problem goes nicely at first; that is, $\ln \ln \ln e^{1/e} = \ln \ln (1/e) \ln e = \ln \ln 1/e = \ln \ln e^{-1} =$

$$\ln (^-1) \ln e = \ln (-1) = ?$$

In the eighteenth century quite a discussion ensued among Euler and other prominent mathematicians as to the resolution of $\ln (^-1)$. Consider the following argument of John Bernoulli (Kline 1972, pp. 409–10):

$$
\begin{aligned}
(^-1)^2 &= 1^2 \\
\ln (^-1)^2 &= \ln (1)^2 \\
2 \ln (^-1) &= 2 \ln 1 \\
\ln (^-1) &= \ln 1 = 0
\end{aligned}
$$

In general, many mathematicians wanted to define the logarithm of a negative number to be the same as the logarithm of its additive inverse. Euler showed that, to define a logarithm of a negative number, one must employ complex numbers.

The flaw in Bernoulli's argument is that $2(\ln x) = \ln x^2$ only when the logarithms exist in the field of real numbers. To overcome a possible flaw in the thinking of some students and to acknowledge the work of Euler, it would be good to point out that logarithms of negative and even complex numbers do exist in the field of complex numbers, but the rules of operation are different.

7. Simplify.

$$e^{i\pi}$$

If you know Euler's relation

$$e^{ix} = \cos x + i \sin x,$$

then it follows that $e^{i\pi} = {}^-1$. But even today, many students are jolted by the fact that a real number raised to an imaginary power could be real. Once Euler discovered how to find logarithms of complex numbers, he was able to evaluate even more novel expressions, like $1i$, $(^-1)i$, and i^i.

8. Find the last two digits of 17^{42}.

Although Euler might have been able to evaluate 17^{42} mentally, using only arithmetic, he would surely have used what is now called Euler's ϕ-function, which gives the number of positive integers less than m that are relatively prime to m. For example $\phi(10) = 4$, $\phi(11) = 10$, and $\phi(12) = 4$. Euler showed that

$$a^{\phi(m)} \equiv 1 (\mathrm{mod}\ m),$$

or that $a^{\phi(m)} - 1$ is divisible by m whenever a and m are relatively prime. To find the last two digits of 17^{42}, it is helpful to know that $\phi(100) = 40$ and $17^{40} \equiv 1 (\mathrm{mod}\ 100)$. Thus 17^{42} has the same last two digits as 17^2, or 8 and 9. Euler developed formulas for calculating $\phi(m)$, such as

$$\phi(P) = P - 1$$

and

$$\phi(P^n) = P^n - P^{n-1},$$

whenever P is a prime. The value of $\phi(m)$ can also be calculated by simply counting the positive integers less than m that are relatively prime to m. Problems involving number theory were Euler's favorites, and he wrote extensively in this area.

Closing Remarks

Euler was neither too proud to discuss the simple problems in his textbooks nor too content to tackle the unsolved problems of his day. A gifted linguist, Euler published his results in Latin, French, German, or Russian as circumstances dictated. Many of his terms and symbols have become standard. These include e, π, i, and $f(x)$. He defined function, logarithms, and trigonometric and other transcendental functions much as we do today. We owe Euler a great debt for his contributions in mathematics, and we should stop at least every century or so to acknowledge this debt.

BIBLIOGRAPHY

Bell, Eric Temple. *Men of Mathematics.* New York: Simon & Schuster, 1937.

Boyer, Carl B. *The History of the Calculus and Its Conceptual Development.* New York: Dover Publishing Co., 1959. (Reprint of *The Concepts of the Calculus*, 1949).

————. *A History of Mathematics.* New York: John Wiley & Sons, 1968.

Eves, Howard. *An Introduction to the History of Mathematics*, rev. ed. New York: Holt, Rinehart & Winston, 1964.

Kline, Morris. *Mathematical Thought from Ancient to Modern Times.* Oxford University Press, 1972.

Struik, Dirk J. *A Concise History of Mathematics*, 3d rev. ed. New York: Dover Publications, 1967.

National Council of Teachers of Mathematics. *Historical Topics for the Mathematics Classroom.* Thirty-first Yearbook. Washington, D. C.: The Council, 1969.

84

Brachistochrone, Tautochrone, Cycloid—Apple of Discord

J. P. PHILLIPS

$\mathcal{T}$HE CURVE generated by the motion of a point on the circumference of a wheel rolling on a flat surface is the cycloid, also known for its mechanical properties as the brachistochrone and the tautochrone, and also described by one mathematics historian as an apple of discord and by others as the Helen of geometry—beautiful, but the source of warfare.

From the moment of its invention, the cycloid generated controversy; indeed, the invention itself is not without dispute [2]. Most sources, but not all [4], agree that it was not known to the Greeks. John Wallis in the seventeenth century credited it to Nicholas of Cusa in the fifteenth century, but he was mistaken. Certainly the curve is described in a book by Charles Bouvelles written in 1501, and he is usually considered the true inventor [1].

The greatest mathematicians of the Scientific Revolution of the seventeenth century were almost obsessed by the striking properties of the cycloid. The list of those who worked on it includes Galileo, Pascal, Roberval, Torricelli (who was said to have died of chagrin at intimations that he had plagiarized its quadrature), Descartes, Fermat, Wren, Wallis, Huygens, Johann Bernoulli, Leibniz, and Newton. This is very nearly the complete galaxy of the stars of the Scientific Revolution, and their

Reprinted from *Mathematics Teacher* 60 (May, 1967): 506–8; with permission of the National Council of Teachers of Mathematics

common interest in the cycloid requires explanation. For that matter, their unceasing squabbles for priority as each new property was discovered constitute an interesting psychological question.

The intense interest in the seventeenth century in mechanics and the mathematics of motion accounts for much of the significance that the cycloid acquired at that time. Much of what was learned about it was produced by the methods of the calculus at a time when calculus had not yet been officially created, and some confusion as to methodology between individual mathematicians was thus inevitable.

Galileo was the first great man to take an interest in the cycloid, and, in what might be called an example of the true spirit of Archimedes, he determined experimentally about 1599 that its area is close to three times that of the generating circle, and went on to suggest that the arch of the cycloid ought to be suitable for building bridges. Such bridges were built later. Galileo also bequeathed an interest in cycloids to his students Torricelli and Viviani, as will be seen.

In the 1630s the French geometer Roberval offered a mathematical proof of the exact equivalence of the area of the cycloidal arch to three times that of the circle, and for his pains was attacked by Descartes for having labored overmuch to produce so small a result. In rebuttal to Descartes's simpler proof, Roberval commented that prior knowledge of the answer to be found had no doubt been of assistance. Descartes met the insult by finding the tangent to the cycloid and challenging Roberval and Fermat to achieve this result for themselves; Fermat did, but Roberval did not or could not.

Since Roberval's results were not published at this time, he became embroiled several years later in further argument about priority of proof, this time with Torricelli, who had almost certainly achieved the area result independently. When Roberval accused him of plagiarism, Torricelli more or less promptly died, and a doubtless over-romantic tradition has attributed his death to dismay at such an accusation. The tangent was again found independently by Viviani, another Italian pupil of Galileo.

Systematic study of the geometry of the cycloid began with Pascal. According to the story, he was suffering from toothache and assorted other pains one night and began to ruminate on the cycloid, at which point the pains abated. Although he had already withdrawn from mathematics to become a religious mystic, Pascal took this as an intimation that God approved of one last mathematical fling, and in eight days he worked out the area of the section produced by any line parallel to the base, the volumes generated by revolution about axis or base, the centers of gravity of these solids, and divers other results. To accomplish these results required the integration of several trigonometric functions, an operation never previously attempted and of course not recognized at the time (1658) as part of the calculus.

Pascal issued a challenge to the mathematicians of Europe to solve these problems also, but only two formal answers were received and neither was good enough to win the prize. One answer came from Wallis, who later rectified a number of errors in it and published it. Informally, Huygens, Wren, and Fermat solved parts of the challenge problem, and Huygens went on to new and different ground.

With Huygens, the primary interest in the cycloid shifts from its mathematical to its mechanical properties. As a purely mathematical accomplishment, Huygens discovered that the evolute of the cycloid is an equal cycloid, but he appears to have been much more interested in applications of his discovery that the cycloid is the tautochrone; that is, a body freely accelerated by gravity along an inverted cycloidal arc requires the same length of time to reach the bottom of the arc no matter how far from the bottom it is to start with. His proposal that pendulum clocks (which he invented) should use a cycloidal motion turned out to offer numerous mechanical difficulties, however.

The most famous discovery concerning the cycloid came in 1696, when Johann Bernoulli undertook to determine the equation of the path down which a particle will fall from one point to another in the shortest time. This was the first formal investigation of the calculus of variations, but is additionally important because Bernoulli saw an opportunity to use it in the controversy then raging between partisans of Newton and Leibniz concerning the invention of the calculus. In correspondence, Leibniz and Bernoulli agreed that only a handful of mathematicians in Europe besides themselves could possibly find the solution—the brachistochrone or cycloid—and that those who could would thereby demonstrate the most thorough mastery of the calculus. The problem was duly submitted as a challenge to mathematicians.

In a celebrated (but poorly documented) performance, Newton is said to have received the problem in the mail after a hard day's work at the mint and to have solved it before the next morning. His solution was published anonymously in the *Philosophical Transactions of the Royal Society,* and Leibniz and Bernoulli were sorely disappointed.

Regrettably, the incident only stirred up Newton's abnormal suspicions that others were trying to claim credit for his work, and several years of underhanded efforts to undermine Leibniz's claim to the calculus followed. The whole episode became a disgrace to the principals and swayed the entire future of mathematics in Europe [3].

BIBLIOGRAPHY

1. Cajori, F. A. *History of Mathematics* (New York: The Macmillan Co., 1919), p. 162.

2. Gunther, S. *Bibliotheca Mathematica* I (1887), 7.

3. More, L. T. *Isaac Newton*. New York: Dover Publications, 1934.

4. Zwikker, C. *The Advanced Geometry of Plane Curves and Their Applications*. New York: Dover Publications, 1963.

85

Gaspard Monge and Descriptive Geometry

LEO GAFNEY

$\mathcal{M}$EANDERING about a second-hand bookstore some time ago, I came upon a musty volume entitled *Géométrie Descriptive* by G. Monge, Paris, 1827 (see Fig. 1). Out of curiosity, I bought the old tome, but gave neither subject nor author another thought until the following summer when I found the name "Monge" sprinkled through a course in differential geometry. He was credited, for example, with having given the first analytical expression of a tangent and with having done pioneering work in the field of partial differential equations. With renewed interest, I dusted off my French dictionary and returned to *Géométrie Descriptive*.

Gaspard Monge

Gaspard Monge was born in Beane, France, in 1746, and died in Paris in 1818. While still in his teens, he constructed a "blueprint" of his home town which won him an appointment to the college of engineers at Mézières. Since he was the son of a peddler, it was a somewhat limited appointment, admitting him not to the school for officer training but to the annex where surveying and drawing were taught. Fortunately for the history of geometry, he was at length permitted to con-

GÉOMÉTRIE

DESCRIPTIVE;

PAR G. MONGE.

CINQUIÈME ÉDITION;
augmentée d'une
THEORIE DES OMBRES ET DE LA PERSPECTIVE,
extraite des papiers de l'auteur,
PAR M BRISSON.
Ancien élève de l'... ... Polytechnique...
Chaussées.

PARIS,
BACHELIER (SUCCESSEUR DE Mme Vve COURCIER),
LIBRAIRE POUR LES MATHÉMATIQUES,
QUAI DES AUGUSTINS.
1827

FIGURE I

struct the plan of a fortress from observed data. Rejecting the cumbersome arithmetic calculations then demanded for such a project, he produced a solution using only ruler and compass. With geometrical methods, remarkable in their elegance and simplicity, Monge at once contributed to the solution of practical military problems of his day

Reprinted from *Mathematics Teacher* 58 (Apr., 1965): 338–44; with permission of the National Council of Teachers of Mathematics and at the same time became the father of a new theoretical study—descriptive geometry.

Due both to the rivalry between French military schools and to the agitated times surrounding the French Revolution, Monge was not permitted to publish or teach his discoveries from 1768, when he was made professor at Mézières, until 1794, when he was elected professor at the short-lived normal school of the Republic. In the intervening years, Monge was not only a teacher but also an active French revolutionary. In 1792 he was minister of naval affairs and director for the manufacture of cannon and powder. These offices, however, did not protect him from the terrorists, and he was forced to flee the guillotine. Returning in 1794, he helped organize the normal school mentioned above.

FIGURE 2

AVERTISSEMENT DE L'ÉDITEUR.

Ce Traité de *Géométrie descriptive* a été composé pour l'usage des Élèves de la première École normale, créée par une loi du 9 brumaire an 3 (30 octobre 1794).

Cette École, qui n'a subsisté que pendant les quatre premiers mois de l'année 1795, et qui était destinée à régénérer l'instruction publique, anéantie en France sous le règne de la terreur, comptait quinze cents Élèves pris dans tous les départemens, parmi les instituteurs et les hommes cultivant les Sciences et les Lettres; elle avait pour professeurs

MM. Lagrange, Laplace .... pour les Mathématiques,
Monge .... pour la Géométrie descriptive,
Haüy .... pour la Physique,
Berthollet .... pour la Chimie,
Daubenton .... pour l'Histoire naturelle,
Thouin .... pour l'Agriculture,
Buache, Mentelle, Volney .... pour la Géographie et l'Histoire,
Laharpe, Garat, Bernardin-de-St.-Pierre, Sicard .... pour la Grammaire, la Littérature et la Morale.

MM. Lacroix et Hachette étaient Professeurs-Adjoints pour la *Géométrie descriptive*. Les leçons orales sur cette science se donnaient, ainsi que pour les autres parties de l'enseignement, à l'amphithéâtre du Jardin des Plantes. On avait disposé, dans les bâtimens de la Sorbonne, de grandes salles de dessin, où les Élèves de l'École normale étaient exercés aux constructions graphiques.

Cette nouvelle édition des leçons de Géométrie descriptive

a

The book which initiated this article begins with an interesting reference to the normal school (see Fig. 2): "This treatise on descriptive geometry was written for the use of students in the first normal school, established by the law of the 9th of Brumaire in the year 3 (30th of October, 1794). This school, which existed only during the first four months of the year 1795, and which was intended to revive public education, annihilated in France under the Reign of Terror, . . . had as professors. . . ." There are two things of note on this page: first, the strange month and year—French Revolution time—and second, the extraordinary mathematics faculty.

Monge continued to support the Republic until its collapse, when he became a zealous partisan of Napoleon and even accompanied Napoleon on his Egyptian campaign, although he did not follow his emperor on the Russian march. With the fall of Napoleon, Monge could not redeem himself once more, and, deprived of his previous honors by Louis XVIII, he died in a relatively degraded state.

Descriptive Geometry

The objectives of descriptive geometry," Monge begins his work, "are two: first, to give an understanding of the methods of representing on a two-dimensional surface objects which in nature have three dimensions. . . . The second objective is to teach the way to determine the forms of objects and to deduce all of the properties resulting from their respective positions."

Monge achieved his first goal with virtually unrevisable success; in striving for the second he opened many doors for future mathematicians.

In a certain sense, Monge begins where the artist Albrecht Dürer (1471–1528) let off. A careful examination of Figures 3 and 4 will reveal the similarity. In each case, a point is projected onto several planes, and these are rotated so as to form a right angle (wall and floor) about the ground line. The method of Monge, therefore, consists chiefly

FIGURE 3

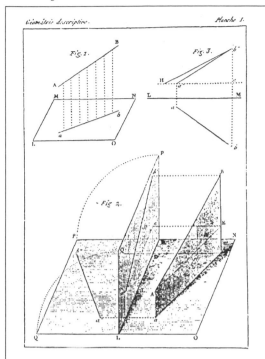

FIGURE 4*

in making orthographic projections of an object on two (or possibly more) planes and establishing a definite relation between the different projections. The two planes are ordinarily the horizontal and a vertical, and the line of intersection is the ground line.

Some of the projections and constructions discussed by Monge, which are accompanied by elegant diagrams, are as follows:

1. Given a point and a line (or plane), to find a line (plane) through the given point parallel (perpendicular) to the given line (plane).

2. Given a point and a surface, to find tangents to the surface from the given point.

3. Given a cylinder or conic, to find the tangent at a particular point.

4. To describe the intersections of surfaces—cylinders, conics, etc.

5. To investigate the curve which is described on a three-dimensional surface when the surface is "sliced" at a point by a plane. (See Figures 46 and 47 in Figure 5. This is actually a projection, but, by considering it on the surface, it is possible to examine tangents, as in Figures 46 and 47, and to discuss the shape of the intersection of two surfaces, as in Figure 49 in Figure 5.)

The surfaces considered by Monge are of two general types. The first is that which can be conceived of as being formed by bending a plane surface without, however, stretching, crumpling, or tearing it. This is called a developable surface. The central problem under this topic is to determine the shape of the plane figure formed by the boundary of a piece of developable surface as the latter is rolled out into a plane. A circular cylinder slit along an element, for example, will become a rectangular figure, and a circular cone slit along an element will become a sector of a circle.

The second type of surface is that which cannot be formed by bending a piece of a plane. A

*D. J. Struik, *Lectures on Analytic and Projective Geometry*, p. 236. Reading, Mass.: Addison-Wesley Publishing Company, 1953. The page is reproduced with permission of the publisher.

FIGURE 5

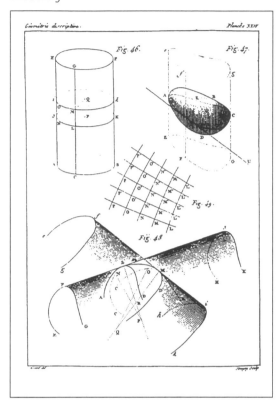

FIGURE 6

Gaspard Monge

(Source: *A Portfolio of Eminent Mathematicians*, edited by David Eugene Smith (Chicago: Open Court Publishing Company, 1908)

sphere and an ellipsoid are two such figures. A practical problem which will throw light on our discussion is that of making a world map. As we know, there are various types of projections, but none of them perfectly represents on a plane the continental figures we find on the sphere of our Earth. The reason is that a sphere cannot be "rolled out smoothly" so as to form a plane surface; it is not a developable surface.

The work of Monge contained in the book we are considering, and which we have outlined above, virtually completes the subject of three-dimensional representation in the plane from the point of view of double projection. Teachers of mechanical or engineering drawing even today will recognize the principles of Monge as forming the backbone of their subject. The actual completion of the subject in the form of a few finishing touches was added by two students of Monge—S. F.

Lacroix and J. N. P. Hachette. Their only really new contribution falls under the topic of shadows.

Two other outstanding pupils of Monge were Charles Dupin, who contributed to the field of differential geometry and after whom the Dupin indicatrix was named, and Victor Poncelet, who, as a prisoner of war in Napoleon's Russian campaign, undertook an investigation of the properties of plane configurations left invariant under projections.

Monge the Teacher

It would be unfair to complete even this brief study of Monge without some discussion of him as a teacher. His biographers acclaim him as one of those very special mathematicians who not only possess the fire of genius but can also ignite it in others. About two-thirds of the way through the book we have been discussing (p. 111), there is evidence of this when Monge takes time out to talk about students and teachers:

If schools have been established . . . where young people can exercise themselves in mathematical constructions and become familiar with the more important phenomena of nature and thus gain most necessary knowledge, a knowledge which will develop their intelligence, and give them the habit and feeling for precision so that these students then will most assuredly contribute to the progress of the national industry, and be guaranteed forever safety from false teachers, and if I were here proposing the basic text for instruction in such schools, the abstract treatment would end right here and we would pass immediately to applications which are more useful and to those which are more frequently used. But we ought not write only for students of the schools; we ought to write also for their professors.

It is possible to conduct ordinary instruction on the level of simple cases which are of use in daily life; but if a student comes just once in his life upon a difficulty which was not treated in class, to whom shall he turn if not to the professor? And what shall the professor answer if he has not familiarized himself with considerations of a more general nature than those which comprise the ordinary content of the subject?

BIBLIOGRAPHY

1. Aleksandrov, O. D., "Curves and Surfaces," *Mathematics, Its Content, Methods and Meaning* (Vol. 1, part 3). Providence, R. I.: American Mathematical Society, 1963.

2. Arago, D. F. J., "Biographie de Gaspard Monge," *Mémoires de l'Acad. d. Sci.*, Paris, XXIV (1853).

3. ———, *Notices Biographiques*, Paris, Vol. 2, 1853, 427–592.

4. Archibald, R. C., "Centers of Similitude and Certain Theorems Attributed to Monge. Were They Known to the Greeks?" *American Mathematical Monthly*, XXII (1915), 6–12.

5. ———, "Historical Note on Centers of Similitude of Circles," *American Mathematical Monthly*, XXIII (1916), 159–61.

6. Aubry, P. V., *Monge, le Savant Ami de Napoleon Bonaparte*. Paris, 1954.

7. Ball, W. W. R., *A Short Account of the History of Mathematics*. New York: Dover Publications, Inc., 1960 (reprinted from 1908 edition).

8. Beumer, M. G., "Gaspard Monge as a Chemist," *Scripta Mathematica*, XIII (1944), 122–23.

9. Bell, E. T., *Men of Mathematics* (Chapter 12). New York: Simon and Schuster, Inc., 1937.

10. Cajori, F., *A History of Mathematics*. New York: The Macmillan Company, 1919.

11. Coolidge, J. L., *A History of Geometrical Methods*. Oxford: The Clarendon Press, 1940.

12. De Launay, L., *Monge, Fondateur de l'École Polytechnique*. Paris, 1933.

13. De Morgan, A., "Monge, Gaspard," *Engl. Cycl.-Biog.*, Vol. 4, 1857.

14. Dupin, C., *Essai Historique sur les Services et les Travaux Scientifiques de Gaspard Monge*. Paris, 1819.

15. Kreyszig, E., *Differential Geometry*. Toronto: University of Toronto Press, 1959.

16. Monge, G., *Géométrie Descriptive*, 5th ed. Paris, 1827.

17. Nielson, N., *Géométrie Français sous la Révolution*, "Monge," 182–90. Copenhagen, 1929.

18. Sanford, V., "Gaspard Monge," *The Mathematics Teacher*, XXVIII (1935), 238–40.

19. Simon, L. G., "The Influence of French Mathematicians at the End of the Eighteenth Century upon the Teaching of Mathematics in American Colleges," *Isis*, XV (1931), 115, 119.

20. Smith, D. E., "Gaspard Monge, Politician," *Scripta Mathematica*, I (1932), 111–22.

21. ———, "Among My Autographs: Monge and the American Colonies" and "Monge the Lesser," *American Mathematical Monthly*, XXVIII (1921), 166, 208–9.

22. Struik, D. J., *Lectures on Analytic and Projective Geometry*. Reading, Mass.: Addison-Wesley Publishing Company, 1953.

23. ———, *Lectures on Classical Differential Geometry*. Reading, Mass.: Addison-Wesley Publishing Company, 1950.

24. ———, "Outline of a History of Differential Geometry," *Isis*, XIX (1933), 92–120; XX (1934), 161–91. Part 4, "Monge and the École Polytechnique," XIX, 113–20.

25. Taton, R., "Gaspard Monge," *Elemente der Mathematik, Beihefte*, IX (1950).

26. ———, *L'Oeuvre Scientifique de Gaspard Monge*. Paris, 1951.

86

Mathematicians of the French Revolution

CARL B. BOYER

$\mathcal{T}$HE EIGHTEENTH century has had the misfortune to come after the seventeenth and before the nineteenth. How could any period which followed the "Century of Genius" and which preceded the "Golden Age" of mathematics be looked upon as anything but an inconsequential interlude? Analytic geometry and the calculus were invented in the seventeenth century; the rise of mathematical rigor and the flowering of geometry are associated with the nineteenth. It is not easy to find comparable developments in the eighteenth century. H. G. Zeuthen wrote a *Geschichte der Mathematik im XVI. und XVII. Jahrhundert* (German ed., Leipzig, 1903), and Felix Klein published *Volesungen über die Entwiklung der Mathematik im 19. Jahrhundert* (2 vols., Berlin, 1926–1927)—both unusually scholarly works—but who ever heard of a history of mathematics in the eighteenth century? Not that the period is a complete void, for the veriest tyro knows that in the very middle of the century which began with the closing years of Newton and ended with the early years of Gauss there lived the prolific Euler—a man for whom calculation was as easy and natural as breathing is for us. But one does not look to the latter part of the eighteenth century for significant trends or discoveries in mathematics. This is in marked contrast to what is true in other fields. For Americans the date 1776 was

decisive; in France the year 1789 was crucial. Nor was the Age of Revolutions confined to the sphere of politics. The Industrial Revolution has changed the whole fabric of Western society, and the thermotic revolution during the same years laid the foundations of modern chemistry. Can it be possible that mathematics during these stirring events was enjoying a Rip Van Winkle nap, uninfluenced by the world about it? It is the purpose of this paper to show that this was far from the case—that the mathematicians of France at the time of the Revolution not only contributed handsomely to the fund of knowledge, but that they were in large measure responsible for the chief lines of development in the explosive proliferation of mathematics during the succeeding century. We shall even be tempted to add to the already impressive list of revolutions of the time two more, a "geometrical revolution" and an "analytical revolution." In developing this argument considerable use has been made of Niels Nielsen's *Géomètres français sous la révolution* (Copenhague, 1929); but whereas Nielsen has catalogued alphabetically the contributions of some four score French mathematicians, it is the object of this paper to furnish a synthesis of the mathematical milieu of the time with emphasis upon but a handful of individuals.

Every age is inclined to think of itself as one of revolution—a period of tremendous change. But almost every age of rapid change has been preceded by a long period in which preparations for the revolution are made, sometimes consciously,

Reprinted from *Scripta Mathematica* 25 (Spring, 1960): 11–31; with permission of *Scripta Mathematica* and Yeshiva University Press.

more often unconsciously; and in this respect the French political and mathematical revolutions were far from exceptional. Among the heralds of the French Revolution were Voltaire, Rousseau, D'Alembert, and Diderot—not one of whom lived to see the fall of the Bastille (Voltaire and Rousseau died in 1778; D'Alembert died in 1783; and Diderot a year later)—and their associate Condorcet, who fell a victim in the holocaust which he helped to father. In mathematics six figures who were to show the way—Monge, Lagrange, Laplace, Legendre, Carnot, and Condorcet—were to be in the midst of the turmoil, and it is primarily about these men that we shall build our account.

The half dozen mathematicians who will be the center of our attention were almost of an age: Lagrange, the oldest, was born in 1736; Condorcet was born in 1743; Monge in 1746; Laplace in 1749; Legendre in 1752; Carnot, the youngest, was born in 1753. With the exception of Condorcet, who died a suicide in prison, these mathematicians all lived to be septuagenarians, and one, Legendre, an octogenarian. And it is a comfort to us to note, in these days of surveys proving that it is mostly the young men who contribute significantly to mathematics, that the chief influence of our five heroes came during their forties, fifties, and sixties. But before we tell about their role as mature leaders, let us look at their early training.

Today we take it as a matter of course that anyone who has unusual ability and interest in mathematics can follow his inclinations and find a place for himself in the scholarly world. In France of the eighteenth century, however, this was far from the case. Universities were not the mathematical foci that they are today; and one is hard put to it to name even one eighteenth-century mathematician at, say, the University of Paris. During the fourteenth century Paris had been one of the mathematical centers of the world, the other being at Oxford; but it had long since lost this position. It was always behind the times: when Europe turned to Cartesianism, Paris clung to Peripatetic Scholasticism; and when most of the scientific world had turned to Newtonionism, Paris fought a rear-guard action for Cartesianism. Most of the French mathematicians of the time were associated not with the universities, but with either the church or the military; others found royal patronage or became private teachers. Only families of some means or position were able to provide their sons with mathematical training, generally through a church or military school. Lagrange, the only one of our group who was not strictly a Frenchman, was born of a once prosperous family who provided for him an education at the college in Turin. As a young man he became professor of mathematics in the military academy of Turin, but later he found royal patrons in Frederick the Great and Louis XVI. Condorcet's family included influential members in the cavalry and the church, and hence his education presented no problem. At Jesuit schools and later at the Collège de Navarre he made an enviable reputation in mathematics; but instead of becoming a captain of cavalry, as his family had hoped, he lived the life of a scholar in much the same sense as Voltaire, Diderot, and D'Alembert. The third of our sextet, Gaspard Monge, was the son of a poor tradesman. However, through the influence of a lieutenant colonel who had been struck by the boy's ability, Monge was permitted to attend some courses at the École Militaire de Mézières; and he so impressed those in authority that he soon became a member of the teaching staff—the only one of our group of six who was primarily a teacher, and as we shall presently see, perhaps the most influential mathematics teacher since the days of Plato. Laplace, too, was born without wealth; but, like Monge, he found influential friends who saw that he got an education—again in a military academy. And for a while Laplace, like Monge, taught mathematics in the school in which he had been educated. Legendre was another member of the sextet who experienced no difficulty in securing an education; but even he was not a university teacher in the strict sense, although for five years he taught in the École Militaire at Paris. The youngest of our group,

Lazare Carnot (who is not to be confused with his son, Sadi Carnot), was sufficiently above bourgeois standing to be permitted to enter the École Militaire at Mézières, where Monge was one of his teachers. Upon graduation Carnot entered the army, although, lacking a title, he could not, under renewed rule of the *ancien régime*, aspire to a rank above that of captain. This must have rankled in his mind—as it did in the case of so many others that the proverb arose at examination time that "the competent were not noble and the noble were not competent." The economic wastefulness of the government may have been the immediate cause of the French Revolution, but it was far from the only one. The enormous waste of human resources was also an important factor, and symptomatic of this was the failure at first of the men of our group to win positions commensurate with their ability. Is it any wonder that not one of the six expressed regret later when the old order passed away? But before we turn to the events which brought about the new order of things, let us glance at the mathematics education our sextet inherited.

In the early eighteenth century France had taken the lead in the exploitation of two capital inventions of the preceding century—analytic geometry and the calculus. Neither of these subjects, however, resembled closely those now given in our textbooks. Representative of the forms in which they appeared in France throughout the eighteenth century are the two works of L'Hospital—*Traité des sections coniques* (Paris, 1707) and *Analyses des infiniment petits* (Paris, 1696). The author of these two books was not a teacher; but no other textbooks could rival his during the eighteenth century. This is only one instance of the observation we shall frequently make that the good teacher and the good textbook writer are not generally the same person. L'Hospital was not responsible in any significant measure for the material in his books; he cribbed from others—in true textbook tradition—and not even the rule on indeterminate forms which bears his name was his own discovery. But he did give the subjects a much-needed

conventional arrangement which enabled the novitiate to digest the new subjects more easily. Nevertheless, neither of L'Hospital's texts would satisfy an Anglo-Saxon instructor of today, for two deficiencies in particular would be immediately apparent. First, there were no lists of exercises to be worked out by the student; more importantly, very little attention was paid to fundamental principles—and this at a time when the new disciplines were badly in need of a sound foundation. Even apart from such criticisms of the L'Hospital textbooks, they would scarcely fit into modern courses. The *Conics* in particular would be a poor substitute for our analytic geometries, for there are no generalities on coordinates, no analytic proofs of elementary theorems, no formulas for slope, distance, angle, no consideration of the straight line, no sketching of higher plane curves, and nothing on problems in three dimensions. L'Hospital's calculus comes somewhat closer to modern material, but here too one finds substantial differences: there are no rules for the differentiation of transcendental functions, and L'Hospital never got around to writing a complementary book on the integral calculus.

L'Hospital's works were not the greatest textbooks of the century—they were simply the most popular. The book which, in didactic quality, stands out above all others of the eighteenth century was Euler's *Introductio in analysin infinitorum* (2 vols., Lausanne, 1748). This added to analytic geometry the sketching of higher plane curves, both algebraic and transcendental; and, together with Euler's treatises on the calculus, it made the logarithmic and trigonometric functions a traditional part of college courses in analysis. Even on the lower level Euler ultimately had a deep influence, for trigonometry became less and less the geometry of half-chords and more and more the study of the now familiar goniometric functions, with the full recognition of their periodicity. The Euler function-concept approach, however, took hold but slowly. The textbooks of L'Hospital reappeared in just as many editions after Euler's *Introductio* as before, whereas the *Introductio* itself did not appear in another edition until almost half a century later. The medium through which the influence of Euler was most felt was that of the multi-volume compendium—the *Cours complet de mathématiques*—which made its appearance at about the time that our sextet was studying its mathematics.

Of the mathematical encyclopedias of the later eighteenth century the most successful, judging from repeated editions, was that by Étienne Bézout, instructor in the school at Mézières which both Monge and Carnot attended. The *Cours de mathématique* of Bézout appeared in six volumes at Paris from 1764–1769, and a second edition began to appear only a year after the completion of the first. During the first third of the nineteenth century it still was a very influential work, especially in America where parts of it appeared in English translation at West Point and other academies. Bézout's *Cours* was intended to cover virtually the entire spectrum of mathematical instruction from arithmetic through the applications of the calculus. His preface claimed that he presupposed of his readers nothing beyond the names of the numbers. Even the techniques involved in the arithmetic operations on integers are explained at length in the first volume, together with some short-cuts in calculation and the proof by nines. The whole of the first volume is devoted to arithmetic techniques. The second volume contains the elements of geometry and trigonometry. The geometry is a thoroughly emasculated structure, with little in the way of proof. Following the definition of parallel lines as those which never meet no matter how far one imagines them prolonged, the theorem on the equality of corresponding angles formed by a transversal is "proved" by the simple observation that the parallel lines, not having any inclination with respect to each other, must necessarily have the same inclination with respect to the transversal. There are no formal proofs of theorems, but angle-measure looms large—as one should expect in a volume intended for future navigators—as does also solid mensuration. The definition of trigonometry is interesting for its emphasis: the subject is defined as "a part of geometry which teaches how to determine or calculate three of the six parts of a triangle [*triangle rectiligne*] from the knowledge of three other parts, when this is possible." There is nothing here of Euler's ratios or the periodic function approach. Anticipating the needs of students of navigation, Bézout hurried on to spherical trigonometry.

Part III of Bézout's *Cours* covers algebra and its application to arithmetic and geometry: and here too the author holds to the belief that the study of mathematics "is less the accumulation of a great number of propositions than the acquisition of the spirit of research and discovery." Admitting that the chief object of his volume on algebra is to prepare students for the study of mechanics, Bézout apologizes for spending some time on applications to arithmetic and geometry, justifying the digression by the need for developing facility in algebra. One sees here, almost a century and a half since the appearance of Descartes's *Géométrie*, that analytic geometry was indeed a step-child in mathematics. One had to apologize for teaching it; and perhaps one reason for this situation was that the purpose of the subject was not what we know it today. The object of "the application of algebra to geometry," as inherited from Descartes, was the geometrical construction of algebraic quantities—such as the roots of the equation $4u^3 - 3r^2u - cr^2 = 0$. Here, too, the influence of Euler was slow in making itself felt.

The fourth part of Bézout's *Cours* is the *raison d'être* of the program—the principles of mechanics—but the author necessarily had to introduce this with a substantial development of the differential and integral calculus, including differential equations. This is one place where the ubiquitous influence of Euler is really felt, especially in the analysis of transcendental functions. The emphasis given to mechanics and to the closing section on navigation is in keeping with the use of the *Cours de mathématiques* as a text in a military academy. The mathematical preeminence of France (and, indeed, of continental Europe as a whole) in the eighteenth century is, in fact, based in large measure on the application of analysis to mechanics as taught in technical schools, and it was under this influence that the mathematicians of the French Revolution had been brought up. This is in marked contrast to the situation in England, which remained a stronghold of synthetic geometry. One should naturally expect the contrast in mathematical spirit to become sharper during the Revolution, for France had greater need for technical training and England became more thoroughly isolated from the continent. But let us look more closely at the mathematical situation in France.

Everyone of the six men we have named as the mathematical leaders during the Revolution had produced abundantly before 1789. Lagrange had published his *Méchanique analytique* (1788), as well as frequent papers on algebra, analysis, and geometry. Condorcet, perhaps the most interesting of the six because of the breadth of his interests, had published *De calcul intégral* as early as 1765, and *Essai sur l'application de l'analyse à la probabilité des décisions rendues à la pluralité des voix* in 1785. A firm believer in the perfectibility of man, a basic tenet of the Philosophes, Condorcet was the only one of our six who can be said to have played an anticipatory role in the events leading to 1789. (It is ironic to note that of our mathematical sextet the one who did most to bring about the Revolution was the only one to lose his life through it, although two others, Carnot and Monge, were not always safe from the guillotine.) Monge had contributed numerous mathematical articles to the *Mémoires* of the *Académie des Sciences*. Inasmuch as he succeeded Bézout as examiner for the School of the Marine, Monge was urged by those in authority to do what Bézout had done—write a *Cours de mathématiques* for the use of candidates. Monge however, was interested in teaching and research rather in writing textbooks, and he completed only one volume of the project—*Traité élémentaire de statique* (Paris, 1788). He was attracted not only to both pure and applied mathematics but also to

physics and chemistry. In particular, he participated with Lavoisier in experiments, including those on the composition of water, which led to the chemical revolution of 1789. Through his numerous activities Monge had become, at the time of the revolution, one of the best known of French scientists. In fact, his reputation as a physicist and chemist was perhaps greater than that as a mathematician, for his geometry had not been properly appreciated. His chief work, the *Géométrie descriptive*, had not been published because his superiors felt that it was in the interests of national defense to keep it confidential. (Classified material is not, you see, a monopoly of the mid-twentieth century.) Laplace and Legendre were regular contributors to learned periodicals, and Carnot by 1786 had published a second edition of his *Essai sur les machines en général* as well as some verses and a work on fortifications.

In looking closely at the achievements of the six men with whom we are concerned, one is struck by a lack of utilitarian motive in their work. Carnot's *Essai* would appear, from the title, to be most technically oriented, but a glance at the book will show that it deals with broad principles, not with technology. The *Mécanique* of Lagrange likewise is concerned with a postulational treatment of the subject, far removed from criteria of practicability. The beauty of Lagrange's work is apparent not to the engineer but to the pure mathematician; even in the more elementary portions of his work there is an aesthetic quality. It is primarily to him that we owe such compact results as

$$\frac{1}{2!} \begin{vmatrix} x_1 & y_1 & 1 \\ x_2 & y_2 & 1 \\ x_3 & y_3 & 1 \end{vmatrix}$$

for the area of a triangle and

$$\frac{1}{3!} \begin{vmatrix} x_1 & y_1 & z_1 & 1 \\ x_2 & y_2 & z_2 & 1 \\ x_3 & y_3 & z_3 & 1 \\ x_4 & y_4 & z_4 & 1 \end{vmatrix}$$

for the volume of a Tetrahedron.

The work of Lagrange on the volume of a tetrahedron looks pretty, but inconsequential; yet it contained an idea which was to become, through the educational reforms of the Revolution, of considerable importance. As Lagrange expressed it, "I flatter myself that the solutions which I am going to give will be of interest to geometers as much for the methods as for the results. These solutions are purely analytic and can even be understood without figures." And true to his promise, there is not a single diagram throughout the work. Monge, too, seems somehow to have come to the conclusion, whether independently or not one cannot tell, that one should avoid the use of diagrams. Even Carnot seems vaguely to have felt somewhat the same way, and his *Essai*, antedating the *Mécanique* of Lagrange, contains not a single diagram.

Laplace, of all the members of our sextet, came closest to being an applied mathematician; but even in his case one must interpret the phrase in a very broad sense. After all, how "practical" in those days was the theory of probability or celestial mechanics? One can safely conclude that, in spite of their education in predominantly technical schools, the great figures in mathematics just before the Revolution had shown remarkable "purity" of interest. Was this to be lost during the national emergency?

The fall of the Bastille in 1789 found our six men divided into two categories: the three L's (Lagrange, Laplace, and Legendre) took no active part in shaping the political events which were to follow; the other three (Carnot, Condorcet, and Monge) welcomed the changed outlook and played definite roles in revolutionary activities. Men from both groups, however, played roles in at least one mathematical project during the Revolution.

The reform of the system of weights and measures is an especially appropriate example of the way in which mathematicians patiently persisted in their efforts in spite of confusion and political difficulties. As early in the Revolution as 1790 Talleyrand proposed the reform of weights and measures. The problem was referred to the Académie des Sciences, in which a committee, of which Lagrange and Condorcet were two of the members, was established to draw up a proposal. Legendre should have been a member, for he had achieved quite a reputation for his triangulation of France; but revolutionary politics seem to have been responsible for his being overlooked. The Committee agreed on a decimal system, although there appear to have been some earnest supporters of a duodecimal scheme. Lagrange firmly supported the decimalists against the duodecimalists, for he was not greatly impressed by the argument about divisibility. (He is reported to have almost regretted not adopting as a base for the system some *prime* number, such as eleven; but it has been suggested that he may have done this simply to obstruct the duodecimalists.) As is well known, the Committee considered two alternatives for the basis of the new system. One was the length of the pendulum which should beat seconds. The equation for the pendulum being $T = 2\pi\sqrt{l/g}$, this would make the standard length g/π^2. But the Committee was so impressed by the accuracy with which Legendre and others had measured the length of a terrestrial meridian that in the end the meter was defined to be the ten-millionth part of the distance between the equator and the pole. The resulting metric system was ready in most respects in 1791, but there was confusion and delay in establishing it. The National Convention in 1793 suppressed the Académie des Sciences, while the Jardin des Plantes was greatly expanded. This inconsistency seems to have been the result of political forces. The Académie was led by older and more conservative men, the Jardin by younger scientists who were eager in their support of the new government. There was, moreover, quite a cult of Robespierre which represented a back-to-nature attitude derived in part from Rousseau. Evidently there was in France an attitude toward physical science something like Goethe's belligerency toward Newtonian physics. The Jardin des Plantes represented "safe" science, that of the Académie was suspect. The closing of the Académie was a blow to mathematics; but the Convention continued the Committee on Weights and Measures, although it purged the Committee of some members, such as Lavoisier, and enlarged it by adding others, including Monge. At one point Lagrange was very nearly lost to the Committee, for the provincially minded Convention had banned foreigners from France; but Lagrange was specifically exempted from the decree, and he remained to serve as head of the Committee. Still later the Committee was placed under the Institut National which had replaced the Académie des Sciences; and Lagrange, Laplace, Legendre, and Monge all served on the Committee at this stage. By 1779 the work of the Committee had been completed, and the metric system as we have it today became a reality. It will be noted that five of our group of six revolutionary mathematicians took active part in this project, only Carnot being unconnected with it; but we shall find that Carnot was engaged in many other essential activities, both political and mathematical. The metric system is, of course, one of the more tangible mathematical results of the Revolution; but in terms of the development of our subject it cannot be compared in significance with other results we shall point to.

Condorcet was, of all the members of our group, the only one who can be said to have worked to bring about the Revolution. A physiocrat, a philosophe, and an encyclopedist, he belonged to the circle of Voltaire and D'Alembert. He was a capable mathematician who had published books on probability and the integral calculus; but he was also a restless visionary and idealist who was interested in anything related to the welfare of mankind. He, like Voltaire, had a passionate hatred of injustice; and although he held the title of marquis, he saw so many inequities in the *ancien régime* that he wrote and worked toward reform. With implicit faith in the perfectibility of mankind and belief that education would eliminate vice, he argued for free public education, something which we in America now take for granted. He was, in short, what would today be known as a liberal.

Like our own Thomas Jefferson, he believed that error, like truth, had a right to freedom. The extent of his inspiration can well be gauged by the fact that an impressive tome of 891 pages was published by Franck Alengry in 1904 with the title *Condorcet: Guide de la révolution française*. This book, however, is on his social and legal philosophy, rather than on his mathematics. Condorcet is perhaps best remembered mathematically as a pioneer in social mathematics, especially through the application of probability and statistics to social problems. When, for example, conservative elements (including the Faculty of Medicine, as well as the Faculty of Theology) attacked those who advocated innoculation against small pox, Condorcet (along with Voltaire and Daniel Bernoulli) came to the defence of variolation. He argued the case through statistics, just as we do today in justifying vaccination against polio; but his approach did not meet with general approval. Even the mathematician D'Alembert was not convinced, schooled as he had been in an absolutistic attitude toward mathematics.

With the opening of the Revolution, Condorcet's thoughts turned almost exclusively to administrative and political problems. The educational system had collapsed under the effervescence of the Revolution, and Condorcet saw that that was the time to try to introduce the reforms he had in mind. He presented his plan to the Legislative Assembly, of which he became President; but agitation over other matters precluded serious consideration of it. Condorcet published his scheme in 1792, but the provision for free education became a target of attack. Instead of adopting constructive measures, the extremists added to the problems by abolishing the schools of law, theology, medicine, and arts. Not until years after his death did France achieve Condorcet's ideal of free public education. Condorcet boldly denounced the Septembrists, and was ordered arrested for his pains. He sought hiding, and during the long months of concealment he composed the celebrated *Sketch for a historical picture of the progress of the human mind*, indicating nine steps in the rise of mankind from a tribal stage to the founding of the French Republic, with a prediction of the bright tenth stage which he believed the Revolution was about to usher in. Shortly after completing this work (in 1794), believing that his presence endangered the lives of his hosts, he left his hiding place. Promptly recognized as an aristocrat, he was arrested; and the following morning he was found dead on the floor of his prison, presumably a suicide.

Condorcet had been sympathetic to the moderate Gironde wing of the Revolution. (As is well known, extremists often are more bitter against the middle of the political spectrum than they are against the opposite end.) Monge was plebian and an important member of the more radical Jacobin Club; but he, too, was to have some trouble. Monge was an enthusiastic partisan and joined patriotic organizations—as did also Vandermonde and Meusnier. He was assigned a role in the reform of weights and measures ordered by the Constituent Assembly in 1790, but his post as examiner for the navy kept him from Paris for a couple of years. On his return to the city in 1792 he was named Ministre de la Marine, apparently on the suggestion of Condorcet. (It had been Condorcet who earlier had secured for him the chair of hydraulics created by Turgot and had aided in his election to the Académie des Sciences. Condorcet had been offered the post of Minister of the Navy, but he declined the responsibility, recognizing the handicap of his aristocratic title.) It was in his capacity as minister that to Monge fell the task of signing the official record of the trial and execution of the king. The French fleet, however, was so poorly organized and so ineffectual that Monge was unable to accomplish anything significant, and within a year he demanded that he be replaced. He nevertheless remained active in politics and governmental operations. He was a prominent member of the Jacobin Club, serving as vice-president; and he devoted an enormous amount of energy to meeting the needs for gunpowder of the revolutionary

arsenal. At the instance of the Committee of Public Safety he published also a *Description de l'art de fabriquer les canons*. Throughout the Revolution Monge sometimes found himself in a precarious position, for he was too liberal for the conservatives and too conservative for the extremists. At one point, after 9 Thermidor, Monge felt that he had to leave Paris to escape the guillotine; but he was soon back again, zealous for public service.

More important for the future of mathematics were the efforts of Monge, after the crisis of foreign invasion had subsided, to establish a school for the preparation of engineers. As Condorcet had been the guiding spirit in the Committee on Instruction, so Monge was the leading advocate of institutions of higher learning. The result was the formation in 1794 of a Commission of Public Works, of which Monge was an active member, charged with the establishment of an appropriate institution. The school was the famous École Polytechnique which took form so rapidly that students were admitted in the following year. At all stages of its creation the role of Monge was essential, both as administrator and as teacher. It is gratifying to note that the two functions are not incompatible, for Monge was eminently successful in both. He was even able to overcome his reluctance to write textbooks, for in the reform of the mathematics curriculum the need for suitable books was acute. Monge found himself lecturing on two subjects both essentially new to a university curriculum. The first of these was known as stereotomy, now more commonly called descriptive geometry. Monge gave a concentrated course in the subject to four hundred students, and a manuscript outline of the syllabus survives. This shows that the course was of wider scope, both on the pure and the applied side, than is now usual. Besides the study of shadow, perspective, and topography, attention was paid to the properties of surfaces, including normal lines and tangent planes to these, and to the theory of machines. Among the problems set by Monge, for example, was that of determining the curve of intersection of two gauche surfaces each of which is generated by a line which moves so as to intersect three skew lines in space. Another was the determination of a point in space equidistant from four lines. Such problems point up a change in mathematical education which was sponsored primarily by the French Revolution. As long ago as the Golden Age of Greece it had been pointed out by Plato that the state of solid geometry was deplorable. Euclid, schooled by the pupils of Plato, had done something about this, climaxing his *Elements* with the proof of the famous theorem that there are five and only five regular solids. Archimedes followed with his mensuration of the volumes of the sphere, cone, and paraboloid of revolution, and Pappus added his theorems relating centers of gravity to areas and volumes of solids. But the decline in mathematics had hit solid geometry much harder than it had plane geometry. One who could not cross the *pons asinorum* could scarcely be expected to reach the study of three dimensions. The inventors of analytic geometry, Descartes and Fermat, had been well aware of the fundamental principle of solid analytic geometry—that every equation in three unknowns represented a surface, and conversely—but they had not taken steps to develop it. One can say that whereas the seventeenth century was the century of curves—the cycloid, the limaçon, the catenary, the lemniscate, the equiangular spiral, the hyperbolas, parabolas, and spirals of Fermat, the pearls of Sluse, and many others—the eighteenth was the century which really began the study of surfaces. Coolidge, in an article in the *American Mathematical Monthly* a few years before he died, called attention to the slow progress made in solid analytic geometry before the days of Euclid. It was Euler who called attention to the quadric surfaces as a family analogous to the conics. Whereas the conics had been exhaustively treated by Apollonius, before the middle of the eighteenth century the general hyperboloid of one sheet and the hyperbolic parabolid had not been identified. Euler's *Introductio* in a sense established the subject of solid analytic geometry (although one must

perforce mention Clairaut as a precursor); but Euler was not a proselytizer and hence his subject found no place in the school curriculum. One reason may have been that, like Descartes, he did not begin with the simplest rectilinear cases. We have seen that Lagrange, influenced perhaps by his calculus of variations, manifested interest in problems in three dimensions and emphasized their analytic solution. He was first, for example, to give the formula

$$D = \frac{ap + bq + cr - d}{\sqrt{a^2 + b^2 + c^2}}$$

for the distance D from a point (p, q, r) to the plane $ax + by + cz = d$. But Lagrange did not have a geometer's heart, nor did he have enthusiastic disciples. Monge, by contrast, was a specialist in geometry—almost the first since Apollonius—as well as a superior teacher and a curriculum builder. (Parenthetically it may be mentioned that Monge had two brothers who also were Professors of Mathematics, thus putting the name of Monge in a class with that of the Bernoullis and the Cassinis as designating a family of mathematicians). The rise of solid geometry consequently was due in large part to the mathematical and revolutionary activities of Gaspard Monge. Had he not been politically active, the École Polytechnique might never have come into being; and had he not been an inspiring teacher, the revival of geometry of three dimensions might not have taken place.

It should be pointed out that the École Polytechnique was not the only school created at the time. The École Normale had been hastily opened to some fourteen or fifteen hundred students less carefully selected than at the École Polytechnique, and it boasted a mathematical faculty of high calibre, Monge, Lagrange, Legendre, and Laplace being among the instructors. It was the lectures of Monge at the École Normale that finally were published in 1799 as his *Géométrie descriptive;* but administrative difficulties made the school short-lived.

Descriptive geometry was not the only contribution of Monge to three-dimensional mathematics, for at the École Polytechnique he taught also the course in "application of analysis to geometry." Just as the abbreviated title "analytic geometry" had not yet come into general use, so also there was no "differential geometry"; but the course given by Monge was essentially an introduction to the latter field. Here, too, no textbook was available, and so Monge found himself compelled to write up and print his *Feuilles d'analyse* (1795) for the use of students. Here the analytic geometry of three dimensions really came into its own; and it was this course, required of all students at the École Polytechnique, which formed the prototype of the present program in solid analytic geometry. Students, however, evidently found the course difficult, for the lectures skimmed very rapidly over the elementary forms of the line and plane, the bulk of the material being on the applications of the calculus to the study of curves and surfaces in three dimensions. Monge ever was reluctant to write textbooks on the elementary level, or to organize material which was not primarily his own. However, he found collaborators ready to edit material which he included in his course; and so in 1802 there appeared in the *Journal de l'École Polytechnique* an extensive mémoire by Monge and Hachette on *Application d'algèbre à la géométrie.* The first theorem in this is typical of a more elementary approach to the subject. It is the well-known eighteenth-century generalization of the Pythagorean theorem: The sum of the squares of the projections of a plane figure upon three mutually perpendicular planes is equal to the square of the area of the figure. By Monge and Hachette the theorem is proved just as in modern courses; and, in fact, the whole volume could serve without difficulty as a text in the twentieth century. Equations for transformations of axes, the usual treatment of lines and planes, the determination of the principal planes of a quadric are treated fully. The one thing that might be missed is the use of determinants, for this, despite the anticipation

by Lagrange, was the work of the nineteenth century.

Among the results first given by Monge are two theorems which bear his name: (1) The planes drawn through the midpoints of a tetrahedron perpendicular to the opposite edges meet at a point M (which has since been called the Monge point of the tetrahedron). M turns out to be the midpoint of the segment joining the centroid and the circumcenter. (2) The locus of the vertices of the trirectangular angle whose faces are tangent to a given quadric surface is a sphere (known as the Monge sphere or director sphere of the quadric).

As we have already indicated, Monge possessed a quite unusual combination of talents, for he was at once a capable administrator, an imaginative research mathematician, and an inspiring teacher. The one trait of a pedagogue he might have had, but lacked, was that of a textbook compiler. But if Monge here showed a deficiency, it was more than made up for by his young and eager students. One can say with no fear of contradiction, that the pupils of Monge let loose a spate of elementary textbooks on analytic geometry such as has never been equalled—not even in our own day, deluged as we are with new books. If one is to judge from the sudden appearance of so many analytic geometries beginning with 1798, a revolution had taken place in mathematical instruction. Analytic geometry, which for a century and more had been overshadowed by the calculus, suddenly achieved a recognized place in the schools; and this "analytical revolution" can be credited primarily to Monge. Between the years 1798 and 1802 four elementary analytic geometries appeared from the pens of Lacroix, Puissant, Lefrançois, and Biot, all directly inspired by the lectures at the École Polytechnique. And Polytechnicians were responsible for about again as many in the next decade. Most of these were eminently successful, appearing in many editions. The volume by Biot achieved a fifth edition in less than a dozen years; that by Lacroix, student and colleague of Monge, appeared in twenty-five editions within ninety-nine years! Perhaps we should speak instead of the "textbook revolution" for Lacroix's other textbooks were almost as spectacularly successful, his *Arithmetic* and his *Geometry* appearing in 1848 in the 20th and 16th editions, respectively. The 20th edition of his *Algebra* was published in 1859, and the 9th edition of his calculus in 1881. And these figures do not include translations into other languages.

Monge is known to most readers as a founder of modern pure geometry. Through Poncelet and other *anciens élèves* of the École Polytechnique pure geometry did undergo a glorious renaissance, largely through the inspiration of Monge; but it may be well for us to dwell for a moment on an aspect of Monge's work which is less well known. Virtually without exception, the textbook writers ascribe the inspiration for their work to Monge, although Lagrange occasionally is mentioned also. Lacroix most clearly expressed the point of view as follows.

> In carefully avoiding all geometric constructions, I would have the reader realize that there exists a way of looking at geometry which one might call *analytic geometry*, and which consists in deducing the properties of extension from the smallest possible number of principles by purely analytic methods, as Lagrange has done in his mechanics with regard to the properties of equilibrium and movement.

Lacroix held that algebra and geometry "should be treated separately, as far apart as they can be; and that the results in each should serve for mutual clarification, corresponding, so to speak, to the text of a book and its translation." Lacroix pointed to Lagrange's work on the tetrahedron as an instance of this point of view, but he believed that Monge "was the first one to think of presenting in this form the application of algebra to geometry." (Delambre likewise ascribed to Monge the "resurrection of the alliance of algebra and geometry.") His section on solid analytic geometry Lacroix admitted to be almost entirely the work of Monge. Perhaps teachers today can take satisfaction in the

thought that analytic geometry as presented by Fermat and Descartes, a lawyer and a philosopher, respectively, remained ineffectual, and only when it was given a new form by genuine pedagogues—Monge and those of his students who in turn became teachers at the École Polytechnique—did it show vitality.

It is interesting to note that Lacroix declined to use the name "analytic geometry" as a title for this textbook, and edition after edition carried the ponderous title *Traité élémentaire de trigonométrie rectiligne et sphérique et application de l'algèbre à la géométrie*. Although the phrase "analytic geometry" had appeared every now and then during the eighteenth century, it seems first to have been used as the title of a textbook by Lefrançois in an edition of his *Essais de Géométrie* of 1804, and by Biot in an 1805 edition of his *Essais de Géométrie Analytique*, the latter of which, translated into English as well as other languages was used for many years at West Point. We need not look in detail at the contents of the texts of Lacroix, Lefrançois, Biot, and others, for they resemble very closely the books you and I have used.

We have dwelt at such length on the work of Monge that one may get the impression that he was the outstanding figure of the Revolution. This is far from the truth, for the mathematician whose name was on the tongue of every Frenchman during the Revolution was not Monge but Carnot. It was Carnot, who, when the success of the Revolution was threatened by confusion within and invasion from without, organized the armies and led them to victory. As ardent a republican as Monge, he nevertheless shunned all political cliques. He had a high sense of intellectual honesty and tried to be impartial in reaching decisions. After investigation he absolved the loyalists of the charge that they had mixed powdered glass in flour for the army; but he felt bound by conscience to vote for the death of the king. (The American Tom Paine, sometimes regarded in this country as dangerously radical, voted *against* the execution of the king.) But reasoned impartiality is difficult to maintain in times of crisis, and Robespierre, whom Carnot had antagonized, threatened that Carnot would lose his head at the first military disaster. Had Carnot been merely a mathematician and a politician, like Monge and Condorcet, he might well have gone to the guillotine. But Carnot had won the admiration of his countrymen for his remarkable military successes; and when a voice in the Convention proposed his arrest, the deputies spontaneously rose to his defense, acclaiming him the "Organizer of Victory." Hence it was instead the head of Robespierre which fell, and Carnot survived to take an active part in the École Polytechnique. Carnot was greatly interested in education at all levels, even though he seems never to have taught a class.

Carnot led a charmed political life until 1797. He had gone from the National Assembly to the Legislative Assembly, to the National Convention, to the powerful Committee of Public Safety, to the Council of Five Hundred and the Directorate. In 1797, however, he refused to join a partisan coup d'etat and was promptly ordered deported. His name was stricken from the roles of the Institut and his chair of geometry was voted unanimously to General Bonaparte. Even Monge, fellow republican and mathematician, approved the intellectual outrage. About the only thing that can be said in extenuation of his action is that Monge seems to have been mesmerized by Napoleon. Monge followed his idol through thick and thin, his devotion being such that he became literally sick every time Napoleon lost a battle. This is in contrast to Carnot, who initially was responsible for Bonaparte's rise to power through his appointment to the Italian campaign. But Carnot did not hesitate to oppose the Frankenstein he had created, although it nearly cost him his life. Mathematically, his proscription turned out to be a good thing, for it gave him an opportunity, while in exile, to complete a work which had been on his mind for some time. One should expect that a man engaged, as was Carnot, in affairs of enormous practical exigency, would tend to think in terms of immediate practicality. Trajectories would appear to be a more likely subject for study than abstract ideas. But the work which Carnot had been planning during his politically busy days was, *mirabile dictu*, the *Réflexions sur la métaphysique du calcul infinitésimal* which appeared in 1797. This was not a work on applied mathematics; it came closer to philosophy than physics, and in this respect it adumbrated the period of rigor and concern for foundations so typical of the following century. Carnot's *Refléxions* became very popular and ran through a number of editions in several languages, proving that even in times that try men's souls pure mathematics finds many devotees.

Carnot was not the only one of our Revolutionary group who felt the need for greater rigor in mathematics. We have mentioned the lamentable state of geometry as portrayed by Bézout's *Cours de mathématiques*. This prompted Legendre, who was, after all, primarily an analyst, to revive some of the intellectual quality of Euclid. The result was the *Éléments de géométrie* which appeared in 1794, the very year of the Terror. Here, too, one sees the very antithesis of what generally is regarded as practical. As Lengendre says in the preface, his object is to present a geometry which shall satisfy the "spirit." It is indeed heartening to note that at least a number of mathematicians were able to resist the anti-intellectual forces of the day, and it should serve to inspire us to do likewise in times of crisis. The result of Legendre's efforts was a remarkable successful textbook—one of the mathematical products of the Rèvolution which had pervasive influence, for twenty editions appeared within the author's lifetime. Legendre wrote that his object was "to compose a very rigorous element" of geometry; but the author did not wax pedantic to the point of making a fetish of rigor at the expense of clarity.

Often we are inclined to think of American mathematics as influenced primarily by German scholarship, for a generation ago one went to Goettingen to be in touch with the foremost scholars in the field. We are prone to forget that during much of the nineteenth century it was French mathematics which dominated American teaching, and this was primarily through the work of the very men we have been considering. Textbooks by Lacroix, Biot, and Lagrange were published in America for use in the schools; but perhaps the most influential of all was the geometry of Legendre. *Davies' Legendre* became almost a synonym for geometry in America. As late as 1885 Dean Van Amringe of Columbia wrote in the preface of another edition, "It is believed that in clearness and precision of definition, in general simplicity and rigor of demonstration, in orderly and logical development of the subject, and in compactness of form, *Davies' Legendre* is superior to any work of its grade for the general training of the logical powers of pupils, and for their instruction in the great body of elementary geometric truth."

But if Carnot and Legendre were disciples of clear and rigorous thought, Lagrange was the high priest of the cult. At the height of the Terror Lagrange had thought seriously of leaving France; but just at this critical point the École Normale and the École Polytechnique were established, and Lagrange was invited to lecture on analysis. Lagrange seems to have welcomed the opportunity to teach, although it had been many years since he had done lecturing at Turin. In the interval he had been under the patronage of the nobility, but during the Revolution he did not take sides. Perhaps this was the result of political apathy, possibly it was due to Lagrange's mental depression at the time. At all events, his appointment to the newly established schools awoke him from his lethargy. The new curriculum called for new lecture notes, and these Lagrange supplied in the form of a classic in mathematics. His *Théorie des fonctions analytiques* appeared in the same year as Carnot's *Métaphysique,* and together they make 1797 a banner year for the rise of rigor. Lagrange's function theory, which developed some ideas he had presented in a paper about twenty-five years earlier, certainly was not useful in the narrower

sense, for the notation of the differential was far more expeditious and suggestive than the Lagrangian derived function, from which our name derivative comes. The whole motive of the work was not to try to make the calculus more utilitarian, but to make it more logically satisfying. The key idea is easy to describe. The function $f(x) = 1/(1 - x)$ when expanded by long division, yields the infinite series $1 + 1x + 1x^2 + 1x^3 + \ldots\ldots + 1x^n + \ldots$. If the coefficient of x^n is multiplied by $n!$, Lagrange called the result the value of the nth derived function of $f(x)$ for the point $x = 0$—with suitable modification for expansions of functions about other points. Lagrange thought that through this device he had eliminated the need for limits or infinitesimals; but alas, there are flaws in his fine scheme. Not every function can be so expanded; and, moreover, the question of the convergence of the infinite series brings back the need for the limit concept. But the work of Lagrange during the Revolution can be said to have had a broader influence through the initiation of a new subject which has ever since been the center of attention in mathematics—the theory of functions.

We have said little so far about Laplace, who in his day was regarded more highly as a mathematician than was Lagrange. There are two reasons for our relative neglect. In the first place, Laplace took virtually no part in revolutionary activities. He seems to have had a strong sense of intellectual honesty in science, but in politics he was completely without convictions. This does not mean that he was timid, for he seems to have associated freely with those of his scientific colleagues who were suspect during the period of crisis. It is said that he too would have been in danger of the guillotine except for his contributions to science, but this statement seems to be questionable. He played a role in the Committee on Weights and Measures, but this was not of great significance. He naturally was a professor at the École Normale and the École Polytechnique, but unlike Monge and Lagrange, he did not publish lecture notes. His publications were primarily on celestial mechanics, in which he stands preeminent in the period since Newton. Laplace did have one fling at political administration some years later, for Napoleon, a great admirer of men of science, appointed him Minister of the Interior—a post which Carnot likewise had held for a while under Napoleon. But it is well known that Laplace, unlike Carnot, showed no aptitude for the office. A second reason for our failure to emphasize the work of Laplace is that it did not have the immediate and persistent influence that can be traced to others in our group. He represents in a sense the end of an era rather than the beginning of a new period—although one must make some exception in the case of his work in probability.

We shall terminate our account with the date 1799, for at that time Napoleon seized power and one can regard the period of the Revolution as over. This was far from the end of the activity of the five survivors in our group, for every one of them continued to make contributions to mathematics, and some to politics as well. Honors came to all of them, Monge, Carnot, and Lagrange being named counts of the empire and Laplace achieving the title of marquis. Of our group of six, only Legendre seems never to have borne a title. Mathematically our story has a happy ending, for our scholars were able to continue their work until the end. Politically, however, at least two of our heroes were to suffer defeat. Carnot and Monge had strong political convictions, and both of them voted for the death of Louis XVI. Carnot, the more consistent of the two, ever was opposed to dictators, and in 1804 he was the only Tribune with sufficient courage and conviction to vote against naming Napoleon emperor. Yet later when he felt that the welfare of France demanded it, Carnot willingly served under Napoleon, both in the army and in governmental administration. Monge, on the other hand, supported his idol from the revolutionary corporal to the despotic emperor. He and Fourier accompanied Bonaparte on the Italian and Egyptian campaigns, and it was Monge who executed the distasteful task of determining what works of art were to be brought back to Paris as war booty. Following the restoration of the French monarchy, Carnot was forced to seek exile in Magdeburg, and Monge was banished and stripped of all his honors, including his place in the École Polytechnique and the Institut National. The turn in events was accepted courageously by Carnot, who continued his scholarly activities, but it broke the spirit of Monge, who died shortly afterward. Lagrange had died a few years before the Napoleonic crisis. Legendre seems to have remained politically neutral throughout the changes, for he was shy and retiring; but he produced a steady stream of publications on elliptic integrals and the theory of numbers, as well as contributions to other parts of mathematics. Toward the end of his life he, too, suffered politically. Because he resisted the move of the government to dictate to the Académie des Sciences, he was deprived of his pension. Laplace, on the other hand, made peace with each regime as it came along, including in editions of his works glowing tributes to whichever side happened to be in power. Posterity, as a result, has admired Laplace for his mathematics while deploring his political maneuvering.

It is now over a century and a half since the days of which we have been speaking, and we can look back on the period dispassionately. Are there any lessons to be drawn from our survey? One that can be drawn is that the things that really count in mathematics, and that have lasting influence, are not those which immediate practicality dictates. Even in times of crisis it is things of the "spirit" (in the French sense) which count most, and this spirit is perhaps best imparted by great teachers. But perhaps more important than this is the moral that, like Carnot, one should never lose heart, no matter how disillusioning the political or intellectual outlook may be.

87

The *Ladies' Diary* . . . Circa 1700

TERI PERL

RECENTLY I learned of the existence of a journal for women, published in England throughout the eighteenth century, which was largely devoted to problems and puzzles in mathematics. It was called the *Ladies' Diary,* and it was fascinating in that it revealed mathematics to be a proper, pleasant, and possible pastime for gentlewomen (fig. 1).

The *Ladies' Diary* was published annually from 1704 to 1841. Questions presented in one issue were answered in the following year's issue. The *Ladies' Diary* was an almanac(k) as well as a collection of practical exercises involving many branches of contemporary mathematics. During its day, it contained contributions by almost all mathematicians of eminence in England, among them Thomas Simpson (of Simpson's rule, 1710–1761), its editor from 1754 to 1760. In all, this journal may be said to have reflected the progress of mathematical science in England in its time. I. Grattan-Guinness, editor of *Annals of Science,* compares the *Ladies' Diary,* in level and purpose, to *Scientific American* today.*

According to Thomas Leybourn, writing about the *Ladies' Diary,*

the diary was projected and begun in the year 1704 by Mr. John Tipper, who conducted it until 1713. It does not appear that the improvement of Mathematical Science was a particular object with the ingenious projector: indeed the law, which the first

FIGURE I

* Grattan-Guinness, 1974; personal communication.

Special thanks are due to Cecily Young Tanner, daughter of mathematicians William and Grace Chisholm Young, who told me about the existence of the diaries in London during the summer of 1974. Mrs. Tanner is herself a mathematician, now retired professor of mathematics at Imperial College, London.

Reprinted from *Mathematics Teacher* 70 (Apr., 1977): 354–58; with permission of the National Council of Teachers of Mathematics contributors imposed on themselves, of not only proposing but also of answering all questions in rhyme, was not favourable to the development of Mathematical genius. [Leybourn 1817]

A look at specific issues of the diaries shows them to have been small—about ten centimeters by fifteen centimeters in size. In the preface to the 1727 issue, the editor

> has the Happiness in a New Path to meeth with no competitors, for the Pleasure he takes in endeavouring to employ Some Spare Hours of the Fair-Sex in a Study Innocent, Useful, and Diverting: And is always willing to shew that Veneration for them whilst they are pleas'd to contribute their Performances, for Illustrating the DIARY, but publishing to the WORLD the HONOUR and Merit of the British FAIR.

The 1727 issue contained "an article by Mr. Lock on the Mathematicks." It also contained some letters to the "Author," some enigmas and answers (we could call these *word puzzles*), and some "Arithmetical Quests."

The first part of each issue was an almanac. In this section relevant items were collected, month by month, page by page. The layout of each page had several distinct features (fig. 2). First, the day and hour of the appearance of the new moon and full moon were given. This was followed by a listing of information associated with each day of that month; for example, the time of sunrise, important birthdates, the length of day and night in hours and minutes, the end of Oxford and Cambridge terms (March 25,F), moonrises, and holidays. At the bottom of each page was a listing of unusual weather for the same month the previous year. Dates of freezing days, windy days, stormy days, great rains as well as ordinary rains were to be found here.

In the earlier issues, many women's names appear as proposers, as well as answerers, of mathematics problems. One name appears frequently, referred to by Mr. Leybourn as "the ingenious" Mrs. Barbara Sidway. Here Mrs. Sidway solves a

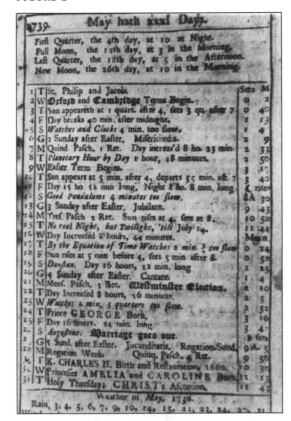

problem referred by a Mr. William Howney, and then goes on to propose one of her own. Mr. Howney's problem goes like this:

> A General who had served the King successfully in his wars, asked as a reward for his services, a farthing for every different file of ten men each, which he could make with a body of 100 men. The King, thinking the request a very moderate one, readily assented. Pray what sum would it amount to?

Mrs. Sidway answered,

$$\frac{91}{1} \times \frac{92}{2} \times \frac{93}{3} \times \frac{94}{4} \times \frac{95}{5} \times \frac{96}{6} \times \frac{97}{7}$$
$$\times \frac{98}{8} \times \frac{99}{9} \times \frac{100}{10}$$

= 17310309456440 farthings
= 180315723501.9s.3d.

The following theorem is proposed to answer all questions of this nature: The number of combinations of m given things, all different from each other, taken by a given number n at a time, is equal to

$$\frac{m}{1} \times \frac{m-1}{2} \times \frac{m-2}{3} \times \frac{m-3}{4} \times \text{etc.}$$

to n terms.

Mrs. Sidway in turn asks readers to solve the question of how "from a given cone to cut the greatest cylinder possible."

Four men presented solutions in the following year's Diary. One of these was a straightforward geometric solution, which I have included here. The other is interesting because it involves *fluxions,* which was the British term for the derivative. Simpson, in his *New Treatise on Fluxions* (1737), defined a fluxion thus: "The magnitude by which any flowing quantity would be uniformly increased in a given portion of time with the generating celerity at any proposed position or instant (was [were]) it from thence to continue invariable) is the fluxion of the said quantity at that position or instant" (Kline 1972, p. 427).

The geometric solution was as follows:

Let ACB be the given cone [fig. 3], DE the diameter, and PQ the altitude of the greatest inscribed cylinder; then the solidity of the cylinder is $DE^2 \times .7854 \times PQ$, therefore $DE^2 \times PQ$ must be a maximum. But DE has to AP a given ratio, namely that which the diameter of the base has to the altitude,

and therefore when $DE^2 \times PQ$ is the greatest, $AP^2 \times PQ$ will also be the greatest: and this is the case when AP is double of PQ; hence the cone must be cut as a third of its altitude from the base.

The solution by fluxions is as follows:

Put $AQ = a$, $BC = b$, and $X = AP$; then by similar triangles $a : b :: X : \frac{b}{a} X = DE$; whence $DE^2 \times PQ = \frac{b^2}{a^2} \times X^2 \times (a - X)$; therefore, leaving out the constant factor $\frac{b^2}{a^2}$, $X^2(a - X)$ must be a maximum. Taking fluxions $2aX\dot{X} - 3X^2\dot{X} = 0$, and reducing, $X = \frac{2}{3} a$, whence $a - X = PQ = \frac{1}{3} a$, as before. [Leybourn 1817]

Before going on to other examples of problems that appeared in the diaries, I have a few comments on some different facets of the magazine. I found the advertisements in the *Ladies' Diary* particularly interesting. The following is a sample of advertisements that appeared in the first half of the eighteenth century.

Astronomy (new publication) by C. Leadbetter. Teacher of the Mathematicks. Printed by J. Wilcox in Little-Britain, and T. Heath near the Fountain-Tavern in the Strand.

Artificial Teeth set in so firm as to eat with them, and so exact as not to be distinguished from Natural; not to be taken out at night . . . greatly helpful to Speech. Also Teeth clean'd or drawn by J. Watts and Sam Rutter, Operators, who apply themselves wholly to the said business, and live in Raquel-Court, Fleet Street, London.

One particular advertiser appears in the 1727 *and* 1728 issues:

Steel Spring and other sorts of Trusses for Ruptures at the Navel, Groin or elsewhere being very Easie and Effectual to Old or Young. Also Bag Trusses for fixed Tumours, and Instruments for the Lame or Crooked, by P. Bartlett at the Golden Ball in St. Paul's Churchyard near Cheapside, London. Persons in the Country, sending their bigness, and which side the Rupture is on, may be supply'd with the Trusses, and proper Directions. His Mother, Mrs. M. Bartlett, at the Golden Ball against St. Bride's Lane in Fleet Street, is Skilful in this Busi-

FIGURE 3

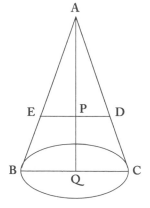

ness to her own Sex, who also makes Sturdy Stockings, Knee and Ankle-pieces very useful for swelt'd limbs.

A look at the diaries for 1776 and 1777 showed advertisements no longer appearing in the magazine. These later issues contain questions that seem different from those in women's magazines of today only in the style of language used. Here are a few sample "queries" I noted.

Ye learned, pray say, why people deceased are always interr'd with their heads to the west?

Whether there is anything that will prevent the cramp when people are swimming; and if there is, what?

On what principle is it that ladies' turn down an empty tea-cup in the bottom of a fruit pye, to prevent the syrrup from boiling out in baking?

A rebus from one of the later issues went like this:

If you a bullock set before
A shallow river's brink,
You may a city's name explore
I' the time you let him drink.

A problem proposed by a Miss Ann Nichols in 1761 is a good example of the rhyme format that was encouraged in the *Ladies' Diary*. Answered by several people in 1762, one writer pointed out an ambiguity in the problem as originally stated. Miss Nichols wrote—

Old John, who had in credit liv'd,
Tho' now reduc'd, a sum receiv'd:
This lucky hit's no sooner found,
Than clam'rous duns came swarming round:
To th' landlord, baker, many more,
John paid in all pounds ninety-four.
Half what remain'd, a friend he lent,
On Joan and self, one-fifth he spent;
And when of all these sums bereft,
One-tenth o' th' sum receiv'd had left.
Now shew your skill, ye learned fair,
And in your next that sum declare.

A Mr. Tho. Sadler replied—

If X be supposed equal the whole sum John received, then will

or

$$\frac{X-94}{2} + \frac{X-94}{5} + \frac{X}{10} + 94$$

$$\frac{8X + 282}{10} = X,$$

per the conditions of the question: whence $X = 141$ £ the sum required. Mr. George Salmon, late of Mr. B. Donn's school, and several others, observe, that this question is ambiguous, it being doubtful whether $^1/5$ of the whole sum received or only $^1/5$ of what remained after the 94 £ was paid, was the part thereof spent on Joan and himself, and accordingly find the whole sum received to be either 141 or 235 pounds.

Here is an interesting puzzle problem from a later issue. Proposed in 1791 and answered in 1792, by which time the problems and solutions were all by men, this problem involves discovering the word described in the riddle by using a mapping from numbers to letters. Two solutions are included.

What's *high* in esteem,
As the ladies now deem,
To declare, from below, condescend:
In Diary next year
Pray let it appear,
'Twill oblige your poetical friend.
$XYZ = 160$; $X^2 + Y^2 + Z^2 = 465$; $X^3 + Y^3 + Z^3 = 8513$.

The answer by a Mr. Joseph Garnett was as follows:

By the question X, Y, and Z are to be whole numbers, and from the third equation no one can be above 20, and some one must be more than 14. Now from the first equation the three roots must evidently be some divisors of 160, which are 1, 2, 4, 5, 8, 10, 16, and 20, among which there are only two above 14, viz. 16 and 20, therefore the numbers are either

$$16 \begin{cases} 10 \text{ and } 1, \\ 5 \text{ and } 2 \end{cases} \text{ or } 20 \begin{cases} 8 \text{ and } 1 \\ 4 \text{ and } 2 \end{cases}$$

of which 20, 8, and 1 are the only ones that will answer the conditions, and the word is HAT.

Another solution to this problem came from a Mr. J. Holt of Manchester:

> Because no word can be formed without a vowel, the value of one of the unknown quantities must answer to a vowel, and be such that the product of the other two in the first equation be a composite number; but 1 and 5 only have these properties. Now if 5 be substituted for X in the second equation, then $Y^2 + Z^2 = 440$: but no two perfect squares whatever will make this number; therefore $X = 1$, consequently $Y^2 + Z^2 = 464$, $YZ = 160$; from this latter equation $Z = 160 \div Y$, which value of Z substituted in the former, it becomes $Y^2 + 25600 \div Y^2 = 464$, or $Y^4 - 464\,Y^2 = -25600$, the roots of which quadratic equation are 8 and 20, which are the values of Y and Z. Hence the required word is HAT.

The problems I've included here are rather straightforward and relatively simple. I had very little time to explore the diaries this first time around, and I tended to skip problems that looked too complex. Of particular interest to me was that whereas the early issues of the *Ladies' Diary* contained many questions and solutions contributed by women, the later ones seem to have fewer and fewer contributions from women. I wonder why this was so. Perhaps this effect was a reflection of the attitude of the editor at a particular time. Perhaps it was a reflection of a larger change in the role of women in society in general. It might be interesting to explore this further. I found the existence of the diaries fascinating in itself and definitely worth a longer look. I hope to get back to them one day, and I hope that others will also consider them worth looking at more carefully.

REFERENCES

Kline, Morris. *Mathematical Thought from Ancient to Modern Times.* New York: Oxford University Press, 1972.

The Ladies' Diary: or, the Woman's Almanack. British Museum Library, Stanford University Library (fragments).

Leybourn, Thomas. *The Mathematical Questions, Proposed in the "Ladies' Diary," and Their Original Answers Together with Some New Solutions, 1704 to 1816.* 4 vols. London: J. Mawman, 1817.

HISTORICAL EXHIBIT 7.5

Women in Mathematics

In both ancient classical and traditional societies, evidence for the participation of women in mathematics is particularly scarce. It is known that the Pythagoreans (6th century B.C.) permitted women to attend some of their ceremonies; but to what extent these women were actually involved with mathematics remains unclear. History recognizes Hypatia (ca. A.D. 370–415) of Alexandria as the first, known, woman mathematician. She was the daughter of the Greek mathematician, Theon, acknowledged for his commentaries on Ptolemy's *Almagst* and Euclid's *Elements*. While Hypatia also distinguished herself as a commentator, writing on Diophantus's *Arithmetica* and Appollonius's *Conic Sections*, she was renowned for her knowledge of mathematics, medicine, and philosophy. Hypatia, as the leader of the Neo-Platonic school of philosophy in Alexandria, was a proponent of "pagan-science" and, as such, was singled out by a fanatical Christian mob and murdered.

With the Christian homogamy of Europe that followed the fall of the Roman Empire, mathematics became the prerogative of churchmen, academicians, and, later, merchants. Thus, by associating mathematics with male-dominated professions, Western traditions excluded women from major mathematical pursuits for many centuries. However, this has not necessarily been the case in non-Western societies. Recent research by Professor Michael Closs* of the University of Ottawa reveals that, during the classical period of their civilization (A.D. 300–900), the Mayan peoples of Mesoamerica employed female mathematician-scribes. An illustration from a painted pottery vessel of this period depicts such a female mathematician at work.

Women's involvement with mathematics is culturally bound and, in the contemporary world, is experiencing a rapid reorientation.

---

* Michael P. Closs, "I Am a Kahal; My parents were Scribes," *Research Reports on Ancient Maya Writing* (May, 1992), pp. 7–22.

SUGGESTED READINGS FOR PART VII

Baron, M. E. *The Origin of the Infinitesimal Calculus.* Oxford: Pergamon Press, 1969.

Boyer, C. B. *This History of Calculus and Its Conceptual Development.* New York: Dover Publications, 1949.

————. *History of Analytic Geometry.* New York: Scripta Mathematics, 1956.

Dyksterhius, E. J. *The Mechanization of the World Picture.* Oxford: Oxford University Press, 1961.

Edwards, C. H. *The Historical Development of the Calculus.* New York: Springer-Verlag, 1979.

Fauvel, John, et al., eds. *Let Newton be!* Oxford: Oxford University Press, 1988.

Meyer, R. W. *Leibniz and the Seventeenth Century Revolution.* Translated by J. P. Stern. Cambridge: Bowes and Bowes, 1952.

Taylor, E. G. R. *The Mathematical Practitioners of Tudor and Stuart England.* Cambridge: The University Press, 1954.

Toeplitz, Otto. *The Calculus: A Genetic Approach.* Chicago: University of Chicago Press, 1963.

Vrooman, Jack. *René Descartes, A Biography.* New York: C. P. Putnam's Sons, 1970.

PART VIII

THE SEARCH FOR CERTAINTY

*F*OLLOWING THE ADVENT OF CALCULUS in the seventeenth century came a period of experimentation and mathematical exploitation. For the next hundred years, mathematicians revelled in the use of the powerful new analytic methods supplied to them by Newton and Leibniz. Under a prevailing spirit of mathematical adventurism, many results were obtained the implications of which were not fully understood at the time. As the mathematical climate of Europe was rapidly changing, the social and intellectual climate in general was also in a state of flux. The French Revolution (1789) and the following Napoleonic period (1804–1815) encouraged a new critical outlook on the sciences, education, and the relation of these institutions to society. Science was becoming democratized. Europe, as a whole, was beginning to experience the impact of industrialization and the rise of technological innovations. Mathematical answers were sought to an increasing number of technological and sociological questions. The realm of mathematics moved from the comfortable isolation of royal courts and elite academies to the lecture halls of universities and research institutes, which were more demanding testing grounds for mathematical ideas. Such ideas now had to be effectively communicated to large receptive audiences and attention was paid to producing adequate textbooks for this task. No more could a single individual possess a universal knowledge of mathematics—the subject had become too broad and complex. The different areas of mathematics came to be classified as either "pure" or "applied" and the constrained specialist appeared on the scene.

The development of mathematics in the nineteenth century underwent alternate phases of consolidation, rigorization, and fragmentation. Outstanding questions from Greek antiquity were resurrected, examined, and, in some cases, put to rest. By the 1880s, the three classical Greek problems—duplication of the cube, angle trisection, and the quadrature of the circle—were proven to be unsolvable under the restriction of ruler and compass construction. Constructibility of regular polygons was also explored using modern techniques and the class of constructible polygons extended. The nagging question of the independence of Euclid's fifth postulate was contested and the postulate was found to be independent within the Euclidean system. Thus, alternate theories of "parallelism" were permitted giving rise to non-Euclidean geometries. Strange new mathematical worlds were appearing.

A reorganization of the calculus by such mathematicians as Augustin Cauchy (1789–1848), Karl Weierstrass (1815–1897), and Bernard Bolzano (1781–1848) amended the haphazard advances of the eighteenth century and set calculus on a sound logical foundation. These efforts to strengthen mathematics often led to more perplexing and intriguing questions and problems, such as: "What is a function?" "How could the concept of continuity be formalized?" "What was the structure of the real numbers?" Number systems were given special attention. Both geometric and analytic interpretations were provided for the complex numbers. Building upon complex number theory, the Irish mathematician William Hamilton (1805–1865) sought to develop meaningful mathematical operations in a four-dimensional space. His efforts resulted in the discovery of quaternions and noncommutative mathematical operations. Algebra was no longer merely an "arithmetic of unknowns" but became a study of the structure of mathematics itself. The German mathematician Georg Cantor (1845–1918) grappled with the concept of infinity (a concept that had defeated Zeno and his fellow sophists) and produced workable theories of classification and manipulation for transfinite numbers. Mathematics was changing so fast that mathematicians were compelled to ask the ultimate question— "What is mathematics?"

Towards the turn of the century, three schools of mathematical thought arose to confront this question. The first of these, formalism, insisted that all mathematics could be treated symbolically and depended on the use of a rigorous axiomatic approach to formulate concepts; new theories then emerged from a sequence of deductive steps from the original axioms. The validity of formalism rested on the consistency of the basic axiomatic system—if the system was consistent and used correctly then correct mathematics would result. Formalism depended on logic but not as the primary means of understanding mathematics. In contrast, a second philosophical school of thought, logicism, contended that mathematics was simply logic. The chief proponents of this theory were the British philosophers Bertrand Russell and Alfred North Whitehead, who published their views in the monumental work *Principia Mathematica* (1910–1913). Finally, the third school of thought was championed by the Dutch mathematician L. E. J. Brouwer, who felt that mathematics was independent of both language and logic. He considered mathematical concepts mental constructs formed by a finite sequence of intuitive steps that could be traced back to the natural numbers. The idea that mathematics is basically intuitive is called intuitionism. Each of these schools eventually encountered difficulties with its doctrines. For example, in 1931 a brilliant young German mathematician, Kurt Gödel, demonstrated that consistency could never be adequately proven within a given formal system, thus casting doubt on the very basis of formalism. To this day, no school of thought has definitively answered the question, "What is mathematics?"

Although a state of uncertainty exists to plague mathematicians, both the nature and extent of that uncertainty is being reduced. In particular, the advent and continuing development of electronic computers and electronic computing techniques are rapidly pushing the frontiers of mathematics into new realms. In 1976, Kenneth Appel and Wolfgang Haken of the University of Illinois used 1000 hours of computer time to devise 1,936 configurations to prove the Four-Color-Conjecture. This work marks the first instance of a computer derived proof in mathematics. The largest known prime number of 65,087 digits has been discovered by computer (1989) as well as a billion decimal place estimate for the value of π. While the usefulness of such knowledge is debatable, its accomplishment testifies to the ongoing challenges of mathematics and how they are being met.

88

Did Gauss Discover That, Too?

RICHARD L. FRANCIS

$\mathcal{A}$ REMARKABLE FEATURE of the history of mathematics is the enormous interval that separates the posing of certain problems and their ultimate resolution. Also remarkable is the long chain of attempted solutions that spanned the many centuries. Examples of this phenomenon are the three famous problems of antiquity, which were resolved only in the modern era.

These challenges in geometric construction were restricted to the Euclidean tools of the unmarked straightedge and compass. They are generally identified as the problems of trisecting an angle, duplicating a cube, and squaring a circle (Fig. 1). Each problem had been the cause of roughly two millennia of mathematical mystery and activity.

The angle-trisection problem requires the construction of an angle whose measure is one-third that of a given angle. Duplication of a cube, also a famous problem, necessitates the construction of the edge of a cube having twice the volume of a given cube. The third famous problem of antiquity, namely, squaring a circle, requires constructing a square having the same area as a designated circle. As all such constructions are to be done by the use of the Euclidean instruments only, it follows that the allowable curves of construction are the line and the circle.

Though posed in the early Greek era of demonstrative geometry (roughly the fifth century

B.C.), virtually all historical treatments of the three famous problems identify the nineteenth century as the time of solution. Thus, the demonstration of the impossibility of squaring a circle is attributed to C. F. Lindemann in 1882, and the impossibility of angle trisection, as well as of cube duplication, is attributed to mathematicians in the 1830s. The 1830s were characterized in large measure by the concentrated pursuit of a theory of

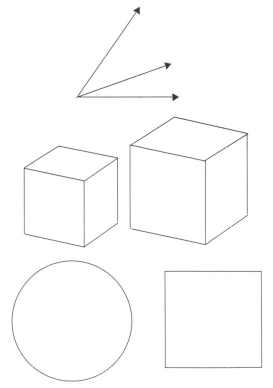

FIGURE I

Famous problems of antiquity

Reprinted from *Mathematics Teacher* 59 (April, 1986): 288–93; with permission of the National Council of Teachers of Mathematics.

equations; further, this period saw the tragic death of the famed algebraist Evariste Galois in 1832.

That the mathematical community generally regards the trisection problem as having been solved in the fourth decade of the nineteenth century is both interesting and perplexing, especially in light of the frequent credit given to the German mathematician Carl Friedrich Gauss for a far-reaching result that makes the problem of angle trisection solvable as a trivial corollary. What was this far-reaching result, attributed to Gauss forty years earlier, and, moreover, was this result his actual work or that of a subsequent mathematician?

Gauss discovered in 1796, at age nineteen, the steps whereby the regular seventeen-sided polygon could be constructed with a straightedge and a compass. This accomplishment was a bright jewel in the long history of geometry in that it overcame limitations on construction that had existed from the time of the ancient Greeks. In the few years after his discovery, notably in his *Disquisitiones Arithmeticae* of 1801, Gauss further identified a class of numbers for which regular polygons with such side numbers were constructible. He implied that a regular polygon of an odd number of sides, say n, can be constructed if n is a prime of the type $(2^p + 1)$ or a product of *distinct* primes all of this form. The only primes of this form known today are 3, 5, 17, 257, and 65 537. Thus note the possibility of constructing a regular polygon of 3×5 sides (known even to the ancients) or that of a regular polygon of, say, 3×17 sides, a revelation stemming from the Gaussian breakthrough. Note also that this construction standard merely expresses a suffcient condition.

Though Gauss is properly acknowledged as having established a constructibility criterion for certain regular polygons, does his proof also include both necessary and sufficient parts? Did he merely establish that all regular odd-sided polygons of the form described are constructible, or did he go a step farther and show (what we actually know today) that these are the *only* regular odd-sided polygons that can be constructed?

Though history overwhelmingly credits Gauss with the sufficiency of the constructiility standard, some serious probing is necessary concerning his disposition of the matter of necessity. Basically, the historical question centers on "Gaussian conjecture" versus "Gaussian proof."

Let us digress and suppose that Gauss's proof actually encompassed both necessary and sufficient conditions of constructibility. Then the impossibility of angle trisection would have followed immediately. Very simply, the regular nonagon would be classified as nonconstructible since the number of sides is the product of two *like* primes of the form $(2^p + 1)$; that is, $9 = 3 \times 3$. In such a case, the central angle of the regular nonagon, namely 40°, would accordingly be nonconstructible, though the 120° angle is easily constructed. Quite obviously, since a 120° angle cannot be trisected, it

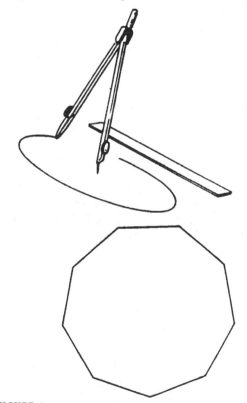

FIGURE 2

Did Gauss really prove that the regular nine-sided polygon is not constructible?

585

thus would follow that no general method of angle trisection existed (Fig. 2).

In our thinking about the problem, a disturbing question arises. If the impossibility of angle trisection is so obviously a consequence of Gauss's work between 1796 and 1801, why was it not boldly publicized and lauded then, as was his breakthrough with the construction of a regular heptadecagon? This problem of trisection had been, throughout the previous centuries, the ultimate mathematical challenge. Can it be maintained that this mathematical gem was somehow overlooked? Or can it be that Gauss merely conjectured or asserted, without proof, the necessity of the constructibility criterion for regular polygons.

Unfortunately, historical accounts give varying answers to the question of the scope of Gauss's work in this regard. Further, more credit is given to Gauss than he is actually due in most historical treatments. *The paradoxical part of this historical dilemma is that the acclamation of Gauss as the demonstrator of the necessary and sufficient polygonal criterion is not consistently joined with an implied solution by Gauss of the impossibility of angle trisection.* Almost always in the history of mathematics, credit for establishing the impossibility of trisecting the general angle is pinpointed much later than a supposedly Gaussian construction would suggest.

To give some indication of the widespread recognition of Gauss as the complete solver of the regular-polygon-construction problem, I shall present several quotations concerning history, number theory, geometry, and general education:

1. From Bell (1937):

The young man proved that a straightedge and compass construction of a regular polygon having an odd number of sides is possible when, and only when, that number is either a prime Fermat number or is made up by multiplying together different Fermat primes. . . . It was this discovery, announced on June 1, 1796. . . . His name was Gauss.

2. Newman (1956) stated essentially the same thing as Bell.

3. From Burton (1976):

In the seventh and last section of *Disquisitiones Arithmeticae*, Gauss proved that a regular polygon of n sides is so constructible if and only [if] either

$$n = 2^k \text{ or } n = 2^k p_1 p_2 \cdots p_r,$$

where $k \geq 0$ and $p_1, p_2 \ldots, p_r$ are distinct Fermat primes.

4. From Boyer (1968):

In the *Disquisitiones*, Gauss answered the question in the affirmative, showing that a regular polygon of n sides can be constructed with Euclidean tools if and only if the number $n \ldots$

5. From *Encyclopaedia Britannica* (1978):

Gauss developed a criterion based on number theory by which it can be decided whether a regular polygon with any given number of corners can be geometrically constructed.

6. From Coxeter (1961):

The question was completely answered by Gauss (1777–1855) at the age of nineteen. Gauss found that a regular n-gon can be so constructed if and only if the odd prime factors of n are distinct Fermat primes. . . . In view of Gauss's discovery, we may say that it has been known since 1796 that the classical trisection problem can never be solved.

These quotations seem fairly typical of the general tendency to ascribe to Gauss a complete disposition of the polygonal construction problem.

But did Gauss actually prove the entirety of such a theorem, namely, that a regular polygon of n sides is constructible *if and only if* $n = 2^k$, $k > 1$, or

$$n = 2^k p_1 p_2 p_3 \cdots p_r,$$

where $k > 0$ and the p_is are distinct primes of the form $(2^m = 1)$? Though the criterion as stated is true, it is virtually a historical certainty that Gauss did not establish the necessity of the standard.

Again, consider the literature that focuses special attention on the actual works of Gauss in this area:

1. From Klein et al. (1962):

Now the implication . . . is not correct, as Pierpont interestingly set forth in his paper "On An Undemonstrated Theorem of the *Disquisitiones Arithmeticae.*" That is, Gauss did not give a proof of the "impossibility" referred to.

2. From Archibald (1949):

He [Gauss] was later able to show that when the Fermat number . . . is prime, it represents the number of sides of a regular polygon constructible with ruler and compasses. He did not prove, however, what is a fact, namely that these are the only polygons, the number of whose sides is prime, which are so constructible.

3. From Buhler (1981):

He [Gauss] adds that it would be futile to try to divide the circle in the cases $n = 7, 11, 13, 19, . . .$ but that the confines of *Disquisitiones Arithmeticae* would not permit him to prove this fact. None of Gauss's papers contains any indication of a proof of this last statement.

4. From Smith (1929), which quotes Gauss in the *Disquisitiones Arithmeticae* showing the extent of its main elaboration on the problem:

Or briefly, it is necessary that n should involve neither any odd prime factor which is not of the form $2^m + 1$, nor even any prime factor of the form $2^m + 1$ more than once.

Note: Gauss then lists the "38 such values of n" less than 300; none of this discussion is accompanied by a proof.

5. From Kline (1972):

The fact that the Gauss condition is necessary was first proved by Pierre L. Wantzel (1814–1848), Jour. de Math. [*Journal de Mathématique*], 2, 1837, 366–372.

6. From Kazarinoff (1968):

Carl Friedrich Gauss published in 1801 in his *Disquisitiones Arithmeticae* exquisite arithmetic yield-ing the constructibility of regular p-gons for p a prime of the form $1 + 2^k$. Closing his discussion, he asserted he had proved that no other regular p-gons were constructible. Gauss never published a proof of this assertion, nor did he ever outline one in his correspondence or notes.

7. See also Pierpont (1895).

Some serious questioning in the mathematical community, even in Gauss's lifetime, concerned the extent of his actual demonstration of the polygonal criterion. *Did he really have a proof or didn't he?* Niels Henrik Abel (1802–29) sought (in 1826), from his teacher, Berndt Michael Holmboe (1795–1850), a resolution of the "mystery" surrounding Gauss's theory on the division of the circle into equal parts (cyclotomy). Perhaps the mystery was essentially one of vague rumors and the casual distortion or metamorphosis of mathematical conjecture into a mathematical proof.

In this day of swift communication, a worldwide mathematical community, and an abundance of research journals, such confusion about a current problem's status is hard to understand. However, rumor, reinforced by many years of acceptance and enthusiastic exaggeration, dies hard. So firmly imbedded in modern mathematical thought is the recognition of Gauss as the complete solver of the polygonal construction problem that the original state of achievement becomes obscured.

It is hoped that an attempt at perspective can put the matter in a better light. It appears, on careful reflection, that the recognition of Gauss as the demonstrator of the entire polygonal-construction criterion is actually less appropriate than the highly questionable acclaim accorded Fermat for his "last theorem." Even Gauss considered Fermat's "theorem" to be unproved in spite of Fermat's claim. Because of his obsession with precision and polish, Gauss was very quick to distinguish between conjecture and proof. Perhaps in the same spirit of Gaussian inquiry the question of *who proved what* should be raised by the historically minded.

The primary historical concern is the discrepancy between the general acknowledgement of Gauss as the complete solver of the regular-polygon-construction problem (ca. 1801) and the nearly unanimous judgment of historians that Pierre Laurent Wantzel (1814–1848) solved the angle-trisection problem in 1837. Consider several of the typical references to Wantzel's achievement and the notable year 1837, which symbolized the first breakthrough with the long-standing problems of antiquity:

1. From Kline (1972):

And Wantzel in the paper of 1837 showed that the general angle could not be trisected nor could a given cube be doubled.

2. From Eves (1983):

A Chronological Table . . . 1837—trisection of an angle and duplication of a cube proved impossible.

3. From Archibald (1949):

The Nineteenth Century and later . . . Trisection of an angle; duplication of cube proved impossible (1837).

4. From Kazarinoff (1968):

In my opinion, Professors Gauss, Klein, Pierpont, and Archibald should each have credited Pierre L. Wantzel (1814–1848), who proved the impossibility of constructing non-Gaussian regular *n*-gons with ruler and compass alone in 1837 in the very article where he solved the problems of angle trisection and cube duplication.

5. From Adler (1957):

The degree of the equation that belongs to the problem of trisecting an angle happens to be 3. Since 3 is not a power of 2, the construction is impossible. The proof of this result was first published by Wantzel in 1837.

6. See Wantzel (1837) and Cajori (1918).

7. From Bell (1945):

These two [trisection and duplication] were not settled till P. L. Wantzel (A.D. 1814–1848, French) in 1837 obtained necessary and sufficient conditions for the solution of an algebraic equation with rational coefficients to be geometrically constructible in the manner specified.... Thus the problems were proved to be impossible.

8. From Coxeter (1969):

[It] can be so constructed if the odd prime factors. . . . In view of Wantzel's theorem, we may say that it has been known since 1837.

Many landmark events occurred in 1837. It was the year Queen Victoria ascended to the British throne. Andrew Jackson was ending an eventful eight years of public service as the seventh president of the United States. In mathematics, two of the famous problems of antiquity (angle trisection and cube duplication) were resolved. Such mathematical achievements, though the work of the twenty-three-year-old Pierre Wantzel, must obviously rest indirectly on the monumental efforts of Wantzel's predecessors, not excluding Gauss. Other giants on whose shoulders Wantzel likely stood include Paolo Ruffini, Niels Henrik Abel, and Evariste Galois.

That the famous trisection construction was shown to be impossible must have been disturbing and disappointing to those not so well versed in mathematics. It was an outcome out of line with centuries of thought in which the construction, though difficult, was merely considered elusive and awaiting a discovery by someone somewhere. Such pronouncements of impossibility, however, were increasingly being made. Note, for example, Abel's demonstration, at about the same time that the general equation of degree 5 or more *could not* be solved radically in terms of its coefficients, that is, in a manner resembling the quadratic formula, wherein a substitution into a radical expression involving coefficients yields all solutions.

In this last area of the non-Euclidean geometries, Gauss is seemingly accorded less recognition than he is actually due, a situation somewhat reversed from his overrecognition in connection with the polygonal-construction standard. Again, if Gauss is credited with establishing both the necessity and sufficiency of the constructibility criterion for regular polygons, then he must be credited also with the immediate, nearly trivial corollary of the impossibility of angle trisection. A disturbing inconsistency in the general historical record seems to give him credit for the first achievement but not the latter. The probable truth suggests he should be given credit for neither. Did Gauss thus discover these, too? *Not likely!*

BIBLIOGRAPHY

Adler, Irving. *The Impossible in Mathematics.* Washington, D.C.: National Council of Teachers of Mathematics, 1957.

Archibald, Raymond C. "Outline of the History of Mathematics." *American Mathematical Monthly* 56 (1949).

Bell, E. T. *The Development of Mathematics.* New York: McGraw-Hill Book Co., 1945.

———. *Men of Mathematics.* New York: Simon & Schuster, 1937.

Boyer, Carl B. *A History of Mathematics.* New York: John Wiley & Sons, 1968.

Buhler, W. K. *Gauss, A Biographical Study.* New York: Springer-Verlag, 1981.

Burton, D. M. *Elementary Number Theory.* Boston: Allyn & Bacon, 1976.

Cajori, Florian. "Pierre Laurent Wantzel." *Bulletin of the American Mathematical Society* 24 (1918): 339–47.

Coxeter, H. S. M. *Introduction to Geometry.* New York: John Wiley & Sons, 1961, 1969.

Encyclopaedia Britannica. "Carl Friedrich Gauss." Vol. 7, 966–68, 1978.

Eves, Howard W. *An Introduction to the History of Mathematics.* New York: CBS College Publishing, 1983.

Hallerberg, Arthur E. "Squaring the Circle—for Fun and Profit." *Mathematics Teacher* 71 (April 1978): 247–55.

Infeld, Leopold. *Whom the Gods Love: The Story of Evariste Galois.* Reston, Va.: National Council of Teachers of Mathematics, 1978.

Kazarinoff, N. D. "On Who First Proved the Impossibility of Constructing Regular Polygons with Ruler and Compass Alone." *American Mathematical Monthly* 75 (1968).

Klein, F., W. F. Sheppard, P. A. Macmahon, and L. J. Mordell. *Famous Problems and Other Monographs.* Bronx, N.Y.: Chelsea Publishing Co., 1962.

Kline, Morris. *Mathematical Thought from Ancient to Modern Times.* New York: Oxford University Press, 1972.

Newman, J. R. *The World of Mathematics.* Vol. 1. New York: Simon & Schuster, 1956.

Pierpont, James. "On an Undemonstrated Theorem of the *Disquisitiones Arithmeticae.*" *Bulletin of the American Mathematical Society* 2 (1895):77–83.

Smith, D. E. *A Source Book in Mathematics.* New York: McGraw-Hill Book Co., 1929.

Wantzel, Pierre L. "Recherches sur les moyens de reconnaître ai un problème de géométrie peut se résoudre avec la règle et le compas." *Journal de Mathématique* 2 (1837): 366–72.

89

Niels Henrik Abel

ROBERT W. PRIELIPP

Niels henrik abel (1802–1829) is without a doubt the greatest mathematician Norway has ever produced. Though he died before reaching the age of twenty-seven, struck down by tuberculosis, he left a bountiful legacy for mathematics. At the time of his death the world was really just beginning to awaken to his genius. A young man who was well acquainted with poverty, Niels Henrik's family life was not always of the best. Neither his father, who died in 1820, nor his mother was a saint (for a very readable account of Abel's life see [2]), and the family was usually quite heavily in debt. Abel's studies were financed by his professors and by the Norwegian government. He was granted a fellowship to travel and study in Europe, and the year and a half which he spent in Germany, Italy, and France was one of the happiest periods in his life. When Abel returned to Norway, he could only obtain a temporary position. The last years of his life were beset by many difficulties.

Almost every student of beginning algebra is familiar with the quadratic formula. By 1600 Cardan and Ferrari had shown how to solve any cubic equation or quartic equation. The seventeenth and eighteenth centuries witnessed innumerable futile attempts to solve the general equation of fifth degree by radicals. While still in school Abel believed he had found such a solution, but he discovered an error before publication. (At the age

Niels Henrik Abel

of sixteen Galois was to repeat the same mistake, believing for a very short time that he had done what cannot be done. This is just one of the many striking parallels in the careers of Abel and Galois.) Shortly thereafter Abel proved that the equation

$$y^5 - ay^4 + by^3 - cy^2 + dy - e = 0$$

was not solvable by radicals; that is, that y cannot be expressed in terms of a, b, c, d, and e by the use of addition, subtraction, multiplication, and division and extraction of roots only, a finite number

Reprinted from *Mathematics Teacher* 62 (Oct., 1969): 482–84; with permission of the National Council of Teachers of Mathematics.

of times. This paper was published at Oslo in 1824 at Abel's expense. In order to save on printing costs, he had to give the paper in a very sketchy form, which in some places affects the lucidity of his reasoning (for the complete text see [3]). It was this proof which opened the road to the modern theory of equations, including group theory and the solution of equations by means of transcendental functions.

Abel proposed for himself the problem of finding all equations solvable by radicals, and succeeded in solving all equations which had commutative groups (see [1] for an explanation of this terminology). These equations are now called Abelian equations.

Clearly Niels Henrik Abel made a vital contribution to the theory of groups. Listed below are some of his other great achievements.

1. Abel wrote a study of the binomial series, considered to be one of the classics of function theory. It contains the principles of convergent series with special applications to power series.

2. While traveling in Europe, Abel stopped in Berlin. He had the good fortune of making the acquaintance of A. L. Crelle who became his lifelong friend. Inspired by Abel, Crelle founded the important *Journal für die reine und angewandte Mathematik* (*Journal for Pure and Applied Mathematics*), many of the early volumes of which are filled by papers by Abel. These papers concern such topics as equation theory, functional equations, integration in finite form, and problems from theoretical mechanics.

3. During the spring of 1826 Abel completed his *Mémoire sur une propriété générale d'une classe très-étendue de fonctions transcendentes* (*Memoir on a General Property of a Very Extensive Class of Transcendental Functions*), which he himself considered his masterpiece. In this he gives a theory of integrals of algebraic functions—in particular, the result known as Abel's theorem, that there is a finite number, the genus, of independent integrals of this nature. This forms the basis for the later theory of Abelian integrals and Abelian functions. The memoir was submitted to the Paris Academy of Sciences. Legendre and Cauchy were appointed as referees. Somehow Cauchy took the memoir home and mislaid it, and both he and Legendre forgot all about it.

4. The theory of elliptic functions (whose discovery had eluded the great Legendre) was developed with great rapidity in competition with K. G. J. Jacobi.

Two of Abel's greatest recognitions came posthumously. Just two days after Abel's death Crelle wrote to say that Abel would be appointed to the professorship of mathematics at the University of Berlin. Together with Jacobi, Abel was awarded the Grand Prix of the French Academy for 1830.

Perhaps Oystein Ore provided us with the best way to conclude this article when he wrote:

Mathematicians, in their characteristic manner, have erected many monuments to Abel, more durable than bronze. The custom prevails of marking new results and great ideas by the names of their originators. Today, anyone who reads advanced mathematical texts will find Abel's name perpetuated in numerous branches of his science: Abelian theorems in abundance, Abelian integrals, Abelian equations, Abelian groups, and Abelian formulas. In short, there are few mathematicians whose names are associated with so many concepts of modern mathematics. Niels Henrik would have been greatly surprised at his own importance.

REFERENCES

1. Lieber, Lillian R., and Lieber, Hugh Gray. *Galois and the Theory of Groups: A Bright Star in Mathesis*. Brooklyn: The Galois Institute of Mathematics and Art, 1956.

2. Ore, Oystein. *Niels Henrik Abel, Mathematician Extraordinary*. Minneapolis: University of Minnesota Press, 1957.

3. Smith, David Eugene. *A Source Book in Mathematics*. Vol. I. New York: Dover Publications, 1959, pp. 261–66.

90

The Parallel Postulate

RAYMOND H. ROLWING
AND MAITA LEVINE

$\mathcal{E}$UCLID'S FAMOUS PARALLEL postulate was responsible for an enormous amount of mathematical activity over a period of more than twenty centuries. The failure of mathematicians to prove Euclid's statement from his other postulates contributed to Euclid's fame and eventually led to the invention of non-Euclidean geometries.

Before Euclid's time, various definitions of parallel lines had been considered by the Greeks and then discarded. Among them were "parallel lines are lines everywhere equidistant from one another" and "parallel lines are lines having the same direction from a given line." But these early definitions were sometimes vague or contradictory. Euclid tried to overcome these difficulties by his definition, "Parallel lines are straight lines which, being in the same plane and being produced indefinitely in both directions, do not meet one another in either direction," and by his fifth postulate, "Let it be postulated that, if a straight line falling on two straight lines makes the interior angles on the same side less than two right angles, the two straight lines, if produced indefinitely, meet on that side on which the angles are less than two right angles."

The statements of Euclid's first four assumptions are: "Let the following be postulated: (1) To draw a straight line from any point to any point. (2) To produce a finite straight line continuously in a straight line. (3) To describe a circle with any center and distance. (4) That all right angles are

equal to one another." All of his assumptions fall into one of two categories. The first is the set of "self-evident" facts concerning plane figures. An example of such an assumption is that "a straight line is the shortest distance between two points." The second category deals with concepts beyond the realm of actual experience. For example, Euclid stated that "a straight line must continue undeviatingly in either direction without end and without finite length." Since it is impossible to experience things indefinitely far off, anything that is said about events there is speculation, not self-evident truth. The fifth postulate falls into this latter category.

The complicated nature of the fifth postulate led numerous mathematicians to believe that it could be proved using the remaining postulates, and, therefore, ought to be a theorem rather than a postulate. Even Euclid might have supported this viewpoint since he did succeed in proving the converse of the postulate. One of the first geometers who attempted to prove a statement equivalent to Euclid's parallel postulate was Posidonius, in the first century B.C. He had defined parallel lines as lines that are coplanar and equidistant. A second early important attempt to prove the parallel postulate was made by Claudius Ptolemy of Alexandria, in the second century.

In the fifth century, Proclus, who had studied mathematics in Alexandria and taught in Athens, worked extensively on the problem of proving Euclid's fifth postulate. He succeeded in showing that the postulate could be proved if the following statement could be established: If L_1 and L_2 are

Reprinted from *Mathematics Teacher* 62 (Dec., 1969): 665-69; with permission of the National Council of Teachers of Mathematics.

any two parallel lines and L_3 any line distinct from and intersecting L_1, then L_3 intersects L_2.

In his argument Proclus used the phrase "distance between parallels." Thus, he assumed that parallel lines are everywhere equidistant, and this assumption is logically equivalent to the fifth postulate. In effect, Proclus assumed what he was trying to prove.

The most elaborate attempt to prove the parallel postulate, and the one most significant for the further development of geometry, was made by the Italian priest, Girolamo Saccheri, who taught mathematics at the University of Pavia. His significant work, published in 1733, was entitled *Euclides ab omni naevo vindicatus sive conatus geometricus quo stabiliuntur prima ipsa geometriae principia*. In this treatise, Saccheri tried to free Euclid from all error, including the supposed error of assuming Postulate V.

One of Saccheri's contributions was the introduction of the Saccheri quadrilateral. The construction of this quadrilateral is as follows. At the endpoints of a segment $\overline{AB}$ construct congruent segments $\overline{AC}$ and $\overline{BD}$, each perpendicular to $\overline{AB}$ and draw $\overline{CD}$. Saccheri tried to prove, on the basis of the first four postulates, that $m \triangle ACD = m \triangle BCD$. He reasoned that if P is the midpoint of $\overline{AB}$ and if Q is the midpoint of $\overline{CD}$ then rt. $\triangle CAP \cong$ rt. $\triangle DBP$, whence $m \triangle ACP = m \triangle BDP$ and $CP = DP$. Then $\triangle CPQ \cong \triangle DPQ$, so $m \triangle PCQ = m \triangle PDQ$, and, therefore, $m \triangle ACD = m \triangle BCD$. This proof does not depend upon the parallel postulate. Saccheri called $\triangle ACD$ and $\triangle BDC$ the *summit angles* of the quadrilateral and he formu-

lated the following three possibilities, which are exhaustive and pairwise mutually exclusive: (1) The summit angles are right angles. (2) The summit angles are obtuse angles. (3) The summit angles are acute angles. These possibilities are generally called the *right angle hypothesis,* the *obtuse angle hypothesis,* and the *acute angle hypothesis,* respectively. Saccheri succeeded in proving that if any of these hypotheses is valid for one Saccheri quadrilateral, it is valid for every quadrilateral of the same type. He proved further that the fifth postulate is a consequence of the right angle hypothesis. And, by asssuming that a straight line is infinitely long, he showed that the obtuse angle hypothesis is self-contradictory.

Disposing of the acute angle hypothesis presented some difficulty, so Saccheri argued intuitively that the "hypothesis of the acute angle is absolutely false, because it is repugnant to the nature of a straight line." Actually, no logical contradiction can be deduced from the acute angle hypothesis, for it gives rise to a new geometry.

So, while Saccheri was looking for a proof of the parallel postulate, he discovered a new world, the world of "absolute geometry," whose theory is independent of the question of the parallel postulate. Included in "absolute geometry" are the theorems concerning congruent triangles, the inequalities involving the measures of the sides and angles of a triangle, and a set of theorems about the Saccheri quadrilateral. The quadrilateral $ABCD$ is a Saccheri quadrilateral if $\triangle A$ and $\triangle B$ are right angles and $AC = BD$. Among the theorems that can be proved are: (1) The diagonals of a Saccheri quadrilateral are always congruent. (2) In any Saccheri quadrilateral, the upper base angles are congruent, i.e., $\triangle C \cong \triangle D$. (3) In any Saccheri quadrilateral, the upper base is congruent to or longer than the lower base, i.e., $CD \geq AB$.

After Saccheri's time, many mathematicians pursued the problem of trying to prove the parallel postulate from the first four postulates. In 1766, J. H. Lambert, a Swiss geometer, showed that Saccheri's obtuse angle hypothesis is consistent

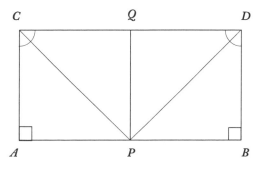

with spherical geometry. In many cases those who attacked the problem worked with statements that are logically equivalent to the fifth postulate rather than with the statement of Euclid. Legendre (1752–1833) tried to prove the following alternative to Euclid's postulate: There exists a triangle in which the sum of the measures of the three angles is equal to the sum of the measures of two right angles. He presented a proof of the fact that the sum of the measures of the angles of a triangle cannot be greater that 180°, but he failed to supply a proof of the fact that the sum cannot be less than 180°. At any rate, his alleged proof of the latter rests on assumptions which are equivalent to the theorem he was trying to establish.

In 1809, Bernhard Friedrich Thibaut tried to demonstrate the existence of a triangle with the property that the sum of the measures of the angles is equal to 180°. His argument was based on the assumption that every rigid motion can be resolved into a rotation and an independent translation. This assumption, however, is equivalent to Postulate V.

In 1813, John Playfair copied Thibaut's argument and tried to correct Thibaut's errors. His attempt was unsuccessful, but it is his alternate statement of the fifth postulate that is now best known and most frequently quoted: "Through a given point, not on a given line, only one parallel can be drawn to the given line." Playfair's statement is surely simpler and more direct than Euclid's.

Karl Friedrich Gauss of Göttingen, Germany, studied the theory of parallels for thirty years. His work resulted in the formulation of a non-Euclidean geometry. In a letter to his friend Franz Adolf Taurinus, dated November 8, 1824, he stated: "The assumption that the angle sum [of a triangle] is less that 180° leads to a curious geometry, quite different from ours but thoroughly consistent, which I have developed to my entire satisfaction. The theorems of this geometry appear to be paradoxical, and, to the uninitiated, absurd, but calm, steady reflection reveals that they contain nothing at all impossible." However, Gauss wrote only a short account of his geometry, which was pub-

lished in 1831. He apparently shrank from the controversy in which a treatise on the new geometry would have involved him. Along with other eminent mathematicians he was influenced by the authority of the German philosopher Immanuel Kant, who had died in 1804. Kant's doctrine emphasized that Euclid's geometry was "inherent in nature." Although Plato had said merely that God geometrizes, Kant asserted, in effect, that God geometrizes according to Euclid's *Elements*.

Gauss was the forerunner of the group of geometers who, instead of trying to prove Euclid's parallel postulate, replaced it by a contradiction of it and thereby invented a new geometry. In 1823, Janos Bolyai replaced Postulate V with the statement: "In a plane two lines can be drawn through a point parallel to a given line and through this point an infinite number of lines may be drawn lying in the angle between the first two and having the property that they will not intersect the given line." Bolyai's first disappointment came when he learned that Gauss had worked on the same problem for thirty years and had achieved the same results. He was disappointed a second time when he discovered that Nikolai Ivanovich Lobachevsky, a professor at the University of Kasan in Russia, had also invented the new geometry and published an account of it in 1829. Bolyai's work was published in 1833 as a twenty-six-page appendix to a semiphilosophical two-volume treatise on elementary mathematics written by his father.

Lobachevsky is now accorded most of the credit for the invention of the new geometry, and his name is the one usually attached to it. His replacement of Euclid's fifth postulate was stated as follows: "Through a point P not on a line there is more than one line which is parallel to the given line." The rest of Euclid's postulates were preserved. A frequently used model of Lobachevskian geometry is the Poincaré model. A detailed description of this model can be found in *Elementary Geometry from an Advanced Standpoint*, by Moise. Of course, any theorems of Euclid's geometry which do not depend on the parallel postulate are valid in

Lobachevsky's geometry. On the other hand, the following theorems are examples of statements which are quite different from the corresponding theorems of Euclidean geometry.

1. *No quadrilateral is a rectangle; if a quadrilateral has three right angles, the fourth angle is acute.*

2. *The sum of the measures of the angles of a triangle is always less than 180°.*

3. *If two triangles are similar, they are congruent.*

Thus, through the discoveries of Gauss, Bolyai, and Lobachevsky, mathematicians recognized the existence of more than one consistent geometry. Leonard M. Blumenthal summarized the significance of the work of these three geometers:

> When the culture was ripe for it, three men, Gauss, Bolyai, and Lobachevsky (a German, a Hungarian, and a Russian) arose in widely separated parts of the learned world, and working independently of one another, created a new geometry. It would be difficult to overestimate the importance of their work. A significant milestone in the intellectual progress of mankind had been passed.

The first period in the history of non-Euclidean geometry, in the opinion of Felix Klein, ended with Lobachevsky's research. This period was characterized by the use of synthetic methods. Riemann, Helmholtz, Lie, and Beltrami were the representatives of the second period in the history of non-Euclidean geometry. Their work involved using the tools of differential geometry.

G. F. B. Riemann is credited with the development of another non-Euclidean geometry, in 1854, which can be realized on a sphere. He began by studying Euclid's postulate that a straight line has infinite length. Discarding this assumption, he invented a geometry in which all lines have finite length. Euclid's first postulate was replaced by the statement: "A straight line is restricted in length and without endpoints." And the parallel postulate was replaced by the statement: "through a point in a plane there can be drawn in the plane no line which does not intersect a given line not passing

through the given point." The remainder of Euclid's postulates were retained. In Riemannian geometry, the following theorems can be proved:

1. *Two perpendiculars to the same line intersect.*

2. *Two lines enclose an area.*

3. *The sum of the measures of the angles of a triangle is greater that 180°.*

4. *If two sides of a quadrilateral are congruent and perpendicular to a third side, the figure is not a rectangle, since two of the angles are obtuse.*

An important contribution to the study of non-Euclidean geometry was made by the Italian mathematician, Eugenio Beltrami. His paper, published in 1868, gave the final answer to the question of the consistency of the new geometries. Bolyai and Lobachevsky had suspected that extending their investigations to three-dimensional space might reveal inconsistencies. Beltrami's paper gave an interpretation of plane non-Euclidean geometry as the geometry of geodesics on a certain class of surfaces in Euclidean space. Therefore, the new geometries must be as consistent as Euclidean geometry.

Gauss was the first to use the term "non-Euclidean" geometry, and Felix Klein first gave to the new geometries the names currently used to describe them. He called Lobachevsky's geometry *hyperbolic*, Riemann's geometry *elliptic*, and Euclid's geometry *parabolic*. This terminology arose from the projective approach to non-Euclidean geometry developed by Klein and Arthur Cayley.

Non-Euclidean geometry not only widened the scope of geometric knowledge; it also stimulated discussions concerning what is a geometry and what is "mathematical truth." As recently as the 1820s, geometry had been thought of as an idealized description of the spatial relations of the world in which we live. Euclid held this viewpoint on the meaning of geometry, and chose for his postulates statements that had their roots in everyday experience. A geometric statement was then regarded as true if it correctly described nature, and false if it did not. But Beltrami's proof that Euclid's geometry was not the only consistent geometry forced mathematicians to abandon the idea that geometric truth involved a description of nature.

David Hilbert called the invention of non-Euclidean geometry "the most suggestive and notable achievement of the last century." And Heath claimed that one of the cornerstones on which Euclid's greatness as a mathematician rests was his parallel postulate, for "when we consider the countless successive attempts made through more than twenty centuries to prove the postulate, many of them by geometers of ability, we cannot but admire the genius of the man who concluded that such a hyopothesis, which he found necessary to the validity of his whole system of geometry, was really undemonstrable."

BIBLIOGRAPHY

Archibald, Raymond Clare. "Outline of the History of Mathematics, Part II," *American Mathematical Monthly,* LVI (January 1949), 21–26.

Bell, E. T. *The Development of Mathematics.* 1st ed. New York: McGraw-Hill Book Co., 1940. Pp. 304–7.

Blumenthal, Leonard M. *A Modern View of Geometry.* San Francisco: W. H. Freeman & Co., 1961. Pp. 4–17.

Cajori, Florian. *A History of Mathematics,* 2d ed. New York: Macmillan Co., 1919. Pp. 48, 306.

Eves, Howard, and Newsom, Carroll V. *An Introduction to the Foundations and Fundamental Concepts of Mathematics.* Rev. ed. New York: Holt, Rinehart & Winston, 1965. Pp. 58–79.

Merriman, Gaylord M. *To Discover Mathematics.* New York: John Wiley & Sons, 1942. Pp. 144–51.

Moise, Edwin. *Elementary Geometry from an Advanced Standpoint.* Reading, Mass.: Addison-Wesley Publishing Co., 1963. Pp. 115–31.

Sanford, Vera. *A Short History of Mathematics.* Boston: Houghton Mifflin Co., 1930. Pp. 276–81.

91

Saccheri, Forerunner of Non-Euclidean Geometry

SISTER MARY OF MERCY FITZPATRICK

*F*EW PEOPLE CONTEST the statement that Lenin, the father of the Russian Revolution, effected the most drastic social upheaval of our modern times. Nor do they argue whether or not the monuments erected or the volumes published to commemorate this feat are in proportion to the enormity of the achievement. These facts are accepted without question or comment. Might it not occur to us to search for a foundation for the old saying that "history repeats itself," in this particular case? And if a search were made, would we not all be astonished to find that no parallel seems to be evident? For just a century earlier, Lenin's countryman, Nikolai Ivanovich Lobachevski (1793–1856), who had brought about the most famous mathematical revolution of all times, received no honors, but was, instead, removed from his post as rector and professor of mathematics at the University of Kazan. No explanation whatsoever was given to this dedicated mathematician who had served the university well, and who according to Edna E. Kramer in her book, *The Main Stream of Mathematics*, was becoming renowned throughout Europe.

Today, Lobachevski's achievement, the discovery of a new and logical non-Euclidean geometry, is universally hailed, not only by the mathematicians, but also by laymen as well. But what, perhaps, is not so well known is the fact that, just as in the case of calculus the propitious mo-

ment for discovery was the age of Newton and Leibniz, so in the matter of Lobachevskian geometry the way was clearly well paved by scores of mathematicians who had attempted to prove Euclid's Fifth Postulate. One of the most notable of these mathematicians, chiefly because of the sincerity of his work and his proven ability as a logician, was a man named Girolamo Giovanni Saccheri (1667–1733), with whom this paper is concerned.

Girolamo Giovanni Saccheri was born at San Remo, Italy, on the night of September 4, 1667. From his earliest years he showed extreme precociousness and a spirit of inquiry, but little else is known of his childhood and early adolescence. In March, 1685, he entered the Order of the Society of Jesus in Genoa. Having completed his noviitate in 1690, he was sent by his superiors to the Collegio di Brera in Milan to teach grammar and, at the same time, to study theology and philosophy. It was here that he met the famous professor of mathematics, the Jesuit Father Tomasso Ceva, for whose brother, Giovanni, the famous Ceva theorem of college geometry is named. Here, too, he became acquainted with the important mathematicians of his day, among them Viviani, who called himself the last pupil of Galileo.

Father Ceva introduced the young scholastic, Saccheri, to the reading of Euclid's *Elements* in the edition of Clavius in which the Fifth Postulate of Euclid is listed as Axiom 13. This assignment apparently became a lifetime pursuit. It seems rather

Reprinted from *Mathematics Teacher* 57: (May, 1964): 323–32; with permission of the National Council of Teachers of Mathematics.

597

strange, however, that Saccheri at this time seems to have paid little, if any, attention to the contemporary discoveries of Newton and Leibniz, although he had analyzed the work of another contemporary, John Wallis (1616–1703), and had pointed out flaws in the latter's proof of the Fifth Postulate.

In 1694, Saccheri was sent to teach philosophy and polemic theology in the Collegio dei Gesuiti of Turin, at which post he remained for a period of three years. As a result of his studies and teaching during the three years, he produced a little book, *Logica demonstrativa*, which, according to his biographer, Alberto Pascal, deserves to be better known than it is by mathematicians and logicians.

The first edition of the *Logica demonstrativa*, which appeared in 1697, did not carry the author's name, but was published under the cover of a thesis by Count Gravere, one of Saccheri's students. There has been much speculation about why Saccheri did not allow his name to appear at this time. Some think that it was because of his standing; he had not yet attained the rank of professor, and as a result, even though his genius was evident, the work would not receive the acclaim it deserved. Others say that his religious superiors did not permit him to publish the work under his own name. Whatever the reason, the book reappeared four years later when, armed with the rank of professor and a change of residence, Saccheri's reputation was soaring in university circles. He was now acclaimed as a brilliant teacher and, incidentally, a man of such remarkable memory that, according to Vera Sanford in *A Short History of Mathematics*, he could play three games of chess at the same time without seeing any of the boards.

A third edition of the *Logica demonstrativa* appeared in 1735, two years after Saccheri's death. Were it not for Giovanni Vailati, who in 1903 brought this work to the attention of the world in his own small publication, it might well have been cast into oblivion. In the *Logica demonstrativa*, the only existing copy of which is preserved in the Stadtbibliothek of Cologne, Saccheri lays down the clear distinction between what he calls *definitiones quid nominis* and *definitiones quid rei*, or between *nominal* and *real* definitions. The former are only intended to explain the meaning that is to be attached to a given term, whereas the latter, besides declaring the meaning of a word, affirm at the same time the existence of the thing defined or, in geometry, the possibility of constructing it. The nominal definition becomes a real definition by means of a *postulate* or an affirmative answer to the question of whether or not the thing exists. In this work Saccheri showed his absorption in the powerful method of *reductio ad absurdum*, an idea he later applied to an investigation of Euclid.

When Saccheri, a full-fledged member of the Society of Jesus, was appointed professor of mathematics at the University of Pavia, he joined the ranks of those mathematicians who had already set out to discover flaws in Euclid's fundamental postulates. The well-known postulate of parallel lines, was not, it was felt, sufficiently obvious. The Greek followers of Euclid had made attempts to prove it; the Arabians, when they acquired the Greek mathematics, also found the parallel axiom unsatisfactory. No one doubted that this was a necessary truth, but many felt that there should be some way of deducing it from the other and simpler axioms of Euclid.

As almost every high school student of geometry knows today, Euclid organized into a deductive system the fundamental geometry known in his day. He selected a few simple geometric facts as a basis and sought to demonstrate that all remaining facts were logical consequences of these. For his basic facts, which he called axioms or postulates, he gave no proof, but used them as the foundation of his system. The five geometric statements selected by Euclid as the basis of his deductive treatment of geometry are introduced as follows:

Let the following be postulated:

1. To draw a straight line from any point to any point.

2. To produce a finite straight line.

3. To draw a circle with any center and distance.

4. That all right angles are equal to one another.

5. That if a straight line falling on two straight lines makes the interior angles on the same side less than two right angles, the two straight lines if produced indefinitely meet on that side on which are the angles less than two right angles.

It is largely to this last (fifth) postulate that Euclid owes his greatness, and yet it is the one which has been the basis of the sharpest attacks on his system. The other four postulates are simple and in sharp contrast to the complicated nature of the statement of the Fifth Postulate. It is not surprising, then, that mathematicians even in Euclid's own time suggested that it should be a theorem rather than a postulate and initiated the efforts to demonstrate it as a logical consequence of the first four. To review these efforts is beyond the scope of this paper, since it is Saccheri's work that claims our attention.

Saccheri, as was stated earlier, had developed an interest in the method of *reductio ad absurdum*. Hence he set out to develop the consequences of denying Euclid's parallel axiom while retaining all the others. In this way he expected to develop a geometry which should be self-contradictory, since he had no doubt that the parallel axiom was a necessary truth.

To arrive at a contradiction in the most convenient form, Saccheri employed a figure found in his Clavius (1574), the birectangular quadrilateral. In his "Indicis loco" of *Euclides vindicatus* he states:

1. In Propositions I and II of the First Book two principles are established, from which in Propositions III and IV is proved that interior angles at the straight joining the extremities of equal perpendiculars erected toward the same parts (in the same plane) from two points of another straight, as base, not merely are equal to each other, but besides are either right or obtuse or acute according as that join is equal to, or less, or greater than the aforesaid base; and inversely.

2. Hence occasion is taken to distinguish three different hypotheses, one of right angle, another of

obtuse, a third of acute: about which, in Propositions V, VI, and VII, is proved that any of these hypotheses is always alone true if it is found true in any one particular case.

Let us examine Propositions I, II, III, and IV. Proposition I states:

If two equal straights (Figure 1) *AC, BD* make with the straight *AB* angles equal toward the same parts: I say that the angles at the join *CD* will be mutually equal.

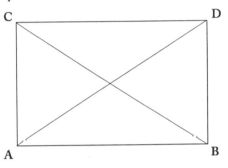

FIGURE I

Saccheri's proof is very simple:

Join *AD, CB*. Then consider the triangles *CAB, DBA*. It follows (Eu. I 4) that the bases *CB, AD* will be equal. Then consider the triangles *ACD, BDC*. It follows (Eu. I 8) that the angles *ACD, BDC* will be equal.

Q.E.D.

Proposition II states:

Retaining the uniform quadrilateral *ABCD*, bisect the sides *AB, CD* (Figure 2) in the points *M* and *H*. I say the angles at the join *MH* will then be right.

Proof: Join *AH, BH*, and likewise *CM, DM*. Because in this quadrilateral the angles *A* and *B* are taken equal and likewise (from the preceding proposition) the angles *C* and *D* are equal; it follows (Eu. I 4, noting the equality of the sides) that in the triangles *CAM, DBM*, the bases *CM, DM* will be equal; and likewise, in the triangles *ACH, BDH*, the bases *AH* and *BH*. Therefore, comparing the triangles *CHM, DHM*, and in turn the triangles *AMH, BMH*, it follows (Eu. I 8) that we have mutually equal, and therefore right, the angles at the points *M* and *H*.

Q.E.D.

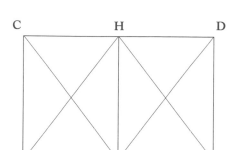

FIGURE 2

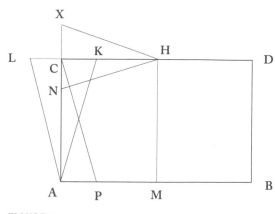

FIGURE 3

Proposition III states:

If two equal straights (Figure 3) *AC, BD,* stand perpendicular to any straight *AB*: I say the join *CD* will be equal to, or less than, or greater than *AB,* according as the angles at *CD* are right, or obtuse, or acute.

Saccheri divides this proof into three parts as follows:

1. Each angle *C* and *D* being right; suppose, if it were possible, either one of these, as *DC,* greater than the other *BA.*

Take in *DC* the piece *DK* equal to *BA,* and join *AK.* Since therefore on *BD* stand perpendicular the equal straights *BA, DK,* the angles *BAK* and *DKA* will be equal (Prop. I). But this is absurd; since the angle *BAK* is by construction less than the assumed right angle *BAC;* and the angle *DKA* is by construction external, and therefore (Eu. I 16) greater than the internal and opposite *DCA* which is supposed right. Therefore neither of the aforesaid straights *DC, BA,* is greater than the other, whilst the angles at the join *CD* are right; and therefore they are mutually equal.

Q.E.D. (primo loco)

2. But if the angles at the join *CD* are obtuse, bisect *AB* and *CD* in the points *M* and *H,* and join *MH.*

Since therefore on the straight *MH* stand perpendicular (Prop. II) the two straights *AM, CH,* and at the join *AC* is a right angle at *A,* the straight *CH* (Prop. I) will not be equal to this *AM,* since a right angle is lacking at *C.*

But neither will it be greater: otherwise in *HC* the piece *KH* being assumed equal to this *AM,* the angles at the join *AK* will be (Prop. I) equal.

But this is absurd, as above. For the angle *MAK* is less than a right: and the angle *HKA* is (Eu. I 16) greater than an obtuse, such as the internal and opposite *HCA* is supposed.

It remains therefore, that *CH,* whilst the angles at the join *CD* are taken obtuse, is less than this *AM*; and therefore *CD,* double the former, is less than *AB,* double the latter.

Q.E.D. (secundo loco)

3. Finally, however, if the angles at the join are acute, *MH* being constructed perpendicular as before (Prop. II), we proceed thus. Since on the straight *MH* stand perpendicular two straights *AM, CH,* and at the join *AC* is a right angle at *A,* the straight *CH* will not be equal to this *AM* (as above), since the angle at *C* is not right. But neither will it be less otherwise, if in *HC* produced *HL* is taken equal to this *AM,* the angles at the join *AL* will be (as above) equal.

But this is absurd. For the angle *MAL* is by construction greater than the assumed right *MAC*; and the angle *HLA* is by construction internal, and opposite, and therefore less than (Eu. I 16) the external *HCA,* which is assumed acute.

It remains, therefore, that *CH,* whilst the angles at the join *CD* are acute, is greater than this *AM,* and therefore *CD,* the double of the former, is greater than *AB,* the double of the latter.

Q.E.D. *(tertio loco)*

Saccheri's general conclusion, therefore, was:

It is established that the join *CD* will be equal to, or less than, or greater than this *AB,* according as the angles at the same *CD* are right, or obtuse, or acute.

Proposition IV is the converse of Proposition III. It is stated as follows:

But inversely (Figure 3) the angles at the join *CD* will be right, or obtuse, or acute, according as the straight *CD* is equal, or less, or greater than the opposite *AB.*

Proof: For if the straight *CD* is equal to the opposite *AB,* and nevertheless the angles at it are either obtuse or acute; now these such angles prove it (Prop. III) not equal, but less, or greater than the opposite *AB;* which is absurd against the hypothesis.

The same uniformly avails in regard to the remaining cases. It holds therefore that the angles at the join *CD* are either right, or obtuse, or acute, according as the straight *CD* is equal to, or less than, or greater than the opposite *AB.*

Q.E.D.

Saccheri called the angles at *C* and *D* the *summit* angles of the quadrilateral, and, as we see from the first four propositions, noted the three following possibilities:

1. The summit angles are right (the right-angle hypothesis).
2. The summit angles are obtuse (the obtuse-angle hypothesis).
3. The summit angles are acute (the acute-angle hypothesis).

In Propositions V, VI, and VII, Saccheri proved that if one of these hypotheses is true for one of his quadrilaterals, it is true for every such quadrilateral. Using the Postulate of Archimedes, which includes implicitly the infinitude of the straight line, Saccheri showed, in Proposition VI, that the Fifth Postulate is a consequence of the right-angle hypothesis, and finally, in Proposition XIV, that the obtuse-angle hypothesis is self-contradictory.

It now remained for Saccheri to dispose of the acute-angle hypothesis, which he explored in the hope of finding a contradiction. Obtaining many results that seemed strange because they differed to a great degree from those that had been established by use of the Fifth Postulate, he never succeeded in finding the desired contradiction. Since he found it impossible to cast out the acute-angle hypothesis on purely logical grounds and so "prove" the Fifth Postulate, Saccheri concluded in Proposition XXXIII of his famous book, *Euclides ab omni naevo vindicatus (Euclid Freed of Every Flaw),* that the "hypothesis of the acute angle is absolutely false because repugnant to the nature of the straight line." For this proof he trusted to intuition and to faith in the validity of the Fifth Postulate rather than to logic, and relied on five Lemmas spread over sixteen pages of his masterpiece, "in which are contained five fundamental axioms relating to the straight line and circle, with their correlative postulates."

It is highly improbable that Saccheri was satisfied with his investigation, for he was an able logician. Did he really expect to find a flaw in Euclid's fundamental postulates, or did he have a preconceived idea that no defect was to be found in his idol? Why the particular title, *Euclid Freed of Every Flaw?* Vailati hints that Saccheri was already convinced before beginning his proofs that he would find no contradiction, and that the sole purpose of his work was to "free Euclid." He points out that Saccheri already, in his youth, had arrived at the idea that the characteristic property of the most fundamental propositions in every demonstrative science is precisely their indemonstrability, except by assuming as hypothesis the falsity of the very proposition to be proved, and in showing how also, when taking this hypothesis as point of departure, one arrives at the conclusion that the proposition in question is true.

Vailati goes on to say, "It was in hopes of reaching in this way a proof of the Parallel Postulate, namely, deducing it from the very hypothesis of its falsity, that Saccheri pushed on in the investigation of the consequences flowing from the other two alternative hypotheses, to which the negation of the Parallel Postulate gave rise, attaining thus results fitting to carry on in their sequence to a discovery far more important than what he had in mind to reach, namely to the discovery of a wholly new geometry of which the old is only a simple particular case."

"In this regard," continues Vailati, "his position is not unlike that of his fellow countryman, Columbus, who precisely in the hopes of reaching by a new way regions already known, was led to the discovery of a new continent."

Dr. Withers, in his book, *Euclid's Parallel Postulate*, published in 1908, attributes Saccheri's conclusion not to a logical but to a psychological difficulty. He states: "We have seen that the assumption (acuti) worried Saccheri very profoundly in his heroic efforts to 'vindicate Euclid.' It was not a logical but a psychological or experiential difficulty which caused Saccheri to reject the logical conclusions to which his own labors clearly and inevitably pointed; and it was certainly the same sort of difficulty which caused the immediate rejection, by himself and subsequent mathematicians, of the assumption (obtusi)."

Although Saccheri's work, *Euclid Freed of Every Flaw*, failed in its aim, it is of great importance. In it the most determined effort had been made on behalf of the Fifth Postulate, and the fact that Saccheri did not succeed in discovering any contradictions among the consequences of the *hypothesis of the acute angle* could hardly help but suggest the question as to whether a consistent logical geometrical system could be built upon the hypothesis, and thus the Euclidean Postulate be impossible of demonstration.

The publication of Saccheri's *Euclides ab omni naevo vindicatus*, in 1733, attracted some attention. Mention is made of it in two early histories of

mathematics, that of J. C. Heilbronner (Leipzig, 1742), and that of Montucla (Paris, 1758). It was also noted by G. S. Klügel, in his dissertation (Göttingen, 1763). But then it seems to have been forgotten until it was accidentally rediscovered by the Jesuit Father Angelo Manganotti in 1889. It is not generally known whether Father Manganotti had previous knowledge of Saccheri and his work or if his first acquaintance with him was through Beltrami, who in 1889 named Saccheri as "the Italian precursor of the Hungarian Bolyai and the Russian Lobachevski."

Following the rediscovery of Saccheri's work, it was translated from the Latin into English by Dr. George Bruce Halsted of the University of Texas, into German by Engel and Stäckel, and into Italian by Boccardini. But with all of this, there were hundreds of mathematicians who knew nothing of this publication. It might be noted here that even Sir Thomas Heath, in his famous three-volume *Euclid's Elements* (Cambridge, 1908), referred to Saccheri's work as a Latin translation of Euclid.

What influence did Saccheri exert on the geometers of the eighteenth century? It is difficult to say. However, Roberto Bonola, in his *Non-Euclidean Geometry* (1906), shows that Saccheri had a very decided influence on a few of the leading mathematicians of that period. The first was Johann Heinrich Lambert (1728–1777), a Swiss mathematician who quotes, in his *Theorie der Parallellinien*, a dissertation of G. S. Klügel (1739–1812) in which the work of the Italian geometer Saccheri is carefully analyzed. Part of Lambert's investigation clearly resembles that of Saccheri. His fundamental figure is a quadrilateral with three right angles. The *three* hypotheses are made concerning the nature of the third angle, and in the treatment the author does not depart from the Saccheri method.

Adrien Marie Legendre (1752–1853), a French mathematician, also displayed knowledge of Saccheri's work. In his *Eléments de géométrie*, his investigations of the theory of parallels were much like Saccheri's, and the results he obtained were to a large extent the same. He chose, however, to place emphasis upon the angle-sum of a triangle and proposed three hypotheses in which the sum of the angles is, in turn, equal to, greater than, and less than two right angles, hoping to be able to reject the last two. Unconsciously assuming the straight line infinite, he was able to eliminate the geometry based on the second hypothesis by proving the following theorem: The sum of the angles of a triangle cannot be greater than two right angles.

Even though Legendre added nothing new to the materials and results obtained by Saccheri and Lambert, yet the simple, straightforward style of his proofs brought him a large following and helped to create an interest in these ideas just at a time when geometers were on the threshold of a great discovery. Some of his proofs, on account of their elegance, are of permanent value. In spite of his many attempts, he was never able to dispose of the third hypothesis, which, as Gauss remarked, was "the reef on which all the wrecks occurred."

So far, the term "non-Euclidean geometry" had not been used, and it remained for Gauss to bring that expression into our mathematical language. Carl Friedrich Gauss (1777–1855), like Saccheri and Lambert before him, had attempted to prove the truth of Euclid's Fifth Postulate by assuming its falsity. Contrary to popular opinion, he did not recognize the existence of a logically sound non-Euclidean geometry by intuition or a flash of genius, but rather he spent many laborious years before he had overcome the inherited prejudice against it. How much influence was exerted on Gauss by Saccheri is best expressed by Segre in his "Congetture intorno alla influenzadi Saccheri sulla formazione della geometria non euclidea" (1903). He remarks that both Gauss and Wolfgang Bolyai, while students at Göttingen, the former from 1795 to 1798 and the latter from 1796 to 1799, were interested in the theory of parallels. It is therefore possible that, through their professors Kästner and Seyffer, who were both deeply versed in this subject, they had become familiar with both Saccheri's *Euclides vindicatus* and Lambert's *Theorie*

der Parallellinien. There is no confirmation of this, however; it is only a conjecture.

In one of his letters to his friend Schumacher, Gauss mentioned a "certain Schweikart." The one referred to was Ferdinand Karl Schweikart (1780–1859), who from 1796 to 1798 was a student of law at Marburg. As he was keenly interested in mathematics he took advantage of the opportunity while at the university to listen to various lectures on the subject, and particularly to those of J. K. F. Hauff, who was somewhat of an authority on the theory of parallels. Schweikart's interest in this theory developed to such an extent that in 1807 there appeared his only published work of a mathematical nature, *Die Theorie der Parallellinien nebst dem Vorschlage ihrer Verbannung aus der Geometrie.*

In this book Schweikart mentioned both Saccheri and Lambert, but the contents were in no way novel and the style was on quite conventional lines. Doubtless his acquaintance with the work of Saccheri and Lambert affected the character of his later investigations. Eleven years later, having discovered a new order of ideas, Schweikart developed a geometry independent of Euclid's hypothesis. So, in 1818, he sent the following memorandum to Gauss for the latter's opinion: ". . . There are two kinds of geometry—a geometry in the strict sense, the Euclidean; and an Astral Geometry. . . . ." The "astral geometry," as discovered by Schweikart, as well as Gauss's "non-Euclidean geometry," corresponds exactly to Saccheri's system for the hypothesis of the acute angle. And since Schweikart in his *Theorie* of 1807 mentioned the work of both Saccheri and Lambert, we can easily assume the influence exerted upon him by these earlier works.

It should not surprise us that the discovery of non-Euclidean geometry was not made by one person, but independently by several in different parts of the world. We know from the history of mathematics that this has happened on more than one occasion, and undoubtedly will happen again. Slowly but surely, the efforts of such men as Saccheri, Lambert, and others, in the investigation of the Fifth Postulate directed the speculations of geometers to the point where the discovery was imminent.

Gauss, whose courage did not match his genius, failed to complete his discoveries. He did, however, keep in touch with the Hungarian mathematician, Wolfgang Bolyai (1775–1856), who had been his fellow-student in Göttingen. The latter's son, János, in 1823 developed a new system of geometry on the "acute angle hypothesis," in which those theorems of Euclid that are independent of the Fifth Postulate still operated, but in which the others were replaced by such amazing conclusions as this: In a plane, instead of one line, two lines can be drawn parallel to a given line, and through this point, an infinite number of lines may be drawn lying in the angle between the first two and having the property that they will not intersect the given line.

Amazing as this is in contrast to Euclidean geometry, it is equally logical. The younger Bolyai's discovery was published in 1832 as a supplement to a work by his father, to the great delight of Gauss, who stated that this publication relieved him of all responsibility to complete his own work. But, as we know today, Bolyai's work was almost completely ignored, and hence, soon forgotten.

Meanwhile, in Russia, a little-known mathematician named Nikolai Lobachevski (1793–1856) was lecturing on the "acute angle geometry" at the University of Kazan. When the German version of his work was published in 1840, Gauss was again enthusiastic, but Bolyai declared that Lobachevski had been strongly influenced by his own publication of 1832. It is now well known that Lobachevski's work had been presented in Russian at least eleven years prior to the German publication. If any influence was exerted on Lobachevski, it was probably through one of his teachers, Johann Martin Christian Bartels (1769–1836), a friend of Gauss who went to Kazan in 1807 and became professor of mathematics at that university.

Some notes of Lobachevski, that were written during the period 1815–1817, show his attempts at the proof of the Fifth Postulate, but his investi-gations seemed to resemble those of Legendre. Whether this was a result of an influence passed on to him from the works of Gauss or even Saccheri, or whether the resemblance is purely coincidental, is difficult to assume. One thing, however, is certain: his discoveries were not favorably received, and the story of his treatment at the hands of the government is given at the beginning of this paper.

Neither Bolyai nor Lobachevski gave any consideration to the "obtuse angle hypothesis." But Georg Friedrich Bernhard Riemann (1826–1866), a student of Gauss in Göttingen, began with a study of the postulate that a straight line is infinitely long, the assumption that had convinced Saccheri that the "obtuse angle hypothesis" is untenable. Discarding this postulate, Riemann developed a geometry in which all lines are of finite length, any pair of lines intersect if they lie in the same plane, and the sum of the angles of a triangle is greater than two right angles.

Reimann, in his work, called attention to the true nature and significance of geometry, and did much to free mathematics from the handicap of tradition. In fact, his work dealt almost entirely with generalities, and was suggestive in nature, so as to leave opportunities for detailed investigations to be carried on later by his successors in the field of non-Euclidean geometry.

Earlier in this paper it was stated that the term "non-Euclidean Geometry" was first used by Gauss to denote that geometry which arises on replacing Euclid's Fifth Postulate by its negation, and keeping unaltered all the remaining postulates that Euclid either explicitly formulated or which enter implicitly into his development of geometry. In this sense, there is only one non-Euclidean geometry, namely, Saccheri's geometry of the "acute-angle hypothesis." But Felix Klein, in 1871, gave a different nomenclature to the existing geometries. That which had been investigated by Saccheri, Gauss, Bolyai, and Lobachevski, he called "hyperbolic geometry"; that of Riemann, he called "elliptic geometry"; while to Euclid's he gave the name of "parabolic geometry." These names were suggested by the fact that a straight line contains two infinitely distant points under the *hypothesis of the acute angle*, none under the *hypothesis of the obtuse angle*, and only one under the *hypothesis of the right angle*. The characteristic postulate of Euclidean geometry states that: "Through a given point, one and only one line can be drawn which is parallel to a given line." On the other hand, the distinctive feature of hyperbolic plane geometry is the assumption that an infinite number of parallels can be drawn to a line through a point. An investigation of the supposition that no line can be drawn through a point parallel to a given line brings us to the hypothesis of the obtuse angle of Saccheri, which geometry he ruled out, because he assumed that straight lines are infinite. It was Riemann who first pointed out the importance of distinguishing between the ideas of *boundlessness* and *infinitude* in connection with the concepts of space. Hence, the transition from Euclidean to elliptic geometry is not a very simple task, and elliptic geometry's characteristic postulate, "Two straight lines always intersect with each other," is not only incompatible with the Euclidean postulate which it replaces, but also with others.

So, at last, slowly but surely, the obstinate puzzle of the Fifth Postulate was being solved. One wonders why this should have taken so long and why the chain reaction set off by Saccheri had not fissioned more violently. But one must remember that during this period the philosophy of Kant (1724–1804), which treated space not as empirical, but as intuitive, dominated the situation. Space was regarded as something existing in the mind, not as a concept resulting from external experience. It required courage at that time to recognize that geometry was an experimental science once it was applied to physical space, and that its postulates and their consequences need only be accepted if convenient, and if they agree reasonably well with experimental data. It was the domination of the Kantian philosophy which dampened Gauss's courage and which influenced the public against the discoveries of Bolyai and Lobachevski.

Gradually a change of viewpoint arrived and the new discovery led to the complete destruction of the Kantian space concept and the revelation of the true distinction between concept and experience, as well as their interrelation. Now that the term "non-Euclidean geometry" has become almost a household expression, many of us may find it difficult to reconcile the attitude of mathematicians of greater and lesser stature over the period of transition. Again, this attitude should not be too much of a shock to us. The history of scientific discovery teaches that every radical change in its separate departments does not suddenly alter the convictions and the presuppositions upon which investigators and teachers have for a considerable time based the presentation of their subjects. Were Saccheri to return today to this new world, he would probably be the first to acknowledge that one step would have brought him to the full realization of the discovery that stared him in the face as he made his heroic attempt to free his idol, Euclid, from every blemish. When he died on October 25, 1733, at the age of sixty-six, it is doubtful if he had any intimation of the fire he had ignited, and of the future generations who would hail him as "the forerunner of non-Euclidean geometry."

BIBLIOGRAPHY

Bell, Eric T. *Mathematics, Queen and Servant of Science.* New York: McGraw-Hill Book Co., Inc., 1951.

———. *Men of Mathematics.* New York: Simon and Schuster, Inc., 1937.

———. *The Development of Mathematics.* New York: McGraw-Hill Book Co., Inc., 1945.

———. *The Magic of Numbers.* New York: McGraw-Hill Book Co., Inc., 1946.

Blumenthal, Leonard M. *A Modern View of Geometry.* San Francisco: W. H. Freeman & Co., Publishers, 1961.

Bonola, Roberto. *Non-Euclidean Geometry* (1906). Translated by H. S. Carslaw, Sydney, 1911. New York: Dover Publications, 1955.

Cajori, Florian. *A History of Mathematics.* New York: The Macmillan Company, 1938.

Eves, Howard. *An Introduction to the History of Mathematics,* revised edition. New York: Holt, Rinehart & Winston, Inc., 1964.

Hofmann, Joseph E. *Classical Mathematics.* New York: Philosophical Library, Inc., 1959.

Kattsoff, Louis O. "The Saccheri Quadrilateral," *The Mathematics Teacher,* LV (December, 1962): 630-36.

Kline, Morris. *Mathematics in Western Culture.* New York: Oxford University Press, Inc., 1953.

Kramer, Edna E. *The Main Stream of Mathematics.* New York: Oxford University Press, Inc., 1953.

Miller, George A. *Historical Introduction to Mathematical Literature.* New York: The Macmillan Company, 1916.

Newman, James R. (ed.) *The World of Mathematics,* Vol. III. New York: Simon and Schuster, Inc., 1956.

Saccheri, Girolamo G. *Euclides ab omni naevo vindicatus* (1733). Translated by George B. Halsted. Chicago: Open Court Publishing Co., 1920.

Sanford, Vera. *A Short History of Mathematics.* Boston: Houghton Mifflin Company, 1930.

Smith, David E., *History of Mathematics,* Vol. II. Boston: Ginn & Company, 1925.

————. *A Source Book in Mathematics.* New York: McGraw-Hill Book Co., Inc., 1929.

Wolfe, Harold E. *Introduction to Non-Euclidean Geometry.* New York: Holt, Rinehart & Winston, Inc., 1945.

92

The Changing Concept of Change: The Derivative from Fermat to Weierstrass

JUDITH V. GRABINER

$\mathcal{S}$OME YEARS AGO while teaching the history of mathematics, I asked my students to read a discussion of maxima and minima by the seventeenth-century mathematician, Pierre Fermat. To start the discussion, I asked them, "Would you please define a relative maximum?" They told me it was a place where the derivative was zero. "If that's so," I asked, "then what is the definition of a relative minimum?" They told me, *that's* a place where the derivative is zero. "Well, in that case," I asked, "what is the difference between a maximum and a minimum?" They replied that in the case of a maximum, the second derivative is negative.

What can we learn from this apparent victory of calculus over common sense?

I used to think that this story showed that these students did not understand the calculus, but I have come to think the opposite: they understood it very well. The students' answers are a tribute to the power of the calculus in general, and the power of the concept of derivative in particular. Once one has been initated into the calculus, it is hard to remember what it was like *not* to know what a derivative is and how to use it, and to realize that people like Fermat once had to cope

Reprinted from *Mathematics Magazine* 56 (Sept., 1983): 195–206; with permission of the Mathematical Association of America.

with finding maxima and minima without knowing about derivatives at all.

Historically speaking, there were four steps in the development of today's concept of the derivative, which I list here in chronological order. The derivative was first *used*; it was then *discovered*; it was then *explored and developed*; and it was finally *defined*. That is, examples of what we now recognize as derivatives first were used on an ad hoc basis in solving particular problems; then the general concept lying behind these uses was identified (as part of the invention of the calculus); then many properties of the derivative were explained and developed in applications both to mathematics and to physics; and finally, a rigorous definition was given and the concept of derivative was embedded in a rigorous theory. I will describe the stops, and give one detailed mathematical example from each. We will then reflect on what it all means—for the teacher, for the historian, and for the mathematician.

The Seventeenth-Century Background

Our story begins shortly after European mathematicians had become familiar once more with Greek mathematics, learned Islamic algebra, synthesized the two traditions, and struck out on their own. François Vieta invented symbolic algebra in 1591; Descartes and Fermat independently invented analytic geometry in the 1630s. Analytic geometry meant, first, that curves could be represented by equations; conversely, it meant also that every equation determined a curve. The Greeks and Muslims had studied curves, but not that many—principally the circle and the conic sections plus a few more defined as loci. Many problems had been solved for these, including finding their tangents and areas. But since any equation could now produce a new curve, students of the geometry of curves in the early seventeenth century were suddenly confronted with an explosion of curves to consider. With these new curves, the old Greek methods of synthetic geometry were no longer sufficient. The Greeks, of course, had known how to find the tangents to circles, conic sections, and some more sophisticated curves such as the spiral of Archimedes, using the methods of synthetic geometry. But how could one describe the properties of the tangent at an arbitrary point on a curve defined by a ninety-sixth degree polynomial? The Greeks had defined a tangent as a line which touches a curve without cutting it, and usually expected it to have only one point in common with the curve. How then was the tangent to be defined at the point $(0, 0)$ for a curve like $y = x^3$ (Figure 1), or to a point on a curve with many turning points (Figure 2)?

The same new curves presented new problems to the student of areas and arc lengths. The Greeks had also studied a few cases of what they called "isoperimetric" problems. For example, they asked: of all plane figures with the same perimeter, which one has the greatest area? The circle, of course, but the Greeks had no general method for solving all such problems. Seventeenth-century mathematicians hoped that the new symbolic algebra might somehow help solve all problems of maxima and minima.

Thus, though a major part of the agenda for seventeenth-century mathematicians—tangents, areas, extrema—came from the Greeks, the subject matter had been vastly extended, and the solutions would come from using the new tools: symbolic algebra and analytic geometry.

Finding Maxima, Minima, and Tangents

We turn to the first of our four steps in the history of the derivative: its *use,* and also illustrate some of the general statements we have made. We shall look at Pierre Fermat's method of finding maxima and minima, which dates from the 1630s [8]. Fermat illustrated his method first in solving a simple problem, whose solution was well known: *Given a line, to divide it into two parts so that the product of the parts will be a maximum.* Let the

FIGURE 1

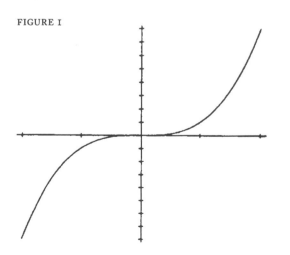

FIGURE 2

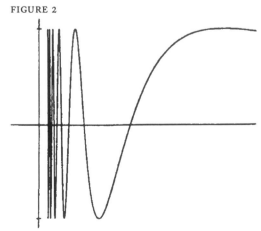

608

FIGURE 3

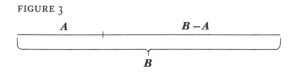

length of the line be designated B and the first part A (Figure 3). Then the second part is $B - A$ and the product of the two parts is

$$A(B - A) = AB - A^2. \qquad (1)$$

Fermat had read in the writings of the Greek mathematician Pappus of Alexandria that a problem which has, in general, two solutions will have only one solution in the case of a maximum. This remark led him to his method of finding maxima and minima. Suppose in the problem just stated there is a second solution. For this solution, let the first part of the line be designated as $A + E$; the second part is then $B - (A + E) = B - A - E$. Multiplying the two parts together, we obtain for the product

$$BA + BE - A^2 - AE - EA - E^2 =$$
$$AB - A^2 - 2AE + BE - E^2. \qquad (2)$$

Following Pappus's principle for the maximum, instead of two solutions, there is only one. So we set the two products (1) and (2) "sort of" equal; that is, we formulate what Fermat called the pseudo-equality:

$$AB - A^2 = AB - A^2 - 2AE + BE - E^2.$$

Simplifying, we obtain

$$2AE + E2 = BE$$

and

$$2A + E = B.$$

Now Fermat said, with no justification and no ceremony, "suppress E." Thus he obtained

$$A = B/2,$$

which indeed gives the maximum sought. He concluded, "We can hardly expect a more general method." And, of course, he was right.

Notice that Fermat did not call E infinitely small, or vanishing, or a limit; he did not explain why he could first divide by E (treating it as non-zero) and then throw it out (treating it as zero). Furthermore, he did not explain what he was doing as a special case of a more general concept, be it derivative, rate of change, or even slope of tangent. He did not even understand the relationship between his maximum-minimum method and the way one found tangents; in fact he followed his treatment of maxima and minima by saying that the same method—that is, adding E, doing the algebra, then suppressing E—could be used to find tangents [8, p. 223].

Though the considerations that led Fermat to his method may seem surprising to us, he did devise a method of finding extrema that worked, and it gave results that were far from trivial. For instance, Fermat applied his method to optics. Assuming that a ray of light which goes from one medium to another always takes the quickest path (what we now call the Fermat least-time principle), he used his method to compute the path taking minimal time. Thus he showed that his least-time principle yields Snell's law of refraction [7] [12, pp. 387–390].

Though Fermat did not publish his method of maxima and minima, it became well known through correspondence and was widely used. After mathematicians had become familiar with a variety of examples, a pattern emerged from the solutions by Fermat's method to maximum-minimum problems. In 1659, Johann Hudde gave a general verbal formulation of this pattern [3, p. 186], which, in modern notation, states that, given *a polynomial of the form*

$$y = \sum_{k=0}^{n} a_k x^k,$$

there is a maximum or minimum when

$$\sum_{k=1}^{n} k a_k x^{k-1} = 0.$$

Of even greater interest than the problem of extrema in the seventeenth century was the

FIGURE 4

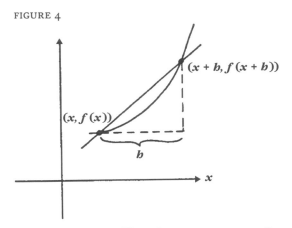

finding of tangents. Here the tangent was usually thought of as a secant for which the two points came closer and closer together until they coincided. Precisely what it meant for a secant to "become" a tangent was never completely explained. Nevertheless, methods based on this approach worked. Given the equation of a curve

$$y = f(x),$$

Fermat, Descartes, John Wallis, Isaac Barrow, and many other seventeenth-century mathematicians were able to find the tangent. The method involves considering, and computing, the slope of the secant,

$$\frac{f(x+h)-f(x)}{h},$$

doing the algebra required by the formula for $f(x + h)$ in the numerator, then dividing by h. The diagram in Figure 4 then suggests that when the quantity h vanishes, the secant becomes the tangent, so that neglecting h in the expression for the slope of the secant gives the slope of the tangent. Again, a general pattern for the equations of slopes of tangents soon became apparent, and a rule analogous to Hudde's rule for maxima and minima was stated by several people, including René Sluse, Hudde, and Christiaan Huygens [3, pp. 185–186].

By the year 1660, both the computational and the geometric relationships between the problem of extrema and the problem of tangents were clearly understood; that is, a maximum was found by

computing the slope of the tangent, according to the rule, and asking when it was zero. While in 1660 there was not yet a general concept of derivative, there was a general method for solving one type of geometric problem. However, the relationship of the tangent to other geometric concepts—area, for instance—was not understood, and there was no completely satisfactory definition of tangent. Nevertheless, there was a wealth of methods for solving problems that we now solve by using the calculus, and in retrospect, it would seem to be possible to gereralize those methods. Thus in this context it is natural to ask, how did the derivative as we know it come to be?

It is sometimes said that the idea of the derivative was motivated chiefly by physics. Newton, after all, invented both the calculus and a great deal of the physics of motion. Indeed, already in the Middle Ages, physicists, following Aristotle who had made "change" the central concept in his physics, logically analyzed and classified the different ways a variable could change. In particular, something could change uniformly or nonuniformly; if nonuniformly, it could change uniformly-nonuniformly or nonuniformly-nonuniformly, etc. [3, pp. 73–74]. These medieval classifications of variation helped to lead Galileo in 1638, without benefit of calculus, to his successful treatment of uniformly accelerated motion. Motion, then, could be studied scientifically. Were such studies the origin and purpose of the calculus? The answer is no. However plausible this suggestion may sound, and however important physics was in the later development of the calculus, physical questions were in fact neither the immediate motivation nor the first application of the calculus. Certainly they prepared people's thoughts for some of the properties of the derivative, and for the introduction into mathematics of the concept of change. But immediate motivation for the general concept of derivative—as opposed to specific examples like speed or slope of tangent—did not come from physics. The first problems to be solved, as well as the first applications, occurred in mathematics, especially geometry (see [1, chapter 7]; see also [3; chapters 4–5], and, for Newton, [17]). The concept of derivative then developed gradually, together with the ideas of extrema, tangent, area, limit, continuity, and function, and it interacted with these ideas in some unexpected ways.

Tangents, Areas, and Rates of Change

In the latter third of the seventeenth century, Newton and Leibniz, each independently, invented the calculus. By "inventing the calculus" I mean that they did three things. First, they took the wealth of methods that already existed for finding tangents, extrema, and areas, and they subsumed all these methods under the heading of two general concepts, the concepts which we now call *derivative* and *integral*. Second, Newton and Leibniz each worked out a notation which made it easy, almost automatic, to use these general concepts. (We still use Newton's $\dot{x}$ and we still use Leibniz's dy/dx and $\int y\,dx$.) Third, Newton and Leibniz each gave an argument to prove what we now call the Fundamental Theorem of Calculus: the derivative and the integral are mutually inverse. Newton called our "derivative" a *fluxion*—a rate of flux or change; Leibniz saw the derivative as a ratio of infinitesimal differences and called it the *differential quotient*. But whatever terms were used, the concept of derivative was now embedded in a general subject—the calculus—and its relationship to the other basic concept, which Leibniz called the integral, was now understood. Thus we have reached the stage I have called *discovery*.

Let us look at an early Newtonian version of the Fundamental Theorem [13, sections 54–5, p. 23]. This will illustrate how Newton presented the calculus in 1669, and also illustrate both the strengths and weaknesses of the understanding of the derivative in this period.

Consider with Newton a curve under which the area up to the point $D = (x, y)$ is given by z (see Figure 5). His argument is general: "Assume any

FIGURE 5

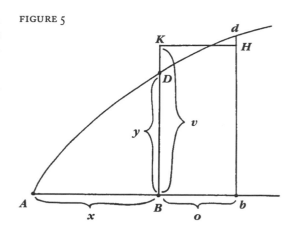

relation betwixt x and z that you please"; he then proceeded to find y. The example he used is

$$z = \frac{n}{m+n}\,ax^{(m+n)/n};$$

however, it will be sufficient to use $z = x^3$ to illustrate his argument.

In the diagram in Figure 5, the auxiliary line bd is chosen so that $Bb = o$, where o is not zero. Newton then specified that $BK = v$ should be chosen so that area $BbHK =$ area $BbdD$. Thus $ov =$ area $BbdD$. Now, as x increases to $x + o$, the change in the area z is given by

$$z(x + o) - z(x) = x^3 + 3x^2o + 3xo^2 + o^3 - x^3 = 3xo^2o + 3xo^2 + o^3,$$

which, by the definition of v, is equal to ov. Now since $3x^2o + 3xo^2 + o^3 = ov$, dividing by o produces $3x^2 + 3ox + o^2 = v$. Now, said Newton, "If we suppose Bb to be diminished infinitely and to vanish, or o to be nothing, v and y in that case will be equal and the terms which are multiplied by o will vanish: so that there will remain . . . "

$$3x^2 = y.$$

What has he shown? Since $(z(x + o) - z(x))/o$ is the rate at which the area z changes, that rate is given by the ordinate y. Moreover, we recognize that $3x^2$ would be the slope of the tangent to the curve $z = x^3$. Newton went on to say that the argument can be reversed; thus the converse holds too. We see that derivatives are fundamentally involved in areas as well as tangents, so the concept of derivative helps us to see that these two problems are mutually inverse. Lebniz gave analogous arguments on this same point (see, e.g. [16, pp. 282–284]).

Newton and Leibniz did not, of course, have the last word on the concept of derivative. Though each man had the most useful properties of the concept, there were still many unanswered questions. In particular, what, exactly, is a differential quotient? Some disciples of Leibniz, notably Johann Bernoulli and his pupil the Marquis de l'Hospital, said a differential quotient was a ratio of infinitesimals; and after all, that is the way it was calculated. But infinitesimals, as seventeenth-century mathematicians were well aware, do not obey the Archimedean axiom. Since the Archimedean axiom was the basis for the Greek theory of ratios, which was, in turn, the basis of arithmetic, algebra, and geometry for seventeenth-century mathematicians, non-Archimedean objects were viewed with some suspicion. Again, what is a fluxion? Though it can be understood intuitively as a velocity, the proofs Newton gave in his 1671 *Method of Fluxions* all involved an "indefinitely small quanity o," [14, pp. 32–33] which raises many of the same problems that the o which "vanishes" raised in the Newtonian example of 1669 we saw above. In particular, what is the status of that little o? Is it zero? If so, how can we divide by it? If it is not zero, aren't we making an error when we throw it away? These questions had already been posed in Newton's and Leibniz's time. To avoid such problems, Newton said in 1687 that quantities defined in the way that $3x^2$ was defined in our example were the *limit* of the ratio of vanishing increments. This sounds good, but Newton's understanding of the term "limit" was not ours. Newton in his *Principia* (1687) described limits as "ultimate ratios"—that is, the value of the ratio of those vanishing quantities just when they are vanishing. He said, "Those ultimate ratios with which quantities vanish are not truly

the ratios of ultimate quantities, but limits towards which the ratios of quantities decreasing without limit do always converge; and to which they approach nearer than by any given difference, but never go beyond, nor in effect attain to, till the quantities are diminished in infinitum" [15, Book I, Scholium to Lemma XI, p. 39].

Notice the phrase "but never go beyond"—so a variable cannot oscillate about its limit. By "limit" Newton seems to have had in mind "bound," and mathematicians of his time often cite the particular example of the circle as the limit of inscribed polygons. Also, Newton said, "nor . . . attain to, till the quantities are diminished in infinitum." This raises a central issue: it was often asked whether a variable quantity ever actually reached its limit. If it did not, wasn't there an error? Newton did not help clarify this when he stated as a theorem that "Quantities and the ratios of quantities which in any finite time converge continually to equality, and before the end of that time approach nearer to each other than by any given difference, become ultimately equal" [15, Book I, Lemma I, p. 29]. What does "become ultimately equal" mean? It was not really clear in the eighteenth century, let alone the seventeenth.

In 1734, George Berkeley, Bishop of Cloyne, attacked the calculus on precisely this point. Scientists, he said, attack religion for being unreasonable; well, let them improve their own reasoning first. A quantity is either zero or not; there is nothing in between. And Berkeley characterized the mathematicians of his time as men "rather accustomed to compute, than to think" [2].

Perhaps Berkeley was right, but most mathematicians were not greatly concerned. The concepts of differential quotient and integral, concepts made more effective by Leibniz's notation and by the Fundamental Theorem, had enormous power. For eighteenth-century mathematicians, especially those on the Continent where the greatest achievements occurred, it was enough that the concepts of the calculus were understood sufficiently well to be applied to solve a large number of problems, both in mathematics and in physics. So, we come to our third stage: *exploration and development.*

Differential Equations, Taylor Series, and Functions

Newton had stated his three laws of motion in words, and derived his physics from those laws by means of synthetic geometry [15]. Newton's second law stated: *"The change of motion [our 'momentum'] is proportional to the motive force impressed, and is made in the direction of the [straight] line in which that force is impressed"* [15, p. 13]. Once translated into the language of the calculus, this law provided physicists with an instrument of physical discovery of tremendous power—because of the power of the concept of the derivative.

To illustrate, if F is force and x distance (so $m\dot{x}$ is momentum and, for constant mass, $m\ddot{x}$ the rate of change of momentum), then Newton's second law takes the form $F = m\ddot{x}$. Hooke's law of elasticity (when an elastic body is distorted the restoring force is proportional to the distance [in the opposite direction] of the distortion) takes the algebraic form $F = -kx$. By equating these expressions for force, Euler in 1739 could easily both state and solve the differential equation $m\ddot{x} + kx = 0$ which describes the motion of a vibrating spring [10, p. 482]. It was mathematically surprising, and physically interesting, that the solution to that differential equation involves sines and cosines.

An analogous, but considerably more sophisticated problem, was the statement and solution of the partial differential equation for the vibrating string. In modern notation, this is

$$\frac{\partial^2 y}{\partial t^2} = \frac{T \partial^2 y}{\mu \partial x^2},$$

where T is the tension in the string and μ is its mass per unit length. The question of how the solutions to this partial differential equation behaved was investigated by such men as d'Alembert,

Daniel Bernoulli, and Leonhard Euler, and led to extensive discussions about the nature of continuity, and to an expansion of the notion of function from formulas to more general dependence relations [10, pp. 502–514], [16, pp. 367–368]. Discussions surrounding the problem of the vibrating string illustrate the unexpected ways that discoveries in mathematics and physics can interact ([16, pp. 351–368] has good selections from the original papers). Numerous other examples could be cited, from the use of infinite-series approximations in celestial mechanics to the dynamics of rigid bodies, to show that by the mid-eighteenth century the differential equation had become the most useful mathematical tool in the history of physics.

Another useful tool was the Taylor series, developed in part to help solve differential equations. In 1715, Brook Taylor, arguing from the properties of finite differences, wrote an equation expressing what we would write as $f(x + h)$ in terms of $f(x)$ and its quotients of differences of various orders. He then let the differences get small, passed to the limit, and gave the formula that still bears his name: the Taylor series. (Actually, James Gregory and Newton had anticipated this discovery, but Taylor's work was more directly influential.) The importance of this property of derivatives was soon recognized, notably by Colin Maclaurin (who has a special case of it named after him), by Euler, and by Joseph-Louis Lagrange. In their hands, the Taylor series became a powerful tool in studying functions and in approximating the solution of equations.

But beyond this, the study of Taylor series provided new insights into the nature of the derivative. In 1755, Euler, in his study of power series, had said that for any power series,

$$a + bx + cx^2 + dx^3 + \dots,$$

one could find x sufficiently small so that if one broke off the series after some particular term—say x^2—the x^2 term would exceed, in absolute value, the sum of the entire remainder of the series

[6, section 122]. Though Euler did not prove this—he must have thought it obvious since he usually worked with series with finite coefficients—he applied it to great advantage. For instance, he could use it to analyze the nature of maxima and minima. Consider, for definiteness, the case of maxima. If $f(x)$ is a relative maximum, then by definition, for small h,

$$f(x - h) < f(x) \text{ and } f(x + h) < f(x).$$

Taylor's theorem gives, for these inequalities,

$$f(x - h) = f(x) - h \frac{df(x)}{dx} + h^2 \frac{d^2f(x)}{dx^2} - \ldots$$
$$< f(x) \tag{3}$$

$$f(x + h) = f(x) + h \frac{df(x)}{dx} + h^2 \frac{d^2f(x)}{dx^2} + \ldots$$
$$< f(x). \tag{4}$$

Now if h is so small that $h\,df(x)/dx$ dominates the rest of the terms, the only way that both of the inequalities (3) and (4) can be satisfied is for $df(x)/dx$ to be zero. Thus the differential quotient is zero for a relative maximum. Furthermore, Euler argued, since h^2 is always positive, if $d^2f(x)/dx^2 \neq 0$, the only way both inequalities can be satisfied is for $d^2f(x)/dx^2$ to be negative. This is because the h^2 term dominates the rest of the series—unless $d^2f(x)/dx^2$ is itself zero, in which case we must go on and think about even higher-order differential quotients. This analysis, first given and demonstrated geometrically by Maclaurin, was worked out in full analytic detail by Euler [6, sections 253–254], [9, pp. 117–118]. It is typical of Euler's ability to choose computations that produce insight into fundamental concepts. It assumes, of course, that the function in question has a Taylor series, an assumption which Euler made without proof for many functions; it assumes also that the function is uniquely the sum of its Taylor series, which Euler took for granted. Nevertheless, this analysis is a beautiful example of the exploration and development of the concept of the differential quotient of first, second, and nth orders—a development which completely solves the problem of characterizing maxima and minima, a problem which goes back to the Greeks

Lagrange and the Derivative as a Function

Though Euler did a good job analyzing maxima and minima, he brought little further understanding of the nature of the differential quotient. The new importance given to Taylor series meant that one had to be concerned not only about first and second differential quotients, but about differential quotients of any order.

The first person to take these questions seriously was Lagrange. In the 1770s, Lagrange was impressed with what Euler had been able to achieve by Taylor-series manipulations with differential quotients, but Lagrange soon became concerned about the logical inadequacy of all the existing justifications for the calculus. In particular, Lagrange wrote in 1797 that the Newtonian limit-concept was not clear enough to be the foundation for a branch of mathematics. Moreover, in not allowing variables to surpass their limits, Lagrange thought the limit-concept too restrictive. Instead, he said, the calculus should be reduced to algebra, a subject whose foundations in the eighteenth century were generally thought to be sound [11, pp. 15–16].

The algebra Lagrange had in mind was what he called the algebra of infinite series, because Lagrange was convinced that infinite series were part of algebra. Just as arithmetic deals with infinite decimal fractions without ceasing to be arithmetic, Lagrange thought, so algebra deals with infinite algebraic expressions without ceasing to be algebra. Lagrange believed that expanding $f(x + h)$ into a power series in h was always an algebraic process. It is obviously algebraic when one turns $1/(1 - x)$ into a power series by dividing. And Euler had found, by manipulating formulas, infinite power-series expansions for functions like sin x, cos x, e^x. If functions like those have power-series expansions, perhaps everything could be reduced to algebra. Euler, in his book *Introduction to the Analysis of the Infinite (Introductio in analysin infinitorum*, 1748), had studied infinite series, infinite products, and infinite continued fractions by what he thought of as purely algebraic methods. For instance, he converted infinite series into infinite products by treating a series as a very long polynomial. Euler thought that this work was purely algebraic, and—what is crucial here—Lagrange also thought Euler's methods were purely algebraic. So Lagrange tried to make the calculus rigorous by reducing it to the algebra of infinite series.

Lagrange stated in 1797, and thought he had proved, that any function (that is, any analytic expression, finite or infinite) had a power-series expansion:

$$f(x + h) = f(x) + p(x)h + q(x)h^2 + r(x)h^3 + \dots, \qquad (5)$$

except, possibly, for a finite number of isolated values of x. He then defined a new function, the coefficient of the linear term in h which is $p(x)$ in the expansion shown in (5) and called it the *first derived function* of $f(x)$. Lagrange's term "derived function" (*fonction dérivée*) is the origin of our term "derivative." Lagrange introduced a new notation, $f'(x)$, for that function. He defined $f''(x)$ to be the first derived function of $f'(x)$, and so on, recursively. Finally, using these definitions, he proved that, in the expansion (5) above, $q(x) = f''(x)/2$, $r(x) = f'''(x)/6$, and so on [11, chapter 2].

What was new about Lagrange's definition? The concept of *function*—whether simply an algebraic expression (possibly infinite) or, more generally, any dependence relation—helps free the concept of derivative from the earlier ill-defined notions. Newton's explanation of a fluxion as a rate of change appeared to involve the concept of motion in mathematics; moreover, a fluxion seemed to be a different kind of object than the flowing

quantity whose fluxion it was. For Leibniz, the differential quotient had been the quotient of vanishingly small differences; the second differential quotient, of even smaller differences. Bishop Berkeley, in his attack on the calculus, had made fun of these earlier concepts, calling vanishing increments "ghosts of departed quantities" [2, section 35]. But since, for Lagrange, the derivative was a function, it was now the same sort of object as the original function. The second derivative is precisely the same sort of object as the first derivative; even the nth derivative is simply another function, defined as the coefficient of h in the Taylor series for $f^{(n-1)}(x + h)$. Lagrange's notation $f'(x)$ was designed precisely to make this point.

We cannot fully accept Lagrange's definition of the derivative, since it assumes that every differentiable function is the sum of a Taylor series and thus has infinitely many derivatives. Nevertheless, that definition led Lagrange to a number of important properties of the derivative. He used his definition together with Euler's criterion for using truncated power series in approximations to give a most useful characterization of the derivative of a function [9, p. 116, pp. 118–121]:

$$f(x + h) = f(x) + hf'(x) + hH, \text{ where } H \text{ goes to zero with } h.$$

(I call this the *Lagrange property of the derivative*.) Lagrange interpreted the phrase "H goes to zero with h" in terms of inequalities. That is, he wrote that,

> *Given D, h can be chosen so that $f(x + h)$ $- f(x)$ lies between $h(f'(x) - D)$ and (6) $h(f'(x) + D)$.*

Formula (6) is recognizably close to the modern delta-epsilon definition of the derivative.

Lagrange used inequality (6) to prove theorems. For instance, he proved that a function with positive derivative on an interval is increasing there, and used that theorem to derive the Lagrange remainder of the Taylor series (9, pp. 122–127], [11, pp. 78–85]. Furthermore, he said, considerations like inequality (6) are what make possible applications of the differential calculus to a whole range of problems in mechanics, in geometry, and, as we have described, the problem of maxima and minima (which Lagrange solved using the Taylor series remainder which bears his name [11, pp. 233–237]).

In Lagrange's 1797 work, then, the derivative is defined by its position in the Taylor series—a strange definition to us. But the derivative is also *described* as satisfying what we recognize as the appropriate delta-epsilon inequality, and Lagrange applied this inequality and its *n*th-order analogue, the Lagrange remainder, to solve problems about tangents, orders of contact between curves, and extrema. Here the derivative was clearly a function, rather than a ratio or a speed.

Still, it is a lot to assume that a function has a Taylor series if one wants to define only *one* derivative. Further, Lagrange was wrong about the algebra of infinite series. As Cauchy pointed out in 1821, the algebra of finite quantities cannot automatically be extended to infinite processes. And as Cauchy also pointed out, manipulating Taylor series is not foolproof. For instance, e^{-1/x^2} has a zero Taylor series about $x = 0$, but the function is not identically zero. For these reasons, Cauchy rejected Lagrange's definition of derivative and substituted his own.

Definitions, Rigor, and Proofs

Now we come to the last stage in our chronological list: *definition*. In 1823, Cauchy defined the derivative of $f(x)$ as the limit, when it exists, of the quotient of differences $(f(x + h) - f(x))/h$ as h goes to zero [4, pp. 22–23]. But Cauchy understood "limit" differently that had his predecessors. Cauchy entirely avoided the question of whether a variable ever reached its limit; he just didn't discuss it. Also, knowing an absolute value when he saw one, Cauchy followed Simon l'Huilier and S.-F. Lacroix in abandoning the restriction that variables never surpass their limits. Finally, though Cauchy, like

Newton and d'Alembert before him, gave his definition of limit in words, Cauchy's understanding of limit (most of the time, at least) was algebraic. By this, I mean that when Cauchy needed a limit property in a proof, he used the algebraic inequality-characterization of limit. Cauchy's proof of the mean value theorem for derivatives illustrates this. First he proved a theorem which states: *if $f(x)$ is continuous on $[x, x + a]$, then*

$$\min_{[x,\, x + a]} f'(x) \leqslant \frac{f(x + a) - f(x)}{a} \leqslant \max_{[x,\, x + a]} f'(x). \quad (7)$$

The first step in his proof is [4, p. 44]:

> Let δ, ε be two very small numbers; the first is chosen so that for all [absolute] values of h less than δ, and for any value of x [on the given interval], the ratio $(f(x + h) - f(x))/h$ will always be greater than $f'(x) - \varepsilon$ and less than $f'(x) + \varepsilon$.

(The notation in this quote is Cauchy's, except that I have substituted h for the i he used for the increment.) Assuming the intermediate-value theorem for continuous functions, which Cauchy had proved in 1821, the mean-value theorem is an easy corollary of (7) [4, pp. 44–45], [9, pp. 168–170].

Cauchy took the inequality-characterization of the derivative from Lagrange (possibly via an 1806 paper of A.-M. Ampère [9, pp. 127–132]). But Cauchy made that characterization into a definition of derivative. Cauchy also took from Lagrange the name derivative and the notation $f'(x)$, emphasizing the functional nature of the derivative. And, as I have shown in detail elsewhere [9, chapter 5], Cauchy adapted and improved Lagrange's inequality proof-methods to prove results like the mean-value theorem, proof-methods now justified by Cauchy's definition of derivative.

But of course, with the new and more rigorous definition, Cauchy went far beyond Lagrange. For instance, using his concept of limit to define the integral as the limit of sums, Cauchy made a good first approximation to a real proof of the Fundamental Theorem of Calculus [9, pp. 171–175], [4, pp. 122–125, 151–152]. And it was Cauchy who

P. Fermat

I. Newton

1637–38

1669

R. Descartes

1684

G. W. Leibniz

L. Euler

1755

A.-L. Cauchy

1797

1823

J.-L. Lagrange

Dates refer to these mathematicians' major works which contributed to the evolution of the concept of the derivative.

1861

K. Weierstrass

not only raised the question, but gave the first proof, of the existence of a solution to a differential equation [9, pp. 158–159].

After Cauchy, the calculus itself was viewed differently. It was seen as a rigorous subject, with good definitions and with theorems whose proofs were based on those definitions, rather than merely as a set of powerful methods. Not only did Cauchy's new rigor establish the earlier results on a firm foundation, but it also provided a framework for a wealth of new results, some of which could not even be formulated before Cauchy's work.

Of course, Cauchy did not himself solve all the problems occasioned by his work. In particular, Cauchy's definition of the derivative suffers from one deficiency of which he was unaware. Given an ε, he chose a δ which he assumed would for any x. That is, he assumed that the quotient of differences converged uniformly to its limit. It was not until the 1840s that G. G. Stokes, V. Seidel, K. Weierstrass, and Cauchy himself worked out the distinction between convergence and uniform convergence. After all, in order to make this distinction, one first needs a clear and algebraic understanding of what a limit is—the understanding Cauchy himself had provided.

In the 1850s, Karl Weierstrass began to lecture at the University of Berlin. In his lectures, Weierstrass made algebraic inequalities replace words in theorems in analysis, and used his own clear distinction between pointwise and uniform convergence along with Cauchy's delta-epsilon techniques to present a systematic and thoroughly rigorous treatment of the calculus. Though Weierstrass did not publish his lectures, his students—H. A. Schwartz, G. Mittag-Leffler, E. Heine, S. Pincherle, Sonya Kowalevsky, Georg Cantor, to name a few—disseminated Weierstrassian rigor to the mathematical centers of Europe. Thus although our modern delta-epsilon definition of derivative cannot be quoted from the *works* of Weierstrass, it is in fact the *work* of Weierstrass [3, pp. 284–287]. The rigorous understanding brought to the concept of the deriva-

tive by Weierstrass is signaled by his publication in 1872 of an example of an everywhere continuous, nowhere differentiable function. This is a far cry from merely acknowledging that derivatives might not always exist, and the example shows a complete mastery of the concepts of derivative, limit, and existence of limit [3, p. 285].

Historical Development versus Textbook Exposition

The span of time from Fermat to Weierstrass is over two hundred years. How did the concept of derivative develop? Fermat implicity used it; Newton and Liebniz discovered it; Taylor, Euler, Maclaurin developed it; Lagrange named and characterized it; and only at the end of this long period of development did Cauchy and Weierstrass define it. This is certainly a complete reversal of the usual order of textbook exposition in mathematics, where one starts with a definition, then explores some results, and only then suggests applications.

This point is important for the teacher of mathematics: the historical order of development of the derivative is the reverse of the usual order of textbook exposition. Knowing the history helps us as we teach about derivatives. We should put ourselves where mathematicians were before Fermat, and where our beginning students are now—back on the other side, before we had any concept of derivative, and also before we knew the many uses of derivatives. Seeing the historical origins of a concept helps motivate the concept, which we—along with Newton and Leibniz—want for the problems it helps to solve. Knowing the historical order also helps to motivate the rigorous definition—which we, like Cauchy and Weierstrass, want in order to justify the uses of the derivative, and to show precisely when derivatives exist and when they do not. We need to remember that the rigorous definition is often the end, rather than the beginning, of a subject.

The real historical development of mathematics—the order of discovery—reveals the creative mathematician at work, and it is creation that makes doing mathematics so exciting. The order of exposition, on the other hand, is what gives mathematics its characteristic logical structure and its incomparable deductive certainty. Unfortunately, once the classic exposition has been given, the order of discovery is often forgotten. The task of the historian is to recapture the order of discovery: not as we think it might have been, not as we think it should have been, but as it really was. And this is the purpose of the story we have just told of the derivative from Fermat to Weierstrass.

This article is based on a talk delivered at the Conference on the History of Modern Mathematics, Indiana Region of the Mathematical Association of America, Ball State University, April 1982; earlier versions were presented at the Southern California Section of the M. A. A. and at various mathematics colloquia. I thank the *Mathematics Magazine* referees for their helpful suggestions.

REFERENCES

[1] Margaret Baron, *Origins of the Infinitesimal Calculus*, Pergamon, Oxford, 1969.

[2] George Berkeley, *The Analyst, or a Discourse Addressed to an Infidel Mathematician*, 1734. In A. A. Luce and T. R. Jessop, eds., *The Works of George Berkeley*, Nelson, London, 1951 (some excerpts appear in [16, pp. 333–338]).

[3] Carl Boyer, *History of the Calculus and its Conceptual Development*, Dover, New York, 1959.

[4] A.-L. Cauchy, *Résumé des leçons données a l'école royale polytechnique sur le calcul infinitésimal*, Paris, 1823. In *Oeuvres complètes d'Augustin Cauchy*, Gauthier-Villars, Paris, 1882– , series 2, vol. 4.

[5] Pierre Dugac, *Fondements d'analyse*, in J. Dieudonné, *Abrégé d'histoire des mathématiques, 1700–1900*, 2 vols., Hermann, Paris, 1978.

[6] Leonhard Euler, *Institutiones calculi differentialis*, St. Peterburg, 1755. In *Operia omnia*, Teubner, Leipzig, Berlin, and Zurich, 1911– , series 1, vol. 10.

[7] Pierre Fermat, *Analysis ad refractiones*, 1661. In *Oeuvres de Fermat*, ed., C. Henry and P. Tannery, 4 vols., Paris, 1891–1912; Supplement, ed. C. de Waard, Paris, 1922, vol. 1, pp. 170–172.

[8] ——, *Methodus ad desquirendam maximam et minimam et de tangentibus linearum curvarum*, *Oeuvres*, vol. 1, pp. 133–136. Excerpted in English in [16, pp. 222–225].

[9] Judith V. Grabiner, *The Origins of Cauchy's Rigorous Calculus*, M. I. T. Press, Cambridge and London, 1981.

[10] Morris Kline, *Mathematical Thought from Ancient to Modern Times*, Oxford, New York, 1972.

[11] J.-L. Lagrange, *Théorie des fonctions analytiques*, Paris, 2nd edition, 1813. In *Oeuvres* de Lagrange, ed. M. Serret, Gauthier-Villars, Paris, 1867–1892, vol. 9.

[12] Michael S. Mahoney, *The Mathematical Career of Pierre de Fermat, 1601–1665*, Princeton University Press, Princeton, 1973.

[13] Isaac Newton, *Of Analysis by Equations of an Infinite Number of Terms* [1669], in D. T. Whiteside, ed., *Mathematical Works of Isaac Newton*, Johnson, New York and London, 1964, vol. 1, pp. 3–25.

[14] ——, *Method of Fluxions* [1671], D. T. Whiteside, ed., *Mathematical Works of Isaac Newton*, vol. 1, pp. 29–139.

[15] ——, *Mathematical Principles of Natural Philosophy*, tr. A. Motte, ed. F. Cajori, University of California Press, Berkeley, 1934.

[16] D. J. Struik, *Source Book in Mathematics, 1200–1800*, Harvard University Press, Cambridge, MA, 1969.

[17] D. T. Whiteside, ed., *The Mathematical Papers of Isaac Newton*, Cambridge University Press, 1967–1982.

HISTORICAL EXHIBIT 8.1

Synopsis of Major Geometries

Founder	Popular Name	Possible Model	Number of Lines Parallel to a Given Line	Sum of the Interior Angles of a Triangle
Euclid (ca. 300 B.C.)	Euclidean	A plane (a)	only one	180°
N. Lobachevsky (1830) J. Bolyai (1832)	Hyperbolic	Poincaré circle (b)	many	less that 180°
G. Riemann	Elliptic	Sphere (c)	none	greater than 180°

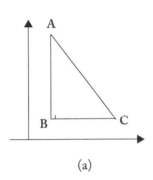

(a)

$\angle A + \angle B + \angle C = 180°$

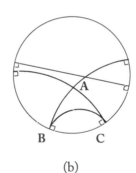

(b)

$\angle A + \angle B + \angle C < 180°$

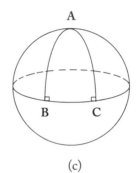

(c)

$\angle A + \angle B + \angle C > 180°$

Poincaré Circle

Given a circle, lines drawn within this circle have the following properties: any line passing through the center of the circle will continue to the circumference as a "straight" line; lines not through the center will be circular arcs orthogonal to the circumference.

93

The Men Responsible for the Development of Vectors

GEORGE J. PAWLIKOWSKI

*I*N ORDER to fully appreciate the study of vectors, one should have some idea of how they came about. This paper is devoted to the men responsible for the development of vectors. Since not all the men responsible for vector development could be discussed here, I have taken a few of the important men in the field and searched for information about them. The following is the result of my study.

Simon Stevin

Simon Stevin was born in Bruges, Belgium, in 1548. He was an engineer, physicist, and mathematician who combined practical sense, theoretical understanding, and originality. His physics led him to the discovery of the hydrostatic paradox. He introduced decimal fractions into arithmetic in *La disme* in 1585. This led to a great improvement in the system of measurement. His most famous work was *De Beghinselen der Weeghconst*. It was in this work, which came out in 1586, that he concerned himself with statics. Stevin investigated statics and determined accurately the force necessary to sustain a body on a plane inclined at any angle to the horizon. He was the first scientist since Archimedes to work with statics and vectors.

Reprinted from *Mathematics Teacher* 60 (Apr., 1967): 393–96; with permission of the National Council of Teachers of Mathematics.

Sir William R. Hamilton

William R. Hamilton was born in Dublin, Ireland, on August 4, 1805. At thirteen, he started to become interested in neutron readings. He studied about the calculus of directed line segments in space and the meaning assigned to their products. By doing this, Hamilton used vector analysis in his study of differential geometry. He studied at Trinity College in Dublin and, when he graduated in 1827, he was appointed as the professor of astronomy.

In 1835, he introduced couples of moments and he used couples of real numbers (*a*, *b*). In 1843, he had a thought that led to the birth of the quaternions. He announced this thought at a meeting of the Royal Irish Academy. Hamilton is remembered most for the quaternions. He developed the quaternions along the lines of quaternion differentials and the linear vector function. The importance of these quaternions is due to their extension of the concept of number. The geometric application of the quaternions is also important. Hamilton expected and anticipated the quaternions would be found useful in physics for giving a mathematical explanation of the universe. However, this was never realized. Hamilton claimed that negative numbers and imaginary numbers were to be treated in the science of order and not in the science of magnitude.

A paper which he presented in 1848 laid the groundwork for linear algebra. In this same paper he introduced *n*-tuples and defined noncom-mutative multiplication for some aspects of what we know today as vectors. In 1853, he published his lectures on the quaternions, and his *Elements of Quaternions* was published in 1866, just one year after his death. This publication dealt with the sum of a scalar and a vector and with linear vector functions. Much of Hamilton's work is still unpublished. He was a brilliant man, fluent in some twelve languages. He used Newton's equations to state his principle which, since it generalizes the principle of least time, is basic to dynamic theory. Hamilton also worked extensively with the theory of fifth-degree algebraic equations.

Hermann Günther Grassmann

Hermann Grassmann was born in Stettin on April 15, 1809. He made his home in Stettin and taught at the gymnasium. Grassmann started with an algebra of points and went on develop his system and theories on vector algebra and the theory of a line vector. In 1844, he wrote *Lineale Ausdehnungslehre* in Euclidean form. In this book, Grassmann developed and worked with a geometry of *n*-dimensions. He used invariant symbolism which we now recognize as a vector. Even though this was quite abstract in his time, it has become an important tool for the twentieth-century mathematician and scientist. He loved mathematics and did much research in it.

Georg Friedrich Bernhard Riemann

Bernhard Riemann was born in the village of Breselenz, Hanover, Germany, on September 17, 1826. He received his early education from his father, who was a Lutheran pastor. His father wanted him to enter the ministry, but Bernhard's love for mathematics won out. He was educated at Göttingen and Berlin under Gauss and Jacobi. He developed a form of non-Euclidean geometry based on a postulate which does not permit parallel lines. His geometry is too complicated to discuss here;

however, it may be said that the parallel displacement of a vector has some of the properties of parallelism that are familiar from elementary Euclidean geometry. Much of Riemann's geometry is concerned with vectors of this nature. His *Gesammelte Mathematische Werke* was posthumously published in 1876. Riemann is known for his theories on differential geometry, contributions to the theory of functions of a complex variable, and the Riemann definite integral—which is the basic form as presented in calculus today. Einstein used much of Riemann's geometry in his work on the theory of relativity. Riemann was a member of the Royal Society. He was a sick man most of his life. He died at an early age on July 20, 1866, of tuberculosis.

Elwin Bruno Christoffel

Elwin Christoffel was a German mathematician who was born in 1829 and died in 1900. He did quite a bit of work with tensor analysis, which is a generalization of vector analysis. In 1870, along with R. Lipschitz, he wrote two papers that dealt with Riemann's theory of quadratic differential forms. It was in these papers that Christoffel introduced the symbols bearing his name. These Christoffel symbols find use in tensor analysis in connection with curvatures of the coordinate system. This type of symbolism was just right for the transformation theory of quadratic differential forms.

Peter Guthrie Tait

Peter Tait was born in Dalkeith, Scotland, on April 28, 1831. He was a mathematician and a physicist who did much research on Sir William R. Hamilton's quaternions. His research material was published in *Elementary Treatise on Quaternions*. Tait also made use of vector analysis in differential geometry and the study of his golf-ball problem. However, his important publications were on the foundation of the kinetic theory of gases. Tait died in the year of 1901.

Josiah Willard Gibbs

J. Willard Gibbs was born in New Haven, Connecticut, on February 11, 1839. His father taught theology at the Yale Divinity School. Gibbs became a student at Yale and stayed there until he received his Ph.D. degree. He was a brilliant student, not only in mathematics, but in the humanities as well. When his studies at Yale terminated, he went abroad to study. In Europe he became acquainted with the world's leading mathematicians. In 1871, he returned to New Haven and started teaching mathematical physics at Yale.

Starting around 1880, he worked to develop geometric algebra into a system of vector algebra suitable for the needs of mathematical physicists. Soon after this date, he privately printed *Elements of Vector Analysis* for his pupils. He was always reluctant about publishing a book, but finally, in 1901, E. B. Wilson's *Vector Analysis* book was published. Dr. Wilson was a student of Gibbs and wrote the book using notes taken from Gibbs's lectures. Gibbs also did work with optics, statistical mechanics, and physical chemistry. He made and used diagrams and models to illustrate his theories. Gibbs adopted the best and simplest system for vectors from Hamilton's work on quaternions and Grassmann's work on geometric algebra. He used the above material in his lecture notes. *The Collected Works of J. Willard Gibbs* (1928) contains all of his published writings. He was a fellow of the Royal Society and a member of the National Academy of Sciences. He died in New Haven, April 28, 1903.

William Kingdon Clifford

William Clifford was born at Exeter, England, in 1845. He was a mathematician and a philosopher. Clifford believed that applied geometry is a part of physics and not a part of pure mathematics. His classification of loci, one year before his death, was an introduction to the study of n-dimensional space in a direction mainly projective. The methods employed in this study have been those of analytic, synthetic, and differential geometry, as well as vector analysis. He was a teacher at Trinity College in Cambridge and at University College in London. Clifford was the first Englishman who actually understood Riemann. With Riemann, he shared a deep interest in the origin of space conceptions. He also developed a geometry of motion for the study of which he generalized Hamilton's quaternions into what he called biquaternions. He wrote *The Common Sense of the Exact Sciences* and a paper titled "On the space theory of matter." This paper came out forty years before Einstein announced his theory of gravitation. Clifford died of tuberculosis in Madena in 1879.

Oliver Heaviside

Oliver Heaviside was born in Camden Town, England. He was a physicist and an electrical engineer who did much with long-distance telephone communication. There are many mathematicians who opposed Heaviside's work because of his methods. His theory was excellent, but he lacked a rigorous mathematical foundation for his work. His *Electrical Papers*, in 1892, and his *Electromagnetic Theory* (1893–1912) are important in the development of the electric theory. Heaviside worked with vectors in England about the same time that Gibbs worked with them in the United States. His work on vectors was a combination of Hamilton's and Tait's, which he altered and adapted to meet his own requirements. As a result of the above, he had a type of vector algebra much like that of Gibbs. Oliver Heaviside was a fellow of the Royal Society. He died in 1925.

Gregorio Ricci-Curbastro

Gregorio Ricci-Curbastro was an Italian mathematician who was born in 1853 and died in 1925. He read the papers that were written by E. Christoffel and R. Lipschitz in 1870 on Riemann's theory of quadratic differential forms.

These papers led Gregorio Ricci-Curbastro, after much study, to his absolute differential calculus. The symbols introduced in the papers by Christoffel and Lipschitz fitted nicely into Ricci-Curbastro's calculus. He helped in the development of the algebra and the calculus of tensors. The above-mentioned tensor analysis is a generalization of vector analysis.

Tullio Levi-Civita

Tullio Levi-Civita was born in 1873 and died in 1941. He studied under Ricci. Along with Ricci and fellow students, he helped to develop absolute differential calculus into the theory of tensors. Tensors provided a unification of many invariant symbolisms. Tullio Levi-Civita and Gergorio Ricci-Curbastro published a systematic account of the research of Ricci in 1901. Levi-Civita was a professor of mathematics at the University of Rome.

BIBLIOGRAPHY

Bezuszka, Stanley J. *Co-op Unit Study Program, Course I.* Boston: College Mathematics Institute Press, 1960.

Cajori, F. *A History of Mathematics.* New York: Macmillan Co., 1919.

Collier's Encyclopedia. New York: Crowell-Collier Publishing Co., 1962.

Newman, James R. (ed.). *The Harper Encyclopedia of Science.* New York: Harper & Row, 1963.

——— (ed.). *The World of Mathematics.* New York: Simon & Schuster, 1956.

Smith, David E. *A Source Book in Mathematics.* New York: McGraw-Hill Book Co., 1929.

Struik, Dirk J. *A Concise History of Mathematics.* New York: Dover Publications, 1948.

624

The Noncommutative Algebra of William Rowan Hamilton

By the early nineteenth century, the study of algebra had evolved from the mere manipulation of symbols to a more abstract investigation of the laws of mathematical operations and how they combined objects. William Rowan Hamilton (1805–1865) was a professor of mathematics at Trinity College, Dublin. In 1833, he devised an algebra for working with number couples. This algebra could be readily applied to complex numbers where a number of the form $a + bi$ could be represented by the ordered couple (a, b). Hamilton sought to extend his algebra further to the consideration of number triples; however, he ran into difficulty. In 1843, he finally extended his theory, but with number quadruples not triples. He called his groupings of four numbers "quaternions." The algebra of quaternions was most unusual—it was not commutative! "Multiplication" with Hamilton's system can be summarized as follows:

Let: $[1, 0, 0, 0]$ be represented by 1,
 $[0, 1, 0, 0]$ by i,
 $[0, 0, 1, 0]$ by j, and
 $[0, 0, 0, 1]$ by k; then the properties of
"multiplication" are illustrated in the Cayley Table below:

*	1	i	j	k
1	1	i	j	k
i	i	-1	k	$-j$
j	j	$-k$	-1	i
k	k	j	$-i$	-1

Thus, the elements heading each column are combined with the elements leading each row to produce the result in the intersection of the column and row, i.e., $j * i = k$, etc.

94

Matrix Theory I: Arthur Cayley—
Founder of Matrix Theory

RICHARD W. FELDMANN, JR.

ALTHOUGH THE TERM "matrix" was introduced into mathematical literature by James Joseph Sylvester[1] in 1850, the credit for founding the theory of matrices must be given to Arthur Cayley, since he published the first expository articles on the subject.

Arthur Cayley[2] was born in Richmond, England, on August 16, 1821. At the age of fourteen he entered King's College in London, and in 1838 he went to Trinity College, from which he graduated with highest honors. He left teaching a few years later, since a permanent position required the taking of religious vows. After practicing law to the extent that it supported him but did not interfere with his mathematical studies, he accepted the newly formed Sadlerian Professorship of Pure Mathematics at Cambridge in 1863. He remained in the capacity until his death on Jaunary 26, 1895, except for the first half of 1882, when he lectured at Johns Hopkins University at the request of Sylvester. Besides his work in matrix theory, Cayley also published copiously in the fields of analytic geometry of n dimensions, determinant theory, linear transformations, and skew surfaces. Arthur Cayley was a voluminous writer, whose collected papers fill twelve large volumes. His most productive period was between 1863 and 1883, when he published about 430 papers.

Reprinted from *Mathematics Teacher* 57 (Oct., 1962): 482–84; with permission of the National Council of Teachers of Mathematics.

Cayley's introductory paper in matrix theory was written in French and published in a German periodical.[3] In this paper, matrices are introduced to simplify the notation which arises in simultaneous linear equations. The set of equations

$$
\begin{aligned}
\xi &= \alpha x + \beta y + \gamma z + \cdots \\
\eta &= \alpha' x + \beta' y + \gamma' z + \cdots \\
\zeta &= \alpha'' x + \beta'' y + \gamma'' z + \cdots
\end{aligned}
$$

is written as

$$
(\xi, \eta, \zeta, \cdots)
= \begin{pmatrix} \alpha, & \beta, & \gamma, & \cdots \\ \alpha', & \beta', & \gamma' z, & \cdots \\ \alpha'', & \beta'', & \gamma'' z, & \cdots \end{pmatrix} (x, y, z, \cdots).
$$

The same article also introduces, although quite sketchily, the ideas of inverse matrix and of matrix multiplication, or "compounding" as Cayley called it.

The above basic properties are expanded in a second expository article[4] which also lists many additional properties of matrices. In this important paper, Cayley works mostly with square matrices with nine elements. He represents the zero matrix,

$$\begin{pmatrix} 0, & 0, & 0 \\ 0, & 0, & 0 \\ 0, & 0, & 0 \end{pmatrix},$$

by "0" and the "matrix unity,"

$$\begin{pmatrix} 1, & 0, & 0 \\ 0, & 1, & 0 \\ 0, & 0, & 1 \end{pmatrix},$$

by "1."

In introducing the algebra of matrices, Cayley defines the addition of two matrices by

$$\begin{pmatrix} a, & b, & c \\ a', & b', & c' \\ a'', & b'', & c'' \end{pmatrix} + \begin{pmatrix} \alpha, & \beta, & \gamma \\ \alpha', & \beta', & \gamma' \\ \alpha'', & \beta'', & \gamma'' \end{pmatrix}$$
$$= \begin{pmatrix} a + \alpha, & b + \beta, & c + \gamma \\ a' + \alpha', & b' + \beta', & c' + \gamma' \\ a'', + \alpha'', & b'' + \beta'', & c'' + \gamma'' \end{pmatrix}$$

and states, without proof, that matrices are commutative[5] and associative under addition. Two types of multiplication are exhibited. The first is designated as "scalar multiplication," that is

$$m \begin{pmatrix} a, & b, & c \\ a', & b', & c' \\ a'', & b'', & c'' \end{pmatrix} = \begin{pmatrix} ma, & mb, & mc \\ ma', & mb', & mc' \\ ma'', & mb'', & mc'' \end{pmatrix}.$$

The second type of multiplication is called "compounding," according to the following scheme:

$$\begin{pmatrix} a, & b, & c \\ a', & b', & c' \\ a'', & b'', & c'' \end{pmatrix} \begin{pmatrix} \alpha, & \beta, & \gamma \\ \alpha', & \beta', & \gamma' \\ \alpha'', & \beta'', & \gamma'' \end{pmatrix}$$

$$= \begin{pmatrix} (a, b, c \big\backslash \alpha, \alpha', \alpha''), & (a, b, c \big\backslash \beta, \beta', \beta''), & (a, b, c \big\backslash \gamma, \gamma', \gamma'') \\ (a', b', c' \big\backslash \alpha, \alpha', \alpha''), & (a', b', c' \big\backslash \beta, \beta', \beta''), & (a', b', c' \big\backslash \gamma, \gamma', \gamma'') \\ (a'', b'', c'' \big\backslash \alpha, \alpha', \alpha''), & (a'', b'', c'' \big\backslash \beta, \beta', \beta''), & (a'', b'', c'' \big\backslash \gamma, \gamma', \gamma'') \end{pmatrix},$$

where $(X, Y, z \big\backslash A, A', A'')$ is a symbol representing $XA + YA' + ZA''$. This form of multiplication is said to be associative, but in general, not commutative.

The nth power of a matrix M is defined as a matrix "compounded" by itself $n-1$ times. It is also noted that $M^n \cdot M^p = M^{n+p}$. Commutative and anticommutative[6] matrices under "compounding" are introduced, using the names of "convertible" and "skew-convertible" matrices.

In Section 17 of this same *Memoir*, Cayley exhibits the inverse of

$$\begin{pmatrix} a, & b, & c \\ a', & b', & c' \\ a'', & b'', & c'' \end{pmatrix}$$

in the form

$$\frac{1}{\nabla} \begin{pmatrix} \partial_a \nabla, & \partial_{a'} \nabla, & \partial_{a''} \nabla \\ \partial_b \nabla, & \partial_{b'} \nabla, & \partial_{b''} \nabla \\ \partial_c \nabla, & \partial_{c'} \nabla, & \partial_{c''} \nabla \end{pmatrix},$$

where ∇ is the determinant of the matrix and $\partial_x \nabla$ is the determinant obtained from ∇ by replacing the element x in ∇ by 1 and all other elements in the row and column containing x by zeros, which, in effect, makes $\partial_x \nabla$ the cofactor of x. When ∇ is zero, the matrix is called "indeterminate"[7] and has no inverse.

Cayley states, "It may be added that the matrix zero is indeterminate; and that the product of two matrices may be zero without either of the factors being zero, if only the matrices are one or both of them indeterminate." This statement is erroneous in that both matrices must be indeterminate.[8]

The transposed matrix is defined by the equation

$$\text{tr} \begin{pmatrix} a, & b \\ c, & d \end{pmatrix} = \begin{pmatrix} a, & c \\ b, & d \end{pmatrix}.$$

Cayley further discusses this matrix by stating that

$$\operatorname{tr}(LMN) = (\operatorname{tr} N)(\operatorname{tr} M)(\operatorname{tr} L).$$

Symmetric and skew-symmetric matrices are defined by $\operatorname{tr} M = M$ and $\operatorname{tr} M = -M$ respectively.

The remainder of the paper discusses the Cayley-Hamilton theorem[9] and methods for finding roots and powers of a matrix. These methods are so vague and cumbersome that no further mention of them will be made.

The final statement in the article refers to rectangular matrices. Cayley notes here that an n by m matrix[10] can be added only to an n by m matrix, and that an n by m matrix can be compounded only by an m by p matrix. He concludes with the observation that the transpose of an n by m matrix is an m by n matrix.

Working with the transformation $\phi(x) = (ax + b)/(cx + d)$ in 1880,[11] Cayley showed that $\phi^n(x)$ has for coefficients the elements of

$$\begin{pmatrix} a, & b \\ c, & d \end{pmatrix}^n.$$

He proved that

$$\begin{vmatrix} a, & b \\ c, & d \end{vmatrix}^n = \left(\frac{1}{\lambda^2 - 1}\right)\left(\frac{a+d}{\lambda+1}\right)^{n-1}$$

$$\left\{(\lambda^{n-1} - 1)\begin{pmatrix} a, & b \\ c, & d \end{pmatrix} + (\lambda^n - \lambda)\begin{pmatrix} -d, & b \\ c, & -a \end{pmatrix}\right\},$$

where λ is determined from

$$(\lambda + 1)^2 / \lambda = (a + d)^2/(ad - bc).$$

An article by A. Buchheim[12] relates that Cayley developed notation to differentiate between right-hand and left-hand division; ab^{-1} and $b^{-1}a$ were represented by

respectively. This notation does not appear in Cayley's published works, and probably comes from his correspondence.

NOTES

1. "Additions to the Articles in the September Number of This Journal 'On a New Class of Theorems . . .' and 'On Pascal's Theorem'," *Philosophical Magazine*, series three, XXXVII (1850), 363–70. Sylvester used the term to represent a rectangular portion of a determinant.

2. More biographical details on Arthur Cayley's life can be found in E. T. Bell, *Men of Mathematics* (New York: 1937); Alexander MacFarlane, *Lectures on Ten British Mathematicians* (New York: 1916); G. B. Halsted, "Arthur Cayley," *American Mathematical Monthly*, II (1895), 102–06.

3. "Remarques sur la notation des fonctions algebraiques," Crelle's *Journal für reine und angewandte Mathematik*, L (1855), 282–85.

4. "Memoir on the theory of matrices," *Philosophical Transactions of the Royal Society of London*, CXLVIII (1858), 17–37.

5. Cayley used the term "convertible."

6. $MN = -NM$.

7. We now use the term *singular*.

8. Consider $AB = 0$, A indeterminate and B not indeterminate. Then $ABB^{-1} = 0$, and $A = 0$.

9. This will be discussed in article III of this series, "The Characteristic Equation."

10. An n by m matrix has n rows and m columns.

11. "On the Matrix $\begin{pmatrix} a, & b \\ c, & d \end{pmatrix}$ and in Connexion Therewith the Function as $(ax + b)/(cx + d)$," Messenger of Mathematics, IX (1880), 104–09.

12. A. Buchheim, "On the Theory of Matrices," *Proceedings of the London Mathematical Society*, XVI (1885), 63–82.

95

Matrix Theory II:
Basic Properties

RICHARD W. FELDMANN, JR.

$\mathcal{A}$FTER CAYLEY PUBLISHED his expository articles on the theory of matrices, many mathematicians published papers expanding Cayley's ideas and developing new concepts, notations, and terminology. Some of these were fruitful and are still used, while others never turned up in print again.

Division

Georg Frobenius[1] proposed that the quotient A/B be used to represent multiplication of A by B^{-1} when $AB^{-1} = B^{-1}A$. He then showed that $(A/B)^{-1}$ is B/A. Since matrix multiplication is not, in general, commutative, a second German mathematician, Kurt Hensel[2], suggested $A\backslash B$ and B/A to represent $A^{-1}B$ and BA^{-1}, respectively. However, both notations were soon dropped and the word *division* dropped from usage in favor of *multiplication by an inverse.*

Absolute Value

The concept of absolute value[3] of a matrix is still seen in modern mathematical literature, but quite infrequently. It was first proposed in 1925 by J. H. M. Wedderburn, who defined the absolute value of a matrix $A = (a_{ij})$, represented by $\lfloor A \rfloor$, to be the square root of

Reprinted from *Mathematics Teacher* 57 (Nov., 1962): 589–90; with permission of the National Council of Teachers of Mathematics.

$$\sum_{i,j=1}^{n} a_{ij}\bar{a}_{ij}$$

where $\bar{a}_{ij}$ is the complex conjugate of a_{ij}. Wedderburn developed some theorems on absolute values. He proved that

$$\lfloor A + B \rfloor \leq \lfloor A \rfloor + \lfloor B \rfloor \text{ and } \lfloor AB \rfloor \leq \lfloor A \rfloor \lfloor B \rfloor.$$

Representing a scalar by λ, he showed

$$\lfloor \lambda A \rfloor = |\lambda| \lfloor A \rfloor \text{ and } \lfloor \lambda I \rfloor = n^{1/2}|\lambda|.$$

The author stated that this concept was foreshadowed in 1909 by I. Schur[4] when he observed that $\lfloor A \rfloor^2$ is the trace of $A\bar{A}^T$, where $\bar{A}^T$ is the transpose of the matrix formed by replacing each element in A by its complex conjugate.

Check for Matrix Multiplication

A fairly simple check for accuracy in multiplying two square matrices, $A = (a_{ij})$ and $B = (b_{ij})$, together was published in 1924 by W. E. Roth.[5] Setting

$$a_i = \sum_{j=1}^{n} a_{ij}, \qquad \alpha_j = \sum_{i=1}^{n} a_{ij}$$

$$b_i = \sum_{j=1}^{n} b_{ij}, \qquad \beta_j = \sum_{i=1}^{n} b_{ij}$$

the sum of the elements in AB is

$$\sum_{k=1}^{n} \alpha_k b_k$$

629

and the sum of the elements in BA is

$$\sum_{k=1}^{n} a_k \beta_k.$$

This result is extended to show that the sum of the elements in A^r is equal to

$$\sum_{i=1}^{n} \sum_{j=1}^{n} a_{ij}^{r-2} \alpha_i a_j.$$

$$| M - rI | = (-1)^n r^n + a_1 r^{n-1} + a_2 r^{n-2} + \cdots + a_n$$

and

$$| A - rI | = (-1)^n r^n + b_1 r^{n-1} + b_2 r^{n-2} + \cdots + b_n,$$

then $\lim a_i = b_i$ for $i = 1, 2, \cdots, n$.

Terminology

As the number of published papers increased, James Joseph Sylvester (1814–1897) introduced a terminology[6] which he hoped would eliminate much of the verbiage necessary when discussing the position of an element in a matrix. He proposed in 1884 that the element in the mth row and nth column be referred to as the element of "latitude" m and "longitude" n.

Limit of a Matrix

Refining some ideas of Giuseppe Peano (1888) and G. A. Bliss (1905), an article[7] by H. B. Phillips in the *American Journal of Mathematics* defined the limit of an n by n matrix $M = (m_{ij})$, whose elements are functions, to be the matrix $A = (a_{ij})$, where $\lim m_{ij} = a_{ij}$. The coefficients in the characteristic function of M were shown to approach, in the limit, the corresponding coefficients in the characteristic function of A. Symbolically, if

REFERENCES

1. "Über lineare Substitutionen und bilineare Formen," Crelle's *Journal für reine und angewandte Mathematik*, LXXXIV (1878), 1–63.

2. "Über den Zusammenhang zwischen den Systemen und ihren Determinanten," Crelle's *Journal für reine und angewandte Mathematik*, CLIX (1928), 246-54.

3. "The Absolute Value of the Product of Two Matrices," *Bulletin of the American Mathematical Society*, XXXI (1925), 304-8.

4. "Über die charakteristischen Wurzeln einer linear Substitution mit einer Anwendung auf die Theorie der Integralgleichungen," *Mathematische Annalen*, LXVI (1909), 488–510.

5. "A Convenient Check on the Accuracy of the Product of Two Matrices," *American Mathematical Monthly*, XXXVI (1929), 37–38.

6. "Lectures on the Principles of Universal Algebra," *American Journal of Mathematics*, VI (1884), 270–86.

7. "Functions of Matrices," *American Journal of Mathematics*, XLI (1919), 266–78.

96

Matrix Theory III:
The Characteristic Equation;
Minimal Polynomials

RICHARD W. FELDMANN, JR.

$\mathcal{T}$HE CHARACTERISTIC EQUATION of a matrix M is found by expanding the determinant of $M - xI$, where I is the identity matrix, and setting it equal to zero. The roots of this equation are the characteristic roots of M.

The first appearance of the characteristic equation in matrix theory was in 1858 in Cayley's basic paper.[1] He does not refer to the equation by this name and used it to prove the theorem now known as the Cayley-Hamilton theorem. However, the terms "characteristic equation" and "characteristic root" are borrowed from papers on determinant theory. The terms "latent equation" and "latent root" were introduced into the literature by J. J. Sylvester. This equation was once called, but now very infrequently, the "secular equation," in honor of Laplace and his research on the secular inequalities of the planets.[2]

A relationship between the coefficients of the characteristic equation and the sums of the determinants of the various-sized principal minors of the matrix was demonstrated by William H. Metzler[3] in 1892. He proved that if the equation is written

$$x^n - m_1 x^{n-1} + m_2 x^{n-2} - \cdots \pm m_{n-1} x \mp m_n = 0,$$

then the determinant of the matrix is m_n, and m_k is the sum of the principal k-rowed minors of the determinant of the matrix.

Thirty-six years later, a German mathematician, Kurt Hensel,[4] published a proof that a nonzero constant term is required for the existence of the inverse matrix, but this is subsumed under Metzler's findings of 1892.

A special case was treated by Georg Frobenius,[5] when he proved that the characteristic function of a nilpotent matrix, i.e., a matrix some positive integral power of which is the zero matrix, is of the form x^r.

Certain special types of matrices were found to have characteristic roots which had predictable properties. Charles Hermite[6] showed that the roots of the characteristic equation of a Hermitian matrix[7] are all real. In 1885, A. Buchheim[8] proved that the characteristic roots of a symmetric[9] matrix with real elements are all real. The fact that the characteristic roots of a real skew-symmetric matrix are pure imaginary was found by Karl Weierstrass[10] in 1870.

A related result, also subsumed under Metzler's findings of 1892, was proven by Henry Taber[11]; he showed that the sum of the characteristic roots is equal to the sum of the diagonal elements (the so-called *trace* or *spur*) of the matrix.

The Cayley-Hamilton theorem states that a matrix satisfies its own characteristic equation; i.e.,

Reprinted from *Mathematics Teacher* 57 (Dec., 1962): 657–59; with permission of the National Council of Teachers of Mathematics.

if $f(x) = 0$ is the characteristic equation of a matrix M, then $f(M) = 0$.

This result was first published by Sir Arthur Cayley in 1858[12]. His original statement of the theorem was, "The determinant, having for its matrix a given matrix less the same matrix considered as a single quantity[13] involving the matrix unity, is equal to zero." Cayley proved this theorem only for second and third order matrices, stating that he felt that further proof was unnecessary. He published the proof for the two-by-two case, which in essence stated

$$\begin{vmatrix} a - \begin{pmatrix} a\,b \\ c\,d \end{pmatrix} & b \\ c & d - \begin{pmatrix} a\,b \\ c\,d \end{pmatrix} \end{vmatrix}$$

$$= \begin{pmatrix} a\,b \\ c\,d \end{pmatrix}^2 - (a+d) \begin{pmatrix} a\,b \\ c\,d \end{pmatrix}$$

$$+ (ad - bc) \begin{pmatrix} a\,b \\ c\,d \end{pmatrix}^0$$

$$= \begin{pmatrix} a^2 + bc & ab + bd \\ ac + cd & d^2 + bc \end{pmatrix}$$

$$- (a+d) \begin{pmatrix} a\,b \\ c\,d \end{pmatrix} + (ad - bc) \begin{pmatrix} 1\,0 \\ 0\,1 \end{pmatrix}$$

$$= \begin{pmatrix} 0\,0 \\ 0\,0 \end{pmatrix}.$$

To reduce the theorem to a compact form, Cayley symbolized a matrix A, considered as a single quantity, by the notation $\bar{A}$, and wrote the theorem

$$\text{Det}\,(\bar{1} \cdot M - \bar{M} \cdot 1) = 0.$$

The first general proof was given by A. Buchheim[14] in the 1884 volume of the *Messenger of Mathematics*. Sixty-three pages later in the same journal, A. R. Forsyth published a proof of the theorem for the three-by-three case.[15] Hamilton's

name is connected with the Cayley-Hamilton theorem because he established the theorem for quaternions in 1853.[16]

Closely related to the characteristic polynomial of a matrix is the minimal polynomial of the matrix. The concept was put forth by Georg Frobenius[17] in 1878. His definition calls the minimal polynomial of a matrix the polynomial of least degree which the matrix satisfies. He stated that it is formed from the factors of the characteristic polynomial and that it is unique.

Seven years later, a pair of articles[18] in the *Comptes Rendus Hebdomadaires des Séances des Académie des Sciences* by Edouard Weyr discussed the problems encountered in trying to find a minimal polynomial. However, the author was unable to shed any light on a method of finding the polynomial. He coined the word "dérogatoire" to describe a matrix whose minimal polynomial has a degree which is less than the degree of the characteristic polynomial.

The first American article on the subject[19] stated that, in general, the degree of the minimal polynomial is not less than the degree of the characteristic polynomial. The author, William H. Metzler, wrote that when this happens the characteristic polynomial is said to "degrade" to a polynomial of lower degree.

In 1904, the German mathematician Kurt Hensel[20] proved that this polynomial is unique and that it divides any other polynomial which the matrix satisfies.

NOTES

1. Sir Arthur Cayley, "A Memoir on the Theory of Matrices," *London Philosophical Transactions*, CXLVIII (1858), 17–37. Also *Collected Works*, Vol. II, 475–96.

2. Mentioned by E. T. Browne in "On the Separation Property of the Roots of the Secular Equation," *American Journal of Mathematics*, LII (1930), 843–50.

3. W. H. Metzler, "On the Roots of Matrices," *American Journal of Mathematics*, XIV (1892), 326–77.

4. Kurt Hensel, "Über den Zusammenhang zwischen den Systemen und ihren Determinenten," Crelle's *Journal für reine und angewandte Mathematik,* CLIX (1928), 246–54.

5. Georg Frobenius, "Über lineare Substitutionen und bilineare Formen," Crelle's *Journal für reine und angewandte Mathematik,* LXXXIV (1878), 1–63.

6. Charles Hermite, "Remarque sur un théorème de M. Cauchy," *Comptes Rendu Hebdomadaires des Séances des Académie des Sciences,* XLI (1855), 181.

7. In a Hermitian matrix, the element a_{ij} is the complex conjugate of a_{ji}.

8. A. Buchheim, "On a Theorem Relating to Symmetrical Determinants," *Messenger of Mathematics,* XIV (1885), 143–44.

9. In a symmetric matrix, the elements a_{ij} and a_{ji} are equal.

10. This fact is taken from T. J. I'A. Bromwich, "On the Roots of the Characteristic Equation of a Linear Substitution," *Acta Mathematica,* XXX (1906), 297–304. The actual article by Weierstrass is referred to only as being on page 134 of Vol. III of his *Gesammte Werke.*

11. Henry Taber, "On the Application to Matrices of Any Order of the Quaternion Symbols S and V," *Proceedings of the London Mathematical Society,* XXII (1891), 67–79.

12. This result is found in Sec. 21 of the article mentioned in Note 1.

13. That is, considered as a unit.

14. A. Buchheim, "Mathematical Notes," *Messenger of Mathematics,* XIII (1884), 62–66.

15. A. R. Forsyth, "Proof of a Theorem by Cayley in Regard to Matrices," *Messenger of Mathematics,* XIII (1884), 139–42.

16. W. R. Hamilton, *Lectures on Quaternions* (Dublin: 1853), p. 566.

17. "Über lineare Substitutionen und belineare Formen," Crelle's *Journal für reine und angewandte Mathematik,* LXXXIV (1878), 1–63.

18. "Sur la théorie des matrices," *Comptes Rendus,* C (1885), 787–89. "Repartition des matrices en espèces et formation de toutes les espèces," *Comptes Rendus,* C (1885), 966–69.

19. "On the Roots of Matrices," *American Journal of Mathematics,* XIV (1892), 326–77.

20. "Theorie der Korper von Matrizen," *Crelle's Journal für reine und angewandte Mathematik,* CXXVII (1904), 116–66.

97

A History of Extraneous Solutions

KENNETH R. MANNING

$\mathcal{T}$HE ASSERTION THAT before the rise of abstract algebra in the mid-nineteenth century, algebra was concerned more with the use of algorithms and rules than with investigations of logical foundations for the subject could well be tested by a detailed study of the treatment of extraneous solutions throughout that period. A treatment of extraneous solutions at a particular time in history would, because of the subtle mathematical principles involved in that treatment, reflect on the state of algebra at least as much as, and perhaps more than, any one topic in mathematics. However, the main purpose of this paper is to reveal the history of extraneous solutions, though reflections on the state of algebra seen from this historical account will be mentioned whenever appropriate.

Background

An *extraneous solution*[1] of an equation is a value for the unknown *x*, which, though introduced in the solution process and satisfying the derived equations, does not satisfy the original equation. The two classes of equations which frequently lead to extraneous solutions are *fractional equations* and *radical equations*.

A *fractional equation* is an equation that contains the unknown variable in the denominator of a fraction. In modern language, a fractional equa-

Reprinted from *Mathematics Teacher* 63 (Feb., 1970): 165–75; with permission of the National Council of Teachers of Mathematics.

tion is a *rational function*, the quotient of two relatively prime polynomials (where the degree of the denominator is greater than zero), set equal to zero. The equation

$$\sum_{i=0}^{n} P_i(x)/Q_i(x) = 0,$$

where each

$$P_i(x) = \sum_{j=0}^{r} \alpha_j x^j,$$

each

$$Q_i(x) = \sum_{j=0}^{t} \beta_j x^j,$$

and degree of

$$(Q_i(x)) \geq 1$$

for at least one *i*, is a fractional equation. A *radical equation* is an equation that contains the unknown variable under a radical sign. In modern language, a radical equation is an *irrational function*—excluding the set of transcendental functions—set equal to zero. The equation,

$$\sum_{i=0}^{n} (j\sqrt{P_i(x)})_i = 0,$$

is a radical equation. Neither fractional equations nor radical equations—unlike polynomial equations—have a degree. Consequently, as it is, the

Fundamental Theorem of Algebra has no governance over the number of roots of such equations.

Correct Methods of Solving Equations

To obtain a true solution of an equation it is sufficient, though never necessary, that there be equivalent equations[2] throughout the solution process, from the original equation with the unknown variable usually in the most complicated form to the final equation usually giving the unknown variable only in terms of known constants. In other words, a true solution may be obtained even if nonequivalent equations do appear in the solution process. However, generally, to obtain precisely *all* the roots of an equation—neither introducing any extraneous solutions nor losing any true solutions—each equation, starting with the original down through the final, must undergo operations that lead only to equivalent equations. Thus a mapping—an operation performed on a particular equation—of one equation into another must be *bijective*.

The two fundamental operations of algebra over the set of real numbers are addition and multiplication. Clearly, any mapping defined as the addition of real numbers is bijective. Therefore, any real number may be added to both members of any equation to produce an equivalent equation. On the other hand, a mapping defined as the multiplication of real numbers is *not* bijective since multiplication by the real number zero has no inverse operation. Therefore, neither multiplying by zero nor attempting to define, and later multiply by, the multiplicative inverse of zero, necessarily produces an equivalent equation. The root, $x = 0$, of a fractional equation is frequently lost when attempting to define, and multiply by, the multiplicative inverse of zero.

Another algebraic operation over the real numbers which is frequently necessary to solve radical equations is raising both members of an equation to an integral power. A mapping defined by raising any real number to an integral power is *not* necessarily bijective, since a real number and its additive inverse can be mapped to the same number, thus forsaking the necessity of injectivity. Hence, a mapping defined by raising to integral powers does not in general produce equivalent equations. Therefore, extraneous solutions can be introduced by such a mapping.

Historical Account of Fractional Equations

There is little or no treatment of fractional equations in the early eighteenth-century algebraic treatises. Neither McLaurin, in *A Treatise of Algebra*, nor Euler, in his *Elements of Algebra*, chose to include fractional equations as either illustrative examples or as problems at the end of the chapter. Among the early eighteenth-century mathematicians who published algebraic treatises, Nicolas Saunderson (1682–1739) was, to the author's knowledge, the only mathematician to incorporate fractional equations into his algebraic treatise, *Elements of Algebra*. Saunderson not only included fractional equations into his algebraic treatise, but he also attempted to solve them. However, from Saunderson's method of solution, one gains no precise knowledge of the extent of Saunderson's awareness of extraneous solutions, since the particular fractional equations chosen do not lead to extraneous solutions when solved by almost any method.

Saunderson takes the equation

$$\frac{42x}{x-2} = \frac{35x}{x-3},$$

and solves as follows:

$$\frac{42x}{x-2} = \frac{35x}{x-3}$$

$$\therefore \quad \frac{42}{x-2} = \frac{35}{x-3}$$

$$\therefore \ 42(x - 3) = 35(x - 2)$$

$$\therefore \ 42x - 126 = 35x - 70$$

$$\therefore \ x = 8.$$

Then, as *part* of the solution process he proceeds in converse to prove equivalence—a concept noticeably absent from other contemporary treatises on algebra—in the following way:

$$x = 8$$

$$\therefore \ 7x = 56$$

$$\therefore \ 42x - 35x = 126 - 70$$

$$\therefore \ 42(x - 3) = 35(x - 2)$$

$$\therefore \ \frac{42}{x - 2} = \frac{35}{x - 3}$$

$$\therefore \ \frac{42x}{x - 2} = \frac{35x}{x - 3}$$

Saunderson concludes that, therefore, $x = 8$ is the root of the equation.[3]

Obviously, $x = 8$ is a root, but equally as obvious, $x = 0$ is also a root. Still more important, Saunderson's solution process failed to yield the root, $x = 0$, as a solution to the original equation. Saunderson attempted to define and multiply by the multiplicative inverse of $x = 0$. Though there were no extraneous solutions introduced in Saunderson's solution process, there was a true solution that was actually lost. Losing true solutions is much more drastic than introducing extraneous solutions, since a root once lost can never be found, whereas an extraneous solution can be rejected on substitution into the original equation.

The loss of true solutions and the introduction of extraneous solutions result from performing illegitimate operations on equations. An awareness of one fault usually accompanies an awareness of the other fault. For this reason, Saunderson perhaps was not aware of the problems of introducing extraneous solutions, just as he was not aware of

the problems of losing true solutions. The best one can do on this point is to conjecture.

Perhaps the earliest printed algebraic treatise discussing any fractional equation with the purpose of demonstrating a problem encountered in solving it, is *Elémens D'Algèbre* by the late eighteenth-century French mathematician, Silvestre François Lacroix (1765–1843). In this treatise Lacroix considers the particular fractional equation,

$$\frac{x^2}{x^2 - 2ax + a^2} = m,$$

and solves as follows:

$$\frac{x^2}{x^2 - 2ax + a^2} = m$$

$$x^2 = a^2 m - 2amx + mx^2$$

$$(1 - m)x^2 + 2amx - a^2 m = 0$$

$$x = - a \left(\frac{m \pm \sqrt{m}}{1 - m} \right).$$

He notes that if $m = 1$, then the denominator becomes zero, in which case the fractional equation is unsolvable. This is the only fractional equation that Lacroix uses to illustrate this point.[4]

Although Lacroix did recognize and discuss this problem in solving a particular fractional equation, he has obtained fallacious results, most ironically because he sought a solution in a most general manner, assuming the fractional equation to be always reducible, or equivalent, to a quadratic equation for every value of m. Lacroix, like many mathematicians who are interested in generalizing, performed all the algebraic operations on the equation while in its literal form. And after obtaining the value for x in terms of literal numbers, only then did he substitute numerical values for the literal numbers in order to determine the numerical value for x in a specific case. If, however, m

= 1 is substituted at the beginning, then the solution follows:

$$\frac{x^2}{x^2 - 2ax + a^2} = m = 1$$

$$x^2 = a^2 - 2ax + x^2$$

$$x = a / 2.$$

Check:

$$\frac{x^2}{x^2 - 2ax + a^2} \rightarrow \frac{(a / 2)^2}{(a - a / 2)(a - a / 2)}$$

$$= \frac{(a / 2)^2}{(a / 2)(a / 2)} = 1.$$

The derived equation is linear when $m = 1$, and not quadratic as Lacroix assumed. The true solution is $a/2$, and neither $0/0$ nor $-2a/0$, as Lacroix obtained.

It is important to realize that Lacroix chose this particular equation because its answer in final form would have a zero denominator when solved by standard methods. He was hardly aware of the *set* of fractional equations and the problems of extraneous solutions in connection with that set. Lacroix saw this particular equation as an example of a particular problem, and proceeded from that viewpoint.

The fractional equation, let alone a discussion of extraneous solutions, was left out of many algebraic treatises of the late eighteenth century through the mid-nineteenth century. Over 1000 elementary and advanced algebraic treatises printed in the nineteenth century were examined by the author to check whether fractional equations were included in these treatises. Most mathematicians, with the exception of Benjamin Peirce (1809–1880), the Harvard mathematician whom we shall discuss later, did not begin to include fractional equations into their treatises until around the 1850s. That fractional equations do not appear in even a substantial minority of these treatises indicates that

the solution of the fractional equation was not in the background of elementary algebra. But still, attention should be given to the isolated use of fractional equations by Peirce.

Benjamin Peirce, in his textbook, *An Elementary Treatise on Algebra*, printed in 1837, was among the first mathematicians to incorporate fractional equations. Peirce translates many physical problems to algebraic equations that result in fractional equations. He makes no distinction in the solution process of fractional equations and polynomial (integral) equations. Peirce does, however, comment on cases in which the physical problems can become meaningless. Usually such cases occur when, in the fractional equation representing the physical situation, a root of a polynomial in a denominator is assumed the answer of the particular physical problem, as Peirce demonstrates with several examples. In such physical situations, the roots of a polynomial in a denominator would be an extraneous solution that would result from the equation if solved by certain methods. As it was in the case of Lacroix, it is very important to decide the status and significance of Peirce's solution procedure for fractional equations: To Peirce every equation represented a particular physical situation, and consequently, he focused on absurdities within that physical situation *rather than* on the processes involved in abstractly solving an equation that was not a specific physical representation of anything.

Fractional equations started to appear in most algebraic treatises around 1850. From around 1850 until around 1880, there was no special treatment in the solution techniques for the set of fractional equations. These equations were included in the treatises among various types of problems. Oftentimes no distinction was made between fractional equations and equations *containing* fractions.[5] The two types of equations were frequently solved in an identical manner. Some mathematicians would demonstrate special artifices to solve certain fractional equations. These artifices would differ from problem to problem; they were not developed from any general mathematical principle.

Extraneous solutions were frequently introduced, and true solutions frequently lost.

It was around 1880 that mathematicians began to treat fractional equations as a special topic. The special treatment of fractional equations at this particular time in history is more than likely related to the history of "division by zero," since fundamental questions about such division were raised in various countries around 1880.

Rudolf Lipochitz explains in 1877 that in dividing F/m by F_1/m_1 where $F_m \neq 0$, it; is not permissible to let $F_1/m_1 = 0$ for the reason that there exists no fraction which, when multiplied, gives F/m. The class concept was stated by Axel Harnack of Dresden in 1881, who used rational numbers in his calculus and from his definition of class he says that the use of zero as a divisor is impossible. Four years later Stolz elaborates upon this definition of number class and excludes zero as a divisor. The students of elementary mathematics accepted this exclusion principle and developed it in their textbooks using the definition of division. . . .[6]

Mathematically, the solution of fractional equations and division by zero are closely related.

In 1882 James E. Oliver, Lucien A. Wait, and George W. Jones, mathematics professors at Cornell University, determined methods of solution of fractional equations in their first edition of *A Treatise on Algebra*. These mathematicians proved the following theorem:

Theor. 2. If both members of an equation be multiplied by any same number, not a function of the unknown elements and not 0 or ∞, the roots of the equation are not changed thereby.[7]

This theorem is especially relevant to the solution of fractional equations, since those solutions usually do require multiplying by the unknown in order to obtain a polynomial equation which can then be solved.

Oliver, Wait, and Jones understand both the problems of introducing extraneous solutions and the problems of losing true solutions.

Note. Unless the reader be sure that every step he has taken is { *valid,* *reversible,* } i.e., that each successive transformed equation is true { whenever only when } the previous ones are true, his results can serve merely to suggest values of the roots for trial. If any step has been { invalid irreversible } the problem may have { fewer other } solutions that he has found. In particular, he must have multiplied by no more factors containing the unknown than were necessary to clear of fractions, and must have { gained lost } no solutions in taking like { powers roots } of both members; or else he must test his results by substituting in the original equation or equations, and say: if they satisfy the equation they are among its true roots; if not, they are strangers introduced in course of the work. The results are to be trusted only after they are tested.[8]

The only fault in this statement is the implication that a fractional equation will be equivalent to the derived polynomial equation if the fractional equation is multiplied by the least common multiple of the denominators. In fact, this is stated explicitly elsewhere in the five-page discussion on fractional equations. Many mathematicians of the decade ardently hold that multiplication by the lowest common denominator introduces no extraneous solutions. The contention is not true. For, consider the equation

$$\frac{x^2 - 3x}{x^2 - 1} + 2 + \frac{1}{x - 1} = 0.$$

Multiplying by the lowest common denominator, $(x^2 - 1)$,

$$x^2 - 3x + 2(x^2 - 1) + 1(x + 1) = 0$$
$$x^2 - 3x + 2x^2 - 2x + 1 = 0$$
$$3x^2 - 2x - 1 = 0$$
$$x = -\tfrac{1}{3}, 1.$$

However, $x = 1$ is an extraneous solution.

Only after reading *Text-Book of Algebra*, by George E. Fisher and Isaac J. Schwatt, mathematics professors at the University of Pennsylvania, was I able to gain insight into what Oliver, Wait, and Jones, and other mathematicians, were aiming for in their contentions. Fisher and Schwatt proved the following principle:

> If both members of a fractional equation, in one unknown number, be multiplied by an integral expression which is necessary to clear the equation of fractions, the integral equation thus derived will be equivalent to the given fractional equation.

Let

$$\frac{N}{D} = 0 \qquad (1)$$

be the given fractional equation when all its terms are transferred to the first member, added algebraically, and the resulting fraction reduced to its lowest terms.

In deriving equation (1) from the given fractional equation, terms were transferred from one member to the other, by Ch. IV., §3, Art. 5; and then only indicated operations were performed. Therefore equation (1) is equivalent to the given fractional equation.

Clearing (1) of fractions, we have the integral equation

$$N = 0. \qquad (2)$$

Any root of (1) reduces N/D to 0. But any value of x which reduces N/D to 0 must reduce N to 0 (Ch. III., §4, Art. 7), and hence is a root of the derived equation. That is, no solution is lost by the transformation.

Any root of the derived equation reduces N to 0. But, since N/D is a fraction in its lowest terms, N and D have no common factor, and therefore cannot both reduce to 0 for the same value of x (Ch. VIII, §4, Art. 2). Consequently, any value of x which reduces N to 0 must reduce N/D to 0. (Ch. III., §4, Art. 6). That is, no root is gained by the transformation.

Therefore the derived integral equation is equivalent to equation (1), and hence to the given fractional equation.[9]

The mathematicians meant to say that if all the fractions in a fractional equation are combined into a single fraction and reduced to lowest terms, then the fractional equation, which now consists of one fraction, may be multiplied by the denominator of the reduced fraction, and will be equivalent to the derived polynomial equation. This is precisely what Fisher and Schwatt proved, though the statement of the theorem is not as precise as the proof of the theorem. The criterion for the equivalence of a fractional equation to a polynomial equation has nothing to do with the lowest common denominator as some mathematicians claimed. To take the previous example, I shall illustrate the correct method of solving a fractional equation.

$$\frac{x^2 - 3x}{x^2 - 1} + 2 + \frac{1}{x - 1} = 0$$

$$\frac{x^2 - 3x}{x^2 - 1} + \frac{2(x^2 - 1)}{(x^2 - 1)} + \frac{1(x + 1)}{(x - 1)(x + 1)} = 0$$

$$\frac{x^2 - 3x + 2x^2 - 2 + x + 1}{x^2 - 1} = 0$$

$$\frac{(3x + 1)(x - 1)}{x^2 + 1} = 0$$

$$\frac{3x + 1}{x - 1} = 0$$

$$3x + 1 = 0$$

$$x = -\tfrac{1}{3}.$$

Throughout the 1890s algebra textbooks discussed fractional equations and the problems of extraneous solutions. Various theorems governing the solution of fractional equations were proved using axioms and fundamental mathematical principles. But in the early part of the twentieth century, textbooks only mentioned the problem of extraneous solutions. The proofs, which were contained in the earlier editions of a text in the 1890s, were frequently left out of later editions of the same text in the early 1900s. Around 1910, James M. Taylor, who wrote an algebraic treatise in the

1890s, wrote an article on the problem of extraneous solutions.[10] Less than a year following, G. A. Miller, who had recently written an article entitled, "Dividing by Zero," wrote yet another article warning against the serious problems of extraneous solutions.[11] Special attention was again brought to the problem in 1931 by Robert E. Bruce,[12] and again in 1966 by Carl B. Allendoerfer.[13]

The algebra textbooks of the twentieth century vary widely in their discussions on fractional and equivalent equations, but few are rivals to the textbooks of the 1890s in presenting logical, accurate, and comprehensible discussions on fractional equations and extraneous solutions.

An Historical Account of Radical Equations

The critical period in the history of radical equations did not begin with the inclusion of radical equations into the treatises as did the critical period in the history of the fractional equation. For as early as the 1860s, radical equations were included in the treatises, but more important, were also treated as a special topic. For two decades following, radical equations posed no problems as to their solution. The crisis in the solution of radical equations occurred exactly when the interpretation of the radical sign changed, around the 1880s. An old interpretation of the radical sign is summarized:

> It should be observed that it is quite immaterial what sign is put before a radical . . . ; for there are *two* square roots of every algebraical expression and we have no symbol which represents one only to the exclusion of the other; so that $+\sqrt{x+1}$ and $-\sqrt{x+1}$ are alike equivalent to $\pm\sqrt{x+1}$; also $x + \sqrt{x+1}$ has the same *two* values as $x \pm \sqrt{x+1}$.[14]

As long as the above interpretation was given to the radical sign, as we shall see later, there was no problem of introducing extraneous solutions in solving radical equations; and hence, there was no crisis in the history of the radical equation.

A few years prior to the previous interpretation, Oliver, Wait, and Jones had already begun, as most mathematicians had by this time, to designate symbols to exclude either the positive or negative square root. The symbol, $+\sqrt{}$, excluded the negative square root, while $-\sqrt{}$ excluded the positive square root. With this new interpretation of the radical sign, it was necessary for Oliver, Wait, and Jones to record the following:

THEOR. 3. *If the two members of an equation be raised to the same integral power, the results are equal; but it is possible that the new equation may have some roots not found in the old one.*

For if P = Q, wherein P or Q or both of them are functions of some unknown element, say x, then

$$P^2 = Q^2, P^3 = Q^3, \ldots, P^n = Q^n, \qquad \text{[II. ax.6]}$$

$$P^2 - Q^2 = 0, P^3 - Q^3 = 0, \ldots, P^n - Q^n = 0,$$

i.e.,

$$(P - Q)(P + Q) = 0, (P - Q)(P^2 + PQ + Q^2)$$

$$= 0, \ldots, (P - Q)(P^{n-1} + P^{n-2}Q \ldots + Q^{n-1}) = 0.$$

But these equations are satisfied either if such values be given the unknown that P − Q = 0, or that

$$P + Q = 0, P^2 + PQ + Q^2 = 0, \ldots, P^{n-1} + P^{n-2}Q \ldots$$
$$+ Q^{n-1} = 0;$$

and in general the roots of the equation P − Q = 0 are not the same as the roots of the equations P + Q = 0, $P^2 + PQ + Q^2 = 0, \ldots, P^{n-1} + P^{n-2}Q + \ldots + Q^{n-1} = 0$.

E.g., if $x = 5$, then $x^2 = 25$, and $x = {}^+5, {}^-5$; but only ${}^+5$ satisfies the original equation and is its root.

So, if $^+\sqrt{(9 - x)} = x - 9$, then $x^2 - 17x + 72 = 0$, and $x = 8, 9$; but 9, not 8, satisfies the original equation, and is its root.

Were that equation $^-\sqrt{(9 - x)} = x - 9$, the root were 8, not 9.[15]

During the early 1890s there was much confusion in the interpretation of the radical sign. However, toward the end of the decade, this confusion decreased. Fisher and Schwatt already had the term, *principal square root*, for the positive square root. They had adopted the modern convention of assuming the principal root if no sign preceded the radical. They discussed the condition of removing the principal root restriction, and consequently revealed the evolution of extraneous solutions of radical equations.

5. But if the restriction to *principal* roots be removed, any irrational equation contains in itself the statements of two or more equations.

E.g., if both *positive* and *negative* square roots be admitted, the equation

$$\sqrt{(x+6)} + \sqrt{(x+1)} = 1$$

is equivalent to the four equations

$$\sqrt{(x+6)} + \sqrt{(x+1)} = 1, \qquad (1)$$
$$\sqrt{(x+6)} - \sqrt{(x+1)} = 1, \qquad (2)$$
$$-\sqrt{(x+6)} + \sqrt{(x+1)} = 1, \qquad (3)$$
$$-\sqrt{(x+6)} - \sqrt{(x+1)} = 1, \qquad (4)$$

in which the roots are limited to *principal* values.

The same rational integral equation will evidently be derived by rationalizing any one of these four equations. Therefore the roots of this rational equation must comprise the roots of these four irrational equations. Consequently, in solving an irrational equation, we most expect to obtain not only its roots but also the roots of the other three equations obtained by changing the signs of the radicals in all possible ways. Some of these equations can be rejected at once as impossible. The roots of the other irrational equations will be the roots of the rational equation. Thus, of the above equations, (1), (3), and (4) can be rejected at once as impossible.

The rational equation derived from any one of the four equations is

$$x + 1 = 4; \text{ whence } x = 3.$$

The number 3 is a root of the one equation not rejected, since

$$\sqrt{(3+6)} - \sqrt{(3+1)} = 1.^{16}$$

The principal square root, beginning in the late 1890s, was thereafter assumed unless a negative sign preceded the radical. As a consequence, the history of radical equations parallels that of fractional equations. Both kinds of equations were discussed as illustrations of equations that may lead to extraneous solutions.

Discussion

In general, any fractional equation can be combined term by term to obtain a single fraction reduced to lowest terms. This fractional equation, now consisting of one reduced fraction, can be multiplied by the denominator of that fraction without destroying equivalence between the fractional equation and the resulting polynomial equation. Thus, a polynomial equation equivalent to the original equation can be obtained. Then the polynomial equation can be solved by various methods. Therefore, there is absolutely no mathematical reason to introduce extraneous solutions in solving fractional equations.

Today, there are two methods that are widely used in solving fractional equations. In the first method, the fractional equation is multiplied, from the outset, by the lowest common denominator of all the fractions in the equation, in order to obtain a polynomial equation. Next, the polynomial is solved. The resulting solutions are substituted back into the original equation to reject extraneous solutions that could have been introduced. In the second method, the roots of the polynomials in the denominators of the fractions in the fractional equation are determined. These roots are the various values which, if substituted for the unknown in the fractional equation, would lead to division by zero. Because of this, these values are, at the outset, excluded as possible roots of the fractional equation. Next, the fractional equation is solved by multiplying by the lowest common denominator. The solutions that do not appear among the excluded values of the unknown are roots of the fractional equation.

There are, however, objections to both of these methods. In the first method where the fractional equation is multiplied by the lowest common denominator at the outset, the derived polynomial equation will have at least as many, and maybe more, solutions as the fractional equation has roots. And therefore, a substitution of the solutions of the derived polynomial equation into the fractional equation would determine the roots of the fractional equation. However, this substitution process, a necessary part of the solution process, is frequently very laborious, and is thus undesirable. This method has another fault in that it leads one to believe that extraneous solutions are inevitably introduced in solving fractional equations. But most drastic, this method does not reveal the exact number of roots of the fractional equation *before* the roots are found. In the second method, it is excessive labor to determine the roots of the polynomials in the denominators. This method, too, leads one to believe that the introduction of extraneous solutions is inherent in any solution of fractional equations. Like the first method, this method reveals no criterion whereby the exact number of roots can be determined *prior to* the actual determination of the values of the roots.

The method whereby the terms of a fractional equation are combined first, etc., is preferable to either of the two methods just discussed. There are no undesirable, though necessary, substitutions to make. Neither is there the problem of determining the roots of the polynomials in the denominators. One is not, by using this method, misled to believe that the introduction of extraneous solutions is inherent in the solution of fractional equations since, in using this method, no extraneous solutions are introduced. But most important, by using this method, one can establish a criterion to determine the exact number of roots of the fractional equation before it is solved, as we shall now see.

By using the method advocated, one can transform every fractional equation into an equivalent polynomial equation. Since, however, the number of roots of a polynomial equation is determined by the *Fundamental Theorem of Algebra*, the number of roots of any fractional equation can be determined from its equivalent polynomial equation. This equivalent polynomial equation is precisely the numerator—after the fractional equation has been combined term by term to obtain a single, reduced fraction—of the single, reduced fraction

equated to zero. Thus the notion of the degree of a fractional equation can be constructed analogous to this notion for polynomial equations, whereby the number of roots of a polynomial equation corresponds to the degree function for that polynomial.[17] The degree of the polynomial in the numerator of the single, reduced fraction can be considered the degree of the original fractional equation. By using this definition of degree, one can determine the number of roots of any fractional equation *prior to* solution, where the number of roots of a fractional equation equals its degree, just as in the case of a polynomial equation. Algorithms can perhaps be found to calculate the degree of the fractional equation while it stands in its original form—before the fractional equation is transformed to an equivalent polynomial equation—by considering the degrees of the numerators and denominators of each fraction with respect to their least common multiples, etc. This paper does not present such a detailed, mathematical algorithm; however, the point that the exact number of roots of a fractional equation can be determined prior to solution is indeed well taken.

On the other hand, the number of roots of a radical equation cannot be determined prior to solution. Because of the restriction of principal square root, the radical equation, after being raised to a power, will not in general be equivalent to the resulting polynomial equation. Extraneous solutions are inevitably introduced in certain cases. A substitution of the solutions into the original radical equation is a necessary part of the solution process. Hence, criteria for a degree function for a radical equation can hardly be constructed.

Conclusion

Before 1830 there was but one isolated discussion of the problems in solving a particular fractional equation. Around 1837 fractional equations were included randomly throughout the treatises of Benjamin Peirce. The entry of fractional equations into Peirce's treatises at this particular time in history may be due to a novel advancement in physics of which fractional equations could have been physical interpretations—but this remains an open question. In any case, fractional equations were thoroughly incorporated into nearly all the algebraic texts after 1850. But it was not until after 1880 that fractional equations were discussed as a special topic, and that mathematically sound, logical solution techniques were developed. A few years earlier, the problem of "division by zero" had been raised in advanced mathematics, and this, most likely, called attention to the problem of solving fractional equations. A discussion of extraneous solutions was a consequence. Around the same time a new interpretation was given the radical sign, whereby a positive or negative square root could be excluded at will. As before, a discussion of extraneous solutions was a consequence. The concepts of equivalence and one-to-one correspondence of certain transformations over the field of real numbers were evoked to explain the nature of extraneous solutions. When, in the various articles printed in mathematical journals, special attention was called to the problem of extraneous solutions in 1910, 1911, 1931, and 1966, the discussions focused on the concepts of equivalence and one-to-one correspondence. Therefore, a history of extraneous solutions has many implications for an historical study of the concepts of equivalence and one-to-one correspondence—both of which I think entered mathematics in the early nineteenth century—from the mid-nineteenth century up to the present time.

As for a modern approach to the solution of fractional equations, the author suggests that a degree function be established for the set of fractional equations. The set of polynomial equations should be treated as a subset of the set of fractional equations. The algebraic studies of rational functions in abstract algebra should be extended in order to parallel the studies of polynomial functions. Such an extension would call attention to, and perhaps elucidate, the problems in the solu-

tion of fractional equations. If then attention is further focused on the problems in the solution of radical equations, there would most likely be a crisis in the history of extraneous solutions.

NOTES

1. Throughout this paper a distinction will be made in the use of the terms, *root* and *solution*. A *root* of an equation is defined as a value that satisfies that equation. A *solution* of an equation is defined as any value obtained through a solution process of that equation. Thus a solution of an equation is not necessarily a root of that equation. A solution of an equation which happens to be a root of that equation will be termed a *true solution* (or just *root*) of the equation.

2. Two equations, $A(x) = 0$ and $B(x) = 0$, are said to be *equivalent* if and only if every root of the equation $A(x) = 0$ is a root of the equation $B(x) = 0$, and *vice versa*.

3. This example is found in Nicholas Saunderson's *Elements of Algebra in Ten Books*, pp. 103–4.

4. This example is found in S. F. Lacroix's *Elémens D'Algèbre*, pp. 174–79.

5. An equation *containing* fractions is an equation in which the coefficients of the unknown quantity are fractions. The equation, $(1/5)x + 1 = 0$, is an equation *containing* fractions, whereas, $1/5x + 1 = 0$ is a fractional equation.

6. H. G. Romig, "Early History of Division by Zero."

7. J. E. Oliver, L.A. Wait, and G. W. Jones, *A Treatise on Algebra*, p. 284.

8. *Ibid.*, pp. 287–88.

9. G. E. Fisher and I. J. Schwatt, *Text-Book of Algebra*, pp. 311–12.

10. See James M. Taylor, "Equations."

11. See G. A. Miller, "Algebraic Equation."

12. See Robert E. Bruce, "Equivalence of Equations in One Unknown."

13. See Carl B. Allendoerfer, "The Method of Equivalence or How to Cure a Cold."

14. Charles Smith, *A Treatise on Algebra*, pp. 119–20.

15. J. E. Oliver, L. A. Wait, and G. W. Jones, *A Treatise on Algebra*, p. 287.

16. G. E. Fisher and I. J. Schwatt, *Text-Book of Algebra*, p. 555.

17. I am aware that a degree function is defined over the field of rational functions as

$$\deg (P(x)/Q(x)) = \deg (P(x)) - \deg (Q(x)).$$

However, this definition is so constructed to determine characteristics of the functions at infinity. Our goal is to determine the number of zeros for such functions; therefore, our definition of degree should be constructed with this goal in mind.

REFERENCES

Allendoerfer, Carl B. "The Method of Equivalence or How to Cure a Cold." *The Mathematics Teacher* 59 (1966): 531–42.

Bruce, Robert E. "Equivalence of Equations in One Unknown." *The Mathematics Teacher* 24 (1931): 238–44.

Fisher, G. E., and I. J. Schwatt. *Text-Book of Algebra.* Norwood, Mass.: Norwood Press, 1898.

Lacroix, S. F. *Elémens D'Algèbre.* 11th ed. Paris, 1815.

Miller, G. A. "The Algebraic Equation." *Monographs on Topics of Modern Mathematics.* Edited by J. W. A. Young, pp. 211–60. 1911.

Oliver, J. E.; L. A. Wait; and G. W. Jones. *A Treatise on Algebra.* Boston: J. S. Cushing & Co., 1882.

Peirce, Benjamin. *An Elementary Treatise on Algebra.* Boston: Monroe & Co., 1837.

Romig, H. G. "Early History of Division by Zero." *American Mathematical Monthly* 31 (1924): 387–89.

Saunderson, Nicolas. *Elements of Algebra in Ten Books.* Cambridge: University Press, 1740.

Smith, Charles. *A Treatise on Algebra.* 2d ed. London: Macmillan and Co., 1890.

Taylor, James M. "Equations." *The Mathematics Teacher* 2 (1910): 135–46.

98

The Evolution of Group Theory

G. H. MILLER

NOW THAT THE VALUE of group theory in theoretical physics, chemistry, electrical engineering, and electronic computing machines is being realized [8], it is of interest to look at the origins of group theory.[1] This paper will be concerned only with a résumé of the significant contributions of earlier investigators through the nineteenth century. There are disagreements among the historical authorities as to who originated certain specific topics in group theory, but for the most part there is a uniformity of opinion.

Early Examples of Groups

Although the abstract concept of group was not considered until centuries later, the additive group modulo 24 was used by the ancient Babylonians and Egyptians in designating the hours of the day. Later civilizations used modulo 12 for the period of daylight and the period of darkness.

Some authorities in group theory [11, 23, 24] are of the opinion that the Egyptian method of handling fractions by the unit fraction method is based on the group concept.

Pre–Nineteenth Century Contributions

There were certain mathematicians who developed concepts in group theory but did not make use of the term "group," as they did not realize the

significance of groups at that time. However, they did provide some of the foundations for group theory.

One of the earliest contributions was made by L. Euler (1707–1783). In Volume 8 (1760–61) of *Novi Commentarii Academiae Petropolitanae,* Euler gave a fundamental generalization of one of Fermat's theorems: The natural numbers which are not greater than m and are relatively prime to m form an Abelian group with respect to multiplication modulo m. The example which he used in his proof is an important example of a finite commutative group.

A decade after Euler's article, Lagrange (1736–1813) wrote "Réflexions sur la Résolution Algébrique des Équations," which was a most influential work in the development of group theory. In this article the important theorem of Lagrange is stated: The order of every permutation group of degree n divides the order of the symmetric group of the same degree. In addition, this same article contained some consideration of the octic group of degree 4, cyclic and noncyclic groups of order and degree 4, and the metacyclic group of order 20 on five letters.

Another important theorem in group theory was developed by P. Ruffini (1765–1822), in which it is shown that the order of a group is divisible by the order of every one of its subgroups, but that there is not always a subgroup in a given group whose order is an arbitrary divisor of the order of the group. In addition, Ruffini developed the classification of permutation groups, although most of his terms are not used by the present-day

Reprinted from *Mathematics Teacher* 57 (Jan., 1964): 26–30; with permission of the National Council of Teachers of Mathematics.

algebraists. These classifications approximated the concepts of intransitive, transitive imprimitive, and transitive primitive.

Nineteenth-Century Contributions

The term "commutative law" was first introduced by F. Servois in 1814 in the journal *Annales de Mathématiques.*

It was the German mathematical genius Karl Friedrich Gauss (1777–1855) who first dealt extensively with the concept of cyclic groups of permutations. He introduced the term "modulus" and the modulus sign, $\equiv$, which are still used today. He considered the addition group and used the zero element as the identity.

Other important additions to group theory were made by two mathematical geniuses who were cut down in the prime of their lives.[2] The first genius, Neils Abel (1802–1829), contributed by utilizing group theory to prove there can be no algebraic solution for a fifth degree equation by showing that the symmetric group of degree five does not contain any subgroup of index 5 except for the symmetric group of degree four. The second mathematician, who made his contribution in twenty short years of life, was Evariste Galois (1811–1832). Despite the fact that his work was not recognized until many years after his death, he was the originator of brilliant applications of finite groups to algebra which now bear his name, Galois theory. He developed the foundations for the solvability of algebraic equations by means of rational operations and extraction of roots. Galois was the first to introduce a term for a group and to use it in the abstract sense as it is used today. He considered such topics as substitution groups and he developed the ideas of quotient groups, invariant subgroups, and right and left cosets. His amazing contributions provided solid and theoretical foundations for a large area of group theory despite the fact that they consisted of less than sixty pages in the journals and were written in many cases in a very short period of time.

The most prolific writer in the middle of the nineteenth century in the area of group theory was Augustin-Louis Cauchy (1789–1857). He considered such topics as circular permutations and the method of transformation of permutation groups, and he dealt with multiply transitive permutation groups of degree 6 and of order 12. In 1844 he published the first work on the transformation of permutations in *Exercises d'analyse.* He anticipated a special case of Sylow's theorem and in addition dealt with the holomorph of the cyclic group of order *n*. He also introduced the important concept of double coset.

To Arthur Cayley (1821–1895) goes the credit for the first recorded attempt to give a definition of an abstract group. This was in 1854 in an article, in the *Philosophical Magazine,* entitled "On the Theory of Groups Depending on the Symbolic Equation $\theta^n = 1$." Although he was the first to offer definition, he was not always consistent in the use of his definition; however, he recognized the necessity for a definition, which many excellent mathematicians did not realize was necessary. He also showed that every finite abstract group is isomorphic to a regular permutation group. He was the first to determine correctly all groups of order 12.

An excellent example of an outstanding mathematician who worked with groups but did not realize the significance of the concept of group is William Rowan Hamilton (1805–1865). In his *Lectures in Quaternions,* published in 1853, the term "associative law" was first introduced into the literature. In this work he made much use of the associative law and emphasized the value of this new law. In 1856 the topic of quaternions was again introduced in a work entitled "A New System of Roots of Unity." The concept of quaternions provided the first example in group theory of a noncommutative group for multiplication. He considered relations between regular solids and groups, and he introduced color groups [23] in the 1899 edition of the *American Journal of Mathematics.*

Toward the latter part of the nineteenth century, many new innovations were made in group theory, as well as many developments of previous concepts.

One of the first specialists in group theory was Marie Evremond Camille Jordan (1838–1923). In 1867 he published his first article on groups of infinite order. He introduced the term "class" of a permutation group and also provided a theorem on the constancy of the factors of composition of a group. In 1870 his book *Traité des Substitutions et des Équations Algébriques* was published. Here he summarized most of his work up to that time. In addition, he introduced the notion of the commutator subgroup of a permutation group but did not develop the theory for commutators. He did some of the original work in what is now called abstract groups. Jordan made the first extensive list of primitive groups up to degree 17. The list, however, was not complete.

The concept of multiply transitive groups was developed by Émile Léonard Mathieu (1835–1890). He established the existence of two five-fold transitive groups of degrees 12 and 24, which are presently known as Mathieu groups. Also, he established the existence of an infinite system of triply transitive groups of degrees $p + 1$ and order $p(p^2 - 1)$.

It was Ludwig Sylow (1832–1918) who continued the work of Galois and developed the eight important Sylow theorems which were reported in *Mathematische Annalen* in 1872. These important theorems are summarized by Hall [7] in modern terminology as follows:

1. If G is of order $n = p^m s$ where p does not divide s and p is a prime, then G contains subgroups of order p^i, where $i = 1, 2, \ldots, m$, and each subgroup of order p^i, where $i = 1, 2, \ldots, m - 1$, is a normal subgroup of at least one subgroup of order p^{i+1}.

2. In a finite group G, the Sylow p-subgroups are conjugate.

3. The number of Sylow p-subgroups of a finite group G is of the form $1 + kp$ and is a divisor of the order of G.

The applications of group theory to the physical universe were promulgated by three men. The first introductory proclamation was made by Felix Klein in 1872. In his *Erlanger Programm* he considered geometry as a series of problems in group theory. These ideas were fully developed in his book *Vorlesungen über das Ikosaeder und die Auflosung der Gleichung vom funften Grade*, which was published in 1884. He can probably be called the initiator of the term "modern mathematics" since he advocated the use of topics of modern mathematics wherever he spoke.

Using some of the basic concepts of Klein, Henri Poincaré (1854–1912), in his "Foundations of Geometry" in the *Monist* in 1898, developed the thesis that all of Euclid's geometry was actually based on the group concept.

Sophius Lie (1842–1899) was the first individual to make use of basic concepts of group theory to solve problems of physics. His text entitled *Theorie de Transformationsgruppen*, which came out in three volumes during 1888–93, was an aid, making use of invariance under continuous transformation groups, in the solution of differential equations problems. This work was produced upon the urging of his good friend Felix Klein. His work was based on his theory of continuous and discontinuous transformation groups.

Richard Dedekind (1831–1916), in the *Bulletin of the American Mathematical Society* (1898), discussed the properties of Hamiltonian groups. These groups were derived from the concepts of Hamilton's quaternions and showed that the miltiplicative group of quaternionic units was a group of order 8. Dedekind also showed that this was an example of a non-Abelian group. He amplified the work done on commutator subgroups.

It was Georg Frobenius (1849–1917) who first introduced the concepts of characteristic element and characteristic subgroups, in the *Quarterly*

Journal of Mathematics in 1896, and he pioneered the work in the area of matrices. He also considered commutator subgroups.

Two other famous Englishmen in group theory were Thomas Penyngton Kirkman (1806–1895) and William Burnside (1852–1927). Kirkman is probably best known for entering his paper on the determination of permutation groups in competition for the grand prize offered by the Paris Academy of Sciences in 1858. Despite the fact that his terminology for group theory was not similar to that used by the mathematicians of his day, he made significant contributions in the theory of groups. This difference in terminology was probably due to his vocation, which was that of a clergyman.

Burnside made several advances in the study of groups of finite order. He considered odd and even permutations and did work on group characteristics. Both Burnside and Frobenius initiated an important area of finite groups known as group representation theory.

The American Contribution

The American contribution to group theory started with the hiring of James Joseph Sylvester (1814–1897) by Johns Hopkins University. Sylvester was a close friend of Arthur Cayley during their days of law practice in England, where they cooperatively developed some basic concepts of group theory and, in particular, the concept of invariance. Sylvester taught courses stressing group theory at Johns Hopkins. Other universities, such as the University of Chicago and Clark University, similarly engaged European mathematicians, among whom Oscar Bolza and Heinrich Maschke were interested in group theory. Under the influence of these men, three Americans received their doctors' degrees in group theory, to become the first in America to obtain degrees in this new area. They were Jacob William Albert Young, "On the Determination of Groups Whose Order Is a Power of a Prime" (1893); Frank Nelson Cole, "A Con-

tribution to the Theory of the General Equation of the Sixth Degree" (1886); and Leonard Eugene Dickson, "The Analytic Representation of Substitution on a Power of a Prime Number of Letters with a Discussion of the Linear Group" (1891).

F. N. Cole (1861–1927) made additional contributions to finite groups in correcting earlier works by Askwith and Cayley, who had omitted one group of degree 7. In addition, he helped to establish that there is only one simple group of order 360 in the list of groups of orders 201 to 500.

Leonard Eugene Dickson (1874–1954) made substantial contributions to group theory. Some of the many areas with which he was concerned were: linear groups and Galois field theory, hypoabelian groups, abstract simple groups, isomorphisms of linear groups, and additional extensions in Galois theory.

Under the inspiration of F. N. Cole, G. A. Miller published extensively on finite groups. Some of the areas he considered in the nineteenth century were: substitution groups on 8, 9, 10 letters; transitive groups of degrees 12, 13, 14, 17; primitive groups of degrees 10, 15; commutator groups; extensions of Sylow's theorem; and Hamilton groups. For his extensive investigations of finite groups he received an award from the Crawkow Academy of Sciences in 1900. This was the only European award ever given to an American in the field of mathematics up until then.

This résumé of the history of group theory shows that practically all of the fundamental concepts of group theory were developed during the nineteenth century. It was during the late nineteenth century that group theory was not only investigated in the realm of pure mathematics but also was applied to the solution of problems in the physical environment. Thus the major ideas of group theory, even though they came in some instances from diverse and seemingly unrelated sources, finally were synthesized by the mathematicians of the nineteenth century into an organized set of concepts and theorems which form the foundations for the theory as we know it today.

NOTES

1. The author is indebted to Professor H. R. Brahana of the University of Illinois for his assistance in the preparation of this article.

2. Abel died of tuberculosis at the age of 26 and Galois was killed in a duel at the age of 20.

BIBLIOGRAPHY

[1] Bell, E. T. *Men of Mathematics*. New York: Simon and Schuster, Inc., 1937.

[2] ———. *The Development of Mathematics*. New York: McGraw-Hill Book Co., Inc., 1940.

[3] Brahana, H. R. "George Abram Miller," *National Academy of Sciences of the U.S.* New York: Columbia University Press, 1957.

[4] Burns, Josephine Elizabeth. "The Foundation Period in the History of Group Theory." Master's Thesis, University of Illinois, 1909.

[5] Cajori, Florian. *A History of Mathematics*. New York: The Macmillan Company, 1919.

[6] Easton, Burton Scott. *The Constructive Development of Group Theory*. Philadelphia: University of Pennsylvania Series in Mathematics, No. 2, 1902.

[7] Hall, Marshall. *The Theory of Groups*. New York: The Macmillan Company, 1959.

[8] Homermesh, Morton. *Group Theory and Its Applications to Physical Problems*. Reading, Mass.: Addison-Wesley Publishing Co., Inc., 1962.

[9] Miller, G. A. "Evolution of the Use of the Modern Mathematical Concept of Group," *Scientific Monthly*, 41:228–33.

[10] ———. "Felix Klein and the History of Modern Mathematics," *Proceedings of the National Academy of Sciences*, 13:611–13.

[11] ———. "Group Theory in the History of Mathematics," *Scientific Monthly*, 47:24–27.

[12] ———. *Historical Introduction to Mathematical Literature*. New York: The Macmillan Company, 1916.

[13] ———. "On a Few Points in the History of Elementary Mathematics," *American Mathematical Monthly*, 16:177–79.

[14] ———. "On Several Points in the Theory of Groups of a Finite Order," *American Mathematical Monthly*, 5:196–97.

[15] ———. "On the History of Finite Abstract Groups," *Journal of the Indian Mathematical Society*, 17:97–102.

[16] ———. "On the History of Several Fundamental Theorems in the Theory of Groups of Finite Order," *American Mathematical Monthly*, 8:213–16.

[17] ———. "Sylvester and Cayley," *Science*, 44:173.

[18] ———. "The Group Theory Element of the History of Mathematics," *Scientific Monthly*, 12:575.

[19] ———. "Twenty-five Important Topics in the History of Mathematics," *Science*, 48:182–84.

[20] Prasad, Ganesh. *Some Great Mathematicians of the Nineteenth Century*, Vol. 1. Benares City, India: S. C. Chatterju, Mahamandal Press, 1933.

[21] ———. *Some Great Mathematicians of the Nineteenth Century*, Vol 12. Benares City, India: S. C. Chatterju, Mahamandal Press, 1934.

[22] Smith, David Eugene. *History of Mathematics*. Vol. 1. Boston: Ginn & Company, 1923.

[23] Speiser, A. *Die Theorie der Gruppen Endlicher Ordnung*. Basel, Switz.: 1956.

[24] Vaerden, Bartel Leendert van der. *Science Awakening*, English translation by Arnold Dresden. Groningen, Neth.: P. Noordhoff, 1954.

99

Sylvester and Scott

PATRICIA C. KENSCHAFT
AND KAILA KATZ

*T*WO BRITISH MATHEMATICIANS, both the victims of discrimination in their native country, were instrumental in stimulating the mathematical research community in the United States. James Joseph Sylvester (1814–1897) and Charlotte Angas Scott (1858–1931) inspired and trained many younger mathematicians here, served as editors of the first continuing mathematical research journal in this country, and contributed substantial research to the early American journals. Both overcame obstacles in their careers and promoted the study of mathematics by young women at a time when it was unpopular to do so.

James Joseph Sylvester

Sylvester was born into a Jewish family in London and attended private Jewish schools until the age of fourteen. After that time he was in educational settings where he often suffered because of anti-Semitism. He attended the University of London, studying under August DeMorgan, but was expelled after five months when it was discovered that he intended to use a knife against one of his tormentors. He then attended the Royal Institution at Liverpool for two years, gaining high distinction in mathematics, though at one point he ran away when harassment by fellow students became unbearable.

James Joseph Sylvester (1814–1897)

In 1831, at the age of seventeen, he matriculated at Cambridge University, but illness prevented him from finishing until 1837. He placed second in the mathematical Tripos examination, a grueling five-day examination required of those who wished to graduate with honors. Unfortunately, in order to receive a degree, a student also had to subscribe to the "Thirty-Nine Articles," religious oaths of allegiance to the Church of England. Not being a member of that church, Sylvester could neither receive his degree, go on for an M.A.

Reprinted from *Mathematics Teacher* 75 (Sept., 1982): 490–94; with permission of the National Council of Teachers of Mathematics.

at Cambridge, nor become a member of the Cambridge faculty. It wasn't until 1872, when Cambridge rescinded these religious tests, that Sylvester was awarded his long overdue degrees. In 1837 he returned to the University of London (whose name had changed to University College) to teach science, where he became a colleague of his former teacher, DeMorgan. After two years there, he left for Ireland. In 1841, Trinity College in Dublin awarded him both the B.A. and the M.A. degrees that his own alma mater would not confer upon him.

Accepting an offer to become professor of mathematics at the University of Virginia in 1841 was a disastrous decision for Sylvester. The Richmond newspapers were against the appointment of the Jewish foreigner. They took every opportunity to tell their readers of their opposition to his appointment as well as that of another foreigner who was Catholic. Some students apparently shared that opposition. An inattentive, hostile student insulted Sylvester on more than one occasion in class, and a faculty committee to which he complained failed to support him. Never one to take abuse lightly, he resigned five months after his arrival. Unable to obtain another teaching position in the States, he returned to London in 1842. He became an actuary, tutored students (including Florence Nightingale) in mathematics in his spare time, and later studied law.

It was in 1850 that he met another lawyer, Arthur Cayley, with whom he collaborated in ground-breaking research into matrices and linear transformations. Together they developed the theory of algebraic invariants, those algebraic expressions that remain invariant or unchanged under linear transformations. Students who study matrices may encounter Sylvester's law of nullity. The nullity of a matrix is the difference between its order and its rank; it is also referred to as the dimension of the kernel. Sylvester's law says that the nullity of the product of two square matrices can never exceed the sum of the nullities of the factors and is never less than the nullity of each

factor (Baker 1904). Sylvester was the first to use several words that are now common in matrix theory, including *invariant, covariant,* and *Jacobian.* Sylvester's research earned for him election as foreign correspondent to the highly prestigious and selective French Academy of Science in 1863.

In 1856 Sylvester became professor of mathematics at the Royal Academy in Woolwich, England. This position ended in 1870 when the school forced him to retire because of his age, although he was only fifty-six years old. The school then tried, unsuccessfully, to swindle him out of part of his pension. Letters to the editor of the *London Times* and a lead article there helped him secure his full pension.

Fortunately, these bitter experiences did not dull his enthusiasm for teaching. His happiest years as a teacher began in September 1876, at the age of sixty-two. The trustees of the Johns Hopkins University, seeking an internationally known mathematician for their new university, offered him a chair, which he accepted. Sylvester's successor in that chair of mathematics at Johns Hopkins described Sylvester as one who was extremely appreciative of the work of others, who was stimulated by his students' questions, and who gave the warmest recognition to any talent or ability displayed by his students.

In 1878 while at Johns Hopkins, he founded the *American Journal of Mathematics* and served as editor for its first five years. During its first ten years he contributed thirty papers to this journal. Although he spent only seven years at Johns Hopkins, he is often credited with being the individual who was most significant in the beginning of mathematical research in the United States. In particular, Thomas Fiske (1905), the founder and president of the American Mathematical Society, said in 1904 in a speech honoring the Society's tenth anniversary, "With the arrival of Professor Sylvester at Baltimore and the establishment of the *American Journal of Mathematics* began the systematic encouragement of mathematical research in America."

In 1883 Sylvester left the United States to accept an invitation from Oxford University to become the Savilian Professor of geometry. He worked there creatively until ill health forced him to retire in 1893, though he continued his mathematical research until his death in 1897.

Charlotte Angas Scott

Charlotte Scott was born in 1858. When she was a teenager, secondary education for women in England was virtually nonexistent. However, her father was president of a college and evidently provided fine tutors for her. On the basis of home study she won a scholarship in 1876 to the recently founded Girton College at Cambridge University, the first college in England open to women (Girton College Records 1876). In 1880 she obtained special permission to take the Tripos examinations with the Cambridge men, and she placed eighth in mathematics in the entire university. Because she was a woman, she could not be present at the award ceremony or even have her name read; but when the name of a young man was read in place of hers, the indignant male students shouted throughout the hall, "Scott of Girton! Scott of

Charlotte Angas Scott (1858–1931)

Girton!" The resulting publicity enabled women to gain the right to take the Tripos exams after an appropriate time of residence at the University and to have their names posted with the men.

Cambridge did not award degrees to women, however, until 1948; so Scott had to retake all her examinations for the University of London. She was awarded a bachelor's degree from that institution in 1882 and a doctorate in 1885. She had no thesis advisor in the modern sense, but evidently Arthur Cayley served in that role. In his obituary she reported that she had attended his lectures at Cambridge for four years beginning in 1880 and that ". . . for the last fourteen years I have been priviledged to know him and experience his kindness" (Scott 1895). He wrote a letter commending her mathematical research, which helped her to obtain a position on the original faculty of Bryn Mawr College immediately after receiving her doctorate (Bryn Mawr College Executive Committee Report 1884).

Thus, Scott too joined the initial faculty of a leading American institution. Her tenure was longer than Sylvester's; after serving as department head for thirty-nine years, she remained on campus without teaching classes an additional year while the last of her seven successful doctoral students completed her degree (Lehr 1974). In 1907, four of the fifteen women with doctorates on the membership list of the American Mathematical Society (AMS) had been Scott's students and a fifth was Scott herself. Many of her students who did not earn doctorates nevertheless attained important positions in higher education in our country.

Scott was a coeditor from 1899 until 1926 of the journal that Sylvester had founded. She had about forty mathematical publications, three of which appeared in that journal before she became coeditor. Although a British citizen until her death, she was a member of the original council of the AMS when it was organized in 1894. After serving for five years as a council member, she became vice-president of the AMS in 1905, and she was the only woman during the first eighty years of the AMS to serve in that position. In his semicentennial celebration speech in 1934, Thomas Fiske, reviewing the first fifty years of the society he had founded, mentioned about thirty men and only one woman, Charlotte Scott (*American Mathematical Society Semicentennial Celebration* 1939).

Scott (1894) wrote a graduate textbook in analytic geometry that was so widely used that it was reprinted by a new publisher in 1924, thirty years after its original appearance (see Bibliography). Through it alone she made a significant impact on mathematical research in our country.

Her own research involved geometrical invariants, quantities that do not change when a figure is modified or the viewpoint is altered. She was especially interested in algebraic figures in two dimensions whose equations were of degree three, four, or five (in both variables); she investigated nodes, cusps, double tangents, and "inflexions" of such curves. One of her most widely quoted papers was a geometric proof of Max Noether's theorem about quantities of the form $Af + Bg = 0$. If $f = 0$ and $g = 0$ are algebraic curves that intersect, then any curve defined by $Af + Bg$ passes through the points of intersection of f and g. Scott's (1899) geometrical proof of the converse (that all curves passing through all points of intersection of f and g must be of the form $Af + Bg$) was published in the *Mathematische Annalen*. It may have been the first mathematical paper written in the United States that received widespread attention in Europe (see Abyankar 1981).

Scott's frequent summer trips across the Atlantic Ocean at a time when such travel was difficult and rare did much to stimulate mathematics in our developing country by maintaining contact with the established world of culture. In 1922 when about 200 people gathered to honor her at Bryn Mawr College, Alfred North Whitehead made a special trip to the United States to attend. His tribute included the following sentences. "A friendship of people is the outcome of personal relations. A life's work such as that of Professor Charlotte Angas Scott is worth more to the world than many anxious efforts of diplomats. She is an example of the universal brotherhood of civilization" (Putnam 1922).

Sylvester and Scott may have been acquainted since they both worked closely with Cayley. Certainly both believed in the importance of women studying mathematics. Sylvester's tutoring of Florence Nightingale during his years as an actuary gives testimony to his belief. Later when a woman named Christine Ladd (later Ladd-Franklin) applied to become a graduate student at the Johns Hopkins University and the trustees debated whether or not to accept a woman, it was Sylvester who insisted that sex should not be a factor in judging mathematical excellence. The parallels in the lives of Sylvester and Scott and in the barriers that they had to overcome to exercise their enthusiasm for teaching and research are striking. The fact that the birth of research mathematics in this country can be largely traced to these two English mathematicians is an important message that bears repeating even now.

BIBLIOGRAPHY

Abyankar, Shreeram S. Telephone conversation on 14 February 1981, with Patricia C. Kenschaft.

American Mathematical Society Annual Register, 1907.

Baker, H. F., ed. *The Collected Mathematical Papers of James Joseph Sylvester*, IV:53. Cambridge:1904–1912, 558.

Ball, W. W. R. *A History of the Study of Mathematics at Cambridge.* Cambridge: At the University Press, 1889.

"The Beginnings of the American Mathematical Society: Reminiscences of Thomas Scott Fiske." *American Mathematical Society Semicentennial Celebration*, 1939.

Bryn Mawr College Executive Committee Report. 7 July 1884. Supplied by Lucy Fisher West, archives librarian.

Cambridge University Commission. *Report of Her Majesty's Commissioners Appointed to Inquire into the*

State, Discipline, Studies, and Revenues of the University and Colleges of Cambridge: Together with the Evidence, and an Appendix. London, Her Majesty's Stationery Office, 1852.

Cartwright, Dame Mary. Interview with Patricia C. Kenschaft, July 1974.

Catalogue of Scientific Papers, Vol. 5. London: Royal Society of London, 1871, 901–4.

Fisher, Charles S. "Death of a Mathematical Theory: A Study in the Sociology of Knowledge." *Archive for History of Exact Sciences* 3(1)(1966): 137–59.

Fiske, Thomas S. "Mathematical Progress in America." *Bulletin of the AMS* 11 (February 1905): 238–46.

Girton College Records. 1876. Supplied by Margaret Gaskell, archives librarian.

Glaisher, J. W. L. "Presidential Address; on the Mathematical Tripos." *Proceedings of the London Mathematical Society* 18 (11 November 1886): 4–38.

Hawkins, Hugh. *Pioneer: A History of the Johns Hopkins University, 1874–1889.* Cornell University Press, 1960.

Jones, E. E. Constance. *Girton College.* Adam and Charles Block, 1913.

Kenschaft, P. C. "Charlotte Angas Scott, 1858–1931."

Association for Women in Mathematics Newsletter 8 (April 1978): 11–12.

Kline, Morris. *Mathematical Thought from Ancient to Modern Times.* New York: Oxford University Press, 1972, Chapter 33.

Lehr, Marguerite. Interview with P. C. Kenschaft, June 1974.

North, John D. "Sylvester, James Joseph," *Dictionary of Scientific Biographies* 13 (1976): 216–22.

Putnam, Emily J. "Celebration in Honor of Professor Scott." *Bryn Mawr Bulletin* 2 (1922):12–14.

Scott, C. A. *An Introductory Account of Certain Modern Ideas and Methods in Plane Analytical Geometry.* London and New York: Macmillan and Co., 1894.

———. "Arthur Cayley, Obituary." *Bulletin of the AMS* 1 (March 1895): 133–41.

———. "A Proof of Noether's Fundamental Theorem." *Mathematische Annalen* 52 (1899): 592–97.

———. *Projective Methods in Plane Analytic Geometry.* New York and London: Chelsea Publishing Co., 1924.

Yates, R. C. "Sylvester at the University of Virginia." *American Mathematical Monthly* 44 (April 1937): 194–201.

100

The Early Beginnings of Set Theory

PHILLIP E. JOHNSON

GEORG CANTOR CREATED and largely developed the theory of sets in approximately the years 1874–1897. In contrast to such developments as the calculus and non-Euclidean geometry, the creation of set theory was, according to all indications, Cantor's alone. Also, set theory was not preceded by a long evolutionary period such as is usually the case with big mathematical breakthroughs. The present article will concern itself primarily with the very earliest set-theoretic works of Cantor, namely, his first two papers in this area.[1]

In view of Cantor's outstanding contributions to mathematics, it is surprising that he did not start out originally to become a mathematician. Cantor's father, a practical-minded, prosperous businessman, wanted Georg to go into the trade or profession of engineering so he could make a living. Georg submitted at first to his father's wishes, but his heart was never in it. Luckily for the mathematical world, his father eventually saw the folly of trying to keep such a brilliant mathematical mind as Georg's tied down to something so mundane as engineering. It may be that Georg would have eventually studied mathematics even without his father's blessing, but certainly his father's consent made things easier for him.[2]

Cantor studied at the University of Berlin under three of the truly great mathematicians of that time: Ernst Eduard Kummer, Karl W. T. Weierstrass, and Leopold Kronecker. He received his Ph.D. degree from Berlin in 1867. His doctoral dissertation, and indeed all his works until the early 1870s, although excellent, gave no hint of the outstanding mathematical originator that he was to become.

The birth of set theory can now be recognized in an 1874 article by Cantor which appeared in *Crelle's Journal*.[3] In this paper he proved the now well known theorem that the set of real algebraic numbers can be put in one-to-one correspondence with the set of natural numbers (positive whole numbers), whereas the same is not true of the set of real numbers.[4] Up to this time different orders of infinity had not been recognized. According to Fraenkel, Cantor himself had first thought that the continuum could be put in one-to-one correspondence with the set of natural numbers.[5] Cantor remarked in the introduction to his article that by combining the two above-mentioned theorems there results a proof of the theorem first proved by Liouville that in each given interval there exist infinitely many transcendental (nonalgebraic) real numbers.[6]

Cantor's next article[7] brought forth considerable opposition, expecially from his former teacher Kronecker. A number of important theorems

Reprinted from *Mathematics Teacher* 63 (Dec., 1970): 690–92; with permission of the National Council of Teachers of Mathematics. This article is adapted from the author's Ph.D. dissertation, "A History of Cantorian Set Theory," George Peabody College for Teachers, 1968, under the direction of Dr. J. Houston Banks.

concerning equivalent sets appeared in this paper.[8] One of the very well known results proved here is that the set of rational numbers can be put in one-to-one correspondence with the set of natural numbers.[9] Another interesting result (even though now obvious) noted by Cantor is something that had been noticed by Bernard Bolzano several years earlier: For infinite sets a set may be equivalent to a (proper) subset of itself; for example, the set of natural numbers is equivalent to the set of even natural numbers.[10, 11] Cantor also conjectured in this paper that the two powers of the rational numbers and the real numbers exhaust all possibilities for infinite subsets of the continuum.[12] Time has shown that he was overly optimistic, in that it has now been shown in the realm of axiomatic set theory that the conjecture is neither provable nor disprovable. Perhaps the result most objected to was Cantor's proof of the independence of the power of the continuum from its number of dimensions. That this result surprised even himself is evident in one of his letters to his close friend Richard Dedekind prior to publication of the proof.[13] It had been commonly assumed that points in two-space cannot be traced back to one-space, yet Cantor's proof said that the set of points in two-space is equivalent to the set of points in one-space. In fact, n-space is equivalent to one-space, and the result can even be expanded to the case of a countable infinity of dimensions.[14] These are some of the results in Cantor's second paper on set theory.

Cantor's second paper on set theory had rough sledding at *Crelle's Journal*, and he never published another paper in that journal. For a while it appeared that the paper would not be published when it was slated to be, apparently because of Kronecker's rejecting the point of view of Cantor's ideas (Kronecker was on the editing staff of the *Journal*).

Following Cantor's two papers that represent the earliest beginnings of set theory was an intensive working period by him in the years 1879–1884, during which he published practically his complete theory of sets. Cantor faced tremendous opposition in gaining recognition of these works. Kronecker and a number of other prominent mathematicians of that time were firmly aligned against his new and strange notions.

Cantor suffered a complete breakdown in the spring of 1884. In retrospect a number of contributing causes can be seen for this, among which were the trouble in getting his important paper of 1878 accepted, the hard struggle to gain recognition for his works of 1879–1884, and the formidable array of influential colleagues against his works. In addition, he was only moderately satisfied with the position that he occupied at Halle and would have preferred the wider field of work offered by the University of Berlin. As long as Kronecker was at Berlin, however, there was little chance that Cantor could get an appointment there.

Although Cantor's mental illness recurred throughout his life, his mental crisis was essentially over at the beginning of 1885; and his confidence in his work, which had been shaken, was reestablished. He published papers in set theory until 1897. Some of this work was particularly noteworthy in crystallizing some of his previous notions.

Hardly had Cantor's work been completed before paradoxes began to appear.[15] Despite the paradoxes and the difficulties in gaining recognition of his work, the creation of set theory is an undeniably important mathematical development. Fortunately Cantor lived long enough to see the beginning of the tremendous impact which his theory was destined to have on the mathematical world and to enjoy the pleasure of the belated recognition that he so much deserved.

NOTES

1. There was one earlier paper that contains some of the rudimentary concepts out of which Cantor's consideration of general point sets grew, but this is not being considered as his first paper on set theory since the paper was really on trigonometric series. The earlier

paper was "Ueber die Ausdehnung eines Satzes aus der Theorie der trigonometrischen Reihen," *Mathematische Annalen* 5 (1872): 123–32.

2. The biographical information about Cantor in this article is chiefly from Abraham Adolf Fraenkel's excellent biography, "Georg Cantor," *Jahresbericht der Deutschen Mathematiker Vereinigung* 39 (1930): 189–266.

3. Georg Cantor, "Ueber eine Eigenschaft des Inbegriffes aller reellen algebraischen Zahlen," *Journal für die reine und angewandte Mathematik* 77 (1874): 258–62.

4. The proofs of these two theorems are readily available in a number of sources and will not be repeated here. The proof (using Cantor's famous diagonal process) usually given of the second theorem mentioned is not the one that Cantor gave at this time but is due to a later work.

5. Fraenkel, "Georg Cantor," p. 237.

6. Ibid., p. 259.

7. Georg Cantor, "Ein Beitrag zur Mannigfaltigkeitslehre," *Journal für die reine und angewandte Mathematik* 84 (1878): 242–58.

8. Two sets are *equivalent* if their elements can be placed in one-to-one correspondence with each other.

9. The proof of this result also is readily available and will not be repeated.

10. Georg Cantor, *Contributions to the Founding of the Theory of Transfinite Numbers*, a translation by Philip E. B. Jourdain of two of Cantor's works published in 1895 and 1897 (New York: Dover Publications, n.d.), p. 41. The book is also provided with an introduction and notes by Jourdain. The introduction deals in part with the work of Cantor from 1870 to 1895 and is an especially helpful source for the material in this article.

11. Cantor used "subset" in the sense that "proper subset" is now used.

12. The *power (cardinal number)* of a set is the collection of all sets that are equivalent to it.

13. Fraenkel, "Georg Cantor," p. 237.

14. Cantor did not use the term "countable" until later. A set is *countable (denumerable)* if it is equivalent to the set of natural numbers.

15. The first published paradox of set theory was the Burali–Forti paradox of 1897, and in the next few years paradoxes began to appear in abundance. The paradoxes have been of considerable importance in motivating study in the foundations of mathematics and in the axiomatizations of set theory by such people as Ernst Zermelo, Fraenkel, John von Neumann, Paul Bernays, and Kurt Gödel.

101

Infinity: The Twilight Zone of Mathematics

WILLIAM P. LOVE

*T*HE CONCEPT OF INFINITY has fascinated the human race for thousands of years. Who among us has never been awed by the mysterious and often paradoxical nature of the infinite? The ancient Greeks were fascinated by infinity, and they struggled with its nature. They left for us many unanswered questions including Zeno's famous paradoxes. The concept of infinity is with us today, and many ideas in modern mathematics are dependent on the infinitely large or the infinitely small. But most people's ideas about infinity are very vague and unclear, existing in that fuzzy realm of the twilight zone.

Our mathematics curriculum includes the study of infinite sets of numbers in algebra and infinite sets of points in geometry, but students rarely understand how these may be related.

Ask your students these questions:

1. Which is larger: the set of all prime numbers or the set of all composite numbers?

2. Which is larger: the set of all rational numbers or the set of all irrational numbers?

3. Which is larger: the number of points in the interval [0, 1] or the number of points in the interval [0, 2]?

4. Which is larger: the number of points in a line or the number of points in a plane?

Reprinted from *Mathematics Teacher* 82 (Apr., 1989): 284–92; with permission of the National Council of Teachers of Mathematics.

Georg Cantor

The photograph of Georg Cantor appears through the courtesy of Ivor Grattan-Guinness.

Most students will answer these questions incorrectly because their intuitive understanding of finite sets does not always apply to infinite sets. In addition, most students are unaware that there are infinitely many different infinities.

Unless some ideas involving mathematical infinities are included in their instruction, students will never develop a true understanding of the various number systems introduced in their algebra classes nor will they understand how these sets relate to points, lines, and planes introduced in their geometry classes.

This article presents in a clear and understandable manner the main ideas concerning the theory of infinity as developed by Cantor and others. It includes most of the important theorems and a sketch of their proofs. Details of proofs may be found in advanced textbooks on set theory.

This material could be used as a reading assignment for interested students, as a topic for presentation at a mathematics club meeting, or as a source of information for teachers. Students might find it an interesting topic for a mathematics fair project. A teacher might challenge students to find their own proofs for some of the easier theorems presented here.

Let us examine infinity more carefully.

Georg Cantor (1845–1918) was one of the first modern mathematicians to seriously examine the infinite. His work was not accepted by everyone, and many considered him insane. He received little recognition in his lifetime and eventually died in an asylum. Today he is considered to be one of the founding fathers of the theories of the infinite, but even now there are some who do not accept his work. He began by looking at properties of sets, especially the cardinality of sets.

Cardinality of Sets

DEFINITION. *The cardinality of a set S is the number of elements in that set. This is denoted by n(S). For example, if F = {fingers on a hand}, then n(F) = 5.*

Sets A and B have the same cardinality if $n(A) = n(B)$. For small finite sets, it is easy to determine if two sets have the same number of elements by simply counting them. But when the sets are large it is not so easy. For example, in a large auditorium, are there more seats or more people? Rather than count them, it is easier to pair each person with one seat and then see if any people or seats are left over. This principle applies to both finite and infinite sets.

DEFINITION. *A one-to-one correspondence between sets A and B is a pairing of the elements of one set with the elements of the other set in such a way that all elements from both sets have exactly one partner.*

This is denoted by $A \leftrightarrow B$.

Thus, if it is possible to find a one-to-one correspondence between set A and set B, then A and B have the same number of elements and have the same cardinality; that is,

$$\text{if } A \leftrightarrow B, \text{ then } n(A) = n(B).$$

In addition,

$$\text{if } A \text{ is a subset of } B, \text{ the } n(A) \le n(B).$$

The First Transfinite Number: Aleph Null

In the 1880s, Cantor applied these ideas to infinite sets. He began by examining the infinite set of natural numbers, denoted by N. He defined the cardinality of the set of natural numbers to be *aleph null*, denoted $\aleph_0$, using the first letter of the Hebrew alphabet, $\aleph$, followed by the subscript zero. Thus, $n(N) = \aleph_0$. This is called a *transfinite* number, being "beyond the finite."

Cantor began to examine various sets of numbers to determine their cardinality relative to the natural numbers. He reasoned that if it was possible to find a one-to-one correspondence between a set S and the set of natural numbers, then they must have the same number of elements; that is,

$$\text{if } S \leftrightarrow N, \text{ the } n(S) = \aleph_0.$$

THEOREM 1. *Cardinality of the set of all even numbers is $\aleph_0$.*

Proof. If evens $\leftrightarrow$ naturals, then $n(\text{even}) = \aleph_0$.

even: 2   4   6   8   10 ... 2k ...
    ↓   ↓   ↓   ↓   ↓   ↓
natural: 1   2   3   4   5   ... k   ...

Therefore,

$$n(\text{even}) = \aleph_0.$$

THEOREM 2. *Cardinality of all nonnegative integral powers of 10 is $\aleph_0$.*

Proof. If powers of 10 $\leftrightarrow$ naturals, then n(powers of 10) = $\aleph_0$.

10^p: 10^0 10^1 10^2 10^3 10^4 ... 10^k ...

natural: 1   2   3   4   5   ... $k+1$ ...

Therefore,

$$n(\text{powers of } 10) = \aleph_0.$$

THEOREM 3. *Cardinality of all prime numbers is $\aleph_0$.*

Proof. If primes $\leftrightarrow$ naturals, then n(primes) = $\aleph_0$.

primes: 2   3   5   7   11   13   ... P_k ...

natural: 1   2   3   4   5   6   ... k ...

Although there is no formula for the pairing, the kth prime is paired with the natural number, k. Therefore,

$$n(\text{primes}) = \aleph_0.$$

THEOREM 4. *Every infinite subset of the natural numbers has cardinality $\aleph_0$.*

The student should recognize the implications of theorem 4. This means that all the following sets have the same number of elements: natural numbers, even numbers, odd numbers, prime numbers, composite numbers, multiples of 5, multiples of 10, perfect squares, perfect cubes, triangular numbers, Fibonacci numbers, and any other infinite subset of the naturals. All these have cardinality $\aleph_0$.

This implication violates one's intuition. How can a proper subset have the same number of elements as the entire set? One of the basic assumptions taught since the time of the ancient Greeks was that "the whole is greater than the

part." This notion is true for finite sets but not true for infinite sets. In 1888, Dedekind proposed this assumption as the basis for a definition of an infinite set.

DEFINITION. *A set S is infinite if and only if S has the same number of elements as one of its proper subsets.*

Cantor next examined sets that were "larger" than the set of natural numbers.

THEOREM 5. *Cardinality of the set of integers is $\aleph_0$.*

Proof. In integers $\leftrightarrow$ naturals, then n(integers) = $\aleph_0$. The positive integers are paired with the even naturals, and the nonpositive integers are paired with the odd naturals.

integers: ... -2 -1 0   1   2   ...

naturals: ...   5   3   1   2   4   ...

That is, whenever $k > 0$, $k \rightarrow 2k$; and when $k \leq 0$, $k \rightarrow -2k + 1$. Therefore,

$$n(\text{integers}) = \aleph_0.$$

Next, Cantor investigated the set of rational numbers. The rationals have the density property so that between any two distinct rationals there is an infinite number of other rationals. Students' intuition will tell them that there are more rational numbers than natural numbers, but Cantor proved that this assumption is false by using an unusual "diagonal process."

THEOREM 6. *Cardinality of the set of rational numbers is $\aleph_0$.*

Proof. If rationals $\leftrightarrow$ naturals, then n(rationals) = $\aleph_0$. Cantor used arrays to list all possible forms a/b. The positive array (see Fig. 1) includes all positive rational numbers, and a similar nonpositive array may be constructed to include all other ratio-

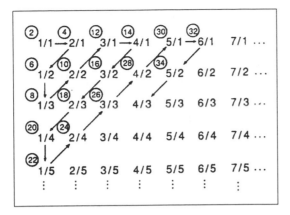

FIGURE I.

Cantor's one-to-one correspondence between the positive rational numbers and the even natural numbers

nal numbers. Cantor established a one-to-one correspondence between the positive array and even numbers as shown in Figure 1. Thus, positive array ↔ evens. In a similar manner, it is possible to construct the nonpositive array such that nonpositive array ↔ odds. Thus, combined arrays ↔ naturals, so n(combined arrays) = $\aleph_0$.

Since 1/2, 2/4, 3/6, . . . all represent the same rational number, we see that the combined arrays contain the set of all rational numbers. Thus,

$$n(\text{rationals}) \leq n(\text{combined arrays}) = \aleph_0.$$

The natural numbers are a subset of the rationals. Therefore,

$$\aleph_0 = n(\text{naturals}) \leq n(\text{rationals}).$$

Thus we have that

$$n(\text{rationals}) \leq \aleph_0,$$

and that

$$\aleph_0 \leq n(\text{rationals}).$$

Therefore,

$$n(\text{rationals}) = \aleph_0.$$

THEOREM 7. *If S is any infinite set whose elements may be uniquely determined by an ordered pair of integers $\{a, b\}$, then cardinality of S is $\aleph_0$.*

The proof is similar to that for theorem 6, since $\{a, b\}$ may replace a/b.

THEOREM 8. *If S is any infinite set whose elements may be uniquely determined by a finite ordered n-tuple $\{a_1, a_2, a_3, \ldots, a_n\}$, then cardinality of S is $\aleph_0$.*

The proof is left to the reader.

Next, Cantor investigated the set of irrational numbers. The irrational numbers consist of algebraic numbers (such as $\sqrt{2}$, $\sqrt{3}$, $\sqrt[3]{5}$) and transcendental numbers (such as π and e).

DEFINITION. *A number, k, is algebraic if and only if k is the solution to some algebraic equation of the form $a_n x^n + a_{n-1} x^{n-1} + \ldots + a_1 x + a_0 = 0$, where all a_i are integers.*

For example, the algebraic equation $x^2 - 2 = 0$ has two solutions: $x_1 = \sqrt{2}$ and $x_2 = -\sqrt{2}$. Thus, both are algebraic numbers. The first root, x_1, may be characterized by the set of coefficients of the equation $(1, 0, -2)$ plus the root number = 1:

$$x_1 \leftrightarrow \{1, 0, -2, 1\}$$

The second root, x_2, may be characterized by the set of coefficients of the equation $(1, 0, -2)$ plus the root number = 2:

$$x_2 \leftrightarrow \{1, 0, -2, 2\}$$

THEOREM 9. *Cardinality of the set of algebraic numbers is $\aleph_0$.*

Proof. Every algebraic number is the solution to some algebraic equation having a finite number of coefficients. Thus, each algebraic number can be uniquely characterized by a finite ordered n-tuple consisting of coefficients plus root number.

Therefore, by theorem 8, the cardinality of the set of all algebraic numbers = $\aleph_0$. An interesting result is observed by using some examples: n(even naturals) = $\aleph_0$; n(odd naturals) = $\aleph_0$; and

$$n(\text{even naturals} \cup \text{odd naturals})$$
$$= n(\text{naturals}) = \aleph_0.$$

This implies that

$$\aleph_0 = \aleph_0 = \aleph_0.$$

In general, any finite number of $\aleph_0$'s added together results in $\aleph_0$.

Cantor wondered if all infinite sets had cardinality $\aleph_0$. He spent three years trying to show that the cardinality of points on a line was $\aleph_0$, but he never succeeded. Instead, he made the most surprising discovery of his life. He found a new infinity.

A Second Transfinite Number: c

THEOREM 10. *Cardinality of points in interval* $(0, 1)$ *is not* $\aleph_0$.

Proof. If $(0, 1) \leftrightarrow N$ is impossible, then $n(0, 1) \neq \aleph_0$. To show this impossibility, Cantor had to

TABLE I

Cantor's Typical Correspondence between Natural Numbers and Decimal Number in (0, 1)

Natural Numbers	Decimal Number in (0, 1)
1	$\leftrightarrow$ 0. $a_1\, a_2\, a_3\, a_4\, a_5\, a_6$ ...
2	$\leftrightarrow$ 0. $b_1\, b_2\, b_3\, b_4\, b_5\, b_6$ ...
3	$\leftrightarrow$ 0. $c_1\, c_2\, c_3\, c_4\, c_5\, c_6$ ...
⋮	⋮
k	$\leftrightarrow$ 0. $k_1\, k_2\, k_3\, k_4\, k_5\, k_6$ ...
⋮	⋮

show that every possible correspondence $(0, 1) \leftrightarrow N$ would not work. To do this, he let each real number in $(0, 1)$ be written as an infinite decimal (such as $0.38 = 0.379\ 999\ 99 \ldots$). Then every such pairing could be shown as in Table 1. For every possible correspondence between N and $(0, 1)$ one can construct a decimal number z in the interval $(0, 1)$ that is not in the table and is not paired with any natural number. Define

$$z = 0.z_1\, z_2\, z_3\, z_4\, z_5 \ldots$$

as follows:

$$z_1 \neq a_1$$

Hence, z is not the number paired with 1.

$$z_2 \neq b_2$$

Hence, z is not the number paired with 2.

$$z_3 \neq c_3$$

Hence, z is not the number paired with 3. And so on. Thus, z is not paired with any natural number in N, so the correspondence $(0, 1) \leftrightarrow N$ is impossible. Therefore, $n(0, 1) \neq \aleph_0$.

Since $n(0, 1)$ is obviously infinite and $n(0, 1) \neq \aleph_0$, there must exist a transfinite cardinal number different from $\aleph_0$ — a second infinity. This discovery shocked the mathematical world. Cantor named the new transfinite number c for the "continuum."

DEFINITION. *Cardinality of points in interval* $(0, 1)$ *is c.*

With Cantor's discovery, two transfinite numbers had been identified: $\aleph_0$ and c. In addition, $\aleph_0$ is the "smallest" infinity and $\aleph_0 < c$. Any set having cardinality $\aleph_0$ is said to be *countably infinite*, and any set having cardinality c is *uncountably infinite*.

THEOREM 11. *Cardinality of points on interval* $[0.1]$ *is c.*

Proof. If $[0,1] \leftrightarrow (0, 1)$, then $n[0,1] = c$. This correspondence is established by three 1–1 onto functions:

$$f \qquad g \qquad h$$
$$[0,1] \leftrightarrow [0,1) \leftrightarrow (0, 1] \leftrightarrow (0, 1)$$

1. First, consider the function h:

Define $h_1(x) = 3/2 - x$ where $x \in (1/2, 1], y \in [1/2, 1)$.

Define $h_2(x) = 3/4 - x$ where $x \in (1/4, 1/2\}, y \in [1/4, 1/2)$.

Define $h_3(x) = 3/8 - x$ where $x \in (1/8, 1/4], y \in [1/8, 1/4)$.

Define $h_i(x) = 3/2^i - x$ where $x \in (1/2^i, 1/2^{i-1}], y \in [1/2^i, 1/2^{i-1})$.

We see that $h(x) = h_1(x) \cup h_2(x) \cup h_3(x) \cup \ldots \cup h_i(x)$ ... defines a one-to-one correspondence $(0, 1] \leftrightarrow (0, 1)$.

2. Second, consider the function g:

Define function $g(x) = 1 - x$. Clearly, $g(x)$ defines a one-to-one correspondence $[0, 1) \leftrightarrow (0, 1]$.

3. Third, consider the function f:

Define the function $f(x)$ such that $f(0) = 0$ and $f(x) = h(x)$ for $x \in (0, 1]$ as defined in part 1 of this proof. Clearly, f defines a one-to-one correspondence $[0, 1] \leftrightarrow [0, 1)$.

4. Thus, we have

$$[0, 1] \leftrightarrow [0, 1) \leftrightarrow (0, 1] \leftrightarrow (0, 1).$$

Hence, $[0, 1] \leftrightarrow (0, 1)$. Therefore, $n[0, 1] = c$.

Next, Cantor showed that all segments have the same number of points.

THEOREM 12. *Cardinality of points on interval* $[a, b]$ *is c.*

Proof. If $[a, b] \leftrightarrow [0, 1]$, then $n[a, b] = c$.

FIGURE 2

Geometric correspondence between points on [0, 1] and [a, b]

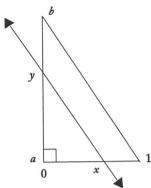

Geometric proof. Construct a right triangle from segments $[0, 1]$ and $[a, b]$ with 0 and a at the right angle. Construct a hypotenuse through 1 and b. Every line parallel to the hypotenuse that intersects the triangle will pair a point x in $[0, 1]$ with exactly one point y in $[a, b]$ as shown in Figure 2. Hence, $[a, b] \leftrightarrow [0, 1]$. Therefore $n[a, b] = c$.

Algebraic proof. Define $f(x) = (x - a)/(b - a)$, where $x \in [a, b]$ and $f(x) \in [0, 1]$. So f defines a one-to-one correspondence $[a, b] \leftrightarrow [0, 1]$. Therefore, $n[a, b] = c$.

COROLLARY. *All segments have the same number of points.*

Next, Cantor showed that lines and segments have the same number of points.

THEOREM 13. *Cardinality of points on line* $(-\infty, +\infty)$ *is c.*

Proof. If $(0, 1) \leftrightarrow (-\infty, +\infty)$, then $n(-\infty, +\infty) = c$.

Geometric proof. The interval $(0, 1)$ may be shaped into a semicircle having center P and at the point 0.5 being tangent to a line. Every ray from P that intersects the semicircle will also intersect the line and thus will pair exactly one $x \in (0, 1)$ with exactly one $y \in (-\infty, +\infty)$ as shown in Figure 3.

FIGURE 3

Geometric correspondence between points on [0, 1] and $(-\infty, +\infty)$

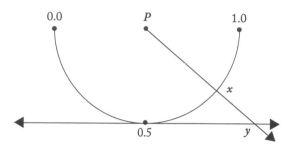

Algebraic proof. We know that

$$(0, 1) \leftrightarrow (-\frac{\pi}{2}, +\frac{\pi}{2})$$

from theorem 12. The function $f(x) = \tan(x)$ defines a correspondence

$$(-\frac{\pi}{2}, +\frac{\pi}{2}) \leftrightarrow (-\infty, +\infty);$$

so

$$(0, 1) \leftrightarrow (-\infty, +\infty).$$

Thus,

$$n(-\infty, +\infty) = c.$$

This means n(real numbers) $= c$.

COROLLARY. *All segments, rays, lines, circles, triangles, squares, rectangles, polygons, and plane curves have the same number of points, namely c.*

COROLLARY. *Cardinality of irrational numbers is c.*

Proof. Since

$$\text{reals} = \text{rational} \cup \text{irrational},$$

we have

$$n(\text{reals}) = n(\text{rationals}) + n(\text{irrationals}).$$

Thus,

$$c = \aleph_0 + n(\text{irrationals}).$$

If n(irrationals) $= \aleph_0$, then we have $\aleph_0 + \aleph_0 = c$, which is false. Thus, n(irrationals $\neq \aleph_0$. This means that n(irrationals) $> \aleph_0$, and assuming the next largest transfinite number is c (i.e., assuming the continuum hypothesis), we conclude that n(irrationals) $= c$.

COROLLARY. Cardinality of transcendental numbers is c.

Proof. Since

$$\text{reals} = \text{algebraic} \cup \text{transcendental},$$

we have

$$n(\text{reals}) = n(\text{algebraic}) + n(\text{transcendental}).$$

Thus,

$$c = \aleph_0 + n(\text{transcendental}).$$

Using the same proof as for the previous corollary, we conclude that n(transcendental) $= c$.

Cantor proved that transcendental numbers existed and that they were more numerous than rational or algebraic numbers.

Next, Cantor examined points in the plane.

Definition. *The unit square is the set of all points in the plane defined by the Cartesian product $(0, 1) \times (0, 1)$. If $P = (x, y)$ is a point in the unit square, then its coordinates may be written in decimal form:*

$$P = (0.x_1 x_2 x_3 \ldots, 0.y_1 y_2 y_3 \ldots)$$

THEOREM 14. *Cardinality of all points in unit square is c.*

Proof. If unit square $\leftrightarrow (0, 1)$, then n(unit square) $= c$. Form a one-to-one correspondence as follows:

$$P = (0.x_1 x_2 x_3 \ldots, 0.y_1 y_2 y_3 \ldots)$$
$$\leftrightarrow n = 0.x_1 y_1 x_2 y_2 x_3 y_3 \ldots$$

For example,

$$P = (0.123\,4 \ldots, 0.987\,6 \ldots)$$
$$\leftrightarrow n = 0.192\,837 \ldots$$

In this manner every point in the unit square is paired with exactly one number in interval $(0, 1)$. Thus, n (unit square) $= c$.

COROLLARY. *Cardinality of points in every plane region having area greater than zero is c.*

COROLLARY. *Cardinality of all points in a plane is c.*

COROLLARY. *Cardinality of all complex numbers is c.*

Proof. There exists a one-to-one correspondence between every complex number $a + bi$ and every point (a, b) in the plane. Thus, n(complex numbers) $= c$.

In a similar manner, Cantor turned his attention to sets of points in three-dimensional space.

DEFINITION. *The unit cube is the set of all points in space defined by the Cartesian product $(0, 1) \times (0, 1) \times (0, 1)$. If $P = (x, y, z)$ is a point in the unit cube, then its coordinates can be written in decimal form:*

$$P = (0.x_1 x_2 x_3 \ldots, 0.y_1 y_2 y_3 \ldots,$$
$$0.z_1 z_2 z_3 \ldots)$$

THEOREM 15. *Cardinality of points in a unit cube is c.*

Proof. If unit cube $\leftrightarrow (0, 1)$, then n(unit cube) $= c$. Form a one-to-one correspondence as follows:

$$P = (0.x_1 x_2 \ldots, 0.y_1 y_2 \ldots, 0.z_1 z_2 \ldots)$$
$$\leftrightarrow n = 0.x_1 y_1 z_1 x_2 y_2 z_2 \ldots$$

Example:

$$P = (0.123 \ldots, 0.456 \ldots, 0.789 \ldots)$$
$$\leftrightarrow n = 0.147\ 258\ 369 \ldots$$

In this manner, every point in the unit cube is paired with exactly one number in $(0, 1)$. Therefore, n(unit cube) $= c$.

COROLLARY. *Cardinality of all points in every geometric solid having volume greater than zero is c.*

COROLLARY. *Cardinality of all points in three-dimensional space is c.*

COROLLARY. *Cardinality of all points in N-dimensional space is c.*

Thus, the same number of points occur in the interval $[0, 1]$ as occur in all of space.

Cantor wondered if $\aleph_0$ and c were the only transfinite numbers, but another was to be discovered.

A Third Transfinite Number: d

Another transfinite number was found to be even larger than either $\aleph_0$ or c.

DEFINITION. Let

$$F = \{\text{all real functions } f \mid f: (0, 1) \to (0, 1)\}.$$

THEOREM 16. *Cardinality of F does not equal c.*

Proof. If $F \leftrightarrow (0, 1)$ is impossible, then $n(F) \neq c$. One must show that every 1–1 correspondence between F and $(0, 1)$ is impossible. Every such correspondence would pair each z in $(0, 1)$ with some function f_z in F as shown:

$$z \in (0, 1) \quad \leftrightarrow \quad f_z \in F$$
$$0.25 \in (0, 1) \quad \leftrightarrow \quad f_{0.25} \in F$$
$$0.38 \in (0, 1) \quad \leftrightarrow \quad f_{0.38} \in F$$
$$\ldots$$

For every possible correspondence between $(0, 1)$ and F, it is always possible to define a function $g \in F$ that is not in the table and is not paired with any real number in $(0, 1)$. For g and f_z to be different, they have to be unequal at one or more points. For example, if $g(x) \neq f_{0.25}(x)$ at $x = 0.25$ then g and $f_{0.25}$ are different. Define g as follows:

$$g(x) \neq f_z(x) \text{ at } x = z$$

for every $z \in (0, 1)$. Hence, g is different from every function in the table at one point and so is not paired with any number in $(0, 1)$.

Therefore, $F \leftrightarrow (0, 1)$ is impossible and $n(F) \neq c$.

Since F has more elements than $(0, 1)$, then $n(F) > c$ and is a larger transfinite number.

DEFINITION. *Cardinality of $F = \{f \mid f: (0, 1) \rightarrow (0, 1)\}$ is d.*

Thus, there exist at least three transfinite cardinal numbers:

$$\aleph_0 < c < d$$

COROLLARY. *Cardinality of $\{f \mid f: R \rightarrow R\}$ is d.*

In the plane exist infinitely many different circles, squares, rectangles, polygons, lines, rays, segments, curves, and figures. Infinitely many sets of points are possible in the plane.

THEOREM 17. *Cardinality of all sets of points in a plane is d.*

COROLLARY. *Cardinality of all sets of points in space is d.*

Infinitely Many Transfinite Numbers

If three transfinite numbers exist, it is a natural question to ask if more exist also. The answer is yes, more do exist.

Cantor proved a remarkable theorem comparing the cardinality of a set S and the cardinality of the set of all subsets of S. He showed that if $n(S) = A$, then n(all subsets of S) $= 2^A$ and $A < 2^A$. Therefore, if the cardinality of the natural numbers is $\aleph_0$, the cardinality of the set of all subsets of naturals must be $2^{\aleph_0}$. He called this new number *aleph 1*, $\aleph_1$. Since there exists a set having cardinality $\aleph_1$, then the cardinality of all subsets of it would be $2^{\aleph_1}$, which he called *aleph 2*, $\aleph_2$. Continuing in this way, Cantor showed that an infinite number of different transfinite numbers exists. He named them all alephs and formed the sequence

$$\aleph_0 < \aleph_1 < \aleph_2 < \aleph_3 < \aleph_4 < \cdots .$$

How do the transfinite numbers c and d fit with all the alephs? Cantor believed that $\aleph_1 = c$ and $\aleph_2 = d$,

but he was not able to prove it. This belief was called the *continuum hypothesis*. An equivalent form of the continuum hypothesis states that no transfinite numbers occur between $\aleph_0$ and c. A generalized continuum hypothesis assumes that the sequence of alephs are the only transfinite numbers.

Many mathematicians tried to prove the continuum hypothesis, and it was considered one of the most important unsolved problems in mathematics in 1900. The question of the continuum hypothesis has an interesting history and was not resolved until 1963 by Paul Cohen. It is now accepted as an axiom in modern set theory. However, by rejecting this axiom a new system can be created called *non-Cantorian set theory*.

Summary

1. All the following sets have cardinality $\aleph_0$:
   - Natural numbers, even numbers, and odd numbers
   - Multiples of 3, 4, 5, 6, and so on
   - Squares, cubes, and powers
   - Primes and composites
   - All infinite subsets of naturals
   - Integers, rational numbers, and algebraic numbers

2. All the following sets have cardinality c (or $\aleph_1$):
   - Transcendental numbers and irrational numbers
   - Real numbers and complex numbers
   - Points on all segments, rays, and straight lines
   - Points on every circle, polygon, or plane curve
   - Points on every unit square, polygonal region, or plane region
   - Points on every plane
   - Points in every sphere, cube, or geometric solid
   - Points in three-dimensional space or n-dimensional space

3. All the following sets have cardinality d (or $\aleph_2$):

- All real-valued functions $f: (0, 1) \rightarrow (0, 1)$
- All real-valued functions $g: R \rightarrow R$
- All possible sets of points in the plane
- All possible sets of points in three-dimensional space.

Conclusion

By learning about infinity, our students will understand a number of ideas fundamental to mathematics:

1. That properties that hold true for finite sets may not necessarily hold true for infinite sets

2. That for infinite sets, a set and a proper subset may have the same number of elements

3. That there is not just one infinity but rather infinitely many different infinities

4. That there is a fundamental difference between the infinite sets of the integers, natural numbers, and rational numbers as compared to the infinite sets of the irrational, real, and complex numbers.

BIBLIOGRAPHY

Abian, Alexander. *The Theory of Sets and Transfinite Arithmetic*. Philadelphia: W. B. Saunders Co., 1965.

Cantor, George. *Contributions to the Founding of the Theory of Transfinite Numbers*. New York: Dover Publications, 1915.

Cohen, Paul J., and Reuben Hersh. "Non-Cantorian Set Theory." In *Mathematics in the Modern World*. San Francisco: W. H. Freeman & Co., 1968.

Dauben, Joseph W. *Georg Cantor*. Cambridge, Mass.: Harvard University Press, 1979.

Davis, Paul J., and Reuben Hersh. *Mathematical Experience*. Boston: Birkhäuser Boston, 1981.

Drake, Frank R. *Set Theory—an Introduction to Large Cardinals*. Amsterdam: North Holland Publishing Co., 1974.

Hallett, Michael. *Cantorian Set Theory and Limitation in Size*. Oxford: Clarendon Press, 1984.

Kinsolving, May R. *Set Theory and the Number System*. Scranton, Pa.: International Textbook Co., 1967.

Lin, Shwu-Yeng, T. *Set Theory—an Intuitive Approach*. Boston: Houghton-Mifflin Co., 1974.

Maor, Eli. *To Infinity and Beyond: A Cultural History of the Infinite*. Boston: Birkhäuser Boston, 1987.

102

Irrationals or
Incommensurables V:
Their Admission to the
Realm of Numbers

PHILLIP S. JONES

We have seen in earlier notes[1] that incommensurable quantities were first discovered in geometric situations and were proved to exist by using the theory of evens and odds. As roots of equations, they were represented by symbols with which operations were performed, and they were approximated by rational numbers. Viete's and Descartes's literal symbolism helped to free these numbers from what in one sense was a too close association with geometric and physical magnitude. One the other hand Fermat's and Descartes's analytic geometry produced a need to associate a number with every point of a line—a feat impossible without the recognition of irrational numbers as abstract entities on a par with other numbers. This was a concept significantly different from the Greek idea of incommensurable magnitudes for whose ratios Eudoxus derived a treatment.

We might note parenthetically, however, that one French geometer, Legendre (1752–1833), in writing an elementary geometry to replace Euclid as a school text, proved propositions on similitude by applying algebraic-numerical reasoning to lit-

eral symbols which represented lengths. He thus turned the tables on the Greeks, who used lengths to represent numbers. But Legendre did not give a clear or modern exposition of irrationals.

A second motivation which we can now see would drive mathematicians to seek to clarify the idea of the irrational was the development of the calculus with its limits and problems of continuity. Augustin Cauchy, one of the prime movers in the rigorization of the calculus, in his *Cours d'Analyse* of 1821 treated irrationals as essential quantities which were familiar to all. He remarked that an irrational number could be the limit of a sequence of rationals. Although this is essentially the idea used later by Cantor, who *defined* irrationals as the limits of sequences of rationals, Cauchy's remark was logically incomplete since he did not give a definition of irrational numbers. Bernard Bolzano (ca. 1817) and C. F. Gauss were also thinking along this line.[2]

The persons responsible for the two chief theories of irrationals in use today are the Germans Georg Cantor (1845–1918) and J. W. R. Dedekind (1831–1916), both of whom published their first papers on this topic in 1872. Cantor's approach is somewhat similar to that propounded by the Frenchman C. Meray in 1869. His work is also

Reprinted from *Mathematics Teacher* 49 (Nov., 1956): 541–43; with permission of the National Council of Teachers of Mathematics.

similar, but less so, to the development of the German scholar, Karl Weierstrass, whose lectures of 1865–1866 on this topic were expanded and published by several of his students.

Cantor worked with sequences of rational numbers. Through definitions, he created a new number field out of the raw materials (rational numbers) which were on hand and understood. This process of extending old concepts or creating new ideas from old materials has happened repeatedly in the history of mathematics.

Cantor called a *fundamental sequence* any sequence of rational numbers $(a_1, a_2, a_3, \ldots, a_m, \ldots, a_n)$ which satisfied the condition that for every positive $\in$ there was an $N_\in$ such that $|a_m - a_n| < \in$ for all $m \geq N_\in$, $n \geq N_\in$. Such sequences satisfy Cauchy's condition for convergence, and hence are sometimes today called *Cauchy, regular* or *convergent.* Perron calls this "the criterion of Bolzano-Cauchy-Cantor."

Cantor considered that every sequence of this type represented a *real* number. It is fairly easy to show that this new set of numbers, defined and represented by limits of regular sequences, contains within it at least one sequence for each rational number. For example the sequence, (1.9, 1.99, 1.999, …, …) formed in the obious way corresponds to 2, while (.6, .66, .666, … …) corresponds to 2/3.

Further, the sum of two regular sequences $(a_1, a_2, a_3, \ldots, a_n, \ldots) + (b_1, b_2, b_3, \ldots, b_n, \ldots)$ is defined to be $(a_1 + b_1, a_2 + b_2, a_3 + b_3, \ldots, a_n + b_n \ldots)$. This sum can be shown not only to possess the usual desirable properties of associativity and commutativity, but also demonstrates that the sum of the sequences representing two rationals will turn out to be a sequence representing that rational which is the sum of the two original rationals.

After similar definitions of the product of two sequences, it can be shown that if sequences of rationals are thought of as elements of a set, then this set, with the operations noted above, forms a number *field.* That is, addition and multiplication are closed, associative, and commutative; there is

an identity element with respect to each operation; each element, except possibly zero, has an additive and multiplicative inverse; and multiplication is distributive with respect to addition.

Not only is the set of all regular sequences a field, but, as suggested above, there is a subset of these sequences which can be identified with the set of all rational numbers. Further, this subset also satisfies the field axioms in such a way that the result of adding or multiplying two rationals corresponds to the sequence which would result from adding or multiplying the sequences associated with the two original rationals. This isomorphism of the set of rationals with a subset of the set of regular sequences shows that the set of sequences is an *extension field* of the field of rationals.

Of course, just as there are many rationals corresponding to the integer 2 (e.g., 2/1, 4/2, 6/3, etc.), so there are many sequences corresponding to the rational 2/1. For example, in addition to that cited above, both (1, 3/2, 7/4, 15/8, …) and (2, 2, 2, 2, …) are such sequences. All of these are equal, however, under the definition of equality for regular sequences.

Before proceeding to a brief discussion of Dedekind's approach to irrationals, we should note several results related to Cantor's work. For example, the final solution of the ancient Greek problem of squaring the circle (with compasses and unmarked straight edge) was ultimately achieved only when Lindemann showed pi to be a transcendental number. Hence the problem is impossible of solution since ruler and compasses can construct only certain types of algebraic irrationals.

Euler had shown e to be irrational in 1737, long before Cantor; and Lambert had then shown pi to be irrational in 1767. However, some irrationals, such as the $\sqrt{2}$, are constructible, others such as $\sqrt[3]{2}$, cannot be so constructed, and in particular none of that class of irrationals called "transcendental"[3] can be so constructed.

Although their existence had long been suspected, it was not until 1844 that Liouville proved that there are such things as transcendental numbers. Even then he did not produce a single particular example. In 1871 Cantor proved that there is an infinite number of transcendentals. To do this, his theory of transfinite numbers and infinite aggregates was needed. This theory showed as well that, although the transcendentals and rationals are both infinite in number, there are actually more transcendentals than there are rationals or even algebraic numbers.

The first number to be proven transcendental was *e*. Hermite demonstrated this in 1873. Lindemann's proof of the transcendence of pi followed in 1882.

In spite of the large number of transcendentals proven to exist, it has been and still is a difficult problem actually to identify them. In fact, the seventh of the famous twenty-three unsolved problems presented by David Hilbert in 1900 was to prove the transcendence of a class of numbers, of which $2^{\sqrt{2}}$ and e^{π} were examples. In 1929, the Russian, A. Gelfond, proved that e^{π} is transcendental, and in 1934 he gave a complete solution for Hilbert's problem, but the testing of particular numbers for transcendence is still not easy.[4]

Dedekind's extension of the rationals also defined new numbers by use of sets of the "old" rational numbers. However, instead of discussing convergent sequences, he talked of partitions of the set of all rationals into two sets, such that every number of the second set was greater than every member of the first. Every such partition was called a *schnitt*, or a *cut*, and by definition was identified with a real number.

In some cuts there is a last element in the first set, or a first element in the second. Cuts of this type are identified with the rational numbers. But there are also cuts that have no last element in the first set, nor first in the second. Such cuts may also be regarded as defining real numbers, but in these cases the numbers correspond to irrationals.

For example, if the first set includes all the rationals whose square is less than 2, it has no last element, and the second set containing all rationals whose square is greater than 2 has no first element.

These two sets are a partition of the set of all rational numbers which defines the $\sqrt{2}$.

After defining addition, multiplication, and equality of cuts, they too can be shown to be a field which has a subset that is isomorphic to the rationals. Hence the set of cuts is an extension field of the rationals.

In teaching, there are two other related topics that ought to be mentioned with the study of incommensurables and irrationals. First, every rational number can be written as a periodic or repeating decimal fraction, while irrationals correspond to infinite nonrepeating decimals: Second, "quadratic surds" correspond to repeating continued fractions. The first fact should be the topic for an interesting discussion in many secondary school classes because the period of the decimal representation of rationals is associated with the use of the base 10, while the correspondence itself may be associated with geometric progressions in second-year algebra. Continued fractions are a less common topic which merit more attention as enrichment material associated with rationalizing radical expressions and the Euclidean algorism.

NOTES

1. P. S. Jones. "Irrationals or Incommensurables I, II, III, IV." *The Mathematics Teacher*, XLIX (February, March, April, October, 1956), pp. 123–127, 187–191, 282–285, 469–471).

2. Oskar Perron, *Irrationalzahlen* (1939), p. 55 ff. Although it has not been quoted directly, much use in this series has also been made of notes of J. Molk's translation exposition of A. Pringsheim's "Nombres Irrationnels et Notion de Limite" in Tome I, vol. 1 of *Encyclopédie des Sciences Mathématiques Pures et Appliquées* (Paris, 1904).

3. Transcendental numbers are those which cannot be the root of any algebraic equation. Algebraic equations are all those which can be written in the form $a_n x^n + a_{n-1} x^{n-1} + \ldots ax + a = 0$ where the exponents are integers and the a_i are rational numbers.

4. Harry Pollard, *The Theory of Algebraic Numbers* (Mathematical Association of America, 1950) pp. 42–46; Einar Hille, "Gelfond's Solution of Hilbert's Seventh Problem," *American Mathematical Monthly*, XLIX (December, 1942), 654–661.

103

The Genesis of Point Set Topology:
from Newton to Hausdorff

JEROME H. MANHEIM

$\mathcal{P}$ARADOXES OF THE INFINITE, discussed by the Greeks, became urgent problems with the codification of calculus. Mathematicians of the seventeenth and eighteenth centuries sought to refer analysis to arithmetic in a way that would allow them to retain their reliance on the visual sense, i.e., on geometrical representation. During this period, when Euclidean geometry was the final truth-referent in mathematics, such an approach was clearly indicated. Berkeley's valid objections (1734) to the processes of Newton and Leibniz were not supplemented with a constructive program. The only possible constructive program, the arithmetization of analysis, would have been most unlikely as long as geometry occupied the dominant role. The works of the early analysts are viewed as the point of departure for systematizing problems which led to the development of point set topology.

The movement towards abstraction and generalization in mathematics is a measure of progress towards the notion of a topological space. In this respect a first debt is owed to Euler (1755) and other formalists, men who sought to work beyond the confines of geometry. But eighteenth-century formalism was a failure because it developed without adequate attention being paid to questions of convergence, and thereby yielded results which were demonstrably false.

Ultimately analysis depended upon a satisfactory theory of limits. Recognition of this fact is attributable to D'Alembert (ca. 1750), while credit for its implementation at what might be called the "naive level" belongs to Cauchy (1821). Mathematicians came to realize that an acceptable structure of analysis made the same demands upon rigor as did geometry and this gave rise to two separate, conflicting, lines of inquiry. The one typified by D'Alembert (1754), Gauss (1812), and Cauchy (1821) considered the limit concept as basic, while men such as Maclaurin (1742), Landen (1764), and Lagrange (1813) sought to achieve rigor by eliminating limit considerations from analysis. Cauchy, in particular, believed that the limit concept, by replacing the vague notion of infinitesimals, would be the instrument which would banish geometric intuition from analysis. Lagrange applied his skill in an effort to achieve a rigorous construction within the confines of the discrete. Although each school achieved a measure of success, neither fulfilled its ambition. Those who sought to employ the limit concept failed to make a valid definition. In essence their definition of a limit depended upon the prior existence of the same limit. Alternatively, the would-be arithmetization implicitly employed the limit notion.

Reprinted from *Mathematics Teacher* 59 (Jan., 1966): 36–41; with permission of the National Council of Teachers of Mathematics. Originally published as Section 7.1 of the final chapter of Jerome H. Manheim, *The Genesis of Point and Set Topology* (Oxford: Pergamon Press; and New York: the Macmillan Company, 1964). Permission to reproduce this section was kindly granted by Pergamon Press.

While the movement towards arithmetization had its beginnings in attempts to rigorize calculus, a major impetus was given to this tendency by the growing realization that the possibility of geometrical representation was inadequate as a base for mathematical truth. Problems that arose as a result of the analysis of the differential equation of the vibrating string caused new consideration to be given to Euler's contention of the inadequacy of geometry. Euler himself continued to play the role of formalist in his considerations of the string problem. This time, however, his manipulations were coupled with exceptional insight into the physical situation. Thus, while Euler (1750) identified a function with its graph, he felt that truth was more apt to be uncovered by formal processes than by reliance on the visual sense. D'Alembert's (1749) notion of function was indeed more general than Euler's, but when limited by the then-current belief that the graph of all functions was a proper subset of all graphs, it became a very restrictive concept. From the point of view of the development of topology, the importance of both these views of functions rests in their destruction.

The broadening of the function concept had its roots in Daniel Bernoulli's solution of the vibrating string problem (1750). Had Bernoulli's argument for a series solution been more mathematical and less speculative, an extension of the notion of function clearly would have been indicated. Instead, Euler gave a counter argument (1755) which was intended to demonstrate the absurdity of Bernoulli's claim to generality, an argument which preserved Euler's concept of function. Lagrange came close to validating Bernoulli's result and deriving the (Fourier) series expansion for very general functions (1759). He failed largely because of his single-minded purpose to prove the correctness of Euler's conclusions.

Fourier, in the course of his investigations on the theory of heat (1807), made the claim: "But herein we have dealt with a single case only of a more general problem, which consists in developing any function whatever in an infinite series of sines or cosines of multiple arcs."[1] Since to Fourier a function included arbitrary representations as well as those given by formulas, this statement initiated numerous inquiries intended to establish the exact reach of its validity. These investigations gave rise to the study of point set theory.

The first rigorous study of Fourier series was undertaken by Dirichlet in 1823. Here, for the first time, sufficiency conditions were soundly proved. This study confirmed a certain generality for Fourier representability and furthermore demonstrated, by counter-example, that the inclusiveness asserted by Fourier (1807) was false. It was in the course of these inquiries that Dirichlet (1837) extended the function concept, specifically the notion of continuous function, to include very general correspondences.

Riemann (1854) recognized the desirability of establishing necessity conditions that a function be Fourier representable. Since the coefficients of such an expansion are found by integration, Riemann first entered into an investigation of the definite integral, extending the meaning to certain functions which do not satisfy the piecewise continuity requirement laid down by Cauchy (1821). The fact that certain infinite sets of discontinuities did not destroy the integrability property, while others did, was a first step towards a classification of such sets. Riemann, however, limited his generalization of the integral concept to the study of Fourier series and did not pursue the set-theoretical consequences. Many of his results were employed by later researchers who sought to establish conditions for the uniqueness of the Fourier representation.

The relative status of geometric and analytic representation was not demonstrably disturbed by the introduction of infinite point sets into Fourier series considerations. The heirs of Fourier believed they had established the equal potency of the two forms of representation and were satisfied that this was the final determination. That there might be functions analytically expressible whose graphs would not be sketched was not even imagined.

Bernard Bolzano (1817) saw the necessity of proving the Fundamental Theorem of Algebra without reference to geometric intuition. His demonstration ultimately depended upon establishing the sufficiency of Cauchy's convergence criterion. Cauchy's supposed analytic proof of this theorem (1821) did not suppress, but simply altered, the role of geometric intuition. Bolzano's failure was logical in nature.

Bolzano gave a method for constructing a continuous curve which was not differentiable at an everywhere dense set of points (ca. 1830), but he did not publish his example. It was unlikely, even had it been published, that it would have had any significant effect on the mathematical community; since no formula was given. Bolzano's curve would not have been considered a function.

An example of a continuous function with an everywhere dense set of points of non-differentiability was given by Riemann in his 1854 *Habilitationsschrift*. But this example (which by all criteria then in effect was surely a function) failed to excite much interest in either Riemann or his distinguished audience, and remained unpublished until 1867.

The method of Riemann was generalized by Hankel (1870) and Cantor (1882). The importance of pathological functions, as they affect the foundations of analysis, resulted from a discovery by Weierstrass (1861). His remarkable function, which was everywhere continuous and nowhere differentiable, impressed upon mathematicians the necessity of restructuring all of analysis without reference to spatial intuition. This recognition was a clear vindication of the need expressed by Bolzano half a century earlier. The programs which began as an attempt to rescue analysis from the onslaught of pathological functions led to a new discipline, point set topology, a subject ideally suited to the study of just such functions.

To rebuild the foundations of analysis it was necessary to obtain a satisfactory definition of irrational numbers from the rationals. Four mathematicians were associated with this process.

Charles Meray (1869) saw that the definition of an irrational number as the limit of a certain sequence of rational numbers, as given by Cauchy in 1821 and in use since then, must rest upon geometric intuition if it was to avoid circularity. In Meray's reconstruction of the real numbers he begins with the notion of *progressive variable*. A progressive variable v is a quantity (there is no connotation of measurement in this use of the term) that receives an infinite set of values $\{v_n\}$, these values being taken from the domain of the rationals. Since the rational numbers are given, there is no difficulty in assigning a meaning to *rational limit*. It follows that if V is a rational limit of $\{v_n\}$ then, for n sufficiently large and for arbitrary $\in > 0$, $|v_{n+p} - v_n| < \in$ for all p. If the inequality still holds when there does not exist a rational limit, the sequence is said to have a *fictitious limit*, and these fictitious limits are called *fictitious numbers*. The fictitious numbers thus created are the irrationals.

Weierstrass (ca. 1870) defined a *numerical quantity* as a set of elements subject to the conditions that each element may occur only a finite number of times and the number of times each element occurs is known. The value assigned to a numerical quantity with a finite number of elements is the sum of its elements. After defining the meaning to be attached to the sum of finitely many numerical quantities, Weierstrass extends the definition to include infinitely many such objects. This allows consideration to be given to the sum of a convergent infinite set of (rational) numerical quantities. Those sums that are not rational define the irrationals.

Cantor (1872) defines an *elementary series* as a sequence of rational numbers $\{a_i\}$ such that a_n and $a_n - a_{n+r} \rightarrow 0$ with $1/n$. Such a sequence may or may not have a rational limit. But whether or not it has such a limit the series serves to define a real number. If there is no rational limit, the real number so defined is called irrational.

Dedekind (ca. 1858) partitioned the rationals into two nonempty subsets A and B such that every rational is in A or B and such that every member of A is less than every member of B. Respecting these sets and the phrases "has a last element," "has a first element," there are four permutations:

1. *A* has a last element and *B* has a first element.
2. *A* has no last element but *B* has a first element.
3. *A* has a last element but *B* has no first element.
4. *A* has no last element and *B* has no first element.

Situation (1) is impossible by virtue of the density property of the rationals. Numbers (2) and (3) serve to define the rational which is the first element of *B* or the last element of *A* respectively. The fourth situation, which does not identify a rational, is used by Dedekind to define the irrationals.

There were numerous objections, both to arithmetization *per se* and to the methods employed by the arithmetizers. But the majority of mathematicians became convinced that analysis was firmly embedded in the integers. With reliance upon spatial intuition eliminated as a necessary component of analytic investigations, it became possible to study hitherto proscribed topics. When abstract spaces made their appearance some years later, no significant objections could be raised. This generalized notion of spaces, however, depended also on a theory of point sets.

Point set considerations developed from attempts to establish conditions under which the representation of a function by a trigonometric series is unique. The need to establish uniqueness arose from recognition of the role of uniform convergence in Fourier series investigations. Abel (1826) had given an example of a convergent sequence of continuous functions which failed to converge to a continuous function. This example had little impact on the development of mathematics until Seidel related this phenomenon to uniform convergence in 1848. Weierstrass demonstrated that if a function had a series expansion which converged uniformly, then the sum of the integrated terms represented the integral of the function. In 1870 Heine noted that the proof that a function had a unique Fourier expansion depended upon termwise integration and hence tac-

itly assumed uniform convergence. This observation inaugurated a sequence of investigations designed to establish the conditions under which the Fourier expansion was unique, a question of practical as well as theoretical importance. The initial results of point set theory were a by-product of these investigations.

The first attempt to rescue Fourier series from what appeared as the destructive influence of uniform convergence was due to Heine (1870). He showed that if a function satisfied the Dirichlet conditions in $(-\pi, \pi)$ its Fourier series is uniformly convergent in any open interval if the function is continuous throughout the interval and at the end points.[2] Four years later Paul du Bois-Reymond showed that continuity of a function was insufficient to guarantee the uniform convergence of its series so that, even for a continuous function, there may be more than one trigonometric series.

The uniqueness problem engaged the interest of Georg Cantor. He began his researches by considering the most general trigonometric series, i.e., a series whose coefficients are not restricted to the integral form. In his papers, which started to appear when he was only twenty-five years old (1870), he established several important theorems. One of these results (1872) showed that an infinite set of points of discontinuity was not, of itself, sufficient to destroy uniqueness, a result complementary to that of du Bois-Reymond (1876). In the course of proving this theorem Cantor found it necessary to distinguish between various types of infinite point sets. Distinctions made by earlier investigators (with the exception of du Bois-Reymond) had invariably been made on the basis of isolating a subset (not necessarily proper) of the rationals from the reals. This type of classification failed to suit Cantor's needs and, what is more surprising, under his arrangement into *kinds* both the rationals and the reals were of the same kind.

Cantor saw that a theory of sets had implications for all branches of mathematics and, as had been anticipated by Dirichlet, that the set concept was more fundamental than that of function. The subsequent development of naive set theory is due almost exclusively to Cantor (1872 *et seq.*). Most mathematicians insisted, contra Cantor, that infinity was only approachable and consequently not subject to an arithmetic. Eventually the importance of Cantor's approach was recognized and, while one group of mathematicians was concerned with eliminating the antinomies of set theory through axiomatization, another reached for generalization.

Throughout the evolution of the function concept geometry had been under attack, but some mathematicians refused to abandon this subject simply because it lost its supremacy to analysis. These men sought to apply concepts which were essentially nongeometric to the classification of geometries. A first attempt was made by Riemann in 1854. Eighteen years later Felix Klein, in his famous *Erlanger Programm,* considered that he had brought order to the classification of geometries via the theory of groups. But the *Erlanger Programm* failed to recognize that a space, by itself, can be an object of study, independent of the nature of the transformations of its elements. The emergence of point set topology demonstrated the inadequacy of Klein's view.

The study of sets of objects other than points began in 1833 with Ascoli's paper on sets of curves. Shortly thereafter (1887) Volterra and Arzela studied sets of functions, researches which gave rise to Functional Analysis. Also in 1887 Hadamard proposed that investigations be made of restricted function sets. Borel (1903) suggested a way of extending the notion of *nearness,* and therefore *limit,* to sets of lines and planes. In this way he generalized many of the concepts originally developed for point sets by Cantor. Fredholm's work of the same year led to the notion of a set of functions as a set of points. About this time Hilbert (1901) considered the set of functions for which the Bolzano-Weierstrass theorem held true.

The generalization by abstraction of the set concept was made by Fréchet. In his 1906 thesis he considered first an arbitrary set of elements E and an operation which maps E into some numerically determined value. The generality Fréchet afforded to the set of relations involving the points

meant that the relations were not necessarily explainable in terms of the group concept. Fréchet's work, based upon limit, did not, however, achieve the generality of a topological space.

Riesz based his treatment of sets on the concept of condensation point (1908), hoping thereby to avoid the limitations of Fréchet's study, but he did not develop many of the consequences. In 1913 Hermann Weyl proposed a development of general two-dimensional manifolds from the point of view of neighborhoods. For all practical purposes Weyl was an axiom away from the notion of a topological space, an axiom of separation.

The appearance in 1914 of Hausdorff's *Grundzüge der Mengenlehre* marks the emergence of point set topology as a separate discipline. A *space* became merely a set of points and a set of relations involving these points, and a geometry was simply the theorems concerning the space. Hausdorff began his development of topology with a small noncategorical set of neighborhood axioms. After deriving many of the properties of the general space thus defined he progressively introduced new axioms, ultimately deducing metric spaces and, finally, particular Euclidean spaces.

Looking backwards, point set topology developed from arithmetization and set theory. The latter owes its origin to problems in Fourier series representation while the former is principally indebted to the introduction of pathological functions into analysis. Both these problems, in turn, had their roots in the notion of function, a notion which for long sought to preserve the dominant role of geometry. The evolution of the function concept began with the codification of calculus, when the apparant paradoxes of the infinitesimal processes became serious problems.

NOTES

1. Joseph Fourier, *The Analytical Theory of Heat,* trans. Alexander Freeman (New York: Dover, 1955), p. 168.

2. Horatio Scott Carslaw, *Introduction to the Theory of Fourier's Series and Integrals* (New York: Dover, 1930), p. 13., notes that it was shown later that a Fourier series could be integrated termwise even if the series did not converge.

104

The Four-Color Map Problem, 1840–1890

H. S. M. COXETER

Möbius

August Ferdinand Möbius (1790–1868), a pupil of Gauss, made important contributions to many branches of mathematics. In particular, he was one of the first people to conceive the idea of four-dimensional space. His student, R. Baltzer[1] remembered, after forty-five years, a lecture during which Möbius had told the following story: "There was once a king in India who had a large kingdom and five sons. In his will he decreed that, after his death, the kingdom should be divided among his sons in such a way that the territory of each should have a common boundary line (not merely a point) with the territories of the remaining four. How was the kingdom divided?" In the next lecture, the students confessed that they had tried in vain to solve the problem. Möbius laughed and remarked that he was sorry they had wearied themselves, because the proposed partition is impossible. Baltzer quotes a proof of the impossibility, given by Möbius's friend Weiske.

De Morgan, Cayley, and Tait

Ten years after the death of Möbius, at a meeting of the London Mathematical Society (June, 1878), the eminent English mathematician Arthur Cayley raised the following question: "Has a solution been given of the statement that in colouring a map of a country, divided into counties, only four distinct colours are required, so that no two adjacent counties should be painted in the same colour?" Cayley, whose collected papers now fill thirteen large volumes, had to confess that he knew no proof himself.

At a meeting of the Royal Society of Edinburgh (March, 1880) P. G. Tait remarked, "Some years ago, while I was still working at knots, Professor Cayley told me of De Morgan's statement that four colours had been found by experience to be sufficient for the purpose of completely distinguishing from one another the various districts on a map. I had previously shown that if an even number of boundaries meet at each point on the diagram, two colours (as on a chessboard) will suffice for the purpose. But in a map, boundaries usually meet in threes. I replied to Professor Cayley that I thought the proof might be made to depend upon the obvious proposition that not more than four points in a plane can be joined two and two by nonintersecting lines. (Here points are made to stand for districts. When two such points are joined by a line they must have different colour-titles.) I did not at the time pursue the subject, as I found that it was more complex than it appeared at first. Mr. Kempe's paper in *Nature* (Feb. 26, 1880) has recalled my attention to the subject." Tait then gave a sketch of his own treatment, which is described in a well-known book, Rouse Ball's *Mathematical Recreations and Essays.*[2]

Reprinted from *Mathematics Teacher* 52 (Apr., 1952): 283; with permission of the National Council of Teachers of Mathematics.

NOTES

1. R. Baltzer, "Eine Erinnerung an Möbius und seinen Freund Weiske," *Bericht über die Verhandlungen der Sächsischen Akademie der Wissenschaften zu Leipzig. Math.-Nat. Kl.* 37 (1885), 1–6, 1.

2. W. W. Rouse Ball, *Mathematical Recreations and Essays,* rev. H. S. M. Coxeter (London: Macmillan & Co., 1959), 224–28.

105

The Origin of the
Four-Color Conjecture

KENNETH O. MAY

CONSIDERING THE FAME and tender age of the four-color conjecture,[1] our knowledge of its origins is surprisingly vague. A well-known tradition appears to stem from W. W. R. Ball's *Mathematical Recreations and Essays,* whose first edition appeared in 1892. There it is said that "The problem was mentioned by A. F. Möbius in his lectures in 1840 . . . but it was not until Francis Guthrie communicated it to De Morgan about 1850 that attention was generally called to it. . . . It is said that the fact had been familiar to practical mapmakers for a long time previously."[2] In spite of repetition by later writers, this tradition does not correspond to the facts.

In the first place, there is no evidence that mapmakers were or are aware of the sufficiency of four colors. A sampling of atlases in the large collection of the Library of Congress indicates no tendency to minimize the number of colors used. Maps utilizing only four colors are rare, but those that do so usually require only three. Books on

cartography and the history of mapmaking do not mention the four-color property, though they often discuss various other problems relating to the coloring of maps.

If cartographers are aware of the four-color conjecture, they have certainly kept the secret well. But their lack of interest is quite understandable. Before the invention of printing it was as easy to use many as to use few colors. With the development of printing, the possibility of printing one color over another and of using such devices as hatching and shading provided the mapmaker with an unlimited variety of colors. Moreover, the coloring of a geographical map is quite different from the formal problem posed by mathematicians because of such desiderata as coloring colonies the same as the mother country and the reservation of certain colors for terrain features, e.g., blue for water. The four-color conjecture cannot claim either origin or application in cartography.

To support his statement about Möbius, Ball refers to an article by Baltzer, a former student of Möbius and the editor of his collected works.[3] However, as has been pointed out by H. S. M. Coxeter,[4] this article shows merely that Weiske communicated to Möbius a puzzle whose solution amounted to the claim that it is impossible to have five regions each having a common boundary with every other. Baltzer witnesses that in 1840 Möbius presented this puzzle to a class and laughingly revealed its impossibility in the next lecture. But nothing was published on the problem, and its

Reprinted from *Mathematics Teacher* 60 (May, 1967): 516–19; with permission of the National Council of Teachers of Mathematics. This article is based on work done during the tenure of a Science Faculty Fellowship from the National Science Foundation. It was presented to the Minnesota Section of the Mathematical Association of America on November 3, 1962, and to the Midwest Junto of the History of Science Society on April 5, 1963. It appeared in *Isis,* LVI (1965, 3, No. 185), pp. 346–48, and is republished here with the kind permission of that journal. The author is indebted to S. Schuster, G. A. Dirac, J. Dyer-Bennet, Oystein Ore, and H. S. M. Coxeter for discussion and suggestions.

source remained unknown until Baltzer discovered Weiske's communication among the Möbius papers more than forty years later, when the four-color conjecture was already well known. Baltzer found no evidence that Möbius worked on the problem, and it is not mentioned in the collected works published in 1885–1887. Evidently Weiske's puzzle and its mention by Möbius were fruitless and had no historical link with the origin of the four-color conjecture.

The statement that attention was generally called to the problem about 1850 appears also to be not entirely accurate. Indeed, the first printed reference to it appeared in 1878, when the *Proceedings of the London Mathematical Society* reported Cayley's question as to whether the conjecture had been proved.[5] Interest was immediate, and a long series of partial solutions and pseudosolutions began with the papers of Kempe[6] and Tait.[7] Cayley, Kempe, and Tait attributed the problem vaguely to Augustus De Morgan.

More precise information was supplied by the physicist Frederick Guthrie in a communication to the Royal Society of Edinburgh in 1880. He wrote,

> Some thirty years ago, when I was attending Professor De Morgan's class, my brother, Francis Guthrie, who had recently ceased to attend them (and who is now professor of mathematics at the South African University, Cape Town), showed me the fact that the greatest necessary number of colors to be used in coloring a map so as to avoid identity of color in lineally contiguous districts is four. I should not be justified, after this lapse of time, in trying to give his proof, but the critical diagram was as in the margin. [The diagram shows four regions in mutual contact.]

> With my brother's permission I submitted the theorem to Professor De Morgan, who expressed himself very pleased with it ; accepted it as new; and as I am informed by those who subsequently attended his classes, was in the habit of acknowledging whence he had got his information.

> If I remember rightly, the proof which my brother gave did not seem altogether satisfactory to himself; but I must refer to him those interested in the subject. I have at various intervals urged my brother

to complete the theorem in three dimensions, but with little success. . . . [8]

A hitherto overlooked letter from De Morgan to Sir William Rowan Hamilton permits us to pinpoint events more precisely than could Frederick Guthrie, after the passage of almost thirty years. On October 23, 1852, De Morgan wrote as follows:

> A student of mine asked me today to give him a reason for a fact which I did not know was a fact, and do not yet. He says, that if a figure be anyhow divided, and the compartments differently coloured, so that figures with any portion of common boundary *line* are differently coloured—four colours may be wanted, but no more. Query cannot a necessity for five or more be invented? As far as I see at this moment, if four *ultimate* compartments have each boundary line in common with one of the others, three of them inclose the fourth, and prevent any fifth from connexion with it. If this be true, four colours will colour any possible map, without any necessity for colour meeting colour except at a point.

> Now, it does seem that drawing three compartments with common boundary, two and two, you cannot make a fourth take boundary from all, except inclosing one. But it is tricky work, and I am not sure of all convolutions. What do you say? And has it, if true, been noticed? My pupil says he guessed it in colouring a map of England. The more I think of it, the more evident it seems. If you retort with some very simple case which makes me out a stupid animal, I think I must do as the Sphynx did. If this rule be true, the following proposition of logic follows:

> If *A, B, C, D*, be four names, of which any two might be confounded by breaking down some wall of definition, then some one of the names must be a species of some name which includes nothing external to the other three.[9]

On October 26, Hamilton replied: ". . . I am not likely to attempt your 'quaternion of color' very soon." Evidently the response of others to the conjecture was equally passive. Even Francis Guthrie published nothing on it, though he lived until 1899 and produced a book and several papers on other topics.

Apparently it was on October 23, 1852, that Frederick Guthrie communicated his brother's conjecture to De Morgan, who did not bother to explain to Hamilton the indirect nature of the communication.[10] Probably De Morgan, in telling others about the problem, mentioned that it had occurred to Guthrie while coloring a map, and this gave rise to the tradition linking the conjecture with the experience of cartographers.[11]

On the basis of the data given here we can replace tradition by the following account of the origin of the four-color conjecture. It was not the culmination of a series of individual efforts, but flashed across the mind of Francis Guthrie, a recent mathematics graduate, while he was coloring a map of England. He attempted a proof, but considered it unsatisfactory—thus showing more critical judgment than many later workers. His brother communicated the conjecture, but not the attempted proof, to De Morgan in October 1852. The latter recognized the essentially combinatorial nature of the problem, gave it some thought, and tried without success to interest other mathematicians in attempting a solution. He communicated it to his students, giving due credit to Guthrie, and to various mathematicians, one of whom revived the problem almost thirty years later and launched it on its erratic course.

Rarely is a mathematical invention the work of a single individual, and assigning names to results is generally unjust. In this case, however, it would seem that the four-color conjecture belongs uniquely to Francis Guthrie and could fairly be called Guthrie's problem.[12]

NOTES

1. In nontechnical terms the four-color conjecture is usually stated as follows: Any map on a plane or the surface of a sphere can be colored with only four colors so that no two adjacent countries have the same color. Each country must consist of a single connected region, and adjacent countries are those having a boundary line (not merely a single point) in common. The conjecture has acted as a catalyst in the branch of mathematics known as combinatorial topology and is closely related to the currently fashionable field of graph theory. More than half a century of work by many (some say all) mathematicians has yielded proofs only for special cases (up to 35 countries in 1940). The consensus is that the conjecture is correct but unlikely to be proved in general. It seems destined to retain for some time the distinction of being both the simplest and most fascinating unsolved problem in mathematics.

2. W. W. Rouse Ball, *Mathematical Recreations and Essays*, rev. H. S. M. Coxeter (London: Macmillan & Co., 1959), p. 223.

3. R. Baltzer, "Eine Erinnerung an Möbius und seinen Freund Weiske," *Bericht über die Verhandlungen der Sächsischen Akademie der Wissenschaften zu Leipzig. Math.-Nat. Kl.* XXXVII (1885), 1–6.

4. H. S. M. Coxeter, "The Four-Color Map Problem," *The Mathematics Teacher*, LII (April 1952), 283–89.

5. A. Cayley, "On the Colouring of Maps," *Proceedings of the London Mathematical Society*, IX (1878), 148. A four-line report of Cayley's query in the session of June 13. See also a note with the same title by Cayley in the *Proceedings of the Royal Geographical Society*, I (1879, N.S.), 259–61.

6. A. B. Kempe, "On the Geographical Problem of Four Colors," *American Journal of Mathematics* II (1879), 193–204.

7. P. G. Tait, "On the Colouring of Maps," *Proceedings of the Royal Society of Edinburgh*, X (1880), 501–3, 729.

8. Frederick Guthrie, "Note on the Colouring of Maps," *Proc. Roy. Soc. Edinburgh*, X (1880), 727–28.

9. R. P. Graves, *Life of Sir William Rowan Hamilton* (Dublin, 1889), III, 423.

10. Biographical data on the Guthrie brothers supports the date indicated by De Morgan's letter. Indeed, Francis took his B.A. at University College London in 1850 and his LL.D. in 1852, whereas his younger brother Frederick was a student there in 1852 and, after a period in Germany, received his bachelor's degree in 1855.

11. In his note of April 1879, referred to in Note 5 above, Cayley begins: "The theorem that four colors are sufficient for any map, is mentioned somewhere by the late Professor De Morgan, who refers to it as a theorem known to map-makers." I have been unable to locate this "somewhere" and suspect that the communication was verbal. The same vague reference to De Morgan appears in a note in *Nature*, XX (1879), 275, and in Kempe's article cited in Note 6.

12. It is so called in E. Lucas, *Récréations mathématiques* (Paris, 1894).

Topological Diversions

Topology is the study of the properties of geometric shapes which do not change when the shapes themselves are deformed through stretching or twisting. Size and distance (length) are irrelevant in topology. Topological transformations such as twisting can result in some interesting objects that possess unusual features. A. F. Möbius (1790–1868) discovered a geometric configuration containing only one side and one surface edge. It is called a Möbius strip in honor of him. A Möbius strip can easily be constructed by taking a thin strip of paper, twisting it once, and joining the two ends together:

In 1882, the German mathematician Felix Klein extended the concept of a Möbius strip into a "bottle" which had no inside nor outside or edges. An anonymous verse celebrates this event:

> A mathematician named Klein
> Thought the Möbius band was devine.
> Said he: "If you glue
> The edges of two,
> You'll get a weird bottle like mine."

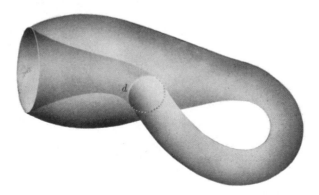

106

Meta-Mathematics and the Modern Conception of Mathematics[1]

MORTON R. KENNER

*I*T IS PERHAPS PERTINENT to begin this article with a few words indicating why it is proper that we who are concerned with mathematics education should also be concerned with such things as meta-mathematics. In this journal, it is hardly necessary to belabor the well-known facts that secondary school and college enrollments have swelled and are continuing to swell, that secondary and undergraduate curricula are being altered, revised, and experimented with. It should also be obvious, from the pages of this journal, that as the facilities of our colleges become more and more overtaxed and as our high school populations overflow into junior colleges—in many cases identical to the high school—our secondary teachers, and hence our secondary teacher training program, of necessity, will have to become cognizant of changes which have occurred in mathematics somewhat beyond and apart from the classically conceived secondary level. These facts create many problems about which we are all aware. I do not intend to try to solve any of them in this article. But, whatever the solutions to these new problems will be—and there will undoubtedly be many solutions—it seems to me that all of these solutions will share at least one thing in common. All of these solutions will in some sense recognize the contemporary view of the abstract nature of mathematics. They will all recognize that mathematics is a good deal more

than how much or how many, a good deal more than do this and don't do that.

Sharing this contemporary view of the nature of mathematics does not necessarily mean that secondary mathematics must be conceived of and taught entirely abstractly—although it may well mean that. But it does mean that we must ourselves, as teachers, know many of the consequences of believing that mathematics is abstract. It should be unnecessary to remind ourselves that it is a poor teacher indeed who teaches everything he knows. In fact, what so often makes excellent teaching excellent is that which we know but which we do not teach. Understanding what meta-mathematics is, to a certain extent, is equivalent to understanding what we mean when we say that mathematics is abstract. In the twentieth century, this latter understanding is certainly an obligation of all mathematics teachers.

The following discussion will attempt to sketch first some of the background necessary to an understanding of what meta-mathematics is, and second to discuss intuitively the nature of the subject and how it is related to abstractive tendencies in mathematics.

Mathematics and Abstraction

But what do we really mean when we say that mathematics is abstract? It is quite probable that we may mean two essentially different things. On the one hand, we may mean by abstractness the

Reprinted from *Mathematics Teacher* 51 (May, 1958): 350–57; with permission of the National Council of Teachers of Mathematics.

sort of thing which enables us to discuss a property (or properties) of a host of particular mathematical systems at the same time. We can—and do—talk about an abstract group, for example, knowing that in some sense we are also talking about certain aspects of the set of symmetries of a triangle, or about certain aspects of the set of rational numbers under addition, or about certain aspects of many other mathematical systems that are quite different in many ways but yet share the one property of being models or representations of an abstract group. This type of abstractness—the type which enables us to deal so efficiently with so many mathematical systems simultaneously by considering universally conceived properties shared by all of them—is an important part of current mathematics. To many, it represents in a nutshell what we mean when we say mathematics is abstract.

There is, however, another equally important meaning of abstractness. This other meaning of abstractness is most easily associated with the intuitive awareness which we display in recognizing that mathematics is independent—in some sense—of empirical knowledge. When we feel that the theorems of mathematics are not verified by experience or are not made true by experience or are not about experience, we are sensing this other meaning of abstractness. This is the side of abstractness which led to Mr. Russell's now famous quip about our ignorance. The evolution of this type of abstractness is common knowledge in the field of geometry where the development of non-Euclidean geometry, from at least one point of view, purports to show that geometry is not a physical science. This, indeed, is one of the major lessons that we learn in studying the development of non-Euclidean geometry. But we should not forget that this evolution took place also in the field of algebra.[2] As late as 1770, Euler still believed that mathematics was the science of magnitudes. Number, for Euler, was conceived as being in the first and last instance an answer to the question "how many?" and in cases of extensive measure, "how much?"

At about this same date, William Frend, in his book, *The Principles of Algebra*, printed in 1796, refused [sic!] to accept either negative or imaginary numbers. Frend defended his position with the following words—words undoubtedly strange to our twentieth-century ears. He said,

> The ideas of number are the clearest and most distinct in the human mind; the acts of the mind upon them are equally simple and clear. There cannot be confusion in them. . . . But numbers are divided into two sorts, positive and negative; and an attempt is made to explain the nature of negative numbers by allusion to book-debts and other arts. Now, when a person cannot explain the principles of a science without a reference to metaphor, the probability is that he has never thought accurately upon the subject. A number may be greater or less than another; it may be added to, taken from, multiplied into, and divided by another number; but in other respects it is intractable; though the whole world should be destroyed, one will be one, a three will be a three; and no art whatever can change their nature. You may put a mark before one, which it will obey: it submits to being taken away from another number greater than itself, but to attempt to take it away from a number less than itself is ridiculous. Yet this is attempted by algebraists, who talk of a number less than nothing, of multiplying a negative number into a negative number and thus producing a positive number, of a number being imaginary. Hence they talk of two roots to every equation of the second degree . . . they talk of solving an equation which requires two impossible roots to make it soluble: they can find out some impossible numbers, which, being multiplied together produce unity. This is all jargon, at which common sense recoils.[3]

It was not until the middle of the nineteenth century that Peacock, Gregory, De Morgan, and Boole, in a series of books and papers, laid the basis for the modern abstract conception of algebra. The first of these men, Peacock, in his *Treatise on Algebra* of 1830, distinguished between arithmetic and algebra. He believed that in arithmetic, the symbols A and B represent or stand for integers or fractions. In algebra, on the other hand,

Peacock felt that these signs need not represent integers and, in general, do not do so. Similarly, Peacock felt that in arithmetic the + and − signs represent or stand for the familiar addition or subtraction, whereas in algebra, only the formal properties of possible operations are relevant to this symbolic algebra. Peacock argued that when formally defined operations are symbolically denoted, certain expressions or forms are obtained which are equivalent to other forms in virtue of the rules of the systems of symbols. The discovery of these equivalent forms, he felt, constituted the principal business of algebra.

Gregory further clarified this point in 1838, when he pointed out that symbolic algebra is the science concerned not with the combinations of operations defined by their nature (i.e., what they are or do), but concerned solely with the laws of combination to which these operations are subject.[4]

De Morgan, in 1839, took a further step and divided the algebra of Peacock and Gregory into two different types of algebra—that which De Morgan called technical algebra and that which he called logical algebra.[5] Technical algebra for De Morgan is the art of using symbols under regulations which, when this part of the subject is considered independently of the other, are proscribed as the definitions of the symbols. Technical algebra then is a formal procedure, an operation on signs or symbols obeying prescribed rules. Logical algebra, on the other hand, is the science which investigates the method of giving meaning to the primary symbols and of interpreting symbolical results. Logical algebra then is similar to the problem of finding concrete examples or models for the system spelled out by the technical algebra. Thus, in a certain sense, we might say that De Morgan's technical algebra is an example of the second type of abstractness discussed above, while logical algebra is an example of the first type.

The important point to be noted here is that the validity of deduction in a technical algebra cannot depend upon the material interpretation of the symbols of the calculus. Technical algebra was algebra in which symbols were used independently of the meaning which logical algebra might give to them later. To adopt this viewpoint of technical algebra means—at least to those concerned with the axiomatizing of branches of mathematics—that it must be possible to regard the symbols of a system completely divorced from any meaning subsequently or earlier put upon them.

This point is a crucial one, for suppose that the meaning of the word "line" or the meaning of the word "point" were really necessary to the axiomatization of Euclidean geometry. This would then mean that our axioms did not fully prescribe what is to be meant by line or by point. This would mean further that part of the meaning of point and line must lie outside of the axioms.

Hilbert recognized this crucial point. His *Foundations of Geometry* was in large part an attempt to formalize Euclidean geometry by giving no meaning to the thought things "point" and "line" which were not *completely* stipulated by the axioms. Hilbert, of course, did much more, but his axiomatizing of Euclidean geometry is a landmark in the development of mathematics primarily because it develops a complete (and very rich) mathematical system without utilizing any meaning apart from that given by the axioms. His work was undoubtedly motivated by historical precedent, he undoubtedly was thinking of a specific model; within the system, however, the "thought things" were given meaning only by the axioms.

The Price of Abstractness

But if we are to look only to the axioms, only to the system, how is it possible to know that our system is not contradictory? That is, how do we know that our axioms and the logical rules we assume to deduce theorems from them are not such that at some time it will be possible to deduce a theorem and its contradictory? To deduce, say, both $1 = 1$ and $1 \neq 1$?

Before considering the question of how we can know that our system is noncontradictory, let us pause to consider the consequences of being able to deduce both a theorem and its contradictory. According to the rules of logic which we use in mathematics, it follows that a proposition and its negation imply any proposition. In symbols: $(P$ and $\sim P) \rightarrow Q$. Thus, if we could deduce P and also could deduce $\sim P$ from our axioms using the accepted logical canons, it would follow that any statement expressible in the symbolism would also be a true theorem.

Now it might seem that the way out of this difficulty is trivial. All we need do is exhibit a statement which is expressible in the symbolism of the system. Show that this statement is false, that is, show that it is not a true statement and cannot be true and hence not a theorem, and we would be home free. For, clearly, if we could deduce a statement and its contradictory, every statement would be true. We have exhibited one which is not true. Hence, clearly it must be impossible to deduce a statement and its contradictory, i.e., our system is not contradictory. We shall see presently that this would indeed be a way out—but it is far from trivial. Let us return to the question raised above concerning our ability to know whether or not a system is contradictory.

Before the nineteenth century, sense intuition had served as a guide in telling us that our axioms were not contradictory. Let us see precisely what this means. When, say, Euclid formulated certain axioms for plane geometry, he felt that these axioms were statements—true and eternal statements—about the geometry of the real world. Thus, the physical world was really a model for the axioms. Now, if our axioms were contradictory, that is, if they could give rise to a pair of contradictory propositions, then our model would have to give rise to a pair of contradictory intuitions. But since Euclid held—with obvious justification—that our sense experience of space was not contradictory, and further, since he held that this space was an exact model of the axioms, it followed that our axioms could not be contradictory. When, however, sense experience was no longer considered to be a valid guide for our technical algebra or techni-

cal geometry—in De Morgan's sense—we could not use sense experience as a model to test the validity of our axioms. New methods were necessary. If our system was to be entirely self-contained (independent of meanings attached to symbols), it was necessary to find a way to insure that no pair of contradictory propositions could arise—i.e., could be deduced.

In his *Foundations of Geometry,* Hilbert to some extent bypassed this problem. In that work, he used the well-known techniques of interpreting geometric statements algebraically—that is, Hilbert used algebra as a model for the geometry and showed how all of the geometric statements could be consistently interpreted in terms of algebraic elements and relations.[6] This, in a sense, is of course what we do in analytical geometry. But clearly, showing that all geometric statements are capable of algebraic interpretation does not establish the fact that the geometric axioms are noncontradictory. All it can possibly do is demonstrate that *if* algebra is noncontradictory, then geometry is also noncontradictory.

It is important to recognize that interpreting the axioms by means of a model (however well known) does not necessarily strike at the heart of the deduction problem. We referred earlier to the fact that $(P$ and $\sim P) \rightarrow Q$. We said then that if it were posssible to demonstrate the existence of a statement which could only be false, our axioms would be noncontradictory. For otherwise, every statement would be true. But clearly such a task is far from trivial. It is not enough to show that a given statement is false, it must also be shown that its denial *cannot* be deduced from our axioms. It is thus necessary to analyze carefully the logical procedures by which theorems are arrived at. We must, in other words, analyze our methods of proof including both original axioms and our logic of deduction. Only in this way will it be possible to prove that from a set of axioms (shown perhaps to be noncontradictory through models) it is not possible to deduce both a proposition and its negation as true theorems.

Hilbert was forced to consider this question in his paper of 1905, "On the Foundations of Logic and Arithmetic."[7] As we remarked above, Hilbert had used algebra (and hence elementary arithmetic) in establishing the consistency of geometry. That is, Euclidean geometry was noncontradictory if algebra was. But when he took up the question of arithmetic itself, there was no model to turn to. Hilbert stated this in the following way:

> Arithmetic is indeed designated as a part of logic and it is customary to pre-suppose in founding arithmetic the traditional principles of logic. But on the attentive consideration, we become aware that in the usual exposition of the laws of logic certain fundamental conceptions of arithmetic are already employed. For example, the concept of the aggregate, and in part also the concept of number. We fall thus into a vicious circle and therefore to avoid paradoxes a partly simultaneous development of the laws of logic and arithmetic is requisite.[8]

Thus, since certain fundamental aspects of arithmetic are already employed in logic, the simultaneous development of both is necessary.

It is clear from the above quotation that the paradoxes involving the theory of aggregates or set theory had just caused great concern in the mathematical world. It would take us far afield to analyze the various types of paradoxes resulting from an uncritical use of the concept of set—in particular, infinite sets.[9] Suffice it to remark that Hilbert felt that the set concept, so fundamental in logic and hence the foundations of arithmetic, must be subjected itself to a new type of critical analysis.[10]

Let us pause for a moment to sum up what we have already discussed. The development of abstraction in mathematics has led to the viewpoint that mathematical assertions, or mathematical systems, as mathematics have nothing in common with sense experience. Technical algebra, from the point of view of De Morgan and subsequently, is algebra in which meanings to be imparted to symbols must be specified by the axiom system without any reference outside the system. Mathematical symbols must derive their entire meaning

from the postulates of a system itself. If they do not, we have not completely axiomatized it. But having abandoned sense experience as a criterion in settling mathematical questions, it is no longer possible to establish the consistency—or noncontradictoriness—of our postulate systems by asserting that sense models of them give rise to no contradicting sense intuitions.[11] The sense models tell us nothing whatsoever about our formal system. But mathematics, if it is to lay claim to perfect rigor, must not give rise to contradictions. When we ask our students to accept certain sets of postulates for geometry or for arithmetic, we do so with the faith that these postulates do *not* give rise to contra-dictions. It is important to show that this faith is well founded.

Meta-Mathematics

To subject our axiom systems and our logic simultaneously to a critical analysis, Hilbert created a new discipline, the discipline of meta-mathematics. In a moment we shall consider more carefully what this new discipline was. Let us now see what the program was which Hilbert delineated for it.[12]

1. Meta-mathematics must enumerate all symbols used in mathematics and logic. These Hilbert called the fundamental symbols.

2. Meta-mathematics must denote uniquely all combinations of these symbols which occur as meaningful propositions in classical mathematics. These Hilbert called formulas.

3. Meta-mathematics must yield processes of construction which enable us to set up or to arrive at all formulas which correspond to the demonstrable assertions of classical mathematics.

4. Meta-mathematics must demonstrate in a finite combinatorial way that these formulas which correspond to calculable arithmetic can be demonstrated in accordance with 3 (above) if and only if the actual calculations of the mathematical assertions corresponding to the formulas results in the validity of the assertions.

It should be noted that item 4 is in essence a demand for a finite proof of consistency—i.e., of noncontradictoriness—of classical mathematics. But what sort of discipline can meta-mathematics be? What sort of discipline will be capable of enumerating all symbols, of enumerating all formulas of classical mathematics, of yielding a process of construction which will arrive at such formulas, and finally of giving an absolute proof of consistency?

The first thing about such a discipline which I hope is becoming obvious is that the *object* which meta-mathematics studies is mathematics itself. Let us see what this means by examining statements of mathematics and statements of meta-mathematics. A statement of mathematics is:

For every *x*, if *x* is a prime and if *x* > 2, then *x* is odd.

Statements in meta-mathematics would be:

"*x*" is a numerical variable.
"2" is a numerical constant.
"Prime" is a predicate expression.
">" is a binary predicate.

Note that when meta-mathematics says "'*x*' is a numerical variable," it is referring to the symbol—sign—"*x*" as it is used in the mathematical statement. When it says that "'2' is a numerical constant," it is referring to the symbol—sign—"2" as it is used in the mathematical statement.

Or as another example, consider the mathematical statement

2 + 3 = 5.

A statement in meta-mathematics would be:

"2 + 3 = 5" is a formula.

And when meta-mathematics says that "'2 + 3 = 5' is a formula," it is referring to how the symbols or signs have been put together and are used in mathematics. Thus, meta-mathematics is a subject whose object of study is mathematics.

The meta-mathematician operates with three separate and distinct subject matters. First, there is the informal mathematical system such as the natural numbers, or the points and lines and their relations in plane Euclidean geometry. This subject matter is informal and intuitive. Second, there is the formal axiomatic treatment of the informal mathematical system. This, you will recall, is motivated by the informal mathematical systems but if the axiomatic treatment of this informal mathematical system is to be successful, the formal system must itself be completely independent of the informal system. This is, of course, an immediate consequence of saying that mathematics is abstract, or that mathematical truths are not dependent upon sense experience. And finally, there is the meta-mathematics. This is a discipline which itself must be carefully developed. It is to a certain extent also informal, but the informalities present in meta-mathematics are of a different type. They are informalities in the sense of establishing criteria for permissible procedures. For example, no infinite operations are permissible.

From the point of view of the meta-mathematics, the formal system is independent of any meaning which may be given to it by an informal mathematical system. Meta-mathematics sees only the symbols, how they are grouped together, what is the logical pattern of passing from one grouping of symbols to another grouping of symbols. At the meta-mathematical level, we may speak of this as deduction, but in the formal system, we must regard it solely as a symbolic process. From the point of view of the informal mathematical system, we may regard this process as the means of arriving at truths about our interpreted objects; but from the meta-mathematics, it appears solely as a procedure which either follows rules and hence can be called "deduction" or does not follow these rules and hence cannot be so called.

It seems proper, at this point, to present a brief example of meta-mathematics which may give more meaning to the preceding discussion.[13] Assuming that we had listed the symbols to be permitted and also the operation of putting them side by side, i.e., operation of juxtaposition, we might then proceed as follows to define "term."

1. 0 is a *term*.
2. A variable is a *term*.
3. If s and t are *terms*, the $(s) + (t)$ is a *term*.
4. If s and t are *terms*, then $(s) \cdot (t)$ is a *term*.
5. If s is a *term*, then $(s)'$ is a *term*.
6. The only terms are those given by 1–5.

THEOREM: $((c)') = (a)$ is a *term*.
PROOF: By 2 a and c are terms.
       By 5 $(c)'$ is a term.
       By 3 $((c)') + (a)$ is a term.

We could now define "formula."

1. If s and t are terms, the $(s) = (t)$ is a *formula*.
2. If A and B are formulas, then $(A) \supset (B)$ is a *formula*.
3. If A and B are formulas, then (A) & (B) is a *formula*.
4. If A and B are formulas, then $(A) \vee (B)$ is a *formula*.
5. If A is a formula, then $\sim (A)$ is a *formula*.
6. If x is a variable and A is a formula, then $\forall_x (A)$ and $\exists_x (A)$ are *formulas*.
7. The only *formulas* are those given by 1–6.

THEOREM:
$(\exists_c ((((c)') + (a)) = (b))) \supset (\sim ((a) = (b)))$ is a formula.
  PROOF: By the terms obtained and by 1, $(a) = (b)$ and $(((c)') + (a)) = (b)$ are formulas.
      By 5 $\sim ((a) = (b))$ is a formula.
      By 7 $\exists_c ((((c)') + (a)) = (b))$ is a formula.
      By 2 $(\exists_c ((((c)') + (a)) = (b))) \supset (\sim ((a) = (b)))$ is a formula.

These two simple examples should indicate the procedure by which meta-mathematics proceeds to "construct" the formulas of classical mathematics.

I should like to conclude by calling your attention to what Hilbert hoped to do and how. We have already seen how the fundamental problem is one of knowing that we cannot ever reach a contradiction in the processes utilized in arithmetic.

(This, I emphasize again, is a consequence of our accepting the view that mathematics may be of reality but is not about reality.) Hilbert hoped to do this by showing that in formal mathematics we utilize certain symbols. Let us list all of them. In formal mathematics, we group them together. Let us explicitly—in our meta-mathematics—state how symbols can be grouped together. In formal mathematics, we use such concepts as substitution and deduction. Let us explicitly—in our meta-mathematics—indicate what is a permissible substitution and deduction. In formal mathematics we state what a contradiction is. Let us explicitly—in our meta-mathematics—state what combinations of symbols are contradictions. In our formal mathematical system we would like to believe that no contradictions can arise. Let us explicitly—in our meta-mathematics—prove, with finite means, that with the explicitly given symbols, and the explicitly stated formulas, and the explicitly formalized logic, it is not possible to obtain a formula which is a combination of symbols that we have called a contradiction in our meta-mathematics.

The real problem then is to be able to prove that we cannot arrive at a combination of symbols called a contradiction. The machinery to do this is meta-mathematics. It is the machinery by means of which we analyze the possible consequences of our moves with symbols. Hilbert did not succeed; that is, he was unable to prove that contradictions—certain combinations of symbols—cannot occur. At least he was unable to do this for arithmetic, i.e., the natural numbers. Such a proof of freedom from contradiction of number theory has still not been given. There is some evidence—the theorems of Kurt Gödel, for example—that it may be impossible. This may not be necessarily so. As Professor Kleene has remarked, all this may do is "pose a challenge to the meta-mathematician to bring to bear methods of finitary proof more powerful than those commonly used in elementary number theory." [14]

Meta-mathematics, then—the science created to subject classical mathematics to formal criticism—has not succeeded in giving us an absolute proof of noncontradiction. But it has, nevertheless, put in our hands powerful tools for analyzing the formal structures which we call mathematics. It has enabled us at the very least to look hard at our symbols and the operations on them with no regard to the meanings which informal mathematics puts upon them. If we really believe that modern mathematics is an abstract science, then we cannot avoid the consequences. And these consequences are precisely that we must then subject to critical analysis what we are, in fact, doing when we say we are engaged in mathematics—and this is what meta-mathematics aims to do.

NOTES

1. Delivered, with slight revisions, on March 29, 1957, at the Annual Meeting of the National Council of Teachers of Mathematics, Philadelphia.

2. For the discussion immediately following of the historical development of abstractness in algebra, I am indebted to the article by Ernest Nagel, "'Impossible Numbers': A Chapter in the History of Modern Logic," *Studies in the History of Ideas* (New York: Columbia University Press, 1935), II, 425–475.

3. *Ibid.*, 436, quoted by Nagel.

4. Duncan F. Gregory, "On the Real Nature of Symbolical Algebra," *Edinburgh Philosophical Transactions*, XIV, Part I (1838), 208.

5. A. De Morgan, "On the Foundations of Algebra," *Transactions of the Cambridge Philosophical Society*, VII (1839), 173.

6. See D. Hilbert, *The Foundations of Geometry* (trans. by E. J. Townsend) (La Salle, Illinois: Open Court Publishing Co., 1950, reprint of 1902 edition), pp. 25–36.

7. D. Hilbert, "On the Foundations of Logic and Arithmetic" (trans. by G. B. Halstead), *The Monist*, XV (1905), 338–352.

8. *Ibid.*, 340.

9. See for example, A. Church, "The Richard Paradox," *American Mathematical Monthly*, 41 (1934), 356–361.

10. It should be noted that Hilbert abandoned his early enthusiasm for the point of view of the Logistic School (Frege-Russell).

11. This argument is especially compelling when dealing with infinite systems when sense intuition often abandons us completely.

12. The following discussion is taken from J. Von Neumann, "Die formalistische Grundlegung der Mathematik," *Erkenntnis*, II (1931), 116 ff.

13. The following discussion is indebted to S. C. Kleene, *Introduction to Meta-Mathematics* (New York: D. Van Nostrand Co., Inc., 1952), pp. 72 ff.

14. *Ibid.*, 211–212.

107

Intuition and Logic in Mathematics

HENRI POINCARÉ

I

It is impossible to study the works of the great mathematicians, or even those of the lesser, without noticing and distinguishing two opposite tendencies, or rather two entirely different kinds of minds. The one sort are above all preoccupied with logic; to read their works, one is tempted to believe they have advanced only step by step, after the manner of a Vauban who pushes on his trenches against the place besieged, leaving nothing to chance. The other sort are guided by intuition and at the first stroke make quick but sometimes precarious conquests, like bold cavalrymen of the advance guard.

The method is not imposed by the matter treated. Though one often says of the first that they are *analysts* and calls the others *geometers*, that does not prevent the one sort from remaining analysts even when they work at geometry, while the others are still geometers even when they occupy themselves with pure analysis. It is the very nature of their mind which makes them logicians or intuitionalists, and they cannot lay it aside when they approach a new subject.

Nor is it education which has developed in them one of the two tendencies and stifled the other. The mathematician is born, not made, and

Reprinted from *Mathematics Teacher* 62: (Mar., 1969): 205–12; with permission of the National Council of Teachers of Mathematics. Originally published in Henri Poincaré, *The Foundations of Science,* trans. George Bruce Halsted ("Science and Education," I [New York and Lancaster: The Science Press, 1929]), 210–22.

it seems he is born a geometer or an analyst. I should like to cite examples, and there are surely plenty; but to accentuate the contrast I shall begin with an extreme example, taking the liberty of seeking it in two living mathematicians.

M. Méray wants to prove that a binominal equation always has a root, or, in ordinary words, that an angle may always be subdivided. If there is any truth that we think we know by direct intuition, it is this. Who could doubt that an angle may always be divided into any number of equal parts? M. Méray does not look at it that way; in his eyes this proposition is not at all evident and to prove it he needs several pages.

On the other hand, look at Professor Klein: he is studying one of the most abstract questions of the theory of functions: to determine whether on a given Riemann surface there always exists a function admitting of given singularities. What does the celebrated German geometer do? He replaces his Riemann surface by a metallic surface whose electric conductivity varies according to certain laws. He connects two of its points with the two poles of a battery. The current, says he, must pass, and the distribution of this current on the surface will define a function whose singularities will be precisely those called for by the enunciation.

Doubtless Professor Klein well knows he has given here only a sketch; nevertheless he has not hesitated to publish it; and he would probably believe he finds in it, if not a rigorous demonstration, at least a kind of moral certainty. A logician would have rejected with horror such a conception, or rather he would not have had to reject it, because in his mind it would never have originated.

Again, permit me to compare two men, the honor of French science, who have recently been taken from us, but who both entered long ago into immortality. I speak of M. Bertrand and M. Hermite. They were scholars of the same school at the same time; they had the same education, were under the same influences; and yet what a difference! Not only does it blaze forth in their writings; it is in their teaching, in their way of speaking, in their very look. In the memory of all their pupils these two faces are stamped in deathless lines; for all who have had the pleasure of following their teaching, this remembrance is still fresh; it is easy for us to evoke it.

While speaking, M. Bertrand is always in motion; now he seems in combat with some outside enemy, now he outlines with a gesture of the hand the figures he studies. Plainly he sees and he is eager to paint, this is why he calls gesture to his aid. With M. Hermite, it is just the opposite, his eyes seem to shun contact with the world; it is not without, it is within he seeks the vision of truth.

Among the German geometers of this century, two names above all are illustrious, those of the two scientists who founded the general theory of functions, Weierstrass and Riemann. Weierstrass leads everything back to the consideration of series and their analytic transformations; to express it better, he reduces analysis to a sort of prolongation of arithmetic; you may turn through all his books without finding a figure. Riemann, on the contrary, at once calls geometry to his aid; each of his conceptions is an image that no one can forget, once he has caught its meaning. . . .

Among our students we notice the same differences; some prefer to treat their problems "by analysis," others "by geometry." The first are incapable of "seeing in space," the others are quickly tired of long calculations and become perplexed.

The two sorts of minds are equally necessary for the progress of science; both the logicians and the intuitionalists have achieved great things that others could not have done. Who would venture to say whether he preferred that Weierstrass had never written or that there had never been a Riemann? Analysis and synthesis have then both their legitimate roles. But it is interesting to study more closely in the history of science the part which belongs to each.

II

Strange! If we read over the works of the ancients we are tempted to class them all among the intuitionalists. And yet nature is always the same; it is hardly probable that it has begun in this century to create minds devoted to logic. If we could put ourselves into the flow of ideas which reigned in their time, we should recognize that many of the old geometers were in tendency analysts. Euclid, for example, erected a scientific structure wherein his contemporaries could find no fault. In this vast construction, of which each piece however is due to intuition, we may still today, without much effort, recognize the work of a logician.

It is not minds that have changed, it is ideas; the intuitional minds have remained the same; but their readers have required of them greater concessions.

What is the cause of this evolution? It is not hard to find. Intuition can not give us rigor, nor even certainty; this has been recognized more and more. Let us cite some examples. We know there exist continuous functions lacking derivatives. Nothing is more shocking to intuition than this proposition which is imposed upon us by logic. Our fathers would not have failed to say: "It is evident that every continuous function has a derivative, since every curve has a tangent."

How can intuition deceive us on this point? It is because when we seek to imagine a curve we can not represent it to ourselves without width; just so, when we represent to ourselves a straight line, we see it under the form of a rectilinear band of a certain breadth. We well know these lines have no width; we try to imagine them narrower and narrower and thus to approach the limit; so we do in a certain measure, but we shall never attain this limit. And then it is clear we can always picture these two narrow bands, one straight, one curved, in a position such that they encroach slightly one upon the other without crossing. We shall thus be led, unless warned by a rigorous analysis, to conclude that a curve always has a tangent.

I shall take as second example Dirichlet's principle on which rest so many theorems of mathematical physics; today we establish it by reasoning very rigorous but very long; heretofore, on the contrary, we were content with a very summary proof. A certain integral depending on an arbitrary function can never vanish. Hence it is concluded that it must have a minimum. The flaw in this reasoning strikes us immediately, since we use the abstract term *function* and are familiar with all the singularities functions can present when the word is understood in the most general sense.

But it would not be the same had we used concrete images, had we, for example, considered this function as an electric potential; it would have been thought legitimate to affirm that electrostatic equilibrium can be attained. Yet perhaps a physical comparison would have awakened some vague distrust. But if care had been taken to translate the reasoning into the language of geometry, intermediate between that of analysis and that of physics, doubtless this distrust would not have been produced, and perhaps one might thus, even today, still deceive many readers not forewarned.

Intuition, therefore, does not give us certainty. This is why the evolution had to happen; let us now see how it happened.

It was not slow in being noticed that rigor could not be introduced in the reasoning unless first made to enter into the definitions. For the most part the objects treated of by mathematicians were long ill-defined; they were supposed to be known because represented by means of the senses or the imagination; but one had only a crude image of them and not a precise idea on which reasoning could take hold. It was there first that the logicians had to direct their efforts.

So, in the case of incommensurable numbers. The vague idea of continuity, which we owe to intuition, resolved itself into a complicated system of inequalities referring to whole numbers.

By that means the difficulties arising from passing to the limit, or from the consideration of infinitesimals, are finally removed. Today in analysis only whole numbers are left or systems, finite or infinite, of whole numbers bound together by a net of equality or inequality relations. Mathematics, as they say, is arithmetized.

III

A first question presents itself. Is this evolution ended? Have we finally attained absolute rigor? At each stage of the evolution our fathers also thought they had reached it. If they deceived themselves, do we not likewise cheat ourselves?

We believe that in our reasonings we no longer appeal to intuition; the philosophers will tell us this is an illusion. Pure logic could never lead us to anything but tautologies; it could create nothing new; not from it alone can any science issue. In one sense these philosophers are right; to make arithmetic, as to make geometry, or to make any science, something else than pure logic is necessary. To designate this something else we have no word other than *intuition*. But how many different ideas are hidden under this same word?

Compare these four axioms: (1) Two quantities equal to a third are equal to one another; (2) if a theorem is true of the number 1 and if we prove that it is true of $n + 1$ if true for n, then will it be true of all whole numbers; (3) if on a straight the point C is between A and B and the point D between A and C, then the point D will be between A and B; (4) through a given point there is not more than one parallel to a given straight.

All four are attributed to intuition, and yet the first is the enunciation of one of the rules of formal logic; the second is a real synthetic a priori judgment, it is the foundation of rigorous mathematical induction; the third is an appeal to the imagination; the fourth is a disguised definition.

Intuition is not necessarily founded on the evidence of the senses; the senses would soon become powerless; for example, we can not represent to ourselves a chiliagon, and yet we reason by intuition on polygons in general, which include the chiliagon as a particular case.

You know what Poncelet understood by the *principle of continuity*. What is true of a real quantity, said Poncelet, should be true of an imaginary quantity; what is true of the hyperbola whose asymptotes are real should then be true of the ellipse whose asymptotes are imaginary. Poncelet was one of the most intuitive minds of this century; he was passionately, almost ostentatiously, so; he regarded the principle of continuity as one of his boldest conceptions, and yet this principle did not rest on the evidence of the senses. To assimilate the hyperbola to the ellipse was rather to contradict this evidence. It was only a sort of precocious and instinctive generalization which, moreover, I have no desire to defend.

We have then many kinds of intuition: first, the appeal to the senses and the imagination; next, generalization by induction, copied, so to speak, from the procedures of the experimental sciences; finally, we have the intuition of pure number, whence arose the second of the axioms just enunciated, which is able to create the real mathematical reasoning. I have shown above by examples that the first two can not give us certainty; but who will seriously doubt the third, who will doubt arithmetic?

Now in the analysis of today, when one cares to take the trouble to be rigorous, there can be nothing but syllogisms or appeals to this intuition of pure number, the only intuition which can not deceive us. It may be said that today absolute rigor is attained.

IV

The philosophers make still another objection: "What you gain in rigor," they say, "you lose in objectivity. You can rise toward your logical ideal only by cutting the bonds which attach you to reality. Your science is infallible, but it can only remain so by imprisoning itself in an ivory tower and renouncing all relation with the external world. From this seclusion it must go out when it would attempt the slightest application."

For example, I seek to show that some property pertains to some object whose concept seems to me at first indefinable, because it is intuitive. At first I fail or must content myself with approximate proofs; finally I decide to give to my object a precise definition, and this enables me to establish this property in an irreproachable manner.

"And then," say the philosophers, "it still remains to show that the object which corresponds to this definition is indeed the same made known to you by intuition; or else that some real and concrete object whose conformity with your intuitive idea you believe you immediately recognize corresponds to your new definition. Only then could you affirm that it has the property in question. You have only displaced the difficulty."

That is not exactly so; the difficulty has not been displaced, it has been divided. The proposition to be established was in reality composed of two different truths, at first not distinguished. The first was a mathematical truth, and it is now rigorously established. The second was an experimental verity. Experience alone can teach us that some real and concrete object corresponds or does not correspond to some abstract definition. This second verity is not mathematically demonstrated, but neither can it be, any more than can the empirical laws of the physical and natural sciences. It would be unreasonable to ask more.

Well, is it not a great advance to have distinguished what long was wrongly confused? Does this mean that nothing is left of this objection of the philosophers? That I do not intend to say; in becoming rigorous, mathematical science takes a character so artificial as to strike every one; it forgets its historical origins; we see how the questions can be answered, we no longer see how and why they are put.

This shows us that logic is not enough; that the science of demonstration is not all science and that intuition must retain its role as complement, I was about to say as counterpoise or as antidote of logic.

I have already had occasion to insist on the place intuition should hold in the teaching of the mathematical sciences. Without it young minds could not make a beginning in the understanding of mathematics; they could not learn to love it and would see in it only a vain logomachy; above all, without intuition they would never become capable of applying mathematics. But now I wish before all to speak of the role of intuition in science itself. If it is useful to the student it is still more so to the creative scientist.

V

We seek reality, but what is reality? The physiologists tell us that organisms are formed of cells; the chemists add that cells themselves are formed of atoms. Does this mean that these atoms or these cells constitute reality, or rather the sole reality? The way in which these cells are arranged and from which results the unity of the individual, is not it also a reality much more interesting than that of the isolated elements, and should a naturalist who had never studied the elephant except by means of the microscope think himself sufficiently acquainted with that animal?

Well, there is something analogous to this in mathematics. The logician cuts up, so to speak, each demonstration into a very great number of elementary operations; when we have examined these operations one after the other and ascertained that each is correct, are we to think we have grasped the real meaning of the demonstration? Shall we have understood it even when, by an effort of memory, we have become able to repeat this proof by reproducing all these elementary operations in just the order in which the inventor had

arranged them? Evidently not; we shall not yet possess the entire reality; that I know not what, which makes the unity of the demonstration, will completely elude us.

Pure analysis puts at our disposal a multitude of procedures whose infallibility it guarantees; it opens to us a thousand different ways on which we can embark in all confidence; we are assured of meeting there no obstacles; but of all these ways, which will lead us most promptly to our goal? Who shall tell us which to choose? We need a faculty which makes us see the end from afar, and intuition is this faculty. It is necessary to the explorer for choosing his route; it is not less so to the one following his trail who wants to know why he chose it.

If you are present at a game of chess, it will not suffice, for the understanding of the game, to know the rules for moving the pieces. That will only enable you to recognize that each move has been made conformably to these rules, and this knowledge will truly have very little value. Yet this is what the reader of a book on mathematics would do if he were a logician only. To understand the game is wholly another matter; it is to know why the player moves this piece rather than that other which he could have moved without breaking the rules of the game. It is to perceive the inward reason which makes of this series of successive moves a sort of organized whole. This faculty is still more necessary for the player himself, that is, for the inventor.

Let us drop this comparison and return to mathematics. For example, see what has happened to the idea of continuous function. At the outset this was only a sensible image, for example, that of a continuous mark traced by the chalk on the blackboard. Then it became little by little more refined; ere long it was used to construct a complicated system of inequalities, which reproduced, so to speak, all the lines of the original image; this construction finished, the centering of the arch, so to say, was removed, that crude representation which had temporarily served as support and which was afterward useless was rejected; there remained only the construction itself, irreproachable in the eyes of the logician. And yet if the primitive image had totally disappeared from our recollection, how could we divine by what caprice all these inequalities were erected in this fashion one upon another?

Perhaps you think I use too many comparisons; yet pardon still another. You have doubtless seen those delicate assemblages of silicious needles which form the skeleton of certain sponges. When the organic matter has disappeared, there remains only a frail and elegant lace-work. True, nothing is there except silica, but what is interesting is the form this silica has taken, and we could not understand it if we did not know the living sponge which has given it precisely this form. Thus it is that the old intuitive notions of our fathers, even when we have abandoned them, still imprint their forms upon the logical constructions we have put in their place.

This view of the aggregate is necessary for the inventor; it is equally necessary for whoever wishes really to comprehend the inventor. Can logic give it to us? No; the name mathematicians give it would suffice to prove this. In mathematics logic is called *analysis* and analysis means *division, dissection*. It can have, therefore, no tool other than the scalpel and the microscope.

Thus logic and intuition have each their necessary role. Each is indispensable. Logic, which alone can give certainty, is the instrument of demonstration; intuition is the instrument of invention.

VI

But at the moment of formulating this conclusion I am seized with scruples. At the outset I distinguished two kinds of mathematical minds, the one sort logicians and analysts, the others intuitionalists and geometers. Well, the analysts also have been inventors. The names I have just cited make my insistence on this unnecessary.

Here is a contradiction, at least apparently, which needs explanation. And first, do you think

these logicians have always proceeded from the general to the particular, as the rules of formal logic would seem to require of them? Not thus could they have extended the boundaries of science; scientific conquest is to be made only by generalization.

In one of the chapters of *Science and Hypothesis,* I have had occasion to study the nature of mathematical reasoning, and have shown how this reasoning, without ceasing to be absolutely rigorous, could lead us from the particular to the general by a procedure I have called *mathematical induction*. It is by this procedure that the analysts have made science progress, and as we examine the detail itself of their demonstrations, we shall find it there at each instant beside the classic syllogism of Aristotle. We, therefore, see already that the analysts are not simply makers of syllogisms after the fashion of the scholastics.

Besides, do you think they have always marched step by step with no vision of the goal they wished to attain? They must have divined the way leading thither, and for that they needed a guide. This guide is, first, analogy. For example, one of the methods of demonstration dear to analysts is that founded on the employment of dominant functions. We know it has already served to solve a multitude of problems; in what consists then the role of the inventor who wishes to apply it to a new problem? At the outset he must recognize the analogy of this question with those which have already been solved by this method; then he must perceive in what way this new question differs from the others, and thence deduce the modifications necessary to apply to the method.

But how does one perceive these analogies and these differences? In the example just cited they are almost always evident, but I could have found others where they would have been much more deeply hidden; often a very uncommon penetration is necessary for their discovery. The analysts, not to let these hidden analogies escape them, that is, in order to be inventors, must, without the aid of the senses and imagination, have a direct sense of what constitutes the unity of a piece of reasoning, of what makes, so to speak, its soul and inmost life.

When one talked with M. Hermite, he never evoked a sensuous image, and yet you soon perceived that the most abstract entities were for him like living beings. He did not see them, but he perceived that they are not an artificial assemblage and that they have some principle of internal unity.

But, one will say, that still is intuition. Shall we conclude that the distinction made at the outset was only apparent, that there is only one sort of mind and that all the mathematicians are intuitionalists, at least those who are capable of inventing?

No, our distinction corresponds to something real. I have said above that there are many kinds of intuition. I have said how much the intuition of pure number, whence comes rigorous mathematical induction, differs from sensible intuition to which the imagination, properly so called, is the principal contributor.

Is the abyss which separates them less profound than it at first appeared? Could we recognize with a little attention that this pure intuition itself could not do without the aid of the senses? This is the affair of the psychologist and the metaphysician, and I shall not discuss the question: But the thing's being doubtful is enough to justify me in recognizing and affirming an essential difference between the two kinds of intuition; they have not the same object and seem to call into play two different faculties of our soul; one would think of two searchlights directed upon two worlds strangers to one another.

It is the intuition of pure number, that of pure logical forms, which illumines and directs those we have called *analysts*. This it is which enables them not alone to demonstrate, but also to invent. By it they perceive at a glance the general plan of a logical edifice, and that too without the senses appearing to intervene. In rejecting the aid of the imagination, which, as we have seen, is not always infallible, they can advance without fear of deceiving themselves. Happy, therefore, are those who can do without this aid! We must admire them; but how rare they are!

Among the analysts there will then be inventors, but they will be few. The majority of us, if we wished to see afar by pure intuition alone, would soon feel ourselves seized with vertigo. Our weakness has need of a staff more solid, and, despite the exceptions of which we have just spoken, it is none the less true that sensible intuition is in mathematics the most usual instrument of invention.

Apropos of these reflections, a question comes up that I have not the time either to solve or even to enunciate with the developments it would admit of. Is there room for a new distinction, for distinguishing among the analysts those who above all use pure intuition and those who are first of all preoccupied with formal logic?

M. Hermite, for example, whom I have just cited, can not be classed among the geometers who make use of the sensible intuition; but neither is he a logician, properly so called. He does not conceal his aversion to purely deductive procedures which start from the general and end in the particular.

108

The Three Crises in Mathematics: Logicism, Intuitionism, and Formalism

ERNST SNAPPER

THE THREE SCHOOLS, mentioned in the title, all tried to give a firm foundation to mathematics. The three crises are the failures of these schools to complete their tasks. This article looks at these crises "through modern eyes," using whatever mathematics is available today and not just the mathematics which was available to the pioneers who created these schools. Hence, this article does not approach the three crises in a strictly historical way. This article also does not discuss the large volume of current, technical mathematics which has arisen out of the techniques introduced by the three schools in question. One reason is that such a discussion would take a book and not a short article. Another one is that all this technical mathematics has very little to do with the philosophy of mathematics, and in this article I want to stress those aspects of logicism, intuitionism, and formalism which show clearly that these schools are founded in philosophy.

Logicism

This school was started in about 1884 by the German philosopher, logician, and mathematician, Gottlob Frege (1848–1925). The school was rediscovered about eighteen years later by Bertrand Russell. Other early logicists were Peano and

Reprinted from *Mathematics Magazine* 52 (Sept., 1979): 207–16; with permission of the Mathematical Association of America.

Russell's coauthor of *Principia Mathematica*, A. N. Whitehead. The purpose of logicism was to show that classical mathematics is part of logic. If the logicists had been able to carry out their program successfully, such questions as "Why is classical mathematics free of contradictions?" would have become "Why is logic free of contradictions?" This latter question is one on which philosophers have at least a thorough handle and one may say in general that the successful completion of the logicists' program would have given classical mathematics a firm foundation in terms of logic.

Clearly, in order to carry out this program of the logicists, one must first, somehow, define what "classical mathematics" is and what "logic" is. Otherwise, what are we supposed to show is part of what? It is precisely at these two definitions that we want to look through modern eyes, imagining that the pioneers of logicism had all of present-day mathematics available to them. We begin with classical mathematics.

In order to carry out their program, Russell and Whitehead created *Principia Mathematica* [10] which was published in 1910. (The first volume of this classic can be bought for $3.45! Thank heaven, only modern books and not the classics have become too expensive for the average reader.) *Principia*, as we will refer to *Principia Mathematica*, may be considered as a formal set theory. Although the formalization was not entirely complete, Russell and Whitehead thought that it was and planned to use it to show that mathematics can be reduced to logic. They showed that all classical mathematics, known in their time, can be derived from set theory and hence from the axioms of *Principia*. Consequently, what remained to be done, was to show that all the axioms of *Principia* belong to logic.

Of course, instead of *Principia*, one can use any other formal set theory just as well. Since today the formal set theory developed by Zermelo and Fraenkel (ZF) is so much better known than *Principia*, we shall from now on refer to ZF instead of *Principia*. ZF has only nine axioms and, although several of them are actually axiom schemas, we shall refer to all of them as "axioms." The formulation of the logicists' program now becomes: Show that all nine axioms of ZF belong to logic.

This formulation of logicism is based on the thesis that classical mathematics can be defined as the set of theorems which can be proved within ZF. This definition of classical mathematics is far from perfect, as is discussed in [12]. However, the above formulation of logicism is satisfactory for the purpose of showing that this school was not able to carry out its program. We now turn to the definition of logic.

* * *

In order to understand logicism, it is very important to see clearly what the logicists meant by "logic." The reason is that, whatever they meant, they certainly meant more than classical logic. Nowadays, one can define classical logic as consisting of all those theorems which can be proven in first order languages (discussed below in the section on formalism) without the use of nonlogical axioms. We are hence restricting ourselves to first order logic and use the deduction rules and logical axioms of that logic. An example of such a theorem is the law of the excluded middle which says that, if p is a proposition, then either p or its negation $\neg p$ is true; in other words, the proposition $p \vee \neg p$ is always true where $\vee$ is the usual symbol for the inclusive "or."

If this definition of classical logic had also been the logicists' definition of logic, it would be a folly to think for even one second that all of ZF can be reduced to logic. However, the logicists' definition was more extensive. They had a general concept as to when a proposition belongs to logic, that is, when a proposition should be called a "logical proposition." They said: *A logical proposition is a proposition which has complete generality and is true in virtue of its form rather than its content.* Here, the word "proposition" is used as synonymous with "theorem."

For example, the above law of the excluded middle "$p \vee \neg p$" is a logical proposition. Namely, this law does not hold because of any special content of the proposition p; it does not matter whether p is a proposition of mathematics or physics or what have you. On the contrary, this law holds with "complete generality," that is, for any proposition p whatsoever. Why then does it hold? The logicists answer: "Because of its form." Here they mean by form "syntactical form," the form of $p \vee \neg p$ being given by the two connectives of everyday speech, the inclusive "or" and the negation "not" (denoted by $\vee$ and $\neg$, respectively.)

On the one hand, it is not difficult to argue that all theorems of classical logic, as defined above, are logical propositions in the sense of logicism. On the other hand, there is no *a priori* reason to believe that there could not be logical propositions which lie outside of classical logic. This is why we said that the logicists' definition of logic is more extensive than the definition of classical logic. And now the logicists' task becomes clearer: It consists in showing that all nine axioms of ZF are logical propositions in the sense of logicism.

The only way to assess the success or failure of logicism in carrying out this task is by going through all nine axioms of ZF and determining for each of them whether it falls under the logicists' concept of a logical proposition. This would take a separate article and would be of interest only to readers who are thoroughly familiar with ZF. Hence, instead, we simply state that at least two of these axioms, namely, the axiom of infinity and the axiom of choice, cannot possibly be considered as logical propositions. For example, the axiom of infinity

says that there exist infinite sets. Why do we accept this axiom as being true? The reason is that everyone is familiar with so many infinite sets, say, the set of the natural numbers or the set of points in Euclidean 3-space. Hence, we accept this axiom on grounds of our everyday experience with sets, and this clearly shows that we accept it in virtue of its content and not in virtue of its syntactical form. In general, when an axiom claims the existence of objects with which we are familiar on grounds of our common everyday experience, it is pretty certain that this axiom is not a logical proposition in the sense of logicism.

And here then is the first crisis in mathematics: Since at least two out of the nine axioms of ZF are not logical propositions in the sense of logicism, it is fair to say that this school failed by about 20% in its effort to give mathematics a firm foundation. However, logicism has been of the greatest importance for the development of modern mathematical logic. In fact, it was logicism which started

mathematical logic in a serious way. The two quantifiers, the "for all" quantifier $\forall$ and the "there exists" quantifier $\exists$ were introduced into logic by Frege [5], and the influence of *Principia* on the development of mathematical logic is history.

It is important to realize that logicism is founded in philosophy. For example, when the logicists tell us what they mean by a logical proposition (above), they use philosophical and not mathematical language. They have to use philosophical language for that purpose since mathematics simply cannot handle definitions of so wide a scope.

The philosophy of logicism is sometimes said to be based on the philosophical school called "realism." In medieval philosophy "realism" stood for the Platonic doctrine that abstract entities have an existence independent of the human mind. Mathematics is, of course, full of abstract entities such as numbers, functions, sets, etc., and according to Plato all such entities exist outside our mind. The mind can discover them but does not create them. This doctrine has the advantage that one can accept such a concept as "set" without worrying about how the mind can construct a set. According to realism, sets are there for us to discover, not to be constructed, and the same holds for all other abstract entities. In short, realism allows us to accept many more abstract entities in mathematics than a philosophy which had limited us to accepting only those entities the human mind can construct. Russell was a realist and accepted the abstract entities which occur in classical mathematics without questioning whether our own minds can construct them. This is the fundamental difference between logicism and intuitionism, since in intuitionism abstract entities are admitted only if they are man made.

Excellent expositions of logicism can be found in Russell's writing, for example [9], [10], and [11].

Intuitionism

This school was begun about 1908 by the Dutch mathematician, L. E. J. Brouwer (1881–1966). The intuitionists went about the foundations of mathematics in a radically different way from the logicists. The logicists never thought that there was anything wrong with classical mathematics; they simply wanted to show that classical mathematics is part of logic. The intuitionists, on the contrary, felt that there was plenty wrong with classical mathematics.

By 1908, several paradoxes had arisen in Cantor's set theory. Here, the word "paradox" is used as synonymous with "contradiction." Georg Cantor created set theory, starting around 1870, and he did his work "naively," meaning nonaxiomatically. Consequently, he formed sets with such abandon that he himself, Russell, and others found several paradoxes within his theory. The logicists considered these paradoxes as common errors, caused by erring mathematicians and not by a faulty mathematics. The intuitionists, on the other hand, considered these paradoxes as clear indications that classical mathematics itself is far from perfect. They

felt that mathematics had to be rebuilt from the bottom on up.

The "bottom," that is, the beginning of mathematics for the intuitionists, is their explanation of what the natural numbers 1, 2, 3, … are. (Observe that we do not include the number zero among the natural numbers.) According to intuitionistic philosophy, all human beings have a primordial intuition for the natural numbers within them. This means in the first place that we have an immediate certainty as to what is meant by the number 1 and, secondly, that the mental process which goes into the formation of the number 1 can be repeated. When we do repeat it, we obtain the concept of the number 2; when we repeat it again, the concept of the number 3; in this way, human beings can construct any *finite* initial segment 1, 2, …, n for any natural number n. This mental construction of one natural number after the other would never have been possible if we did not have an awareness of time within us. "After" refers to time and Brouwer agrees with the philosopher Immanuel Kant (1724–1804) that human beings have an immediate awareness of time. Kant used the word "intuition" for "immediate awareness" and this is where the name "intuitionism" comes from. (See Chapter IV of [4] for more information about this intuitionistic concept of natural numbers.)

It is important to observe that the intuitionistic construction of natural numbers allows one to construct only arbitrarily long *finite* initial segments 1, 2, …, n. It does not allow us to construct that whole closed set of all the natural numbers which is so familiar from classical mathematics. It is equally important to observe that this construction is both "inductive" and "effective." It is inductive in the sense that, if one wants to construct, say, the number 3, one has to go through all the mental steps of first constructing the 1, then the 2, and finally the 3; one cannot just grab the number 3 out of the sky. It is effective in the sense that, once the construction of a natural number has been finished, that natural number has been constructed in its entirety. It stands before us as a completely finished mental construct, ready for our study of it. When someone says, "I have finished the mental construction of the number 3," it is like a brick-layer saying, "I have finished that wall," which he can say only after he has laid every stone in place.

We now turn to the intuitionistic definition of mathematics. According to intuitionistic philoso-phy, mathematics should be defined as a mental activity and not as a set of theorems (as was done above in the section on logicism). It is the activity which consists in carrying out, one after the other, those mental constructions which are inductive and effective in the sense in which the intuitionistic construction of the natural numbers is inductive and effective. Intuitionism maintains that human beings are able to recognize whether a given men-tal construction has these two properties. We shall refer to a mental construction which has these two properties as a *construct* and hence the intuitionistic definition of mathematics says: *Mathematics is the mental activity which consists in carrying out constructs one after the other.*

A major consequence of this definition is that all of intuitionistic mathematics is effective or "con-structive" as one usually says. We shall use the adjec-tive "constructive" as synonymous with "effective" from now on. Namely, every construct is construc-tive, and intuitionistic mathematics is nothing but carrying out constructs over and over. For instance, if a real number r occurs in an intuitionistic proof or theorem, it never occurs there merely on grounds of an existence proof. It occurs there because it has been constructed from top to bottom. This implies for example that each decimal place in the decimal ex-pansion of r can in principle be computed. In short, all intuitionistic proofs, theorems, definitions, etc., are entirely constructive.

Another major consequence of the intuitionistic definition of mathematics is that mathematics can-not be reduced to any other science such as, for instance, logic. This definition comprises too many mental processes for such a reduction. And here, then, we see a radical difference between logicism and intuitionism. In fact, the intuitionistic attitude

toward logic is precisely the opposite from the logicists' attitude: According to the intuitionists, whatever valid logical processes there are, they are all constructs; hence, the valid part of classical logic is part of mathematics! Any law of classical logic which is not composed of constructs is for the intuitionist a meaningless combination of words. It was, of course, shocking that the classical law of the excluded middle turned out to be such a meaningless combination of words. This implies that this law cannot be used indiscriminately in intuitionistic mathematics; it can often be used, but not always.

* * *

Once the intuitionistic definition of mathemat-ics has been understood and accepted, all there remains to be done is to do mathematics the intuitionistic way. Indeed, the intuitionists have developed intuitionistic arithmetic, algebra, analy-sis, set theory, etc. However, in each of these branches of mathematics, there occur classical theo-rems which are not composed of constructs and, hence, are meaningless combinations of words for the intuitionists. Consequently, one cannot say that the intuitionists have reconstructed all of clas-sical mathematics. This does not bother the intu-itionists since whatever parts of classical math-ematics they cannot obtain are meaningless for them anyway. Intuitionism does not have as its purpose the justification of classical mathematics. Its purpose is to give a valid definition of math-ematics and then to "wait and see" what math-ematics comes out of it. Whatever classical math-ematics cannot be done intuitionistically simply is not mathematics for the intuitionist. We observe here another fundamental difference between logicism and intuitionism: The logicists wanted to justify all of classical mathematics. (An excellent introduction to the actual techniques of intuition-ism is [8].)

Let us now ask how successful the intuitionistic school has been in giving us a good foundation for mathematics, acceptable to the majority of math-ematicians. Again, there is a sharp difference between the way this question has to be answered in the present case and in the case of logicism. Even hard-nosed logicists have to admit that their school so far has failed to give mathematics a firm foundation by about 20%. However, a hard-nosed intuitionist has every right in the world to claim that intuitionism has given mathematics an entirely satisfactory foundation. There is the meaningful definition of intuitionistic mathematics, discussed above; there is the intuitionistic philosophy which tells us why constructs can never give rise to contradictions and, hence, that intuitionistic mathematics is free of contradictions. In fact, not only this problem (of freedom from contradiction) but all other problems of a foundational nature as well receive perfectly satisfactory solutions in intuitionism.

Yet if one looks at intuitionism from the outside, namely, from the viewpoint of the classical mathematician, one has to say that intuitionism has failed to give mathematics an adequate foundation. In fact, the mathematical community has almost universally rejected intuitionism. Why has the mathematical community done this, in spite of the many very attractive features of intuitionism, some of which have just been mentioned?

One reason is that classical mathematicians flatly refuse to do away with the many beautiful theorems that are meaningless combinations of words for the intuitionists. An example is the Brouwer fixed point theorem of topology which the intuitionists reject because the fixed point cannot be constructed, but can only be shown to exist on grounds of an existence proof. This, by the way, is the same Brouwer who created intuitionism; he is equally famous for his work in (nonintuitionistic) topology.

A second reason comes from theorems which can be proven both classically and intuitionistically. It often happens that the classical proof of such a theorem is short, elegant, and devilishly clever, but not constructive. The intuitionists will of course reject such a proof and replace it by their own constructive proof of the same theorem. However, this constructive proof frequently turns out to be about ten times as long as the classical proof and

often seems, at least to the classical mathematician, to have lost all of its elegance. An example is the fundamental theorem of algebra which in classical mathematics is proved in about half a page, but takes about ten pages of proof in intuitionistic mathematics. Again, classical mathematicians refuse to believe that their clever proofs are meaningless whenever such proofs are not constructive.

Finally, there are the theorems which hold in intuitionism but are false in classical mathematics. An example is the intuitionistic theorem which says that every real-valued function which is defined for *all* real numbers is continuous. This theorem is not as strange as it sounds since it depends on the intuitionistic concept of a function: A real-valued function f is defined in intuitionism for all real numbers only if, for every real number r whose intuitionistic construction has been completed, the real number $f(r)$ can be constructed. Any obviously discontinuous function a classical mathematician may mention does not satisfy this constructive criterion. Even so, theorems such as this one seem so far out to classical mathematicians that they reject any mathematics which accepts them.

These three reasons for the rejection of intuitionism by classical mathematicians are neither rational nor scientific. Nor are they pragmatic reasons, based on a conviction that classical mathematics is better for applications to physics or other sciences than is intuitionism. They are all emotional reasons, grounded in a deep sense as to what mathematics is all about. (If one of the readers knows of a truly scientific rejection of intuitionism, the author would be grateful to hear about it.) We now have the second crisis in mathematics in front of us: It consists in the failure of the intuitionistic school to make intuitionism acceptable to at least the majority of mathematicians.

It is important to realize that, like logicism, intuitionism is rooted in philosophy. When, for instance, the intuitionists state their definition of mathematics, given earlier, they use strictly philosophical and not mathematical language. It would, in fact, be quite impossible for them to use mathematics for such a definition. The mental activity which is mathematics can be defined in philosophical terms but this definition must, by necessity, use some terms which do not belong to the activity it is trying to define.

Just as logicism is related to realism, intuitionism is related to the philosophy called "conceptualism." This is the philosophy which maintains that abstract entities exist only insofar as they are constructed by the human mind. This is very much the attitude of intuitionism which holds that the abstract entities which occur in mathematics, whether sequences or order-relations or what have you, are all mental constructions. This is precisely why one does not find in intuitionism the staggering collection of abstract entities which occur in classical mathematics and hence in logicism. The contrast between logicism and intuitionism is very similar to the contrast between realism and conceptualism.

A very good way to get into intuitionism is by studying [8], Chapter IV of [4], [2], and [13], in this order.

Formalism

This school was created in about 1910 by the German mathematician David Hilbert (1862–1943). True, one might say that there were already formalists in the nineteenth century since Frege argued against them in the second volume of his *Grundgesetze der Arithmetik* (see the book by Geach and Black under [5], pages 182–233); the first volume of the *Grundgesetze* appeared in 1893 and the second one in 1903. Nevertheless, the modern concept of formalism, which includes finitary reasoning, must be credited to Hilbert. Since modern books and courses in mathematical logic usually deal with formalism, this school is much better known today than either logicism or intuitionism. We will hence discuss only the highlights of formalism and begin by asking, "What is it that we formalize when we formalize something?"

The answer is that we formalize some given *axiomatized* theory. One should guard against confusing axiomatization and formalization. Euclid *axiomatized* geometry in about 300 B.C., but formalization started only about 2200 years later with the logicists and formalists. Examples of axiomatized theories are Euclidean plane geometry with the usual Euclidean axioms, arithmetic with the Peano axioms, ZF with its nine axioms, etc. The next question is: "How do we formalize a given axiomatized theory?"

Suppose then that some axiomatized theory T is given. Restricting ourselves to first order logic, "to formalize T" means to choose an appropriate first order language for T. The vocabulary of a first order language consists of five items, four of which are always the same and are not dependent on the given theory T. These four items are the following: (1) A list of denumerably many variables—who can talk about mathematics without using variables? (2) Symbols for the connectives of everyday speech, say $\neg$ for "not," $\wedge$ for "and," $\vee$ for the inclusive "or," $\rightarrow$ for "if then," and $\leftrightarrow$ for "if and only if"—who can talk about anything at all without using connectives? (3) The equality sign $=$; again, no one can talk about mathematics without using this sign. (4) The two quantifiers, the "for all" quantifier $\forall$ and the "there exist" quantifier $\exists$; the first one is used to say such things as "*all* complex numbers have a square root," the second one to say things like "*there exist* irrational numbers." One can do without some of the above symbols, but there is no reason to go into that. Instead, we turn to the fifth item.

Since T is an axiomatized theory, it has so called "undefined terms." One has to choose an appropriate symbol for every undefined term of T and these symbols make up the fifth item. For instance, among the undefined terms of plane Euclidean geometry, occur "point," "line," and "incidence," and for each one of them an appropriate symbol must be entered into the vocabulary of the first order language. Among the undefined terms of arithmetic occur "zero," "addition," and "multiplication," and the symbols one chooses for them are of course 0, +, and ×, respectively. The easiest theory of all to formalize is ZF

since this theory has only one undefined term, namely, the membership relation. One chooses, of course, the usual symbol $\in$ for that relation. These symbols, one for each undefined term of the axiomatized theory T, are often called the "parameters" of the first order language and hence the parameters make up the fifth item.

Since the parameters are the only symbols in the vocabulary of a first order language which depend on the given axiomatized theory T, one formalizes T simply by choosing these parameters. Once this choice has been made, the whole theory T has been completely formalized. One can now express in the resulting first order language L not only all axioms, definitions, and theorems of T, but more! One can also express in L all axioms of classical logic and, consequently, also all proofs

one uses to prove theorems of T. In short, one can now proceed entirely within L, that is, entirely "formally."

But now a third question presents itself: "Why in the world would anyone want to formalize a given axiomatized theory?" After all, Euclid never saw a need to formalize his axiomatized geometry. It is important to ask this question, since even the great Peano had mistaken ideas about the real purpose of formalization. He published one of his most important discoveries in differential equations in a formalized language (very similar to a first order language) with the result that nobody read it until some charitable soul translated the article into common German.

Let us now try to answer the third question. If mathematicians do technical research in a certain

704

branch of mathematics, say, plane Euclidean geometry, they are interested in discovering and proving the important theorems of the branch of mathematics. For that kind of technical work, formalization is usually not only no help but a definite hindrance. If, however, one asks such foundational questions as, for instance, "Why is this branch of mathematics free of contradictions?", then formalization is not just a help but an absolute necessity.

It was really Hilbert's stroke of genius to understand that formalization is the proper technique to tackle such foundational questions. What he taught us can be put roughly as follows. Suppose that T is an axiomatized theory which has been formalized in terms of the first order language L. This language has such a precise syntax that it itself can be studied as a *mathematical* object. One can ask for instance: "Can one possibly run into contradictions if one proceeds entirely formally within L, using only the axioms of T and those of classical logic, all of which have been expressed in L?" If one can prove mathematically that the answer to this question is "no," one has there a mathematical proof that the theory T is free of contradictions!

This is basically what the famous "Hilbert program" was all about. The idea was to formalize the various branches of mathematics and then to prove *mathematically* that each one of them is free of contradictions. In fact if, by means of this technique, the formalists could have just shown that ZF is free of contradictions, they would thereby already have shown that all of classical mathematics is free of contradictions, since classical mathematics can be done axiomatically in terms of the nine axioms of ZF. In short, the formalists tried to create a mathematical technique by means of which one could prove that mathematics is free of contradictions. This was the original purpose of formalism.

* * *

It is interesting to observe that both logicists and formalists formalized the various branches of mathematics, but for entirely different reasons.

The logicists wanted to use such a formalization to show that the branch of mathematics in question belongs to logic; the formalists wanted to use it to prove mathematically that that branch is free of contradictions. Since both schools "formalized," they are sometimes confused.

Did the formalists complete their program successfully? No! In 1931, Kurt Gödel showed in [6] that formalization cannot be considered as a mathematical technique by means of which one can prove that mathematics is free of contradictions. The theorem in that paper which rang the death bell for the Hilbert program concerns axiomatized theories which are free of contradictions and whose axioms are strong enough so that arithmetic can be done in terms of them. Examples of theories whose axioms are that strong are, of course, Peano arithmetic and ZF. Suppose now that T is such a theory and that T has been formalized by means of the first order language L. Then Gödel's theorem says, in nontechnical language, "No sentence of L which can be interpreted as asserting that T is free of contradictions can be proven formally within the language L." Although the interpretation of this theorem is somewhat controversial, most mathematicians have concluded from it that the Hilbert program cannot be carried out: Mathematics is not able to prove its own freedom of contradictions. Here, then, is the third crisis in mathematics.

Of course, the tremendous importance of the formalist school for present-day mathematics is well known. It was in this school that modern mathematical logic and its various offshoots, such as model theory, recursive function theory, etc., really came into bloom.

Formalism, as logicism and intuitionism, is founded in philosophy, but the philosophical roots of formalism are somewhat more hidden than those of the other two schools. One can find them, though, by reflecting a little on the Hilbert program.

Let again T be an axiomatized theory which has been formalized in terms of the first order language L. In carrying out Hilbert's program, one has to talk about the language L as one object, and while doing this, one is not talking within that safe language L itself. On the contrary, one is talking about L in ordinary, everyday language, be it English or French or what have you. While using our natural language and not the formal language L, there is of course every danger that contradictions, in fact, any kind of error, may slip in. Hilbert said that the way to avoid this danger is by making absolutely certain that, while one is talking in one's natural language about L, one uses only reasonings which are absolutely safe and beyond any kind of suspicion. He called such reasonings "finitary reasonings," but had, of course, to give a definition of them. The most explicit definition of finitary reasoning known to the author was given by the French formalist Herbrand ([7], the footnote on page 622). It says, if we replace "intuitionistic" by "finitary":

> By a finitary argument we understand an argument satisfying the following conditions: In it we never consider anything but a given finite number of objects and of functions; these functions are well defined, their definition allowing the computation of their values in a univocal way; we never state that an object exists without giving the means of constructing it; we never consider the totality of all the objects x of an infinite collection; and when we say that an argument (or a theorem) is true for all these x, we mean that, for each x taken by itself, it is possible to repeat the general argument in question, which should be considered to be merely the prototype of these particular arguments.

Observe that this definition uses philosophical and not mathematical language. Even so, no one can claim to understand the Hilbert program without an understanding of what finitary reasoning amounts to. The philosophical roots of formalism come out into the open when the formalists define what they mean by finitary reasoning.

We have already compared logicism with realism, and intuitionism with conceptualism. The philosophy which is closest to formalism is "nominalism." This is the philosophy which claims that abstract entities have no existence of any kind, neither outside the human mind as maintained by realism, nor as mental constructions within the human mind as maintained by conceptualism. For nominalism, abstract entities are mere vocal utterances or written lines, mere names. This is where the word "nominalism" comes from, since in Latin *nominalis* means "belonging to a name." Similarly, when formalists try to prove that a certain axiomatized theory T is free of contradictions, they do not study the abstract entities which occur in T but, instead, study that first order language L which was used to formalize T. That is, they study how one can form sentences in L by the proper use of the vocabulary of L; how certain of these sentences can be proven by the proper use of those special sentences of L which were singled out as axioms; and, in particular, they try to show that no sentence of L can be proven and disproven at the same time, since they would thereby have established that the original theory T is free of contradictions. The important point is that this whole study of L is a strictly syntactical study, since no meanings or abstract entities are associated with the sentences of L. This language is investigated by considering the sentences of L as meaningless expressions which are manipulated according to explicit, syntactical rules, just as the pieces of a chess game are meaningless figures which are pushed around according to the rules of the game. For the strict formalist "to do mathematics" is "to manipulate the meaningless symbols of a first order language according to explicit, syntactical rules." Hence, the strict formalist does not work with abstract entities, such as infinite series or cardinals, but only with their meaningless names which are the appropriate expressions in a first order language. Both formalists and nominalists avoid the direct use of abstract entities, and this is why formalism should be compared with nominalism.

The fact that logicism, intuitionism, and formalism correspond to realism, conceptualism, and nominalism, respectively, was brought to light in Quine's article, "On What There Is"([1], pages 183–196). Formalism can be learned from any modern book in mathematical logic, for instance [3].

Epilogue

Where do the three crises in mathematics leave us? They leave us without a firm foundation for mathematics. After Gödel's paper [6] appeared in 1931, mathematicians on the whole threw up their hands in frustration and turned away from the philosophy of mathematics. Nevertheless, the influence of the three schools discussed in this article has remained strong, since they have given us much new and beautiful mathematics. This mathematics concerns mainly set theory, intuitionism and its various constructivist modifications, and mathematical logic with its many offshoots. However, although this kind of mathematics is often referred to as "foundations of mathematics," one cannot claim to be advancing the philosophy of mathematics just because one is working in one of these areas. Modern mathematical logic, set theory, and intuitionism with its modifications are nowadays technical branches of mathematics, just as algebra or analysis, and unless we return directly to the philosophy of mathematics, we cannot expect to find a firm foundation for our science. It is evident that such a foundation is not necessary for technical mathematical research, but there are still those among us who yearn for it. The author believes that the key to the foundations of mathematics lies hidden somewhere among the philosophical roots of logicism, intuitionism, and formalism and this is why he has uncovered these roots, three times over.

Excellent literature on the foundations of mathematics is contained in [1] and [7].

REFERENCES

[1] P. Benacerraf and H. Putnam, *Philosophy of Mathematics*, Prentice-Hall, 1964.

[2] M. Dummett, *Elements of Intuitionism*, Clarendon Press, Oxford, England, 1977.

[3] H. B. Enderton, *A Mathematical Introduction to Logic*, Academic Press, 1972.

[4] A. A. Fraenkel, Y. Bar-Hillel, and A. Levy, *Foundations of Set Theory*, North-Holland, Amsterdam, Netherlands, 1973.

[5] G. Frege, *Begriffschrift*, in *Translations from the Philosophical Writings of Gottlob Frege* by P. Geach and M. Black, Basil Blackwell, Oxford, England, 1970. Also in [7] pp. 1–82.

[6] K. Gödel, On formally undecidable propositions of *Principia Mathematica* and related systems, in [7] pp. 596–616.

[7] J. van Heijenoort, *From Frege to Gödel*, Harvard Univ. Press, Cambridge. Available in paperback.

[8] A. Heyting, *Intuitionism, An Introduction*, North-Holland, Amsterdam, Netherlands, 1966.

[9] B. Russell, *Principles of Mathematics*, 1st ed. (1903) W. W. Norton, New York. Available in paperback.

[10] B. Russell and A. N. Whitehead, *Principia Mathematica*, 1st ed. (1910) Cambridge Univ. Press, Cambridge, England. Available in paperback.

[11] B. Russell, *Introduction to Mathematical Philosophy*, Simon and Schuster, New York, 1920. Available in paperback.

[12] E. Snapper, What is mathematics?, *Amer. Math. Monthly*, no. 7, 86 (1979) 551–557.

[13] A. S. Troelstra, *Choice Sequences*, Oxford Univ. Press, Oxford, England, 1977.

109

Kurt Gödel,
Mathematician and Logician

GERALD E. LENZ

On 15 January 1978 the *New York Times* reported the death of Kurt Gödel, mathematician and logician. On 20 January, *Time* magazine carried a brief summary of Gödel's work. It is not often that the demise of a mathematician receives attention from such popular publications.

Who was Kurt Gödel and what were his contributions to mathematics? Why should he have been described (*New York Times*, 22 January 1978) as the "discoverer of the most significant mathematical truth of this century, incomprehensible to laymen, revolutionary for philosophers and logicians?"

Born in what is now Czechoslovakia in 1906, Gödel received his Ph.D. from the University of Vienna in 1930 and became a naturalized U.S. citizen in 1948. His most famous contribution to the foundations of mathematics was made when he was only twenty-five years old, in 1931, while still at the University of Vienna.

His work at that time involved the much sought after proof that there can never be any contradictions in mathematics. A demonstration of the consistency of mathematics (i.e., the impossibility of proving contradictory statements if the rules or axioms are followed) was of growing importance because mathematics had moved further away from the concrete world of experience in the late nineteenth and early

Kurt Gödel
Sketch courtesy of Christopher Munoz

twentieth centuries. The existence of hyperbolic geometry, as demonstrated by Gauss, Lobachevski, and Bolyai, presented the mathematics community with a geometric system that could not easily be reconciled with the more intuitive Euclidean geometry. This shift from the mathematics of the familiar (where consistency was not questioned) to more abstract kinds of mathematics motivated the search for consistency proofs. The German mathematician David Hilbert was one of the leading advocates of such proofs. Demonstrations of what can be proved about the formal language of mathematics are termed metamathematical.

Gödel's amazing discovery was that no such proof can ever be constructed. His results are based on the development of mathematics given by Bertrand Russell and Alfred North Whitehead in *Principia Mathematica*, but they also hold for related systems.

Reprinted from *Mathematics Teacher* 73 (Nov., 1980): 612–14; with permission of the National Council of Teachers of Mathematics.

What Gödel showed was that no axiom system for mathematics as we know it is powerful enough to lead to a proof of its own consistency. Such a proof would require that additional axioms be added to the system. The consistency of this new, larger set of axioms would then be in doubt.

Gödel's 1931 results go further. In his 1931 paper, entitled *On Formally Undecidable Propositions of Principia Mathematica and Related Systems I*, he demonstrated that if mathematics is consistent, then it is not complete. That is, if there are no contradictions in mathematics, then there exist mathematical statements that can be neither proved nor disproved. Such statements are said to be formally undecidable.

Thus at the age of twenty-five Kurt Gödel showed that the axiomatic method in mathematics has a fatal weakness. His work is important not only because of what it says about the foundations of mathematics but also because of the philosophical significance of his results for the analysis of all knowledge.

Of course there are many formally *decidable* propositions in mathematics. The task of determining which propositions are formally decidable within a given system is important in current research in mathematical logic.

On 20 September 1938 Kurt Gödel married Adele Porkert. Gödel, who had been a visiting member of the Institute for Advanced Study in Princeton, New Jersey, as early as 1933, became a permanent member of the Institute in 1946. His application for U.S. citizenship was supported by Albert Einstein, who had been Gödel's colleague at the Institute. Einstein had received a lifetime appointment to the Institute in 1933, the year he came to the U.S. to escape Hitler's Germany. Throughout the years Gödel received many awards and honorary degrees, including the United States's highest recognition in science, the National Medal of Science, presented by Gerald Ford in 1975.

One of Gödel's greatest achievements occurred, however, in 1939. We shall quickly review some of the information necessary to understand those results.

Two sets are in one-to-one correspondence if there is some way to match all the elements of both sets such that each element in one is matched to one and only one element from the other. Any two finite sets having the same number of elements are in one-to-one correspondence. However, for infinite sets the situation is more interesting.

For example, the set of all integers is in one-to-one correspondence with the set of even integers, and the set of integers is also in one-to-one correspondence with the rational numbers. However, the integers are not in one-to-one correspondence with the real numbers. In a sense the reals constitute a set that is "too big" to be in one-to-one correspondence with the integers. However, the reals are in one-to-one correspondence with all the points on a Euclidean circle and with all the complex numbers.

This approach to the comparison of infinite sets is due to Georg Cantor, who published it in the late nineteenth century. It was Cantor himself who raised the question concerning the existence of a set that is "too large" to be in one-to-one correspondence with the integers but "too small" to be in one-to-one correspondence with the reals. The assertion that there is no such set became known as the continuum hypothesis. In 1900 Hilbert placed the proof of the continuum hypothesis first in a list of the most important problems facing mathematicians in the twentieth century.

The problem proved to be difficult indeed. It was not until Gödel's 1939 work that even a partial solution was available. Gödel demonstrated that the continuum hypothesis is consistent with the other axioms of set theory. In other words, it is impossible to disprove Cantor's hypothesis. This remarkable result is reminiscent of the situation of the parallel axiom in Euclidean geometry. It can be demonstrated that the parallel axiom cannot be disproved using the other axioms for geometry. However, in Euclidean geometry it is also known that the parallel axiom cannot be proved as a theorem.

Whether or not the continuum hypothesis could be proved remained an open question until 1963. In that year Paul Cohen, who had been a fellow at the Institute for Advanced Study at Princeton from 1959 to 1961, proved that the continuum hypothesis is independent of the other axioms of set theory. In other words it cannot be proved as a theorem using only the other axioms. Thus the status of the continuum hypothesis in set theory is analogous to that of the parallel axiom in geometry.

Gödel is responsible for many other discoveries. In 1939 he showed that the axiom of choice is also independent of the other axioms of set theory. (The axiom of choice says that if one has an infinite collection of sets, it is always possible to form a new set consisting of one element from each of the given sets.) Earlier, Gödel had shown that first-order logic is complete. He also worked in physics, notably with Einstein's field equation. His work there led to the "rotating universe" model, which allows (at least in theory) for the possibility of an agent influencing the past.

The man who accomplished all this was somewhat of a private person despite the fact that he had honorary degrees from Princeton, Yale, and Harvard and was the recipient of the first Albert Einstein Award for achievement.

Kurt Gödel died of heart disease at the Princeton Medical Center at 12:50 P.M. on Saturday, 14 January 1978. He is survived by Adele Gödel, his wife.

The Bibliography contains a list of several sources for those interested in learning more about the work of Gödel. One of the best of those (Nagel and Newman 1958) states:

> Gödel's findings thus undermined deeply rooted preconceptions and demolished ancient hopes that were being freshly nourished by research on the foundations of mathematics. But his paper was not altogether negative. It introduced into the study of foundation questions a new technique of analysis comparable in its nature and fertility with the algebraic method that René Descartes introduced into geometry. This technique suggested and initiated new problems for logical and mathematical investigation. It provoked a reappraisal, still under way, of widely held philosophies of mathematics, and of philosophies of knowledge in general. [p. 6–7]

BIBLIOGRAPHY

Cohen, Paul J. *Set Theory and the Continuum Hypothesis.* New York: W. A. Benjamin, 1966.

Cohen, Paul J., and Reuben Hersh. "Non-Cantorian Set Theory." *Scientific American,* December 1967, pp. 104–16.

Crossley, J. N., C. J. Nash, C. J. Brickhill, J. C. Stillwell, and N. H. Williams. *What Is Mathematical Logic?* London: Oxford University Press, 1972.

Flint, Peter B. "Kurt Gödel, 71, Dies." *New York Times,* 15 January 1978.

Gödel, Kurt. *On Formally Undecidable Propositions of Principia Mathematica and Related Systems, I.* Reprinted in English in *From Frege to Gödel: A Source Book in Mathematical Logic, 1879–1931,* edited by Jean Van Heijenoort. Cambridge, Mass.: Harvard University Press, 1967.

———. "What Is Cantor's Continuum Problem?" *American Mathematical Monthly* 54 (October 1947): 515–25.

Heijenoort, J. Van. "Gödel's Theorem." In *Encyclopedia of Philosophy.* New York: Crowell Collier & Macmillan, 1967.

Kleene, Stephen C. "The Work of Kurt Gödel." *Journal of Symbolic Logic* 41 (1976): 761–78.

"Milestones." *Time,* 30 January 1978.

Nagel, Ernest, and James R. Newman. *Gödel's Proof.* New York: New York University Press, 1958.

———. "Gödel's Proof." *Scientific American,* June 1956, pp. 71–86.

"Prisoners of Logic and Law—a Certain Genius." *New York Times,* 22 January 1978.

Rosser, Barkley. "An Informal Exposition of Proofs of Gödel's Theorems and Church's Theorem." *Journal of Symbolic Logic* 4 (1939): 56–61.

Sweenburne, Richard. *Space and Time.* New York: St. Martin's Press, 1968.

Whitehead, A. N., and Bertrand Russell. *Principia Mathematica.* Cambridge: At the University Press, 1910, 1912, 1913.

Wiebe, Richard. "Gödel's Theorem (Part I)." *Two Year College Journal of Mathematics* 6 (May 1975): 13–17.

———. "Gödel's Theorem (Part II)." *Two Year College Journal of Mathematics* 6 (September 1975): 4–7.

Thinking the Unthinkable:
The Story of Complex Numbers
(with a Moral)

ISRAEL KLEINER

*T*HE USUAL DEFINITION of complex numbers, either as ordered pairs (a, b) of real numbers or as "numbers" of the form $a + bi$, does not give any indication of their long and tortuous evolution, which lasted about three hundred years. I want to describe this evolution very briefly because I think some lessons can be learned from this story, just as from many other such stories concerning the evolution of a concept, result, or theory. These lessons have to do with the impact of the history of mathematics on our understanding of mathematics and on our effectiveness in teaching it. But more about the moral of this story later.

Birth

This story begins in 1545. What came earlier can be summarized by the following quotation from Bhaskara, a twelfth-century Hindu mathematician (Dantzig 1967):

> The square of a positive number, also that of a negative number, is positive; and the square root of a positive number is two-fold, positive and negative; there is no square root of a negative number, for a negative number is not a square.

Jerome Cardan (1501–1576)

From *A Portfolio of Eminent Mathematicians*, ed. David Eugene Smith (Chicago: Open Court, 1896)

In 1545 Jerome Cardan, an Italian mathematician, physician, gambler, and philosopher, published a book entitled *Ars Magna* (The great art), in which he described an algebraic method for solving cubic and quartic equations. This book was a great event in mathematics. It was the first major achievement in algebra since the time, 3000 years earlier, when the Babylonians showed how to solve quadratic equations. Cardan, too, dealt with quadratics in his book. One of the problems he proposed is the following (Struik 1969):

> If some one says to you, divide 10 into two parts, one of which multiplied into the other shall produce . . . 40, it is evident that this case or question is impossible. Nevertheless, we shall solve it in this fashion.

Reprinted from *Mathematics Teacher* 81 (Oct., 1988): 583–92; with permission of the National Council of Teachers of Mathematics. The author would like to acknowledge financial assistance from the Social Sciences and Humanities Research Council of Canada.

Cardan then applied his algorithm (essentially the method of completing the square) to $x + y = 10$ and $xy = 40$ to get the two numbers $5 + \sqrt{-15}$ and $5 - \sqrt{-15}$. Moreover, "putting aside the mental tortures involved" (Burton 1985), Cardan formally multiplied $5 + \sqrt{-15}$ by $5 - \sqrt{-15}$ and obtained 40. He did not pursue the matter but concluded that the result was "as subtle as it is useless" (NCTM 1969). Although eventually rejected, this event was nevertheless historic, since it was the first time ever that the square root of a negative number was explicitly written down. And, as Dantzig (1985) has observed, "the mere writing down of the impossible gave it a symbolic existence."

In the solution of the cubic equation, square roots of negative numbers had to be reckoned with. Cardan's solution for the cubic $x^3 = ax + b$ was given as

$$x = \sqrt[3]{\frac{b}{2} + \sqrt{\left(\frac{b}{2}\right)^2 - \left(\frac{a}{3}\right)^3}}$$

$$+ \sqrt[3]{\left(\frac{b}{2}\right) - \sqrt{\left(\frac{b}{2}\right)^2 - \left(\frac{a}{3}\right)^3}},$$

the so-called Cardan formula. When applied to the historic example $x^3 = 15x + 4$, the formula yields

$$x = \sqrt[3]{2 + \sqrt{-121}} + \sqrt[3]{2 - \sqrt{-121}}.$$

Although Cardan claimed that his general formula for the solution of the cubic was inapplicable in this case (because of the appearance of $\sqrt{-121}$), square roots of negative numbers could no longer be so lightly dismissed. Whereas for the quadratic (e.g., $x^2 + 1 = 0$) one could say that no solution exists, for the cubic $x^3 = 15x + 4$ a real solution, namely $x = 4$, does exist; in fact, the two other solutions, $-2 \pm \sqrt{3}$, are also real. It now remained to reconcile the formal and "meaningless" solution

$$x = \sqrt[3]{2 + \sqrt{-121}} + \sqrt[3]{2 - \sqrt{-121}}$$

of $x^3 = 15x + 4$, found by using Cardan's formula, with the solution $x = 4$, found by inspection. The task was undertaken by the hydraulic engineer Rafael Bombelli about thirty years after the publication of Cardan's work.

Bombelli had the "wild thought" that since the radicands $2 + \sqrt{-121}$ and $2 - \sqrt{-121}$ differ only in sign, the same might be true of their cube roots. Thus, he let

$$\sqrt[3]{2 + \sqrt{-121}} = a + \sqrt{-b}$$

and

$$\sqrt[3]{2 - \sqrt{-121}} = a - \sqrt{-b}$$

and proceeded to solve for a and b by manipulating these expressions according to the established rules for real variables. He deduced that $a = 2$ and $b = 1$ and thereby showed that, indeed,

$$\sqrt[3]{2 + \sqrt{-121}} + \sqrt[3]{2 - \sqrt{-121}}$$

$$= (2 + \sqrt{-1}) + (2 - \sqrt{-1}) = 4$$

(Burton 1985). Bombelli had thus given meaning to the "meaningless." This event signaled the birth of complex numbers. In his own words (*Leapfrogs* 1980):

> It was a wild thought, in the judgement of many; and I too was for a long time of the same opinion. The whole matter seemed to rest on sophistry rather than on truth. Yet I sought so long, until I actually proved this to be the case.

Of course, breakthroughs are achieved in this way—by thinking the unthinkable and daring to present it in public.

The equation $x^3 = 15x + 4$ represents the so-called irreducible case of the cubic, in which all three solutions are real yet they are expressed (by Cardan's formula) by means of complex numbers. To resolve the apparent paradox of cubic equations exemplified by this type of equation, Bombelli developed a calculus of operations with complex numbers. His rules, in our symbolism, are $(\pm 1)i =$

$\pm i$, $(+i)(+i) = -1$, $(-i)(+i) = +1$, $(\pm 1)(-i) = \mp i$, $(+i)(-i) = +1$, and $(-i)(-i) = -1$. He also considered examples involving addition and multiplication of complex numbers, such as $8i + (-5i) = +3i$ and

$$(\sqrt[3]{4 + \sqrt{2i}})(\sqrt[3]{3 + \sqrt{8i}}) = \sqrt[3]{8 + 11\sqrt{2i}}.$$

Bombelli thus laid the foundation stone of the theory of complex numbers.

Many textbooks, even at the university level, suggest that complex numbers arose in connection with the solution of quadratic equations, especially the equation $x^2 + 1 = 0$. As indicated previously, the cubic rather than the quadratic equation forced the introduction of complex numbers.

Growth

Bombelli's work was only the beginning of the saga of complex numbers. Although his book *L'Algebra* was widely read, complex numbers were shrouded in mystery, little understood, and often entirely ignored. Witness Simon Stevin's remark in 1585 about them (Crossley 1980):

> There is enough legitimate matter, even infinitely much, to exercise oneself without occupying oneself and wasting time on uncertainties.

Similar doubts concerning the meaning and legitimacy of complex numbers persisted for two and a half centuries. Nevertheless, during that same period complex numbers were extensively used and a considerable amount of theoretical work was done. We illustrate this work with a number of examples.

As early as 1620, Albert Girard suggested that an equation of degree n may have n roots. Such statements of the fundamental theorem of algebra were, however, vague and unclear. For example, René Descartes, who coined the unfortunate word "imaginary" for the new numbers, stated that although one can imagine that every equation has as many roots as is indicated by its degree, no (real)

numbers correspond to some of these imagined roots.

The following quotation, from a letter in 1673 from Christian Huygens to Gottfried von Leibniz in response to the latter's letter that contained the identity

$$\sqrt{1 + \sqrt{-3}} + \sqrt{1 - \sqrt{-3}} = \sqrt{6},$$

was typical of the period (Crossley 1980):

> The remark which you make concerning . . . imaginary quantities which, however, when added together yield a real quantity, is surprising and entirely novel. One would never have believed that
>
> $$\sqrt{1 + \sqrt{-3}} + \sqrt{1 - \sqrt{-3}}$$
>
> make $\sqrt{6}$ and there is something hidden therein which is incomprehensible to me.

Gottfried Wilhelm von Leibniz (1646–1716)

From *A Portfolio of Eminent Mathematicians*, ed. David Eugene Smith (Chicago: Open Court, 1896)

Leibniz, who spent considerable time and effort on the question of the meaning of complex numbers and the possibility of deriving reliable results by applying the ordinary laws of algebra to them, thought of them as "a fine and wonderful refuge of the divine spirit—almost an amphibian between being and non-being" (*Leapfrogs* 1980).

Complex numbers were widely used in the eighteenth century. Leibniz and John Bernoulli used imaginary numbers as an aid to integration. For example,

$$\int \frac{1}{x^2 + a^2}\, dx = \int \frac{1}{(x + ai)(x - ai)} dx$$

$$= -\frac{1}{2ai} \int \left(\frac{1}{x + ai} - \frac{1}{x - ai} \right) dx$$

$$= -\frac{1}{2ai}$$

$[\log (x + ai) - \log (x - ai)]$. This use, in turn, raised questions about the meaning of the logarithm of complex as well as negative numbers. A heated controversy ensued between Leibniz and Bernoulli. Leibniz claimed, for example, that log i = 0, arguing that $\log(-1)^2 = \log 1^2$, and hence $2 \log(-1) = 2 \log 1 = 0$; thus $\log(-1) = 0$, and hence $0 = \log(-1) = \log i^2 = 2 \log i$, from which it follows that log i = 0. Bernoulli opted for log i = $(\pi i)/2$; this equation follows from Euler's identity $e^{\pi i} = -1$, which implies that $\log(-1) = \pi i$ and hence that $\log i = \frac{1}{2} \log(-1) = (\pi i)/2$, although this argument is not the one that Bernoulli used. The controversy was subsequently resolved by Leonhard Euler (*Leapfrogs* 1978).

Complex numbers were used by Johann Lambert for map projection, by Jean D'Alembert in hydrodynamics, and by Euler, D'Alembert, and Joseph-Louis Lagrange in incorrect proofs of the fundamental theorem of algebra. (Euler, by the way, was the first to designate $\sqrt{-1}$ by i.)

Euler, who made fundamental use of complex numbers in linking the exponential and trigonometric functions by the formula $e^{ix} = \cos x + i \sin x$, expressed himself about them in the following way (Kline 1972):

> Because all conceivable numbers are either greater than zero, less than zero or equal to zero, then it is clear that the square root of negative numbers cannot be included among the possible numbers.... And this circumstance leads us to the concept of such numbers, which by their nature are impossible and ordinarily are called imaginary or fancied numbers, because they exist only in the imagination.

Even the great Carl Friedrich Gauss, who in his doctoral thesis of 1797 gave the first essentially

Leonhard Euler (1707–1783)

From *A Portfolio of Eminent Mathematicians*, ed. David Eugene Smith (Chicago: Open Court, 1908)

correct proof of the fundamental theorem of algebra, claimed as late as 1825 that "the true metaphysics of $\sqrt{-1}$ is elusive" (Kline 1972).

It should be pointed out that the desire for a logically satisfactory explanation of complex numbers became manifest in the latter part of the eighteenth century, on philosophical, if not on utilitarian, grounds. With the advent of the Age of Reason in the eighteenth century, when mathematics was held up as a model to be followed, not only in the natural sciences but in philosophy as well as political and social thought, the inadequacy of a rational explanation of complex numbers was disturbing.

The problem of the logical justification of the laws of operation with negative and complex numbers also beame a pressing pedagogical issue at, among other places, Cambridge University at the turn of the nineteenth century. Since mathematics was viewed by the educational institutions as a paradigm of rational thought, the glaring inadequacies in the logical justification of the operations with negative and complex numbers became untenable. Such questions as, "Why does $2 \times i + i = 2$?" and "Is $\sqrt{ab} = \sqrt{a}\sqrt{b}$ true for negative a and b?" received no satisfactory answers. In fact, Euler, in his text of the 1760s on algebra, claimed $\sqrt{-1}\sqrt{-4} = \sqrt{4} = +2$ as a possible result. Robert Woodhouse opined in 1802 that since imaginary numbers lead to right conclusions, they must have

Karl Friedrich Gauss (1777–1855)

From *A Portfolio of Eminent Mathematicians*, ed. David Eugene Smith (Chicago: Open Court, 1908)

a logic. Around 1830 George Peacock and others at Cambridge set for themselves the task of determining that logic by codifying the laws of operation with numbers. Although their endeavor did not satisfactorily resolve the problem of the complex numbers, it was perhaps the earliest instance of "axiomatics" in algebra.

By 1831 Gauss had overcome his scruples concerning complex numbers and, in connection with a work on number theory, published his results on the geometric representation of complex numbers as points in the plane. Similar representations by the Norwegian surveyor Caspar Wessel in 1797 and by the Swiss clerk Jean-Robert Argand in 1806 went largely unnoticed. The geometric representation, given Gauss's stamp of approval, dispelled much of the mystery surrounding complex numbers. In the next two decades further development took place. In 1833 William Rowan Hamilton gave an essentially rigorous algebraic definition of complex numbers as pairs of real numbers. (To Hamilton the complex number (a, b) consisted of a pair of "moments of time," since he had earlier defined real numbers, under Immanuel

Kant's influence, as "moments of time.") In 1847 Augustin-Louis Cauchy gave a completely rigorous and abstract definition of complex numbers in terms of congruence classes of real polynomials modulo $x^2 + 1$. In this, Cauchy modeled himself on Gauss's definition of congruences for integers (Kline 1972).

Maturity

By the latter part of the nineteenth century all vestiges of mystery and distrust of complex numbers could be said to have disappeared, although a lack of confidence in them persisted among some textbook writers well into the twentieth century. These authors would often supplement proofs using imaginary numbers with proofs that did not involve them. Complex numbers could now be viewed in the following ways:

1. Points or vectors in the plane
2. Ordered pairs of real numbers
3. Operators (i.e., rotations of vectors in the plane)
4. Numbers of the form $a + bi$, with a and b real numbers
5. Polynomials with real coefficients modulo $x^2 + 1$
6. Matrices of the form
$$\begin{bmatrix} a & b \\ -b & a \end{bmatrix},$$
with a and b real numbers
7. An algebraically closed, complete field (This is an early twentieth-century view.)

Although the preceding various ways of viewing the complex numbers might seem confusing rather than enlightening, it is of course commonplace in mathematics to gain a better understanding of a given concept, result, or theory by viewing it in as many contexts and from as many points of view as possible.

The foregoing descriptions of complex numbers are not the end of the story. Various developments in mathematics in the nineteenth century enable us to gain a deeper insight into the role of

William Rowan Hamilton (1805–1865)

complex numbers in mathematics and in other areas. Thus, complex numbers offer just the right setting for dealing with many problems in mathematics in such diverse areas as algebra, analysis, geometry, and number theory. They have a symmetry and completeness that is often lacking in such mathematical systems as the integers and real numbers. Some of the masters who made fundamental contributions to these areas say it best: The following three quotations are by Gauss in 1801, Riemann in 1851, and Hadamard in the 1890s, respectively:

> Analysis . . . would lose immensely in beauty and balance and would be forced to add very hampering restrictions to truths which would hold generally otherwise, if . . . imaginary quantities were to be neglected. (Birkhoff 1973)

> The origin and immediate purpose of the introduction of complex magnitudes into mathematics lie in the theory of simple laws of dependence between variable magnitudes expressed by means of operations on magnitudes. If we enlarge the scope of applications of these laws by assigning to the variables they involve complex values, then there appears an otherwise hidden harmony and regularity. (Ebbinghaus 1983)

> The shortest path between two truths in the real domain passes through the complex domain. (Kline 1972)

The descriptions of such developments are rather technical. Only the barest of illustrations can be given:

(1) In algebra, the solution of polynomial equations motivated the introduction of complex numbers: Every equation with complex coefficients has a complex root—the so-called fundamental theorem of algebra. Beyond their use in the solution of algebraic polynomial equations, the complex numbers offer an example of an algebraically closed field, relative to which many problems in linear algebra and other areas of abstract algebra have their "natural" solution.

(2) In analysis, the nineteenth century saw the development of a powerful and beautiful branch of mathematics, namely complex function theory. We have already seen how the use of complex numbers gave us deeper insight into the logarithmic, exponential, and trigonometric functions. Moreover, we can evaluate real integrals by means of complex function theory. One indication of the efficacy of the theory is that a function in the complex domain is infinitely differentiable if once differentiable. Such a result is, of course, false in the case of functions of a real variable (e.g., $f(x) = x^{4/3}$).

(3) The complex numbers lend symmetry and generality in the formulation and description of various branches of geometry, for example, Euclidean, inversive, and non-Euclidean. Thus, by the introduction of ideal points into the plane any two circles can now be said to intersect at two points. This idea aids in the formulation and proof of many results. As another example, Gauss used the complex numbers to show that the regular polygon of seventeen sides is constructible with straightedge and compass.

(4) In number theory, certain Diophantine equations can be solved neatly and conceptually by the use of complex numbers. For example, the equation $x^2 + 2 = y^3$, when expressed as $(x + \sqrt{2}i)(x - \sqrt{2}i) = y^3$, can readily be solved, in integers, using properties of the complex domain consisting of the set of elements of the form $a + b\sqrt{2}i$, with a and b integers.

716

(5) An elementary illustration of Hadamard's dictum that "the shortest path between two truths in the real domain passes through the complex domain" is supplied by the following proof that the product of sums of two squares of integers is again a sum of two squares of integers; that is,

$$(a^2 + b^2)(c^2 + d^2) = u^2 + v^2,$$

for some integers u and v. For,

$$
\begin{aligned}
(a^2 + b^2)(c^2 + d^2) \\
&= (a + bi)(a - bi)(c + di)(c - di) \\
&= [(a + bi)(c + di)][(a - bi)(c - di)] \\
&= (u + vi)(u - vi) \\
&= u^2 + v^2.
\end{aligned}
$$

Try to prove this result without the use of complex numbers and without being given the u and v in terms of a, b, c, and d.

In addition to their fundamental uses in mathematics, some of which were previously indicated, complex numbers have become a fixture in science and technology. For example, they are used in quantum mechanics and in electric circuitry. The "impossible" has become not only possible but indispensable.

The Moral

Why the history of mathematics? Why bother with such "stories" as this one? Edwards (1974) puts it in a nutshell:

> Although the study of the history of mathematics has an intrinsic appeal of its own, its chief raison d'être is surely the illumination of mathematics itself.

My colleague Abe Shenitzer expresses it as follows:

> One can *invent* mathematics without knowing much of its history. One can *use* mathematics without knowing much, if any, of its history. But one cannot have a mature appreciation of mathematics without a substantial knowledge of its history.

Such appreciation is essential for the teacher to possess. It can provide him or her with insight, motivation, and perspective—crucial ingredients in the making of a good teacher. Of course, whether this story has succeeded in achieving these objectives in relation to the complex numbers is for the reader to judge. However, beyond the immediate objective of lending insight, this story and others like it may furnish us with a slightly better understanding of the nature and evolution of the mathematical enterprise. It addresses such themes or issues as the following:

(1) *The meaning of number in mathematics.* Complex numbers do not fit readily into students' notions of what a number is. And, of course, the meaning of number has changed over the centuries. This story presents a somewhat better perspective on this issue. It also leads to the question of whether numbers beyond the complex numbers exist.

(2) *The relative roles of physical needs and intellectual curiosity as motivating factors in the development of mathematics.* In this connection it should be pointed out that the problem of the solution of the cubic, which motivated the introduction of complex numbers, was *not* a practical problem. Mathematicians already knew how to find approximate roots of cubic equations. The issue was to find a theoretical algebraic formula for the solution of the cubic—a question without any practical consequences. Yet how useful did the complex numbers turn out to be! This is a recurring theme in the evolution of mathematics.

(3) *The relative roles of intuition and logic in the evolution of mathematics.* Rigor, formalism, and the logical development of a concept or result usually come at the end of a process of mathematical evolution. For complex numbers, too, first came *use* (theoretical rather than practical), then *intuitive understanding*, and finally *abstract justification*.

(4) *The nature of proof in mathematics.* This question is related to the preceding item. But although (3) addresses the evolution of complex numbers in its broad features, this item deals with local questions of proof and rigor in establishing various results about complex numbers (cf., e.g., the derivation of the value of logi by von Leibniz and Bernoulli). One thing is certain—what was acceptable as a proof in the seventeenth and eighteenth centuries was no longer acceptable in the nineteenth and twentieth centuries. The concept of proof in mathematics has evolved over time, as it is still evolving, and not necessarily from the less to the more rigorous proof (cf. the recent proof, by means of the computer, of the four-color conjecture). Philip Davis goes a step further in outlining the evolution of mathematical ideas (Davis 1965):

> It is paradoxical that while mathematics has the reputation of being the one subject that brooks no contradictions, in reality it has a long history of successful living with contradictions. This is best seen in the extensions of the notion of number that have been made over a period of 2500 years. From limited sets of integers, to infinite sets of integers, to fractions, negative numbers, irrational numbers, complex numbers, transfinite numbers, each extension, in its way, overcame a contradictory set of demands.

(5) *The relative roles of the individual and the environment in the creation of mathematics.* What was the role of Bombelli as an individual in the creation of complex numbers? Cardan surely had the opportunity to take the great and courageous step of "thinking the unthinkable." Was the time perhaps not ripe for Cardan, but ripe for Bombelli about thirty years later? Is it the case, as John Bolyai stated, that "mathematical discoveries, like springtime violets in the woods, have their season which no human can hasten or retard" (Kline 1972)? This conclusion certainly seems to be borne out by many instances of independent and simultaneous discoveries in mathematics, such as the geometric representation of complex numbers by

George Pólya (1887–1985)
Courtesy of Birkhäuser Boston

Wessell, Argand, and Gauss. The complex numbers are an interesting case study of such questions, to which, of course, we have no definitive answers.

(6) *The genetic principle in mathematics education.* What are the sources of a given concept or theorem? Where did it come from? Why would anyone have bothered with it? These are fascinating questions, and the teacher should at least be aware of the answers to such questions. When and how he or she uses them in the classroom is another matter. On this matter George Pólya (1962) says the following:

> Having understood how the human race has acquired the knowledge of certain facts or concepts, we are in a better position to judge how the human child should acquire such knowledge.

Can we not at least have a better appreciation of students' difficulties with the concept of complex numbers, having witnessed mathematicians of the first rank make mistakes, "prove" erroneous theorems, and often come to the right conclusions for insufficient or invalid reasons?

Some Suggestions for the Teacher

Let me conclude with some comments on, and suggestions for, the use of the history of mathematics in the teaching of mathematics, in particular with reference to complex numbers. Many of the points are implicit in the preceding story.

(1) I first want to reiterate what I view as the major contribution of this story for the teacher. Pólya (1962) puts it very well:

> To teach effectively a teacher must develop a feeling for his subject; he cannot make his students sense its vitality if he does not sense it himself. He cannot share his enthusiasm when he has no enthusiasm to share. How he makes his point may be as important as the point he makes; he must personally feel it to be important.

The objective of my story, then, is to give the teacher some feeling for complex numbers, to imbue him or her with some enthusiasm for complex numbers.

When it comes to suggestions for classroom use, it cannot be overemphasized that these are only suggestions. The teacher, of course, can better judge when and how, at what level, and in what context to introduce and relate historical material to the discussion at hand. The introduction of historical material can, however, convey to the student the following important lessons, which are usually not imparted through the standard curriculum.

(2) Mathematics is far from a static, lifeless discipline. It is dynamic, constantly evolving, full of failures as well as achievements.

(3) Observation, analogy, induction, and intuition are the initial and often the more natural ways of acquiring mathematical knowledge. Rigor and proof usually come at the end of the process.

(4) Mathematicians usually create their subject without thought of practical applications. The latter, if any, come later, sometimes centuries later. This point relates to "immediate relevance" and to "instant gratification," which students often seek from any given topic presented in class.

(5) We must, of course, supply the student with "internal relevance" when introducing a given concept or result. This point brings us to the important and difficult issue of motivation. To some students the applications of a theorem are appealing; to others, the appeal is in the inner logical structure of the theorem. A third factor, useful but often neglected, is the source of the theorem: How did it arise? What motivated mathematicians to introduce it? With complex numbers, their origin in the solution of the cubic, rather than the quadratic, should be stressed. Cardan's attempted division of ten into two parts whose product is forty reinforces this point. How much further one continues with the historical account is a decision better made by the teacher in the classroom, bearing in mind the lessons that should be conveyed through this or similar historical material.

(6) Historical projects deriving from this story about complex numbers can be given to able students as topics for research and presentation to, say, a mathematics club. Possible topics are the following:

(a) The logarithms of negative and complex numbers.

(b) What is a number? That is, discuss the evolution of various number systems and the evolution of our conception of what a number is.

(c) Hypercomplex numbers (e.g., the quaternions). Their discovery is another fascinating story.

(d) Gauss's congruences of integers and Cauchy's congruences of polynomials. The latter lead to a new definition (description) of complex numbers.

(e) An axiomatic characterization of complex numbers (see (7) under the heading "Maturity"). In this connection we ought to discuss the notion of characterizing a mathematical system, and thus the concept of isomorphism. (Cf. the various equivalent descriptions of complex numbers discussed previously.)

(7) Many elementary and interesting illustrations of Hadamard's comment demonstrate that "the shortest path between two truths in the real

719

domain passes through the complex domain." We are referring to elementary results from various branches of mathematics, results whose statements do not contain complex numbers but whose "best" proofs often use complex numbers. One such example was given previously. Some others by Cell (1950), Jones (1954), and the NCTM (1969) can be found in the Bibliography.

BIBLIOGRAPHY

Birkhoff, Garrett. *A Source Book in Classical Analysis.* Cambridge, Mass.: Harvard University Press, 1973.

Burton, David M. *The History of Mathematics.* Boston: Allyn & Bacon, 1985.

Cell, John W. "Imaginary Numbers." *Mathematics Teacher* 43 (December 1950):394–96.

Crossley, John N. *The Emergence of Number.* Victoria, Australia: Upside Down A Book Co., 1980.

Dantzig, Tobias, *Number—the Language of Science.* New York: The Free Press, 1930, 1967.

Davis, Philip J. *The Mathematics of Matrices.* Waltham, Mass.: Blaisdell Publishing Co., 1965.

Ebbinghaus, Heinz-Dieter, et al. *Zahlen.* Heidelberg: Springer-Verlag, 1983.

Edwards, Charles H. *The Historical Development of the Calculus.* New York: Springer-Verlag, 1974.

Flegg, Graham. *Numbers—Their History and Meaning.* London: Andre Deutsch, 1983.

Jones, Phillip S. "Complex Numbers: An Example of Recurring Themes in the Development of Mathematics—I–III." *Mathematics Teacher* 47 (February, April, May 1954): 106–14, 257–63, 340–45.

Kline, Morris. *Mathematical Thought from Ancient to Modern Times.* New York: Oxford University Press, 1972.

Leapfrogs: Imaginary Logarithms. Fordham, Ely, Cambs., England: E. G. Mann & Son, 1978.

Leapfrogs: Complex Numbers. Fordham, Ely, Cambs., England: F. G. Mann & Son, 1980.

McClenon, R. B. "A Contribution of Leibniz to the History of Complex Numbers." *American Mathematical Monthly* 30 (November 1923): 369–74.

Nagel, Ernest. "Impossible Numbers: A Chapter in the History of Modern Logic." *Studies in the History of Ideas* 3 (1935): 429–74.

National Council of Teachers of Mathematics. *Historical Topics for the Mathematics Classroom.* Thirty-first Yearbook. Washington, D.C.: The Council, 1969.

Pólya, George. *Mathematical Discovery.* New York: John Wiley & Sons, 1962.

Sondheimer, Ernest, and Alan Rogerson. *Numbers and Infinity—an Historical Account of Mathematical Concepts.* New York: Cambridge University Press, 1981.

Struik, Dirk J. *A Source Book in Mathematics, 1200–1800.* Cambridge, Mass.: Harvard University Press, 1969.

Windred, G. "History of the Theory of Imaginary and Complex Quantities." *Mathematical Gazette* 14 (1930?): 533–41.

111

The Development of Modern Statistics

DALE E. VARBERG

First Lecture

THAT AREA OF STUDY which we now call statistics has only recently come of age. While its origins may be traced back to the eighteenth century, or perhaps earlier, the first really significant developments in the theory of statistics did not occur until the late nineteenth and early twentieth centuries, and it is only during the last thirty years or so that it has reached a full measure of respectability. It was antedated by the theory of probability and has its roots embedded in this subject. In fact, any serious study of statistics must of necessity be preceded by a study of probability theory, for it is in the latter subject that the theory of statistics finds its foundation and fountainhead.

The word *statistik* was first used by Gottfried Achenwall (1719–1772), a lecturer at the University of Göttingen.[1] He is sometimes referred to as the "Father of Statistics"—perhaps mistakenly, since he was mainly concerned with the description of interesting facts about his country.

Our English word "statistic" means different things to different people. To the man on the

street, statistics is the mass of figures that the expert on any subject uses to support his contentions—it's "what you use to prove anything by." To the more sophisticated person, the word may evoke some notion of the procedures which are used to condense and interpret a collection of data, such as the computing of means and standard deviations. But to the practitioner of the craft, statistics is the art of making inferences from a body of data, or, more generally, the science of making decisions in the face of uncertainty.

Statisticians concern themselves with answering such questions as: Is this particular lot of manufactured items defective? Is there a connection between smoking and cancer? Will Kennedy win the next election? In answering these questions, it is necessary to reason from the specific to the general, from the sample to the population. Therefore, any conclusions reached by the statistician are not to be accepted as absolute certainties. It is, in fact, one of the jobs of the statistician to give some measure of the certainty of the conclusions he has drawn.

It should not be inferred from this lack of certainty that the mathematics of statistics is nonrigorous. The mathematics that forms the basis of statistics stems from probability theory and has a firm axiomatic foundation and rigorously proved theorems.

If we conceive of statistics as the science of drawing inferences and making decisions, it is

Reprinted from *Mathematics Teacher* 57 (Apr., 1963): 252–57; 57 (May, 1963): 344–48; with permission of the National Council of Teachers of Mathematics. This is the text of two lectures on the history of statistics given by Professor Varberg at a National Science Foundation Summer Institute for High School Mathematics Teachers held at Bowdoin College during the summer of 1962. Notes for these lectures were taken by Alvin K. Funderburg.

appropriate to date its beginnings with the work of Sir Francis Galton (1822–1911) and Karl Pearson (1857–1936) in the late nineteenth century. Starting here, modern statistical theory has developed in four great waves of ideas, in four periods, each of which was introduced by a pioneering work of a great statistician.[2]

The first period was inaugurated by the publication of Galton's *Natural Inheritance* in 1889. If for no other reason, this book is justly famous because it sparked the interest of Karl Pearson in statistics. Until this time, Pearson had been an obscure mathematician teaching at University College in London. Now the idea that all knowledge is based on statistical foundations captivated his mind. Moving to Gresham College in 1890 with the chance to lecture on any subject that he wished, Pearson chose the topic: "the scope and concepts of modern science." In his lectures he placed increasingly stronger emphasis on the statistical foundation of scientific laws and soon was devoting most of his energy to promoting the study of statistical theory. Before long, his laboratory became a center in which men from all over the world studied and went back home to light statistical fires. Largely through his enthusiasm, the scientific world was moved from a state of disinterest in statistical studies to a situation where large numbers of people were eagerly at work developing new theory and gathering and studying data from all fields of knowledge. The conviction grew that the analysis of statistical data could provide answers to a host of important questions.

An anecdote, related by Helen Walker,[3] of Pearson's childhood illustrates in a vivid way the characteristics which marked his adult career. Pearson was asked what was the first thing he could remember. "Well," he said, "I do not know how old I was, but I was sitting in a high chair and I was sucking my thumb. Someone told me to stop sucking it and said that unless I do so the thumb would wither away. I put my two thumbs together and looked at them a long time. 'They look alike to

me,' I said to myself, 'I can't see that the thumb I suck is any smaller than the other. I wonder if she could be lying to me.'"

We have here in this simple story, as Helen Walker points out, "rejection of constituted authority, appeal to empirical evidence, faith in his own interpretation of the meaning of observed data, and finally imputation of moral obliquity to a person whose judgment differed from his own." These were to be prominent characteristics throughout Pearson's whole life.

This first period, then, was marked by a change in attitude toward statistics, a recognition of its importance by the scientific world. But, in addition to this, many advances were made in statistical technique. Among the technical tools invented and studied by Galton, Pearson, and their followers were the standard deviation, correlation coefficient, and the chi square test.

About 1915, a new name appeared on the statistical horizon, R. A. Fisher (1890–). His paper of that year on the exact distribution of the sample correlation coefficients ushered in the second period of statistical history and was followed by a whole series of papers and books which gave a new impetus to statistical inquiry. One author has gone so far as to credit Fisher with half of the statistical theory that we use today. Among the significant contributions of Fisher and his associates were the development of methods appropriate for small samples, the discovery of the exact distributions of many sample statistics, the formulation of logical principles for testing hypotheses, the invention of the technique known as analysis of variance, and the introduction of criteria for choice among various possible estimators for a population parameter.

The third period began about 1928 with the publication of certain joint papers by Jerzy Neyman and Egon Pearson, the latter a son of Karl Pearson. These papers introduced and emphasized such concepts as "Type II" error, power of a test, and confidence intervals. It was during this period that industry began to make widespread application of statistical techniques, especially in connection with quality control. There was increasing interest in taking of surveys with consequent attention to the theory and technique of taking samples.

We date the beginning of the fourth period with the first paper of Abraham Wald (1902–1950) on the now often used statistical procedure—sequential sampling. This paper of 1939 initiated a deluge of papers by Wald, ended only by his untimely death in an airplane crash when at the height of his powers. Perhaps Wald's most significant contribution was his introduction of a new way of looking at statistical problems, what is known as statistical decision theory. From this point of view, statistics is regarded as the art of playing a game, with nature as the opponent. This is a very general theory, and, while it does lead to formidable mathematical complications, it is fair to say that a large share of present-day research statisticians have found it advantageous to adopt this new approach.

Having given this brief bird's-eye view of statistical history, we move to a discussion of some of the most basic of statistical concepts. For this purpose it will be convenient to refer to a table showing the heights and weights of twelve people

FIGURE I

Table of Heights and Weights

Individual	X	Y
1	60	110
2	60	135
3	60	120
4	62	120
5	62	140
6	62	130
7	62	135
8	64	150
9	64	145
10	70	170
11	70	185
12	70	160

FIGURE 2

Frequency Diagram

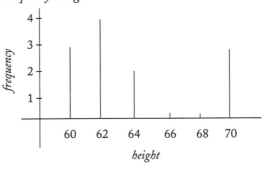

(Fig. 1). The height X is shown in inches; the weight Y is shown in pounds.

To get some feeling for such a collection of data, it is clearly desirable to display the data pictorially. William Playfair (1759–1823) of England is usually given credit for introducing the idea of graphical representation into statistics. His writings, mostly on economics, were illustrated with extremely good graphs, histograms, bar diagrams, etc. In our problem, the data is most simply represented by means of what is called a frequency diagram.

We have shown such a diagram for the height X (Fig. 2). A similar diagram for Y would be easy to construct. While such pictures do help our intuition, we need more than this if we are to treat the data mathematically. We need mathematical measures which describe the data precisely.

Among the most important of such measures are the measures of central tendency. The earliest of these, actually dating back to the Greeks, is the arithmetic mean μ, which for a discrete variable X, such as we have in our example, is defined by

$$\mu_X = (1/n) \sum_{i=1}^{n} x_i.$$

Here x_i denotes a value of the variable X, and n is the size of the population. In our example, the mean μ_X of the heights is 63.83; the mean μ_Y of the weights is 141.67.

To understand the significance of this concept, we rewrite the definition in the form

$$\mu_X = (1/n) \sum x_j f_j.$$

Here f_j stands for the frequency of occurrence of the value x_j and the summation extends over the distinct values of the variable X. Consider now a weightless rod on which there is a scale running through the range of the variable X, and suppose that at x_j is attached a mass of size f_j/n. This gives a system of total mass 1, which will have μ_X as its center of mass, that is, the system will balance on a fulcrum placed at μ_X. In the case of the heights, the system would look as in Figure 3. This interpretation of the mean will be helpful later when we consider the notion of a continuously distributed variable.

While the concept is probably quite old, it was not until 1883 that the median was introduced into statistics by Galton as a second measure of central tendency.[4] The median is simply the middle value of the distribution in the case of an odd number of values and is the average of the two middle values otherwise. The median height in our example is 62.

Another measure of central tendency is the mode, introduced by Karl Pearson around 1894. The mode is the most frequently occurring value, if there is one. In the case where two or more values occur with equal frequency, the mode is not well defined. In the example, the mode of heights is again 62.

If the distribution of a variable X is exactly symmetrical, i.e., if its frequency diagram is exactly symmetrical about a vertical line, then the mean, median, and mode (if there is a mode) will agree. The reader should be able to convince himself that the converse is false by constructing a non-symmetrical distribution for which the mean, median, and mode agree.

FIGURE 3

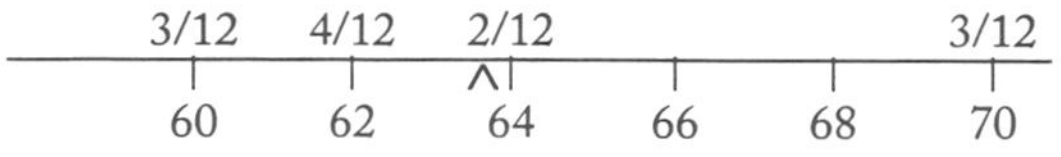

3/12	4/12	2/12			3/12
60	62	64	66	68	70

For most purposes, certainly for theoretical purposes, the mean is the most useful measure of central tendency, although admittedly it may take much calculation to get it. The median does, however, have a property which is sometimes advantageous. It is not as subject to distortion due to a few extreme values. For example, if in the table of heights of twelve persons, one of the 70-inch persons were exchanged for a 90-inch person, the mean would be changed considerably while the median would be unaffected.

We next consider measures of dispersion, i.e., measures of how the data spreads out about the mean. Perhaps the first such measure was the probable error introduced by Bessel in 1815 in connection with problems in astronomy. Most commonly used today is the standard deviation σ, this terminology due to Karl Pearson (1894). It is defined for a discrete variable X by

$$\sigma_X = \left[(1/n \sum_{i=1}^{n} (x_i - \mu_X)^2 \right]^{1/2}.$$

Inspection of this formula reveals that σ tends to be large when the data is widely dispersed, small when the data clusters about the mean.

To introduce the next notion, which is correlation, we refer back to the table of heights and weights (Fig. 1). Inspection of the data reveals that these two variables are somehow related. Even common sense tells us that tall people should generally weigh more than short people. Graphically, this relationship can be portrayed by means of what is called a scatter diagram, this being merely a plot of the data in the Cartesian plane (see Fig. 4). The relationship, if linear, will be indicated by a tendency of the points to simulate a straight line.

In the late nineteenth century, Sir Francis Galton asked whether such a relationship between two sets of data could be measured, and he introduced the notion of correlation. It was Karl Pearson, however, who gave us our present coefficient of correlation ρ defined by

FIGURE 4

Scatter Diagram

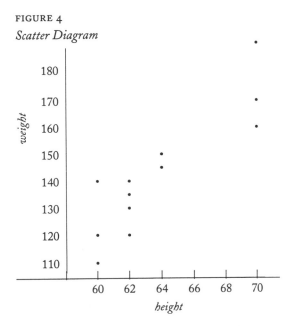

$$\rho = (1/n\sigma_X\sigma_Y \sum_{i=1}^{n} (x_i - \mu_X)(y_i - \mu_Y).$$

It is a matter of simple algebra to show that ρ ranges between −1 and +1. A value of zero indicates no linear relationship; +1 indicates that the data lies on a straight line of positive slope; −1 means that the data lies on a line of negative slope; values near ±1 suggest a strong linear relationship, while values near zero are characteristic of little such relationship. In our example, r is about 0.9. It should be emphasized that r is a measure of linear relationship. The data may lie on a circle, in which case $|\rho|$ will be very small. However, in this case the variables would certainly be related, albeit not linearly related.

Sir Francis Galton, prominent in our discussion thus far, was a cousin of Charles Darwin and did some statistical work for him. His interest in correlation has already been mentioned—it was his writings on this subject which turned the brilliant Karl Pearson toward the study of statistics. But Galton will be remembered most vividly by teachers because he was the first to suggest the use of the normal curve in connection with problems of grading.

The normal curve, which actually dates back at least to Abraham De Moivre in 1733, is a highly useful concept to statistics. It is determined by the equation

$$f(x) = (1/\sqrt{2\pi}\ \sigma) \exp[-(x-\mu)^2/2\sigma^2].$$

Here μ and σ are parameters which turn out to be the mean and standard deviation. The normal curve is often thought of roughly as any "bell shaped" curve. However, this is inaccurate, for other functions, such as $g(x) = [\pi(1+x^2)]^{-1}$, also have graphs which are bell shaped and yet lack completely the qualities which make the normal curve so useful. While the definition of the normal curve given above may appear complicated, from the point of view of the mathematician it is one of the simplest and best behaved of all curves. Figure 5 pictures a special normal curve.

If the area under the normal curve from $-\infty$ to $+\infty$ were to be calculated by integration, it would be found to be 1. Approximately two-thirds of this area lies between points one standard deviation to the left and one standard deviation to the right of the mean. The probability that a normal variable assumes values on any interval $a \leq x \leq b$ is equal to the area above this interval and under the corresponding normal curve. Areas under the normal curve for various intervals are tabulated in any standard book of mathematical tables.

Earlier, in the discussion of discrete distribution, it was shown that the mean could be

FIGURE 5

The Normal Curve ($\mu = 1$, $\sigma = 1$)

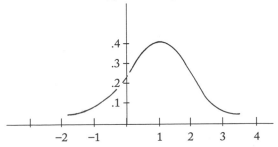

interpreted as the center of mass of a system of discrete masses of total mass 1. The normal distribution described above is an example of a continuous distribution. Reasoning by analogy, we may associate with the normal distribution an idealized continuous rod of mass 1 running indefinitely far in both directions with density varying according to the function f which determines the normal curve. From calculus, the center of mass μ of such a rod would be given by

$$\mu = \int_{-\infty}^{\infty} xf(x)dx.$$

This is, in fact, the formula which we use to define the mean of a continuous distribution. Perhaps surprisingly, not every continuous distribution has a mean, for the above integral may fail to converge. This is the case, for example, for the Cauchy distribution, determined by the equation $g(x) = [\pi(1+x^2)]^{-1}$, as the reader may verify.

Recalling the formula in the discrete case, it is natural to define the standard deviation for a continuous distribution by

$$\sigma = \left[\int_{-\infty}^{\infty} (x-\mu)^2 f(x)dx\right]^{1/2}.$$

It is a matter of a fairly simple integration to check that if these formulas are used to calculate the mean and standard deviation for the normal distribution, they turn out to be the two parameters μ and σ respectively.

Second Lecture

In attempting to answer questions concerning large populations, it becomes necessary, from a practical standpoint, to work with *samples* from the population. The population parameters, such as the mean μ and the standard deviation σ are then unknown. Now assuming that a sample is properly selected (how to do this is an important statistical problem),

it should be possible to get good estimates of the population parameters from the sample. R. A. Fisher has introduced criteria for judging whether a sample statistic is a good estimator for a population parameter. We shall not discuss these criteria. Suffice it to say that if $(x_1, x_2, \ldots, x_n)$ denotes a sample from a population with mean μ and standard deviation σ, then the sample mean $\bar{x}$ and sample standard deviation σ, defined by

$$\bar{x} = (1/n) \sum_{i=1}^{n} x_i$$

and

$$s = \left[(1/n) \sum_{i=1}^{n} (x_i - \bar{x})^2\right]^{1/2}$$

respectively, turn out to satisfy most of the criteria that Fisher has given.[5]

Now if we were to take many samples from a population and compute $\bar{x}$ for each of them, we would get many different values, but presumably these values would tend to cluster about the population mean μ. Looked at this way, $\bar{x}$ is a variable which is distributed in some fashion. This raises an important question. Given a certain distribution of the population variable, how is $\bar{x}$, the sample mean, distributed? A theorem, which we state without proof, partially answers this question.

THEOREM. *If the population variable is normally distributed with mean μ and standard deviation σ, then $\bar{x}$ is normally distributed with the same mean μ but with standard deviation $\bar{\sigma} / \sqrt{n}$, n being the size of the sample.*

FIGURE 6

Normal distributions of X and x for n = 10, 20

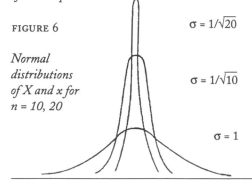

$\sigma = 1/\sqrt{20}$

$\sigma = 1/\sqrt{10}$

$\sigma = 1$

Recalling the significance of the standard deviation, we conclude, not unexpectedly, that as the sample size is made larger and larger the values of $\bar{x}$ tend to cluster more and more closely about μ. This is illustrated pictorially in Figure 6.

A grave limitation arises in the use of this theorem since it is doubtful whether any population variables in real life are exactly normally distributed and many are not even approximately so. We are saved by the most famous theorem in probability theory and one of the most famous in all of mathematics. It is known as the central limit theorem. One form of it is the following:

THEOREM. *If the population variable is distributed in any fashion whatever so long as it has a mean μ and a standard deviation σ, then $\bar{x}$ is approximately normally distributed with mean μ and standard deviation $\sigma/\sqrt{n}$. The word approximately is used in the sense that the distribution of $\bar{x}$ approaches nearer and nearer to the normal distribution as n becomes larger and larger.*

The central limit theorem has a long history. It was proved in its first form about 1733 by Abraham De Moivre for the case of tossed pennies where there are only two possible outcomes. The form which we have stated is due to J. W. Lindeberg in 1922.[6] In recent years, Russian mathematicians have generalized this theorem to its absolute extreme, giving necessary and sufficient conditions for $\bar{x}$ to have the normal distribution as its limiting distribution.

To show one of the uses that the statistician makes of the central limit theorem, consider the following typical problem taken from a well known textbook by Hoel.[7] "A manufacturer of string has found from past experience that samples of a certain type of string have a mean breaking strength of 15.6 pounds with a standard deviation of 2.2 pounds. A time-saving change in the manufacturing process of this string is tried. A sample of 50 pieces is then taken, for which the mean breaking strength turns out to be only 14.5 pounds. On the basis of this sample, can it be concluded that the new process has had a harmful effect on the strength of the string?"

A statistician views the problem this way. It is necessary to test the hypothesis H_0 that $\mu = 15.6$ against the alternative hypothesis H_1 that $\mu < 15.6$. It will be assumed that X, the breaking strength, still has a standard deviation of 2.2, although the possibility certainly exists that the change in manufacturing process has changed the standard deviation. We may now appeal to the central limit theorem, for, no matter how X is distributed, we know that $\bar{x}$ is approximately normally distributed with mean μ and standard deviation $\sigma/\sqrt{n}$, or equivalently that $z = (\bar{x} - \mu)\sqrt{n}/\sigma$ is approximately normally distributed with mean 0 and standard deviation 1. Now if a standard normal table is consulted, we find that a value of 14.5 for $\bar{x}$ is so far away from 15.6 that under hypothesis H_0 the probability of a value this small occurring is only .0002. It seems safe therefore to reject H_0 and accept H_1. The tables also reveal that, if we follow the usual statistical procedure of rejecting H_0 whenever a value occurs which is so small that under H_0 it would occur only 5 per cent of the time, we should reject H_0 for a value of $\bar{x}$ less than 15.09. Any value less than 15.09 falls in the so-called critical region (see Fig. 7).

We mention again the possibility of error due to the assumption that the standard deviation σ has not changed under the new process. Actually, σ is no longer a known parameter. We can, however, calculate s, the sample standard deviation. In 1908, a chemist, William Gosset, writing under the pseudonym of "Student" discovered the distribution of the variable $t = (\bar{x} - \mu)\sqrt{n-1}/s$ (note σ is replaced by s). He showed that if X is normally distributed, the t has the "Student's t" distribution with $n - 1$ degrees of freedom. This distribution is of sufficient importance so that it is tabulated in

FIGURE 7

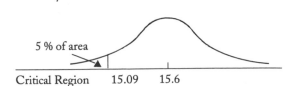

5 % of area

Critical Region   15.09   15.6

most statistical tables. The requirement that X be normally distributed is a stringent one. However, even if X is only approximately normally distributed, statisticians have found it advantageous to use the t distribution when σ is unknown, especially if n is small. If n is large, the difference between s and σ may be expected to be small; hence there is less need to consult the t distribution tables.

We have credited Gosset with the discovery of the t distribution in 1908. However, it should be pointed out that Gosset's results were really guesses and were not rigorously verified until about 1926 by R. A. Fisher, a bright young statistician who had since 1915 begun to play an increasingly influential role.

Fisher is said to have been a precocious child,[8] having mastered such subjects as spherical trigonometry at an early age. He was attracted to the physical sciences and received his B.A. in astronomy from Cambridge in 1912. The theory of errors of astronomy eventually led to his interest in statistical problems. We have picked the date 1915 for his introduction to the statistical world because in that year he published a paper on the distribution of the sample correlation coefficient. This paper initiated a study of the exact distribution of various sample statistics, at which Fisher has been eminently successful. In this study he has been guided by his tremendous geometrical intuition, often arriving at results which have later been proved correct only by the concentrated effort of some of the world's top-flight mathematicians.

Fisher has made many other contributions. Previously, we mentioned his introduction of criteria for judging whether a sample statistic is a good estimator for a population parameter. These include the concepts of consistency, efficiency, and sufficiency, which took form in an impressive memoir of 1921. Also falling in this category is his introduction of maximum likelihood estimators.

In 1919, Fisher left his job of teaching mathematics in public schools to work at the Rothamsted Agricultural Experiment Station. Here he developed sampling techniques and randomization

procedures which are now used all over the world. Two of his books, *Statistical Methods for Research Workers*, published in 1925, and *Design of Experiments*, published in 1935, have had a tremendous influence. The second chapter of this latter book is included in the *World of Mathematics*.[9] In this very charming article, Fisher tells of a woman who claims that she can tell whether or not the milk in her tea was poured before or after the tea was poured, and goes on to describe an experiment designed to prove or disprove her claim.

The theory of testing and estimation received a fresh start with the work of Jerzy Neyman and E. S. Pearson that began about 1928. This resulted in a reformulation and modification of many of the ideas originally introduced by Fisher. To illustrate one of their contributions, we refer back to the problem of the string manufacturer. Our conclusion was that, if a sample led to an $\bar{x}$ which was less than 15.09, the hypothesis H_0 was to be rejected (see Fig. 7). Neyman and Pearson raised questions of the following variety. Why do we take as critical region the region to the left of 15.09? Why do we not choose a "two tailed" critical region with two and one-half per cent of the area at the extreme left and two and one-half per cent of the area at the the extreme right of the distribution? What criteria should be used in choosing critical regions? Must intuition be used or can sound mathematics be brought into the picture? There resulted the formulation of a table which revealed that two distinct types of error were possible, and which Neyman and Pearson named Type I and Type II errors (see Fig. 8).

FIGURE 8

Types of error in testing hypotheses

	H_0 True	H_1 True
$\bar{x} \geq 15.09$ Accept H_0	Correct decision	Type II error
$\bar{x} < 15.09$ Accept H_1	Type I error	Correct decision

Neyman and Pearson were able to sum up their findings in this guiding principle: *Among all tests (critical regions) possessing the same Type I error, choose one for which the size of the Type II error is as small as possible.* While the application of this principle is reasonably complicated, the influence of Neyman and Pearson has made this principle and the related notion of power function important statistical concepts, and a general mathematical theory dealing with this kind of problem has been developed.

No modern account of the history of statistics would be complete without mention of the name of Abraham Wald. Beginning about 1939 and continuing until his death in 1950 in an airplane crash when at the height of his powers, Wald profoundly influenced the direction of statistical theory. His followers are among the leaders of the field today.

Wald was born in Romania of orthodox Jewish ancestry.[10] Because of his religion, he was denied certain educational privileges and found it necessary to study on his own. A clue to his mathematical ability is contained in the fact that as a result of his own studies he was able to make valuable suggestions concerning Hilbert's *Foundations of Geometry,* suggestions which were incorporated in the seventh edition of that book.

Wald later attended the University of Vienna and received his Doctor's Degree after taking only three courses. At this period of time in the political history of Austria, he was denied the privilege of academic work and had to accept a private position helping a banker broaden his knowledge of higher mathematics. He became interested in the theory of economics. Later he became a close associate of the economist, Morgenstern, who collaborated with John von Neumann in the field of game theory.

Wald came to the United States during the dark days before World War II, during which his parents and sisters were eventually to lose their lives in a gas chamber. His interest in economics led him to study statistics, and he soon became an outstanding theoretical statistician. Perhaps his most important con-

tribution was the introduction of a new way of looking at statistics—statistical decision theory. From this point of view, statistics is regarded as the art of playing a game with nature as the opponent. Much of the recent theoretical study has been in this direction and even elementary textbooks are now beginning to adopt this point of view.

Abraham Wald made many other important contributions, but space allows us to cite only one—sequential analysis. Although perhaps not original with him, the theory was certainly developed by him. The technique was judged to be so important in the way of minimizing sampling procedures in manufacturing processes that it was highly classified by the military during World War II.

We shall illustrate the idea of sequential analysis in connection with the problem of quality control in industry. Before the introduction of sequential methods, the standard procedure was to take a fixed size sample from each lot of manufactured items and then accept or reject the lot on the basis of the number of defectives in the sample. This procedure ignores the fact that information about the lot can be obtained from the rate at which defective items occur in the sampling process.

In sequential sampling, three possible states which may occur during the sampling process are recognized: (1) early appearance of defective items

FIGURE 9

Sequential sampling

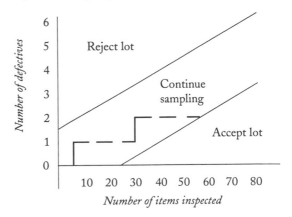

leading to a quick decision to reject the lot, or (2) early appearance of a lack of defective items leading to a quick decision to accept the lot, or (3) inconclusive evidence in which case the sampling process should be continued. This is illustrated pictorially in Figure 9 for a particular example. The guidelines separating the three regions are determined according to the Type I and Type II errors which are allowed. In this example, a decision to *accept* would be made after examining the sixtieth item.

From the diagram, it is clear that while it may be possible to make a quick decision to accept or reject, it is also possible to stay in the intermediate region for a long time or perhaps even indefinitely. Wald, however, showed that, with probability 1, a decision to accept or reject will be made in a finite number of steps. Actual experience has shown that sequential sampling usually results in a saving of about fifty per cent in the cost of sampling when compared with the traditional fixed-sample-size procedures.

NOTES

Editor's note: Notes 5–10 were originally numbered 1–6. Likewise, Figures 6–9 were originally numbered 1-4.

1. Walker, Helen M., *Studies in the History of Statistical Method* (Baltimore: The William and Wilkins Company, 1929). The book has been of great value in the preparation of this lecture, especially in connection with statistical history before 1900.

2. In dividing the history of statistics into four periods, we are following Helen M. Walker, "The Contributions of Karl Pearson," *Journal of the American Statistical Association*, LIII (1958), 11–22.

3. *Ibid.*, p. 13.

4. Actually, Gustav Fechner had employed this measure under the name *der Centralwerth* in 1874 and had given a description of its properties. Galton's use of the concept appears to go back as early as 1869, but the name *median* is first used by him in 1883.

5. In order to be unbiased, n should be replaced by $n-1$ in the formula for s.

6. J. W. Lindeberg, "Eine neue Herleitung des Exponentialgesetzen in der Wahrscheinlich-keitsrechnung," *Mathematisch Zeitschrift*, XV (1922), 211–25.

7. Paul G. Hoel, *Introduction to Mathematical Statistics*, 2nd Edition (New York: John Wiley and Sons, Inc., 1954), p. 103.

8. For a discussion of the life and contributions of Fisher, see Harold Hotelling, "The Impact of R. A. Fisher on Statistics," *Amer. Stat. Assoc.*, XLVI (1951), 35–46.

9. James R. Newman, ed., *The World of Mathematics*, Vol. 3 (New York: Simon and Schuster, 1956), pp. 1512–21.

10. Karl Menger, "The Formative Years of Abraham Wald and his Work in Geometry," *Ann. Math. Stat.*, 23:1 (1952), 14–20.

A Chronological Outline of the Evolution of Computing Devices

SINCE VERY EARLY TIMES, humans have sought to simplify the tasks of numerical record keeping and computation by using physical devices. These devices have varied greatly in scope and context from the adaption of personal body parts such as fingers and toes for simple 1:1 tallying to the use of inanimate objects, notched sticks, and knotted cords for numerical record keeping to the operation of complex mechanical and electrical machines for scientific calculations. Today's high-speed digital computers can trace their origins to the finger manipulations of our ancient ancestors. The path of this evolution is outlined in the list of accomplishments and names of individuals given below.

Date	Accomplishment or Event
?	Use of hands and fingers to communicate numerical facts
ca. 30,000 B.C.	Tally bones recovered from European sites
ca. 8000 B.C.	Clay tokens used in Babylonia for numerical record keeping
ca. 600 B.C.	Abacus used in Classical Greece
ca. 500 B.C.	Computing rods and counting board introduced in China
ca. A.D. 600	With the collapse of Imperial Rome, use of the column abacus dominates European computing. In European setting evolves into a line abacus computing table which remains in use until time of late Renaissance.
ca. A.D. 1400	Quipu used by Inca people of pre-Columbian America
A.D. 1614	John Napier develops logarithms, invents Napier's rods for carrying out multiplication
1620	Edmund Gunter develops logarithmic scale basis for slide rule capable of performing four basic operations
1623	Wilhelm Schickard invents computing machine that can perform four operations
1642	Blaise Pascal builds gear-driven computer that can perform addition and subtraction with six-digit numbers
1671	Gottfried Wilhelm Leibniz refines design of gear computer to include "stepped cylinders" allowing for operation of multiplication and division by repeated additions or subtraction.
1673	Sir Samuel Morland invents multiplying machines in England
1805	Joseph Marie Jacquard develops punch card input for textile looms
1820	Thomas de Colmar standardizes design for mechanical computing machines

Date	Accomplishment or Event
1830	Charles Babbage conceives of great computing engines capable of 26-digit computations. Babbage's designs incorporate specifications of modern digital computers, i.e., input, processing unit, output.
1875	Frank Baldwin obtains American patent for popular calculating machine
1941	Konrad Zuse developes Z3, a relay calculator possessing 64 word memory
1944	Automatic Sequence Controlled Calculator (ASCC) built at Harvard
1945	Electronic Numerical Integrator and Computer (ENIAC) begins operation at University of Pennsylvania, contains 18,000 vacuum tubes, performs 360 multiplication/sec.
	John von Neumann develops Electronic Discrete Variable Calculator (EDVAC) at University of Pennsylvania.
1947	Transistor developed at Bell Laboratories
1951	U.S. Census Bureau accepts delivery of Remington Rand UNIVAC 1. The computer contains 5000 vacuum tubes and performs 1000 calculations/sec.
1953	Magnetic core memory introduced into computers
1957	Fortran programming language introduced
1959	Concept of integrated circuits conceived by Robert Noyce
1960	Cobol language introduced
1964	IBM 360 marketed, employs binary addressing, introduces cheap feasible time-sharing and virtual memory.
	Basic language introduced
1968	First Ph.D. in computer science awarded at University of Pennsylvania
1969	UNIX operating system introduced
	Edgar Codd proposes relational database model to IBM
	Intel develops microprocessor
1970	Floppy disc introduced
1971	Pascal language introduced
	First pocket calculators appear
1975	Microcomputers marketed
1976	Cray 1 supercomputer becomes operational
	Kenneth Appel and Wolfgang Haken resolve 4-color conjecture using a computer
1980	Ada language introduced
1985	The Connection Machine developed by Thinking Machines Corporation, a highly parallel supercomputer possessing 65,536 processors
1988	Computer networking well established

112

ENIAC: The First Computer

A A R O N   S T R A U S S

*T*HE ELECTRONIC DIGITAL COMPUTER is a war baby. The high-speed calculating and data-processing machines have led to the development of new ideas and methods in pure mathematics; they have made possible technological progress leading to interplanetary space travel; and they administer activities of daily life from traffic control and department store sales to industrial payrolls and income tax collection. The computer is truly a major contribution of American mathematics, science, and technology. Yet the stimulus for the production of ENIAC—as so often has been true throughout history—was the demand for ever more sophisticated military methods.

The First World War introduced the first long-range weapons of twentieth-century warfare: cannons that could fire at targets miles away and planes that could drop bombs on targets from miles above. But the gunners and bombardiers who used these weapons needed detailed information for aiming the artillery or bombs. Gunners needed to know angles of elevation for various combinations of target distance and crosswind conditions. Bombardiers needed directions on release time as a function of altitude and air speed. The new weapons received only limited use in World War I. But in the early stages of World War II,

long-distance artillery and bombers played a major role in the fighting.

In 1942, the United States Army sent a young mathematician, Herman H. Goldstine, to the Moore School of Engineering at the University of Pennsylvania to supervise the preparation of firing and bombing tables. Computations required for accurate tables were enormous, and the Moore School was chosen for the project because it had been the home of a great deal of research and development on computing machines (mostly analog computers) during the 1930s. To calculate the trajectory of a typical projectile fired from a large cannon could take over 750 multiplications and many other calculations. These computations would require twelve man-hours using desk calculators, an hour or two using the fastest digital computers available, or at least twenty minutes using the fastest analog computers. Thus hours, or even days, were required to furnish a firing table for just one type of gun. Goldstine and two engineers, John Mauchly and J. P. Eckert, proposed to the Army that the Moore School develop a digital computer that would be at least 100 times faster than the fastest computer then available. The Army officially accepted the proposal in June 1943. By 1946, a computer called ENIAC (Electronic Numerical Integrator and Computer) had been constructed with the capability of performing the trajectory calculations mentioned above in less than ten seconds. More important, the design was soon improved so that subsequent models were a thousand times faster than ENIAC.

Reprinted from *Mathematics Teacher* 69 (Jan., 1976): 66–72; with permission of the National Council of Teachers of Mathematics.

Author's note: I thank Jack Goldhaber, Jack Minker, Chris Rapalus, and Werner Rheinboldt for their helpful suggestions concerning this article.

Historical Evolution of the Computer Concept and Technology

Though ENIAC was produced in the United States through efforts directed by Goldstine, Mauchly, and Eckert, the project built on important ideas of many mathematicians and engineers—from both earlier and contemporary times. The calculating capabilities of ENIAC and its successors were undoubtedly dreamed of by mathematicians for centuries. But the main outline of the computer story is given by the following chronology.

1623:	Calculator of Wilhelm Schickard (Germany).
1642:	Calculator of Blaise Pascal (France).
1673:	Calculator of Gottfried Leibniz (Germany).
1805:	Automated loom of Joseph Marie Jacquard (France).
1822–62:	Calculators of Charles Babbage (England).
1853:	Calculator of Pehr Georg Scheutz (Sweden).
1874:	Calculator of Martin Wiberg (Sweden).
1890:	Use of punch card tabulators in U.S. Census.
1929+:	IBM Computing Bureau at Columbia University.
1937+:	IBM Computer project at Harvard University.
1937+:	Computer project at Bell Telephone Laboratories.
1937+:	Electronic computer ideas of John V. Atanasoff (Iowa).
1943:	Contract to build ENIAC (Philadelphia).
1944:	John von Neumann joins ENIAC project.
1945:	Planning of EDVAC and von Neumann's "First Draft."
1946:	ENIAC finished and operating.
1947:	Conversion of ENIAC into a stored program computer.
1946–58:	Institute for Advanced Study computer project (Princeton).
1951:	Remington Rand UNIVAC.
1953:	IBM 701.

The first calculators of Schickard, Pascal, and Leibniz were hand operated—that is, mechanical. They were about the size of a television set. In principle, they operated like the little plastic adding machines that some people use when shopping in supermarkets, except that some also did multiplication and division. These calculators came before the Industrial Revolution, and they showed that at least this one human activity of calculating could be mechanized. But the calculators were slow and certainly not widely used.

Many fundamental principles of modern computers were suggested in the work of an Englishman, Charles Babbage (1792–1871). He proposed using steam power (the "modern" energy source in those days) to run the calculator and using punched cards for entering data into the machine (an idea apparently acquired from Jacquard, who automated weaving by using punched cards to guide weaving patterns in the looms). Babbage and his wife, Lady Lovelace (daughter of Lord Byron), described programs for their machines in ways that influenced programs developed 100 years later. Unfortunately, nineteenth-century technology could not match the ideas Babbage produced, and he never completely finished building a calculator.

Most of Babbage's ideas were developed for what he called an "Analytical Engine." A much simpler calculator described by Babbage, called a "Difference Engine," was actually built by Scheutz and later improved by Wiberg. In fact, Wiberg used his machine to calculate tables of logarithms, which he then entered in the Philadelphia Exhibition of *1876*!

The first connection between the United States and calculating machines seems to have been with the use of tabulators in the 1890 U.S. Census. Herman Hollerith developed a tabulating system in his doctoral work at Columbia University during the 1880s, improved that system, and eventually used it to tabulate the entire 1890 census within one month after the returns were in, an unprecedented feat. Hollerith's tabulators read punch cards and were electromechanical—that is, they used electric power to throw relays and switches. They were about the size of a dining-room hutch cabinet, or a bedroom chest of drawers. Not only did these machines mark the entrance of the United States into the calculating-machine field, but they were the first to use many of Babbage's advanced ideas and the first to be used for an important, large-scale commercial venture. The use of Hollerith machines spread. For example, in the 1920s, L. J. Comrie (England) used such a machine to calculate tables of the positions of astronomical objects.

One of the early manufacturers of Hollerith machines, the Tabulating Machine Company, later became IBM. Between the two world wars, IBM helped set up calculating laboratories at two American universities. The laboratory at Harvard operated under H. H. Aiken and produced the Mark I, a giant, electromechanical calculator of complex design, similar to another developed at Bell Telephone Laboratory. Both were outmoded by the ENIAC, but they served to introduce many people at IBM, Harvard, and Bell Labs into the computing field. The IBM-Harvard Mark I was about ten feet high, three feet deep, and over fifty feet long. These electromechanical calculators operated in principle like an electric desk calculator or a supermarket cash register, except that they read punch cards, stored numbers, took square roots, and were more complex.

Why weren't these giant electromechanical computers good enough? They were too slow; they could not furnish the data needed by the war effort. The problem was that it took too long for an electric current to throw a relay. Vacuum tubes, on the other hand, could be charged or discharged thousands of times faster than a switch could be thrown. Vacuum tubes are electronic; switches are electromechanical. The modern, quiet, department-store cash registers that click instead of ring are examples of electronic adding and filing machines. The new pocket calculators made by Texas Instruments, Hewlett-Packard, and others are examples of electronic (solid state) calculating machines. However, thousands of tubes were needed for an electronic computer, and much design work was necessary to make the computer reliable enough to

be useful. This was a "hardware" problem. There were also "software" problems of how to communicate with the machine. Pioneering work on hardware problems was done by Atanasoff, whose ideas influenced Mauchly just before Mauchly went to the Moore School in 1941. This brings us to the Moore School/U.S. Army contract to build the ENIAC, which, incidentally, was about the size of the IBM-Harvard Mark I.

About a year after the ENIAC project commenced, the Moore School was visited by one of the most brilliant and versatile mathematicians of this century, John von Neumann. Von Neumann was born and educated in Europe and had emigrated to America in the 1930s to accept a professorship at Princeton and then at Princeton's Institute for Advanced Study. He became interested in the ENIAC project, joined it in the summer of 1944, and made fundamental contributions to both the hardware and software areas. By the fall of 1944, most of the work of designing the ENIAC was finished. This left the designers relatively free to design a second computer while the ENIAC was being constructed. They had many improvements in mind; for example, they wanted to reduce the number (18,000) of vacuum tubes required by the ENIAC. This new machine was called EDVAC (Electronic Discrete Variable Computer), and von Neumann wrote his ideas in a paper called "First Draft of a Report on the EDVAC." Although this paper was not intended for publication, it was widely circulated and was credited by a number of others for showing them how to design a computer.

It is hard to explain, and harder to overstate, the magnitude of von Neumann's contribution. Perhaps his most fundamental contribution was that he described exactly what a computer is, or ought to be. He said it should have a central arithmetic part that is actually a calculating machine, capable of performing the basic arithmetic functions. It should have a control part, which determines how the machine deals with its orders, and in what kind of sequence it obeys its instructions. It should have the ability to read data from the outside world (input) and to inform the outside world of its results (output). And it should have a memory, capable of storing numbers for further use and for storing instructions. In fact this feature of storing its instructions is perhaps what distinguishes a computer from a calculator. Indeed, von Neumann went on to develop the concept of the self-modifying, stored program: some of the instructions tell the computer how to modify some of its other instructions.

After the war the ENIAC/EDVAC group split up. Mauchly and Eckert formed a computer company that eventually produced the UNIVAC, while von Neumann and Goldstine headed a project to continue research work on new computers at the Institute for Advanced Study. Later von Neumann became a consultant to IBM and Goldstine went there full time. The Smithsonian Institution in Washington, D.C., now has an exhibit of these early computers and their calculator predecessors. But the contemporary offspring of ENIAC are in evidence throughout our daily lives.

The Computer and America

In what sense is the computer an *American* contribution to (or from) mathematics? First the ENIAC was the world's first electronic digital computer, and it was designed and built in the United States. Second, von Neumann's "First Draft" was perhaps the most influential single paper written on designing computers. Third, the virtual explosion of computers on the scene in the 1950s was composed of American-made computers.

A more difficult question is, What were the particular features of the United States that caused the first computer to be built here? Most of the earlier developments (before 1890) occurred elsewhere. In the history of computers, two factors stand out: (1) the interplay of industry, government, and academia; and (2) the existence in the 1930s and 1940s of fast-growing electrical and electronic technology and big electrical companies.

Regarding (1): In addition to the IBM-Harvard example mentioned earlier and the lowering of the

barriers separating industry, universities, and government during the war, it is interesting to note that the Institute for Advanced Study computer project was financed variously by the Office of Naval Research, Army Ordnance, the Atomic Energy Commission, the Radio Corporation of America, and Princeton University. This type of interplay has come to be the topic of much controversy in recent years.

Regarding (2): The idea of a computer was not totally new. What was new was the sophistication of electronics necessary to achieve the speed and the reliability to make the whole thing worthwhile. Vacuum tubes (electronic) are much faster than relays (electromechanical). For example, the whole field of numerical meteorology was made possible by the speed of modern computers. A model was run on the Institute for Advanced Study computer in 1953 and yielded a twenty-four-hour weather forecast in six minutes. That model took twenty-four hours to run on the ENIAC and, of course, hopelessly longer on its predecessors. The technology was quite highly developed in the United States, as can be exemplified by the electrical, telephone, telegraph, and radio networks, and by big companies such as General Electric, the Radio Corporation of America, and Western Electric.

Some work was done on computers in Germany during the period 1937–45, especially by Konrad Zuse. In fact, Zuse actually built several electromechanical computers and achieved some success in designing a compiler. (A compiler is a language close to a human one, and the computer itself is programmed to translate this language into its own special code, which it then follows.) Why didn't the Germans continue? Here we arrive at another reason why computers developed in the United States: it was the only major industrial nation to escape physical devastation during World War II.

Speculation on the causes of historical events is a hazardous occupation in which one is inevitably limited by incomplete information. In fact, controversy over who should get credit for inventing the computer is manifest in several pending patent suits between early contributors to the field. Nonetheless, the factors outlined above seem to have had a clear and important role in making the computer an American contribution to mathematics and technology.

The Computer and Mathematics

The development of the computer was sparked by problems in applied mathematics. But the design of effective computers was also facilitated in important ways by theoretical ideas from mathematics, and the availability of modern computing power has stimulated growth of entirely new branches of pure and applied mathematics.

As von Neumann conceived of the computer, it should consist of processor, control, input/output, and memory—operations of which should be guided by a stored program. Mathematics provides the theoretical models of the electronic components that performed these functions. In early computers, the easiest and most reliable way to use electronics to represent numbers was by using, not the *amount* of current, but simply the *direction* of flow—positive or negative, on or off; a switch could be open or closed. Thus it was a natural to use the binary system (base 2) of numeration, rather than the decimal. Most computers after the ENIAC used binary arithmetic, although the ENIAC itself was a decimal machine. Binary arithmetic is an example of Boolean algebra, which, together with other aspects of formal symbolic logic, was the main tool of pure mathematics used in the design of early computers. More recently, concepts from lattice theory, numerical analysis, and combinatorics have become fundamental tools in the design of computers and their software.

The individual aspects of a computer design provide no striking new insight into mathematics, but combined in an electronic system, they provide a calculating and logical tool with incredible speed. The significance of this speed has been stated well by others:

> But why are these great speeds so important? Are they really necessary or have they been devised only to keep the new machine busy?" Let us start the

discussion by stating categorically that without the speed made possible only by electronics our modern computerized society would have been impossible: machines that can do as much as ten or twenty or thirty or even a hundred humans are very important but do not revolutionize modern society. They are extremely valuable and help greatly to ease the burden on humans, but they do not make possible an entirely new way of life ... clearly efforts such as are typified by numerical weather prediction would be quite impossible without electronic computers This is then the whole point of the modern machines. It is not simply that they expedite highly tedious, burdensome, and lengthy calculations being done by humans or electromechanical machines. It is that they make possible what could never be done before! The electronic principle did much more than free men from the loss of hours like slaves in the labor of calculating: it also enabled them to conceive and execute what could not ever be done by men alone. It is what made possible putting men on the moon—and bringing others back safely from an abortive trip there. [Goldstine 1972, pp. 145 47]

Another argument that continually arises is that machines can do nothing that we cannot do ourselves, though it is admitted that they can do many things faster and more accurately. The statement is true, but also false. It is like the statement that, regarded solely as a form of transportation, modern automobiles and aeroplanes are no different than walking. One can walk from coast to coast of the U.S. so that statement is true, but is it not also quite false? Many of us fly across the U.S. one or more times each year, once in a while we may drive, but how few of us ever seriously consider walking more than 3000 miles? The reason the statement is false is that it ignores the order of magnitude changes between the three modes of transportation: we can walk at speeds of around 4 miles per hour, automobliles travel typically around 40 miles per hour, while modern jet planes travel at around 400 miles per hour. Thus a jet plane is around two orders of magnitude faster than unaided human transportation, while modern computers are around six orders of magnitude faster than hand computation. It is common knowledge that a change by a single order of magnitude may produce fundamentally new effects in most fields of technology; thus the change by six orders of magnitude in computing have produced many fundamentally new effects that are being simply ignored when the statement is made that computers can only do what we could do for ourselves if we wished to take the time. [Hamming 1964, pp. 1–2]

These order-of-magnitude changes in the speed with which traditional mathematical calculations can be carried out have enhanced the applicability of classical ideas. The computer can quickly calculate tables of sines and cosines, logarithms, square roots, definite integrals, statistical parameters, or roots of equations. When the first computer was completed in 1946, it found immediate use in the Los Alamos thermonuclear project, but also in more peaceful pursuits like studying the distribution of prime numbers.

But the computer's ability to cope with the many calculations required in the solution of complex system probelms has also played a fundamental role in creating important new branches of mathematics. For instance, whereas any reasonable linear programming application is a situation involving dozens of variables subject to equally numerous linear constraints, the calculations required to solve the system are prohibitively complex in the absence of programmable computer power. The application of computers to space travel or weather prediction stimulates further research in the mathematics of these areas. The tremendous number of calculations involved creates "round-off error" problems that have become the basis of a new mathematical specialty called numerical stability. Problems in computer design have led to the study of topics like the theory of algorithms and the mathematics of grammar and language. Areas of mathematics that had been investigated and forgotten have gained new life because of the needs and potential of computers. For instance, numerical analysis, combinatorics, graph theory, network theory, and automata theory are now prominent topics for pure and applied mathematical research. The ability of computers to store and retrieve large amounts of information has prompted extensive investigation of their potential as both medium and manager of instruction in all areas of education.

One of the chief differences between a calculator and a computer is that the latter can be programmed in advance, the program itself can be stored in the computer, and the program can modify itself, just as it can modify numbers. In 1966, almost twenty years after the first stored program computer, it was proved mathematically that self-modifying programs are more powerful than non-self-modifying ones; another, yet undoubtedly not the last, example of mathematics "created" by the computer.

Reflections

The chronicle of developments that produced the first computer in America, with major contributions by Americans, can be viewed simply as a story of impressive mathematical and scientific achievement. But like any thoughtful reading of history, that story raises broader questions about the motivation, process, and impact of progress in our discipline: Why does it seem to take a war to produce scientific advances that could just as well come about without wars? The computer is not the only war baby that has had enormous practical payoff in peaceful pursuits. Radar and rockets were also both products of World War II; and over two thousand years ago Archimedes produced many of his startling mathematical results in response to problems of national defense.

A discussion of the American contribution to any field raises the question, Who is an American? One who is native born? A citizen? A resident? In acknowledging the American contribution to a discipline, where is the dividing line between justifiable pride on the one hand and a kind of America-can-do-everything nationalism on the other? Big Business and the cooperative efforts of government, industry, and universities can produce good (such as the computer) or evil (such as destruction of the environment). How should mathematicians and laymen deal with this problem?

Many further references to specific parts of the computer story appear in Goldstine's fine book, *The Computer from Pascal to von Neumann* (1972), which served as the principal source for the historical development contained in this article. More extensive discussion of software developments may be found in the second-to-last chapter of Goldstine and the first few pages of Rosen (1967).

The texts listed under student references highlight history as well as future implications of computer technology.

BIBLIOGRAPHY

For Teachers

Goldstine, Herman H. *The Computer from Pascal to von Neumann.* Princeton: Princeton University Press, 1972.

Hamming, R. W. "Impact of Computers." *American Mathematical Monthly* 72, pt. 2 "Computers and Computing," (January 1964): 1–7.

Richtmyer, R. D. "The Post-War Computer Development." *American Mathematical Monthly* 72, pt. 2 "Computers and Computing," (January 1964): 8–14.

Rosen, Saul, ed. *Programming Systems and Languages.* New York: McGraw-Hill Book Co., 1967.

von Neumann, John. *The Computer and the Brain.* New Haven: Yale University Press, 1958.

For Students

Eames, Charles, and Ray Eames. *A Computer Perspective.* Cambridge, Mass.: Harvard University Press, 1973.

Feingold, Carl. *Introduction to Data Processing.* Dubuque, Iowa: Wm. C. Brown, 1971.

Kemeny, John G. *Man and the Computer.* New York: Charles Scribner's Sons, 1972.

Martin, James, and Adrian R. D. Norman. *The Computerized Society.* Englewood Cliffs, N.J.: Prentice-Hall, 1970.

Marxner, Ellen. *Elements of Data Processing.* Albany, N.Y.: Litton Educational Publishing, 1971.

Miller, Boulton B. *Computers: A User's Introduction.* Edwardsville, Ill.: Bainbridge, 1974.

Orilia, Lawrence, William M. Fuori, and Anthony D'Arco. *Introduction to Computer Operations.* New York: McGraw-Hill Book Co., 1973.

Sanders, Donald H. *Computers in Society.* New York: McGraw-Hill Book Co., 1973.

Shelly, Gary B., and Thomas J. Cashman. *Introduction to Flowcharting and Computer Programming Logic.* Fullerton, Calif.: Anaheim Publishing Co., 1972.

113

Dynamical Systems:
Birkhoff and Smale

JAMES KAPLAN AND AARON STRAUSS

$\mathcal{B}$EGINNING PERHAPS with Newton in the seventeenth century many brilliant thinkers have attempted to express the observable behavior of the physical world in the form of relatively simple mathematical formulas and equations. Frequently these equations are "differential equations," which relate the position of some object (or objects) to its velocity, acceleration, temperature changes, energy, external forces, and so on. At first, differential equations were used by Newton to describe the motions of both celestial and earthbound objects under the influence of various gravitational forces. His famous equation $F = ma$, the second of Newton's three laws, is really a differential equation. Later, differential equations were used to describe the behavior of certain mechanical systems such as springs, pendulums, and engines; then to describe properties of electrical networks; and recently they were applied to describe some aspects of the twentieth-century fields of atomic physics, solid state electronics, biology, economics, and space travel.

In the late nineteenth century the superb French mathematician Henri Poincaré (1854–1912) initiated a study of differential equations from a more abstract point of view. In this way he hoped to unify the existing theories and shed some light on certain problems in celestial mechanics that had

been puzzling scientists at that time. He studied systems in which the particles moved according to two or three basic mathematical "laws." These laws were general enough to include most of the specific situations already mentioned. These systems came to be known as *dynamical systems.*

Today the study of dynamical systems is an active, exciting field of mathematical research. Perhaps more than any other field, it combines in a basic way the three principal mathematical areas of analysis, algebra, and topology and uses this combination to obtain direct applications to the problems in the physical world mentioned above. Since the time of Poincaré, important theoretical advances in dynamical systems have been made by several American mathematicians. This article focuses on two of them—George David Birkhoff (1884–1944) and Stephen Smale (1930–).

We shall begin with some examples to give the reader more of a feeling for dynamical systems, briefly discuss the accomplishments of Birkhoff and Smale, and finally describe in more detail one major advance made by each.

Examples

Imagine that a sheet of paper is placed over a bar magnet. If a tiny iron ball bearing is placed on the sheet, held at rest for a moment, and then released, the bearing will move under the influence of the magnetic force. If the bearing is first dipped in ink and then released on the paper, it will leave a path

Reprinted from *Mathematics Teacher* 69 (Oct., 1976): 495–501; with permission of the National Council of Teachers of Mathematics.

FIGURE I

Four particles and their integral curves

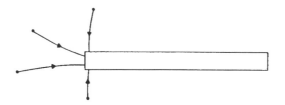

FIGURE I

Four particles and their integral curves

as it moves. If the bearing is again dipped in ink and placed down at exactly the same initial spot, it will have exactly the same motion, that is, it will follow its previous path. If the bearing is redipped and placed at different spots, different paths will be traced (Fig. 1). This is a primitive example of a dynamical system. In the language of dynamical systems, the bearing is a "particle," its movement is the "motion of the particle," each path is a "trajectory," or "integral curve," or "path," and each path is "unique" in the sense that a bearing placed twice in the same spot will follow the same path both times. The collection of all the paths (although we actually draw only a representative sample) is called the "phase portrait" of the dynamical system.

Now suppose that black iron filings are sprinkled on the paper and that the paper is gently tapped to allow the filings to settle into a pattern (Fig. 2). The pattern of black curves will look very much the same as the phase portrait of paths obtained previously. The difference is that no particle traced out the black curves. All the black

FIGURE 2

Black curves representing a vector field

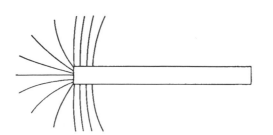

curves were formed simultaneously and directly from the lines of magnetic force. In the language of dynamical systems, the black curves are the "vector field"; they show the "external force" that will produce the motion of particles. The particles will move "along" or "smoothly through" the vector field and trace out paths. Primitive examples similar to this include the motion of a Ping-Pong ball in a mountain stream and the path of a drop of black paint that is dropped into a pail of white paint that in turn is being stirred slowly and uniformly.

Uncontrived examples are more complicated because it is necessary to include velocity as part of the description of a particle's position. (The next two examples require the use of analytic geometry; the reader who has not learned this subject is advised to skip the next two paragraphs.)

Suppose a pendulum swings, and at time t its angle with the vertical is denoted by $x(t)$, or simply x (Fig. 3). The motion is represented by paths in a two-dimensional space. The horizontal axis is x, the vertical axis is the velocity of x, denoted $\dot{x}$. The

FIGURE 3

Simple pendulum

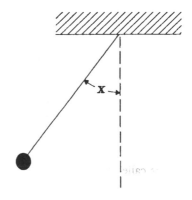

point $(2,0)$ means $x = 2$, $\dot{x} = 0$; point $(1, 1)$ means $x = 1$, $\dot{x} = 1$, and so on. If there is no air resistance, the pendulum has a periodic motion. This is represented by a closed curve around the origin (e.g., a circle) in the x, $\dot{x}$ plane. This is called a periodic

FIGURE 4

Path of a pendulum

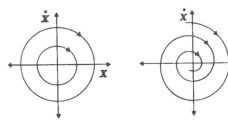

(a) no air resistance   *(b) some air resistance*

path (see Fig. 4a). If there is air resistance, the pendulum swings back and forth, but the amplitude (maximum angle) steadily decreases, and the pendulum approaches a state of rest. This is represented in the x, $\dot{x}$ plane as a spiral contracting toward the origin (see Fig. 4b). The system without air resistance is called a *conservative* dynamical system. This name comes from the fact that in such a system energy is conserved. In the system with air resistance, energy is lost as the pendulum slows down.

In the earth-moon-sun system of three mutually dependent objects, six dimensions are needed, three for the position of each body and three for the velocity of each body. This system seems to be, for all practical purposes, conservative.

In studying any dynamical system the principal goal is to predict the future behavior of the system from information that we have about it at present. This is far from easy; in fact, for a conservative system of three bodies, each of which exerts a force on the other two and in which no simplifying assumptions are made, this goal has not yet been achieved. This problem is called the "three-body problem," and mathematicians have been trying to solve it for hundreds of years. Through attempts to solve this and other problems, many results have been obtained that, although they do not solve the problem, help us to understand dynamical systems. We can now turn to a discussion of the lives of Birkhoff and Smale and to the results they obtained.

Birkhoff and Smale

George David Birkhoff, who was intellectually a disciple of Poincaré, was unquestionably one of the world's premier mathematical intellects. In addition to his work on dynamical systems, which occupied most of his life, he published papers on such diverse problems as the four-color problem, relativity theory, number theory, and even aesthetics. His collected works comprise about a hundred memoirs, nearly three thousand pages, and he also published several books.

One of the first native-born, American-educated American mathematicians, Birkhoff was born in Michigan in 1884 and received his doctorate (classical ordinary differential equations) at the University of Chicago in 1907. He held positions at the University of Wisconsin and Princeton and, for most of his career, Harvard. The high quality of his work was recognized worldwide. In 1918 he received the Querini-Stampalia Prize of the Royal Institute of Science, Arts, and Letters of Venice; in 1923 he received the Bôcher Prize of the American Mathematrical Society; he was awarded the annual prize of the American Association for the Advancement of Science in 1926; he received the Pontifical Academy of Sciences prize in 1933. He was also a member of the National Academy of Sciences, the American Academy of Arts and Sciences, the American Philosophical Society, president of the American Mathematical Society (1925), and president of the American Association for the Advancement of Science (1937).

Stephen Smale received his doctorate (modern differential topology) at the University of Michigan in 1956. He has since held positions at the University of Chicago, the Institute for Advanced Study at Princeton, Columbia University, and is presently at the University of California at Berkeley. Smale's research is on differential topology and global analysis (dynamical systems from a global, or topological, point of view). He has applied his mathematical techniques to problems in electrical circuit theory, economics, and, more recently, biology. In the beginning of his career Smale worked on, among other things, the famous Poincaré conjecture, which proposed that if an object has certain specific properties that spheres have, then that object must in fact be a sphere. This problem in the field of algebraic topology represents an area other than dynamical systems in which both Smale and Poincaré did fine work. Smale solved the Poincaré conjecture in higher dimensions (while sunbathing on a beach, he once said), and for this feat he was awarded the Veblen Prize of the American Mathematical Society in 1964 and the Fields Medal of the International Mathematical Union in 1966. A Nobel Prize is not awarded in mathematics; the Fields Medal is probably its equivalent. It was at this point that Smale began his work on dynamical systems.

Obviously, it would be impossible in a single short article to describe in any detail even a small fraction of the work of two creative and prolific mathematicians of the stature of Birkhoff and Smale. Instead, we shall content ourselves with trying to describe intuitively two mathematical ideas that are important in dynamical systems theory, one of which is associated with each man—the "ergodic theorem" with Birkhoff and "structural stability" with Smale. In a sense, both of these ideas were developed to overcome a common problem in dynamical systems: a lack of completely accurate (i.e., error free) information. In order to understand these ideas we must examine in a little more detail the notion of a dynamical system. For convenience, we shall restrict ourselves to two-dimensional systems. In what follows, the reader who is unfamiliar with analytic geometry should keep in mind the example of the sheet of paper over the bar magnet in Figures 1 and 2.

Dynamical Systems

Consider a particle p that moves in the usual x,y-plane under the influence of certain forces. Suppose that corresponding to each point q in the plane there is associated an arrow $F(q)$ whose tail

FIGURE 5

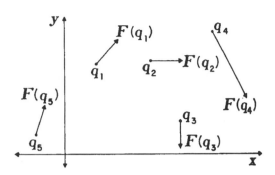

is located at q (see Fig. 5). The direction of the arrow indicates the direction of the external force on a particle located at q, and the length of the arrow denotes the force's magnitude. The collection of all such arrows (or *vectors*) in the plane is called a *vector field*. A particle placed at any point q will be moved by the force $F(q)$, and at each subsequent location it will be acted upon by the force associated with that new location. The path followed by a particle under the influence of the vector field is called an *integral curve*, and the collection of all possible integral curves (which is the same as all possible motions of a particle whose motion is determined by the vector field) constitutes the *phase portrait* of the dynamical system (see Fig. 6).

Recall that our goal in the study of dynamical systems is to be able to predict future behavior from present information. Thus it seems clear that a knowledge of the complete phase portrait of our system is sufficient for our purpose. Given the current location of a particle, we need only follow the integral curve through that point to see where the particle will be at any future instant.

It is at this point that we encounter one of the fundamental problems of science, alluded to earlier: the difficulty of measurement. We can never determine *exactly* the forces (or integral curves) in our vector field, just as we can never determine exactly the current position of a moving particle. We are limited by the accuracy of our measuring devices. Worse, the small errors made when taking

FIGURE 6

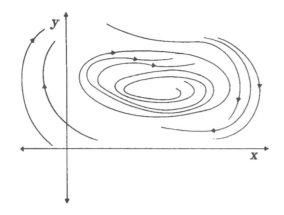

our initial measurements will tend to become magnified as we try to make predictions further and further into the future. In a sense, Birkhoff's ergodic theorem and the notion of structural stability that has come to be associated with Smale are attempts to resolve in different ways the problem of imperfect information.

How would you attempt to resolve the problem if you were faced with it? For instance, suppose you measured the length of a rod on three separate occasions and found its length to be 40.1, 39.9, and 40.0 centimeters. What value would you use for the length of the rod? Probably, you would assume it to be 40.0 centimeters, which is the average of the different measurements.

Now let's return to our dynamical system. Consider again a particle p moving in the x,y-plane. Suppose we wish to know the value of a particular quantity (such as the energy) that depends on the location of p. Such quantities are sometimes called *observables* of the system. We might ask, "What will be the value of the observable at some future time t?" As we have indicated, if we never know the exact position of p, we shall never know the true value of the observable. Instead, by analogy with the previous discussion, we might ask, "As a particle p moves, what will be the average value of our observable?" Such considerations serve as motivation for the subject of *ergodic theory*.

The average value of our observable as the particle moves along an integral curve is an example of a *time average*, since the values of the observable we are averaging are computed at each future time. There is another kind of average, too—a *space average*. We could consider all possible locations of the particle p at one instant of time (say, the present) and then average the value of the observable corresponding to each of these locations.

Perhaps an example can clarify this distinction. Suppose we want to determine the average energy possessed by a molecule of air in a certain room. Here, energy is the observable. There are two ways to do this. One method is to follow a molecule forever, record its energy, and compute its average energy over its lifetime. This is a *time average*. The second method is to compute the energies of every molecule of air in the room and find the average value of all those energies. This is a *space average*. Birkhoff's ergodic theorem states that, under certain technical assumptions, for conservative dynamical systems (systems in which energy is neither gained nor lost) the space average of any observable equals its time average. Notice how this achieves our fundamental goal of predicting future behavior from present information. We can use the present average value of the observable over all of space to predict its average future value. Provided we are willing to settle for average values of conservative systems, we can resolve the problem of imperfect information.

Another attempt to resolve this same problem is the notion of structural stability. To understand this idea, let's first imagine that the phase portrait of our dynamical system is drawn on a flat rubber sheet. Any new phase portrait that is obtainable from the old one simply by stretching or pulling the rubber sheet is called *topologically equivalent* to it (see Fig. 7). Topologically equivalent phase portraits exhibit roughly the same behavior. Notice that we are not permitted to cut a hole in the rubber, nor can we fold it over. Figure 4 contains

FIGURE 7

Topologically equivalent phase portraits

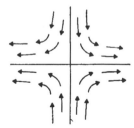

two phase portraits that are *not* topologically equivalent.

With this in mind, let's return for a moment to the picture of our vector field (see Fig. 8). Imagine that we alter each of the arrows by a "little bit." We can change its length or its direction. In so doing we obtain a new vector field that, in turn, will generate a new phase portrait. We say that our original dynamical system is *structurally stable* if every new dynamical system obtained by "slight" modification of the vector field has a phase portrait that is topologically equivalent to the old one. This idea was originally used by Andronov, a Soviet engineer, and Pontryagin, a Soviet mathematician, but it has since become associated with Smale because of his outstanding work on it.

If we are willing to restrict ourselves to structurally stable systems, then imperfect information is not crucial, since our lack of exact knowledge will result in a slightly modified dynamical system that is topologically equivalent to the original one.

FIGURE 8

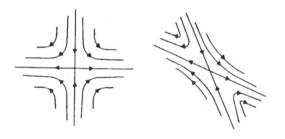

Thus we can use our approximation to predict future behavior, secure in the knowledge that our results will be approximately correct. "Most" dynamical systems in the plane are, in fact, structurally stable. The same is unfortunately not true in higher dimensions.

Reflections

Needless to say, for reasons of space, we are forced to omit from our discussion several other American mathematicians who also made important contributions to dynamical system theory. Furthermore, this field is by no means exclusively American. We have concentrated on the American contributions because of the Bicentennial.

Our chief regret is that we cannot refer the reader to an elementary, yet complete, exposition of dynamical systems. The subject is an advanced one, requiring a lot of knowledge from many areas of mathematics. We can only hope that this article has served to give the reader at least a taste of the flavor of this rich subject.

Finally, we cannot resist a comment about the roles of these two mathematical giants, Birkhoff and Smale, in the development of mathematics in America. Prior to Birkhoff, Americans had to study in Europe to learn significant mathematics. Birkhoff was one of the "new breed," American born and American educated. This American tradition is now firmly established, verified by the biographies of Smale and his contemporaries and even from the fact that many European mathematicians now come to the United States to complete part of their education. But in another sense, a tradition seemed to end with Birkhoff's death— the tradition that mathematicians did fundamental and important work in areas other than pure mathematics. We may be witnessing a return to the tradition of Poincaré and Birkhoff, for Stephen Smale has obtained and continues to obtain important and interesting results in electrical engineering, biology, and economics.

REFERENCES

Birkhoff, George David. "What Is the Ergodic Theorem?" *American Mathematical Monthly* 49 (1942): 222–26.

———. *Collected Mathematical Papers.* American Mathematical Society, 1950.

Gillespie, C. C., ed. *Dictionary of Scientific Biography,* vol. 2. "G. D. Birkhoff." New York: Charles Scribner's Sons, 1970.

Halmos, P. *Ergodic Theory.* New York: Chelsea Publishing Co., 1956.

Morse, M. "George David Birkhoff and His Mathematical Work." *Bulletin of the American Mathematical Society* 52 (1946): 357–91.

Smale, S. "Differentiable Dynamical Systems." *Bulletin of the American Mathematical Society* 73 (1967): 747–817.

———. "What Is Global Analysis?" *American Mathematical Monthly* 76 (1969): 4–9.

114

Fermat's Last Theorem: 1637–1988

CHARLES VANDEN EYNDEN

$\mathcal{A}$ROUND 1637 the French jurist and amateur mathematician Pierre de Fermat wrote in the margin of his copy of Diophantus's *Arithmetic* that he had a "truly marvelous" proof that the equation $x^n + y^n = z^n$ has no solution in positive integers if $n > 2$. Unfortunately the margin was too narrow to contain it. In 1988 the world thought that the Japanese mathematician Yoichi Miyaoka, working at the Max Planck Institute in Bonn, West Germany, might have discovered a proof of this theorem. Such a proof would be of considerable interest because no evidence has been found that Fermat ever wrote one down, and no one has been able to find one in the 350 years since. In fact Miyaoka's announcement turned out to be premature, and a few weeks later articles reported holes in his argument that could not be repaired.

Reports of the latest supposed proof of Fermat's last theorem, as the result has come to be called, seem to have been triggered by a talk given by Miyaoka in Bonn to about three dozen mathematicians on 26 February 1988. Those present started calling friends about his apparent proof, and eventually word reached the newspapers and weekly magazines.

When Miyaoka's proof fell through, members of the mathematical community expressed recriminations, fearing that the public would judge that

mathematicians could not tell what was a proof and what was not. Some blamed reporters for rushing into print too soon, whereas others blamed mathematicians for talking to reporters before the proof was confirmed. Another view was that any publicity was better than none. At least mathematics was shown to be a living subject, practiced by human beings capable of mistakes like anyone else.

The Early History of Fermat's Equation

The mathematical historian Eric Temple Bell (1961) says that Fermat is the only amateur to reach the first rank in mathematics. Certainly he would have had difficulty achieving recognition in modern times, since he never published a mathematical work of any kind under his name. In fact only one number-theoretic proof by Fermat survives. His famous marginal note was found by his son Samuel, who went through Fermat's letters and unpublished works with an eye toward their posthumous publication.

Probably one reason Fermat's problem has survived is that it is possible to prove it for particular exponents. Thus early workers on the problem might have come up with something publishable, if not a general proof. Fermat himself probably had a proof for $n = 4$, and Gauss succeeded for the more difficult case of $n = 3$. Many famous mathematicians settled other cases. The proof for $n = 5$

Reprinted from *Mathematics Teacher* 82 (Nov., 1989): 637–40; with permission of the National Council for Teachers of Mathematics.

was first given by Legendre in 1825, using an idea of Dirichlet. The latter's settling of $n = 14$ in 1832 was superseded by Lamé's 1839 proof that Fermat's equation is impossible for $n = 7$.

Proving Fermat's last theorem for a given exponent n also settles it for any multiple of n. For example, knowing that

$$x^7 + y^7 = z^7$$

is impossible in positive integers also covers the case $n = 14$, since if we had

$$X^{14} + Y^{14} = Z^{14},$$

then

$$x = X^2, y = Y^2, z = Z^2$$

would satisfy the first equation. Thus, to prove Fermat's last theorem in general, it suffices to prove that

$$x^n + y^n = z^n$$

is impossible in positive integers for $n = 4$ and for any value of n that is an odd prime. We can also assume that the greatest common divisor of x, y, and z is 1, since, if d were a common factor, then the equation could be divided by d^n. Such a solution is said to be *primitive*. It is easily seen that for a primitive solution x, y, and z are relatively prime in pairs.

The case $n = 4$ is the only one for which a short proof is known. This proof depends on the integral solutions (called *Pythagorean triples*) to the equation $x^2 + y^2 = z^2$. The general solution to this equation was known by Diophantus and perhaps even by the Babylonians, who before 1600 B.C. at least had a way of generating Pythagorean triples, including 4961, 6480, and 8161 (Neugebauer 1952). The theorem needed is the following. Proofs can be found in Niven and Zuckerman (1980) and Vanden Eynden (1987).

THEOREM. *Suppose the positive integers x, y, and z form a primitive Pythagorean triple. Then one of x and y is even and the other odd. If x is even, then relatively prime positive integers u and v exist such that $x = 2uv$, $y = u^2 - v^2$, and $z = u^2 + v^2$.*

We shall use this theorem to show that Fermat's last theorem holds in the case $n = 4$. Suppose we have a primitive solution to $x^4 + y^4 = z^4$ in positive integers. Setting $w = z^2$ yields the equation

$$x^4 + y^4 = w^2, \tag{1}$$

and so x^2, y^2, w is a primitive Pythagorean triple. Then, assuming x^2 is even, by the theorem just stated, relatively prime positive integers u and v exist such that

$$x^2 = 2uv, y^2 = u^2 - v^2, \text{ and } w = u^2 + v^2.$$

From the second of these equations we have $v^2 + y^2 = u^2$. Since we know that y is odd, we can apply the theorem again to find relatively prime positive integers s and t such that

$$v = 2st, y = s^2 - t^2, \text{ and } u = s^2 + t^2.$$

Note that $x^2 = 2uv = 2(s^2 + t^2)(2st)$, and so

$$(x/2)^2 = st(s^2 + t^2).$$

(Recall that x was even.) Since s and t are relatively prime, the three factors on the right of this equation are relatively prime in pairs. For, if the prime p divides s, for example, then it cannot divide t and so does not divide $s^2 + t^2$ either. It follows from the unique factorization of integers into primes that each of s, t, and $s^2 + t^2$ must itself be a square, say

$$s = x_1^2, \qquad t = y_1^2,$$

and

$$s^2 + t^2 = w_1^2,$$

where x_1, y_1, and w_1 are relatively prime positive integers. But from these equations we have

$$x_1^4 + y_1^4 = w_1^2, \qquad (2)$$

which is of the same form as equation (1). Notice that

$$w_1 \le w_1^2 = s^2 + t^2 = u \le u^2 < u^2 + v^2 = w.$$

Thus from (1) we have derived another case of the sum of two fourth powers being a square, with the squared integer smaller than the one we started with.

This example illustrates a type of argument due to Fermat, from the one proof he left, known as the *method of infinite descent*. Starting with equation (2), we could similarly find positive integers x_2, y_2, and w_2 such that

$$x_2^4 + y_2^4 = w_2^4,$$

with $w_2 < w_1$. Continuing in this way we get an infinite sequence of positive integers

$$w > w_1 > w_2 > w_3 > \cdots > 0.$$

Since this result is obviously impossible, so is the original equation

$$x^4 + y^4 = z^4,$$

and Fermat's last theorem has been proved for the case $n = 4$.

The Work of Kummer

The most significant contribution toward a proof of Fermat's last theorem was made by the German mathematician Ernst Eduard Kummer. An often-repeated story says that in the 1840s Kummer thought he had a proof based on doing arithmetic in certain domains of *algebraic integers*. These are complex numbers that behave in many ways like ordinary integers. Gauss already used such numbers in his proof of the case $n = 3$, namely, the complex numbers of the form $a + bp$, where a and b are ordinary integers and p is a root of the equation

$x^2 + x + 1 = 0$. The sum and product of such numbers are of the same form, and concepts of divisibility and primality can be defined for them. The story goes that Kummer's proof assumed that an element of any such domain can always be written as the product of primes in a unique way, a property that in fact holds in some, but not all, cases. Only when Kummer submitted his proof to Dirichlet was his mistake discovered.

Recent research by Harold M. Edwards has thrown doubt on the foregoing story. (Most of the historical material in this article is based on Edwards's excellent book on Fermat's last theorem [1977].) Its source seems to be a talk given in 1910, more than sixty years after the alleged incident, by Hensel, who had gotten it from a third party who was not a mathematician. Kummer's supposed proof has never been found, and in general the story is not supported by contemporary documents.

Records of the Paris Academy from 1847 show that both Lamé and Cauchy believed that factorization in algebraic number domains was unique and that a proof of Fermat's last theorem could be based on this idea. Both realized that unique factorization would have to be proved, however, to make the proof complete. In fact, Kummer had published examples of domains where unique factorization did not hold in an obscure journal three years before, and no written evidence has been found that he had ever assumed that such domains did not exist.

What Kummer did prove was that Fermat's last theorem held for all "regular" primes, which will be defined shortly. The power of Kummer's result is indicated by the fact that the smallest prime that is not regular is 37. Thus the cases $n = 3, 5, 7, 11, 13, 17, 19, 23, 29$, and 31 (and many others) are disposed of all at once. In fact the only primes less than 100 that are not regular are 37, 59, and 67.

The regular primes can be characterized in the following way. First we define the *Bernoulli numbers* $B_0, B_1, B_2 \ldots$ by the equation

$$\left(B_0 + \frac{B_1 x}{1!} + \frac{B_2 x^2}{2!} + \cdots\right)$$

$$\cdot \left(\frac{x}{1!} + \frac{x^2}{2!} + \frac{x^3}{3!} + \cdots\right) = x.$$

By multiplying the expressions on the left as if they were polynomials and equating coefficients of like powers of x, the numbers B_0, B_1, ... can be computed recursively. The reader may want to check that $B_0 = 1$, $B_1 = -1/2$, $B_2 = 1/6$, $B_3 = 0$, and $B_4 = -1/30$. It turns out that B_i is always a rational number, and $B_i = 0$ for values of i that are odd and greater than 1. We say that a prime p is *regular* when it does not divide the numerator of any of the Bernoulli numbers B_2, B_4, ... , B_{p-3}. For example, 7 is regular because it does not divide the numerators of B_2 or B_4.

Unfortunately neither Kummer nor anyone since has been able to prove the existence of infinitely many regular primes. The existence of infinitely many primes that are not regular has been proved.

Results after Kummer

Although many theorems have been proved since Kummer on the subject of Fermat's last theorem, most cannot be stated simply. Many amateurs were encouraged to work on the problem by a prize of 100 000 marks offered for a proof (but not a disproof) of the theorem in the will of Paul Wolfskehl in 1909. The proof had to be accepted as valid by the German academy of sciences in Göttingen. Inflation after World War I reduced the value of the prize to almost nothing, but it is now worth about $5500. The Mathematics Institute of the University of Göttingen still receives three or four "proofs" per month aiming at the prize.

In modern times computers have been turned loose on the problem. Samuel Wagstaff and

Jonathan Tanner (1987) have used computer techniques to prove the theorem for all exponents $n \leq$ 150 000. A recent theoretical advance was the proof of Gerd Faltings that for any fixed value of n, Fermat's equation can have only a finite number of primitive solutions.

Methods of proof are now applied that go far beyond anything Fermat could have imagined. These new approaches make it increasingly difficult to believe that Fermat really had a proof.

BIBLIOGRAPHY

Bell, Eric Temple. *The Last Problem*. New York: Simon & Schuster, 1961.

Dembart, Lee. "Scientists Buzzing—Fermat's Last Theorem May Have Been Proved." *Los Angeles Times*, 8 March 1988, 3, 23.

Dickson, Leonard Eugene. *History of the Theory of Numbers*. Vol. 2. New York: Chelsea Publishing Co., 1952.

Edwards, Harold M. *Fermat's Last Theorem*. New York: Springer-Verlag, 1977.

Hilts, Philip J. "Famed Math Puzzle May Have Been Solved." *Washington Post*, 9 March 1988, 1, 4.

Kolata, Gina. "Progress on Fermat's Famous Math Problem." *Science* 235 (27 March 1987): 1572–73.

Mordell, Louis J. *Three Lectures on Fermat's Last Theorem*. Cambridge: Cambridge University Press, 1921.

Neugebauer, Otto. *The Exact Sciences in Antiquity*. Princeton University Press, 1952.

Niven, Ivan, and Herbert S. Zuckerman. *An Introduction to the Theory of Numbers*. 4th ed. New York: John Wiley & Sons, 1980.

Ribenboim, Paulo. *13 Lectures on Fermat's Last Theorem*. New York: Springer-Verlag, 1979.

Tanner, Jonathan W., and Samuel S. Wagstaff, Jr. "New Congruences for the Bernoulli Numbers." *Mathematics of Computation* 48 (1987): 341–50.

Vanden Eynden, Charles. *Elementary Number Theory*. New York: Random House, 1987.

EPILOGUE:

$\mathcal{F}$ermat's $\mathcal{L}$ast $\mathcal{T}$heorem, 1993

AN EPILOGUE is usually an afterthought added to a written work to reflect on its contents or, at a later instance, to modify some of its conclusions. In a sense, it stands apart from the textual message itself. In the case of this epilogue, it is very much a part of the textual message of just how mathematics works.

When I completed the compilation of *From Five Fingers to Infinity* in 1991, I purposefully ended with Vanden Eynden's article on Fermat's Last Theorem, the most famous unsolved problem in the history of mathematics. His article highlights some of the uncertainty in mathematics: an unsolved problem, an almost correct solution, a continued search for a solution, and illustrates the persistent nature of mathematicians in striving to solve the problem of Fermat's Last Theorem. In this mission, the work of one mathematician supplies a basis for a latter colleague to carry on the quest. Teamwork is evident. The research process of seeking a solution to a difficult mathematical problem is like a relay-race with each participant carrying the baton a little closer to the victory line before handing it off to his or her teammate. Well, the race to solve Fermat's Last Theorem appears to have been won!

On June 23, 1993, Dr. Andrew Wiles, of Princeton University, delivered the third of a series of conference lectures at the Newton Institute, Cambridge University. Dr. Wiles's lecture series was entitled "Modular Forms, Elliptic Curves, and Galois Representations" and described his research findings associating the theory of elliptic curves, that is, curves designated by equations of the form $y^2 = x^3 + ax + b$, where a and b are integers, with Fermat's Last Theorem. Wiles proved a conjecture which originated in 1955 with the Japanese mathematician Yutaka Taniyama and was later modified by Andre Weil (1968) and Goro Shimura (1971). It is known as the Shimura-Taniyama-Weil conjecture and associates the behavior of elliptic curves with another class of curves known as modular curves. In 1986, Kenneth Ribet of the University of California found a link between elliptic curves and Fermat's Last Theorem. He showed that if a counterexample existed to Fermat's Last Theorem then it would be possible to construct an elliptic curve that was not modular. Building upon Ribet's theories and employing methods developed by still other researchers, Wiles established the Shimura-Taniyama-Weil conjecture for a class of elliptic curves including some relevant to proving Fermat's Last Theorem. In accomplishing this feat, he, in principle, proved Fermat's Last Theorem and established solution techniques that will greatly advance number theory research.

In commenting upon Andrew Wiles's achievement, Barry Mazur, a number theorist at Harvard University who himself had been involved in research on the subject, noted that although the problem was an ancient one solved with modern techniques, still it

was answered strictly in its own terms. That's rather amazing right
there. It underscores how stable mathematics is through the centuries—how mathematics is one of humanity's long continuous
conversations with itself.

Wiles's victory certainly testifies to his personal genius and dogged persistence—he
worked on the problem for seven years. He deserves the laurels of the victor but, in a
larger sense, his accomplishment is also a victory for the mathematical community and
demonstrates international cooperation and interdependence in the field of mathematical research. In developing his solution scheme, Andrew Wiles employed theories from
many branches of mathematics: crystalline cohomology, Galois representations, L-
functions, modular forms, deformation theory, Gorenstein rings . . . and relied on
research findings from colleagues in France, Germany, Italy, Japan, Australia, Colombia,
Brazil, Russia, and the United States. Wiles, himself, is a British citizen working at an
American university. This relay race has been won.

—Frank Swetz

SUGGESTED READINGS FOR PART VIII

Barker, Stephen F. *Philosophy of Mathematics*. New York: Prentice Hall, 1964.

Black, Max. *The Nature of Mathematics: A Critical Survey*. New York: Humanities Press, 1950.

Courant, Richard, and Robbins, Herbert. *What is Mathematics?* New York: Oxford University Press, 1941.

Dauben, Joseph Warren. *Georg Cantor: His Mathematics and Philosophy of the Infinite*. Princeton, N.J.: Princeton University Press, 1979.

Davis, Philip. *The Mathematical Experience*. Boston: Birkhauser, 1981.

Goldstine, H. H. *The Computer from Pascal to Von Neumann*. Princeton, N.J.: Princeton University Press, 1972.

Gratton-Guinness, I. *The Development of the Foundations of Mathematical Analysis from Euler to Riemann*. Cambridge, Mass.: MIT Press, 1970.

Greenberg, Marven. *Euclidean and Non-Euclidean Geometry: Development and History*. San Francisco: W. H. Freeman, 1974.

Hall, Tord. *Carl Friedrick Gauss*. Translated by A. Froderberg. Cambridge, Mass.: MIT Press, 1970.

Johnson, P. E. *A History of Set Theory*. Boston: Prindle, Weber & Schmidt, 1972.

Kline, Morris. *Mathematics: The Loss of Certainty*. New York: Oxford University Press, 1979.

Meschkowski, Herbert. *Ways of Thought of Great Mathematicians*. Translated by John Dyer-Bennet. San Francisco: Holden-Day, 1964.

Nový, Luboš. *Origins of Modern Algebra*. Leyden, The Netherlands: Noordhoff International Publishing, 1973.

Singh, Jagjit. *Great Ideas of Modern Mathematics: Their Nature and Use*. New York: Dover Publications, 1959.

Styazhkin, N. I. *History of Logic from Leibniz to Peano*. Cambridge, Mass.: MIT Press, 1969.

The search for certainty continues . . .

FRANK SWETZ
July 1, 1994

Magic Pentagram
by Lee Sallows,
Nijmegen,
The Netherlands.

A General Bibliography

Throughout this book, references and Suggested Reading lists on specific areas of the history of mathematics have been given. This bibliography is offered as a supplement to that information and is intended for those readers who might wish to extend and strengthen their knowledge on the history of mathematics.

JOURNALS

While journals, *per se,* do not usually find their way into a general bibliography, the inclusion of articles comprising this book indicates that many worthwhile contributions on the history of mathematics can be found in journals. The journals suggested below, although professional or academic in nature, are semi-popular in outlook and are extremely readable; as such they can frequently be found in comprehensive libraries.

Historia Mathematica. (International journal of the history of mathematics) New York: Academic Press, Inc.

The Mathematics Magazine. Washington, D.C.: The Mathematical Association of America.

The Mathematics Teacher. Reston, VA: The National Council of Teachers of Mathematics (NCTM).

BOOKS ON THE HISTORY OF NUMBERS AND NUMERATION

Dantzig, Tobias. *Number the Language of Science.* New York: Macmillan, 1954.

Flegg, Graham. *Numbers: Their History and Meaning.* New York: Schocken Books, 1983.

Freitag, H. T., and A. H. Freitag. *The Number Story.* Reston, VA: NCTM, 1960.

Ifrah, Georges. *From One to Zero: A Universal History of Numbers.* New York: Viking Penguin, 1985.

Menninger, Karl. *Number Words and Number Symbols: A Cultural History of Numbers.* Cambridge, MA: MIT Press, 1977.

Smeltzer, Donald. *Man and Number.* New York: Collier Books, 1962.

Smith, D. E., and J. Ginsburg. *Numbers and Numerals.* Reston, VA: NCTM, 1958.

GENERAL SURVEY BOOKS

Boyer, Carl B., and Uta Merzback. *A History of Mathematics.* New York: John Wiley & Sons, 1989.

Burton, David M. *The History of Mathematics: An Introduction.* Boston: Allyn and Bacon, Inc., 1985.

Cajori, F. *A History of Mathematics.* New York: Chelsea, 1985.

Eves, Howard. *An Introduction to the History of Mathematics.* Philadelphia, PA: Saunders, College Publishing, 1983.

Fauvel, John, and Jeremy Gray, eds. *The History of Mathematics: A Reader.* Milton Keynes, England: The Open University, 1987.

Katz, Victor. *A History of Mathematics: An Introduction.* New York: Harper Collins, 1993.

Kline, Morris. *Mathematical Thought from Ancient to Modern Times.* 3 vols. New York: Oxford University Press, 1972.

Scott, J. F. *A History of Mathematics: From Antiquity to the Beginning of the Nineteenth Century.* New York: Barnes & Noble, 1969.

Struik, D. J. *A Concise History of Mathematics.* New York: Dover Publications, 1987.

SELECTIVE AND INTERPRETATIVE WORKS

Aaboe, Asger. *Episodes from the Early History of Mathematics.* Washington, D.C.: Mathematical Association of America, 1978.

Beckmann, Petr. *A History of π.* New York: St. Martin's Press, 1971.

Bell, E. T. *Men of Mathematics.* New York: Simon and Schuster, 1965.

Bergamini, David. *Mathematics.* Alexandria, VA: Time-Life Books, 1970.

Berggren, J. L. *Episodes in the Mathematics of Medieval Islam.* New York: Springer-Verlag, 1986.

Calinger, Ronald, ed. *Classics of Mathematics.* Oak Park, IL: Moore Publishing Co., 1982.

Campbell, Douglas M., and John Higgins. *Mathematics: People, Problems, Results.* 3 vols. Belmont, CA: Wadsworth, 1984.

Campbell, P., and L. Grinstein. *Women of Mathematics.* New York: Greenwood Press, 1987.

Dantzig, Tobias. *The Bequest of the Greeks.* New York: Charles Scribner's Sons, 1955.

Davis, P., and R. Hersh. *The Mathematical Experience.* Boston: Birkhäuser, 1981.

Devlin, Keith. *Mathematics: The New Golden Age.* Harmondsworth, England: Penguin Books, 1988.

Dilke, O.A.W. Mathematics and Measurement. Berkeley, CA: University of California Press, 1987.

Dörrie, H. *100 Great Problems of Elementary Mathematics: Their History and Solution.* New York: Dover Publications, 1965.

Dunham, William. *Journey Through Genius: The Great Theorems of Mathematics.* New York: Wiley, 1990.

Eves, Howard. *Great Moments in Mathematics (Before 1650).* Washington, D.C.: The Mathematical Association of America, 1981.

———. *Great Moments in Mathematics (After 1650).* Washington, D.C.: The Mathematical Association of America, 1982.

Gillings, Richards J. *Mathematics in the Time of the Pharaohs.* New York: Dover Publications, 1982.

Grattan-Guinness, Ivor, ed. *From Calculus to Set Theory 1630–1910.* London: Duckworth, 1980.

Hofstadter, D. *Gödel, Escher and Bach: An Eternal Golden Braid.* New York: Vintage Books, 1979.

Hogben, Lancelot. *The Wonderful World of Mathematics.* New York: Doubleday and Company, 1968.

———. *Mathematics for the Millions.* London: Merlin, 1989.

Hollingdale, Stuart. *Makers of Mathematics.* Harmondsworth, England: Penguin Books, 1989.

Hooper, Alfred. *Makers of Mathematics.* New York: Vintage Books, 1948.

Karpinski, L. *The History of Arithmetic.* Chicago: Rand McNally, 1925.

Kline, Morris. *Mathematics in Western Culture.* New York: Oxford University Press, 1953.

———. *Mathematics: A Culture Approach.* Reading, MA: Addison-Wesley, 1962.

———. *Mathematics: The Loss of Certainty.* New York: Oxford University Press, 1983.

———. *Mathematics and the Search for Knowledge.* New York: Oxford University Press, 1985.

———, ed. *Mathematics in the Modern World.* San Francisco: W. H. Freeman and Co., 1979.

———, ed. *Mathematics an Introduction to its Spirit and Use.* San Francisco: W. H. Freeman and Co., 1979.

Kramer, Edna E. *The Mainstream of Mathematics.* Greenwich, CA: Fawcett, 1964.

———. *The Nature and Growth of Modern Mathematics.* New York: Hawthorn, 1970.

Li Yan and Du Shiran. *Chinese Mathematics: A Concise History.* New York: Oxford University Press, 1987.

Maor, Eli. *To Infinity and Beyond: A Cultural History of the Infinite.* Boston: Birkhäuser, 1987.

Meschkowski, H. *Ways of Thought of Great Mathematicians.* San Francisco: Holden-Day, 1964.

Midonick, H. O., ed. *The Treasury of Mathematics.* New York: Philosophical Library, 1965.

Moffatt, Michael, ed. *The Ages of Mathematics.* 4 vols. New York: Doubleday Inc., 1977.

Morgan, Bryan, *Men and Discoveries in Mathematics.* London: John Murry, 1972.

Neugebauer, Otto. *The Exact Sciences in Antiquity.* New York: Dover Publications, 1969.

Newman, J.R., ed. *The World of Mathematics.* 4 vols. New York: Simon and Schuster, 1956.

Osen, Lynn M. *Women in Mathematics.* Cambridge, MA: MIT Press, 1974.

Perl, Terri. *Math Equals: Biographies of Women Mathematicians.* Menlo Park, CA: Addison-Wesley, 1978.

Resnikoff, H. L., and R. O. Wells. *Mathematics in Civilization.* New York: Dover Publications, 1985.

Rowe, David, and John McCleary, eds. *The History of Modern Mathematics.* 2 vols. San Diego, CA: Academic Press, 1989.

Smith, David E. *History of Mathematics.* 2 vols. New York: Dover Publications, 1958.

Steen, Lynn Arthur. *Mathematics Today: Twelve Informal Essays.* New York: Springer-Verlag, 1978.

Stillwell, John. *Mathematics and Its History.* New York: Springer-Verlag, 1969.

Struik, D. J., ed. *A Source Book in Mathematics, 1200–1800*. Princeton, N.J.: Princeton University Press, 1986.

Swetz, Frank. *Capitalism and Arithmetic: The New Math of the 15th Century*. La Salle, IL: Open Court, 1987.

Swetz, Frank, and T. I. Kao. *Was Pythagoras Chinese? An Examination of Right Triangle Theory in Ancient China*. University Park, PA: The Pennsylvania State University Press, 1977.

Turnbull, Herbert. *The Great Mathematicians*. New York: New York University Press, 1969.

van der Waerden, B. L. *Geometry and Algebra in Ancient Civilizations*. New York: Springer-Verlag, 1983.

———. *A History of Algebra: From al-Khowarizmi to Emmy Noether*. New York: Springer-Verlag, 1985.

Zaslavsky, Claudia. *Africa Counts: Number and Pattern in African Culture*. Boston: Prindle, Weber & Schmidt, 1973.

Zeintarn, M. *The History of Computing*. Framingham, MA: C.W. Communications, 1981.

757

Index